Fat Mimetics for Food Applications

Fat Mimetics for Food Applications

Edited by

Miguel Ângelo Parente Ribeiro Cerqueira
International Iberian Nanotechnology Laboratory
Braga, Portugal

Lorenzo Miguel Pastrana Castro
International Iberian Nanotechnology Laboratory
Braga, Portugal

WILEY

This edition first published 2023

Registered Offices
John Wiley & Sons, Inc., 111 River Street, Hoboken, NJ 07030, USA
John Wiley & Sons Ltd, The Atrium, Southern Gate, Chichester, West Sussex, PO19 8SQ, UK

For details of our global editorial offices, customer services, and more information about Wiley products visit us at www.wiley.com.

Wiley also publishes its books in a variety of electronic formats and by print-on-demand. Some content that appears in standard print versions of this book may not be available in other formats.

A catalogue record for this book is available from the Library of Congress

Hardback ISBN: 9781119780014; ePub ISBN: 9781119780038; ePDF ISBN: 9781119780021;
oBook ISBN: 9781119780045

Cover image: © ThomasVogel/Getty Images
Cover design by Wiley

Set in 9.5/12.5pt STIXTwoText by Integra Software Services Pvt. Ltd., Pondicherry, India

To our families for their love and support.

Contents

Foreword

We underestimated the global demand for (solid) fats with both food and non-food applications, which resulted in the current supply shortage. Food demands remain unsatisfied because the volume of (solid) fats is insufficient to meet the demands of sustainability, health, and cost.

In the 1950s, we thought that butter could be a long-term fat solution, but its global consumption has slightly decreased given consumers' concerns about saturated fats and cholesterol in their diet. To address these concerns, food processors replaced butter with margarine in the 1960s. Margarines were initially processed with partially hydrogenation vegetable oils. Unfortunately, the partial hydrogenation process produces unhealthy artificial trans fats. After the 1990s, partial hydrogenation was replaced by an interesterification of a solid fat (mainly from palm oil) with a liquid vegetable oil (rapeseed, soybean or sunflower) to produce trans-free margarines.

Palm oil emerged as the leading global source of (solid) fat. Palm kernel oil or palm stearin, has low production costs and is used by leading palm oil producers. Palm oils are driven by the high oil yields per hectare for oil palm relative to other vegetable oil crops. Well-managed oil palm estates should achieve four to six tons of crude palm oil per hectare; this is nearly five times the amount of oil obtained from a hectare planted with sunflowers.

Sustainability is one potential check on the use of palm oil is sustainability, which has recently come to the forefront. Earlier this century, palm oil earned a negative environmental reputation. As a result, a sustainable palm supply chain was founded in 2004, and many Western EU countries have pledged to source only 100% sustainable palm oil by 2020. France, Italy, and Belgium, and more recently, Spain, Germany and the UK have begun to see palm oil as unhealthy. Health concern is pushing sustainable palm into the background. The debate centers on saturated fat intake, and recently, on 3-MCPD (3-monochloropropane diol) risks associated with palm oil. The food demand for palm oil in Europe has dropped from 3.1 million tonnes in 2013 to 2.2 million tonnes in 2019. So, "exotic fats" are becoming credible alternatives to palm fats.

"Exotic fats" are typically fats such as shea, sal, illipé, kokum, and mango kernel. One of the major drawbacks of "exotic fats" is the high price and poor availability of raw materials. As an illustration, shea is rarely grown on plantations; the nuts are mainly collected in the savannah by local people and are sold in small volumes to processing plants. This makes the supply potentially susceptible to political unrest or to other unexpected events. Moreover, the cost of shea butter was over $3,500 per tonne, while palm (kernel) fat was below $1,000 per tonne during the same period (2019–2021).

Can fat mimetics be the solution? This book answers that question. In any case, to serve as a solution, fat mimetics would need to be efficient, scalable, and cost effective. These are fundamental concerns that scientists should always keep in mind.

Fabrice TURON, PhD
Head of Research
FAT & Associés, France

List of Contributors

Ainaz Alizadeh
Department of Food Science and Technology
Tabriz Branch
Islamic Azad University
Tabriz, Iran

Maria A. Azevedo
Centre of Biological Engineering
University of Minho
Campus de Gualtar
Braga, Portugal

Akhoon Asrar Bashir
ICAR-Central Institute of Post-Harvest
Engineering and Technology (CIPHET)
Ludhiana, Punjab, India

Deepti Bharti
Department of Biotechnology and Medical
Engineering
National Institute of Technology
Rourkela, Odisha, India

Filip Van Bockstaele
Department of Food Technology
Safety and Health
Food Structure and Function (FS&F) Research
Group
Ghent University
Coupure Links, Ghent, Belgium

Mariel Calderón-Oliver
Tecnológico de Monterrey
Escuela de Ingeniería y Ciencias
Mexico

Sonia Calligaris
Department of Agricultural, Food,
Environmental, and Animal Sciences
University of Udine
Udine, Italy

Miguel Ângelo Cerqueira
INL – International Iberian Nanotechnology
Laboratory
Braga, Portugal

Xiao-Wei Chen
Lipid Technology and Engineering
College of Food Science and Engineering
Henan University of Technology
Zhengzhou, People's Republic of China

Poonam Choudhary
ICAR-Central Institute of Post-Harvest
Engineering and Technology (CIPHET)
Ludhiana, Punjab, India

Gabrielli Nunes Clímaco
Department of Food Engineering and
Technology
School of Food Engineering
University of Campinas – UNICAMP
Campinas, SP, Brazil

Rosiane Lopes da Cunha
Department of Food Engineering and
Technology
School of Food Engineering
University of Campinas (UNICAMP)
Campinas, SP, Brazil

Ilkem Demirkesen
Department of Animal Health, Food and Feed Research, General Directorate of Agricultural Research and Policies
Ministry of Agriculture and Forestry
Ankara, Turkey

Koen Dewettinck
Department of Food Technology
Safety and Health
Food Structure and Function (FS&F) Research Group
Ghent University
Coupure Links, Ghent, Belgium

Hector Escalona-Buendía
Departamento de Biotecnología
UAM-Iztapalapa
Mexico

Anne-Laure Fameau
Unité Matériaux et Transformations (UMET) – UMR 8207, INRAE
Team PIHM
Villeneuve d'Ascq, France

Luiz Henrique Fasolin
Department of Food Engineering and Technology
School of Food Engineering
University of Campinas – UNICAMP
Campinas, SP, Brazil

Elliot Paul Gilbert
Australian Centre for Neutron Scattering
Australian Nuclear Science and Technology Organisation
Lucas Heights, NSW, Australia

Australian Institute for Bioengineering and Nanotechnology and Centre for Nutrition and Food Sciences
The University of Queensland
St. Lucia, Brisbane, QLD, Australia

Catarina Gonçalves
International Iberian Nanotechnology Laboratory
Braga, Portugal

Dharani Gopi
National Institute of Food Technology
Entrepreneurship and Management
Thanjavur, Tamil Nadu, India

Andrew J. Gravelle
Department of Food Science and Technology
University of California, Davis
Davis, CA, USA

Buse N. Gürbüz
INL – International Iberian Nanotechnology Laboratory
Braga, Portugal

Doman Kim
Graduate School of International Agricultural Technology
Seoul National University
Pyeongchang-gun
Gangwon-do, South Korea

Institute of Food Industrialization
Institutes of Green Bioscience and Technology
Seoul National University
Pyeongchang-gun
Gangwon-do, South Korea

Yogesh Kumar
ICAR-Central Institute of Post-Harvest Engineering and Technology (CIPHET)
Ludhiana, Punjab, India

Suyong Lee
Department of Food Science & Biotechnology and Carbohydrate Bioproduct Research Center
Sejong University
Seoul, Republic of Korea

Felipe de Andrade Maia
Department of Food Engineering and Technology
School of Food Engineering University of Campinas – UNICAMP Campinas
SP, Brazil

Lara Manzocco
Department of Agricultural, Food, Environmental, and Animal Sciences

University of Udine
Udine, Italy

Alejandro G. Marangoni
University of Guelph
Guelph, ON, Canada

Artur J. Martins
International Iberian Nanotechnology Laboratory
Braga, Portugal

Sofia Melchior
Department of Agricultural, Food, Environmental, and Animal Sciences
University of Udine
Udine, Italy

Behiç Mert
Department of Food Engineering
Middle East Technical University
Ankara, Turkey

Karunairaj Michael
National Institute of Food Technology Entrepreneurship and Management
Thanjavur, Tamil Nadu, India

Reed A. Nicholson
Motif FoodWorks, Inc.
Boston, MA, USA

Paula Kiyomi Okuro
Department of Food Engineering and Technology
School of Food Engineering
University of Campinas (UNICAMP)
Campinas, SP, Brazil

Kunal Pal
Department of Biotechnology and Medical Engineering
National Institute of Technology
Rourkela, Odisha, India

Lorenzo Miguel Pastrana Castro
INL – International Iberian Nanotechnology Laboratory
Braga, Portugal

Ivana A. Penagos
Department of Food Technology
Safety and Health
Food Structure and Function (FS&F) Research Group
Ghent University
Coupure Links, Ghent, Belgium

Carolina Siqueira Franco Picone
Department of Food Engineering and Technology
School of Food Engineering
University of Campinas – UNICAMP
Campinas, SP, Brazil

Aurora Pintor-Jardines
Departamento de Biotecnología
UAM-Iztapalapa
Mexico

Stella Plazzotta
Department of Agricultural, Food, Environmental, and Animal Sciences
University of Udine
Udine, Italy

Bikash K. Pradhan
Department of Biotechnology and Medical Engineering
National Institute of Technology
Rourkela, Odisha, India

Ashish Rawson
National Institute of Food Technology Entrepreneurship and Management
Thanjavur, Tamil Nadu, India

Ana Paula Badan Ribeiro
Department of Food Engineering and Technology
School of Food Engineering
University of Campinas (UNICAMP)
Campinas, SP, Brazil

Michael A. Rogers
Associate Professor and Tier II Canada Research Chair in Food Nanotechnology
University of Guelph
Guelph, Ontario, Canada

Kato Rondou
Department of Food Technology
Safety and Health
Food Structure and Function (FS&F) Research Group
Ghent University
Coupure Links, Ghent, Belgium

Satish Saigiri
Agro-Nanotechnology and Advanced Materials Research Center
Institute of Postharvest and Food Sciences
Agricultural Research Organization
The Volcani Center
Rishon LeZion, Israel

Patricia Severiano-Pérez
Departamento de Alimentos y Biotecnología
Facultad de Química, UNAM
Mexico

Thaís Jordânia Silva
Department of Food Engineering and Technology
School of Food Engineering
University of Campinas (UNICAMP)
Campinas, SP, Brazil

Mayanny Gomes da Silva
Department of Food Engineering and Technology
School of Food Engineering
University of Campinas (UNICAMP)
Campinas, SP, Brazil

Mitra Soofi
Department of Food Science and Technology
Tabriz branch
Islamic Azad University
Tabriz, Iran

Davanam Srikanth
National Institute of Food Technology
Entrepreneurship and Management
Thanjavur, Tamil Nadu, India

Sunil C. K.
National Institute of Food Technology
Entrepreneurship and Management
Thanjavur, Tamil Nadu, India

Xiao-Quan Yang
Laboratory of Food Proteins and Colloids
Department of Food Science and Engineering
South China University of Technology
Guangzhou, People's Republic of China

Preface

The scientific community and the food industry are constantly challenged to meet consumer demands for new food products nutritionally balanced and healthy while maintaining the product's safety, good organoleptic properties, and convenience. One pivotal way to meet those challenges is in replacing fats in processed foods. Fats are known for their health implications and influence on foods' shelf-life and nutritional properties. Replacing fats by reducing the amount of fats or by replacing unhealthy fats with healthy fats has been a challenge in the last 20 years. These challenges are mainly related to the physicochemical properties of fats and their influence on food structure and with organoleptic properties.

This book provides an overview of the strategies used for developing fat mimetics for food appli cations. It will cover all the physical strategies and present the main characterization techniques for studying fat mimetics behaviour. It will grant insight into how fat mimetics in food structure are understood and how fat mimetics affect the all characteristics of food products.

The book is organized into five sections. The first section is the foundation for subsequent sections and covers the demanding for fat mimetics for food applications and the relationship of the structural properties in fat mimetics. Section II is devoted to the primarily materials used in fat mimetics development and to the structures that result from different methodologies and approaches. Section III will cover the methodologies used to characterize the developed replacers not only from a physical and sensorial perspective, but also in terms of how fat replacers behave in the gastrointestinal tract. The examples of the food products will be discussed in section IV, where what has been done in terms of using fat mimetics in food products will be presented; this section aims to cover all possible use of these materials. Section V focuses on the industrial perspective. and will present a future perspective using some real cases for projects within the food industry and some commercial examples.

The information in this book is expected to increase the scientific understanding of fat mimetics. The authors are scientists with different areas of expertise covering principal aspects of food science and technology; some authors also bring expertise from the fields of engineering, chemistry and materials. The diverse range of input will give to the readers a broad overview of the field, boosting their capability to integrate different aspects of fat mimetics for food development.

We expect this book will become a unique source of information for scientists and researchers who are doing research and development with fat mimetics on processed foods.

Acknowledgements

We want to thank all the authors for their contributions and by lending their excellent perspectives into the best information available. A big thank you to Kerry Powell and Rebecca Ralf, editorial project managers, for their help during the editorial processes and always giving the best advice. We also want to specially thank Fabrice Turon for his contribution in the forward of this book.

The authors acknowledge the financial support of the projects: hOLIVEcream – Healthy olive oil based creams enriched with berries for application in bakery and pastry, with the reference n.º NORTE-01–0247-FEDER-046947, co-funded by Norte 2020 – North-Regional Operational Programme under the PORTUGAL 2020 Partnership Agreement, through the European Regional Development Fund (ERDF); IceCare: Cardio-Healthy Functional Ice creams (NORTE-01–0247-FEDER-039927), cofinanced by Norte 2020 and the European Union, through the European Regional Development Fund (ERDF), and BetterFat4Meat: Development of structured fats for to replace animal fats in meat products (POCI-01–0247- FEDER-039718), cofinanced by Compete 2020, Portugal 2020 and the European Union, through the European Regional Development Fund (ERDF).

Editors

Miguel Ângelo Parente Ribeiro Cerqueira

Miguel Ângelo Parente Ribeiro Cerqueira is Staff Researcher in the Food Processing and Nutrition Group at the International Iberian Nanotechnology Laboratory, and his research is focused on developing micro and nanosized bio-based structures for food applications. He works on edible and biodegradable materials for packaging, encapsulation of functional compounds using emergent encapsulation technologies, and structuring gels, such as oleogels.

Miguel holds both a Bachelor and PhD (in Biological Engineering from the University of Minho. He received three scholar merit awards during his Bachelor program, and his PhD thesis was awarded for the best PhD Thesis by the School of Engineering of the University of Minho. During his PhD studies, Miguel attended special projects at the Federal University of Ceará, University of Aveiro, and University College Cork. In 2011, he started as his Postdoctoral Researcher studies at UM and undertook special studies at the University of Vigo, University of Campinas) and Institute of Agrochemistry and Food Technology – CSIC. He authored over 130 scientific articles, published 20 book chapters, received two patents, and has edited three books. Miguel supervised more than 15 students (PhD and MSc). He is co-founder of two start-ups (Improveat, Lda in 2013 and 2BNanoFood, Lda in 2020). In 2014, Miguel won the Young Scientist Award and, in 2016, he was selected as Inaugural Member of the International Academy of Food Science and Technology Early Career Scientist Section both organized by the International Union of Food Science and Technology. In 2018, he joined the list of Highly Cited Researchers from Clarivate Analytics.

Lorenzo Miguel Pastrana Castro

Lorenzo Miguel Pastrana Castro is Chair of the Research Office and Group Leader of the Food Processing and Nutrition Group at the International Iberian Nanotechnology Laboratory (INL). He is also a full professor of Food Science at the University of Vigo and is a visiting professor at the Universidade Federal Rural de Pernanbuco (Brasil) and Universita Catolica del Sacro Cuore (Italy).

In 2015, Lorenzo joined the INL as Head of the Department of Life Sciences, a division that included three research units, namely food, environment and health. At INL, his research is oriented with a multidisciplinary approach integrating methods and concepts of biotechnology, nanotechnology, and mathematical modeling. Lorenzo is currently working in three main research areas: Food structure with an emphasis on 3D printing materials, encapsulation technologies for improving functional foods and food personalization, and active and intelligent food packaging. He is author of more than 200 scientific contributions and has four licenced patents relating the development of new food products and process. He was the PI of over 30 national and European research projects and was the contractor and promoter of two food start-ups.

Lorenzo was also heavily active in promoting the connexion between academia and industry as the director of the Knowledge Transfer Office (2009–2010) at the University of Vigo, the founder of the Galician Agri-Food Technology Platform (2006) and as a current member of the scientific board of the Portugal Foods innovation cluster.

Section I

Introduction to Fat Mimetics

1.1

Why Does the Food Industry Need Fat Mimetics?

Miguel Ângelo Parente Ribeiro Cerqueira and Lorenzo Miguel Pastrana Castro*

INL–International Iberian Nanotechnology Laboratory, Braga, Portugal
** Corresponding author*

1.1.1 The Role of Lipids in Foods and Human Nutrition

The human brain has a high energy demand and thus the brain uses more energy than any other organ. It is also evident when comparing other mammals, or even primates, that humans' distinctive nutritional needs are a consequence of the evolution of humans brain size. Thus, humans require fat-dense diets to supply energy needs. Fatty acids provide more than twice the calories per gram of proteins or carbohydrates; on the other hand, higher amounts of polyunsaturated fatty acids are essential to brain development (Leonard, Snodgrass, and Robertson 2010; Lehner et al. 2021). Since this evolution happened in a fat scarcity environment, humans developed the remarkable ability to detect and metabolize high-fat foods as well as the capacity toward a strong preference for the smell, texture, and taste of lipid-rich foods (Leonard, Snodgrass, and Robertson 2010).

Nowadays, in the diets of people dwelling in developed countries, there is no scarcity of fat anymore, and lipids (fats and oils) are present in the composition of many processed foods where they play important technological, nutritional, and sensorial roles. Their nutritional value is related to essential fatty acid composition and to their role in the absorption of liposoluble vitamins such as vitamins A, D, E, and K. On the other hand, the intake of these molecules increased in recent decades. Consumption of monounsaturated and polyunsaturated lipids is related to health benefits. However, in some western populations, large intakes of saturated and trans fatty acids are responsible for serious health issues linked to non-communicable diseases (NCDs) such as obesity, diabetes, and cardiovascular diseases (Te Morenga and Montez 2017; Astrup et al. 2019). These findings lead to a demand for new solutions that can guarantee the presence of healthy lipids in foods while the foods' functional properties are maintained.

The functionalities of solid fats and liquid oils in foods are very diverse; this diversity comes from the fats' and oils' physical and chemical properties. Their capacity to form emulsions is essential for producing salad dressings, mayonnaise, gravies, and cheese sauces. Emulsions also modify the flavor and texture of many foods; for example, ice cream's creamy texture is the result of an emulsion. The melting point of solid saturated fats or liquid oils determines the temperature behaviour of products such as chocolate or ice creams, respectively. Fats and oils in the shortening process are also responsible for tenderizing baked products. Some optical properties of foods, such as the opaque or glossy appearance of milk, are also due to fat. Fats and oils dissolve and preserve flavors while playing an important role in making foods satisfying or making us feel full. Finally, fats and oils are used to transfer heat and to facilitate crust formation during frying. Those functionalities

Fat Mimetics for Food Applications, First Edition. Edited by Miguel Ângelo Parente Ribeiro Cerqueira and Lorenzo Miguel Pastrana Castro.

are difficult to mimic; thus, the replacement of fats in foods has been a challenge among researchers and the industry. In the last 20 years, there has been an increase in new solutions within the marketplace, and there have been new developments within the scientific community.

1.1.2 Current Status of Fats in the Food Industry

The use of lipids in food formulation is widespread. Lipids are present in ingredients added to foods (normally in minor amounts), or they are added directly to foods in the form of vegetable oils or animal fat. Foods such as margarine, bakery products, chocolates, and spreads are some foods where lipids are used. One of the most commonly used strategies to change lipids' properties is hydrogenation, where the liquid fats are converted into solid and semi-solid fats with the aim of achieving different functionalities. In the past, partially hydrogenated vegetable fats were used to improve the physical properties of food products, however, several studies have shown trans fatty acids (TFAs) that result from the hydrogenation process have harmful effects on human health, showing that partially hydrogenated vegetable fats are one of the primary sources of industrially-produced TFAs in foods. As a result, in 2015, the Food and Drug Administration (FDA) determined that partially hydrogenated oils, the main source of artificial trans fats in processed foods, are not generally recognized as safe (GRAS). They gave the companies until 1 January 2020 to stop distributing food containing partially hydrogenated oils (FDA 2018). Also, the European Commission (EU) limited the use of trans fats, limiting the presence of 2 grams per 100 grams of total fat; in this case, the trans fats naturally occurring in the fat of animal origin are not included (European Commission 2019).

As an alternative, the industry started using saturated fatty acids (SFA), which can be obtained by fractionating oils, such as palm oil. Palm oil is one of the most used sources of fats for foods due to its low price, texture and rheological properties, high stability against oxidation. and good storage time. However, due to the high amount of saturated fats in palm oil, high consumption of palm oil can lead to some health problems. Studies showed that the consumption of saturated fats (except for stearic acid) influences cholesterol levels in the blood and leads to cardiovascular diseases (Zhu, Bo, and Liu 2019). The World Health Organization's current guidelines for a healthy diet mentions that reducing the amount of total fat intake to less than 30% of total energy intake helps prevent unhealthy weight gain in adults; further, the risk of developing non-communicable diseases (NCDs) is lowered by a) reducing saturated fats to less than 10% of total energy intake; b) reducing trans fats to less than 1% of total energy intake; and c) replacing both saturated fats and trans fats with unsaturated and polyunsaturated fats (WHO 2020). The European Union has also promoted healthier foods by allowing food producers to have a nutrition claim on their products' labels. This nutrition claim is related to a food product's particular beneficial nutritional property that can be related to energy or to a specific compound present in the food product. There are eight nutrition claims directly related to fats, namely: 1) low fat, 2) fat-free, 3) low saturated fat, 4) saturated fat-free, 5) source of omega-3 fatty acids, 6) high omega-3 fatty acids, 7) high monounsaturated fat, and 8) high polyunsaturated fat (European Commission 2022).

In addition to the health issues related to the consumption of some fats, there is a huge concern, among the entire food industry, related to the environmental impact of oil production. According to the Food and Agriculture Organization of the United Nations (FAO), global livestock production makes up 14.5% of all anthropogenic (human-caused) emissions, i.e. 7.1 gigatonnes of carbon dioxide (CO_2) equivalent per year (Gerber 2013). In the case of lipids, one of the great examples is palm oil, where some studies revealed the considerable impact of lipid production on deforestation and the release of CO_2 (Meijaard et al., 2020). In this regard, the use of fat mimetics (that can be based on other vegetable oils) can bring some new insights and possibilities to the industry.

Therefore, the removal of hydrogenated fats and the suggested limitation on the consumption of saturated fats has been changing how the food industry looks at lipids. There is a real need for healthier, trans fatty acid-free, stable, and solid-like fats, which maintain their structure at ambient temperatures, assuring a longer shelf-life. One of the solutions are the fat replacers that can be divided into fat substitutes, fat mimetics, or fat extenders, depending on their chemical composition and physical behaviour. Fat mimetics are one of the most studied alternatives, where the use of physically structured oils (oleogels) seems to be one of the most promising routes. This is related to the way they are produced, mostly using only food ingredients and without chemically modifying the lipids, thus presenting some benefits towards the sustainability and clean label trend. But also, other fat mimetics based on proteins and carbohydrates have shown their applicability and are already used in several commercial products.

1.1.3 Food Trends and Fat Mimetics

In 2017, FAO pointed out several food trends (FAO 2018). Some of those are related to the increasing population, the dietary transition in low- and middle-income countries, diet-related diseases and sustainability. Also, consumer engagement has been seen by several stakeholders as a food trend, where a consumer's behaviour and perception have a massive impact on the future direction of the agri-food system and companies' decisions. Consumers are looking for healthy and more natural products, alternative proteins, and products that can be considered clean label and sustainable. Therefore, the industry needs new technologies and strategies to answer those demands. One of the strategies can be fat mimetics; for example, the use of fat mimetics can guarantee the reduction of unhealthy fats or/and their replacement by mono- and polyunsaturated fats and, simultaneously, offer technical functionalities similar to fats.

Plant-based foods and meat alternatives are also food trends that fat mimetics strategies can greatly impact. While proteins and hydrocolloids are already used to develop those products, incorporating oils to develop plant-based foods with a balanced lipid profile is still a challenge. Therefore, using fat mimetics could be a way to tailor the lipid profile of these kinds of products.

Lipid-based fat mimetics mainly use vegetable oils, but it is also possible to use oils obtained from microalgae, insects and even fermentation. The options play a big role in developing more sustainable fat alternatives. Fat mimetics strategies could help overcome technological challenges.

Acknowledgements

The authors acknowledge the financial support of the projects: hOLIVEcream – Healthy olive oil based creams enriched with berries for application in bakery and pastry, with the reference n.º NORTE-01-0247-FEDER-046947, co-funded by Norte 2020 – North-Regional Operational Programme under the PORTUGAL 2020 Partnership Agreement, through the European Regional Development Fund (ERDF); IceCare: Cardio-Healthy Functional Ice creams (NORTE-01-0247-FEDER-039927), co-financed by Norte 2020 and the European Union, through the European Regional Development Fund (ERDF), and BetterFat4Meat: Development of structured fat for use in meat products by replacement of animal fat (POCI-01-0247- FEDER-039718). Co-financed by Compete 2020, Portugal 2020 and the European Union, through the European Regional Development Fund (ERDF).

References

Astrup, A., Bertram, H.C.S., Bonjour, J.-P., de Groot, L.Cp., Otto, M.C., Feeny, E.L., Garg, M.L., Givens, I., Kok, F.J., Krauss, M.K., Lamarche, B., Lecerf, J-M., Legrand, P., McKinley, M., Micha, R., Michalski, M-C., Mozaffarian, D., and Soedamah-Muthu, S.S. (2019). WHO draft guidelines on dietary saturated and trans fatty acids: time for a new approach? *BMJ* 366: l4137. doi: 10.1136/bmj.l4137.

European Comission (2019). Commission Regulation (EU) 2019/649. *Official Journal of the European Union* 17–20. doi: 10.2903/sp.efsa.2018.en-1433.

European Commission (2022). *Nutrition claims*. https://food.ec.europa.eu/safety/labelling-and-nutrition/nutrition-and-health-claims/nutrition-claims_en (accessed 1 June 2022).

FAO (2018). *The Future of Food and Agriculture – Trends and Challenges*. Rome.

FDA (U.S. Food and Drug Administration) (2018). *Trans fat*. https://www.fda.gov/food/food-additives-petitions/trans-fat (accessed 5 June 2022).

Gerber, P.J., Steinfeld, H., Henderson, B., Mottet, A., Opio, C., Dijkman, J., Falcucci, A., and Tempio, G. (2013). *Tackling Climate Change Through Livestock – A Global Assessment of Emissions and Mitigation Opportunities*. Rome: Food and Agriculture Organization of the United Nations (FAO).

Lehner, A., Staub, K., Aldakak, L., Eppenberger, P., Rühli, F., Martin, R., and Bender, N.P.P. (2021). Impact of omega-3 fatty acid DHA and EPA supplementation in pregnant or breast-feeding women on cognitive performance of children: systematic review and meta-analysis. *Nutrition Reviews* 79 (5): 585–598. doi: 10.1093/nutrit/nuaa060.

Leonard, W., Snodgrass, J., and Robertson, M. (2010). Evolutionary perspectives on fat ingestion and metabolism in humans. In: *Fat Detection: Taste, Texture, and Post Ingestive Effects* (ed. J.P. Montmayeur and J. le Coutre), 3–18. CRC Press/Taylor & Francis. doi: 10.1201/9781420067767-c1.

Meijaard, E., Brooks, T.M., Carlson, K.M., Slade, E.M., Garcia-Ulloa, J., Gaveau, D.L., Ser Huay Lee, J., Santika, T., Juffe-Bignoli, D., Struebig, M., Wich, S., Ancrenaz, M., Koh, L.P., Zamira, N., Abrams, J.F., Prins, H.H., Sendashonga, C.N., Murdiyarso, D., Furumo, P.R., ... Sheil, D., Slade, E. (2020). The environmental impacts of palm oil in context. *Nature Plants* 6 (12): 1418–1426. doi: 10.1038/s41477-020-00813-w.

Te Morenga, L. and Montez, J.M. (2017). Health effects of saturated and trans-fatty acid intake in children and adolescents: Aystematic review and meta-analysis. *PloS One* 12 (11): e0186672. doi: 10.1371/journal.pone.0186672.

WHO (World Health Organization) (2020). *Healthy diet*. https://www.who.int/news-room/fact-sheets/detail/healthy-diet (accessed 1 June 2022).

Zhu, Y., Bo, Y., and Liu, Y. (2019). Dietary total fat, fatty acids intake, and risk of cardiovascular disease: A dose-response meta-analysis of cohort studies. *Lipids in Health and Disease* 18 (1). doi: 10.1186/s12944-019-1035-2.

1.2

Overview of the Structure-Property Relationship in Fat Mimetics

Reed A. Nicholson[1] *and Alejandro G. Marangoni*[2,*]

[1]*Motif FoodWorks, Inc., Boston, MA, USA*
[2]*University of Guelph, Guelph, ON, Canada*
**Corresponding author*

1.2.1 Introduction

Food materials described by the term "fat mimetics" have changed over the years (Marangoni et al. 2020). Fat mimetics as we currently understand them are materials that can also be described as structured oils. There are several reasons behind the interest in using fat mimetics in food products, including replacing and reducing trans and saturated fatty acids, respectively, improving the lipid source's sustainability, and replacing animal fats for religious reasons or dietary choice. The concept of fat mimetics pertains to structuring a liquid oil into a solid fat or semi-solid structure that mimics the physical properties and the functionality of solid fats, which are traditionally high in saturated fatty acids. Fat mimetics can be produced through various means that are broadly categorized as (1) indirect oil structuring, (2) oil structuring (oleogelation), and (3) biphasic structured systems, as shown in Figure 1.2.1. These systems were described in greater detail in a recent review by Marangoni and coworkers (2020). In short, indirect oil structuring, through either solvent exchange or colloid templating, involves dispersing a hydrophilic material in water to produce a scaffolding framework. In the case of the solvent exchange method, hydrophilic regions are exposed by heating the material, which allows for the water to be exchanged with oil using a solvent of intermediate polarity. For colloid templating, a continuous water emulsion or foam is created and is subsequently dried via freeze or spray-drying, which produces a porous scaffolding, which is then used to bind oil. Oleogels may be produced through direct oil structuring using either polymer gelators (*i.e.*, ethylcellulose or chitin) or a variety of lipid gelators as monocomponent or mixed systems. In each case, the gelator material is dispersed in oil at high temperatures, which upon cooling forms a gel network. The final category includes a series of emulsions in which aggregation or the crowded packing of the dispersed phase structures the material. Alternatively, one of or both of the continuous and dispersed phases may be structured, or the crystalline interfacial layer can be utilized to structure the system. This chapter will provide an overview of several techniques used to characterize the material properties of fat mimetics and how structural attributes influence these material properties.

Fat Mimetics for Food Applications, First Edition. Edited by Miguel Ângelo Parente Ribeiro Cerqueira and Lorenzo Miguel Pastrana Castro.

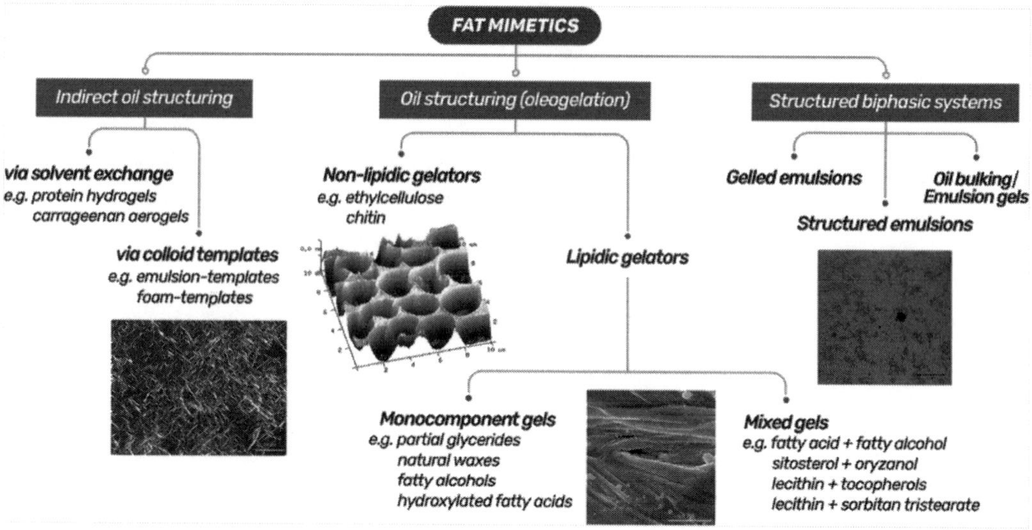

Figure 1.2.1 Schematic diagram of strategies used for fat mimetics. Microstructures of different samples are shown. Left to right: polarized light microscopy image (scale bar = 200 um) of structured oil prepared by foam template (structuring agent: hydroxypropyl methylcellulose); atomic force microscopy image of an oleogel prepared by direct dispersion of a non-lipidic gelator (structuring agent: ethylcellulose); Cryo-scanning electron microscopy image (scale bar = 6 um) of an oleogel prepared with a lipidic gelator (structuring agent: sunflower wax); and confocal microscopy image (scale bar = 25 um) of a structured biphasic system (structured emulsion) prepared by encapsulating tightly packed dispersed oil droplets in crystalline layers of monoacylglycerols (Marangoni et al. 2020, Reproduced from Royal Society of Chemistry).

1.2.2 Rheological Properties

Rheology is the study of the deformation and flow of matter. The rheological properties of a fat or a fat mimetic are determined by the features of the structural network (crystalline fat, polymeric network, etc.) within the liquid oil phase and pertain to the texture and functionality of the material. In turn, rheology is of prime importance to a lipid material's behavior during processing, performance in food systems, and the overall consumer acceptability (Marangoni et al. 2012).

In the case of viscoelastic solids such as fats and fat mimetics, the materials' plastic flow is a matter of interest; however, fats and fat mimetics are often characterized using small amplitude oscillatory shear (SAOS) rheology. For SAOS testing, a lipid material may be subjected to oscillatory shear deformations at strain inputs of $\gamma_0 \approx 0.01$–0.1%, which typically falls within the linear viscoelastic range (LVR). Within this region, the stress response of the material is not dependent on the strain or the oscillatory frequency. At these low strains, the material is not yielding, meaning that structural breakdown is not occurring. As a result, SAOS rheology characterizes the microstructural properties of the material.

Under small deformations, fats demonstrate solid-like linear behavior. The elastic (solid-like) and viscous (liquid-like) characteristics are related to the average storage modulus, G', and loss modulus, G'', respectively. A characteristic property of soft-viscoelastic solids and gels, such as fats and fat mimetics is $G' > G''$. In crystalline fat systems, G' is known to be related to both the solid fat content and the fractal dimension of the network through a power law (Nicholson and Marangoni 2019). This relationship would not hold true for most fat mimetics as they typically contain low levels of solid fat, it is important to note that an ideal fat will display a storage modulus of $1 \times 10^5 - 5 \times 10^6$ Pa (Patel, Nicholson, and Marangoni 2020).

While it is undeniable that the properties characterized by SAOS rheology are important with respect to fats and fat mimetics, the level of strain used in SAOS rheology is substantially lower than the levels these materials would be subjected to during processing and usage. For this reason, two fats may have similar SAOS rheology features but may behave very differently when used in food applications. To appropriately investigate the behavior of a fat or fat mimetic that would be exhibited during relevant deformations, large amplitude oscillatory shear (LAOS) rheology, which examines the stress response at much higher strains, should be employed. It is only at these larger deformations that the material's plastic flow can be examined. Plastic flow gives rise to spreadable material.

The stress response of a viscoelastic material outside of the LVR is non-sinusoidal and is dependent on shear strain input and oscillatory frequency. As a result, the non-linear response is insufficiently reflected by G' and G''. LAOS responses are commonly represented with Bowditch-Lissajous curves wherein the total stress (τ) within each oscillatory cycle is displayed as a function of the strain input for the elastic component or the shear rate input for the viscous component. Bowditch-Lissajous curves in the SAOS region are ellipsoidal with distortion occurring in the LAOS region. The stress response of a material can be quantified numerically using Fourier Transform or Chebyshev polynomial analysis (Ewolt, Hosoi, and McKinley 2008). Furthermore, a Pipkin diagram can be used to display Lissajous plots produced over a range of strain inputs (positioned on the x-axis) and frequencies (positioned on the y-axis). This depiction of the rheological behavior is useful for qualitatively examining the material's nonlinear response to deformation under a variety of conditions. The work of Macias-Rodriguez and Marangoni looked to characterize the differences in the rheological behavior of laminating and all-purpose baking shortenings (Macias-Rodriguez and Marangoni 2016, 2017a, 2017b, 2018; Macias-Rodriguez, Ewolt, and Marangoni 2018). It is important to keep in mind that these two types of shortening are used for very different applications because of differences in their physical properties, which is namely their responses to deformation with laminating shortening undergoing plastic flow rather than brittle fracture as is seen for all-purpose shortening. The authors demonstrated that these two materials only began to display different rheological behaviors when testing was performed in the LAOS region. Ultimately, it was shown that the local viscosities of the two materials at minimum and maximum shear rates differed markedly, with laminating shortening showing a higher local viscosity, which resulted in a more effective energy dissipation during deformation. Furthermore, ultra small-angle x-ray diffraction analysis demonstrated an association between the number of recognizable length scales in a material and its plasticity. Fewer length scales result in a brittle material while more length scales produce a material that undergoes plastic flow behavior when deformed.

While this work highlighting the importance of LAOS rheology in demonstrating the differences in responses to deformation was performed on fat systems rather than fat mimetics, the authors were able to show that various compositions (palm based, soy based, trans-free, trans-containing) were able to produce shortenings with similar LAOS rheology responses. This means that the nano and mesoscale structures generated with differing processing parameters are more important to the response to deformation of the material than the specific molecular composition. This gives hope that if the nano and mesoscale structure of the network within a fat mimetic were to be tailored in the correct fashion, there is potential for proper plastic flow behavior to be achieved in the fat mimetic. However, for this to be achieved, emphasis need to be put on experimentation with deformations outside of the linear regime that bring about and characterize either plastic flow behavior or brittle fracture. As such, LAOS rheology needs to become more commonly implemented in the study of fat mimetics.

1.2.3 Large Deformation Testing

In many cases, rheological characterization of fat mimetic systems is difficult or impractical to perform due to the length of time necessary for structuring the system in order to stabilize at various temperatures of interest. Additionally, transfer of fat mimetic systems onto the rheometer could potentially produce artifacts. This is particularly true in cases involving weak systems containing low gelator concentrations. For this reason, large deformation testing is frequently implemented to test the mechanical properties of fat mimetics with hardness being a matter of particular interest. Hardness may be tested using penetration tests with a cylindrical probe or a cone penetrometer wherein the sample is contained within a cup or tube. Testing hardness through this methodology can be very useful as it allows one to evaluate hardness at gelator concentrations close to the critical gelation concentration; it also allows one to determine the dependence of the material hardness on gelator concentration. Since sample preparation for hardness testing is performed separately from the testing instrument, the effects of gelation temperature, gelation time, post-gelation temperature treatments, and long-term monitoring can be conveniently tested. This provides valuable information regarding changes in the structural network and its effects on the large deformation properties that are often most beneficial at low gelator concentrations. For samples that have a higher degree of structure, the material may be allowed to gel within a mold, producing a free-standing material that can then be subjected to parallel plate compression testing, such as texture profile analysis (TPA). When TPA is performed, the values for several other parameters (fracturability, cohesiveness, resilience, and springiness) are obtained, which can provide useful structural insights. The hardness of fat mimetics is certainly a useful mechanical property, and much of the food application work surrounding fat mimetics performed thus far has taken these measurements into consideration. However, the overall network strength should not be considered independently from the other structural properties when designing fat mimetics for food applications. Since spreadability is of the utmost importance for these materials, one needs to be able to differentiate between plastic flow and brittle fracture. Fracturability, as measured by TPA and other compression tests, highlights a material's brittle fracture; however, since these tests require that the material in question be self-standing, this measurement can only be performed on fat mimetics containing high gelator concentrations.

Recently the use of back extrusion has been expanded to examine the spreadability of fat mimetics (Gravelle et al. 2017a). When back extrusion is performed, a force-deformation profile is collected. In the work of Gravelle et al. (2017a), the region of steady-state flow within this profile was fitted with a second order polynomial and the root mean square deviation (RMSD) was calculated, representing the "brittleness factor". Simply put, a jagged pattern (higher RMSD) corresponds with a brittle material, while a smooth pattern (lower RMSD) corresponds with a material that undergoes plastic flow (*i.e*, is more spreadable). Ethylcellulose oleogels are known to be quite brittle, as are many other fat mimetics, such as fatty acid + fatty alcohol mixtures. An oleogel prepared with 8% (w/w) stearyl alcohol-stearic acid at a ratio of 7:3 had a brittleness factor of 2.38 ± 0.08. When ethylcellulose was incorporated at 4% (w/w), the brittleness factor dropped to 0.25 ± 0.07, while also reducing the hardness of the material. The brittleness factor increased to 1.79 ± 0.50 when the ethylcellulose concentration was increased to 6% (w/w). Still, it was lower than that of the stearyl alcohol-stearic acid system in the absence of ethylcellulose, even though this material had a hardness that was much greater than that of the 8% (w/w) stearyl alcohol-stearic acid oleogel. Furthermore, the ability of the stearyl alcohol-stearic acid to plasticize ethylcellulose was confirmed through an examination of the thermal properties of the ethylcellulose. Addition of stearyl alcohol-stearic acid reduced the onset glass transition temperature and caused the glass transition to occur over a broader temperature range. This research not only introduced an efficient method of quantifying differences in the brittleness of fat mimetics, but it also demonstrated the potential of using hybrid gelator systems for improving spreadability. While plasticity is attributed to weakened intermolecular interactions and an increased free volume of the system (Chee 1985; Gravelle et al. 2017a), the work of Macias-Rodrigues and Marangoni that was highlighted earlier in this chapter demonstrated that materials underwent plastic flow under deformation when more length scales were present. This would be the case with hybrid gelator systems, and it is the case with triacylglycerol systems containing triacylglycerol species melting at an array of temperatures. Furthermore, if the incorporation of surfactant molecules (fatty acids and fatty alcohols) into ethylcellulose oleogels can improve the plasticity of these materials, then there would be a reason to believe that it may be beneficial to incorperate certain other surfactant molecules. This may also be the case if certain lipid gelator molecules were included in the fat mimetic systems produced through indirect means of oil structuring, so long as they interact with the structural scaffolding in a positive manner. Even in the event that no interaction is occurring, introducing structuring materials at a different length scale may still be beneficial.

Back extrusion was used to demonstrate plastic flow more recently by Nicholson and Marangoni (Nicholson and Marangoni 2020, 2021). In this work, back extrusion was performed on margarines prepared with glycerolysis structured oil systems along with a commercially available margarine product and a commercially available butter, each at their intended use temperatures. Although the force-deformation profiles were not fitted with second order polynomials and a RMSD was not calculated, it was qualitatively obvious that the jaggedness of the margarines prepared with glycerolysis structured oil systems was very low, much like that of the commercial butter and margarine products. This thereby demonstrated that the margarines prepared with glycerolysis structured oil systems underwent plastic flow rather than brittle fracture. Furthermore, a glycerolysis structured oil system is another example of a lipid system containing structuring material of varying length scales. The presence of mixed crystals of monoacylglycerols, diacylglycerols, and triacylglycerols may be what to what the plastic nature is attributed.

1.2.4 Microstructure

With triacylglycerol systems, the birefringence of the fat crystals makes light microscopy and polarized light microscopy useful tools for investigating the microstructure of the material. Many lipidic gelators also share this characteristic. As such, the microstructures of fat mimetics prepared from a variety of materials have been examined in this fashion. These include waxes (Blake et al. 2014; Blake and Marangoni 2015a, 2015b, 2015c), hydroxylated fatty alcohols (Rogers, Wright, and Marangoni 2008), fatty acids + fatty alcohols (Gandolfo, Bot, and Flöter 2004; Blach et al. 2016; Gravelle et al. 2017a, 2017b), monoacylglycerols (Ojijo et al. 2004; Kesselman and Shimoni 2007), monoacylglycerol + diacylglycerol mixtures (Nicholson and Marangoni 2020, 2021), and several others. Minimal sample preparation means that the size and morphology of crystal and polycrystal aggregates can be viewed quickly and easily, which allows for the investigation of fat mimetics prepared with a range of gelator concentrations and under differing processing conditions to be performed efficiently.

As pointed out in a recent review by Flöter et al. (2021), structural hypotheses need to be supported with other experimental techniques (Flöter et al. 2021). While the need for alternative experimental methods to support hypotheses is true in all areas of science for a variety of reasons, this is particularly true with light microscopy. Among several limitations of light microscopy, the simple fact that light microscopy only provides a two-dimensional view of the crystalline material means that the structural make up cannot be fully discerned through this method alone. In addition, crystallization conditions and sample preparation may lead to artifacts. Crystallizing samples in bulk before transferring them to the microscope slide may affect the crystal morphology, while crystallizing the material between the slide and coverslip causes the crystals to orient themselves with their edges being aligned with the glass surfaces. As a result of these limitations, many structural differences will remain uncharacterized without a three-dimensional view. Given this, cryogenic scanning electron microscopy (cryo-SEM) is frequently used to characterize the microstructure of fat mimetic systems. Blake and Marangoni (2015a) used cryo-SEM to demonstrate that rice bran wax, sunflower wax, and candelilla wax crystallize in a platelet morphology (Blake and Marangoni 2015a). This differed from their previous observation of needle-like structures for rice bran wax and sunflower wax and of spherulitic structures for candelilla wax when viewed under light microscopy (Blake et al. 2014). These structural differences were observerable due to both the crystallization conditions to which samples were subjected for viewing under light microscopy and with the fact that only the two-dimensional view of the crystals were obtained. To reduce the contact area with the hydrophilic glass surface, the hydrophobic wax crystals align themselves so that the edge of the platelets is oriented toward the slide and coverslip. This same phenomenon was observed for stearyl alcohol + stearic acid mixtures (Blach et al. 2016). Cryo-SEM again elucidated that the needle-like crystals observed under light microscopy were platelets observed two-dimensionally along their edge. These works demonstrate the effectiveness of cryo-SEM for thoroughly examining crystal morphology in lipidic gelator fat mimetic systems. Further, cryo-SEM can also provide insights into the structuring material's surface characteristics. Additionally, the use of cryo-SEM goes beyond lipid gelator systems. Structural scaffolding of fat mimetics produced from the non-lipidic gelators ethylcellulose and chitin have been viewed using cryo-SEM (Zetzl, Marangoni, and Barbut 2012; Zetzl et al. 2014; Nikiforidis and Scholten 2015). The porous structures of ethylcellulose viewed with cryo-SEM were confirmed by atomic force microscopy (Zetzl et al. 2014). Meanwhile, the structural network of chitin oleogels were further visualized with confocal laser scanning microscopy (CLSM) (Nikiforidis and Scholten 2015). CLSM is a useful microscopy tool as it has a great depth resolution, which means there is limited sample disruption. Reconstruction of two-dimensional sections' images provides a three-dimensional structural view. Selective phase staining is useful in multicomponent systems, such as in food materials containing fat mimetics or fat mimetics containing a lipid phase and a protein phase. Confocal Raman microscopy is useful for investigating the composition and structure of fat spreads and dispersions. Confocal Raman microscopy may also be especially useful for characterizing the structural properties of structured biphasic systems if multivariate curve resolution were used to decompose the Raman spectra of these fat mimetic systems to observe solid fat and liquid oil distribution (Van Dalen et al. 2017; Martens et al. 2018).

The influence of processing conditions, such as shear fields and temperature gradients, on the crystalline network of triacylglycerol networks is well reported. These findings were recently documented in a review of structure in fat and oil systems (Nicholson and Marangoni 2019). At higher cooling rates or lower crystallization temperatures, which produces smaller crystals, nucleation is favoured over crystal growth. Shear increases heat and mass transfer along with the frequency of collisions between molecules. Shear induced crystal breakage can also change the size of microcrystalline elements; fragments may form nuclei for growth of future crystals. Shear and higher cooling rates also promote the formation of mixed crystals. The effects of shear on the crystal network properties are dependent on the shear rates applied (Maleky and Marangoni 2011; Maleky, Smith, and Marangoni 2011; Acevedo, Block, and Marangoni 2012; Acevedo and Marangoni 2014). These effects on network structure would also be expected with fat mimetics produced using lipidic gelators. While light microscopy certainly has limitations, it remains a useful tool for investigating the structural network features of lipidic gelator systems. Based on light microscopy images, the box-counting fractal dimension can be determined, thereby quantifying the homogeneity of mass distribution within the crystal network. For two-dimensional images of a crystal network, the closer the box-counting fractal dimension is to the maximum obtainable value of 2, the more uniformly distributed is the mass within the network. The percentage or fraction of pixels indicative of the solid crystalline material may be subtracted from the overall pixels of a micrograph to determine the pore area fraction of a network. A greater amount of solid crystalline materials are indicated by a lower pore area fraction. Blake and Marangoni (2015b) showed that crystallizing rice bran wax, sunflower wax, and candelilla wax in peanut oil at faster cooling rates (5°C/min vs 1°C/min) resulted in smaller crystals (Blake and Marangoni 2015b). When oleogels were prepared with a wax concentration of less than 5%, the higher cooling rate also resulted in an increased box-counting fractal dimension and a decreased pore area fraction. Furthermore, under slow cooling (1.5°C/min), shear rates of 50 s^{-1} resulted in smaller mean crystal lengths for rice bran wax and sunflower wax systems (Blake and Marangoni 2015c). In the case of the rice bran wax crystal networks, the application of shear increased the box-counting fractal dimension and decreased the pore area fraction. Interestingly, shear increased the crystal lengths within the candelilla wax system. In glycerolysis-structured oil systems, shear has the potential to alter crystal size and network properties. When a cottonseed glycerolysis product was crystallized under shear on a stainless-steel tempering table, the resulting crystal network was continuous and contained very small crystals (Nicholson et al. 2022). In triacylglycerol systems, altering crystallization temperature and the application of shear are known to improve the texture and plasticity, leading to improved functionality and sensory attributes (Nicholson and Marangoni 2019). These improvements are due, in part, to the induced changes in the microstructure of the crystal network. The referenced studies on wax and glycerolysis structured systems highlight the ability of these common lipid processing practices to alter the network microstructure of fat mimetics prepared with lipidic gelators. Usage of these microstructure altering techniques creates the potential for improving fat mimetic systems' functionality.

For crystalline triacylglycerol systems, the storage modulus (G′) is related to crystalline mass distribution in space through the power law: $G' \propto (SFC/100)^m$, where SFC is the solid fat content (the amount of crystalline mass) and m is dependent on the fractal dimension and on whether the system is in the strong-link regime or weak-link regime (Nicholson and Marangoni 2019). Similarly, the relationship between the hardness and microstructure of stearyl alcohol + stearic acid mixtures was described by Blach et al. (2016) with a power-law function (Blach et al. 2016). Blach and coworkers showed that while systems structured with pure stearic acid and pure stearyl alcohol could not be described in this manner, the hardness of the systems that were structured by mixtures of the two surfactants was due to an increase in the crystalline mass along with this mas' distribution within the network. This relationship was represented by the equation: $H = \gamma(SFC/100)^n$, where H represented the hardness at 10°C measured by back extrusion testing, where SFC represented the solid fat content, and where n represented the scaling exponent corresponding with the crystalline mass distribution within space. The pre-exponential factor γ was related to the structuring material's properties and to the size of the structural units. A similar model was also successfully used to describe mineral wax crystal networks as closed-cell cellular solids and determined the Young's modulus for the links between spherical oil filled cells (Miyazaki and Marangoni 2014). The size of the cells was shown to be responsible for differences in the paraffin and polyethylene wax-oil mixtures' mechanical strength. Furthermore, Hwang et al. (2011) demonstrated dependence of the crystal size and firmness of sunflower wax oleogels on the cooling rate (Hwang et al. 2011). Higher cooling rates decreased the crystal size, leading to firmer materials, much like what is often expected with triacylglycerol systems.

1.2.5 Oil Binding Capacity

Fat mimetics are composed of a liquid oil entrapped within a scaffolding of gelator material, whether it be a crystalline lipidic material or a non-lipidic material or an emulsion system in which a structured system component holds the oil phase within; for the fat mimetic to behave like a solid fat, the liquid oil must remain in place. The oil binding capacity (OBC) is therefore a critical property of fat mimetics with implications on storage, processing, and food material quality. The OBC of fat mimetics are typically examined with centrifugation testing or by determining the weight gained by a segment of filter paper when the fat mimetic is placed on it, but it may also be examined with compression testing (with or without the sample again being placed on filter paper). Flöter et al. (2021) describes errors associated with some of these OBC testing methods. Among them, OBC determined through centrifugation testing is a function of the density differences between the liquid oil and the scaffolding. In addition, the filter paper methods underestimate OBC due to solid material adhering to the filter paper. While one would hope OBC testing methods would yield comparable results, Flöter et al. (2021) established that this was not the case as the centrifugation method showed results that differed from the filter paper method three different gelator systems were compared. Comparability between these two methods could be improved by including a mesh underneath the fat mimetic in the container during the centrifugation procedure. With this recent highlighting of differences between OBC monitoring methods, it is certainly important to take care when choosing the appropriate measurement method.

Blake et al. (2014) measured the OBC of rice bran wax, sunflower wax, candelilla wax, and carnauba wax crystal networks at 40°C with the aid of filter paper-lined funnels (Blake et al. 2014). Exuded oil was collected at various time intervals over the course of 24 h. This allowed for the oil binding to be classified as either short-term oil binding or long-term oil binding, which helped differentiate between the portions of the oil that were unbound or strongly bound. The oil binding properties were then related to various microstructural elements of the wax crystal networks. As would be expected, an increase in the amount of crystalline material, and similarly, a decrease in the porosity, resulted in a lower level of fast-leaking oil, lower rates of oil loss, and a greater overall OBC. This work also showed that as the size of the crystals in the network became smaller, the amount of oil that was strongly bound to the crystals increased, resulting in a lower oil loss throughout the experimental procedure as well as lower rates of both fast and slow oil loss. This was attributed to the increased crystal surface area, which allowed for more oil adsorption. Furthermore, the box-counting fractal dimension was also related to oil loss parameters. A more even distribution of crystalline material within the network resulted in lower rates of oil loss and higher overall OBC. OBC dependence on the crystal network's box-counting fractal dimension was also shown for glycerolysis structured cottonseed oil systems wherein differing glycerolysis reaction conditions resulted in variations within the species of monoacylglycerol and diacylglycerol species produced (Nicholson and Marangoni 2020). The network that displayed the highest OBC also possessed the highest box-counting fractal dimension, even when the same amount of crystalline material was present. Figure 1.2.2 shows the microstructures observed under polarized light microscopy for glycerolysis structured cottonseed oil systems following 24 h glycerolysis

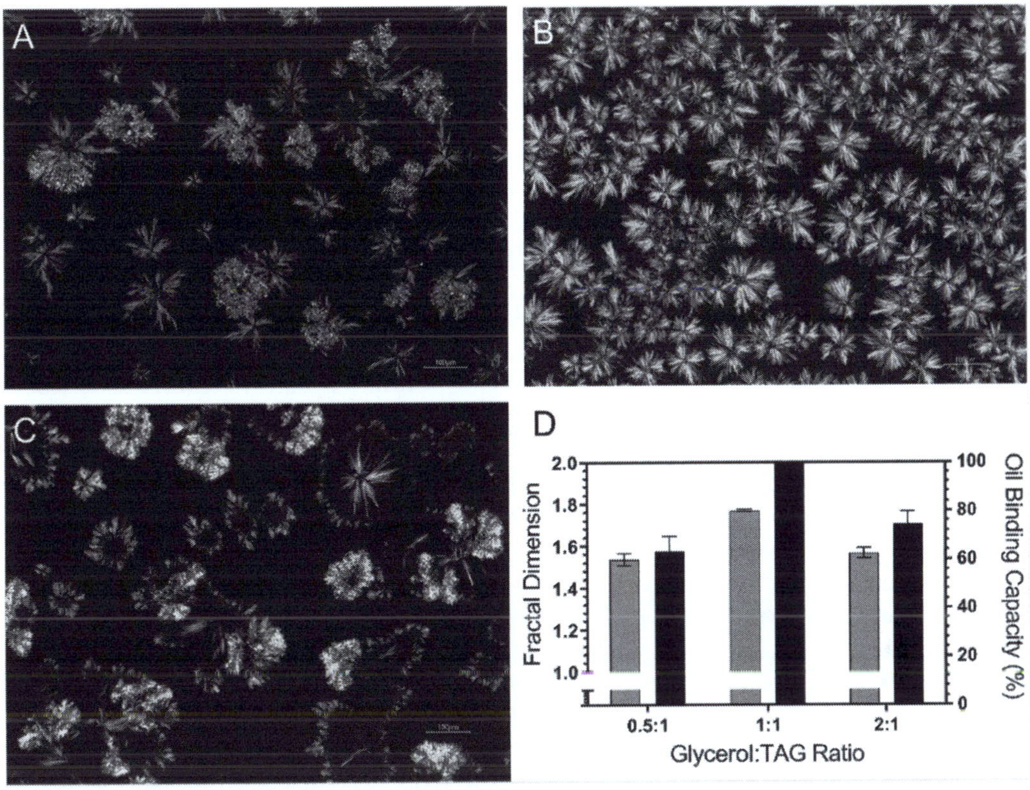

Figure 1.2.2 Polarized light microscopy images of glycerolysis structured cottonseed oil systems reacted for 24 h in the presence of (A) 0.5:1, (B) 1:1, and (C) 2:1 glycerol:triacylglycerol (TAG) (mol:mol) ratios. (D) Box-counting fractal dimension (grey bars) and oil binding capacity (%) (black bars) for each of the networks.

reactions in the presence of 0.5:1 (Figure 1.2.2A), 1:1 (Figure 1.2.2B), and 2:1 (Figure 1.2.2C) glycerol:triacylglycerol (mol:mol) ratios. The three respective systems possessed box-counting fractal dimensions of 1.54, 1.77, and 1.57 and showed OBCs of 63.4%, 100%, and 74.5% under the subjected centrifugation conditions.

When the effect of cooling rate on the OBC of rice bran wax, sunflower wax and candelilla wax networks in peanut oil was investigated, it was determined that a faster cooling rate (5°C/min vs 1°C/min) increased the OBC for systems containing wax concentrations of less than 5% (Blake and Marangoni 2015b). The increased OBC was the result of the smaller crystals formed, of the reduced pore area fraction, and of the increased box-counting fractal dimension, representing a more homogenous distribution of crystalline mass. When shear ($50\ s^{-1}$) was applied during slow cooling (1.5°C/min) of rice bran wax in peanut oil, the microstructural changes that were induced (*i.e.*, smaller mean crystal lengths, increased box-counting fractal dimension, decreased pore area fraction) increased the OBC of this network. Unfortunately, applying shear in this instance decreased the OBC of the sunflower wax and candelilla wax networks because the surface area for oil absorption was reduced. Nonetheless, this work highlights that common techniques used to engineer the microstructural and textural properties of triacylglycerol systems may also be successfully used in some lipidic gelator fat mimetic systems to improve the OBC. Furthermore, ethylcellulose oleogel networks are notorious for possessing a low OBC, especially following deformation. This could be improved by including lipidic gelator molecules, which may offer a solution to network brittleness issues common to these fat mimetic materials. Including lipidic gelator molecules wherein the network microstructure can be manipulated through processing techniques could prove useful in fat mimetics prepared through indirect oil structuring.

1.2.6 Conclusions and Next Steps

Each of the characteristics discussed in this chapter are important when evaluating a fat mimetic and its potential in food applications from a stability and bulk textural perspective. Thus far, the use of tribology has been more or less absent from fat mimetics studies and understandably so, as food system tribology has only been a mainstream research topic for the past decade or so. In addition, the early years of fat mimetic's research consisted mainly of determining whether or not a material could actually gel and be useful as a fat mimetic. Only recently have researchers studying fat mimetics begun to effectively tailor the textural and rheological properties for food applications. As the fat mimetics field moves forward, it is important to characterize the friction and wear properties of the fat mimetics and the food systems containing them in order to gain a better understanding of how these materials affect the food products' mouthfeel properties. It is necessary that this research compliments textural and rheological characterizations to understand how a fat mimetic system affects the entire eating experience. Each of the various materials used in fat mimetics construction has molecular structures and molecular moieties that differ from those in traditional triacylglycerol systems. While traditional triacylglycerol systems do contain minor components, such as monoacylglycerols, diacylglycerols, phospholipids, and sterols, these molecules are found in much lower quantities than in fat mimetic systems. It is likely that these molecules will interact with saliva and oral surfaces differently than triacylglycerols do, thereby influencing the lubrication and mouthcoating properties of the lipid phase. The lubricating properties of lipids both reduce the friction coefficient in the boundary lubrication regime and decrease the entrainment speed at which the system enters the hydrodynamic lubrication regime (Nicholson et al. 2022). Changes to a Stribeck curve in the presence of lipids are depicted in Figure 1.2.3.

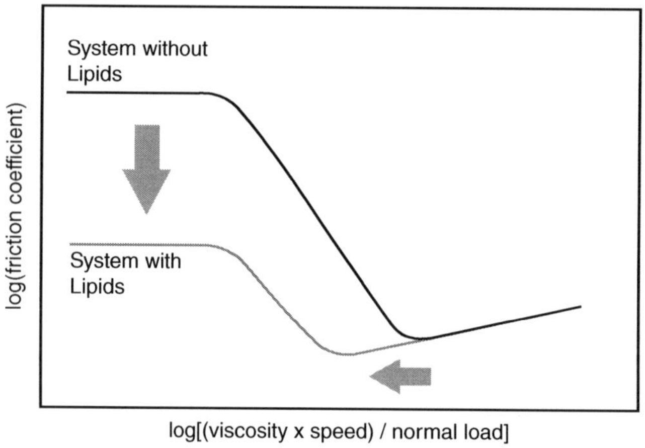

Figure 1.2.3 Illustration of a Stribeck curve demonstrating the frictional properties of a system without lipids (black line) and a system with lipids (grey line). Arrows have been included to highlight the shifts in the Stribeck curve due to the presence of lipids. Note: Stribeck curves can be broken into three lubrication regimes: (1) boundary lubrication, (2) mixed lubrication, and (3) hydrodynamic lubrication.

Presently, the effects that many of the molecules providing structure in fat mimetics have on lubrication behaviors are uncertain. Furthermore, sensory testing frequently involves panelists consuming only a small quantity of a food sample; this is especially the case with experimental food products that are still a long way from entering the market. These types of sensory tests are likely not enough for differences in the lubrication and mouthcoating properties to be fully grasped. For a fat mimetic to be effectively used as a replacement for crystalline fat systems, it is important that all aspects of the fat mimetic system be non-detrimental to all phases of oral processing. As such, tribological differences between fat mimetic systems and triacylglycerol systems need to be characterized.

References

Acevedo, N.C., Block, J.M., and Marangoni, A.G. (2012). Unsaturated emulsifier-mediated modification of the mechanical strength and oil binding capacity of a model edible fat crystallized under shear. *Langmuir* 28 (46): 16207–16217.

Acevedo, N.C. and Marangoni, A.G. (2014). Functionalization of non-interesterified mixtures of fully hydrogenated fats using shear processing. *Food and Bioprocess Technology* 7: 575–587.

Blach, C., Gravelle, A.J., Peyronel, F., Weiss, J., Barbut, S., and Marangoni, A.G. (2016). Revisiting the crystallization behavior of stearyl alcohol: Stearic acid (SO: SA) mixtures in edible oil. *RSC Advances* 6: 81151–81163.

Blake, A.I., Co, E.D., and Marangoni, A.G. (2014). Structure and physical properties of plant wax crystal networks and their relationship to oil binding capacity. *Journal of the American Oil Chemists' Society* 91: 885–903.

Blake, A.I. and Marangoni, A.G. (2015a). Plant wax crystals display platelet-like morphology. *Food Structure* 3: 30–34.

Blake, A.I. and Marangoni, A.G. (2015b). The use of cooling rate to engineer microstructure and oil binding capacity of wax crystal networks. *Food Biophysics* 10: 456–465.

Blake, A.I. and Marangoni, A.G. (2015c). The effect of shear on the microstructure and oil binding capacity of wax crystal networks. *Food Biophysics* 10: 403–415.

Chee, K.K. (1985). Prediction of glass transition temperatures of plasticized polymers. *European Polymer Journal* 21 (1): 29–31.

Ewolt, R.H., Hosoi, A.E., and McKinley, G.H. (2008). New measures for characterizing nonlinear viscoelasticity in large amplitude oscillatory shear. *Journal of Rheology* 52: 1427–1458.

Flöter, E., Wettlaufer, T., Conty, V., and Scharfe, M. (2021). Oleogel-their applicability and methods of characterization. *Molecules* 26: 1673.

Gandolfo, F.G., Bot, A., and Flöter, E. (2004). Structuring of edible oils by long-chain FA, fatty alcohols and their mixtures. *Journal of the American Oil Chemists' Society* 81: 1–6.

Gravelle, A.J., Blach, C., Weiss, J., Barbut, S., and Marangoni, A.G., Barbut, S., Marangoni, A.G. (2017a). Structure and properties of an ethylcellulose and stearyl alcohol/stearic acid (EC/SO:SA) hybrid oleogelator system. *European Journal of Lipid Science and Technology* 119. 1700069.

Gravelle, A.J., Davidovich-Pinhas, M., Barbut, S., and Marangoni, A.G. (2017b). Influencing the crystallization behavior of binary mixtures of stearyl alcohol and stearic acid (SOSA) using ethylcellulose. *Food Research International (Ottawa, Ont.)* 91: 1–10.

Hwang, H.S., Kim, S., Singh, M., Winkler-Moser, J.K., and Liu, S.X. (2011). Organogel formation of soybean oil with waxes. *Journal of the American Oil Chemists' Society* 89 (4): 639–647.

Kesselman, E. and Shimoni, E. (2007). Imaging of oil/monoglyceride networks by polarizing near-field scanning optical microscopy. *Food Biophysics* 2: 117–123.

Macias-Rodriguez, B. and Marangoni, A.G. (2016). Rheological characterization of triglyceride shortenings. *Rheologica Acta* 55: 767–779.

Macias-Rodriguez, B.A., Ewolt, R.H., and Marangoni, A.G. (2018). Nonlinear viscoelasticity of fat crystal networks. *Rheologica Acta* 57: 251–266.

Macias-Rodriguez, B.A. and Marangoni, A.G. (2017a). Bakery shortenings: structure-mechanical function relations. *Applied Rheology* 27: 33410–33417.

Macias-Rodriguez, B.A. and Marangoni, A.G. (2017b). Rheological differences between commercial fat types quantified using non-linear rheological techniques. *Journal of Food Engineering* 212: 87–96.

Macias-Rodriguez, B.A. and Marangoni, A.G. (2018). Linear and nonlinear rheological behavior of fat crystal networks. *Critical Reviews in Food Science and Nutrition* 58: 2398–2415.

Maleky, F. and Marangoni, A.G. (2011). Thermal and mechanical properties of cocoa butter crystallized under an external laminar shear field. *Crystal Growth & Design* 11 (6): 2429–2437.

Maleky, F., Smith, A.K., and Marangoni, A.G. (2011). Laminar shear effects on crystalline alignments and nanostructure of a triacylglycerol crystal network. *Crystal Growth & Design* 11 (6): 2335–2345.

Marangoni, A.G., Acevedo, N., Maleky, F., Co, E., Peyronel, F., Mazzanti, G., Quinn, B., and Pink, D. (2012). Structure and functionality of edible fats. *Soft Matter* 8: 1275–1300.

Marangoni, A.G., van Duynhoven, J.P.M., Acevedo, N., Nicholson, R.A., and Patel, A.R. (2020). Advances in our understanding of the structure and functionality of edible fats and fat mimetics. *Soft Matter* 16: 289–306.

Martens, K.J.A., van Dalen, G., Heussen, P.C.M., Voda, A., Nikolaeva, T., and van Duynhoven, J.P.M. (2018). Quantitative structural analysis of fat crystal networks by means of raman confocal imaging. *Journal of the American Oil Chemists' Society* 95: 259–265.

Miyazaki, Y. and Marangoni, A.G. (2014). Structural-mechanical model of wax crystal networks-a mesoscale cellular solid approach. *Materials Research Express* 1: 025101.

Nicholson, R.A., Gravelle, A.J., Baier, S.K., and Marangoni, A.G. (2022). Oral processing and consumer perception: lipid-based systems. In: *Oral Processing and Consumer Perception: Biophysics, Food Microstructures and Health* (ed. B. Wolf, S. Bakalis, and J. Chen), 187–213. Cambridge, UK: Royal Society of Chemistry.

Nicholson, R.A. and Marangoni, A.G. (2019). Food structure development in oil and fat systems. In: *Handbook of Food Structure Development* (ed. F. Spyropoulos, A. Lazidis, and I. Norton), 115–133. Cambridge, UK: Royal Society of Chemistry.

Nicholson, R.A. and Marangoni, A.G. (2020). Enzymatic glycerolysis converts vegetable oils into structural fats with the potential to replace palm oil in food products. *Nature Food* 1: 684–692.

Nicholson, R.A. and Marangoni, A.G. (2021). Lipase-catalysed glycerolysis extended to the conversion of a variety of edible oils into structural fats. *Current Research in Food Science* 4: 163–174.

Nikiforidis, C.V. and Scholten, E. (2015). Polymer oleogelation with chitin and chitin nanocrystals. *RSC Advances* 5: 37789–37799.

Ojijo, N.K.O., Kesselman, E., Shuster, V., Eichler, S., Eger, S., Neeman, I., and Shimoni, E. (2004). Changes in microstructural, thermal, and rheological properties of olive oil/monoglyceride networks during storage. *Food Research International (Ottawa, Ont.)* 37: 385–393.

Patel, A.R., Nicholson, R.A., and Marangoni, A.G. (2020). Applications of fat mimetics for the replacement of saturated and hydrogenated fat in food products. *Current Opinion in Food Science* 33: 61–68.

Rogers, M.A., Wright, A.J., and Marangoni, A.G. (2008). Crystalline stability of self-assembled fibrillar networks of 12-hydroxystearic acid in edible oils. *Food Research International (Ottawa, Ont.)* 41: 1026–1034.

Van Dalen, G., van Velzen, E.J.J., Heussen, P.C.M., Sovago, M., van Malssen, K.F., and van Duynhoven, J.P.M. (2017). Raman hyperspectral imaging and analysis of fat spreads. *Journal of Raman Spectroscopy* 48: 1075–1084.

Zetzl, A.K., Gravelle, A.J., Kurylowicz, M., Dutcher, J., Barbut, S., and Marangoni, A.G. (2014). Microstructure of ethylcellulose oleogels and its relationship to mechanical properties. *Food Structure* 2: 27–40.

Zetzl, A.K., Marangoni, A.G., and Barbut, S. (2012). Mechanical properties of ethylcellulose oleogels and their potential for saturated fat reduction in frankfurters. *Food & Function* 3: 327–337.

Section II

Materials and Methods Used for the Production of Fat Mimetics

2.1

Natural Wax-Based Oleogels for Food Application

Bikash K. Pradhan[1], *Satish Saigiri*[2], *Deepti Bharti*[1], *Doman Kim*[3,4], *and Kunal Pal*[1,*]

[1] *Department of Biotechnology and Medical Engineering National Institute of Technology, Rourkela, India*
[2] *Agro-Nanotechnology and Advanced Materials Research Center, Institute of Postharvest and Food Sciences, Agricultural Research Organization, The Volcani Center, Rishon LeZion, Israel*
[3] *Graduate School of International Agricultural Technology, Seoul National University, Pyeongchang-gun, Gangwon-do, South Korea*
[4] *Institute of Food Industrialization, Institutes of Green Bioscience and Technology, Seoul National University, Pyeongchang-gun, Gangwon-do, South Korea*
[*] *Corresponding author*

2.1.1 Introduction

Presently, much of the world's population eats processed food products; this increase in consumption is owed to significant advances in food processing technologies within the last few decades. These advances make it more affordable to distribute food products to a large population. Many traditional food items are being replaced with processed foods even though conventional food products are healthier than processed foods (Wertheim-Heck and Raneri 2019). A reason for this replacement is due to saturated fats' abilities to tailor texture, mouthfeel, and other organoleptic properties (Temkov and Mureşan 2021). Substituting saturated fats is challenging even though saturated fats are the reason processed foods are considered unhealthy. Regular saturated fat consumption can lead to heart diseases, to various inflammatory reactions, and to increased oxidative stress (Mirsafaei et al. 2020; Juul, Vaidean, and Parekh 2021). However, with the advances in communication systems, especially the widespread availability of the internet, people have become more health-conscious, and a significant portion of the population has made drastic changes to their dietary habits. Hence, many are looking for healthy food products devoid of saturated fats.

In the past, hydrogenated vegetable oils were used to replace saturated animal fats. Hydrogenated vegetable oils allow food manufacturers to develop food products with suitable textures and tastes (Issaoui and Delgado 2019). Though hydrogenated, vegetable oils were initially promoted as healthy alternatives to animal-derived saturated fats; however, these artificially produced fats were later deemed unhealthy (Sadiqa and Noemana 2019). This shift was related to the incomplete hydrogenation of the fats during the hydrogenation process. Because of this process, the remaining double bonds were capable of undergoing isomerization. Isomerization generated trans fats. Trans fats were found to not be suitable for human consumption because they harmed the human body. Regular trans fats consumption significantly increases serum low-density lipoprotein (LDL) cholesterol levels while reducing high-density lipoprotein (HDL) cholesterol levels (Shah and Thadani 2019). Accordingly, consuming trans fats has been banned in several countries.

As mentioned above, the food processing industries need solid fats for improving the texture and the mouthfeel of food products. Hence, these industries are looking for an alternative to hydrogenated and animal-derived saturated fats. For this reason, structuring the vegetable oils by oleogelation has received a great deal of attention from researchers. Oleogelation refers to converting oils into semi-solid formulations using a network of solid components (Bharti et al. 2021). The solid component in the oleogels may be categorized into natural waxes (e.g., mango butter, cocoa butter,

Fat Mimetics for Food Applications, First Edition. Edited by Miguel Ângelo Parente Ribeiro Cerqueira and Lorenzo Miguel Pastrana Castro.

kokum butter, beeswax, candelilla wax, and rice bran wax), polysaccharides (e.g., ethylcellulose), and fluid-filled structures made up of emulsifiers (e.g., Tween 80 and Span 80). Among these solid components, someone has primarily explored natural waxes for oleogelation. The oleogels developed with the natural waxes are self-standing viscoelastic semi-solid formulations, which are stable for a longer duration and hence have a longer shelf life (Sahu et al. 2021). It is possible to tailor the physical and textural properties of these oleogels quite easily. Further, because of the thermos-reversible property of the gelator, the oleogels of natural waxes are thermo-reversible. Oleogels help retain polyunsaturated fatty acids (PUFA) beneficial to human health.

At first, the mechanism of oleogelation by natural waxes will be briefly discussed. Thereafter, a scientometric analysis of the Web of Science database will be conducted. This analysis allows us to identify the most commonly explored natural waxes for developing oleogels. The chemical composition and the properties of the waxes will then be discussed. Lastly, applications for natural wax-derived oleogels in the food industry will be reviewed.

2.1.2 Mechanism of Oleogelation

Natural waxes have the potential to form oleogels at a low concentration, usually ranging from 1–10% (w/w) (Manzoor et al. 2022). Waxes are solid at room temperature, display hydrophobic tendencies; they typically have a high melting point. The formation of oleogels involves establishing a three-dimensional network from the wax molecules wherein the oil gets trapped and creates a structured arrangement. The stages of structuring are similar to the structuring stages of traditional solid fats, which have been used for many years. The similarity lies in the stages of fast nucleation, growth, aggregation, and complex network formation (Patel and Dewettinck 2016); yet there is a difference in the crystal morphologies of solid fats.

In many cases, a lower number of waxes is required to form a three-dimensional fat network; this can cause an efficient oil binding in oleogels. Oleogels can be formulated by dispersing waxes in vegetable oil (Figure 2.1.1) (Manzoor et al. 2022). The minimum amount of wax required to form oleogels in the vegetable oil is the critical gelation concentration (CGC). The dispersion occurs at a temperature higher than the melting point of the wax (T_m), followed by cooling to a lower temperature. The initiation of fat crystal growth promotes the formation of a highly crystalline phase that interacts to form a fat network. Waxes primarily contain linear molecular units; therefore, crystal growth occurs either in 1-dimension or 2-dimensions, which gives an appearance of a needle/fiber or of plate-like fat structures, respectively (Patel 2015). Fibrous structure formation initiates from a single nucleation point and is followed by the fat crystal's unidirectional growth. Crystal growth in lateral and perpendicular directions often yields plate-like morphology in the case of monoglycerides (Sein, Verheij, and Agterof 2002). The process also involves fibers branching at specific points. The phenomenon of crystallographic mismatch branching (CMB) results in the self-assembled fibrillar network of waxes (Cabrera, Rojas, and Moreno 2020). CMB involves the nucleation of a daughter fiber at the parent fiber, giving rise to a fork-like arrangement. The self-assembly of gelator molecules inside the vegetable oil is directed by hydrogen bonding, π–π stacking, van der Waals forces, and hydrophobic forces. The cooling rate is again a crucial aspect that determines the morphology and arrangement of the oleogels' crystalline network. A faster cooling rate does not provide sufficient time for the nuclei of wax molecules to grow from a metastable to a stable form, which may produce smaller fat crystals. The formation of oleogels can be confirmed via the inverted tube method, which checks the absence of oil flow under gravity.

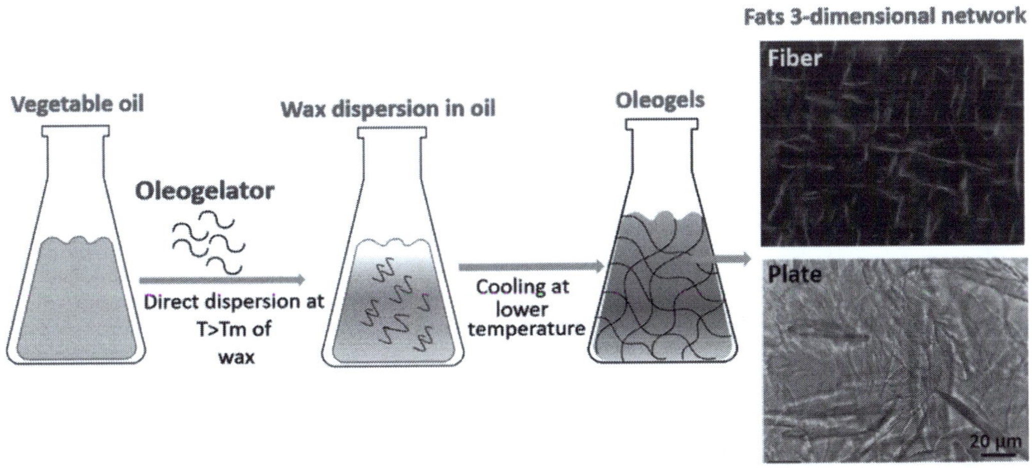

Figure 2.1.1 Schematics of oleogel formation using direct dispersion of wax molecules in vegetable oil (microstructure depicting the plate morphology has been obtained from (Uvanesh et al. 2016, from John Wiley & Sons) under the license no. 5320810894677).

Chemically, waxes are considered heterogeneous materials, comprising fatty alcohols, fatty acids, and hydrocarbon chains. The proportions of the resulting constituents differ among natural waxes. These constituents significantly impact gelation behavior and crystal morphology. The CGC, in the case of oleogels, depends upon the oil's polarity, which predicts the nature of wax-wax and wax-oil interactions (Doan et al. 2018). The polarity of the vegetable oil can be confirmed through the ratio of saturated to unsaturated fatty acids, chain length, and triglyceride conformation. Therefore, modulation in the oil's polarity can result in oleogels with modified mechanical and structural properties forming. Some external factors that can affect gelation are the rate of cooling, shear rate, time, and temperature involved in the crystallization process (Doan et al. 2018). Altogether, internal and external factors can affect the gelation of oleogels, stability, and storage.

2.1.3 Bibliography Meta-Analysis

The bibliography meta-analysis was carried out to identify the commonly exploited waxes for preparing oleogels. It was performed using the "Web of Science core collection" database on 12 February 2022. For the analysis, the keyword string of "oleogel* (Topic) and wax* (Topic) and food (Topic)" was searched. The search results returned 86 articles, 21 review articles, 4 early access, and 1 proceeding paper. Then, the publications under the "articles" category were selected for analysis. The publication details were exported as a text file and analyzed in VosViewer software. Co-occurrence analysis based on keywords was carried out using the software. Herein, the keywords that appeared at least five times were selected. Subsequently, from the pool of keywords, the word oleogel and the names of the natural waxes were selected to generate a "Network Visualization" plot (Figure 2.1.2). Figure 2.1.2 demonstrates that the most commonly used natural waxes include rice bran wax, candelilla wax, and beeswax. In the next section, we briefly discuss the properties of these waxes.

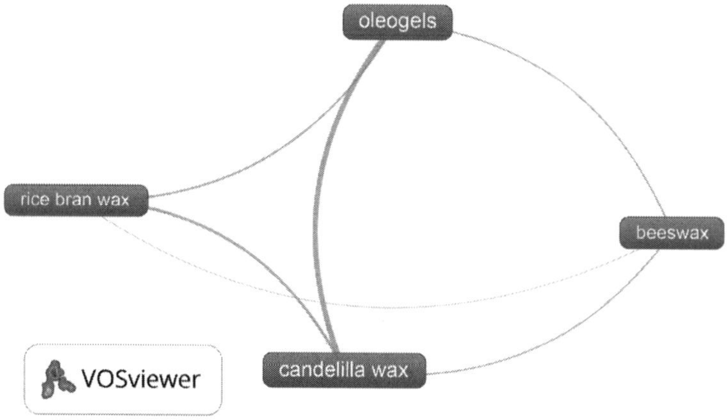

Figure 2.1.2 Network Visualization plot of the commonly explored natural waxes to develop oleogels.

2.1.4 Natural Waxes

2.1.4.1 Candelilla Wax

Candelilla wax (CW) is a plant wax extracted from candelilla shrubs. These shrubs, namely *Euphorbia antisyphilitica* and *Euphorbia Cerifera*, mostly grow in northern Mexico and southern Texas, two arid regions in the United States of America (Aranda-Ledesma et al. 2022). The plant is leafless, and the stems are coated with wax to protect the shrubs against the extreme hot and cold desert climate conditions by acting as a moisture barrier. CW also protects the plant from insects and microbes. CW is easily obtained by boiling the stems in water or by extracting using benzene (Kowalczyk and Baraniak 2014). Refined wax is yellow or brown with a complex mixture of fatty acids and hydrophobic organic compounds.

2.1.4.1.1 Chemical Composition of CW

The chemical composition of CW and other waxes is presented in Table 2.1.1. Hydrocarbons comprised of long-chain *n*-alkanes (C_{26}–C_{33}) constitute the major fraction of CW (Marangoni and Garti 2018). Hentriacontane ($C_{33}H_{64}$) is a major component of all the available *n*-alkanes (Toro-Vazquez et al. 2007; Doan et al. 2017). Two types of esters, namely, C_{41}:m/z 607 and C_{39}:m/z 579, are predominantly present in CW (Moreau et al. 2018; Aranda-Ledesma et al. 2022). Pentacyclic triterpenoids occupy the bulk of alcohols. Free fatty acids having 16–30 carbons are generally present in CW. Among the various free fatty acids, melissic acid (C_{30}, ~ 31%) and palmitic acid (C_{16}, ~ 18%) are present in higher quantities (Doan et al. 2017; Aranda-Ledesma et al. 2022).

2.1.4.1.2 Physico-Chemical Properties of CW

CW contains a high quantity of saturated hydrocarbons, fatty acids, esters, and alcohols. Given this, CW appears hard and brittle. Also, the melting point (MP) of CW ranges between 67 and 69°C. In addition to the melting point, properties such as saponification value (SV: amount (mg) of KOH needed to saponify one gram of fat (wax here)), acid value (AV), and iodine value (IV)) are also greatly influenced by the chemical composition of wax. A comparative account of these properties corresponding to waxes is in Table 2.1.2. CW has very low SV due to the presence of unsaponifiable compounds such as long-chain fatty acids, hydrocarbons, higher aliphatic alcohols, and sterols.

Table 2.1.1 Chemical composition of natural waxes (Adapted from Kowalczyk and Baraniak 2014; Doan et al. 2017; Vali et al. 2005; Fratini et al. 2016).

	Percentage, wt. %		
Chemicals	**CW**	**RBW**	**BW**
Hydrocarbons	50–57	1–3	12–16
Esters	28–29	84–85	50–67
Alcohols and sterols	12–14	2–3	1–2
Free fatty acids	7–9	3–3.5	12–14
Moisture	0.5–1	0.5–1	–
Inorganic residues	0.7	0.5–1	–
Exogenous residues	–	–	6

CW: candelilla wax, RBW: Rice bran wax, BW: beeswax

Table 2.1.2 Physical properties of natural waxes (Adapted from Doan et al. 2017; Dassanayake et al. 2009; Fratini et al. 2016).

Physical properties	CW	RBW	BW
SV (mg KOH/g wax)	35–136	80–90	85–100
AV (mg KOH/g wax)	12–18	3–8	7–36
IV	14–27	8–15	7–16
MP (°C)	67–69	78–81	61– 66

CW: candelilla wax, RBW: Rice bran wax, BW: beeswax, SV: Saponification value, AV: acid value, IV: Iodine value, MP: melting point

This can also be interpreted as CW matter being highly unsaponifiable due to its lower ester content than of other waxes. Similar to SV, IV (mass of iodine in grams consumed by 100 g of chemical substance (wax here)) is also low for CW. Such a low IV indicates the presence of few unsaturated fatty acids in CW, which can be correlated to the high stability of CW. The stability of a chemical substance is evaluated by measuring the AV (amount (mg) of KOH needed to neutralize one gram of chemical substance (wax here)). AV of CW is also very low, indicating that it is a very stable material against the hydrolysis of its components. This feature is crucial and provides ideal properties to CW for various applications in the food industry. Having such stable and suitable physical properties has enabled CW to become a successful gelator for preparing oleogels for food applications.

2.1.4.2 Rice Bran Wax

Rice bran wax (RBW) is a natural plant wax from rice (*Oryza sativa*) bran, a byproduct of rice milling. Rice is largely consumed in East Asia and is a food staple in this region's countries. Accordingly, East Asia is the major supplier of rice bran oil (RBO) and rice bran wax. During RBO extraction from rice bran, RBW is also obtained. RBO wax concentration is about 3–4% of the total lipids (Saunders 1985). While refining crude oil, RBW is extracted at the dewaxing step, in which fat is cooled down to separate wax by either filtration or centrifugation (Saunders 1985).

2.1.4.2.1 Chemical Composition of RBW

Though RBW is a complex material with various chemical compounds, it is considered a mono-component material since it predominantly consists of wax esters. The quality and content of RBW in RBO vary with the extraction conditions, such as extraction temperature, the solvent used, source, and history of bran. All the reports have consistently confirmed that wax esters occupy the majority fraction in RBW (Vali et al. 2005; Doan et al. 2017). A generalized composition of RBW is given in Table 2.1.1. Wax esters in RBW constitute more than 80% of the total compounds. In another report of RBW profiling, wax esters were found to be even higher, more than 90% of the whole composition (Blake, Co, and Marangoni 2014). The wax contains both aliphatic and aromatic esters. It is reported that the wax contains 40% aliphatic esters, 13% ω-hydroxy esters, and 8% cinnamic aliphatic diesters (Dassanayake et al. 2009). The free fatty acid content of RBW is less than 5 %. The majority of these free fatty acids have 16 to 32 carbon atoms. Among the various free fatty acids, lignoceric acid (C_{24}) is present as the highest fraction (40%). In addition to the fat-related compounds, RBW also contains 4–6% resins, which complicates its extraction/purification from oil (Vali et al. 2005). Resins are responsible for the dark reddish-brown color and the unpleasant odor of crude RBW.

2.1.4.2.2 Physico-Chemical Properties of RBW

RBW composition is complex, although, not in the same manner of CW. It influences the physico-chemical properties. A detailed account of the properties of RBW is given in Table 2.1.2. The presence of alkanes, long-chain fatty acids, fatty alcohols, and their esters has resulted in a high melting point of 78–81 °C. Higher saturated fatty acid esters and other long-chain compounds add to the unsaponifiable matter. The wax has low SV (80–90), which can be associated with the presence of a higher quantity of saturated fatty acid esters. Similar to SV, AV and IV are also very low for RBW, indicating that the wax is stable against hydrolysis and rancidity. The lack of unsaturated fatty acids attributes to the stability of the wax. The stability offered by RBW has allowed its use in the food, cosmetic, pharmaceutical, polymer, and leather industries. In the past decade, RBW has been used as a molecular gelator for structuring vegetable oils to obtain oleogels. A relatively high melting point allows RBW to crystallize at ambient temperatures quickly. This results in fine dispersion in vegetable oils, which forms a homogeneous solution and eventually stable oleogels (Morales-Rueda et al. 2009). Thus, synthesized oleogels are being applied as substitutes for traditional shortenings and related foods (Zhao et al. 2020).

2.1.4.3 Beeswax

Beeswax (BW) is a natural wax produced in beehives by the genus Apis honeybees. Wax is secreted from the abdominal glands of bees to build honeycomb cells for honey storage, larval and pupal protection, and hive maintenance. Virgin beeswax, immediately secreted, is white; it later becomes yellow after mixing with pollen, propolis, and larval debris. After the honey is harvested, wax is recovered from the comb as a byproduct. Whole combs are crushed or pressed to recover about 10–15% of wax per kilogram of honey, which is much higher than yields from frame beehive keeping, which are only 1–2%. Higher purity can be achieved by melting combs in boiled water where yellow wax gets separated from the comb residue.

2.1.4.3.1 Chemical Composition of BW

BW is produced worldwide, but the largest global supplier is India. Depending on the bee species, the diet of the bee, the place of production (Asia, Africa, Europe), and the age of wax, the quality, and composition of BW vary (Tinto, Elufioye, and Roach 2017). The general composition of BW is given in Table 2.1.1. BW is a complex material with esters of fatty acids, fatty alcohols, hydrocarbons, free fatty acids, alcohols, and other substances. BW consists of hydrocarbons where a carbon chain length of 27 to 33 are predominantly present. The most common hydrocarbons in BW include tricosane, pentacosane, heptacosane, hentriacontane, and nonacosane. Free fatty acids and free fatty alcohols exist with chain lengths of C_{24}–C_{32} and C_{28}–C_{35}, respectively (Moreau et al. 2018). Lignoceric acid (C_{24}) is the most prominent fatty acid in BW (Moreau et al. 2018). Compared to these compounds, the composition of wax esters is more complex as these wax ester compositions are available as linear monoesters, diesters, triesters, hydroxy monoesters, acid esters, and complex hydroxy polyesters, and acid polyesters (Hepburn 1986). Amongst all wax esters, linear monoesters exist in a higher proportion (35%), followed by diesters (14%). Linear monoesters and hydroxymonoesters are present with chain lengths ranging from C_{40}–C_{48} and are mainly derived from palmitic, 15-hydroxypalmitic, and oleic acids (Münstedt and Bogdanov 2009). 15-hydroxypalmitic acid or corresponding diols also form complex esters by attaching to another fatty acid molecule through their hydroxyl groups. A peculiar feature of BW is unsaturated hydrocarbons and fatty esters along with saturated compounds. In addition to significant major components, BW does possess unidentified exogenous residues for about 6% as pollen, propolis, and pieces of floral components (Hepburn 1986).

2.1.4.3.2 Physico-Chemical Properties of BW

BW is a solid, yellow-colored, brittle wax with a melting point of 61 to 66 °C. Identical to CW and RBW, BW also has lower SV, AV, and IV (Table 2.1.2). Unsaturated fatty acids and wax esters did not significantly enhance AV and IV, indicating that BW is stable at room temperature. But the presence of a lower melting point and of unsaturation suggest that the strength and stability of BW are less than CW and RBW. A combination of BW and CW can result in a stable eutectic product with greater hardness than before due to similar melting profiles and compositions. The instability of BW is also evident from the change in its crystallization pattern with time during storage. The change in crystallization increases the elasticity and stiffness of wax. In other words, the stability of wax changes during storage. This indicates that the hardness of BW changes with time and temperature. BW exhibits higher elasticity at low temperatures, while plasticity increases with higher temperatures.

2.1.5 Applications of the Natural Wax-Based Oleogels

Natural wax-based oleogels are advantageous due to their unique low melting point, higher gel strength, lower cost, and higher oil binding capacity (Hwang and Winkler-Moser 2020). It has been reported that the gel strength of an oleogel can be enhanced by adding natural waxes. Hence, blending a smaller amount of natural wax with other oleogelators can help to achieve a higher gel strength. Also, natural waxes possess additional health benefits because they lower the serum cholesterol level. Wax-based oleogels can substitute saturated fat in some fat-based food formulations in the food industry. Shortenings, creams, whipped cream, and margarine are typical examples of these formulations (Doan et al. 2016). In this section, some applications of previously selected natural wax-based oleogels in food applications are discussed.

2.1.5.1 Candelilla Wax

As per European Food Safety Authority (EFSA), Candelilla wax (CW) is mainly used as a glazing agent and texturizer (Aranda-Ledesma et al. 2022). However, it has other potential applications as a food additive and a clouding agent, and it can be added to baking mixes as surface finisher. CW can produce a strong oleogel when used at a lower concentration (1.5%). On the contrary, a comparatively higher concentration of hard fat is needed for gel formation (Sahu et al. 2021). CW is less expensive (~8 USD per kg) and is more readily available than other low-molecular-mass organic gelling agents, fats, butters, etc. Also, the wax poses a lower health risk than conventional fats. It has also been reported that the odor of the wax is perceptible during baking (Aranda-Ledesma et al. 2022). Due to the advantages mentioned above, CW has potential applications in various food applications. In their research, Jang et al. (2015) used a lower concentration of CW-oleogels (3 wt% and 6 wt%) as an alternative to the shortening agent in cookie dough. They found that the cookie dough prepared using CW-based oleogels showed high spreadability. On the contrary, the cookies showed a reduced snapping property. In another study, the CW-based oleogel was blended in cookie dough at varying ratios: 30:70 and 60:40. They found the CW-based dough's high shear resistance and hardness compared to the dough prepared without wax (Mert and Demirkesen 2016a). In their research, Lim et al. (2017) replaced the shortening agent in muffin batter with a wax-based oleogel prepared using grape seed oil. They found an increased specific gravity of the prepared batter that significantly affected the muffin volume after baking. Apart from baking, the CW-based oleogel has also been employed in noodle-making. Oh et al. (2020) incorporated the CW-based oleogel in two concentrations (8% and 16%) into the noodle dough (Oh and Lee 2020). Compared to the control (without oleogel), the oleogel-based noodles were less dense and showed an increased pore size distribution. There was a consequent reduction in the cooking loss with a corresponding increase in water absorption. Table 2.1.3 summarizes recently published articles that used CW-based oleogels in various food applications. Though the prior research in the field suggests many advantages of CW, there are some limitations associated with its use. CW-oleogel-based dough shows better shear resistance and hardness than the dough prepared with pristine waxes. However, the properties are not better than the dough prepared without any wax or wax-based oleogels. This is because the CW-oleogels lose their solid-like structure in response to the shear forces that occur during dough-mixing. The CW-based oleogel cannot completely replace saturated fat during cookie dough preparation. Also, adding CW in a whipped cream formulation increases the droplet immobilization capacity which reduces coalescence. Hence, partial replacement ($\leq$ 10%) of oil/butter with the CW-based oleogel is suggested for achieving more acceptable properties.

2.1.5.2 Rice Bran Wax

Among the widely available waxes, the rice bran wax showed potential for oleogelation as it can structure oil at a reduced concentration as low as 0.5–1 wt%. The high oil binding property is due to the plant-based wax's low polarity and long-chain length. The advantage of using rice bran wax is its easy availability and low cost, making it favorable for food manufacturers and consumers. In their study, Zhao et al. (2020) have prepared oleogel with either expeller pressed (EP) corn oil or refined corn oil at different concentrations of rice bran wax. They compared the two formulations' physical, thermal, rheological, and crystallization properties. The oleogel was used in cookie making and showed comparable characteristics to control. Thus, it can be used as a shortening agent in baking. In a similar study, the authors compared the properties of two rice bran wax-based

Table 2.1.3 Candelilla wax-based oleogel in different food applications.

Food application	CW conc. (% wt)	Oil used	Change in the characteristics of the final food product	References
Cookie	1.05	Soyabean oil and Sunflower oil	The addition of oleogel significantly increased the final product's volume and improved adhesiveness and creaminess.	(Da Silva, Fernandes, and Arellano 2021)
Wheat sponge cake	5	Canola oil	The butter substitution by oleogel decreased the cake's hardness and increased the specific volume, and starch digestibility.	(Alvarez-Ramirez et al. 2020)
Sweet pan bread	–	Rice bran oil	No significant microstructural change was observed even after replacing the up to 75% of the butter with oleogel. However, adding oleogel increased its nutritional value by reducing the % of unsaturated fat.	(Jung et al. 2020)
Shortcrust pastry	1–5	Sunflower oil	No significant microstructural change and hardness properties were found.	(Tavernier et al. 2018)
Biscuits	3 and 6	Oielic rapeseed oil	The level of unsaturated fatty acids in the prepared oleogel-based cookies was significantly high compared to the shortening cookies.	(Jang et al. 2015)
Muffins	8	Grapeseed oil	The muffin batter with wax-based oleogel as a shortening agent shows less elasticity and lower viscosity, and an enhanced shear-thinning behavior. Also, it significantly increased the muffin's volume by increasing the specific gravity of the prepared batter.	(Lim et al. 2017)
Margarines	2.6	Sunflower oil	Adding CW-based oleogel to the margarine makes it more spreadable and increases its oxidative stability. Also, the taste, texture, and melting point have been enhanced.	(Da Silva et al. 2018)
Margarines	3–7	Soybean oil	The formulation showed a slightly higher melting point and firmness than that of conventional margarine. However, the dropping point and the solid fat content were found closer to commercially available spread.	(Hwang and Winkler-Moser 2020)
Cookies	2.5	Sunflower oil	The cookies prepared with the CW-oleogel show a higher spread ratio and slightly harder texture than commercial shortening.	(Mert and Demirkesen 2016b)
Muffins	10	Sunflower oil	CW-oleogel-based muffins showed higher oxidative stability, harder texture, and higher specific gravity than the control.	(Jeong, Lee, and Oh 2021)
Maize Tortillas	5	Canola oil	A reduced hardness and increased tensile strength were observed in the tortilla after the CW-oleogel was incorporated. Further, in the in vitro study, the tortilla containing oleogel had a slow and readily digestible starch fraction compared to the control.	(Vernon-Carter et al. 2020)

CW: candelilla wax

oleogels prepared using either extra virgin olive oil or sunflower oil. The prepared samples showed pseudoplastic behavior (Zhao et al. 2020). Wanwanut et al. (2017) have explored the health effects of rice bran wax and rice bran oil oleogel by studying its lipid digestibility compared to margarine in rats. Findings show that RBW-based oleogel can decrease the serum and liver triacylglycerol (TAG) level by 30%. Hence, the authors suggested oleogel as a healthy substitute for margarine. Pandolsook et al. (2017) replaced margarine in cookie making with a rice bran wax-based oleogel at different levels (Pandolsook and Kupongsak 2017). A significant change in the physico-chemical characteristics, including the prepared cookie's hardness and color, was observed. However, the diameter, width, and spread remained unaffected with no significant change. Table 2.1.4 represents the lists of potential applications of rice bran wax-based oleogel in food applications. Though a lot of research has reported a substantial change in the oleogel after the RBW addition, much research contradicts these findings.

2.1.5.3 Beeswax

Beeswax (BW) is one of the most commonly used structures that forms agents during oleogelation. It is a complex mixture of esters, alcohols, fatty acids, hydrocarbons, and other chemical components (Sarkisyan et al. 2021). In the past, BW has been efficiently used in various food-based applications. It includes a coating or glazing agent and a texturizer in the chewing gum base (Martins et al. 2016). Recently, it has been applied primarily as a shortening agent in various fat-based foods. In their research, Yilmaz, Öğütcü, and Yüceer (2015) employed a BW-hazelnut oil-based oleogel

Table 2.1.4 Rice bran wax – based oleogel in different food applications.

Food application	RBW conc. (% wt)	Oil used	Change in the characteristics of the final food product	References
Ice cream	10	Oleic sunflower oil	The ice cream with RBW oleogel showed a higher overrun than ice cream using liquid oil. The addition of the oleogel also significantly improved the texture and appearance of ice cream.	(Zulim Botega et al. 2013)
Cheese	10	Soybean oil	The oleogel-based cheese had more volatile samples in the oxidation study than in the control study.	(Park, Bemer, and Maleky 2018)
Sausage	2.5 and 10	Soybean oil	RBX showed potential in replacing the pork fat in frankfurter-type sausages with a significantly high lipid oxidation.	(Wolfer et al. 2018)
Sausage	2.5 and 10	Soybean oil	No significant change was observed after RBX-oleogel to the bologna sausages. However, an increase in the fatty acid profile was found in the oleogel-based sausage.	(Tarté et al. 2020)
Biscuit	4.5	Sunflower oil	The biscuits prepared with RBW-wax-based oleogel had a higher hardness than the control.	(Principato et al. 2021)
Cheese cream	–	Soybean oil	The RBW-wax-based cheese cream showed high hardness, stickiness, and spreadability compared to the full-fat commercial control.	(Bemer et al. 2016)

RBW: Rice bran wax

during cookie preparation (Yilmaz and Öğütcü 2015). They compared the physical, sensory, and textural properties of the oleogel-based cookies against commercial bakery shortening. The results showed that the oleogel-cookies resemble the control cookies' properties. In a similar study, Yi et al. (2017) compared the physiological, sensory, and oxidative properties of two types of oleogels: carnauba wax-canola oil and beeswax-canola oil (Yi et al. 2017). The evaluation was made at different concentrations of the structure-forming agents within the oleogel (0–15%). The study suggested replacing the oleogels with the commercially available solid fat (Yi et al. 2017). In the work by Yilmaz et al. (2015), a vitamin-enriched BW-hazelnut oil-based oleogel was formulated to replace the spreadable fats/margarine in various food applications. The authors observed that the physical and structural properties of these organogels are similar to the wax oleogels. In another study, a BW-virgin olive oil-oleogel (7.7 wt%) was developed that showed more stability in texture, oxidation, and color during storage compared to the control (Öğütcü, Arifoğlu, and Yilmaz 2015). The authors recommended the prepared oleogel as a potential replacement for the spread stock. Table 2.1.5 presents some of the recently reported applications of the BW-based oleogels in food products. It has been reported that the concentration of BW in the oleogel may range from between 0.5 and10%; however, it may exceed this range. Though the critical concentration of BW for most

Table 2.1.5 Beeswax – based oleogel in different food applications.

Food application	BW concentration (% wt)	Oil used	Change in the characteristics of the final food product	References
Pork Burger	11	Olive oil, linseed oil, and fish oil	The addition of BW-based oleogel increased lipid oxidation, but there was no change in the sensory parameter.	(Gómez-Estaca et al. 2019b)
Lipid pates	11	Olive oil, linseed oil, and fish oil	The nutritional content in the oleogel-based pates increased significantly; however, the changes in the sensory parameters were not significant.	(Gómez-Estaca et al. 2019a)
Pattis	10	Rapseed oil	The addition of the BW-based wax improved the cooking loss, fatty acid composition, and nutritional content; however, the oxidative stability and hardness decreased significantly.	(Gao et al. 2021)
Potato stripes	3 and 8	Sunflower oil	The use of oleogel for frying significantly increased the color, smell, crispness, and flavor.	(Aydeniz Guneser, Yilmaz, and Uslu 2021)
Cookies	6	Refined sunflower oil	The RBW-based oleogel is a good alternative to solid fat while making cookies.	(Frolova, Sobolev, and Kochetkova 2021)
Beef burger	5, 7.5, and 10	Soybean oil	Though there was no change in the color of the cooked beef burger after adding oleogel, a significant decrease (11%) in the cooking loss and in the fat absorption (1.6%) was observed.	(Moghtadaei, Soltanizadeh, and Goli 2018)
Sponge cake	3%	Sunflower oil	The oleogel-based sponge cake released two times less fat than the control sample.	(Dikhtyar et al. 2021)

BW: beeswax

oil is 6 wt%, the critical concentration may vary depending on the type of oil used (Frolova et al. 2022). Also, the crystallization and melting temperatures of the prepared oleogel change depending on the content of fatty acid present. These variables influence the crystallization process. A detailed analysis of effect of the type of oil, its composition, and the BW concentration on the food properties is needed for a better understanding on its effect on the physico-chemical properties.

2.1.6 Conclusion

Oleogel is a novel alternative to structure oils without using trans and saturated fats. Its use in diets reduces fat content and adds to food's the health benefits. Wax-based oleogels have wide applications in these fields and are thus applicable in various food industries. The chapter provides comprehensive ideas about the how the aforementioned wax-based oleogels have recently been applied in the food industry. Further, readers can have a thorough idea of the mechanism of oleogelation and the physico-chemical properties of these wax-based oleogels. The information shared in this chapter can act as a reference for researchers working on wax-based oleogels.

References

Alvarez-Ramirez, J., Vernon-Carter, E., Carrera-Tarela, Y., and Garcia-Hernandez, A. (2020). Effects of candelilla wax/canola oil oleogel on the rheology, texture, thermal properties and in vitro starch digestibility of wheat sponge cake bread. *LWT* 130: 109701.

Aranda-Ledesma, N.E., Bautista-Hernández, I., Rojas, R., Aguilar-Zárate, P., del Pilar Medina-Herrera, N., Castro-López, C., Martínez-Ávila, G.C.G. (2022). Candelilla wax: Prospective suitable applications within the food field. *LWT* 159: 113170.

Aydeniz Guneser, B., Yilmaz, E., and Uslu, E.K. (2021). Sunflower oil–beeswax oleogels are promising frying medium for potato strips. *European Journal of Lipid Science and Technology* 123: 2100063.

Bemer, H.L., Limbaugh, M., Cramer, E.D., Harper, W.J., and Maleky, F. (2016). Vegetable organogels incorporation in cream cheese products. *Food Research International* 85: 67–75.

Bharti, D., Kim, D., Cerqueira, M.A., Mohanty, B., Habibullah, S.K., Banerjee, I., and Pal, K. (2021). Effect of biodegradable hydrophilic and hydrophobic emulsifiers on the oleogels containing sunflower wax and sunflower oil. *Gels* 7: 133.

Blake, A.I., Co, E.D., and Marangoni, A.G. (2014). Structure and physical properties of plant wax crystal networks and their relationship to oil binding capacity. *Journal of the American Oil Chemists' Society* 91: 885–903.

Cabrera, S., Rojas, J., and Moreno, A. (2020). Oleogels and their contribution in the production of healthier food products: The fats of the future. *Journal of Food and Nutrition Research* 8: 10.

Da Silva, T.L., Chaves, K.F., Fernandes, G.D., Rodrigues, J.B., Bolini, H.M.A., and Arellano, D.B. (2018). Sensory and technological evaluation of margarines with reduced saturated fatty acid contents using oleogel technology. *Journal of the American Oil Chemists' Society* 95: 673–685.

Da Silva, T.L.T., Fernandes, G.D., and Arellano, D.B. (2021). Development of reduced saturated fat cookie fillings using multicomponent oleogels. *Journal of the American Oil Chemists' Society* 98: 1069–1082.

Dassanayake, L.S.K., Kodali, D.R., Ueno, S., and Sato, K. (2009). Physical properties of rice bran wax in bulk and organogels. *Journal of the American Oil Chemists' Society* 86: 1163–1173.

Dikhtyar, A., Andrieieva, S., Fedak, N., Grinchenko, N., and Pyvovarov, Y. (2021). Determining patterns in the formation of functional-technological properties of a fat-based semi-finished product in the technology of sponge cake products. *Eastern-European Journal of Enterprise Technologies* 6: 114.

Doan, C.D., Patel, A.R., Tavernier, I., de Clercq, N., Raemdonck, K.V., Van de Walle, D., Delbaere, C., and Dewettinck, K. (2016). The feasibility of wax-based oleogel as a potential co-structurant with palm oil in low-saturated fat confectionery fillings. *European Journal of Lipid Science and Technology* 118: 1903–1914.

Doan, C.D., Tavernier, I., Okuro, P.K., and Dewettinck, K. (2018). Internal and external factors affecting the crystallization, gelation and applicability of wax-based oleogels in food industry. *Innovative Food Science & Emerging Technologies* 45: 42–52.

Doan, C.D., To, C.M., De Vrieze, M., De Vrieze, M., Lynen, F., Danthine, S., Brown, A., Dewettinck, K., and Patel, A.R. (2017). Chemical profiling of the major components in natural waxes to elucidate their role in liquid oil structuring. *Food Chemistry* 214: 717–725.

Fratini, F., Cilia, G., Turchi, B., and Felicioli, A. (2016). Beeswax: a minireview of its antimicrobial activity and its application in medicine. *Asian Pacific Journal of Tropical Medicine* 9: 839–843.

Frolova, Y., Sarkisyan, V., Sobolev, R., Makarenko, M., Semin, M., and Kochetkova, A. (2022). The influence of edible oils' composition on the properties of beeswax-based oleogels. *Gels* 8: 48.

Frolova, Y.V., Sobolev, R., and Kochetkova, A. (2021). Comparative analysis of the properties of cookies containing oleogel based on beeswax and its fractions. *IOP Conference Series: Earth and Environmental Science*. IOP Publishing, 012033.

Gao, Y., Li, M., Zhang, L., Wang, Z., Yu, Q., and Han, L. (2021). Preparation of rapeseed oil oleogels based on beeswax and its application in beef heart patties to replace animal fat. *LWT* 149: 111986.

Gómez-Estaca, J., Herrero, A.M., Herranz, B. and Álvarez, M.D. (2019a). Characterization of ethyl cellulose and beeswax oleogels and their suitability as fat replacers in healthier lipid pâtés development. *Food Hydrocolloids* 87: 960–969.

Gómez-Estaca, J., Pintado, T., Jiménez-Colmenero, F., and Cofrades, S. (2019b). Assessment of a healthy oil combination structured in ethyl cellulose and beeswax oleogels as animal fat replacers in low-fat, PUFA-enriched pork burgers. *Food and Bioprocess Technology* 12: 1068–1081.

Hepburn, H., Hepburn, H. (1986). Composition and synthesis of beeswax. *Honeybees and Wax: An Experimental Natural History*, 44–56.

Hwang, H.S. and Winkler-Moser, J.K. (2020). Properties of margarines prepared from soybean oil oleogels with mixtures of candelilla wax and beeswax. *Journal of Food Science* 85: 3293–3302.

Issaoui, M., Delgado, A.M. (2019). Grading, labeling, and standardization of edible oils. *Fruit oils: Chemistry and functionality*, 9–52.

Jang, A., Bae, W., Hwang, H.-S. et al. (2015). Evaluation of canola oil oleogels with candelilla wax as an alternative to shortening in baked goods. *Food Chemistry* 187: 525–529.

Jeong, S., Lee, S., and Oh, I. (2021). Development of antioxidant-fortified oleogel and its application as a solid fat replacer to muffin. *Foods* 10: 3059.

Jung, D., Oh, I., Lee, J., and Lee, S. (2020). Utilization of butter and oleogel blends in sweet pan bread for saturated fat reduction: Dough rheology and baking performance. *LWT* 125: 109194.

Juul, F., Vaidean, G., and Parekh, N. (2021). Ultra-processed foods and cardiovascular diseases: potential mechanisms of action. *Advances in Nutrition* 12: 1673–1680.

Kowalczyk, D. and Baraniak, B. (2014). Effect of candelilla wax on functional properties of biopolymer emulsion films–a comparative study. *Food Hydrocolloids* 41: 195–209.

Lim, J., Jeong, S., Lee, J., Park, S., Lee, J., and Lee, S. (2017). Effect of shortening replacement with oleogels on the rheological and tomographic characteristics of aerated baked goods. *Journal of the Science of Food and Agriculture* 97: 3727–3732.

Limpimwong,W., Kumrungsee, T., Kato, N., Yanaka, N., and Thongngam, M. (2017). Rice bran waxoleogel: A potential margarine replacement and its digestibility effect in ratsfed a high-fat diet. *Journal of Functional Foods* 39: 250–256.

Manzoor, S., Masoodi, F., Naqash, F., Rashid, R. (2022). Oleogels: Promising alternatives to solid fats for food applications. *Food Hydrocolloids for Health* 2: 100058.

Marangoni, A.G. and Garti, N. (2018). *Edible Oleogels: Structure and Health Implications*. Elsevier.

Martins, A.J., Cerqueira, M.A., Fasolin, L.H., Cunha, R.L., and Vicente, A.A., Cunha, R.L. (2016). Beeswax organogels: Influence of gelator concentration and oil type in the gelation process. *Food Research International* 84: 170–179.

Mert, B. and Demirkesen, I. (2016a). Reducing saturated fat with oleogel/shortening blends in a baked product. *Food Chemistry* 199: 809–816.

Mert, B. and Demirkesen, I. (2016b). Evaluation of highly unsaturated oleogels as shortening replacer in a short dough product. *LWT-Food Science and Technology* 68: 477–484.

Mirsafaei, L., Reiner, Ž., Shafabakhsh, R., and Asemi, Z. (2020). Molecular and biological functions of quercetin as a natural solution for cardiovascular disease prevention and treatment. *Plant Foods for Human Nutrition* 75: 307–315.

Moghtadaei, M., Soltanizadeh, N., and Goli, S.A.H. (2018). Production of sesame oil oleogels based on beeswax and application as partial substitutes of animal fat in beef burger. *Food Research International* 108: 368–377.

Morales-Rueda, J.A., Dibildox-Alvarado, E., Charó-Alonso, M.A., Weiss, R.G., and Toro-Vazquez, J.F. (2009). Thermo-mechanical properties of candelilla wax and dotriacontane organogels in safflower oil. *European Journal of Lipid Science and Technology* 111: 207–215.

Moreau, R.A., Harron, A.F., Hoyt, J.L., Powell, M.J., and Hums, M.E. (2018). Analysis of wax esters in seven commercial waxes using C30 reverse phase HPLC. *Journal of Liquid Chromatography & Related Technologies* 41: 604–611.

Münstedt, K. and Bogdanov, S. (2009). Bee products and their potential use in modern medicine. *Journal of ApiProduct and ApiMedical Science* 1: 57–63.

Öğütcü, M., Arifoğlu, N., and Yilmaz, E. (2015). Preparation and characterization of virgin olive oil-beeswax oleogel emulsion products. *Journal of the American Oil Chemists' Society* 92: 459–471.

Oh, I. and Lee, S. (2020). Rheological, microstructural, and tomographical studies on the rehydration improvement of hot air-dried noodles with oleogel. *Journal of Food Engineering* 268: 109750.

Pandolsook, S. and Kupongsak, S. (2017). Influence of bleached rice bran wax on the physico-chemical properties of organogels and water-in-oil emulsions. *Journal of Food Engineering* 214: 182–192.

Park, C., Bemer, H.L., and Maleky, F. (2018). Oxidative stability of rice bran wax oleogels and an oleogel cream cheese product. *Journal of the American Oil Chemists' Society* 95: 1267–1275.

Patel, A.R. (2015). *Alternative Routes to Oil Structuring*. Springer.

Patel, A.R. and Dewettinck, K. (2016). Edible oil structuring: an overview and recent updates. *Food & Function* 7: 20–29.

Principato, L., Sala, L., Garrido, G.D., and Spigno, G. (2021). Effect of dietary fibre and thermal conditions on rice bran wax oleogels for biscuits preparation. *Chemical Engineering Transactions* 87: 49–54.

Sadiqa, Z. and Noemana, A. (2019). Selection of edible oil for cooking. *Journal of Cardiovascular Disease* 15: 1–2.

Sahu, D., Bharti, D., Kim, D., Sarkar, P., and Pal, K. (2021). Variations in microstructural and physico-chemical properties of candelilla wax/rice bran oil–derived oleogels using sunflower lecithin and soya lecithin. *Gels* 7: 226.

Sarkisyan, V., Sobolev, R., Frolova, Y., Malinkin, A., Makarenko, M., and Kochetkova, A. (2021). Beeswax fractions used as potential oil gelling agents. *Journal of the American Oil Chemists' Society* 98: 281–296.

Saunders, R. (1985). Rice bran: Composition and potential food uses. *Food Reviews International* 1: 465–495.

Sein, A., Verheij, J.A., and Agterof, W.G. (2002). Rheological characterization, crystallization, and gelation behavior of monoglyceride gels. *Journal of Colloid and Interface Science* 249: 412–422.

Shah, B. and Thadani, U. (2019). Trans fatty acids linked to myocardial infarction and stroke: What is the evidence? *Trends in Cardiovascular Medicine* 29: 306–310.

Tarté, R., Paulus, J.S., Acevedo, N.C., Prusa, K.J., and Lee, S-L. (2020). High-oleic and conventional soybean oil oleogels structured with rice bran wax as alternatives to pork fat in mechanically separated chicken-based bologna sausage. *LWT* 131: 109659.

Tavernier, I., Doan, C.D., Van Der Meeren, P., Heyman, B., and Dewettinck, K. (2018). The potential of waxes to alter the microstructural properties of emulsion-templated oleogels. *European Journal of Lipid Science and Technology* 120: 1700393.

Temkov, M. and Mureșan, V. (2021). Tailoring the structure of lipids, oleogels and fat replacers by different approaches for solving the trans-fat issue—a review. *Foods* 10: 1376.

Tinto, W., Elufioye, T., and Roach, J. (2017). Waxes. *Pharmacognosy*. Elsevier pp. 443–455.

Toro-Vazquez, J., Morales-Rueda, J., Dibildox-Alvarado, E., Charó-Alonso, M., Alonzo-Macias, M., and González-Chávez, M.M. (2007). Thermal and textural properties of organogels developed by candelilla wax in safflower oil. *Journal of the American Oil Chemists' Society* 84: 989–1000.

Uvanesh, K., Sagiri, S., Senthilguru, K., Pramanik, K., Banerjee, I., Anis, A., Al-Zahrani, S., and Pal, K. (2016). Effect of span 60 on themicrostructure, crystallization kinetics, and mechanical properties of stearicacid oleogels: an in-depth analysis. *Journal of Food Science* 81: E380–E387.

Vali, S.R., Ju, Y.-H., Kaimal, T.N.B., and Chern, Y.-T. (2005). A process for the preparation of food-grade rice bran wax and the determination of its composition. *Journal of the American Oil Chemists' Society* 82: 57–64.

Vernon-Carter, E.J., Alvarez-Ramirez, J., Meraz, M., Bellow-Perez, L.A., and Garcia-Diaz, S. (2020). Canola oil/candelilla wax oleogel improves texture, retards staling and reduces in vitro starch digestibility of maize tortillas. *Journal of the Science of Food and Agriculture* 100: 1238–1245.

Wertheim-Heck, S.C. and Raneri, J.E. (2019). A cross-disciplinary mixed-method approach to understand how food retail environment transformations influence food choice and intake among the urban poor: Experiences from Vietnam. *Appetite* 142: 104370.

Wolfer, T.L., Acevedo, N.C., Prusa, K.J., Sebranek, J.G., and Tarté, R. (2018). Replacement of pork fat in frankfurter-type sausages by soybean oil oleogels structured with rice bran wax. *Meat Science* 145: 352–362.

Yi, B., Kim, M.-J., Lee, S.Y., and Lee, J. (2017). Physico-chemical properties and oxidative stability of oleogels made of carnauba wax with canola oil or beeswax with grapeseed oil. *Food Science and Biotechnology* 26: 79–87.

Yilmaz, E. and Öğütcü, M. (2015). The texture, sensory properties and stability of cookies prepared with wax oleogels. *Food & Function* 6: 1194–1204.

Yilmaz, E., Öğütcü, M., and Yüceer, Y.K. (2015). Physical properties, volatiles compositions and sensory descriptions of the aromatized hazelnut oil-wax organogels. *Journal of Food Science* 80: S2035–S2044.

Zhao, M., Lan, Y., Cui, L., Monono, E., Rao, J., and Chen, B. (2020). Formation, characterization, and potential food application of rice bran wax oleogels: Expeller-pressed corn germ oil versus refined corn oil. *Food Chemistry* 309: 125704.

Zulim Botega, D.C., Marangoni, A.G., Smith, A.K., and Goff, H.D. (2013). The potential application of rice bran wax oleogel to replace solid fat and enhance unsaturated fat content in ice cream. *Journal of Food Science* 78: C1334–C1339.

2.2

Phytosterols and Other Sterols

Artur J. Martins

International Iberian Nanotechnology Laboratory, Braga, Portugal

2.2.1 Introduction

Different classes of gelator molecules have been generating interest in recent years due to their specific nature and to their oil structuring capacity. To meet a particular food product's requirements, oil structuring agents or oleogelators should be capable of producing gel systems with a certain level of tailoring proficiency, which will allow them to actually serve as feasible alternatives as fat substitute drivers. Food engineering perceives that the principles of food formulation should always comprise food safety and quality concepts without disregarding consumer acceptance as a priority when developing a novel food product. It is thus extremely challenging to meet all criteria stating that in order to serve as an agent for oil structuring, these alternatives should be food grade, affordable, nonreactive with other ingredients, versatile, and prone to tailoring features (Flöter et al. 2021). Within the design of novel food products, underestimating these considerations is certainly a recipe for disaster, since new food systems should offer very similar sensory perceptions for consumers in addition to healthier benefits. Saturated fats are the responsible constituents for food's characteristic mouthfeel and critical texture properties (*i.e.*, hardness, plasticity, cohesion, etc.), which translate to foodstuffs as dissimilar as ice creams or meat products (Stortz et al. 2012). Apart from the interesting texture and mechanical properties, which can magnify oleogels' fittingness as fat mimicking materials, reducing saturated fatty acids in common foods has driven research within this field as the search for triacylglycerols-like building blocks has intensified. Triacylglycerols (TAGs: tri-esters comprised of a glycerol bound to three fatty acid molecules) are intrinsically connected to fat crystallization behavior, melting point, and solubility properties, and they serve as the main functional lipid phase in foodstuffs. Here is one of the main issues for the global production of TAGs. Palm oil is the most widely traded vegetable oil, accounting for nearly 60% of worldwide oilseed exports (Carter et al. 2007; Ayompe, Schaafsma, and Egoh 2021). Its main advantage is the significantly higher production per hectare and its distinctive combination of fatty acids; however, the wide use of palm oil to produce triacylglycerols (TAGs) for functional fat/lipid constituents embodies several concerns among food manufacturing processes, since its production has become increasingly controversial because of environmental and social problems (e.g., deforestation, habitat and biodiversity loss, disruption of food chains, and air and water pollution) (Ayompe, Schaafsma, and Egoh 2021). Also, because of the improvement in testing methods, contaminants (e.g., glycidyl fatty acid esters, 3-MCPD, and 2-MCPD) have been detected in palm oil and palm oil fats (as well as in other oils and fats) because of the deodorisation stages of oil refining. As a consequence, the European Food Safety Authority (EFSA) expressed concerns (EFSA Journal 2016) setting the tone

Fat Mimetics for Food Applications, First Edition. Edited by Miguel Ângelo Parente Ribeiro Cerqueira and Lorenzo Miguel Pastrana Castro.

concerning consumers' exposure to such contaminants, managing potential risks and stating levels of exposure among distinct consumer age groups (*i.e.*, formula consumption in infants, and margarine and pastries consumption in consumers ages three and above).

Recently, examples of oleogel applications for the development of novel food products with scalability potential has increased the overall global interest in these promising edible materials. Finding healthy replacements for common TAGs is driving the field. This chapter discusses the suitability of sterols (a subclass of steroid alcohols) to formulate oleogels and their derivative structures for use in developing fat mimetics within food formulation frameworks. Plant sterols, commonly referred to as phytosterols, exhibit a steroid structure composed by a hydroxyl group at position 3 and a side chain at position 17; they can have one or more double bonds along the steroid skeleton (Figure 2.2.1). Structurally, these phytosterols are closely related to stanols (which do not exhibit unsaturated bonds), which are normally considered part of the phytosterol group (Bot 2018).

Plant sterols can be obtained from numerous natural sources, namely vegetable oils, such as rapeseed oil, soybean oil, sunflower oil, and corn oil. Corn and rapeseed are the richest sources of sterols, averaging total sterol contents per 100 g of refined oil of, and 700–750 and 750–800 mg respectively for corn and rapeseed oil (Fernandes and Cabral 2007; Cantrill 2008). There are a few recent proposals involving lipase-assisted procedures; however, on an industrial scale, phytosterols are commonly recovered using a refinement process called *deodistillation* that removes the flavor and odor from the oil (Maniet, Jacquet, and Richel 2019). Sterol recovery comes from the unsaponified fraction of oil distillates; the oil type will influence the phytosterol content obtained, which can range from 8–20%, resultant from 0.3–0.5% of the overall refined oil volume (Cantrill 2008). Phytosterols can also be obtained from crude tall oil, which is a byproduct of wood pulp manufacturing. In this extraction route, the heavy phase (pitch tall oil) has fewer volatiles (a major part of the unsaponifiables and a minor amount of fatty and rosin acids), and the sterol content may range from 5–15%. β-sitosterol is the most abundant plant sterol, and the commercial sources of tall oil and vegetable oil present a β-sitosterol percentage ranging from 36–80% and 43–52% (Cantrill 2008).

In regard to γ-oryzanol, this is a complex molecule containing ferulate, sterol esters, and triterpene alcohols. It is found in rice bran and rice bran oil, being about 20% of the unsaponifiable fraction in rice bran oil (Srikaeo 2014). The γ-oryzanol content in rice bran oil can go slightly above 1% after oil refining, with an overall content of 20% in the unsaponified fraction (Narayan, Barhate, and Raghavarao 2006). In terms of γ-oryzanol recovery, apart from solid-liquid extraction and crystalization/precipitation

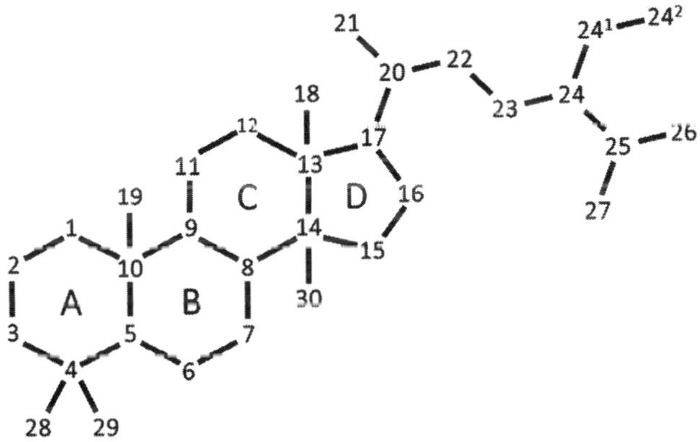

Figure 2.2.1 Steroid skeleton nomenclature. Arjen Bot (2019) / Elsevier.

methods, the supercritical fluid extraction is also an alternative methodology of high interest (Bran, King, and John 2004; Jesus, Grimaldi, and Hense 2010; Claycamp et al. 2014; Kayathi et al. 2021).

Amid the studied sterol-based oleogels, there is a remarkable similarity among the sterol-based oleogels that are commonly produced using the γ-oryzanol-β-sitosterol binary system with regular fat; this is amongst the many interesting features that embodies their suitability for industrial food applications. A specific time-dependent aggregation (that will be discussed later in section 2.2.2.2) at the molecular level will lead the phytosterols/oryzanol content to construct nanotubular conformations used for developing the well-known macro-scale transparent gels (Duffy et al. 2009). The fact that these components are deprived of any of the saturated fatty acids, which account for the crystallization process in the majority of other systems (*i.e.*, natural waxes; mono and diglycerides; and sorbitol tristearates), epitomizes a convenient feature that could facilitate their use in making healthier food products. Given the ability to replace saturated fats with healthier possibilities, also having in mind oleogel technological properties and aptitude to serve as vehicles for bioactives has led food science to meaningfully investigate introducing these oleogels into food matrices (Kouzounis, Lazaridou, and Katsanidis 2017; Franco et al. 2019, 2020; Martins et al. 2019; Martins et al. 2020). This structuring system also seems to positively respond as a two-way ingredient in the sense that it can answer to the need of having an oleogelator that; apart from being food grade, it can be likewise a functional ingredient. Thus, the combination between the above-mentioned gelators meets both purposes—nutritional and technological, since the value is added when it is incorporated into food systems (Lemus et al. 2014). Sterol-based oleogels bear physiological effects on human diet. According to the European Food Safety Authority (EFSA), phytosterols, besides being approved for use in foodstuffs, have been proven to reduce the risk of coronary heart disease and to lower levels of LDL cholesterol (European Food Safety Authority 2012).

2.2.2 γ-Oryzanol-Sterols System

Apart from the above-mentioned matters, this oleogel system is not as easy to comprehend and predict as some other systems based on crystalline or polymeric networks that employ mono- or multi-component systems. When the discussion is around mixed sterol-based oleogels, we are usually referring to the predominantly studied γ-oryzanol-β-sitosterol binary system (Bot and Agterof 2006; Bot, Den Adel, and Roijers 2008; Bot et al 2009; Bot, den Adel, and Heussen 2010; Sawalha et al. 2013; Matheson and Dalkas et al. 2017; Matheson et al. 2017; Dalkas et al. 2018). Because multi-component mixtures of TAGs combined with minor components are the main constituents of edible oils and fats, sterol-based oleogels follow a non-TAG structuring route by means of gelling fibrillar building blocks. The result is a highly stable oleogel that exhibits an elevated binding capacity (Flöter et al. 2021). Normally, a solid fat mixture is characterized by its solid fraction content (*i.e.*, the mass fraction of solid present at a certain temperature); this quality allows for a prediction on the material's selected physical properties (Himawan, Starov, and Stapley 2006). The solid fat content is the gelators' mass within the oleogel formulation, which in this case is namely the γ-oryzanol and β-sitosterol content.

2.2.2.1 Crystallites and Oil Gelation

As mentioned elsewhere, sterols can develop crystallites 3.8 nm thick that can display either hydrated (hemi-hydrate and monohydrate) or anhydrous forms depending on environmental conditions (Bot 2018). As the most abundant plant sterol, β-sitosterol is, therefore, the most studied regarding oleogel formation. Christiansen et al. reported on the ability to change sterol (β-sitosterol) crystal size and shape through supersaturation using different quantities of water and the sterol itself (Christiansen et al. 2002). This is important for digestive matters, specifically for the ability of the bioactive to be adsorbed at the intestinal level, specifically considering the interference on physiological matters like cholesterol absorption. Still focusing solely on β-sitosterol and its crystalline construction, the capacity to form a three-dimensional bilayer structure in an apolar medium was verified. This reveals different ordering for powder and oily slurry, which is a contrast to the long-range ordering observed in an aqueous slurry. This is indicative of a co-crystallization of anhydrous and hemi-hydrate crystals, for the first and a more prevalent monohydrate conformation for the later (Bot, den Adel, and Heussen et al. 2010).

Despite being generally mentioned as if it is a pure molecule, γ-oryzanol composition depends on the strain of rice from which it is extracted because its composition is a mixture of components that account for over 85%, which are the 24-methylene cycloartanyl ferulate (53 – 57%), cycloartenyl ferulate (27 – 3%), campesteryl ferulate (8 – 11%), and β-sitosteryl ferulate (1 – 4%) (Matheson et al. 2017). The lack of prominent Bragg diffractions for γ-oryzanol in an oil medium, suggests a high level of solubility and the crystallization data indicates the absence of a characteristic bilayer conformation that is typical of systems displaying esterified acid moieties' interactions. Despite struggles in wetting the γ-oryzanol powder and the unclear data on γ-oryzanol hydrates (compared to β-sitosterol), two hydration states were still detected for this compound. X-ray diffraction experiments using both γ-oryzanol and β-sitosterol molecules, traced back the crystalline formation in oil, water, or emulsions to the pure compounds (Bot, den Adel, and Heussen et al. 2010).

It was generally known and regarded that independently, neither γ-oryzanol nor β-sitosterol could arrest the oil phase and develop gels by themselves. However, a recent work demonstrated the ability of γ-oryzanol (> 7 wt%) to develop crystals in an oil medium (see Figure 2.2.2), and formulations with γ-oryzanol content above 15 wt% were able to generate self-supporting networks (from platelet crystals a few microns in length), which resulted in in stable oleogels using different types of edible oils (*i.e.*, canola oil, peanut oil, soybean oil, and palm olein) (Wijarnprecha et al. 2021). This structuring performance was strictly attributed to the solidification of the γ-oryzanol ferulates during concurrent mixing prompted under dissimilar cooling rates excluding the possibility that some minor components (also refined γ-oryzanol was tested) could co-adjuvate in this gelation process.

2.2.2.2 γ-Oryzanol-Sterols Mechanism

Apart from monocomponent crystalline conformations, a completely different outcome ensues for mixtures developed under an oil medium. Here, there is a unique crystallographic fingerprint attained as an outcome of the fibrillar self-assembled structure ensuing from the sterols and γ-oryzanol molecular docking (see Figure 4.3). Even though both γ-oryzanol and β-sitosterol are not pure constituents due to the presence of minor components, we consider their oil mixture being a ternary system. Briefly, the analysis of this mechanism at a molecular scale envisions a self-assembly of the γ-oryzanol and β-sitosterol after solubilization in the oil medium, despite the staking of the two molecules. This happens because hydrogen bonds between the hydroxyl groups of β-sitosterol and the carbonyl groups of γ-oryzanol are formed. The combination of these two compounds produces a tubular-shaped arrangement where a set of entangled, unidirectional fibrils grow into a fibrillar network (Martins et al. 2019). These binary fibrils can be grouped in oryzanol-sitosterol or those based on surfactant-like molecules such as tocopherol, lecithin, and monoglycerides (Okuro et al. 2018). The mechanism for forming the gel

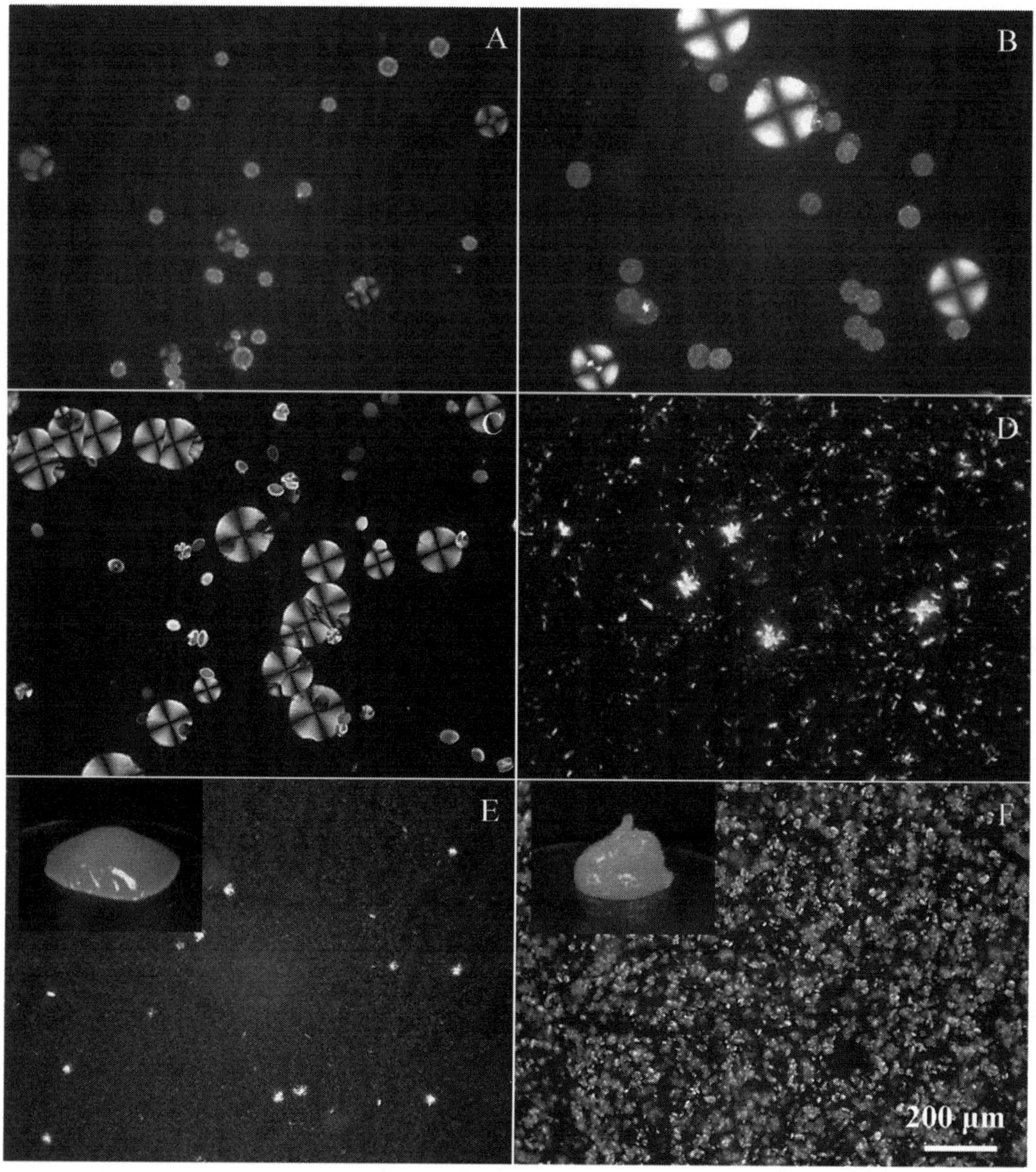

Figure 2.2.2 Polarized light micrographs of γ-oryzanol crystals in oil formed under static (regime A) (A, B, and C) and dynamic conditions (regime D) (D, E, and F) at different γ-oryzanol concentration [5 wt% (A and D), 7 wt% (B and E) and 25 wt% (C and F)] following storage at 5 °C for 1 week. E and F show γ-oryzanol in RBO self-supporting networks. Scale bar: 200 μm. Wijarnprecha et al. (2011), with permission from Elsevier.

was suggested by Bot & Agterof (Bot and Agterof 2006). This proposed helical ribbon assembly mechanism was verified in subsequent reports (Rogers 2011; Dalkas et al. 2018). These curved one-dimensional assembled building blocks are responsible for producing an oleogel with very distinctive characteristics; one of the most interesting is the stability and the translucent nature of said oleogels (Sawalha et al. 2013; Martins et al. 2019). The length of the fibers is theoretically dependent of processing conditions like storage and cooling rates (Martins et al. 2018). Clear gels are only achievable due to the nano size of such building blocks—approximately 7 nm in diameter (reaching up to 10 nm) with the ferulic acid moiety of γ-oryzanol (Scharfe et al. 2019). This tends to be smaller than the wavelength of visible light, which consequently produces see-through solid-like gels (Duffy et al. 2009; Rogers et al. 2014). The high levels of transparency combined with the strong mechanical properties are important aspects that increase the feasibility of this oleogel system toward being applicable in food formulations (Bot, Den Adel, and Roijers 2008; Bot and Veldhuizen et al. 2009). The cropping mechanism of both stearane groups from the sitosterol units (managed by hydrogen bonds in both molecules), is responsible for placing the ferulic acid moiety of the γ-oryzanol outside of the inter-connected structural arrangement (Dalkas et al. 2018). This is responsible for a marginally tilted stacking that would originate the wedged-shaped tubules (Figure 2.2.3).

The path of oleogelation appears to be a multi-step process that consists of a primary formation of fibrils and of a sequential bundling effect that is responsible for the formation of a strong enough complex or network through $\pi - \pi$ and vdW interactions (aromatic ring and methoxy group) (Dalkas et al. 2018). This intra-aggregation can still be affected by an increase in the gelators' total mass, which can materialize for a 1:1 molar ratio where stacking the tubular structures can induce different opacity levels in the oleogels produced (Martins et al. 2019). Knowing this, we can state that with the presence of molecules capable of bridging the available groups within the self-assembly arrangement, the sol-gel transition, optical and mechanical properties will undergo serious modifications. The position of the hydroxyl group is critical because it limits solubility; hence, upon cooling, the gelator molecules undergo phase separation. Phase separation minimizes free surface energy, causing hydroxyl groups to align, which is then responsible for reducing contact with the apolar solvent. This behavior can somehow be tailored for fat mimicking purposes. Considering that macro-properties can be adjusted using molecules (that have a polar nature), and depending on their chemical structure, this will induce a certain level of instability within the fatty system or work in the opposite way. The study of Calligaris et al. showed that the oil phase unsaturation levels (flaxseed > sunflower > extra virgin olive oil) play a role in oleogel hardness and in sol-gel transition temperatures; the study reported lower hardness values are associated with increasing oil viscosity (Calligaris et al. 2014). Something similar was observed by Sharfe et al.; sol-gel transition temperatures were diminished for oils with increasing levels of minor polar components due to increased solvent viscosity and to the anticipated interactions of polar molecules with the

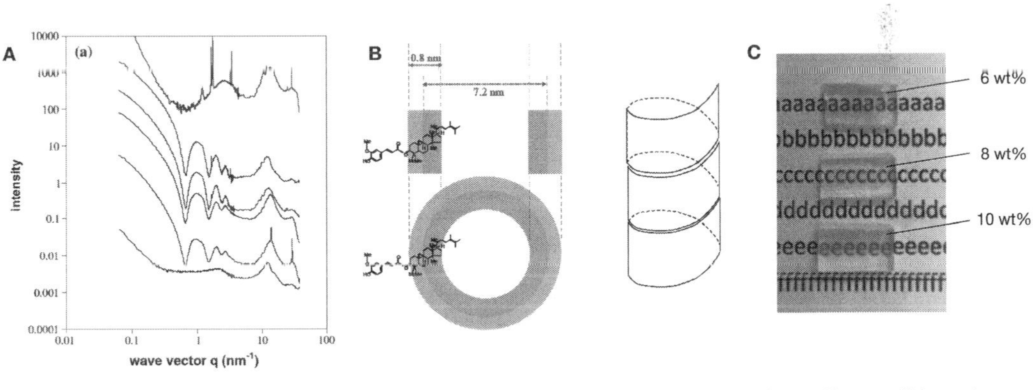

Figure 2.2.3 A) SAXS and WAXS diffractograms for β-sitosterol and γ-oryzanol in sunflower oil in various ratios: an 8% total sterol(ester)s; The β-sitosterol:γ-oryzanol ratio from top to bottom: 100:0; 80:20; 60:40; 40:60; 20:80; 0:100; Reproduced from Bot, A. et al. 2008 / with permission of John Wiley & Sons; B) diagram of the cross-sectional width and fiber length for the supramolecular structure of γ oryzanol:cholesterol hollow tubules, Reproduced from Adel et al. (2010) / with permission of IOP Publishing; C) Visual appearance of oleogels containing extra virgin olive and 5 % of β-sitosterol and γ-oryzanol (2:3). C) Transparency associated with these type of oleogels, produced with increasing concentration of β-sitosterol and γ-oryzanol. Martins et al. (2019) with permission from John Wiley & Sons.

oleogelators (Scharfe et al. 2019). The same authors likely observed that the upsurge in oleogel hardness occurred when low concentrations of tocopheryl acetate were added to the oil mixture (decreasing its effect for formulation with higher concentrations), whereas for samples with oleic acid and monoglycerides a reversed viscoelastic behavior was observed and increased G' and G'' parameters emerged in gels containing tocopheryl acetate and water. This means the oleogels' fibrillar microstructure noticeably changed in the presence of such compounds (Scharfe, Prange, and Flöter 2021). The ratio of the gelling components must be carefully managed or tailored. Guaranteeing a successful oil gelation/structuring leads to a desirable set of properties suitable for the oleogels' end application as described. If the gelators' ratio is tilted toward increased β-sitosterol content, the critical gelation control mass can experience variations; hence, oleogel's end properties will change analogously. The literature review discloses that the presence of an increased β-sitosterol fraction (unbalancing the ratio) during the protocol of oleogel development will decrease gel transparency. This leads to the appearance of crystallized artefacts or dendrites, which results in forming less attractive oleogels. Such variations were previously observed by Martins et al. while working with a commercial (non-pure) mixture of phytosterols. Fostering an unbalanced ratio between γ-oryzanol and β-sitosterol because of the lopsided (increased) presence of other phytosterols developed a different level of affinity toward γ-oryzanol, which resulted in shaping the gelation mechanics and oleogel's final characteristics (e.g., opacity and mechanical properties) (Martins et al. 2019). Given that, the critical oleogelator concentration might represent an important target, since food formulation always thrives around commodity foods' costs. It is advised to use a gelator mass of at least 8% (above critical gelling concentration) to perform successful oil structuring without resorting to subsequently cooling temperatures (Martins et al. 2019). Cooperative gelation embodies one major advantage for these multi-component systems, in the sense that physical properties may easily be tuned through the adjustment of the relative proportions of its individual components; however, such a procedure should be carefully done to avoid gel destabilization.

Considering this binary system, other phytosterols can achieve similar results due to their equivalent molecular construction. The structure of phytosterols (*i.e.*, β-sitosterol) displays a steroidal ring base with a hydroxyl C-3 group and a double bond placed between C-5 and C-6 within the ring; for interconnection with the γ-oryzanol to happen, such connection sites must be available (Rogers 2011; Martins et al. 2019). Plant sterols share a fused polycyclic structure with differing unsaturation positions and numbers. β-sitosterol, stigmasterol, and campesterol are amongst the most common (4-desmethyl) phytosterols, and their metabolic pathway and function have been widely studied. Despite some differences observed in distinctive sterol types, these act more or less the same way when solubilized in the presence of γ-oryzanol. It was observed (Figure 2.2.4) that with provenance from the sterol ring, the increase of double bonds decreased in the tubules' diameter (producing narrow tubules) (Bot et al. 2009).

Disparities in system miscibility, observed for sitosterol rich oleogels, will influence the way that these micro-crystalline bodies or dendrites emerge from the nucleation sites. Similarly, at the formation of either thin fibers comprised of co-aggregates or of macroscopic (hylotropic) crystals dependent on the amount of cholesterol and oryzanol in the mixture were reported (Rogers 2011). It was also recently reported that the number of the nucleation sites for these crystalline bodies (thin and highly ramified) increases when the phytosterol content is higher than oryzanol. This increase of solids concentration reversely affects the crystallization of the samples produced with an unbalanced oleogelators ratio, where thicker and less ramified dendrites are seen. The impact of cooling rates on the hardness was also observed (Martins et al. 2022). All these particularities are responsible for shaping sterol-based oleogels' ultimate features (*i.e.*, binding capacity, hardness, and optical features). Apart from the mentioned plant sterols, the application of the—also naturally occurring—lupeol, α-amyrin, β-amyrin, and cycloartenol (among others), could be considered for future oleogelation. These 4,4-dimethyl phytosterols

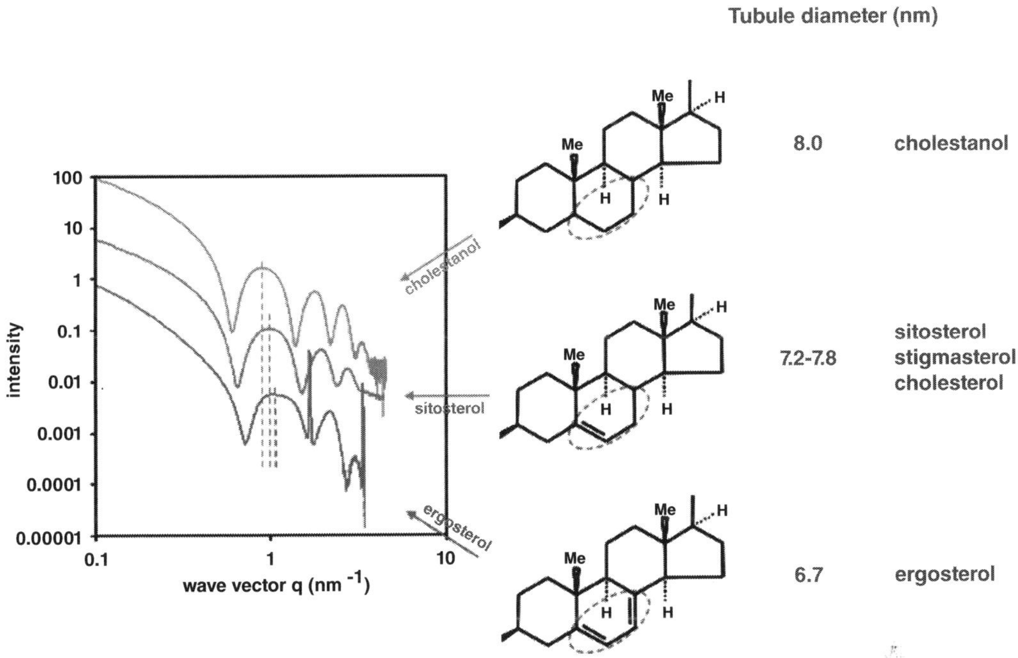

Figure 2.2.4 Effect of sterol ring structure on tubules diameter. The sharp peaks in the ergosterol + oryzanol system are caused by separate crystallization of ergosterol crystals. Reproduced from Adel et al. 2010 / with permission of IOP Publishing.

characteristically exhibit two methyl groups at the aliphatic A-ring's C-4 position, displaying unsaturated (available to bridge) bonds in the ring or the side chain (Zhang et al. 2020). Lupeol is relatively low cost, is commercially available, and is mainly gathered from fruits and vegetables (Saleem 2009), and it appears to be a good candidate for developing oleogel applications, too (Figure 2.2.5).

The reported cholesterol-lowering properties of all the mentioned molecules comprise a very important aspect of their utilization as gel ingredients; however, there is still a research gap on the benefits of 4,4-dimethyl phytosterols when applied as functional foods or nutraceuticals.

Gelation unpredictability is one of the main factors when working with these sterol-based gels. As previously idenftified, this system exhibits an erratic gelation performance deprived of a precise gelling time. Moreover, gelation can be triggered if there is a trivial mechanical disorder during the stabilization period (Bot and Agterof 2006). Such rheopetic nature is observed during the "curing" stage if some type of shear is introduced; nonetheless, it can be argued that the structuring will occur differently if quiescent conditions were to be preserved. This event is possibly a materialization of increased interfibrillar connection points with tubules contacting and forming bundles (Figure 2.2.6A), while system percolation occurs and less molecular ordering is achieved. A linear arrangement of the Avrami relation was used by Matheson et al. as the authors reported on the assembly of β-sitosterol: γ-oryzanol based nanotubules. Calculations considered the temporal deviations of both the G modulus and Tg, where n is the Avrami growth parameter; t is time; K as the constant rate of growth; G (N) is the plateau modulus when gelation is complete:

$$\ln\left(-\ln\left(1-\frac{G'(t)}{G'(\infty)}\right)\right)=\ln(K)+n\ln(t-\mathrm{Tg}) \quad (1)$$

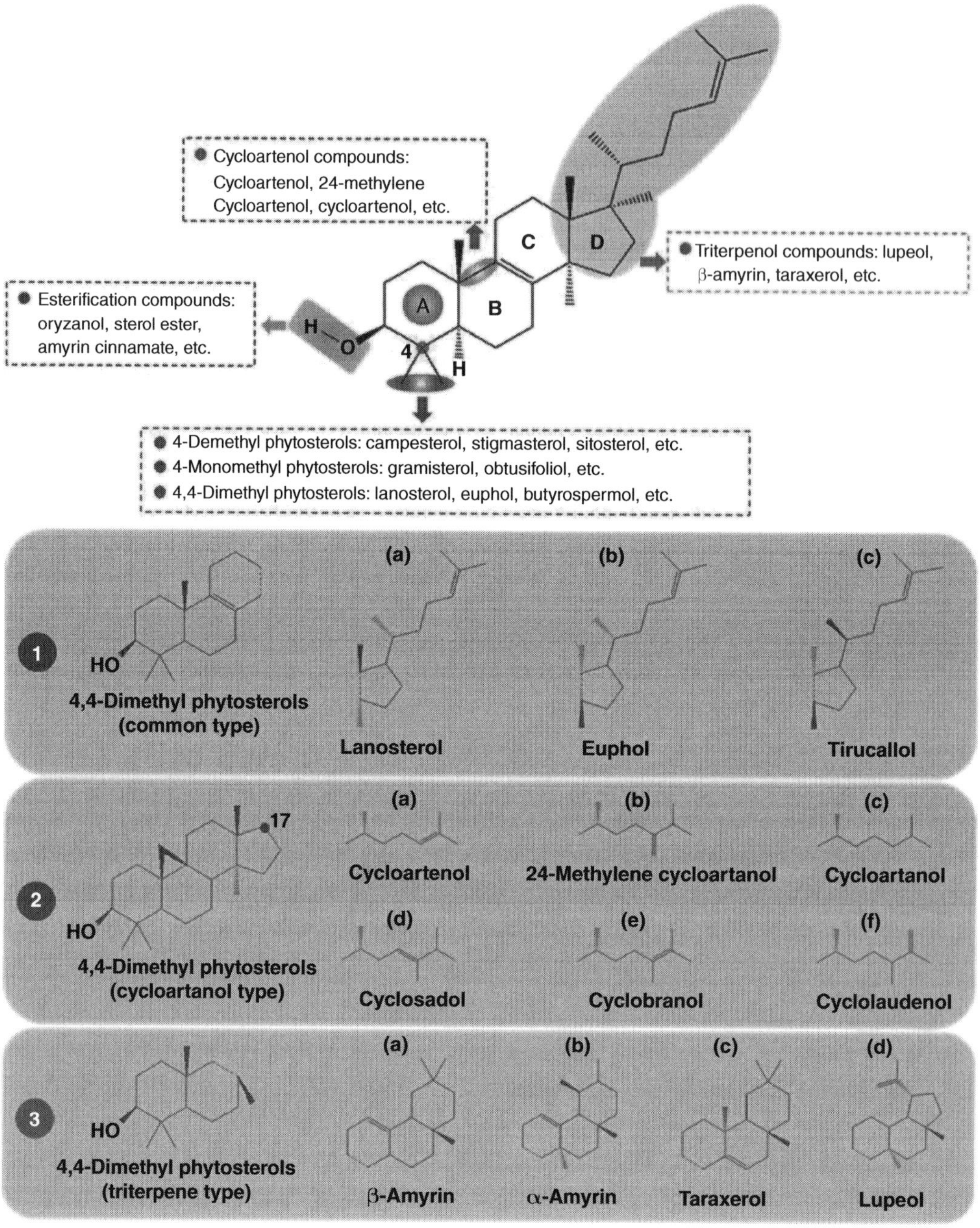

Figure 2.2.5 Summary of the structure–compound relationship of the major 4,4-dimethyl phytosterols. The 4,4-dimethyl phytosterols typically possess two methyl groups at the carbon-4 atom position of the ali-phatic A-ring; Reproduced from RSC, Zhang et al. 2020 / With permission from Royal Society of Chemistry.

Considering this, the authors reported that the nanosized tubules, formed by the molecular docking of γ-oryzanol and β-sitosterol, were associated with the 1-D growth of gelators nuclei, presenting anisotropy, which is similar to a supercooled system ($n < 1$) (Matheson et al. 2017). This also indicates that an increased oleogelator concentration would lead to the uprise of the n parameter (Figure 2.2.6), reflecting the aggregation of nanofibrils, which are kinetically reversible at large scales (Matheson et al. 2017).

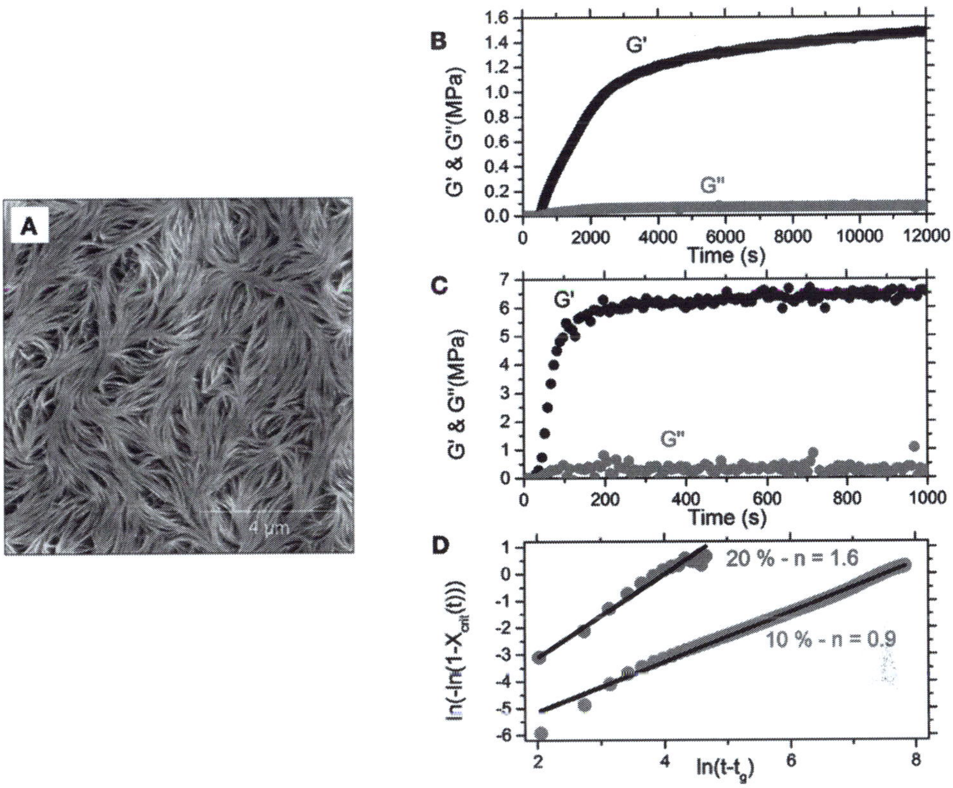

Figure 2.2.6 A) AFM of 20% w/w oleogel sample showing that fibrils aggregate into thick bundles; evolution of the storage (*G′*) and loss (*G″*) modulus for: A) 10% w/w sample and B) 20% w/w sample undergoing the sol –gel transition. C) Plots used to extract n value, along with the linear fitting to the data to yield Avrami parameters of n = 0.9 and n = 1.6. Matheson et al. (2017), with permission from American Chemical Society.

Such observations reveal that the sol – gel transition is not immediate (upon reaching a certain temperature after cooling); the subsequent growth of the viscoelastic parameters suggests that there is an aggretation phenomena after fibril formation. This is consistent with recent observations, using Confocal Raman Spectroscopy, where the molecular arrangements within the system, took a certain amount of time to fully develop into the final conformation (Martins et al. 2022). Therefore, the critical gelation time can vary greatly, depending on the gelators' ratio and concentration, while cooling rates would also effect this. Raman modes attributed to the aromatic ring stretching vibrations C = C started developing in the very beginning of the stabilization period for the oleogels with greater solids percentage; by contrast, oleogels with lower gelator concentrations only exhibited the same behavior after 24 h of curing time. This means that the nanotubular conformation for the first ones is shaped more quickly but that the complete arrangement takes time to develop. Given this, Raman spectra exposed the ways in which the interfibrillar aggregation and stabilization time is associated for different gels. A proven relationship between ferulic acid and the tubular formation was observed. Significant intensity ratio analysis changes showed the occurrence of the inversion of modes 1160 cm^{-1} and 1190 cm^{-1} (I1160/I1190), as represented in Figure 2.2.7A. The initial intensity ratio was effectively equal to the final one for 10 wt% samples, meaning that a faster gelation process occurred. This ratio was also steered by the dissimilar intensity ratios for 8 wt% (between 1 and 3) and 10 wt% (at around 0.5). The intensity ratio of modes 1590 cm^{-1} and 1440 cm^{-1}

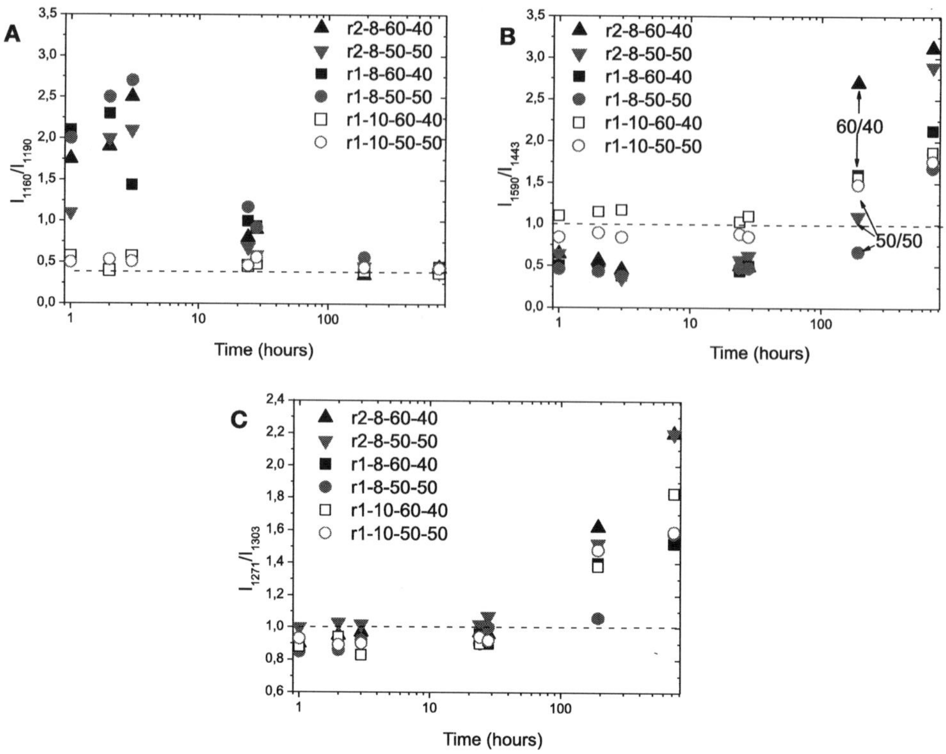

Figure 2.2.7 Raman modes that revealed the highest intensity variation as a function of storage for sterol-based oleogels with 8 and 10 wt% of γ-oryzanol:β-sitosterol (50:50 and 60:40 ratio – being 60:40 the equimolar ratio and 50:50 the one with increased β-sitosterol); 10 wt% (open symbols) and 8 wt% (full symbols); red and black colors represent 50:50 and 60:40 ratios, respectively. xx axis on a logarithmic scale. The horizontal dashed line indicates (A) the final condition reached; (B,C), the value 1 for the intensity ratio (meaning equal intensity). Martins, A.J. et al. 2022 / MDPI / CC BY 4-0.

displayed an intensity inversion, which resulted from the storage time effect (Figure 2.2.7B) from a < 1 value (initially) to a > 1 value (at the end). The β-sitosterol and γ-oryzanol ratio also effected gelation time (Figure 2.2.7b) as the primary gelling stage was reached earlier for the 60:40 samples. Likewise, the time performance observed for I1271/I1303 uncovered the intensity inversion that occurred with the faster cooling rate, which demonstrated a steeper and faster inversion compared with the slower rate (Figure 2.2.7C).

It was also observed how the tubular self-assembled system's micro and nanometric stability is intertwined with aspects like oleogelator ratios and concentrations, as well as cooling rates and storage periods. These particularities granted insights into the tangible gelation point as a matter the molecular stabilization's time scale. This can be decisive for the processing and manufacturing novel products that integrate oleogels in their formulation.

2.2.2.3 Hydrates

One of the interesting aspects about sterol-based gelation is hydrate crystals' formation. Within the oryzanol-sterols system, hydrate forms' formation is responsible for gel disturbance; this is a consequence of the lack of dispersibility of the water droplets in the gel matrix. Studies of emulsion-based gels showed that tubular self-assembled arrangement creation is fostered by reducing water activity and/or by developing emulsions using oils with low polarity. Bot et al. discovered that the presence of hydrates due to emulsification on sterol-based gels leads to extremely soft gels (Bot et al. 2009; Bot, den Adel, and Heussen 2010). As discussed earlier in this chapter, modifications among fibrillar arrangements will likely occur when polar molecules are available (Figure 2.2.8A) in the vicinity of the ferulic acid moieties, since this will foster gel disturbance during sol-gel transition temperatures. Such disturbance in the fibrillar network was investigated by Matheson et al. using molecular dynamic simulations. This investigation explained the consequences of forming hydrogen bonds between water, and the hydroxyl group of sitosterol and the carbonyl group of oryzanol, wherein the gel structure would be compromised (Matheson and Koutsos et al. 2017). Similar behavior was observed as γ-oryzanol and β-sitosterol mixtures were applied in structuring w/o emulsions. The tubular network did not shape completely, or it only partially formed with the tendency to endure disintegration alongside prolonged storage periods. To some degree, this ternary system produced an

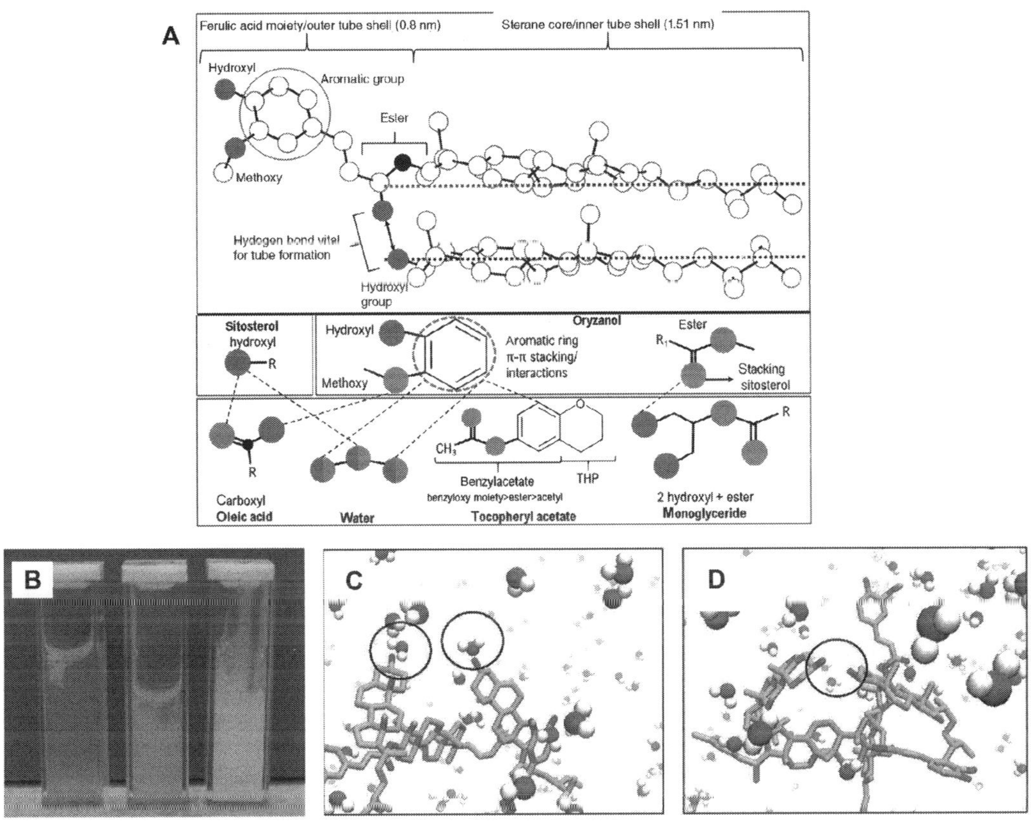

Figure 2.2.8 A) Top: Stacking of γ-oryzanol and β-sitosterol, red indicates main interactions points that contribute to the tube's stability or to form inter-tubule interactions. Bottom: Potential interactions of additives' functional groups with functional groups of sitosterol and oryzanol. Red circles: hydrogen bond donor. Green circles: hydrogen bond acceptor. Red dotted line: π–π interactions and aromatic stacking, graphic does not aim to provide complete overview of all possible interactions; B) sunflower oil sterol-based oleogel organogel (left), a gel from the same batch with added glycerol droplets (center), and an emulsion-based oleogel from the initial batch with added (right); C) and D) show the hydrogen bond network between the hydroxyl group of sitosterols and water molecules and the hydrogen bond between sitosterols. Sitosterol is shown with cyan carbons and the hydrogen bonds are in circles. A) Reproduced from Scharfe, M. et al. 2021 / with permission of John Wiley & Sons; B, C, and D) Reproduced from Matheson et al. (2018), with permission from Royal Society of Chemistry.

emulsion association with reduced viscoelastic (and optical) properties from the original binary gel. Sawalha et al. observed that emulsion-based oleogels with a water activity below 0.9 are exceptionally stable (for at least one year), which displays clear evidence of tubules within its structure (Sawalha et al. 2012). Despite this, the authors showed that by keeping the water activity under 0.9, with the addition of sodium chloride, sterol hydration was prevented, thus maintaining the tubular structuring network. Molecular dynamics simulations indicated that using glycerol as a component in a ternary system (as a substitute for water) induced a successful association with water molecules; when these were maintained at a certain level, the capacity to shape the known tubular structure was increasingly stable (Matheson et al. 2018).

As discussed earlier, γ-oryzanol and β-sitosterol hydrates can develop under distinct environmental conditions. It was reported that the presence of water will interfere with forming the self-assembled fibrillar network, which will produce different types of building blocks. Recently, Pinto et al. developed water-in-oleogel structured emulsions using γ-oryzanol and β-sitosterol. Due to the presence of a dispersed aqueous phase, hydrate crystalline forms (hemi-hydrates and monohydrates) of β-sitosterol developed at the water–oil interface, thus modifying the hydrogen-bonding network commonly generated between the β-sitosterol and oryzanol (Pinto et al. 2022). Hence, the influence of the oleogelator ratio on gel firmness would be more significant for emulsion-based oleogels than for pure oleogels. However, it can highlight the feasibility of structured emulsion-based systems toward the production of fat mimetic like structures, so that oleogel particle-filled systems can be considered good fat substitution candidates in foods. With that in mind, a ternary system composed of an aqueous part, possibly a polymeric-based hydrogel, could act similarly, producing an emulsified composition with some degree of viscoelasticity with an apolar fraction with sterol-based fibrillar building blocks.

2.2.3 Other Combinations Including Sterols

Aging polycrystalline networks with and without hydrates are some facets to consider with co-crystallization methods using sterols and other oleogelators. Recent studies have focused on the combination of phytosterols and other oleogelators, which has escalated the development of numerous complex microstructures that include O/W, W/O, and bi-continuous assemblies (please see table 2.2.1). It is expected that in coming years, further exploration on the combination of oleogelators involving sterols will translate to new scientific findings. Given this, mixtures of sterols and glycerides as well as the development of oleofoams or whipped oleogels, and the development of food bio-inks for 3D printing are explored in this text.

2.2.4 Perspective on the Industrial Applicability

Defining strategies/guidelines related to developing products using sterol-based oleogels is still a conundrum. Difficulties lie in openly identifying the ways in which building blocks behave in different oleogel systems and how their actions will translate in industrial applications where a greater range of elemental interactions makes such identification even harder. For instance, the effects of the gelator molecules on the stabilization of nutraceuticals (lipophilic and hydrophilic), can present consequences for the oleogels/bigels' physical stability due to potential crystal interactions between the bioactives and the sterol molecules during crystalization. It is straightforward to consider that most of the technological and fundamental qualities for fat mimicry, combined with

Table 2.2.1 Gelator combinations including sterols toward oleogel production.

Non-sterol gelator material	Approved for food applications	Rationale/Description	References
Mono and diglycerides of fatty acids	EU (additive E471) and FDA	Oleogel composed of monoglyceride combinations and oil phase; Monoglycerides oleogels and phytosterols mixtures used as 3D printing materials; Production of highly stable whippable oleogels through adsorption of lipidic crystals at the oil-air interface.	(Bin Sintang et al. 2017, 2020; Chen et al. 2019; Zampouni et al. 2021) (Cotabarren, Cruces and Palla 2019, 2020) (Truong, Prakash, and Bhandari 2019)
Sodium caseinate and octenylsuccinate starch	EU (additive E1450) and FDA	Structured flavoring oil-in-water emulsions for controlled release of volatiles) based on the crystallization behavior of β-sitosterol in the presence of emulsifier (sodium caseinate, and octenylsuccinate starch).	(Chen et al. 2016)
Lecithin	EU (additive E 322) and FDA	Highly viscous suspensions and solid-like gels with an aqueous phase); (Oleogels as vehicles for resveratrol and blueberry extracts).	(Matheson and Dalkas et al. 2017) (Qu et al. 2022)
Plant-based saponins (Glycyrrhizic acid, tea saponin)	GRAS status from FDA	Development of self-standing structured emulsion gels/bigels); Development of oleogel-based flavor emulsions and oil powders using ultrasonic cavitation emulsification).	(Li et al. 2021) (Chen et al. 2021)

consumer-preferred properties, usually occur in a hybrid system. On the other hand, overcoming regulatory concerns might present additional challenges because of the synergistic effects among distinct gelators.

With industry interest fostered by academic research, is essential to establish a more direct approach towards exploring food formulations and integrating the most appropriate oleogels and their derivatives. The suitability of oleogels' applications in foods will remain connected to the role of the structuring kinetics; this is a critical point for successful implementation and scalability. State transition in chocolate would be greatly influenced by oleogel structuring kinetics; chocolate manufacturing could be heavily targeted for this technological breakthrough because chocolate is constituted by a continuous phase organized by a blend of fats. In this case, merits could go from developing heat-resistant chocolate (Stortz and Marangoni 2013), preventing surface fat bloom (as a consequence of oil migration) (Wendt et al. 2017), and into substituting expensive and poor yield cocoa butter (Espert et al. 2021). The role of the food matrix is somewhat elemental for effective industrial implementation in that variations on processing environmental conditions will be key for developing distinct food products. For example, lipid phase transformations (e.g., polymorphism) during the processing conditions (Bot and Flöter 2013) will influence products like baked goods (confectioneries, pastries, or spreads), which embody a critical class of products where oleogel structuring kinetics will represent an elementary function (Fayaz et al. 2017; Espert et al. 2021; Selvasekaran and Chidambaram 2021). Oleogel kinetics is expected to play a secondary role in products with a semi-solid phase (emulsified) core ingredient, like ice creams or the so-called processed cold meats (Gao et al. 2021). As noted, the water activity of the food matrix will direct the oleogel applicability since it will affect the fibrillar self-assemblies' maintenance, which come from the oryzanol-sterol system. Our perspective is that using phytosterol-like molecules as gelators in both hybrid and native gelling systems would facilitate the introduction of sterol-based oleogel systems in the industrial food processing. It is known that molecular arrangements within sterol-based oleogels can produce oleogels with different technological properties when compared to those shaped in a water-free environment. The study of hydrate systems with these oleogelators is still underexplored. Future studies should increase knowledge of their applicability as fat replacers with the capability to mimic natural fat (providing similar structural configuration and supplementary quality-wise characteristics) while being nutritionally healthier. The intrinsic lipid content of oleogels could represent a primary challenge when combined with aqueous food matrices; hence, developing emulsion-based systems will surely enable their easy integration, while meaningfully reducing the fat saturation of certain products while expanding their potential toward integrating hydrophilic bioactive compounds. Food matrix compositions can be highly diverse and complex, and the interaction of already formed oleogel structures with a different set of food compositions is yet to be understood. This represents an exciting technological challenge for the future.

2.2.5 Conclusion

Commonly, molecular engineering concepts are applicable toward the design and modulation of the functional and physicochemical properties of oleogel structures and their derivatives, thus improving the targeting of functionality toward very specific applications. The main goal of this chapter was to provide an overview about the suitability of sterol-based oleogels to be used as fat mimicking materials in food products. It is greatly recognized that the most interesting oil structuring results for these kinds of oleogels are obtained by creating nanometric tubular structures using the β-sitosterol and γ-oryzanol combination (in equimolar proportion). This produces transparent and high-mechanically resistant gels. Both β-sitosterol and γ-oryzanol are naturally-sourced materials; they are highly compatible, and their availability is crucial for the material's price cutback and to scale oleogel production for food manufacturing. These structures are highly likely to arrest the oil phase, consuming relatively low amounts of oleogelator material wherein no chemical modifications are induced, which ensures the gelation mechanics' success. Features such as disintegration set-temperature, mechanical properties, and even optical characteristics of this system are highly susceptible to tailoring actions through concentration-effects. When non-equimolar ratios of β-sitosterol and γ-oryzanol are used, crystallites are produced. The role of phytosterols on lowering intestinal cholesterol absorption and of being pivotal elements for faecal removal of cholesterol are certainly advantageous features toward healthier food formulations. Identifying a food grade alternative to γ-oryzanol within this binary system is still unmapped; that progress will lend a different perspective for the feasible applicability of this gelling system and similar ones as well.

References

Ayompe, L.M., Schaafsma, M., and Egoh, B.N. (2021). Toward sustainable palm oil production: The positive and negative impacts on ecosystem services and human wellbeing. *Journal of Cleaner Production* 278: 123914. doi: 10.1016/j.jclepro.2020.123914.

Bin Sintang, M.D. et al. (2017). Oil structuring properties of monoglycerides and phytosterols mixtures. *European Journal of Lipid Science and Technology* 119 (3). doi: 10.1002/ejlt.201500517.

Bin Sintang, M.D. et al. (2020). Modulating the crystallization of phytosterols with monoglycerides inthe binary mixture systems: Mixing behavior and eutectic formation. *Chemistry and Physics of Lipids* Elsevier 230 (April): 104912. doi: 10.1016/j.chemphyslip.2020.104912.

Bot, A. (2018). Phytosterols. *Encyclopedia of Food Chemistry* 225–228. doi: 10.1016/B978-0-08-100596-5.21626-0.

Bot, A. and Agterof, W.G.M. (2006). Structuring of edible oils by mixtures of γ-oryzanol with β-sitosterol or related phytosterols. *JAOCS, Journal of the American Oil Chemists' Society* 83 (6): 513–521. doi: 10.1007/s11746-006-1234-7. Wiley.

Bot, A., den Adel, R., Roijers, E.C., and Regkos, C. (2009). Effect of sterol type on structure of tubules in sterol + γ-oryzanol-based organogels. *Food Biophysics* 4 (4): 266–272. doi: 10.1007/s11483-009-9124-9.

Bot, A., den Adel, R., and Heussen, P.C.M. (2010). Effect of water on self-assembled tubules in β-sitosterol + γ-oryzanol-based organogels. *Journal of Physics. Conference Series* 247 (1): 12025. Available at: http://stacks.iop.org/1742-6596/247/i=1/a=012025.

Bot, A., Den Adel, R., and Roijers, E.C. (2008). Fibrils of γ-oryzanol + β-sitosterol in edible oil organogels. *JAOCS, Journal of the American Oil Chemists' Society* 85 (12): 1127–1134. doi: 10.1007/s11746-008-1298-7.

Bot, A. and Flöter, E. (2013). Application of edible oils. *Edible Oil Processing* 223–249. doi: 10.1002/9781118535202.ch8. Wiley Online Books.

Bot, A., Veldhuizen, Y.S.J., den Adel, R., and Roijers, E.C. (2009). Non-TAG structuring of edible oils and emulsions. *Food Hydrocolloids* 23 (4): 1184–1189. doi: 10.1016/j.foodhyd.2008.06.009. Elsevier Ltd.

Bran, R., King, J.W., and John, D. (2004) (12) United States Patent. 1(12).

Calligaris, S., Mirolo, G., Da Pieve, S., Arrighetti, G., and Nicoli, M.C. (2014). Effect of oil type on formation, structure and thermal properties of γ-oryzanol and β-sitosterol-based organogels. *Food Biophysics* 9 (1): 69–75. doi: 10.1007/s11483-013-9318-z.

Cantrill, R. (2008). *Phytosterols, phytostanols and their esters*. Chemical and Technical Assessment (CTA), Food and Agriculture Organization of the United Nations. Available at: https://www.fao.org/fileadmin/templates/agns/pdf/jecfa/cta/69/Phytosterols.pdf (accessed 29 April 2022).

Carter, C., Finley, J., Fry, J., Jackson, D., and Willis, L. (2007). Palm oil markets and future supply. *European Journal of Lipid Science and Technology* 109 (4): 307–314. doi: 10.1002/ejlt.200600256.

Chen, X.W. et al. (2019). Engineering phytosterol-based oleogels for potential application assustainable petrolatum replacement. *RSC Advances* Royal Society of Chemistry 10 (1): 244–252. doi: 10.1039/c9ra06950j.

Christiansen, L.I., Rantanen, J.T., von Bonsdorff, A.K., Karjalainen, M.A., and Yliruusi, J.K. (2002). A novel method of producing a microcrystalline β-sitosterol suspension in oil. *European Journal of Pharmaceutical Sciences* 15 (3): 261–269. doi: 10.1016/s0928-0987(01)00223-8.

Claycamp, D.L. et al. (2014) (12) United States Patent. 2(12).

Dalkas, G., Matheson, A.B., Vass, H., Gromov, A., Lloyd, G.O., Koutsos, V., Clegg, P.S., and Euston, S.R. (2018). Molecular interactions behind the self-assembly and microstructure of mixed sterol organogels. *Langmuir* 34 (29): 8629–8638. doi: 10.1021/acs.langmuir.8b01208. American Chemical Society.

Duffy, N., Blonk, H.C.G., Beindorff, C.M., Cazade, M., Bot, A., and Duchateau, G.S.M.J.E. (2009). Organogel-based emulsion systems, micro-structural features and impact on in vitro digestion. *JAOCS, Journal of the American Oil Chemists' Society* 86 (8): 733–741. doi: 10.1007/s11746-009-1405-4. Springer-Verlag.

Espert, M., Hernández, M.J., Sanz, T., and Salvador, A. (2021). Reduction of saturated fat in chocolate by using sunflower oil-hydroxypropyl methylcellulose based oleogels. *Food Hydrocolloids* 120 (April). doi: 10.1016/j.foodhyd.2021.106917.

European Food Safety Authority (2012). Scientific Opinion on the substantiation of a health claim related to 3 g / day plant sterols / stanols and lowering blood LDL-cholesterol and reduced risk of (coronary) heart disease pursuant to Article 19 of Regulation (EC). *EFSA Journal* 10 (5): 2693. doi: 10.2903/j.efsa.2012.2693.

Fayaz, G., Goli, S.A.H., Kadivar, M., Valoppi, F., Barba, L., Calligaris, S., and Nicoli, M.C. (2017). Potential application of pomegranate seed oil oleogels based on monoglycerides, beeswax and propolis wax as partial substitutes of palm oil in functional chocolate spread. *LWT* 86: 523–529. doi: 10.1016/j.lwt.2017.08.036.

Fernandes, P. and Cabral, J.M.S. (2007). Phytosterols: Applications and recovery methods. *Bioresource Technology* 98 (12): 2335–2350. doi: 10.1016/j.biortech.2006.10.006.

Flöter, E., Wettlaufer, T., Conty, V., and Scharfe, M. (2021). Oleogels-their applicability and methods of characterization. *Molecules (Basel, Switzerland)* 26: 6. doi: 10.3390/molecules26061673.

Franco, D., Martins, A.J., López-Pedrouso, M., Purriños, L., Cerqueira, M.A., Vicente, A.A., Pastrana, L.M., Zapata, C., and Lorenzo, J.M. (2019). Strategy toward replacing pork backfat with a linseed oleogel in Frankfurter sausages and its evaluation on physicochemical, nutritional, and sensory characteristics. *Foods* 8 (9): 366. doi: 10.3390/foods8090366.

Franco, D., Martins, A.J., López-Pedrouso, M., Cerqueira, M.A., Purriños, L., Pastrana, L.M., Vicente, A.A., Zapata, C., Lorenzo, J.M. (2020). Evaluation of linseed oil oleogels to partially replace pork backfat in fermented sausages. *Journal of the Science of Food and Agriculture* 100 (1): 218–224. doi: 10.1002/jsfa.10025.

Gao, Y., Li, M., Zhang, L., Wang, Z., Yu, Q., and Han, L. (2021). Preparation of rapeseed oil oleogels based on beeswax and its application in beef heart patties to replace animal fat. *LWT* 149 (February): 111986. doi: 10.1016/j.lwt.2021.111986. Elsevier Ltd.

Himawan, C., Starov, V.M., and Stapley, A.G. (2006). Thermodynamic and kinetic aspects of fat crystallization. *Advances in Colloid and Interface Science* 122 (1–3): 3–33. doi: 10.1016/j.cis.2006.06.016.

Jesus, S.P., Grimaldi, R., and Hense, H. (2010). Recovery of γ-oryzanol from rice bran oil byproduct using supercritical fluid extraction. *The Journal of Supercritical Fluids* 55 (1): 149–155. doi: 10.1016/j.supflu.2010.08.004.

EFSA Journal (2016). Risks for human health related to the presence of 3- and 2-monochloropropanediol (MCPD), and their fatty acid esters, and glycidyl fatty acid esters in food. *EFSA Journal* 14 (5). doi: 10.2903/j.efsa.2016.4426.

Kayathi, A., Chakrabarti, P.P., Bonfim-Rocha, L., Cardozo-Filho, L., Bollampalli, A., and Jegatheesan, V. (2021). Extraction of γ-Oryzanol from defatted rice bran using supercritical carbon dioxide (SC-CO2): process optimisation of extract yield, scale-up and economic analysis. *Process Safety and Environmental Protection* 148: 179–188. doi: 10.1016/j.psep.2020.09.067.

Kouzounis, D., Lazaridou, A., and Katsanidis, E. (2017). Partial replacement of animal fat by oleogels structured with monoglycerides and phytosterols in frankfurter sausages. *Meat Science* 130: 38–46. doi: 10.1016/j.meatsci.2017.04.004.

Lemus, C., Angelis, A., Halabalaki, M., and Skaltsounis, A.L. (2014). *γ-Oryzanol. An Attractive Bioactive Component from Rice Bran., Wheat and Rice in Disease Prevention and Health*. Elsevier. doi: 10.1016/B978-0-12-401716-0.00032-5.

Maniet, G., Jacquet, N., and Richel, A. (2019). Recovery of sterols from vegetable oil distillate by enzymatic and non-enzymatic processes. *Comptes Rendus Chimie* 22 (4): 347–353. doi: 10.1016/j.crci.2019.01.004.

Martins, A.J., Lorenzo, J.M., Franco, D., Pateiro, M., Domínguez, R., Munekata, P.E.S., Pastrana, L.M., Vicente, A.A., Cunha, R., and Cerqueira, M.A. (2020). Characterization of enriched meat-based pâté manufactured with oleogels as fat substitutes. *Gels* 6 (2): 1–14. doi: 10.3390/gels6020017.

Martins, A.J., Cerqueira, F., Vicente, A.A., Cunha, R.L., Pastrana, L.M., and Cerqueira, M.A. (2022). Gelation behavior and stability of multicomponent sterol-based oleogels. *Gels* 8 (1): 37. doi: 10.3390/gels8010037.

Martins, A.J., Cerqueira, M.A., Pastrana, L.M., Cunha, R.L., and Vicente, A.A. (2019). Sterol-based oleogels' characterization envisioning food applications. *Journal of the Science of Food and Agriculture* 99 (7): 3318–3325. doi: 10.1002/jsfa.9546. Wiley.

Martins, A.J., Lorenzo, J.M., Franco, D., Vicente, A.A., Cunha, R.L., Pastrana, L.M., Quiñones, J., and Cerqueira, M.A. (2019). Omega-3 and polyunsaturated fatty acids-enriched hamburgers using sterol-based oleogels. *European Journal of Lipid Science and Technology* 121 (11): 1900111. doi: 10.1002/ejlt.201900111. Wiley.

Martins, A.J., Vicente, A.A., Cunha, R.L., and Cerqueira, M.A. (2018). Edible oleogels: An opportunity for fat replacement in foods. *Food & Function* 9 (2): 758–773. doi: 10.1039/c7fo01641g. Royal Society of Chemistry.

Matheson, A., Dalkas, G., Mears, R., Euston, S.R., and Clegg, P.S. (2018). Stable emulsions of droplets in a solid edible organogel matrix. *Soft Matter* 14 (11): 2044–2051. doi: 10.1039/c8sm00169c. Royal Society of Chemistry.

Matheson, A.B., Dalkas, G., Gromov, A., Euston, S.R., and Clegg, P.S. (2017). The development of phytosterol-lecithin mixed micelles and organogels. *Food & Function* 8 (12): 4547–4554. doi: 10.1039/C7FO01271C. Royal Society of Chemistry.

Matheson, A.B., Koutsos, V., Dalkas, G., Euston, S., and Clegg, P. (2017). Microstructure of β-sitosterol:γ-oryzanol edible organogels. *Langmuir* 33 (18): 4537–4542. doi: 10.1021/acs.langmuir.7b00040.

Narayan, A.V., Barhate, R.S., and Raghavarao, K.S.M.S. (2006). Extraction and purification of oryzanol from rice bran oil and rice bran oil soapstock. *JAOCS, Journal of the American Oil Chemists' Society* 83 (8): 663–670. doi: 10.1007/s11746-006-5021-2.

Okuro, P.K., Malfatti-Gasperini, A., Vicente, A.A., and Cunha, R.L. (2018). Lecithin and phytosterols-based mixtures as hybrid structuring agents in different organic phases. *Food Research International* 111 (March): 168–177. doi: 10.1016/j.foodres.2018.05.022. Elsevier.

Pinto, T.C., Martin, A.J., Pastrana, L., Pereira, M.C., and Cerqueira, M.A. (2022) *Water-in-oleogel* emulsion based on γ -oryzanol and phytosterol mixtures: Challenges and its potential use for the delivery of bioactives. 9 (November 2021): 1–9. doi: 10.1002/aocs.12636.

Rogers, M.A. (2011). Co-operative self-assembly of cholesterol and γ-oryzanol composite crystals. *CrystEngComm* 13 (23): 7049. doi: 10.1039/c1ce05818e. The Royal Society of Chemistry.

Rogers, M.A., Strober, T., Bot, A., Toro-Vazquez, J.F., Stortz, T., and Marangoni, A.G. (2014). Edible oleogels in molecular gastronomy. *International Journal of Gastronomy and Food Science* 2 (1): 22–31. doi: 10.1016/j.ijgfs.2014.05.001.

Saleem, M. (2009). Lupeol, a novel anti-inflammatory and anti-cancer dietary triterpene. *Cancer Letters* 285 (2): 109–115. doi: 10.1016/j.canlet.2009.04.033.

Sawalha, H., den Adel, R., Venema, P., Bot, A., Flöter, E., and van der Linden, E. (2012). Organogel-emulsions with mixtures of β-sitosterol and γ-oryzanol: influence of water activity and type of oil phase on gelling capability. *Journal of Agricultural and Food Chemistry* 60 (13): 3462–3470. doi: 10.1021/jf300313f. American Chemical Society.

Sawalha, H., Margry, G., den Adel, R., Venema, P., Bot, A., Flöter, E., and van der Linden, E. (2013). The influence of the type of oil phase on the self-assembly process of γ-oryzanol + β-sitosterol tubules in organogel systems. *European Journal of Lipid Science and Technology* 115 (3): 295–300. doi: 10.1002/ejlt.201100395.

Scharfe, M., Ahmane, Y., Seilert, J., Keim, J., and Flöter, E. (2019). On the effect of minor oil components on β-Sitosterol/γ-oryzanol Oleogels. *European Journal of Lipid Science and Technology* 121 (8): 1–13. doi: 10.1002/ejlt.201800487.

Scharfe, M., Prange, D., and Flöter, E. (2021). The composition of edible oils modifies β-sitosterol/γ-oryzanol oleogels. Part II: addition of selected minor oil components. *JAOCS, Journal of the American Oil Chemists' Society* (March): 1–21. doi: 10.1002/aocs.12556.

Selvasekaran, P. and Chidambaram, R. (2021). Advances in formulation for the production of low-fat, fat-free, low-sugar, and sugar-free chocolates: An overview of the past decade. *Trends in Food Science and Technology* 113 (December 2020): 315–334. doi: 10.1016/j.tifs.2021.05.008. Elsevier Ltd.

Srikaeo, K. (2014). Organic rice bran oils in health. In: *Wheat and Rice in Disease Prevention and Health* (ed. R.R. Watson, V.R. Preedy, and S.B.T.-W. Zibadi R. in D. P. and H.), 453–465. San Diego: Academic Press. doi: 10.1016/B978-0-12-401716-0.00035-0.

Stortz, T.A., Zetzi, A.K., Barbut, S., Cattaruzza, A., and Marangoni, A.G. (2012). Edible oleogels in food products to help maximize health benefits and improve nutritional profiles. *Lipid Technology* 24 (7): 151–154. doi: 10.1002/lite.201200205.

Stortz, T.A. and Marangoni, A.G. (2013). Ethylcellulose solvent substitution method of preparing heat resistant chocolate. *Food Research International* 51 (2): 797–803. doi: 10.1016/j.foodres.2013.01.059. Food Research International.

Wendt, A., Abraham, K., Wernecke, C., Pfeiffer, J., and Flöter, E. (2017). Application of β-Sitosterol + γ-Oryzanol-Structured Organogel as migration barrier in filled chocolate products. *Journal of the American Oil Chemists' Society* 94 (9): 1131–1140. doi: 10.1007/s11746-017-3024-9.

Wijarnprecha, K., Santiwattana, P., Sonwai, S., and Rousseau, D. (2021). Vegetable oil structuring via γ-oryzanol crystallization. *LWT* 139 (May 2020). doi: 10.1016/j.lwt.2020.110564.

Zhang, T., Liu, r., Chang, M., Jin, Q., Zhang, H. and Wang, X. (2020). Health benefits of 4,4-dimethyl phytosterols: An exploration beyond 4-desmethyl phytosterols. *Food & Function* 11 (1): 93–110. doi: 10.1039/c9fo01205b. Royal Society of Chemistry.

2.3

Lecithin

Thaís Jordânia Silva[1], *Paula Kiyomi Okuro*[2], *Mayanny Gomes da Silva*[2], *Ana Paula Badan Ribeiro*[2], *and Rosiane Lopes da Cunha*[2,*]

[1]*Center for Natural Sciences, Federal University of São Carlos (UFSCar), Campus Lagoa do Sino, Buri, SP, Brazil*
[2]*School of Food Engineering, University of Campinas (UNICAMP), Cidade Universitária Zeferino Vaz, Campinas, SP, Brazil*
* *Corresponding author*

2.3.1 Introduction

Fats and oils are ingredients present in many foods, such as margarines, bakery products, chocolates, and spreads (Chaves et al. 2018). For many years, partially hydrogenated vegetable fat was used to improve the physical properties of food products. After the harmful effects of trans fatty acids (TFA) formed in partial hydrogenation on human health were revealed, new technological alternatives were developed (Briggs et al. 2017); however, the alternatives increased saturated fatty acids (SFA) consumption, which increases the risk of cardiovascular diseases. One of the alternatives for replacing trans/saturated fats is fat mimetics such as oleogels and oleogel-based emulsions, which are new fats with a healthier lipid profile (Silva et al. 2021a).

Lecithin is a natural food additive that is widely applied as an emulsifying agent. Lecithin is a compound comprising a mixture of phospholipids obtained from vegetable sources such as soybean, sunflower, rapeseed, or rice, and from animal sources (Bueschelberger et al. 2015; Bot et al. 2021). Lecithin has a 5.5% expected average annual consumption increase between 2020 and 2025. This growth in the lecithin market was influenced by the COVID-19 pandemic, an event that directly affected its production and demand. Specifically, the Food and Drug Administration (FDA) relaxed soy lecithin labeling requirements, exempting it from the list of allergens and rebalancing the lecithin industry; the changes also amplified the demand for natural (Mordor Intelligence 2021) and plant-based ingredients. Many consumers are interested in adopting a plant-based diet motivated by issues involving health, sustainability, and ethics (Miki et al. 2017; Grossmann and McClements 2021). Thus, there is a tendency to consume vegetable derived lecithins.

Many components used to produce fat mimetics are natural, such as proteins, plant sterols, lecithin, tocopherols, and plant waxes (Patel et al. 2020). Considering that it is a natural ingredient, fat mimetic from lecithin could be an innovative alternative in the food industry. Although phospholipids can be used as structuring agents, the actual impact of commercially available phospholipids on bulk fat remains to be studied (Robert et al. 2020). For this, it is necessary to understand their interaction with other food ingredients and potential gelators.

In this chapter, we will look at the prospects for lecithin-based strategies in developing fat mimetics. Initially, the general characteristics of lecithin are presented, as well as its sources. In the

Fat Mimetics for Food Applications, First Edition. Edited by Miguel Ângelo Parente Ribeiro Cerqueira and Lorenzo Miguel Pastrana Castro.

following sections, we will discuss the production process of vegetable lecithins and the possible modifications that can be used to improve physical and chemical properties. The self-assembly structuring mechanism of lecithin is discussed with its dependence on polarity solvent. Finally, we bring some techno-functionalities of lecithin in both the crystallization of fats and the development of oleogels and emulsions. We seek to understand the particularities of lecithin that make it so essential for foods' physical characteristics. Finally, we propose some research paths that deal with lecithin as an alternative natural ingredient to be used in products containing fats.

2.3.2 Lecithin Chemistry

2.3.2.1 Types and Composition

According to Codex Alimentarius, lecithin refers to a predominant mixture of phospholipids (> 50%) obtained from animal or vegetable sources (Krog and Sparsø 2004). Phospholipids are present in cell walls and have essential functions in organisms such as protection, separation, and transport of components. In some cases, phospholipids are also a source of acetylcholine and essential fatty acids (McClements 2016). The phospholipid structure consists of two nonpolar fatty acids esterified at the *sn-1* and *sn-2* positions of glycerol and a phosphate group with a hydrophilic residue attached to the *sn-3* position. Generally, saturated fatty acids are preferentially esterified at the *sn-1* position, while unsaturated fatty acids are esterified at the *sn-2* position. In the *sn-3* position, depending on the functional group attached to the phosphate group, lecithin can have phosphatidylcholine (PC), phosphatidylethanolamine (PE), phosphatidylinositol (PI), or phosphatidylserine (PS) (Figure 2.3.1) (Li et al. 2015b; Chung et al. 2017).

PC contains choline with a negative charge on the phosphate group and a positive charge on the trimethylamine group, providing a hydrophilic characteristic. PI has an inositol moiety and five hydroxyl groups, which make the molecule strongly hydrophilic. PE contains an amino group and

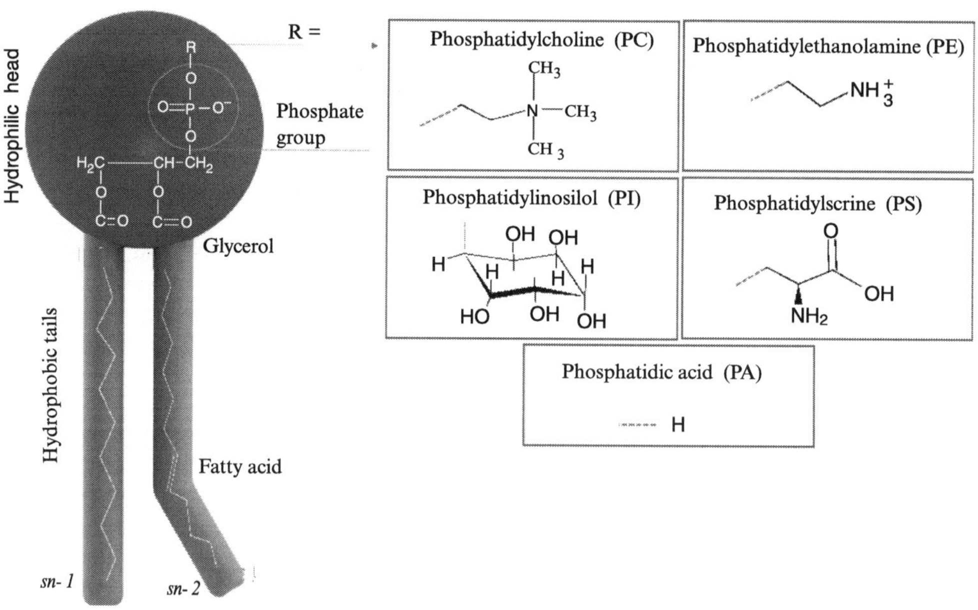

Figure 2.3.1 Chemical structure and classification of the different groups of phospholipids.

a phosphate group; it is poorly hydrated, but it becomes hydrophilic when subjected to acid degumming. PC has oil-in-water (O/W) emulsification-promoting characteristics; PE and PI have water-in-oil (W/O) emulsifying properties, being PE to a lesser extent (Joshi et al. 2006). Phosphatidic acid (PA) is the simplest phospholipid with a hydrogen atom as a functional group; it has two hydroxyl groups that can dissociate. PA can be hydrated and non-hydratable and can be present in lecithin or oil degummed in water. Under acidic pH conditions, PA has a strong hydrophobic character, but if the pH is raised, some of the PA will dissociate and form a monosodium salt (Dijkstra 2017).

Due to its amphiphilic nature, lecithin can be easily adsorbed at the interface, which reduces the interfacial tension with the hydrophobic tail of the fatty acid positioned toward the oily phase while the polar head group is in the aqueous phase (Deng 2021). Therefore, the hydrophilic-lipophilic balance (HLB) of lecithin is determined by its head group and fatty acids. In its natural form, lecithin has low to intermediate HLB values (2–8) and is therefore more suitable for stabilizing W/O systems (low HLB) or forming bilayers or liposomes in aqueous solutions (intermediate HLB values). The hydrolyzed lecithin has an HLB of around 4, but when enriched in the PC fraction, its HLB is approximately 7 and can be dispersed in oily and aqueous phases (Bot et al. 2021). However, lecithin can also be used with other surfactants to improve stability and to form different structures (McClements 2016).

Lecithin can be extracted from soy, eggs, milk, sunflower, rice bran, rapeseed, corn, and other food stuffs (Chen 2015; Chung et al. 2017). Soy lecithin contains approximately 65–75% phospholipids; it is the most widely used lecithin in the food industry given its simple production simple and low cost (Chen 2015). Table 2.3.1 depicts the phospholipid profile and fatty acids composition of main vegetable lecithins according to the Food Chemicals Codex.

Lecithins' composition interferes with its emulsifying property, such as PC acting at the oil-water interface forming a lamellar layer. On the other hand, PE stabilizes W/O emulsions by a hexagonal phase, as does lysophosphatidylcholine in O/W emulsion (Van Nieuwenhuyzen 2015). As noted in Table 1, phospholipid content in vegetable oils varies in composition and properties according to species and between different varieties of the same oil seeds source. Therefore, it is of pivotal importance to know the fingerprint of commercial lecithins to guide their rational application.

2.3.2.2 Technological Manufacture of Lecithin

As mentioned, lecithin is obtainable from both animal and vegetable sources. Vegetable lecithin is a co-product of oil refining, obtained mainly in the degumming stage, followed by drying and cooling operations (Figure 2.3.2). Vegetable lecithin is the most commercially available and used; it has a lot of potential regarding fat mimetic development strategies (Bot et al. 2021).

Degumming is a step the vegetable oils refinement process that aims to eliminate phospholipids associated with problems during the process, such as tank deposition and oil loss. Because of the degumming process, these phospholipids are separated and reused to obtain lecithin. Degumming can be carried out by three main processes: aqueous, acidic, or enzymatic. Traditional gum production occurs through aqueous or acid degumming (Krog and Sparsø 2004). Water (1–3%) is used in aqueous degumming, while usually citric acid is applied in acid degumming; although, acetic, tartaric, or phosphoric acid may also be alternatives in acid degumming. Both processes consist of heating crude oil to around 70°C for a specified time period. In the case of acid degumming, the acid cleaves the bonds of non-hydratable phospholipids (phosphatidic acid salts) and hydrates them; the added water will later be eliminated. The water hydrates the polar lipids for subsequent lecithin gum

Table 2.3.1 Phospholipids and fatty acid composition of lecithin.

	Lecithin Source		
Phospholipid (%)	**Soy**	**Sunflower**	**Rapeseed**
Phosphatidylcholine (PC)	15	16	17
Phosphatidylethanolamine (PE)	11	8	9
Phosphatidylinositol (PI)	10	14	10
Phosphatidic acid (PA)	4	3	4
Fatty acid (%)			
Palmitic (16:0)	16	11	7
Stearic (18:0)	4	4	1
Oleic (18:1)	17	18	56
Linoleic (18:2)	55	63	25
Linolenic (18:3)	7	0	6
Others	1	4	8
Acetone-insoluble matter	62	62	63
Acetone-soluble matter	37	37	36
Food Chemicals Codex			
Acetone-insoluble (%, min.)	50	50	50
Moisture (%, max.)	1.5	1.5	1.5
Toluene insolubles (%, max.)	0.3	0.3	0.3
Acid value (mg KOH/g, max.)	36	36	36
Peroxide value (meq O_2/kg, max.)	100	100	100
Lead (ppm)	1	1	1

Source: Adapted from Van Nieuwenhuyzen W. (2015).

separation by centrifugation. This resulting gum contains water, phospholipids, glycolipids, carbohydrates, some triglycerides, free fatty acids, and carotenoids (Bueschelberger et al. 2015; Bot et al. 2021). Crude lecithin is obtained after careful drying (Bueschelberger et al. 2015; List 2015).

During degumming, the composition of phospholipids affects how efficiently lecithin is obtained. Based on the chemical structure of the phospholipids presented in Section 2.1, PC and PI have highly hydrophilic groups at all pHs. Therefore, PC and PI are easily hydrated during degumming, which increases the extraction yield. On the other hand, PE is poorly hydrated, requiring acidic degumming to protonate its phosphate group, making it more hydrophilic. The hydration of phosphatides is driven by their hydration rate constants, wherein the hydration rate of PC is arbitrarily set at 100; PI, PE and PA have a relative hydration rate of 44, 16, and 8.5, respectively. This parameter helps to explain the technological performance of specific lecithins and the preferential use, for example, from PC-rich sources (Dijkstra 2017).

In degumming, phospholipids are structured as reverse micelles in vegetable oils and hexane-oil mixtures (Mcclements 2020). Factors such as the composition of the medium and the presence of

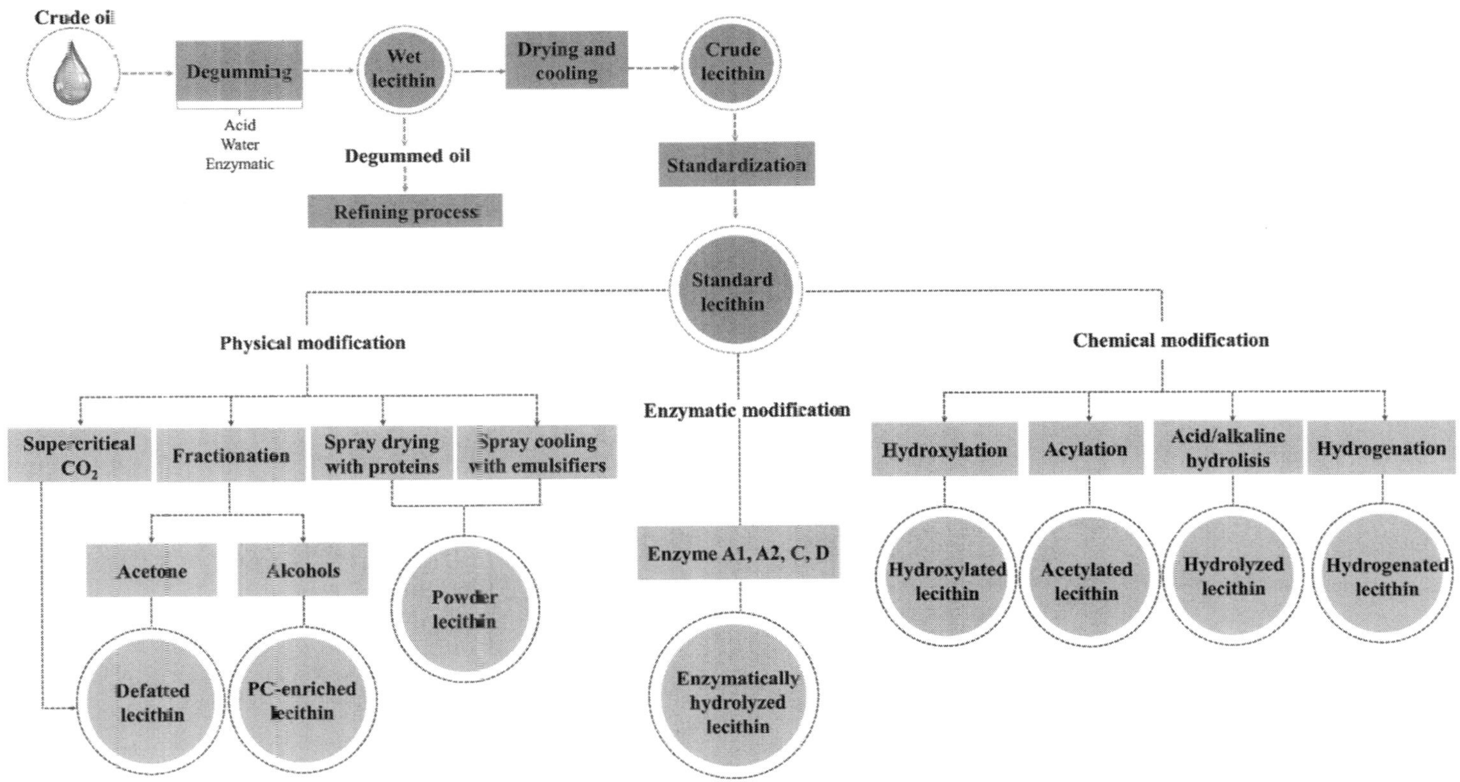

Figure 2.3.2 Schematic representation of the steps involved in lecithin manufacture and modification. Source: Adapted from Miyasaki et al. (2016) and Eot et al. (2021).

moisture strongly affect micelle formation. The greater hydrophobicity of hexane and the presence of moisture result in a decrease of the critical micellar concentration (CMC) (Manjula et al. 2011).

After the degumming step, the gum contains about 70% phospholipids and 30% oil and water. Lecithin is recovered by vacuum drying and steam heating to achieve a moisture content of less than 1%. Excessive drying causes dark-colored lecithin, although, insufficient drying produces a cloudy product (List 2015). Lecithin is stored in suitable containers that stay at 21°C for up to one year (Szuhaj 1983). For food applications, crude lecithin must be subjected to microbiological standardization through hydrogen peroxide treatment. Standardization and required quality may vary according to the country's legislation; quality is mainly a result of efficiency and control in degumming. Lecithin quality is evaluated by appearance (high viscosity or semi-liquid or brown colored powder products), peroxide value (less than 10 meq/Kg), acetone-insoluble substances (not less than 60%), acid value (less than 35 mg of potassium hydroxide per gram), and toluene-insoluble substances (less than 0.3%) (Bueschelberger et al. 2015; Bot et al. 2021). Furthermore, lecithins can have different characteristics depending on their source, variety, region, climate, storage conditions, and other factors (Bueschelberger et al. 2015).

2.3.2.3 Strategies of Lecithin Modification

When used alone, natural lecithin is not suitable for W/O or O/W emulsions stabilization, but it is effective if combined with other emulsifiers (Van Nieuwenhuyzen 1981). Lecithin can be modified to produce more hydrophilic surfactants and lysolecithins, which can stabilize micelles, microemulsions, or O/W emulsions (Chen 2015). A wide range of lecithin can be produced, each with its own characteristics and functionality. Modifications are essential to achieve and to adjust optimal ratios between hydrophilic and lipophilic groups and to ensure optimal surfactant properties (Van Nieuwenhuyzen 2015). Lecithin can be subjected to physical (fractionation, spray drying with proteins, spray cooling with synthetic emulsifiers), chemical (acid and alkali hydrolysis, acetylation, hydroxylation, hydrogenation), and enzymatic modifications as shown in Figure 3.2.2 (Van Nieuwenhuyzen 1981). Chemical modifications involve several methods, but not all are used in food applications because of legal issues that limit the use of these lecithins (Bueschelberger et al. 2015).

2.3.2.3.1 Physical Modification

To obtain lecithin with different functions, physical methods can be used to separate individual phospholipid components. Physical modification involves different fractionation processes that isolate the lecithin fractions, and their properties are influenced by extraction time/temperature and solvent concentration. Spray drying with proteins, spray cooling with synthetic emulsifiers, and membrane technology are also classified as physical modification processes.

Separation of neutral and polar lipids – The separation of neutral (soluble in acetone) and polar lipids—glycolipids and phospholipids—(insoluble in acetones) is also known as de-oiling, which is the main fractionation method. It consists of adding acetone to the gum and then separating the fractions insoluble in acetone, wherein the oil is removed and the phospholipids are concentrated and purified. The final product, after drying, must have less than 25 ppm acetone, more neutral flavors, and greater functionality (Szuhaj 1983; Joshi et al. 2006).

Extraction of supercritical carbon dioxide (CO_2) – Oil-free lecithins are produced with supercritical gases. Under conditions of 40°C and 300 bar, CO_2 has solubility characteristics similar to liquid acetone, dissolving neutral lipids. The advantage of this process is the ease of separating gas and oil without residual oxygen and solvents. Disadvantages are the low oil capacity to dissolve in CO_2, high solvent proportions, and low yield (Teberikler et al. 2001; van Nieuwenhuyzen and Tomás 2008).

Fractionation of oil-free lecithin with alcohols – In fractionation with acetone, the resulting product (insoluble in acetone) contains 35–85% phospholipids. As already mentioned, this product is composed of PC, PE, PI, and PA in different proportions. Each fraction has a different solubility in alcohol with PC being the most soluble among the phospholipids, PI and PA being almost insoluble, and PE being found in both fractions. Therefore, PC is easily extracted with alcohol in a mixing vessel or in countercurrent columns (Joshi et al. 2006). Treatment of oil-free lecithin with 90% ethanol yields a soluble fraction containing mainly PC and PE and the insoluble fraction PI. The soluble fraction is more suitable for O/W emulsions and the insoluble ones for W/O emulsions (van Nieuwenhuyzen 1981; List 2015). Fractionation also encompasses oil removal with acetone to remove triglycerides and to increase phospholipid content (Bot et al. 2021). The composition of acetone-precipitated soy phospholipids is approximately 41% PC, 34% PE, 19% PI, and 6% other phospholipids. Isolated phosphatides can be further fractionated by ethyl alcohol or another short-chain alcohol (Krog and Sparsø 2004). Sunflower lecithin that is enriched with PC by alcohol fractionation and that is enzymatically modified presents the best emulsifying properties against creaming and coalescence destabilization (List 2015; Cabezas et al. 2016).

Spray drying with proteins – In this modification method, lecithin (pasty, oily, viscous, and sticky) is coated with proteins using the spray-drying process. The lecithin resulting from this process is a more easily handled product with different applications and properties (Joshi et al. 2006).

Spray cooling with synthetic emulsifiers – Lecithin is combined with emulsifiers (e.g., mono- and diacylglycerols) or diluted with special oils such as sunflower oils or medium-chain triglycerides. This product has fewer viscous properties than standard lecithin and is sprayed to create flaked or powdered products (Joshi et al. 2006).

2.3.2.3.2 Enzymatic Modification

Specific enzymes, such as phospholipases A1, A2, C, and D, hydrolyze lecithin, result in a product with better emulsification properties in O/W systems. The greater the degree of hydrolysis, the smaller the droplets generated in emulsification; although, process conditions vary according to the enzyme. Each phospholipase has its preferential attack point, which generates different products. Phospholipase A2 splits lecithin at the *sn-2* position, forming lysophospholipids and free fatty acids. Phospholipase A1, in contrast, splits the fatty acid in the *sn-1* position. Phospholipase D increases the PA content of lecithin (Joshi et al. 2006; Bueschelberger et al. 2015; List 2015). On the other hand, enzymes are unable to catalyze the hydrolysis of non-hydratable phosphatides under industrial conditions. Consequently, the enzymatic industrial degumming step must be preceded by chemical degumming to arrive at a degummed oil with a residual phosphorus content low enough to be physically refined (Dijkstra 2010).

2.3.2.3.3 Chemical Modification

Acetylation – In this method, the PE group is acetylated with acetic acid anhydride, resulting in acetyl-PE. The processes' principle is blocking of the "zwitterionic" PE group, which improves its properties. Acetylated lecithins are excellent emulsifiers in O/W systems with an HLB between 9–10; they also have good thermal stability that results in better browning resistance (Joshi et al. 2006; Bueschelberger et al. 2015; List 2015).

Hydroxylation – Hydrolaxation consists of using hydrogen peroxide with acids (e.g., lactic acid), which adds hydroxyl groups in the unsaturated fatty acid's double bond. This chemical reaction does not change the phosphatide group and results in a highly hydrophilic product (HLB around 10) that has improved emulsifying properties (Joshi et al. 2006; Bueschelberger et al. 2015; List 2015). Hydroxylated lecithin is best for O/W emulsions and easily disperses in cold water. It is common to combine acetylation or hydrolysis with hydroxylation and use an acetylated or hydrolyzed lecithin as a raw material for hydroxylation to improve the reaction, which produces an even more hydrophilic lecithin (Van Nieuwenhuyzen 2015).

Hydrogenation – The hydrogenation process has the same principle as the hydrogenation conventionally used to produce plastic fats. The unsaturated fatty acids present in lecithin are added with hydrogens that transform the double bond into a single one, that is, into a saturated fatty acid. Suitable catalysts for this process are palladium and platinum. This modification aim to improve oxidative stability and emulsifying properties (Bueschelberger et al. 2015). This process is preferably carried out with very pure PC fractions because the impurities present in the crude lecithin can interfere with the catalysts, thus affecting the reaction (Van Nieuwenhuyzen 2015).

Hydrolysis by acids and alkalis – Hydrolysis by acids and alkalis consists of separating fatty acids under acidic and alkaline conditions; however, these processes are less specific than enzymatic hydrolysis, which can result in lecithin with an undesirable dark color (Bueschelberger et al. 2015). The products created have improved properties for forming O/W emulsions in the presence of calcium and magnesium ions. Moreover, they have been used to manufacture greases, lubricants, and plastic formulations (Joshi et al. 2006).

The type of lecithin added (native or hydrolyzed) is crucial for its structuring capacity. For example, hydrolyzed lecithin is more hydrophilic compared to native. Hydrolyzed lecithin alters crystal morphology, induces recrystallization, and slows polymorphic transitions. However, the interaction between lecithins and fat crystals is only possible if they are present during the main crystallization step; otherwise, they cannot act as a structuring agent (Delacharlerie et al. 2016).

2.3.2.4 Lecithin Self-Assembly: Dependence of Solvent Medium

Self-assembly of amphiphilic surfactants, such as lecithin, in both polar and nonpolar solutions above the critical micelle concentration (CMC) and Krafft temperature is an important property that can be explored in different applications such as food materials, pharmaceutical (drug delivery), and biotechnology fields (Lehtinen et al. 2017; Kim et al. 2021). Phospholipids self-assembly is highly dependent on the molecular geometry of the phospholipids (*i.e.*, polar head size, head group hydration, and acyl chain unsaturation and length) and on the solution's physicochemical conditions such as polarity, pH, temperature, and ionic strength (Li et al. 2015a; Lombardo et al. 2016). Figure 2.3.3 illustrates the possibilities of structural organization of phospholipids in polar and non-polar media. This ability of phospholipids could underpin formation and stabilization strategies like those found in oleogels and emulsions.

In water, phospholipids molecules can spontaneously assemble into micelles, hexagonal phase, and vesicles with a lipid bilayer organization mainly driven by hydrophobic interactions that are affected by the phospholipids' molecular shape (Li et al. 2015a). It means that, per the balance between attracting forces on the polar headgroups and on repulsive forces between the hydrophobic tails, phases that have a positive curvature (e.g., micellar), no curvature (lamellar), or a negative curvature (e.g., inverted hexagonal) can be established. Transitions between these phases can be triggered by changing the balance of forces by manipulating temperature or hydration, for example (Marrink and Mark 2004).

In micelles, the hydrophobic part is buried in their core, while polar hydrophilic moieties are in contact with the surrounding aqueous phase (Njauw et al. 2013). Depending on the geometry and concentration of the phospholipids without considering the surrounding medium, they can spontaneously assemble into bilayers, similar to biological membranes (Lombardo et al. 2016; Bot et al. 2021). In turn, liposomes are configured when the hydrophilic head groups of the phospholipids

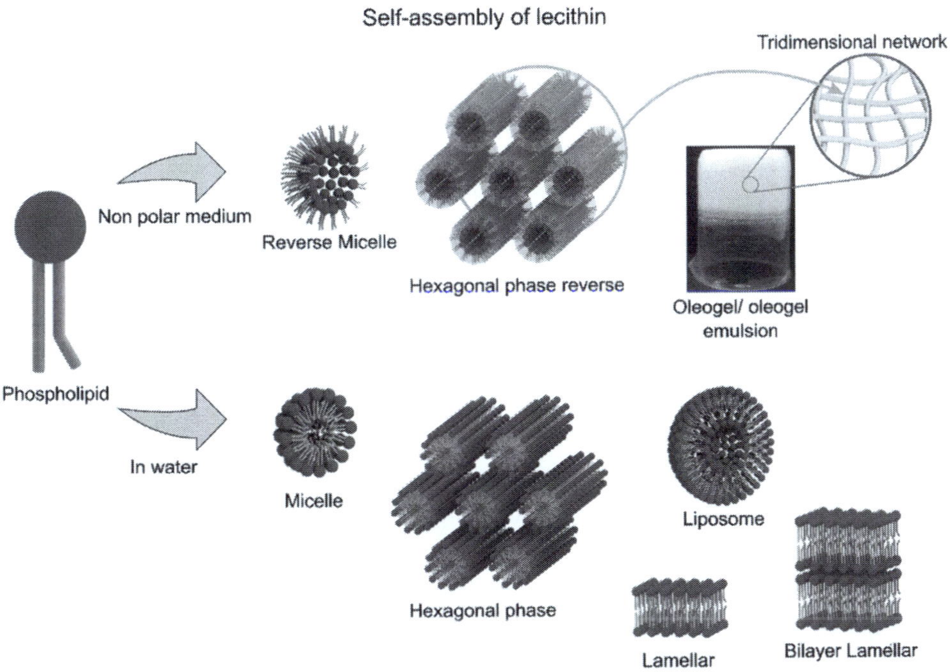

Figure 2.3.3 Schematic illustration of self-assembled structures of the phospholipids in non-polar and polar media.

are positioned toward the aqueous phase; whereas, the hydrophobic tail of each of the monolayers faces each other in the middle of the bilayer vesicle to avoid contact with water molecules (Khorasani et al. 2018). On the other hand, lecithin self-assembly in a nonpolar medium is thermodynamically driven by other interactions. In this case, the polar head of the molecule is oriented toward the micellar structure's interior, and the hydrocarbon tails are oriented toward the organic solvent phase, forming reverse micelles with spheroidal or elliptical shapes (Raghavan 2009; Martínez-Ávila et al. 2019).

In general, the behavior of amphiphilic molecules, including lecithin, in nonpolar environments is much less known than the behavior of their self-assembly in water. In this case, lecithin has a reverse cone shape, and it is capable of forming reverse spherical or ellipsoidal micelles in non-polar solvents (Shchipunov and Hoffmann 1998). Lecithin reverse micelles can grow axially into an elongated cylindrical shape similar to polymer-like strands when a trace amount of water is added (Shchipunov and Hoffmann 1998). In such a framework, hydrogen bonds between water molecules and lecithin lead to changes in molecular geometry. Once the cylinders become long enough, they entangle themselves creating a transient network in which these reverse cylindrical micelles may constantly break and recombine (Shchipunov and Hoffmann 1998; Okuro et al. 2020). On a macroscale edge, such an organization promotes a sharp increase in system viscosity that becomes several times greater compared to pure lecithin in oil (Njauw et al. 2013; Kim et al. 2021). This is the mechanism underlying the formation of lecithin organogels. Other polar additives, such as acids and salts, can replace the role of water addition, unleashing lecithin wormlike micelles and thus creating the 3D network (Imai et al. 2013; Njauw et al. 2013).

Reverse micelles of phospholipids have been extensively investigated in organic solvents such as n-decane and cyclohexane. However, few investigations have dealt with vegetable oils, which are more polar than the aforementioned solvents (Imai et al. 2013; Njauw et al. 2013; Lehtinen et al. 2017; Penttilä et al. 2019). Nevertheless, both experimental results and simulations indicated that the type of solvent strongly influences the aggregation behavior of phospholipids (Lehtinen et al. 2017; Penttilä et al. 2019).

Lecithin self-assembly properties and interactions with other components open up the opportunity for its use in a myriad of strategies to the tune of techno-functionalities. In the next section, we address the prospects of lecithins as crystal habit modifiers and their unique role in developing oleogels and emulsions.

2.3.3 Exploring Techno-Functionalities of Lecithin

2.3.3.1 Conventional Fat: The Role of Lecithin as Crystallization Modifier in Lipid Systems

Lipids are complex systems formed mainly by triacylglycerols (TAGs) and other minor components. The crystallization process and the crystalline organization of TAGs are crucial for technological applications, contributing to a myriad of quality parameters such as texture, flavor, melt-in-mouth sensation, and stability during storage (Metin and Hartel 2005). The crystallization process involves two main steps: nucleation and growth. Nucleation is the initial organization of TAG in the liquid state; this organization forms the nucleus. For nucleation to occur, a supersaturated state or subcooling conditions must be present. Also, there is a maximum of free energy that needs to be overcome. Once these conditions are met, a critically sized nucleus is formed (Himawan et al. 2006). Above this critical size, a stable nucleus forms and grows, while clusters of smaller, non-critical size molecules are dispersed in the liquid lipid phase. This process of spontaneous nucleation is denominated homogeneous nucleation. However, it rarely occurs under typical processing conditions.

In practice, nucleation is usually dominated by a heterogeneous mechanism, in which TAGs' foreign particles or molecules and container walls provide part of the energy required for nucleation, which reduces the energy barrier for nucleation. Nucleation rate is essential for the number and size of crystals formed, their polymorphic shape, and the final distribution of crystalline solids (Marangoni 2004; Lawler and Dimick 2008; Sato 2018). After stable nuclei are formed, crystal growth occurs when TAG molecules present in the liquid phase are incorporated. In this process, molecules from the liquid phase migrate to the crystal's surface where they are rearranged and oriented. A growth unit (single molecule or a group of molecules) migrates to the crystal's surface until it finds a suitable kink site for being incorporated into the network. Growth continues as long as there is a driving force for crystallization and the system comes into equilibrium or until it becomes fully crystallized. In natural fats, the different TAG species form mixed or compound crystals. The probability of two TAGs crystallizing together depends on the similarities or differences in their molecular configuration (such as chain length, degree of unsaturation, or arrangement of fatty acids in the glycerol structure) and on the conditions under which crystallization is carried out (Metin and Hartel 2005; Sato 2018).

Lipid crystallization is largely influenced by the presence of foreign components in the network. In this context, emulsifiers have been reported as influential in the crystallization of lipid materials by modifying the steps of nucleation, crystal growth, and by promoting polymorphic transitions (Figure 2.3.4) (Rizzo et al. 2015; Alfutimie et al. 2016). In this class of materials, lecithins represent an important group of amphiphilic components described as crystallization

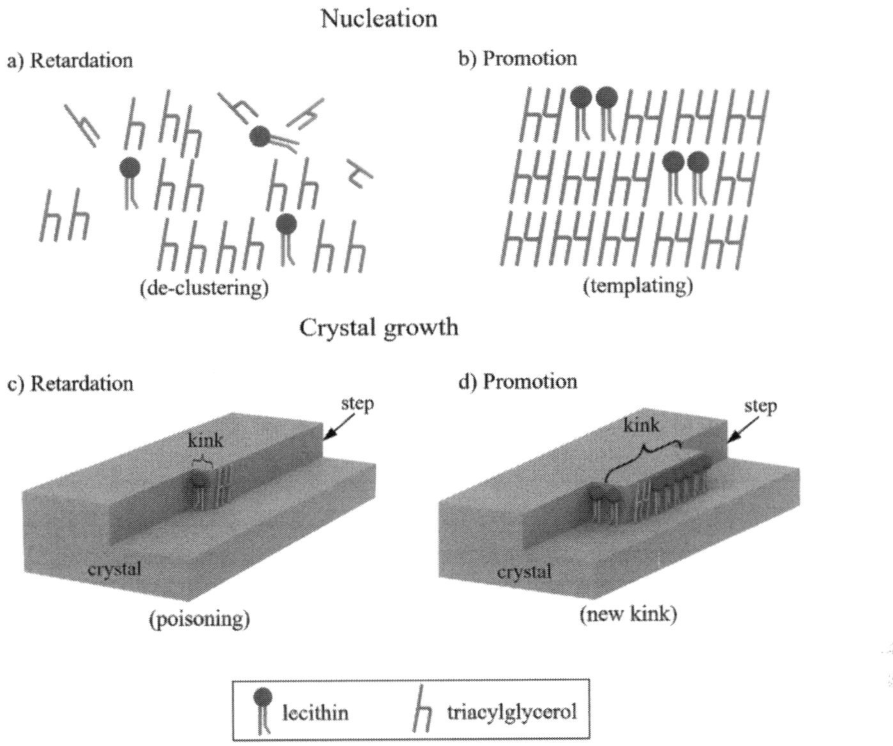

Figure 2.3.4 Possible effects of lecithin on nucleation and crystal growth.

interferents. Despite the wide, industrial use of lecithins, there is still a large knowledge gap related to the precise effect of lecithin as a crystallization modifier in fats (Patel and Dewettinck 2015; Ribeiro et al. 2015). Lecithins can promote different effects on lipid crystallization, depending on their head groups (polar), types of fatty acids in the hydrophobic portion, degree of hydrophobicity, solubility in liquid oil, similarity or dissimilarity of lecithin fatty acids and the fatty acids of the lipid phase, and the added concentration and cooling rate (Garti and Yano 2001; Rigolle et al. 2015; Rizzo et al. 2015).

Some mechanisms have been proposed to explain how emulsifiers such as lecithin interfere in crystallization. The first point is related to the solubility of these interfering components. When the solubility is high or its concentration is below the solubility limit, the emulsifier does not crystallize, which prevents nuclei forming in the lipid phase through attracting molecular interactions that occur between the emulsifier and the lipid molecules (*i.e.*, de-clustering). On the other hand, if the emulsifier has low solubility or is above the solubility limit, it may crystallize before the lipid. In this case, the crystallized emulsifier can act as a crystallization seed, thereby promoting crystallization of the lipid phase (*i.e.*, templating) (Smith et al. 2011; Daels et al. 2015; Smith and Sato 2018). In the crystal growth stage, the emulsifier can delay or modify the direction of growth, blocking the adsorption of TAG molecules in the kinks on the surface of the growing crystal (poisoning). Crystal growth can be promoted when the adsorption of emulsifier molecules on the crystal surface creates new kinks for incorporating TAGs molecules. In addition, absorbing emulsifying molecules on the crystal surface can also alter the interactions between the formed crystals and, thus, modify the crystal lattice, which promotes or delays polymorphic transformations (Garti and Yano 2001; Patel and Dewettinck 2015; Tran et al. 2015; Smith and Sato 2018; Atik et al. 2020).

Some phospholipids such as lyso-PC and PI delay the liquid phase's crystallization (Savage and Dimick 1995). However, adding soy lecithin (2% w/w) reduced cocoa butter's crystallization induction time from 25 min to 4 min. This result was attributed to the PC domain of the soy lecithin used (Simoes et al. 2021). According to Savage and Dimick, phosphatidylcholine molecules exhibit an inverse hexagonal configuration during cocoa butter crystallization. This conformation would be favorable for a co-crystallization between TAGs and PC molecules, which results in reduced crystallization induction times. On the other hand, the effect of soy lecithin in its native form was different from the hydrolyzed effect on the crystallization of lipid systems composed of palm oil and rapeseed oil. Native soy lecithin promoted the growth of well-organized crystals with dense spherulites. In contrast, hydrolyzed soy lecithin provided bigger crystals with a less organized and a less dense fat network. These results were based on the fact that hydrolyzed soy lecithin is more hydrophilic and has a lower affinity for the lipid phase (Delacharlerie et al. 2016).

Other studies have reported that lecithins from different sources such as milk phospholipids, sunflower lecithin, rapeseed lecithin, and even soy lecithin delayed or did not exert an effect on the lipids' crystallization process (Vanhoutte et al. 2002; Dhonsi and Stapley 2006; Lončarević et al. 2013; Cooper et al. 2019). Lecithins' influence on lipid crystallization depends on the similarity between the fatty acids that compose the hydrophobic portion of the phospholipid molecules and the fatty acids of the lipid phase. In general, the greater the dissimilarity between the fatty acids and the higher the molar mass of the emulsifier, the greater the inhibitory potential in the crystallization process (Cerdeira et al. 2005; Smith and Sato 2018). Therefore, the often-divergent results reported in the literature can be explained by the lecithins' naturally diverse composition, which is highly dependent on the agronomic, genetic and environmental parameters of the sources from which they are extracted (Robert et al. 2020). Additionally, use of different instrumental methodologies and various supercooling conditions between studies may contribute to this behavior.

2.3.4 Application of Lecithin in Alternative Oil-Structuring Routes

2.3.4.1 Oleogels

Oleogelation is a promising technology for creating healthier elastic solid fats with mechanical characteristics similar to conventional fats (e.g., consistency and plasticity). Thus, oleogels emerge as a suitable alternative for oil structuring as they are able to retain the fatty acid composition profile of the liquid oil with low SFAs and zero TFAs (Guo et al. 2019, 2020b).

Oleogel formation is a complex process involving supramolecular interactions in which thermodynamic dissolution parameters, such as enthalpy and entropy of gelators in the chosen liquid oils give information about gelation behavior. This concept underpins the creation of "solvent-mediated" gels. In this case, different solvents (diverse affinity of the solute-solvent pair) and/or variations in the crystallization/aggregation/self-assembly properties of the structuring agent can be explored to develop different types of gels (Su et al. 2015; Patel 2017). Fundamentally, lecithin-based oleogels are a representative case of a "solvent-mediated" route as they are typically formed in certain critical ratios of phospholipid(s) in a vegetable oil and trace amounts of water tuning the polarity of the medium (Patel 2017). This is supported by the fact that the formation of lecithin self-assembling structures are driven by a specific solubility balance between lecithin and the solvent medium. The process of forming supramolecular gels from low molecular weight oleogelators such as lecithin involves heating the supersaturated solution followed by cooling it. By decreasing the temperature, amorphous and ordered aggregates are formed, promoting the structural organization of varied length scales. These gels are thoroughly dependent on forming these hierarchically self-assembled structures, and if they are handled properly, gels with tuned properties can be achieved (Su et al. 2015).

Lecithin oleogels were first introduced by Scartazzini and Luisi (1988). The authors observed that upon the addition of critical amounts of water into non-aqueous solutions of naturally-occurring lecithins, there was an increase in the viscosity and the subsequent formation of stable and isotropic gels. As mentioned in Section 2.4, adding water as a co-solvent depends on its ability to alter the self-assembly of phospholipids by establishing H-bonding between adjacent phosphate groups (reduction of the interface curvature). This unleashes a longitudinal aggregation (uniaxial growth) of reverse micelles (known as wormlike micelles) that develop a temporal network capable of immobilizing the liquid oil (Imai et al. 2013; Nikiforidis and Scholten 2014). Thus, the network formed by the reverse micelles is held together by lecithin-water hydrogen bonds and hydrophobic interactions between the organic and aqueous phases (Bodennec et al. 2016). Likewise, relative to adding water, lecithin and α-tocopherol self-assembly has been reported as an efficient strategy for developing structured lipid materials for food application (Nikiforidis and Scholten 2014). Interestingly, tuning the medium polarity for oleogel formation may not be necessary when a combination of gelators is used. It can be seen when lecithin is combined with low melting waxes (Okuro et al. 2018), sucrose esters (Bin Sintang et al. 2017) and ceramide (Guo et al. 2019, 2020a, 2020b).

In fact, lecithin plays a far-reaching role in oleogels' design. There are usually two motivations in applying lecithin to oil structuring: (i) as aforementioned, lecithin is a ubiquitous natural ingredient in the food industry, and (ii) lecithin molecules may self-assemble via noncovalent interactions, such as hydrogen bonds, even in a nonpolar medium (Aguilar-Zárate et al. 2019). However, lecithin is less likely to be applied as the sole oleogelator (single component) due to its inability to form gels in liquid vegetable oils without changing the polarity of the medium or because of the high critical oleogelator concentration in the presence of water (sometimes higher than 30% w/w depending on its type and composition) (Bodennec et al. 2016). Therefore, the technological solution to overcome this limitation is to explore a combination between lecithin and other molecules (herein called mixed system) in which it may act as a main or co gelator in the presence of gelling or non-gelling ingredients.

At least two different classes of molecules are present in a mixed system (or multicomponent), but the mechanism underpinning oil structuring is less obvious. Thus, the assembly pattern of the multicomponent supramolecular oleogel depends on whether the lecithin and another component assemble into various structures (co-assembly or self-sorting). In co-assembly, the two species interact with each other to form structures containing random amounts of both molecules, whereas in self-sorting, the supramolecular structure contains only one gelator (Guo et al. 2020b). Revisiting lecithin properties such as molecular geometry, packing parameter, and hydrophilic-lipophilic balance may be essential for understanding how mixed gels form. Table 2.3.2 illustrates a myriad of studies using lecithin as a primary or co-oleogelator for oil structuring purposes.

The studies summarized in Table 2.3.2 depict the lecithin's role, depending on the corresponding combination. Lecithin seems to be more widely used than it is fundamentally understood, especially when it comes to its self-assembly and interactions in non-polar solvents. Apart from the molecular fingerprint involved in the interactions of lecithin and other species, the mixture's ratio is seemingly quite relevant in dictating the components' supramolecular interactions. For instance, a series of

Table 2.3.2 Strategies using lecithin as either main or co-oleogelator for oil structuring.

Composition					
Lecithin	(An)Other Component(s)	Mixture Ratio/ Continuous Phase	(Supra)Molecular Interactions	Meso-Scale Changes	References
• Deoiled and unsaturated lecithin (20wt.% PC) • Fractionated and fully hydrogenated soybean lecithin (90wt.% PC)	• Ethylcellulose with an average viscosity of 20 cP	• 10 wt.% ethylcellulose with 1 wt.% lecithin/ high oleic canola oil	• Lecithin creates a synergistic effect with the polymeric gel network through hydrogen bond establishment	• Addition of 1 wt.% of soy lecithin increased the shear moduli 10-fold and gel hardness 20-fold for 10 wt.% ethylcellulose oleogels	Aguilar-Zárate et al. (2019)
• Soybean lecithin (molecular weight: 758.1 $g.mol^{-1}$), purity > 98%	• Ceramide III (molecular weight: 584 $g.mol^{-1}$), purity > 92%	• Screening of different concentrations ([oleogelator] = 10 and 20 wt.%) /sunflower oil	• LEC as a crystal habit modifier, altering the crystal morphology of the host gelator (CER) • Crystal morphology was affected by mixture ratio (needle-like, rosette-like and spherulite-shaped crystals)	• Stronger gels ($G' > 10^5$ Pa) were formed with superior oil binding capacity (up to 99.84%) at specific LEC:CER ratios (6:4 and 5:5)	Guo et al. (2020)
• Soybean lecithin, purity > 95%	• Sugar alcohols that contain three (glycerol), four (DL-threitol), five (xylitol), and six (D-sorbitol) -OH groups	• LEC concentration fixed at 80 mM, screening of different sugar alcohols concentrations (0–90 mM)/rice bran oil (among other solvents)	• Structural transition from spherical to cylindrical micelle • Hydrogen bonds between phosphate (PO^{4-}) in lecithin and – OH groups in sugar alcohols	• Sugar alcohols interact with lecithin increasing the viscosity of lecithin organogels • Combination of lecithin and D-sorbitol in rice bran oil prospects for food applications	Kim et al. (2021)

• Sunflower lecithin	• Sucrose esters (monoester (±11%), diester (±15%), triester (±16%) and polyester (±58%))	• Screening of different concentrations, ([oleogelator] = 10 wt.%)/ refined sunflower oil	• LEC modifies the self-assembly of sucrose esters (SEs) by interrupting the extensive hydrogen bonding between their monomers	• 7:3 SEs:LEC combination showed enhanced rheological properties compared to the other ratios, suggesting better molecular ordering induced by LEC	Bin Sintang et al. (2017)
• Oil-free soya lecithin (purified to 99% PC) (+water addition 1 wt.%)	• α-tocopherol	• PC and α-tocopherol ratios of 1:1, 1:2 and 2:1, ([oleogelator] = 25 wt.%) /refined sunflower oil	• α-tocopherol and PC can be used to alter the packing geometry providing supramolecular structures • Combined assembly of α-tocopherol and PC, stabilized by physical interactions	• Formation of reversible oleogels (subjected to shear or temperature increasing to values above 35°C)	Nikiforidis and Scholten (2014)
• Soybean lecithin (Phospholipon® 90G, (96.9 wt.% PC, 1.1 wt.% LPC and tocopherol max. 0.3 wt.%))	• Fruit wax, *Myrica cerifera* (FW, mainly comprised of fatty acids 37 wt.% and fatty alcohols 62.5 wt.%)	• FW:LEC ratios (100:0, 75:25, 50:50, 25:75 and 0:100), ([oleogelator] = 7 wt.%) /refined sunflower oil	• Phospholipid acted as a crystal habit modifier changing the microstructure of wax crystalline structure • Hydrogen bonding between FW and LEC moieties	• Denser structure for FW:LEC (75:25 and 50:50) oleogelator mixture • Mechanical properties, oil binding capacity and thixotropic recovery were enhanced upon adding lecithin	Okuro et al. (2018)

LEC-lecithin; PC-phosphatidylcholine; LPC-lysophosphatidylcholine; CER-ceramide; SE-sucrose ester; FW-fruit wax studies unraveling the interaction between lecithin (LEC) and ceramide (CER) show the complexity of the role of lecithins in oleogelation (Guo et al. 2019, 2020a, 2020b). Initially, the novel idea of water-induced multi-component oleogelation in sunflower oil with lecithin, ceramide, and water was explored (Guo et al. 2019). In this case, water influenced the assembly process of CER and LEC crystals; rather than simply being dispersed in the systems, unleashed gel formation by H-bonding and van der Waals interactions were the primary driving forces (Guo et al. 2019). The authors further investigated that water temperature influenced the interaction between oleogelators, showing that CER and LEC crystallization shifted individual oleogelators from co-assembly (at 20°C) to self-sorting at a higher temperature (95°C) (Guo et al. 2020a). Finally, even with no water addition, CER and LEC showed an outstanding oil gelation capacity. LEC and CER co-assembled in an approximately equal molar ratio, while in other ratios, self-sorting combined with fibers' co-assembly by lateral association. Such behavior highlighted the different underlying microstructures observed depending on the CER:LEC ratios and the temperature when the interactions occur (Guo et al. 2020b).

In addition to oleogels, bigels are also an option using lecithin for developing soft matter for fat replacement. Bigels are biphasic structured systems composed of a mixture of hydrogel and oleogel. In turn, soy lecithin-stearic acid (oleogelator) and whey protein concentrate (hydrogelator) were combined to create semisolid, edible bigels (Bollom et al. 2020).

2.3.4.2 Emulsion Strategies

In the realm of fat mimetics systems, another promising and still under-explored application for lecithins is in the development of structured emulsions. Structured emulsions exhibit higher viscosity than conventional W/O or O/W emulsions and can provide a fat-like semi-solid structure. This characteristic has important implications for the development of low-fat and low-calorie foods (Jimenez-colmenero et al. 2015; Marangoni et al. 2020; Patel et al. 2020). Development of structured emulsions generally involves three main approaches: the use of biopolymer layers to coat the dispersed phase, the incorporation of the emulsion into another dispersed phase with higher viscosity, such as in hydrogels, and the controlled aggregation of droplets from the dispersed phase producing emulsions with viscous or pasty characteristics (McClements 2012; Marangoni et al. 2020).

The synergistic interaction at the interface between lecithin and other components of the emulsion, especially proteins, is the interaction associated with forming structured emulsions with lecithin. Adding 0.5 to 1% (w/v) lecithin caused changes in the secondary structure and induced the partial unfolding of soluble mussel proteins in O/W emulsions, which consequently inducing aggregation and increased turbidity of the systems. This behavior was attributed to hydrophobic interactions between lecithin and mussel proteins (Zou et al. 2020). Hydrophobic interactions between whey protein and soy lecithin have also reportedly been responsible for reducing the gelation time of whey protein (Sun et al. 2018). Synergistic interaction between gelatin and soy lecithin at the interface has been reported as an important mechanism for forming a structured network after 21 days of storage of fish oil-loaded emulsion gels (Zhang et al. 2020b). The interaction between the phospholipids naturally present in krill oil and the casein present in the continuous phase of high internal phase O/W emulsions (O/W-HIPE) promoted a reduction in droplet size, which aggregated and consequently formed a three-dimensional cross-linked network structure in the emulsions that strengthened emulsion gel's properties (Liu et al. 2021). The interaction between bilayers and monolayers of phosphatidylcholine and sodium caseinate produced O/W-HIPE (70% (w/w) fish oil) with a lower viscosity than that of mayonnaise (70–80% fat) but with a higher viscosity than milk (Yesiltas et al. 2019). It was also reported that the interaction between polyglycerol polyricinoleate

(PGPR) and lecithin produced W/O-HIPE with good kinetic stability and with high viscosity because alternating layers formed between PGPR and lecithin at the interface (Okuro et al. 2019).

In the case of oleogel emulsions, these systems were formulated from 20% (w/w) and 30% (w/w) of lecithin and stearic acid mixture as oleogelators with the incorporation of 10% (w/w) and 20% (w/w) of water. Interestingly, oleogel emulsions showed improved mechanical properties compared to the corresponding oleogel formulation with the same structuring agents but with much lower water content (1% w/w) (Gaudino et al. 2019).

2.3.5 Beyond Oil-Structuring Purposes: Role of Lecithin as an Emulsifier and in the Vehiculation of Bioactive Components

Lecithin has been widely studied and applied in emulsified systems because it is a natural amphiphilic product (obtained from egg yolk, milk, sunflower, and soybean among others) with interfacial properties (Deng 2021). Lecithins have varied compositions according to the source, presenting variation both in the polar fraction formed by different phospholipids and in the nonpolar fraction formed by fatty acids (Table 2.3.1) (List 2015; Alhajj et al. 2020). This varied composition and factors such as hydration, pH, presence of salts, concentration, and temperature influence how different structural organizations of lecithin form in the interfacial region of emulsions, such as micelles, lamellar structures or bilayer sheets, and liquid crystals. These structures directly contribute lecithins' ability to form stable emulsions, acting as a mechanical and electrostatic barrier to coalescence (Klang and Valenta 2011; Pichot et al. 2013).

The challenge of using lecithins alone as emulsifiers in O/W emulsions is that different phospholipids have different electrical characteristics depending on the nature of their head groups. For example, PC and PE are zwitterionic, *i.e.*, they exhibit negative, positive, or overall neutral charges depending on the pH solution relative to the pKa values of the ionizable groups. Consequently, the electrical charge on droplet surfaces can vary depending on different types and ratios of phospholipids (McClements et al. 2017). This behavior, combined with the fact that some natural lecithins exhibit a more hydrophobic characteristic, may cause the lecithins to not provide adequate electrostatic repulsion to prevent destabilization phenomena involving droplet aggregation (McClements 2016). An alternative to this problem that thus improves the stability of emulsions is the use of lecithin combined with other types of emulsifiers, such as proteins, polysorbates, and others (Rocchio et al. 2017; Sandoval-Cuellar et al. 2020; Real et al. 2021; Yang et al. 2021). Another approach is the use of hydrolyzed lecithins wherein one of the fatty acids that form the nonpolar fraction of phospholipid molecules is removed. This produces more hydrophilic lecithins, known as lysolecithin, which can form more stable emulsions (Cabezas et al. 2016; El-Abhar et al. 2020; Ma et al. 2020).

Lecithin has been widely applied toward the development of new delivery systems for bioactive compounds, such as microparticles, nanoemulsions, liposomes, and lipid nanoparticles. These delivery systems can provide several benefits such as protection against chemical and physical degradation, masking of off-flavors (such as bitterness or astringency), controlled release of bioactive compounds at target sites—thus increasing their efficacy, and other benefits (McClements et al. 2007; Gasa-Falcon et al. 2020; Pateiro et al. 2021). Several bioactive compounds have been satisfactorily incorporated into delivery systems stabilized with lecithins from different sources, either alone or in association with other emulsifiers. These include olive leaf phenolic compounds (Soleimanifard et al. 2020), antifungal compounds (Zhang et al. 2020a), astaxanthin (Tamjidi et al. 2017, 2018), omega-3 (Lin et al. 2014; Sreedhar et al. 2019; Chen et al. 2020), krill oil (Lin et al. 2020), alpha-lipoic acid (Najafi et al. 2021), and vitamin E (Ozturk et al. 2014).

Soy lecithin and gum arabic were used to stabilize paprika oleoresin nanoemulsions, showing better soy lecithing efficacy for forming nano-sized droplets (d < 150 nm) compared to gum arabic (d < 539.6 nm) (Flores-Andrade et al. 2021). Bioactive compounds in nanoemulsions can increase their solubility and *in vitro* bioaccessibility, as occurred with curcumin (Yang et al. 2021). In addition, chia oil encapsulation efficiency, oxidative stability, and color were improved in microparticles developed with the sunflower lecithin bilayer system compared to microparticles of mono-layered phosphatidylcholine and chia oil in bulk, which were evaluated separately (Julio et al. 2019). On the other hand, a significant increase in particle size was observed in nanostructured lipid carriers stabilized with a mixture of phosphatidylcholine/Tween 80 and loaded with β-sitosterol under acidic conditions (pH = 2). This behavior was attributed to the fact that the electrical charge of phospholipids is dependent on the medium's pH (Soleimanian et al. 2019). Lecithin-coated droplets typically exhibit relatively high negative charge at basic and neutral pH conditions, preventing droplet aggregation due to strong electrostatic repulsion. However, when the pH is reduced, the charge of the droplets becomes progressively less negative; when it passes through zero and becomes slightly positive where is consequent observation of an aggregation phenomena (Ozturk et al. 2014; McClements et al. 2017).

Liposomes are unilamellar and multilamellar vesicles with internal water cores suitable for including hydrophilic components. The structural characteristics and physical properties of lecithin make them great agents for liposome formation. Phospholipids with shorter fatty acid chain lengths can form liposomes with smaller sizes. The presence of unsaturated and saturated or hydrogenated fatty acids plays a fundamental role in the stability and efficiency of micelle encapsulation (Bot et al. 2021). An advantage of soy lecithin is that it has greater stability in liposomes. Liposomes can be used to incorporate bioactive compounds into foods, improve flavor release, and extend the shelf life of products. However, this technique is rarely used in foods due to its low physicochemical stability and sensitivity to environmental conditions such as acidity, ionic strength, temperature, and mechanical forces (Li et al. 2019; Bot et al. 2021; Wang et al. 2019). The technology can also be applied in the combined encapsulation of hydrophilic and lipophilic antioxidant mixtures, demonstrating greater efficiency and stability against oxidation (Van Nieuwenhuyzen 2015).

2.3.6 Food Applications

The functional properties and applications of lecithin are strongly influenced by the physicochemical properties and geometric structure of phospholipids (Bot et al. 2021). Egg lecithins were the first lecithins to be applied in food; however, their high cost encouraged searning for new lecithin sources. Thus, the efficiency of vegetable lecithins, such as soy, sunflower, and rapeseed in food-related applications were considered (Van Nieuwenhuyzen 2015). Lecithin is an ingredient with different functions in foods, such as margarine, bakery shortenings, chocolate, cheese, instant powders, processed meats, dairy products, and supplements (Van Nieuwenhuyzen 1981; Robert et al. 2020; Deng 2021). As a natural product, lecithin is well accepted by consumers (Szuhaj 1983). The main functions of lecithins are summarized in Table 2.3.3. PC-enriched lecithins are widely used in salt-free margarine due to their anti-spattering and emulsification characteristics. There is adidtional evidence that lecithin PC participates in several relevant neurochemical processes, is the precursor and metabolite of acetylcholine, and plays a role in single-carbon metabolism (Tayebati and Amenta 2013). The moieties of lecithin with PA and PI have excellent application in W/O emulsions and in chocolate manufacturing (Joshi et al. 2006).

Table 2.3.3 Lecithins applied to foods and their traditional functionality.

Food	Function
Margarine	Co-emulsifier, anti-spatter, mouthfeel
Chocolate	Viscosity control, adhesiveness reduction, crystallization control, rheology, yield value
Instant Powders	Agglomeration, wetting, dispersibility
Baked Goods	Volume improvement, fat dispersion, anti-staling, firmness, freshness
Cheese	Emulsifier, slice separation agent
Meat and poultry glazes	Browning agent, emulsifier, phosphate dispersant
Dairy products	Emulsifier, wetting agent, anti-spatter and release agent
Spreads and salad products	Emulsifier, crystallization control
Chewing gum base	Rheology, tackiness, brittleness

Source: Adapted from Szuhaj (1983), Van Nieuwenhuyzen (2015).

Recently, fat mimetics have been studied to eliminate TFA, reduce SFA, use natural ingredients, and develop clean label products (Meza et al. 2016; Chaves et al. 2018; Demirci et al. 2020; Park and Maleky 2020; Patel et al. 2020; Silva et al. 2021a). As reported in previous sections, lecithins have been applied in oil structuring strategies, but a complete understanding of their function and further behavior when incorporated into food matrices are still lacking. However, it is a fact that lecithin has a structuring effect, significantly increasing hardness, affecting crystal morphology and polymorphic stability of model fat systems (Delacharlerie et al. 2016). There are still few studies that apply lecithin-based fat mimetics in foods. Lecithin-based oleogels have limited application in foods due to the formation of a weak gel that is sensitive to shear and that can be modified in water (Bodennec et al. 2016). Therefore, lecithin will likely be used in hybrid oleogels, in which it can act synergistically with other oleogelators such as sterols and waxes. Therefore, strategies to form oleogels with greater stability must be developed to facilitate food application.

Lecithin with a high PC fraction facilitates emulsion formation with smaller particle sizes, which expands its application in foods (Bot et al. 2021). However, to delineate the best food application, the chemical composition of lecithin must be very well defined and evaluated. Additionally, the role of lecithin as a fat mimetic in hybrid systems still needs to be better studied. Herein we present the main applications of lecithin in food formulations appearing either as a raw ingredient or as different colloidal systems such as emulsions, oleogels, and liposomes. As depicted in Table 2.3.3, lecithin is an outstanding ingredient in the food industry, having different techno-functionalities. While this is evident, there are plenty of research opportunities addressing lecithin's role as a versatile player in developing fat mimetics. Certainly, this may pave the way for building knowledge that expands the use of lecithin for the texture of foods.

2.3.6.1 Margarines

From a physicochemical point of view, margarines are defined as W/O emulsions containing an at least 80% lipid phase. The texture of margarine is influenced by the network of fat crystals and the stability of the emulsion (Silva, et al. 2021b). Therefore, the crystallization behavior of fats is essential for product quality control (Delacharlerie et al. 2016). Mono- and di-acylglycerols are the main ingredients that stabilize margarines. An emulsifier that stabilizes W/O emulsions has HLB

3–6, but lecithin has a broad HLB (2–8) (Bueschelberger et al. 2015). So why is the addition of lecithin observed? Despite the presence of emulsifying agents, lecithin acts as a co-emulsifier and also has important anti-spattering properties. Lecithin is used to facilitate the formation of small droplets, to stabilize oil droplets, and to reduce interfacial tension (Sun et al. 2020). Lecithin works by forming a phospholipid film that surrounds the water droplets and prevents undesired coalescence. Lecithin behaves differently in salted and unsalted margarine. In salted margarine, natural lecithin performs well. However, in unsalted margarines, calcium and magnesium ions in water and milk proteins inactivate some phospholipids, such as PE by flocculation, requiring a hydrolyzed or choline-enriched lecithin (Van Nieuwenhuyzen 1981). Oleogel emulsions formulated with sunflower wax (2, 3.5, and 5% w/w in high oleic sunflower oil) and soybean lecithin powder (0.12% w/w) in high oleic sunflower oil have potential applications for margarine. The waxes showed an agglomerated structure of platelets, and the lecithin had a stabilizing effect and formed a dense and thin network, as well as a decrease in the coalescence process (Sandoval et al. 2020).

2.3.6.2 Bakery Products

The bakery industry uses lecithins in several products such as breads, cookies, and cakes. Partially hydrolyzed lecithin can increase the volume and freshness of breads. Phospholipids physically bind wheat gluten via H-bridges, increasing elasticity and baking volume as well as improving fermentation tolerance. Complexation of hydrolyzed phospholipids into the amylose α-helix delays the recrystallization of amylose and thus enhances anti-staling, crumb freshness, and shelf-life properties. In pretzel recipes, lecithin acts as a lubricating agent and improves processing speed (Van Nieuwenhuyzen 2015).

In low-calorie cookies, a 30% fat reduction is possible with the addition of lecithin. A low-fat recipe without lecithin results in a sticky dough, but adding lecithin by up to 0.5% reduces stickiness, improves smoothness, and ensures adequate dough handling (Van Nieuwenhuyzen 2015). The flours used to make cookies require low protein to prepare good quality cookies. Therefore, in protein-rich flours, one solution is to prevent gluten development during dough formation by adding additives such as lecithin (Ahmad et al. 2021). Glazing is a coating glaze applied to the surface of foods to produce a homogeneous gloss film that enhances their shine and appearance. Furthermore, it provides a moisture barrier to extend the shelf life of bakery and confectionery products, which could be performed by fat mimetics. Meza et al. (2016) developed glazes with 0, 3, 6, and 9% (w/w) fat mimetic (sunflower oil and soy lecithin) demonstrating that these low-fat enamels have similar properties to a full-fat formulation made with vegetable shortening.

Replacing fats in bakery products is a great challenge, as shortening plays an important role in softening, releasing flavor, influencing plasticity, incorporating air, reducing the elastic nature, and shrinking the dough during molding, among other functions (Demirkesen and Mert 2019). Despite the extensive literature involving applications, no studies evaluated the use of lecithin in oleogels for bakery products, and indeed a recent study review on the development of oleogels in bakery products by Demirkesen and Mert (2019) does not report studies with lecithin. Fat mimetics are evaluated in bakery products with different oils and oleogelators, which demonstrates distinct behaviors. However, problems are reported in oleogels applied to bakery products related to irreversible structural breakage under dough mixing conditions. Thus, oleogel can quickly lose its solid structure as a result of shear forces. Considering that lecithin has emulsifying properties and the potential to interact with some waxes and other oleogelators, studies involving these formulations are still needed in bakery products. Furthermore, taking into account the properties that lecithin has, such as the increase in volume and the freshness of bread, its use as a fat mimetic for application in baking, in combination with another structuring agent, still needs to be evaluated (Demirkesen and Mert 2019).

2.3.6.3 Chocolate

By definition, chocolate is produced with cocoa butter (lipophilic phase) as the continuous phase, which is added to other ingredients that act in the dispersed phase, such as liquid cocoa, sugar, and lecithin, among others. The network is formed by the continuous phase that promotes the final properties of chocolate (Asghar et al. 2017). Lecithin facilitates lubricating sugar and cocoa particles by reducing viscosity, good flow during tempering, molding, or enrobing (Van Nieuwenhuyzen 2015). In addition, lecithins can act as a co-crystallizer of cocoa butter, changing the growth rate, modifying the morphology/number of crystals, or affecting the polymorphic transition (Miyasaki et al. 2016). Lecithins play a fundamental role in reducing the fat bloom in chocolate by retarding the polymorphic transition of cocoa butter and the interaction between sugar and fat. One theory is that a monolayer of emulsifiers adsorbed on fat crystals makes them more polar, while a layer of emulsifiers onto sugar crystals makes them more nonpolar (Dedinaite et al. 1998; Jin and Hartel 2020).

In chocolates, fat mimetics are an alternative to the limited resources of cocoa butter and its high cost. There are still limited fat mimetics studies related to chocolates, but there are already chocolates developed with ethylcellulose and palm kernel oil on the market (Stortz and Marangoni 2013). In addition, chocolate with corn oil-based oleogel and a combination of oleogelators (β-sitosterol with oryzanol, stearic acid, or lecithin) was developed forming three combinations: β-sitosterol: oryzanol, β-sitosterol: stearic acid, and β-sitosterol: lecithin (Sun et al. 2021). The β-sitosterol: oryzanol combination formed strong gels, a denser crystallization network, and better heat resistance compared to the β-sitosterol:stearic acid and the β-sitosterol:lecithin oleogel. However, the β-sitosterol: lecithin oleogel showed a softer texture due to its weaker and more dispersed intermolecular structure, presence of β crystals, and pseudo gel properties. When used in chocolate to replace half of the cocoa butter, the characteristics resembled dark chocolate.

Another interesting combination for oil structuring is sorbitan tri-stearate and lecithin, which form a thermo-reversible and shear-sensitive gel. Individually these components do not form a gel, but when combined, they form a structure with crystals and weak junctions; this weakness is due to lecithin's influence. This characteristic can be adjusted by modifying processing conditions, such as shear rate and cooling rate. Soybean oil was structured as an equimolar mixture (1:1) of sorbitan tri-stearate and lecithin (6%, 8%, or 10% w/w) to assess its effect on oil migration in pralines. This phenomenon is considered the primary cause of chocolate's fat bloom. Oleogels showed time-dependent, shear-thinning properties, indicating that it is possible to prevent the migration of fat fillings. At 40°C, there were not observed solids in the oleogels and did not affect the product's sensory quality; further, structural stability during storage was demonstrated (Si et al. 2016).

Increasing the content of unsaturated triglycerides in oleogels reduces fat bloom. Li et al. (2021) developed chocolate by reducing 50% (w/w) of cocoa butter by three oleogels (monoglyceride stearate, β-sitosterol + lecithin (8:2, w/w), or ethylcellulose oleogel). The oleogel-based chocolate showed thermal and polymorphic stability (β_V crystals) and delayed blooming. Indeed, the lowest whiteness index was observed in chocolate based on oleogel β-sitosterol + lecithin. Emulsifiers can act on chocolate by inhibiting bloom, increasing the melting point of fat, and preventing polymorphic transition. Therefore, in combination with another structuring agent, lecithin is a promising candidate for application in chocolate.

2.3.6.4 Dairy Products

In dairy products, lecithins can positively affect the protein properties in gel formation because they can interact with adsorbed and non-adsorbed protein, thereby increasing protein-protein cross-links and strengthening the gel (Euston and Goff 2019). This effect of lecithin makes it an essential ingredient, and its use is mainly as an ingredient in formulating dairy products and not as a fat mimetic. In addition, phospholipids can affect the milk fat crystallization. A high amount of phospholipids delays crystallization onset, due to their incorporation into the crystal growth site (Vanhoutte et al. 2002). Despite of this, we can find studies that evaluated the fat-reducing content of dairy products using lecithin.

In cheese, fat influences flavor, texture, and processing conditions acting as a plasticizer in the casein matrix, promoting a smooth texture. Li et al. (2019) evaluated the fat replacement in low-fat mozzarella cheese using lecithin. Fat replacers were added to the cheese prior to the addition of chymosin at a level of 20 g maltitol and inulin for kg of cheese and 5 g of lecithin for kg of cheese. Maltitol improved the texture and reduced the hardness of the reduced-fat cheese. Lecithin reduced hardness and increased the degree of a free oil release. Results showed that inulin and lecithin decreased the hydrophobic interaction in low-fat cheese, causing undesirable functional properties in mozzarella cheese (Li et al. 2019). The functionality of tapioca starch and lecithin was also evaluated alone (1% w/w starch or 0.2% w/w lecithin) or in combination (0.5% w/w starch and 0.1% w/w lecithin) as fat mimetics in low-fat feta cheese. The combination of starch and lecithin improved the flavor, texture, and general acceptance of the cheeses (Sipahioglu et al. 1999).

The stability and physical properties of recombined cream (20% fat and 1.5% protein) with added lecithin was evaluated (Zhou et al. 2017). This product consists of anhydrous milk fat and milk protein, which has the disadvantage of low stability. The authors reported that lecithin improved the stability of recombined dairy cream by increasing lecithin concentration, which reduced the mean oil droplet size and the surface protein concentration. This result has been attributed to interactions between lecithin and serum proteins, which are often ignored (Zhou et al. 2017). Lecithin also influences the stability of coffee cream, affecting the protein content of the cream layer after emulsion preparation and reducing the additional adsorption of milk protein at the oil/water interface after sterilization (Meeren et al. 2005). Tekin et al. (2017) developed a low-fat ice cream with double emulsions composed of polyglycerol polyricinoleate, lecithin, and guar gum. In this system, it was possible to reduce the fat content of the ice cream to 2.8% without negatively affecting the quality of the ice cream. Some studies have reported fat substitutes in ice cream (Munk et al. 2018; Botega et al. 2013), but these generally were unsuccessful due to poor sensory quality. Baer et al. (1997) evaluated the effects of mixed emulsifiers (polysorbate 80 with monoacylglycerols and diacylglycerols, 40% (w/w) α-monoacylglycerols, 70% (w/w) α-monoacylglycerols, and lecithin) on low-fat ice cream (2% w/w). Emulsifiers increased the mix's consistency and viscosity and reduced whipping times and ice crystal sizes. Ice cream with lecithin had the lowest flavor score and reduced overrun. The tasters reported soy-like off-taste in the ice creams that had higher lecithin content. However, all emulsifiers improved the body and texture of low-fat ice cream. Although some studies point to the benefits of phospholipids, much remains to be researched regarding mixtures of components incorporating lecithin as a fat substitute in dairy products.

2.3.7 Final Remarks and Perspectives

This chapter highlights the widespread use of lecithins as a versatile ingredient in the food industry. Despite the variety of lecithins applications, it is difficult to find specific data on their properties. The main reason is the variety of lecithins with the most diverse compositions. Based on the lecithin composition (PC, PI, PE, PS, and PA contents) and on the fatty acid profile, lecithins may have distinct properties, such as greater hydrophobicity or hydrophilicity. Furthermore, lecithins structures can be easily modified to improve characteristics such as solubility, emulsifier, and gelling capacity. Even knowing all the structural characteristics of lecithin is not enough to fully predict its behavior in food matrices. Moreover, the functionalities of lecithins are tuned by environmental conditions, such as solvent properties, pH, and temperature. Lecithins can modulate fat crystallization by inducing or slowing it down. However, the effect as crystallization modifiers depends on the fatty acid composition of the lecithin and oil, solubility, degree of hydrophobicity, concentration, and cooling rate. In addition to their impact on the crystallization of hard fats, lecithins can participate in oil structuring by the development of oleogels or emulsions-derived systems. Lecithin alone does not have a good structuring capacity, but in multicomponent systems, when there is a synergistic interaction, it can promote the formation of a denser and more stable network. In oleogels, positive interactions of lecithins with ceramides, ethylcellulose, sucrose ester, waxes, and sterols have been reported. Despite being one of the most used ingredients in the food industry, some mechanisms and interactions of lecithin are not fully understood. Some aspects that can be pointed out in relation to lecithin applied to the development of fat mimetics are: (i) the effect of lecithin on vegetable oils of different compositions; (ii) extensive mapping of lecithin interaction with other components rationalized by solubility balance; (iii) behavior of lecithin in structured or oleogelled emulsions and bigels; and (iv) application and evaluation of lecithin-based fat mimetics in foods products. An extra edge for the techno-functionality perspective would be investigating lecithins' bio-functionalities, and investigating their role in digestion, possibly modulating the phenomena of interfacial lipid digestion and the formation of mixed micelles involved in the bioaccessibility and bioavailability of bioactive compounds.

Acknowledgements

Thais J. Silva and Mayanny G. da Silva thanks CAPES (Finance code 001). Paula K. Okuro thanks FAPESP (grant #2018/20308-3 and #2019/26046-3 São Paulo Research Foundation -FAPESP). Rosiane Lopes Cunha thanks CNPq- (307168/2016-6) for the productivity grant. Ana Paula Badan Ribeiro thanks CNPq- (303429/2018-6) for the productivity grant.

References

Aguilar-Zárate, M., Macias-Rodriguez, B.A., Toro-Vazquez, J.F., and Marangoni, A.G. (2019). Engineering rheological properties of edible oleogels with ethylcellulose and lecithin. *Carbohydrate Polymers* 205: 98–105.

Ahmad, S., Naz, A., Usman, M., and Amjad, A. (2021). Impediment effect of chemical agents (additives) on gluten development in cookie dough. *Journal of Food Science and Technology* 59: 1396–1406.

Alfutimie, A., Al-Janabi, N., Curtis, R., and Tiddy, G.J. (2016). The effect of monoglycerides on the crystallisation of triglyceride. *Colloids and Surfaces A: Physicochemical and Engineering Aspects* 494: 170–179.

Alhajj, M.J., Montero, N., Yarce, C.J., and Salamanca, C.H. (2020). Lecithins from vegetable, land, and marine animal sources and their potential applications for cosmetic, food, and pharmaceutical sectors. *Cosmetics* 7: 1–19.

Lillah, Asghar, A., Pasha, I., Murtaza, G., and Ali, M. (2017). Improving heat stability along with quality of compound dark chocolate by adding optimized cocoa butter substitute (hydrogenated palm kernel stearin) emulsion. *LWT - Food Science and Technology* 80: 531–536.

Atik, D.S., Bölük, E., Toker, O.S., Palabiyik, I. (2020). Investigating the effects of lLecithin-PGPR mixture on physical properties of milk chocolate. *LWT - Food Science and Technology* 129: 109548.

Baer, R.J., Wolkow, M.D., and Kasperson, K.M. (1997). Effect of emulsifiers on the body and texture of low fat ice cream. *Journal of Dairy Science* 80: 3123–3132.

Bin Sintang, M.D., Danthine, S., Patel, A.R., Rimaus, T., Van De Walle, D., and Dewettinck, K.,,,, (2017). Mixed surfactant systems of sucrose esters and lecithin as a synergistic approach for oil structuring. *Journal of Colloid and Interface Science* 504: 387–396.

Bodennec, M., Guo, Q., and Rousseau, D. (2016). Molecular and microstructural characterization of lecithin-based oleogels made with vegetable oil. *RSC Advances* 6: 47373–47381.

Bollom, M.A., Clark, S., and Acevedo, N.C. (2020). Development and characterization of a novel soy lecithin-stearic acid and whey protein concentrate bigel system for potential edible applications. *Food Hydrocolloids* 101: 105570.

Bot, F., Cossuta, D., and O'Mahony, J.A. (2021). Inter-relationships between composition, physicochemical properties and functionality of lecithin ingredients. *Trends in Food Science and Technology* 111: 261–270.

Botega, D.C., Marangoni, A.G., Smith, A.K., and Goff, H.D. (2013). Development of formulations and processes to incorporate wax oleogels in ice cream. *Journal of Food Sicence* 78: 1845–1851.

Briggs, M., Petersen, K., and Kris-Etherton, P. (2017). Saturated fatty acids and cardiovascular disease: Replacements for saturated fat to reduce cardiovascular risk. *Healthcare* 5: 29.

Bueschelberger, H.G., Tirok, S., Stoffels, I., and Schoeppe, A. (2015). Lecithins. In: *Emulsifiers in Food Technology* (ed. V. Norn), 21–60. John Wiley & Sons, Ltd.

Cabezas, D.M., Diehl, B.W.K., and Tomás, M.C. (2016). Emulsifying properties of hydrolysed and low HLB sunflower lecithin mixtures. *European Journal of Lipid Science and Technology* 118: 975–983.

Cerdeira, M., Pastore, V., Vera, L.V., Martini, S., Candal, R.J., and Herrera, M.L. (2005). Nucleation behavior of blended high-melting fractions of milk fat as affected by emulsifiers. *European Journal of Lipid Science and Technology* 107: 877–885.

Chaves, K.F., Barrera-Arellano, D., and Ribeiro, A.P.B. (2018). Potential application of lipid organogels for food industry. *Food Research International (Ottawa, Ont.)* 105: 863–872.

Chen, L. (2015). Emulsifiers as food texture modifiers. In: *Modifying Food Texture: Novel Ingredients and Processing Techniques* (ed. J. Jianshe Chen and A. Rosenthal), 27–49. Elsevier Ltd.

Chen, Y., Ge, H., Zheng, Y., Zhang, H., Li, Y., Su, X., Panpipat, W., Lai, O.M., Tan, C.P., and Cheong, L.Z. (2020). Phospholipid–protein structured membrane for microencapsulation of DHA oil and evaluation of its in vitro digestibility: Inspired by milk fat globule membrane. *Journal of Agricultural and Food Chemistry* 68: 6190–6201.

Chung, C., Sher, A., Rousset, P., Decker, E.A., and McClements, D.J. (2017). Formulation of food emulsions using natural emulsifiers: Utilization of quillaja saponin and soy lecithin to fabricate liquid coffee whiteners. *Journal of Food Engineering* 209: 1–11.

Cooper, Z., Simons, C., and Martini, S. (2019). Retardation of crystallization through the addition of dairy phospholipids. *Journal of the American Oil Chemists' Society* 96: 1205–1218.

Daels, E., Rigolle, A., Raes, K., De Block, J., and Foubert, I. (2015). Monoglycerides, polyglycerol esters, lecithin, and their mixtures influence the onset of non-isothermal fat crystallization in a concentration dependent manner. *European Journal of Lipid Science and Technology* 117: 1745–1753.

Dedinaite, A., Claesson, P.M., Campbell, B., and Mays, H. (1998). Interactions between modified mica surfaces in triglyceride media. *Langmuir* 7463: 5546–5554.

Delacharlerie, S., Petrut, R., Deckers, S., Flöter, E., Blecker, C., and Danthine, S. (2016). Structuring effects of lecithins on model fat systems: A comparison between native and hydrolyzed forms. *LWT - Food Science and Technology* 72: 552–558.

Demirci, M., Lee, C.C., Çavuş, M., and Çağlar, M.Y. (2020). Oleogels for food applications. In: *Biopolymer-Based Formulations: Biomedical and Food Applications* (ed. K. Pal, B. Banerjee, P. Sarkar, D. Kim, W.P Deng, N. K. Dubey, and K. Majumder), 781–811. Elsevier Inc. All.

Demirkesen, I. and Mert, B. (2019). Recent developments of oleogel utilizations in bakery products. *Critical Reviews in Food Science and Nutrition* 60: 2460–2479.

Deng, L. (2021). Current progress in the utilization of soy-based emulsifiers in food applications—A Review. *Foods* 10: 1354.

Dhonsi, D. and Stapley, A.G.F. (2006). The effect of shear rate, temperature, sugar and emulsifier on the tempering of cocoa butter. *Journal of Food Engineering* 77: 936–942.

Dijkstra, A.J. (2010). Enzymatic degumming. *European Journal of Lipid Science and Technology* 112: 1178–1189.

Dijkstra, A.J. (2017). About water degumming and the hydration of non-hydratable phosphatides. *European Journal of Lipid Science and Technology* 119: 1–11.

El-Abhar, M., Mahmoud, G., Hanafy, E., El-Mallah, M.H., and El-Shami, S.M. (2020). Comparative study of modified soy lecithins as oil in water (O/W) emulsifiers. *Egyptian Journal of Chemistry* 63: 0–0.

Euston, S.R. and Goff, H.D. (2019). Emulsifiers in dairy products and dairy substitutes. In: *Food Emulsifiers and Their Applications* (ed. G. Hasenhuettl and R. Hartel), 217–254. Springer, Cham.

Flores-Andrade, E., Allende-Baltazar, Z., Sandoval-González, P.E., Jiménez-Fernández, M., Beristain, C.I., and Pascual-Pineda, L.A. (2021). Carotenoid nanoemulsions stabilized by natural emulsifiers: whey protein, gum Arabic, and soy lecithin. *Journal of Food Engineering* 290: 110208.

Garti, N. and Yano, J. (2001). The roles of emulsifiers in fat crystallization. In: *Crystallization Processes in Fats and Lipid Systems* (ed. N. Garti and J. Yano), 225–264. New York, USA: CRC Press.

Gasa-Falcon, A., Odriozola-Serrano, I., Oms-Oliu, G., and Martín-Belloso, O. (2020). Nanostructured lipid-based delivery systems as a strategy to increase functionality of bioactive compounds. *Foods* 9: 325.

Gaudino, N., Mirzaee, S., Clark, S., and Marangoni, A.G. (2019). Development of lecithin and stearic acid based oleogels and oleogel emulsions for edible semisolid applications. *Food Research International (Ottawa, Ont.)* 116: 79–89.

Grossmann, L. and Mcclements, D.J. (2021). The science of plant-based foods: approaches to create nutritious and sustainable plant-based cheese analogs. *Trends in Food Science and Technology* 118: 207–229.

Guo, S., Song, M., He, X., Yang, F., Cao, Y., Rogers, M., and Lan, Y. (2019). Water-induced self-assembly of mixed gelator system (ceramide and lecithin) for edible oil structuring. *Food & Function* 10: 3923–3933.

Guo, S., Lv, M., Chen, Y., Hou, T., Zhang, Y., Huang, Z., Cao, Y., Rogers, M., and Lan, Y. (2020a). Engineering water-induced ceramide/lecithin oleogels: Understanding the influence of water added upon pre- and post-nucleation. *Food & Function* 11: 2048–2057.

Guo, S., Song, M., Gao, X., Dong, L., Hou, T., Lin, X., Tan, W., Cao, Y., Rogers, M., and Lan, Y. (2020b). Assembly pattern of multicomponent supramolecular oleogel composed of ceramide and lecithin in sunflower oil: Self-assembly or self-sorting? *Food & Function* 11: 7651–7660.

Himawan, C., Starov, V.M., and Stapley, A.G.F. (2006). Thermodynamic and kinetic aspects of fat crystallization. *Advances in Colloid and Interface Science* 122: 3–33.

Imai, M., Hashizaki, K., Taguchi, H., Saito, Y., and Motohashi, S. (2013). A new reverse wormlike micellar system from a lecithin, multivalent carboxylic acid and oil mixture. *Journal of Colloid and Interface Science* 403: 77–83.

Jimenez-Colmenero, F., Salcedo-Sandoval, L., Bou, R., Cofrades, S., Herrero, A.M., and Ruiz-Capillas, C. (2015). Novel applications of oil-structuring methods as a strategy to improve the fat content of meat products. *Trends in Food Science and Technology* 44: 177–188.

Jin, J. and Hartel, R.W. (2020). Accelerated fat bloom in chocolate model systems : Replacement of cocoa powder with sugar particles and the effects of lecithin. *Journal of the American Oil Chemists' Society* 97: 377–388.

Joshi, A., Paratkar, S.G., and Thorat, B.N. (2006). Modification of lecithin by physical, chemical and enzymatic methods. *European Journal of Lipid Science and Technology* 108: 363–373.

Julio, L.M., Copado, C.N., Crespo, R., and Diehl, B.W.K. (2019). Design of microparticles of chia seed oil by using the electrostatic layer-by-layer deposition technique. *Powder Technology* 345: 750–757.

Khorasani, S., Danaei, M., and Mozafari, M.R. (2018). Nanoliposome technology for the food and nutraceutical industries. *Trends in Food Science and Technology* 79: 106–115.

Kim, M.G., Oh, E.J., Jin, K.S. et al. (2021). Effect of sugar alcohols on the reverse self-assembly of lecithin in diverse organic solvents. *Journal of Molecular Liquids* 330: 115670.

Klang, V. and Valenta, C. (2011). Lecithin-based nanoemulsions. *Journal of Drug Delivery Science and Technology* 21: 55–76.

Krog, N.J. and Sparsø, F.V. (2004). Food emulsifiers: Their chemical and physical properties. In: *Food Emulsions* (ed. S.E. Friberg, K. Larsson, and J. Sjoblom). Marcel Dekker, Inc.

Lawler, P.J. and Dimick, P.S. (2008). Crystallization and polymorphism of fats. In: *Food Lipids: Chemistry, Nutrition and Biotechnology* (ed. C.C. Akoh, B. David and D.B. Min), 253. New York, USA: CRC Press.

Lehtinen, O.-P., Nugroho, R.W.N., Lehtimaa, T., Vierros, S., Hiekkataipale, P., Ruokolainen, J., Sammalkorpi, M., and Österberg, M. (2017). Effect of temperature, water content and free fatty acid on reverse micelle formation of phospholipids in vegetable oil. *Colloids Surfaces B Biointerfaces* 160: 355–363.

Li, H., Yu, H., Liu, Y., Wang, Y., Li, H., and Yu, J. (2019). The use of of inulin, maltitol and lecithin as fat replacers and plasticizers in a model reduced-fat mozzarella cheese-like product. *Journal of the Science of Food and Agriculture* 99: 5586–5593.

Li, J., Wang, X., Zhang, T., Wang, C., Huang, Z., Luo, X., and Deng, Y. (2015a). A review on phospholipids and their main applications in drug delivery systems. *Asian Journal of Pharmaceutical Sciences* 10: 81–98.

Li, J., Wang, X., Zhang, T., Wang, C., Huang, Z., Luo, X., and Deng, Y. (2015b). A review on phospholipids and their main applications in drug delivery systems. *Asian Journal of Pharmaceutical Sciences* 10: 81–98.

Li, L., Liu, G., and Lin, Y. (2021). Physical and bloom stability of low-saturation chocolates with oleogels based on different gelation mechanisms. *LWT - Food Science and Technology* 140: 110807.

Lin, X., Wang, Q., Li, W., and Wright, A.J. (2014). Emulsification of algal oil with soy lecithin improved DHA bioaccessibility but did not change overall in vitro digestibility. *Food & Function* 5: 2913–2921.

Lin, Y., Yin, W., and Liu, G. (2020). Development and characterisation of a novel krill oil nanostructured lipid carrier based on 1,3-glycerol distearate. *International Journal of Food Science & Technology* 55: 3493–3502.

List, G.R. (2015). Soybean lecithin: food, industrial uses, and other applications. In: *Polar Lipids: Biology, Chemistry, and Technology* (ed. M.U. Ahmad and X. Xu), 1–33. AOCS Press.

Liu, Y., Fu, D., Bi, A., Wang, S., Li, X., Xu, X., and Song, L. (2021). Development of a high internal phase emulsion of antarctic krill oil diluted by soybean oil using casein as a co-emulsifier. *Foods* 10: 917.

Lombardo, D., Calandra, P., Barreca, D., Magazù, S., and Kiselev, M.A. (2016). Soft interaction in liposome nanocarriers for therapeutic drug delivery. *Nanomaterials* 6: 125.

Lončarević, I., Pajin, B., Omorjan, R., Torbica, A., Zaric, D., Maksimovic, J., and Gajic, J.S. (2013). The influence of lecithin from different sources on crystallization and physical properties of nontrans fat. *Journal of Texture Studies* 44: 450–458.

Ma, Z., Zhao, Y., Khalid, N., Shu, G., Neves, M.A., Koboyashi, I., and Nakajima, M. (2020). Comparative study of oil-in-water emulsions encapsulating fucoxanthin formulated by microchannel emulsification and high-pressure homogenization. *Food Hydrocolloids* 108: 105977.

Manjula, S., Kobayashi, I., and Subramanian, R. (2011). Characterization of phospholipid reverse micelles in nonaqueous systems in relation to their rejection during membrane processing. *Food Research International (Ottawa, Ont.)* 44 (44): 925–930.

Marangoni, A.G. (2004). *Fat Crystal Networks*. New York, USA: Marcel Dekke.

Marangoni, A.G., van Duynhoven, J.P.M., Acevedo, N.C., Nicholson, R.A., and Patel, A.R. (2020). Advances in our understanding of the structure and functionality of edible fats and fat mimetics. *Soft Matter* 16: 289–306.

Marrink, S. and Mark, A.E. (2004). Molecular view of hexagonal phase formation in phospholipid membranes. *Biophysical Journal* 87: 3894–3900.

Martínez-Ávila, M., De la Peña-gil, A., Álvarez-Mitre, F.M., Charó-Alonso, M.A., and Toro-Vazquez, J.F. (2019). Self-assembly of saturated and unsaturated phosphatidylcholine in mineral and vegetable oils. *Journal of the American Oil Chemists' Society* 96: 273–289.

Mcclements, D.J. (2020). Advances in nanoparticle and microparticle delivery systems for increasing the dispersibility, stability, and bioactivity of phytochemicals. *Biotechnology Advances* 38: 107287.

McClements, D.J. (2012). Advances in fabrication of emulsions with enhanced functionality using structural design principles. *Current Opinion in Colloid & Interface Science* 17: 235–245.

McClements, D.J. (2016). *Food Emulsions - Principles, Practices, and Techniques*, 3e. Boca Raton: CRC Press.

McClements, D.J., Bai, L., and Chung, C. (2017). Recent advances in the utilization of natural emulsifiers to form and stabilize emulsions. *Annual Review of Food Science and Technology* 8: 205–236.

McClements, D.J., Decker, E.A., and Weiss, J. (2007). Emulsion-based delivery systems for lipophilic bioactive components. *Journal of Food Science* 72: R109–R124.

Meeren, P.V.D., El-bakry, M., Neirynck, N., and Noppe, P. (2005). Influence of hydrolysed lecithin addition on protein adsorption and heat stability of a sterilised coffee cream simulant. *International Dairy Journal* 15: 1235–1243.

Metin, S. and Hartel, R.W. (2005). Fundamentals of lipid crystallization. In: *Bailey's Industrial Oil Fat Products* (ed. F. Shahidi), 1–31. New York, USA: Wiley.

Meza, B.E., Peralta, J.M., and Zorrilla, S.E. (2016). Rheological characterization of full-fat and low-fat glaze materials for foods. *Journal of Food Engineering* 171: 57–66.

Miki, A.J., Livingston, K.A., Karlsen, M.C., Folta, S.C., and McKeown, N.M. (2017). Using evidence mapping to examine motivations for following plant-based diets. *Current Developments in Nutrition* 4: 1–15.

Miyasaki, E.K., Luccas, V., and Kieckbusch, T.G. (2016). Modified soybean lecithins as inducers of the acceleration of cocoa butter crystallization. *European Journal of Lipid Science and Technology* 117: 1–11.

Mordor Intelligence (2021). Lecithin market - growth, trends, covid-19 impact, and forecasts (2021-2026). *Mordor Intell.* https://www.mordorintelligence.com/industry-reports/lecithin-market (accessed 5 January 2022).

Munk, M.B., Munk, D.M.E., Gustavsson, F., and Risbo, J. (2018). Using ethylcellulose to structure oil droplets in ice cream made with high oleic sunflower oil. *Journal of Food Science* 83: 2520–2526.

Najafi, A., Daghigh Kia, H., and Hamishehkar, H. (2021). Does alpha-lipoic acid–loaded nanostructured lipid carriers improve post-thawed sperm quality and ameliorate apoptosis-related genes of rooster sperm? *Poultry Science* 100: 357–365.

Nikiforidis, C.V. and Scholten, E. (2014). Self-assemblies of lecithin and α-tocopherol as gelators of lipid material. *RSC Advances* 4: 2466–2473.

Njauw, C.W., Cheng, C.Y., Ivanov, V.A., Khokhlov, A.R., and Tung, S.H. (2013). Molecular interactions between lecithin and bile salts/acids in oils and their effects on reverse micellization. *Langmuir* 29: 3879–3888.

Okuro, P.K., Gomes, A., Costa, A.L.R., Adame, M.A., and Cunha, R.L. (2019). Formation and stability of W/O-high internal phase emulsions (HIPEs) and derived O/W emulsions stabilized by PGPR and lecithin. *Food Research International (Ottawa, Ont.)* 122: 252–262.

Okuro, P.K., Malfatti-Gasperini, A.A., Fasolin, L.H., Vicente, A.A., and Cunha, R.L. (2020). Self-organizing structures of phosphatidylcholine in nonaqueous solvents: tailoring gel-like systems. *Journal of Surfactants and Detergents* 23: 725–735.

Okuro, P.K., Tavernier, I., Bin Sintang, M.D., Sintang, M.D.B., Skirtach, A.G., Vicente, A.A., Dewettinck, K., and Cunha, R.L. (2018). Synergistic interactions between lecithin and fruit wax in oleogel formation. *Food & Function* 9: 1755–1767.

Ozturk, B., Argin, S., Ozilgen, M., and McClements, D.J. (2014). Formation and stabilization of nanoemulsion-based vitamin E delivery systems using natural surfactants: Quillaja saponin and lecithin. *Journal of Food Engineering* 142: 57–63.

Park, C. and Maleky, F. (2020). A critical review of the last 10 years of oleogels in food. *Frontiers in Sustainable Food Systems* 4: 1–8.

Pateiro, M., Gómez, B., Munekata, P.E.S., Barba, F.J., Putnik, P., Kovačević, D.B., and Lorenzo, J.M. (2021). Nanoencapsulation of promising bioactive compounds to improve their absorption, stability, functionality and the appearance of the final food products. *Molecules* 26: 1547.

Patel, A.R. (2017). A colloidal gel perspective for understanding oleogelation. *Current Opinion in Food Science* 15: 1–7.

Patel, A.R. and Dewettinck, K. (2015). Comparative evaluation of structured oil systems: Shellac oleogel, HPMC oleogel, and HIPE gel. *European Journal of Lipid Science and Technology* 117: 1772–1781.

Patel, A.R., Nicholson, R.A., and Marangoni, A.G. (2020). Applications of fat mimetics for the replacement of saturated and hydrogenated fat in food products. *Current Opinion in Food Science* 33: 61–68.

Penttilä, P.A., Vierros, S., Utriainen, K., Carl, N., Rautkari, L., Sammalkorpi, M., and Österberg, M. (2019). Phospholipid-based reverse micelle structures in vegetable oil modified by water content, free fatty acid, and temperature. *Langmuir* 35: 8373–8382.

Pichot, R., Watson, R.L., and Norton, I.T. (2013). Phospholipids at the interface: Current trends and challenges. *International Journal of Molecular Sciences* 14: 11767–11794.

Raghavan, S.R. (2009). Distinct character of surfactant gels: A smooth progression from micelles to fibrillar networks. *Langmuir* 25: 8382–8385.

Real, D.A., Hoffmann, S., Leonardi, D., Goycoolea, F., and Salomon, C. (2021). A quality by design approach for optimization of Lecithin/Span® 80 based nanoemulsions loaded with hydrophobic drugs. *Journal of Molecular Liquids* 321: 114743.

Ribeiro, A.P.B., Masuchi, M.H., Miyasaki, E.K., Domingues, M.A.F., Stroppa, V.L.Z., de Olivera, G.M., and Kiechbusch, T.G., AFontenel. (2015). Crystallization modifiers in lipid systems. *Journal of Food Science and Technology* 52: 3925–3946.

Rigolle, A., Gheysen, L., Depypere, F., Landuyt, A., Van Den Abeele, K., and Foubert, I. (2015). Lecithin influences cocoa butter crystallization depending on concentration and matrix. *European Journal of Lipid Science and Technology* 117: 1722–1732.

Rizzo, G., Norton, J.E., and Norton, I.T. (2015). Emulsifier effects on fat crystallisation. *Food Structure* 4: 27–33.

Robert, C., Couëdelo, L., Vaysse, C., and Michalski, M.-C. (2020). Vegetable lecithins: a review of their compositional diversity, impact on lipid metabolism and potential in cardiometabolic disease prevention. *Biochimie* 169: 121–132.

Rocchio, J., Neilsen, J., Everett, K., and Bothun, G.D. (2017). A solvent-free lecithin-Tween 80 system for oil dispersion. *Colloids and Surfaces A: Physicochemical and Engineering Aspects* 533: 218–223.

Sandoval, J.M., Carelli, A., Palla, C., and Baümler, E. (2020). Preparation and characterization of oleogel emulsions: a comparative study between the use of recovered and commercial sunflower waxes as structuring agent. *Journal of Food Science* 85: 2866–2878.

Sandoval-Cuellar, C.E., de Jesus Perea-flores, M., and Quintanilla-Carvajal, M.X. (2020). In-vitro digestion of whey protein- and soy lecithin-stabilized High Oleic Palm Oil emulsions. *Journal of Food Engineering* 278: 109918.

Sato, K. (2018). *Crystallization of Lipids: Fundamentals and Applications in Food, Cosmetics, and Pharmaceuticals*. Hoboken, NJ, USA: John Wiley & Sons.

Savage, C.M. and Dimick, P.S. (1995). Influence of phospholipids during crystallization of hard and soft cocoa butters. *Manufacturing Confectioner* 75: 127–132.

Scartazzini, R. and Luisi, P.L. (1988). Organogels from lecithins. *Journal of Physical Chemistry* 92: 829–833.

Shchipunov, Y.A. and Hoffmann, H. (1998). Growth, branching, and local ordering of lecithin polymer-like micelles. *Langmuir* 14: 6350–6360.

Si, H., Cheong, L.Z., Huang, J., Wang, X., and Zhang, H. (2016). Physical properties of soybean oleogels and oil migration evaluation in model praline system. *Journal of the American Oil Chemists' Society* 93: 1075–1084.

Silva, T.J., Barrera-Arellano, D., and Ribeiro, A.P.B. (2021a). Oleogel-based emulsions: Concepts, structuring agents, and applications in food. *Journal of Food Science* 86: 2785–2801.

Silva, T.J., Barrera-Arellano, D., and Ribeiro, A.P.B. (2021b). Margarines: Historical approach, technological aspects, nutritional profile, and global trends. *Food Research International (Ottawa, Ont.)* 147: 110486.

Simoes, S., Lelaj, E., and Rousseau, D. (2021). The presence of crystalline sugar limits the influence of emulsifiers on cocoa butter crystallization. *Food Chemistry* 346: 128848.

Sipahioglu, O., Alvarez, V. B., and Solano-Lopez, C. (1999). Structure, physico-chemical and sensoryproperties of feta cheese made with tapioca starch and lecithin as fat mimetics. *International Dairy Journal* 9: 783–789

Smith, K.W., Bhaggan, K., Talbot, G., and van Malssen, K.F. (2011). Crystallization of fats: influence of minor components and additives. *Journal of the American Oil Chemists' Society* 88: 1085–1101.

Smith, K.W. and Sato, K. (2018). Effects of foreign and indigenous minor components. In: *Crystallization of Lipids: Fundamentals and Applications in Food, Cosmetics, and Pharmaceuticals* (ed. K. Sato), 263–281. Hoboken, NJ, USA: Wiley.

Soleimanian, Y., Goli, S.A.H., Varshosaz, J., and Maestrelli, F. (2019). β-sitosterol lipid nano carrier based on propolis wax and pomegranate seed oil: Effect of thermal processing, pH, and ionic strength on stability and structure. *European Journal of Lipid Science and Technology* 121: 1800347.

Soleimanifard, M., Sadeghi Mahoonak, A., Ghorbani, M., Heidari, R., and Sepahvand, A. (2020). The formulation optimization and properties of novel oleuropein-loaded nanocarriers. *Journal of Food Science and Technology* 57: 327–337.

Sreedhar, R., Kumar, V.S., Bhaskaran Pillai, A.K., and Mangalathillam, S. (2019). Omega-3 fatty acid based nanolipid formulation of atorvastatin for treating hyperlipidemia. *Advanced Pharmaceutical Bulletin* 9: 271–280.

Stortz, T.A. and Marangoni, A.G. (2013). Ethylcellulose solvent substitution method of preparing heat resistant chocolate. *Frin* 51: 797–803.

Su, M.M., Yang, H.K., Ren, L.J., Zheng, P., and Wang, W. (2015). Solvent-mediated gel formation, hierarchical structures, and rheological properties of organogels. *Soft Matter* 11: 741–748.

Sun, P., Xia, B., Ni, Z.J., Wang, Y., Elam, E., Thakur, K., Ma, Y.L., and Wei, Z.J. (2021). Characterization of functional chocolate formulated using oleogels derived from β-sitosterol with γ-oryzanol/lecithin/stearic acid. *Food Chemistry* 360: 130017.

Sun, X., Zhang, L., Tian, S., and Yang, K. (2020). Phospholipid composition and emulsifying properties of rice bran lecithin from enzymatic degumming. *LWT - Food Science and Technology* 117: 108588.

Sun, X.M., Wang, C.N., and Guo, M.R. (2018). Interactions between whey protein or polymerized whey protein and soybean lecithin in model system. *Journal of Dairy Science* 101: 9680–9692.

Szuhaj, B.F. (1983). Lecithin production and utilization. *Journal of the American Oil Chemists' Society* 60: 306–309.

Tamjidi, F., Shahedi, M., Varshosaz, J., and Nasirpour, A. (2017). Stability of astaxanthin-loaded nanostructured lipid carriers as affected by pH, ionic strength, heat treatment, simulated gastric juice and freeze–thawing. *Journal of Food Science and Technology* 54: 3132–3141.

Tamjidi, F., Shahedi, M., Varshosaz, J., and Nasirpour, A. (2018). Stability of astaxanthin-loaded nanostructured lipid carriers in beverage systems. *Journal of the Science of Food and Agriculture* 98: 511–518.

Tayebati, S.K. and Amenta, F. (2013). Choline-containing phospholipids: relevance to brain functional pathways. *Clinical Chemistry and Laboratory Medicine* 51: 513–521.

Teberikler, L., Koseoglu, S., and Akgerman, A. (2001). Selective extraction of phosphatidylcholine from lecithin by supercritical carbon dioxide/ethanol mixture. *Journal of the American Oil Chemists' Society* 78: 115–120.

Tekin, E., Sahin, S., and Sumnu, G. (2017). Physicochemical, rheological and sensory properties of low-fat ice cream designed by double emulsions. *European Journal of Lipid Science and Technology* 119: 1–32.

Tran, T., Green, N.L., and Rousseau, D. (2015). Spheroidal fat crystals: Structure modification via use of emulsifiers. *Journal of Crystal Growth* 15: 5406–5415.

Van Nieuwenhuyzen, W. (1981). The industrial uses of special lecithins: A review. *Journal of the American Oil Chemists' Society* 58: 886–888.

Van Nieuwenhuyzen, W. (2015). *Production and Utilization of Natural Phospholipids, Polar Lipids: Biology, Chemistry, and Technology*. AOCS Press.

van Nieuwenhuyzen, W. and Tomás, M.C. (2008). Update on vegetable lecithin and phospholipid technologies. *European Journal of Lipid Science and Technology* 110: 472–486.

Vanhoutte, B., Dewettinck, K., Foubert, I., and Vanlerberghe, B. (2002). The effect of phospholipids and water on the isothermal crystallisation of milk fat. *European Journal of Lipid Science and Technology* 104: 490–495.

Wang, X., Liu, L., Xia, S., Muhoza, B., Cai, J., Zhang, X., Duhoranimana, E., and Su, J. (2019). Sodium carboxymethyl cellulose modulates the stability of cinnamaldehyde-loaded liposomes at high ionic strength. *Food Hydrocolloids* 93: 10–18.

Yang, -Q.-Q., Sui, Z., Lu, W., and Corke, H. (2021). Soybean lecithin-stabilized oil-in-water (O/W) emulsions increase the stability and in vitro bioaccessibility of bioactive nutrients. *Food Chemistry* 338: 128071.

Yesiltas, B., Torkkeli, M., Almásy, L., Dudás, Z., Wacha, A.F., Dalgliesh, R., García-Moreno, P.J., Sørensen, A.D.M., Jacobsen, C., and Knaapila, M. (2019). Interfacial structure of 70% fish oil-in-water emulsions stabilized with combinations of sodium caseinate and phosphatidylcholine. *Journal of Colloid and Interface Science* 554: 183–190.

Zhang, S., Song, W., Wu, H., Wang, J., Wang, Y., Zhang, Z., and Lv, H. (2020a). Lecithins-Zein nanoparticles for antifungal treatment: enhancement and prolongation of drug retention in skin with reduced toxicity. *International Journal of Pharmaceutics* 590: 119894.

Zhang, T., Ding, M., Zhang, H., Tao, N., Wang, X., and Zhong, J. (2020b). Fish oil-loaded emulsions stabilized by synergetic or competitive adsorption of gelatin and surfactants on oil/water interfaces. *Food Chemistry* 308: 125597.

Zhou, X., Chen, L., Han, J., Shi, M., Wang, Y., and Zhang, L. (2017). Stability and physical properties of recombined dairy cream: effects of soybean lecithin. *International Journal of Food Properties* 20: 2223–2233.

Zou, H., Zhao, N., Li, S., Sun, S., Dong, X., and Yu, C. (2020). Physicochemical and emulsifying properties of mussel water-soluble proteins as affected by lecithin concentration. *International Journal of Biological Macromolecules* 163: 180–189.

2.4

Mono- and Diglycerides

Sofia Melchior, Stella Plazzotta, Sonia Calligaris, and Lara Manzocco*

Department of Agricultural, Food, Environmental, and Animal Sciences, University of Udine, Udine, Italy
** Corresponding author*

2.4.1 Introduction

Within the last 10 years, scientific and industrial communities have examimed the challenge of modifying foods' lipid profile by reducing the total fat content or by increasing the content of unsaturated fatty acids, while also reducing the saturated and trans fats. These parties' interest is due largely to the burden of diet associated non-communicable diseases. Traditional saturated fats, such as animal fats, tropical oils, margarine, and shortenings are multipurpose ingredients that are found in many food formulas in which they confer peculiar characteristics. Thus, the partial or total replacement of traditional saturated fats is challenging due to the intrinsic contribution of said ingredients to the final food product's processability and to its overall chemical, physical, and sensorial attributes. Reducing fat content or substituting it with liquid oil severely decreases the final food's acceptance. This unacceptable quality strongly motivates fat alternatives' research. The first step is surely the development of novel systems that present specific compositional and structural features. In particular, formulations of fat alternatives should be based on unsaturated fatty acids, eventually associated with water and air, with the final aim of reducing not only the saturated fat content but also the overall caloric density. The structure of such systems is guaranteed by molecules able to organize themselves at different lengths of scales, generating materials with the functional properties of solid fats. In this context, the structuring capacity of mono- and diglycerides allows for the generation of a plethora of different structured systems that can mimic the functionalities of conventional solid fats in different foods. In this chapter, after an initial description of mono- and diglycerides characteristics, the discussion will mainly focus on the different structures that these molecules form in different environments (*i.e.*, water, oil, multiphase, etc.) describing their hierarchical structural organization from a fundamental point of view, moving to factors affecting structure. Finally, relevant food applications will also be discussed.

2.4.2 Monoglycerides and Diglycerides

Monoglycerides (MGs) and diglycerides (DGs) are polar lipid molecules comprising one or two fatty acids esterified to a glycerol molecule, respectively. Due to their chemical characteristics, they exhibit an amphiphilic nature and are the most commonly used polar lipids in several food products, including cakes, bakery products, margarines, frozen desserts, and spreads (Naik et al. 2014;

Fat Mimetics for Food Applications, First Edition. Edited by Miguel Ângelo Parente Ribeiro Cerqueira and Lorenzo Miguel Pastrana Castro.

Nicholson and Marangoni 2018). Based on EU regulation (EC 2008), MGs and DGs are "*quantum satis*" additives (E471) with no need for a numerical acceptable daily intake (ADI), as reported in a recent re-evaluation of EFSA (EFSA 2017). Similarly, in US they are recognised as GRAS by FDI. The wide use of MGs and DGs in foods generates a market estimated to be worth 8.26 billion USD by 2027 (Data Bridge Market Research 2020) accounting for about 70% of the total emulsifier market (Nicholson and Marangoni 2018).

MGs and DGs for commercial applications are usually produced by direct esterification of glycerol with fatty acids or glycerolysis of fats. Considering the two processes, glycerolysis is more affordable than direct esterification, due to the increasing cost of fatty acid and to the quantity of glycerol required (Pouilloux et al. 1999; Hasenhuettl 2008; Lo et al. 2008; Miao and Lin 2018). Both procedures can be chemically conducted using acid, basic, or metal oxide catalysts; they can also be enzymatically conducted with lipase catalysts. Usually, chemical processes are faster, more reproducible, and cost-effective than enzymatic ones even when performed at high temperatures (200–250°C). This condition may be responsible for undesirable side reactions, dark colour by-products, and off flavors (Miao and Lin 2018; Nicholson and Marangoni 2018). Regardless of the method used, thanks to the similar reactivity of the three hydroxyl groups of glycerol, the final product consists in mixtures of MGs, DGs, and triglycerides. Considering the DGs, two possible isomers can be produced: 1,2-DGs with fatty acids esterified in the sn-1 and sn-2 positions and 1,3-DGs with fatty acids linked in the sn-1 and sn-3 positions (Lo et al. 2008; Nicholson and Marangoni 2018). Regarding MGs, they can be divided in 1-MG (or α-MG) and 2-MG (β-MG) (Susi et al. 1961). The MGs' content in the mixture can be increased via distillation obtaining products with high MG levels. With this method, the total MG content reaches 93–97%, with 1-monoester content being at least 90% in commercial distilled MGs (Krog and Sparsø 2004; Barfod and Sparsø 2007; Miao and Lin 2018). Depending on the saturation degree of the fatty acids, distilled MGs can also be classified in saturated and unsaturated MGs (Alfutimie et al. 2014). In general, as the chain length increases, the melting point increases, whereas, MG hydrophilic properties are inversely correlated with the number of carbon atoms. The fatty acid chain length of commercially distillated MGs depends on the source of raw materials (Miao and Lin 2018). Among the wide variety of commercially available distilled MGs, those with mainly palmitic (40–60%) and stearic (60–40%) acids are the best performing ingredients for generating fat substitutes. They are characterized by a melting point higher than 55°C and have the ability to self-assemble in water, oil, or multiphase environments, forming crystalline structures (Krog and Sparsø 1997). Therefore, this chapter's focus will mainly be on distilled saturated MGs. It will be noted when DG-based fat mimetics are considered.

2.4.3 Fat Mimetics Based on Mono- and Diglycerides

MGs, and in some cases DGs, can generate a plethora of plastic viscoelastic materials embedding oil, water, or both oil and water phases. In all cases, gelled structures that appear to be similar to plastic fats are obtained. Gels are defined as a continuous, three-dimensional network of connected molecules or particles (such as crystals, fibrils, etc.) entrapping a large volume of a continuous liquid phase (Abdallah and Weiss 2000). Gels have intermediate characteristics between solids and liquids: the three-dimensional network structure makes gels partially behave as elastic solids, however, molecules are freely flowing in the continuous liquid phase, so that the gel also behaves in some respects as a viscous liquid (Walstra and van Vilet 2008).

Gels can be classified based on the entrapped liquid phase. When the immobilized liquid is water, hydrogels are obtained; whereas, when oil is the solvent, an oleogel is formed. Finally, emulsion-gels were regarded as gels containing both a water and an oil phase.

2.4.3.1 Hydrogels

The behavior of MGs in water has been extensively reviewed in the literature (Krog and Larsson 1968; Sato 2001; Sagalowicz et al. 2006). When mixed with water, MGs' self-assembly in several different structures depending on the MG type, concentration, and temperature. To obtain a MG-hydrogel at room temperature, the MG must be added to an aqueous solution and the mixture must be heated above the melting point of the selected MGs (Krafft temperature, T_c) (Figure 2.4.1).

Water penetrates in the crystal polar regions, and the hydrocarbon chains undergo a solid-liquid transition, resulting in the formation of a lamellar liquid-crystalline phase ($L\alpha$). In the liquid state, the hydrocarbon chains are highly mobile with the molecules that undergo lateral diffusion, rotation, and flip-flop (position change from one side of the bilayer to the other). These phenomena are ascribed to the so-called lyotropic mesomorphism (Lutton 1971; Heertje et al. 2002; Sein et al. 2002; Krog and Sparsø 2004). When the system is further cooled below the T_c, the MG hydrocarbon chains lose their mobility and crystallize in the hexagonally packed L_β phase (α-gel), which results in a hydrogel (Lutton 1971; Sein et al. 2002; Krog and Sparsø 2004). The final hydrogel structure is thus represented by a 3D network of crystalline MG lamellas separated by water (Figure 2.4.1).

The α-gel is metastable; it gradually converts into triclinic packed β-crystals called coagel (Heertje et al. 1998; Sein et al. 2002; Wang and Marangoni 2014). This polymorphic transition leads to an irreversible structural organization with densely packed molecules and with water leaking from the gel, which reduces the water layer's thickness between MG bilayers (Heertje et al. 1998; Sein et al. 2002; Krog and Sparsø 2004; Wang and Marangoni 2014).

The α-gel is able to hold a high air quantity without exhibiting water exudation for weeks. In particular, a stable foam is obtained by whipping the system during cooling below the T_c (Heertje et al. 1998, 2002). Alternatively, the coagel can entrap air, but the foam does not remain stable for long (Heertje et al. 1998). Foams are two-phase media where gas is dispersed throughout a continuous fluid or solid phase (Blázquez et al. 2014). The gas bubbles are retained within the continuous phase due to the surface-active species' presence, which adsorb at the gas-liquid interface, accounting for both foamability and foam stability (Brun et al. 2015). Aqueous foams are surely the most common type of foams, in which gas bubbles are dispersed in an aqueous continuous phase. These foams are stabilized by surfactant molecules, polymers, or particles, which strongly adsorb

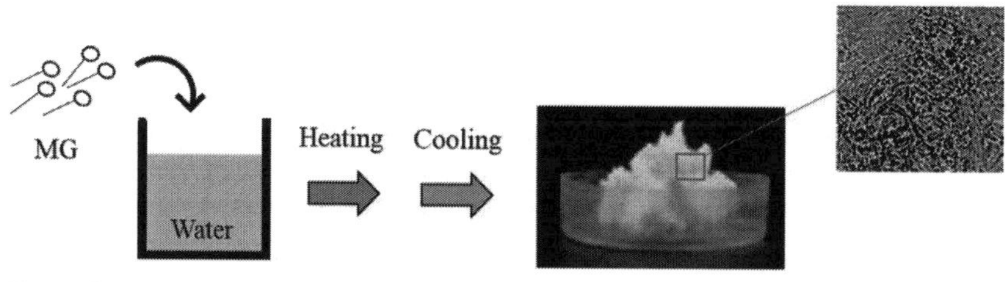

Figure 2.4.1 Schematic representation of the preparation methodology of MG-hydrogels. Macroscopic and microscopic images from Melchior et al. (2021), Royal Society of Chemistry / CC BY NC-3.0.

at the air-water interface, significantly reducing surface tension, bubble coalescence, drainage, and coarsening (Blázquez et al. 2014). The possibility of obtaining hydrogels and their structural characteristics strongly depends on many compositional and environmental factors (Table 2.4.1).

2.4.3.1.1 Effect of Compositional Factors

The final hydrogel structure strongly depends on the MG type. The mixed fatty acid lengths of industrial MGs affect their melting and crystallization properties, as well as their polymorphic behavior, in comparison with the MG with monodisperse fatty acid tails. As a general rule, the melting and crystallizing temperatures of mixed fatty acid MG are lower and show polymorphic behavior more stably than pure components (Krog and Larsson 1968). Based on the literature, 5% MG is required to obtain a structurally stable gel; at lower concentrations, in fact, few poorly connected crystals are formed (Sein et al. 2002).

The most critical issue in MG-hydrogel formation is the swelling behavior that is defined as the ability of MGs to embed water among lamellas. The swelling capacity of the aqueous lamellar phase of commercial MG is limited; thus, introducing a charged group on the surface of the lipid bilayers is a common practice for amplifying the repulsive forces and for increasing their swelling as well (Krog and Sparsø 1997). The most commonly used co-surfactants are soy lecithin, diacetyl tartaric acid ester of MGs and DGs (DATEM), neutralized stearic acid, and sodium stearoyl lactylate (SSL) (Heertje et al. 1998; Sein et al. 2002; Krog and Sparsø 2004). Based on co-surfactant concentration, the swelling can vary from 10–20 Å to several hundred Angstroms in pure MG-water system and commercial MG-water system, respectively (Heertje et al. 1998).

Besides co-surfactants, the swelling behavior also depends on the pH of the aqueous phase. At high pH the acid co-surfactant is ionized, ensuring effective repulsion between bilayers with an increase in lamellar stability. Usually, free fatty acids of commercial MG sources are effective for swelling the lamellar phase at pH around 7. Other surfactants, such as SSL and DATEM, can be used at a lower pH but with lower efficacy. For instance, DATEM is fully ionized at pH 5 and half ionized at pH 3 (Heertje et al. 1998). The pH is usually adjusted with NaOH, but other bases, either strong or weak can also be used (Valoppi et al. 2015a).

The swelling behavior is also affected by the presence of other components, such as salts and proteins (Heertje et al. 1998; Melchior et al. 2021). Ions of NaCl, added as an ingredient in many food formulations, shield the electrostatic repulsive forces induced by anionic surfactants between the bilayers. For this reason, it is essential to select the proper balance between pH and concentrations of anionic species and salt (Heertje et al. 1998). On the other side, proteins can

Table 2.4.1 Compositional and processing factors affecting MG-hydrogel structure.

Factors affecting MG-hydrogel structure		
Compositional factors		MG type and content
		Co-surfactant type and content
		Presence and content of additional components (e.g. salt, proteins)
Processing factors	Setting conditions	Cooling rate
		Shear rate
	Storage conditions	Temperature
		Time

interact with MG, which reduces their availability for networking and results in less swelling. Consequently, the melting point of the hydrogel decreases, and both the gel strength and the rheological properties are lower as compared with the system solely containing the aqueous solution (Valoppi et al. 2015a; Melchior et al. 2021).

2.4.3.1.2 Effect of Processing Factors

The main issue in considering processing parameters is in understanding how these factors affect the transition of α-gel into coagel. This aspect is particularly important since this transition causes syneresis from the matrix during storage.

The cooling rate is reportedly responsible for the hydrogel stability with slow cooling rates displaying higher α-gel stability, which makes sense given that slow cooling rates are associated with longer times to self-assemble into a fully hydrated lamellar structure (Wang and Marangoni 2015). Simultaneously, the application of shear during and after crystallization could reduce the water binding capacity of MG gels by changing their structure at nano and micro level. High stirring conditions led to the formation of a stiffer, more regular and finer coagel network (Heertje et al. 1998).

During storage, both temperature and time affect the rate in which the coagel forms, as well as its stability and properties. At the beginning of the α-gel to coagel transition, small crystals increase their dimensions without affecting hardness. However, after a certain size, they interconnect with each other, resulting in a major hardness increase. At high temperatures, the Ostwald ripening phenomena may occur and the network coarsens causing hardness loss and water exudation (Heertje et al. 1998; Melchior et al. 2021).

2.4.3.2 Oleogels

Oleogels are self-standing, anhydrous, viscoelastic materials structured by a three-dimensional supramolecular network of self-assembled molecules (oleogelators) in edible oils. Oleogelation is an emerging strategy that converts liquid oil into a gel wherein the continuous liquid oil phase is immobilized into a supramolecular network of self-assembled molecules known as oleogelators (Co and Marangoni 2012; Sagiri et al. 2014; O'Sullivan et al. 2016; Patel 2017; Calligaris et al. 2020). Within the last decade, this approach has gotten considerable interest from the academic and industrial sectors due to the high potential of oleogels to replace plastic fat, to inhibit oil migration, and to modulate lipolysis and delivery systems for bioactive compounds (Chen and Terentjev 2009; Heymans et al. 2018).

MG-based oleogels are usually prepared by directly dispersing MGs into the oil phase at temperatures above their melting point, followed by cooling to lower temperatures (Figure 2.4.2).

During cooling, the isotropic fluid forms an inverse lamellar phase, characterized by glycerol heads packed in a 2D hexagonal order (Figure 2.4.2). Further cooling causes the initial formation of a sub-α

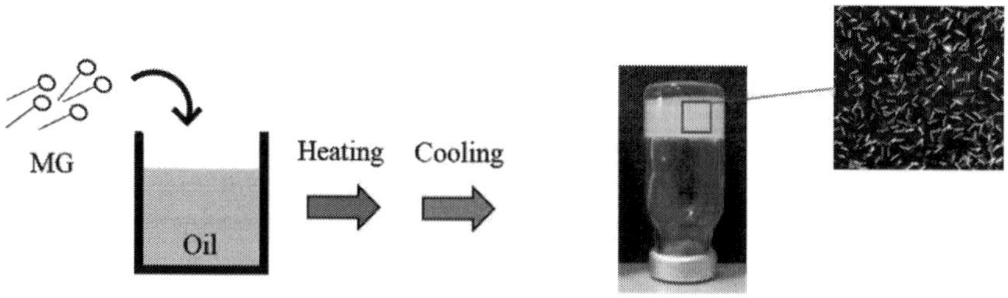

Figure 2.4.2 Schematic representation of the preparation methodology of oleogel. Macroscopic and microscopic images from Da Pieve et al. (2010), Royal Society of Chemistry / CC BY-NC 3.0.

crystalline state with orthorhombic chain packing (Chen and Terentjev 2009; Co and Marangoni 2012; Palla et al. 2014; Heymans et al. 2018). Both the inverse lamellar phase and the sub-α crystalline state are not thermodynamically stable and progressively undergo a polymorphic transition to more stable β-crystals during aging (Ojijo et al. 2004; Co and Marangoni 2012; Palla et al. 2014; Rocha-Amador et al. 2014). These crystalline structures subsequently grow, forming small needle-like crystal morphology, as shown in polarized light microscopy images by many authors (Ojijo et al. 2004). Finally, a continuous three-dimensional network stabilized by covalent interactions like van del Waals forces and hydrogen bonds is formed (Rosen-Kligvasser and Davidovich-Pinhas 2021).

The size, shape and interactions among crystals finally affect the system's macroscopic properties. Several factors could affect such a structure, as reported in Table 2.4.2 The proper selection of these parameters allows the material structure to be steered.

2.4.3.2.1 Effect of Compositional Factors

The MG amount is particularly important since a increase in concentration increases gelling temperature and gel hardness. To obtain gels with rheological properties in the range of plastic fats, MG concentrations higher than 5% (w/w) are generally needed. The increase in MG concentration seems to not modify the nanoscopic gel features, but they progressively enhance the presence of OH groups and hydrophobic interactions needed for gel strengthening (Da Pieve et al. 2011).

The type of oil deeply affected the MG-oleogel structure (Cerqueira et al. 2017; Valoppi et al. 2017; Fasolin et al. 2018; Ferro et al. 2019). Different crystal morphologies were observed depending on the oil type. Needle-like structures were prevalent in cod liver oil (Da Pieve et al. 2010, 2011), sunflower oil (Fayaz et al. 2020), extra virgin olive oil (Alongi et al. 2022), and medium chain triglycerides (MCT) (Fayaz et al. 2021), while larger spherulitic or rosette-like crystals were detected in corn oil (Kesselman and Shimoni 2007) and hazelnut oil (Yılmaz and Öğütcü 2014). To better elucidate the impact of oil type on oleogel formation, Valoppi et al. (2017) analyzed the ability of MGs (10% w/w) to structure different oils (castor, cod liver, corn, extravirgin olive, flaxseed, peanut, and sunflower oil, and a mixture of medium chain triacylglycerols). In general, increasing the oil viscosity was favorable for the gel-network formation. However, a particular behavior was detected with castor oil wherein the gel's ability to take form was reduced by the high solubility of MGs in this matrix (Valoppi et al. 2017). Additionally, the effect of the presence of minor components in oils,

Table 2.4.2 Factors affecting MG-oleogel structure.

Factors affecting MG-oleogel structure		
Compositional factors		MG type and content
		Oil type
		Presence of minor components
		Presence of other oleogelators
Processing factors	Setting conditions	Heating temperature
		Cooling rate
		Shear rate during cooling
		Ultrasonication
	Storage conditions	Temperature
		Time

such as waxes, free fatty acids, oxidation products, antioxidants (e.g., tochopherols, polyphenols), cannot be underestimated (Co and Marangoni 2012; Puşcaş et al. 2021; Alongi et al. 2022).

Recently, the possibility of obtaining mixed MG-oleogels by adding other oleogelators to reinforce the gel strength was studied. The best results in terms of gel strength were obtained by mixing MGs with phytosterols (Bin Sintang et al. 2017; Truong et al. 2019) or waxes (Barroso et al. 2020; Rodríguez-Hernández et al. 2021).

2.4.3.2.2 Effect of Processing Factors

Processing parameters, such as the cooling conditions and especially the cooling rate were reported to be particularly important. In the case of MG-olive oil oleogels, Ojijo et al. (2004) identified 0.6°C/min as the critical cooling rate, above which a stronger gel was formed. Moreover, by increasing the cooling rate, a decrease in the onset temperature of gelation was observed, indicating that gelation required a longer time to begin. However, under these conditions structural development after initiation was faster. This behavior was due to the supercooling effect, which make MGs crystallize at temperatures below their actual crystallization temperatures (Krog 2001). At low cooling rates, orderly lamellar structures were disposed thus promoting nuclei formation on reaching critical size or isothermal temperature (Toro-Vasquez et al. 2001). At very low cooling rates, both α- and β-crystals may be formed; the result is different gelation properties, while at faster cooling rates, the crystallization only occurs in the α-form (Sein et al. 2002; Ojijo et al. 2004). When a high cooling rate was used, the oleogel network was also characterized by small crystals and more branched clusters (Palla et al. 2019).

Oleogels obtained by static and sheared processes exhibit different structures at the nano-, micro-, and macro-level. As observed by Da Pieve et al. (2010), the absence of shearing during crystallization promotes forming junction zones among MG crystals resulting in a strong structure. By contrast, shearing induces the formation of weak oleogels characterized by reduced interactions between small crystal clusters. Stirring speed could also represent a pivotal parameter since it influences the molecular alignment, especially when oils rich in MUFAs and low in PUFAs are used (Rocha-Amador et al. 2014).

The importance of shear conditions was also highlighted by Ojijo et al. (2004). These authors observed a reduced gelation of olive oil-MG mixtures stirred at 55°C due to lamellar bilayer reorientation and multilamellar vesicles' formation that impairs α-crystalline gel phase formation. In general, excessive stirring causes structural disaggregation. For this reason, static cooling is generally preferred to form firmer oleogels.

The effect of applying ultrasonication during cooling was recently studied by the group of Martini and collaborators (da Silva et al. 2019b). Even if few studies are present in the current literature on this topic, results suggest that ultrasound application at a frequency between 20 and 100 kHz could emerge as a strategy to tailor the structure of oleogels. It was reported by da Silva and Danthine (2021) that ultrasonication can modify the crystal morphology of MGs in oleogels with no changes in polymorphic forms.

Finally, MG-oleogel structure reportedly evolved upon storage. Given this, Giacomozzi et al. (2021) detected a significant reduction in the rheological properties and hardness of oleogels made by MGs and by high oleic sunflower oil stored for 3 weeks at 5°C even if oil binding capacity was only slightly decreased. These modifications were attributed to the decrease of the overall interactions within the gel network, generating a weaker crystalline structure. Similarly, Alongi et al. (2022) noted a firmness reduction during 120 days of storage at 20°C in oleogels made of extra virgin olive oil.

Some studies also highlighted the capacity of the oleogel structure to control lipid oxidation during storage. In particular, the oxidation rate in MG-oleogels was reported to be lower than that in the unstructured oil (Yılmaz and Öğütcü 2014; Giacomozzi et al. 2021; Alongi et al. 2022). However, a certain degree of oxidation could be induced by the stressful conditions applied during oleogel preparation (Lim et al. 2017; Giacomozzi et al. 2021; Alongi et al. 2022). Given this, the heating phase should be carefully controlled, reducing the storage time and as the temperature suffered by the product before mixing to the greatest possible extent.

2.4.3.3 From Oleogels to Oleofoams

A further improvement of oleogels is the conversion of oleogels into oleofoams, which is also called oleofoams or whipped oils. Non-aqueous foams are composed of gas bubbles dispersed in a non-aqueous solvent (Fameau and Saint-Jalmes 2017). As compared to aqueous foams, the study on the potentialities of non-aqueous ones is relatively novel; this is mainly due to the low surface activity of surfactants in lipophilic solvents as compared to aqueous ones (Fameau and Saint-Jalmes 2017). Oleofoams can be produced using saturated MG due to the presence of MG crystals both in the bulk oil and at the surface of air bubbles (Gunes et al., 2017). In this case, DG have also been reported as possible structuring molecules. Saturated DGs based on saturated fatty acids with a chain length from C14 to C18 and a purity from 65 to 90% have been used for oleofoam production (Shrestha et al. 2006, 2008; Lei et al. 2020). The air bubbles and the bulk oils are stabilized by packed layers of MGs/DGs crystals in the form of inverse lamellae (Heymans et al. 2018). As a result, soft, aerated, and self-standing materials are obtained (Figure 2.4.3), which could be used not only as low-saturated fats alternatives, but also as air incorporation ingredients, allowing for an overall calorie-intake reduction (Fameau and Saint-Jalmes 2017).

Oleofoams are formed by whipping the pre-formed oleogels (prepared as described in Section 2.4.3.2) or by incorporating air while they are setting (Figure 2.4.3). Air might be incorporated either by a random agitation exerted by a whisk or through injection into the liquid submitted to shear (Brun et al. 2015; Drenckhan and Saint-Jalmes 2015; Gunes et al. 2017; Heymans et al. 2018). MG- and DG-based oleofoams generally show an overrun ranging from 50 to 75%, with bubbles in the range from 10 to 50 μm (Brun et al. 2015; Fameau and Saint-Jalmes 2017). These overrun values are lower than those usually obtained in aqueous foams, where it can reach values as high as 99%. At higher gas volume fractions, gas bubbles likely coalesce or are expelled from the foam (Fameau and Saint-Jalmes 2017). Moreover, these foams have reportedly shown a gel-like behavior, with the storage modulus being higher than the loss modulus until a certain critical stress. This rheological behavior results from interfacial stabilization due to either jamming of the crystalline layers around air bubbles or the formation of a crystalline network in the continuous oil phase (Gunes et al. 2017). Despite the mechanical stress suffered by the system during whipping, the final oleofoam may show G′ and G″ values as higher than the initial oleogels they derive from. This was observed, for example, in oleofoams based on a mixture of MG and DG and rapeseed oil and was attributed to the impact of the rigid

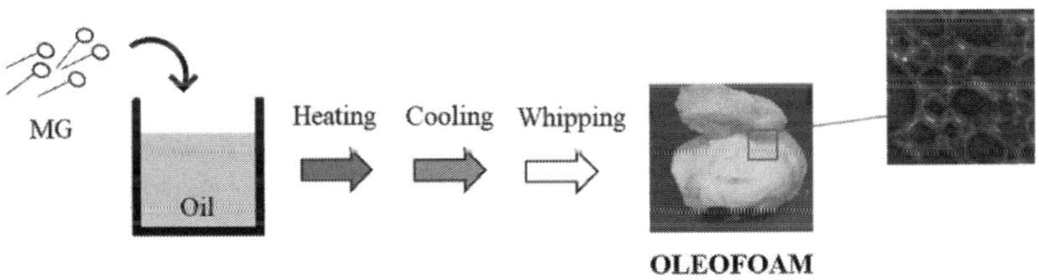

Figure 2.4.3 Schematic representation of the preparation methodology of oleofoams. Macroscopic and microscopic images from Truong et al. (2019), Elsevier.

crystal layers surrounding air bubbles (Brun et al. 2015). This physical effect of the crystalline particles jammed at the bubbles' air-oil interface has been classified as a Pickering-like stabilization mechanism. It provides oleofoams with a high stability against destabilization phenomena, mechanically hindering air transfer and bubble contact (Gunes et al. 2017).

2.4.3.3.1 Effect of Compositional Factors

The MG/DG content in oil was reportedly critical in determining oleofoams' characteristics: the higher the crystalline particles concentration, the higher the foam stability. In a study by Shrestha et al. (2010), diglycerol monomyristate was used to produce an oil oleofoam, which was stable for 20 min at 1% diglycerol monomyristate, and was stable for more than a month in 10% DG systems. Overrun around 55% was obtained only in a 5–20% MG concentration range (Brun et al. 2015; Fameau and Saint-Jalmes 2017). This result can be explained by considering that the overrun was close to the maximum random close packing fraction of monodisperse spheres, above which randomly distributed spheres are compressed against each other. Below and above a certain MG concentration, foaming is not possible due to the oleogel's rheological properties. In particular, according to Gunes et al. (2017) the whipping of weak gels that flow under gravity would not be successful, probably due to the absence of bulk stabilization against drainage and coalescence. Similarly, a G' of the gel higher than 10^8 Pa would have a low air incorporation efficiency due to the gel's solid-like structure.

2.4.3.3.2 Effect of Processing Factors

Beside the factors that need to be controlled during oleogel production (Table 2.4.2), whipping performances should be optimized. The most common device used to prepare oleofoams is a simple kitchen blender. A blender incorporates air bubbles and breaks them down to smaller dimensions as the foaming proceeds (Drenckhan and Saint-Jalmes 2015). Further steps may be added to more accurately control the dimensional distribution of the air bubbles in the oleofoam. For example, Brun et al. (2015) subjected a MG-rapeseed oleofoam to viscous stress under simple shear conditions and obtained the break-up of mother bubbles into daughter ones with a smaller size. An alternative production process is based on depressurization following gas dissolution in oleogel matrix. However, besides being more complex in terms of required reactants (*i.e.*, food grade gas) and equipment, oleofoams have demonstrated poorer characteristics as compared to analogous samples produced by simple whipping. For example, an oleogel of 5% MG in sunflower oil was foamed by depressurization using N_2O (Gunes et al. 2017). The obtained foam had visible oil release and air bubble coalescence upon resting for a few minutes, while the corresponding whipped foam was stable for one month under the same conditions. This was attributed to the larger size and distribution of the air bubbles obtained upon depressurization as compared to those obtained upon whipping (Gunes et al. 2017).

The selection of the whipping time during oleogel setting is particularly critical (Heymans et al. 2018). MG-oleofoams whipped after complete oleogel crystallization showed the highest gel hardness. Presumably, the presence of large crystalline clusters allowed their partial breakage during whipping in smaller particles that could act as efficient building blocks in the formation of a stronger network after whipping (Heymans et al. 2018). However, opposite results were observed in terms of foamability: oleogels whipped during the initial crystallization, thus containing MG α-crystals and low crystal concentration, incorporated air more efficiently as compared to crystalized systems. This was attributed to the low viscosity of partially crystalline systems, allowing more flow and turbulence during whipping and thus a higher air incorporation as compared to fully set oleogels. Moreover, the low number of crystals present in the partially crystalline oleogels is expected to easily and spontaneously accumulate at the oil-air interface, further favoring air capture (Heymans et al. 2018). According to these hypotheses, crystals already present in the set oleogel structure would be mechanically forced to the air-oil interface during whipping. By contrast, in the study of Lei et al. (2020), the higher overrun of DG oleofoams was observed upon full crystallization of the initial oleogels, leading to the formation of β′ crystals. This was attributed to the smaller particle size and, hence, to a higher surface area of these crystals as compared to the other polymorphic forms that are predominant in initial crystallization. It can be thus concluded that these contradictory results could be accounted for the specificity of MGs/DGs used in the foaming experiments. An effect of the crystallization rate was also found: high cooling rates have been reported to be associated with higher oleogel foamability due to the lower crystal dimension; by contrast, the big crystals obtained upon low crystallization rates resulted in a decreased air incorporation ability, since there is more difficultly with jamming at the air bubble surface (Truong et al. 2019). Whipping duration is one additional factor affecting oleofoam structure. Excessive foaming has led to overrun decrease and to uneven bubble size distribution; this is due to MG crystal bridging under shearing and to subsequent bubble merging (Du et al. 2021). As evidenced by the study of Heymans et al. (2018), despite their lower foamability performance, the fully crystallized oleogels formed oleofoams characterized by higher structuration and stability, in terms of rheological behavior and oil leakage, over time, as compared to oleogels foamed when crystallization was still ongoing. This is probably due to the higher physical stability guaranteed by the gelation of bulk oil and not only at the oil-air interface.

Considering storage temperature, increased temperature detrimentally impacts oleofoam stability because the destabilization of the crystalline surface layer occurs when approaching the melting temperature of the MG mixture used to prepare oleogels (Fameau and Saint-Jalmes 2017). It should be noted that the foam stability upon temperature increase can be modulated by selecting MG-based on their chain length: as saturated MG chain length increases, the melting temperature also increases as does oleofoam stability. This was demonstrated in the study of Shrestha et al. (2006), wherein oleofoams stabilized by glycerol monocaprylin (11 carbons) and glycerol monocaprin (13 carbons) showed a stability at room temperature of a few minutes and of several hours, respectively.

2.4.3.4 Gelled Emulsions

It is well known that an emulsion consists at least of two immiscible liquids with one of the liquids being dispersed as small spherical droplets in the other. In the food industry, the two immiscible liquids are usually oil and water, which mainly generate two types of emulsions: water-in-oil (W/O), in which water is the dispersed phase and oil the continuous one, and oil-in water (O/W), in which oil is dispersed in the continuous aqueous phase (Walstra and van Vilet 2008). Water and oil are divided by the interfacial layer in which emulsifiers, also called surfactants, are adsorbed.

Both the dispersed and continuous phase can be modified by adding additional compounds capable of steering the physical properties of the final emulsion. In particular, it is possible to add compounds that can structure both the continuous phase and the dispersed phase, leading to the formation of gelled emulsions. Saturated MG have been demonstrated to be exploitable for the production of O/W and W/O emulsion gels presenting the physical properties of saturated fats.

2.4.3.4.1 Oil-in-Water Gelled Emulsions

MG-based O/W emulsions were developed in 2007 by the research group of Marangoni and coworkers (Batte et al. 2007a, 2007b; Marangoni et al. 2007). These systems are based on the peculiar capacity of MG lamellar bilayers to simultaneously entrap both water and oil phases. In these systems, MGs act as both surfactants, stabilizing the water-oil interface, and structuring agents of the bulk water phase, forming a three-dimensional network.

O/W emulsion gels containing MGs are prepared by heating the oil phase containing the MGs and the co-surfactants at temperatures higher than the T_c (usually 70°C), until the saturated MGs and co-surfactants have completely melted and are evenly dispersed in the oil. Separately, the water phase, opportunely formulated to allow swelling, is similarly heated to 70°C. The oil phase is then poured into the water and homogenized with a high-speed mixer (Figure 2.4.4).

Upon cooling, semi-solid materials with rheological properties typical of gels can be obtained (Batte et al. 2007a, 2007b; Marangoni et al. 2007). In this context, MG crystalline bilayers not only surround and stabilize oil droplets, but they also extend among the droplets, leading to a continuous solid network that structures the continuous water phase (Marangoni et al. 2007). Similar to hydrogels (Section 2.4.3.1), these systems are based the crystalline lamellar phase formation, which occurs when the temperature is decreased below the T_c. This system is unique because the MG lamellar bilayers can create shells entrapping a high quantity of oil (Batte et al. 2007a, 2007b; Marangoni et al. 2007). Also in this case, the α-gel phase is thermodynamically unstable and could convert into the more densely compacted coagel made of plate-like β-crystals.

The structure of MG-emulsion gels can be determined by a wide number of compositional and processing factors that are able to steer the lamellar swelling behavior. These parameters are all reported in Table 2.4.1 with the addition of some considerations reported below.

2.4.3.4.2 Effect of Compositional Factors

By changing the relative proportion of the ingredients, different systems are obtained. In particular, Batte et al. (2007b) and Marangoni et al. (2007) studied the effect of composition in the development of structured O/W emulsions that contained monostearin, oil, water, and stearic acid as co-surfactant. The formation of a structured emulsion with gel-like properties is mainly dependent on the water/monostearin ratio and is less dependent on the system's oil content. In fact, a white gel formed at the water/monostearin ratios between 2 and 20%, in the presence of oil up to 60%. Above this oil content, emulsion gels do not form due to the fact that there is not enough MG to completely stabilize the oil droplets. In a more recent work, the same authors demonstrated that the water content and

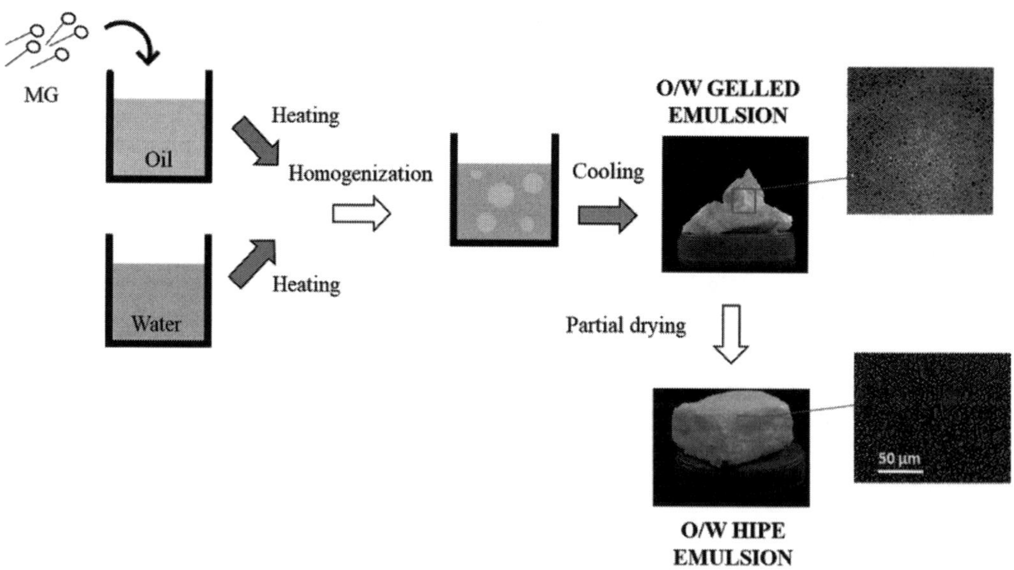

Figure 2.4.4 Schematic representation of the preparation methodology of O/ W gelled emulsions and O/ W HIPE. Macroscopic and microscopic images from Calligaris et al. (2021), John Wiley & Sons.

MGs:water ratio cannot be lower than 25% and 1:5 w/w, respectively, to allow for a proper hydration and sufficient oil encapsulation. Moreover, the overall MG content cannot be lower than 3% (w/w) since, below this concentration, the lamellar bilayers are extremely thin, thus failing to provide the mechanical support required for oil encapsulation (Blake and Marangoni 2015).

The composition of the aqueous phase appears to be one of the principal variables affecting the MG swelling capacity. For instance, the presence of NaCl and/or whey protein greatly hinders the swelling of MG lamellar bilayers by shielding the electrostatic charge of co-surfactants (Batte et al. 2007a; Mao et al. 2014). A similar undesired effect on the structural properties of the MG structured emulsion was also observed when oligosaccharides and hydrocolloids, such as maltose, trehalose, guar gum and xanthan gum, were added; their water binding capacity probably disturbs the structural organization of MG multilayers (Blake and Marangoni 2015). The structure of these emulsions can also be tailored by using specific bases to reach the alkaline pH required for a proper swelling of MG layers. Aqueous solutions of weak bases ($KHCO_3$, $NaHCO_3$, and NH_4HCO_3) or strong bases (NaOH and KOH) were used to formulate MG structured emulsions in the work of Valoppi et al. (2015b). Concentrations of strong bases and weak bases at 1–100 mM were required to obtain gelled emulsions with complex rheological behavior and higher structuration. Valoppi et al. (2015a) noted that the substitution of the aqueous phase with UHT defatted milk also reduced the structuration of the emulsion, which was attributed to the presence of milk proteins and salts and that interfered with the MG network formation.

Also the lipid phase composition affects the MG-emulsion gel properties. In particular, the presence of crystalline fats increases the gel's mechanical properties. This is the case for using sunflower oil or anhydrous milk fats as lipid phases in MG-based emulsions gels prepared by Marino et al. (2017). This was due to the presence of lipid crystals inside the emulsion droplets in conjunction with the crystals of MGs around them. Similar results were obtained by adding fully hydrogenated soybean oil, and palm and palm kernel oil (Blake and Marangoni 2015).

Interestingly, it is possible to reinforce the gel strength by adding molecules that can gel the oil by forming an oleogel inside the emulsion droplets. This approach was applied by Blake and Marangoni (2015) by adding different waxes (rice bran, sunflower, candelilla, and carnauba waxes).

Finally, the presence of surface-active components can affect the physical properties of O/W structured emulsions, due to interactions with MG. A recent study of Du et al. (2021) showed that additing tea saponins to a MG structured O/W emulsion could improve the rheological properties and the stability of the system. However, it was only observed in a specific range (saponins 1% on the water phase, and MGs 1–3% on the oil phase), wherein the saponins and the MGs exerted a synergistic effect in the interface's stabilization. Instead, higher MG concentrations resulted in the destabilization of the emulsion structure, which was attributed to competitive effects between saponins and MGs in the stabilization of the water-oil interface.

2.4.3.4.3 Effect of Processing Factors

All the processing factors reported in Table 2.4.1 could modify the final characteristics of O/W emulsion gels structured with MGs being able to steer, as previously described, the swelling behavior of MGs. However, in this context the homogenization rate was added as an additional factor impacting emulsion structure. In fact, the homogenization rate effects the oil phase dispersion in the continuous water phase: as the homogenization rate increases, the oil droplet size and distribution are reduced. Even if this reduction in particle dimension and polydispersity is associated with a higher physical stability of the emulsion, it also results in a reduction of system structuration, as evidenced by rheological evaluation. This can be attributed to lower networking among multi-layered MGs, since more MGs are required to stabilize the oil droplet surface; this increases as the particle dimension decreases (Batte et al. 2007a). Note that the cooling rate after homogenization should be relatively high (> 6°C/min) to allow fast dispersed oil phase stabilization. A slow cooling rate results in a longer time spent by the material at high temperatures, at which coagulation and flocculation of small globules turn into larger ones occurs.

2.4.3.5 From O/W Gelled Emulsions to High Internal Phase Emulsions (HIPE)

Starting from O/W emulsion gel concepts, it is possible to prepare high internal phase emulsions (HIPE), obtaining a plastic material with a gel strength significantly higher than that of O/W emulsion gels. HIPE are concentrated emulsions characterized by a large volume of the dispersed phase. The volume fraction is above 0.70, which results in the deformation of the dispersed phase droplets into polyhedra, which are separated by thin films in the continuous phase. Similar to dilute emulsions, HIPE are both kinetically and thermodynamically unstable; nevertheless, it is possible to prepare metastable systems that show no change to their properties over long periods of time (Cameron and Sherrington 1996). This can be obtained by accurately selecting surfactants or particles capable of stabilizing the interface and developing rapid and simple approaches to tune the droplet's physical and chemical properties. However, HIPE O/W emulsions based on MGs cannot be prepared by simply increasing the disperse phase fraction (Calligaris et al. 2021) due to the delicate compositional equilibrium that requires specific MG/water/oil ratios. Nevertheless, a HIPE emulsion has been obtained by partially and slowly drying the initial MG structured emulsion in the work of Calligaris et al. (2021) (Figure 2.4.4).

These HIPE emulsions display a gel-like behavior, which is typical of structured systems attributed to closely crowded dispersed oil droplets and to the high packing fraction. In the study of Calligaris et al. (2021), the conversion of the MG O/W structured emulsion into a HIPE by means of partial water removal led to a significant increase in both G′ and to the critical stress of emulsions, due to the increase in both MG content and to the packing fractions of non-spherical particles. It is noteworthy that, starting from an emulsion containing 9% MGs, the rheological properties of the partially-dried HIPE reached those of commercial laminating fats (G′ 4.5×10^5 Pa, critical stress 634 Pa).

In this context, drying temperature and drying extent were crucial processing parameters to avoid system breakdown.

2.4.3.6 Water-in-oil Gelled Emulsions

Water-in-oil (W/O) gelled emulsions can also be prepared by using MG mixtures. These systems can be defined as water-in-oleogel emulsions, since the oil phase is gelled by an oleogelator, which forms a continuous fat crystal network in the oil phase, embedding the aqueous dispersed phase (Wijarnprecha et al. 2021). These systems are quite new, so literature on this topic is limited.

Preparation of these systems is not different from O/W structured emulsions. Also in this case, the oil phase is heated at temperatures high enough to obtain the complete solubilization of the used compounds. The pre-heated water phase is poured in the oil phase, and the system is homogenized and cooled down (Figure 2.4.5) as described for the O/W emulsions. During cooling, the network formed by MG inverse lamellae stabilized the dispersed aqueous phase conferring texture and palatability (Wijarnprecha et al. 2021). In these systems, MGs behave both as emulsifiers, able to stabilize the oil-water interface, and as gelators of the continuous oil phase. In particular, da Silva et al. (2019b) used MGs as emulsifiers to produce W/O structured emulsions in which the oil phase was gelled with a mixture of hard fat based on stearic acid and behenic acid and candelilla wax. Similarly, Toro-Vazquez et al. (2013) used MGs in

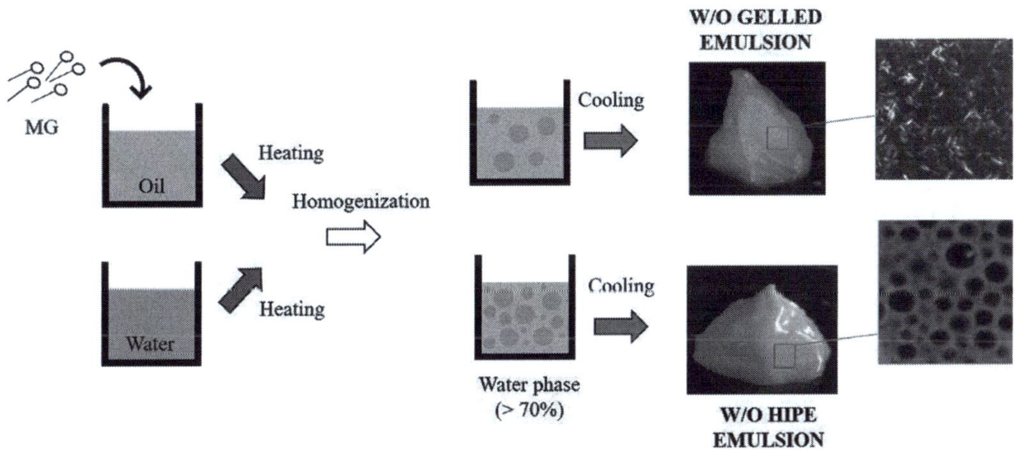

Figure 2.4.5 Schematic representation of the preparation methodology of W/O gelled emulsions and W/O HIPE. Macroscopic and microscopic images from Wijarnprecha et al. (2019), Elsevier, and Liu et al. (2022), Elsevier.

W/O structured systems stabilized by candelilla wax. A recent work by Contreras-Ramírez et al. (2020) focused on developing oleogel-in-water emulsions by using MGs to gel the continuous oil phase, while a lipophilic surfactant (polyglycerol polyricinoleate) was used to stabilize the oil-water interface. Also DGs have recently been used with polyglycerol polyricinoleate to produce these kind of emulsions per Liu et al. (2022).

Water-in-oleogel emulsions, commonly appear as soft self-standing materials, showing a typical gel-like behavior with G′ values in the range from 10^2–10^4 Pa. For example, the emulsions obtained by Liu et al. (2022) using 6% DGs and 30–60% water phase presented G′ values ranging from 10^2–10^4 Pa. Similar rheological values were also shown by an oleogel containing 2% wax, 2% hard fat, 2% MGs, and 94% sunflower oil with an added 1–30% of water (da Silva et al. 2019a). Wijarnprecha et al. (2019) also reported G′ values around 1.6×10^4 Pa for a rice bran wax and for MG-based oleogel system embedding 25% aqueous phase. A G′ higher than 10^5 was obtained by Contreras-Ramírez et al. (2020), who developed water-in-oleogels containing MGs, coconut oil, and a 25% water phase, which was probably due to the high saturated fatty acid content of coconut oil (86%); this contributed to the system's structuration. The stability of W/O gelled emulsions has mainly been measured by the release of structured oil during storage, which also causes the encapsulated water to separate (Ghosh et al. 2015; da Silva et al. 2019a). The stability of these systems is mainly dependent on the continuous oil phase structuration.

Generally speaking, increasing the structural properties of oleogels by adding water droplets is difficult due to interference of the dispersed phase with the crystal network (e.g., crystal size, and 3D spatial organization) (Wijarnprecha et al. 2021). In most cases, in fact, water droplets act as "inactive fillers" decreasing the system's structuration. For example, adding MGs to W/O structured emulsions stabilized by candelilla waxes resulted in an emulsion structuration reduction; this was attributed to the fact that just a fraction of the added MGs was adsorbed at the oil-water interface, while the remaining fraction crystallized in the continuous phase, leading to a decrease in wax crystalline networking (Toro-Vazquez et al. 2013). However, water addition could also improve the structuration of the oleogel system as reported by da Silva et al. (2019a). In this case, adding water at 1% increased the physical properties of oleogels structured by candelilla wax, hard fat, and MGs.

Finally, it should be said that the properties of the dispersed aqueous phase, such as volume fraction, droplet size and distribution, and possible structuration, may affect the emulsion's final structure. However, literature studies are still very scarce. In any case, studies conducted on W/O gelled systems, even if not containing MGs or DGs, suggest that smaller droplet sizes with narrower distributions increase the stability of the emulsion against coalescence and phase separation (Binks and Rocher 2009; Hughes et al. 2009; Wijarnprecha et al. 2021).

2.4.3.7 From W/O Gelled Emulsions to High Internal Phase Emulsions (HIPE)

It is interesting to note that W/O HIPE can also be obtained using MGs and DGs. In this case, HIPE with high water (up to 80%) is obtained, which can be used to fabricate low-calorie foods wherein the lipid fraction is replaced by a structured watery system. Even if the available literature is still scarce, HIPE W/O emulsions structured by DGs that contain polyglycerol polyricinoleate emulsifiers have recently been prepared by Liu et al. (2022). The procedure reported in Figure 2.4.5 was followed by homogenizing the pre-heated oil and water phase, followed by cooling to allow DG crystallization. In the same study, it was found that the gelation of the internal water phase of a DG-structured HIPE by using κ-carrageenan was successful in enhancing the rheological properties and stability of the emulsion, due to the peculiar properties of the used hydrocolloid. The latter, in fact, shows no absorbing properties and therefore does not interfere with the continuous oil phase nor with the water droplet interface, thus avoiding alterations in the DG network.

2.4.4 Food Applications

Table 2.4.3 reports an overview of the literature studies that address applying MG and DG-based fat alternatives in food prototypes. Low-calorie spreads and fillers have been produced by exploiting hydrogels. The addition of these MG-hydrogels allows large water amounts to be intrdouced into the food product without compromising the consistency and mouthfeel. Practical examples are in spreads and whippable non-dairy creams (Heertje et al. 1998, 2002). Based on their different physical properties, hydrogels in the form of coagel better fulfilled the requirements of firmer products, such as spreads, while α-gel hydrogels have been preferred for formulating aerated products such whippable creams (Heertje et al. 1998). In the same product category, oleogels have been proposed to substitute conventional fats with high content of saturated or trans fats. Fayaz et al. (2017) evaluated the ability of MG-pomegranate seed oil oleogel to act as partial palm oil replacer in chocolate spread. The authors observed that MG crystals interacted with palm oil crystals leading to a close three-dimensional network with strong gel properties. Oleogels containing high oleic sunflower oil and MGs were also used as fat replacers in filling creams for sandwich cookies (Palla et al. 2021). Results highlighted that a filler formulated with 260 g/kg of oleogels performed similarly to those observed in traditional products. On the other hand, higher oleogel content was associated with softer filler texture and increased adhesiveness, which may be desirable to guarantee the adhesion between the sandwich cookies. Finally, also oleofoams have been used in aereated chocolate filling (Chisholm et al. 2018). In recent patents, incorporating air bubbles through oleofoams begets specific texture (puffed and aerated) and sensory attributes (soft body and taste), as well as a reduction of the overall calories (Mishima et al. 2016; Lei et al. 2020).

With bakery products in mind, possible applications of MG gels as fat mimetics is among the most widely investigated. In this context, oleogels and emulsion gels appear particularly promising.

For instance, oleogels made by MGs and high oleic sunflower oil were used to replace commercial margarine in muffins (Giacomozzi et al. 2018). Obtained products presented higher specific volume,

Table 2.4.3 Applications of mono- and di-glyceride systems as fat alternatives in different foods.

Application	Fat Mimetic	Substituted fat	Reference
Spreads and creams			
Zero fat spread	Hydrogel	Margarine	Heertje et al. (2002)
Low-fat whippable non-diary cream	Hydrogel	Partial replacement of fat cream	Heertje et al. (2002)
Chocolate spread	Oleogel	Partial replacer of palm oil	Fayaz et al. (2017)
Filling creams	Oleogel	Beef fat	Palla et al. (2021)
Aerated fillings for chocolate, extruded cereal products, biscuits	Oil foams	Shortening	Chisholm et al. (2018)
Bakery			
Muffin	Oleogel	Butter	Giacomozzi et al. (2018)
Cookies	O/W structured emulsion	Shortening	Goldstein and Seetharaman (2011)
White bread	O/W structured emulsion	Palm oil	Manzocco et al. (2012)
Sweet bread	Oleogel	Palm oil	Calligaris et al. (2013)
Sweet bread	O/W structured emulsion	Palm oil	Calligaris et al. (2013)
Sponge cakes, biscuits, laminated pastry	Oil foams	Shortening	Gunes et al. (2016)
Laminated products (croissants and Danishes)	O/W structured emulsion	Palm oil and commercial shortening	Blake and Marangoni (2015)
Puff pastry	O/W HIPE	Laminating palm margarine	Calligaris et al. (2021)
Bakery products	O/W structured emulsions	traditional shortenings, spreads and creams	https://www.coasun.com
Dairy			
Ice cream	O/W structured emulsions	Milk cream	Calligaris et al. (2018)
Semi-solid low-fat cheese, Mozzarella cheese	Hydrogel	Milk fat	Heertje et al. (2002)

similar hardness values, and a more connected and homogeneous crumb structure compared with the traditional product. Interestingly, reformulated muffins also exhibited a higher capacity for oil retention compared with traditional ones after 10 days of storage. O/W MG structured emulsions have also been successfully applied as fat substitutes in different baked goods with performances depending on the final product. In the case of cookies, control samples prepared with all-purpose shortenings displayed superior quality attributes as compared to those reformulated with the MG emulsion. It is difficult to mimic the friability of cookies with a water-containing fat replacer (Goldstein and Seetharaman 2011). By contrast, in the case of white and sweet bread, no specific functional gaps limiting the use of MG structured emulsions were highlighted. Bread samples prepared with MG emulsions showed, in fact, quality characteristics comparable to the control bread prepared with palm oil. This was attributed to the MG crystallized lamellas' capacity to favor oil spreading over flour, which replaces the typical functionality of triacylglycerol crystal networks (Manzocco et al. 2012; Calligaris et al. 2013). Moreover, the reformulation not only led to an 81% reduction of saturated fat content, but it also decreased bread staling during storage (Manzocco et al. 2012). The same authors also used MG-oleogels as a palm oil alternative. In this case, however, the system showed performances more poorly than that with the MG emulsion, since less leavened and firmer bread was yielded. This result was probably reflective of strong MG engagement in oil networking, thus reducing MG interactions (Calligaris et al. 2013). Due to its excellent performances in baked goods, the O/W MG emulsion has been patented as a fat mimetic for bakery products and is currently industrially used as healthier alternative for saturated fats under the commercial name Coasun (https://www.coasun.com). A still open issue in fat substitution in the bakery sector is that of laminating fats, *i.e.*, fats used to produce laminated products, such as puff pastry and croissants. An ideal laminating fat substitute should be plastic enough to withstand the lamination process. To achieve this aim, the rheological properties of laminating fats should be similar to the fat properies of dough and, their melting temperature should be higher than that of the lamination process to avoid the fat melting and to prevent dough layers from coming in contact (Blake and Marangoni 2015). Such peculiar rheological properties have been obtained by adding 7.5% waxes to the formulation of MG O/W emulsions (Blake and Marangoni 2015). These gelled emulsions performed well in the baking trials of laminated products (Danishes and croissants). Since the use of waxes is limited by label constraints, possibly associated with sensory defects (unpleasant waxy taste), an alternative strategy to increase the rheological response of the MG emulsion was developed by Calligaris et al. (2021). In this case, the O/W MG gelled emulsion was subjected to partial drying, leading to a HIPE emulsion containing an 80% lipid phase. The suitability of the obtained HIPE as a laminating fat was then demonstrated by producing puff pastry.

MG fat replacers were also applied in the dairy sector. In particular, patented hydrogel formulations have been proposed to produce low-calorie dairy products, including spreadable cheese and Mozzarella cheese (Heertje et al. 2002). Calligaris et al. (2018) used O/W gelled emulsion containing milk as the aqueous phase to develop a low-saturated fat ice cream. The latter showed only minor modifications in terms of quality characteristics in comparison to the control sample containing milk cream as fat phase.

2.4.5 Novel Functionalities of MG and DG Fat Mimetics

Beside the reduction of saturated fat and/or overall caloric density of food products, the specific structure of MG-based fat mimetics offers additional possibilities in the development of novel functionalities further contributing to the enhancement of the healthy profile of foods. In this regard, Truong et al. (2019) reported the possibility of developing MG-based oleofoams to deliver phytosterols. The latter did not influence the oleofoams' physical properties and stability, which was attributed to the phytosterols' ability to co-crystallize with MG, thus avoiding weakening the MG network.

The co-presence of water and oil in DG-based water-in-oleogel HIPE systems has been exploited to encapsulate both lipophilic (β-carotene) and hydrophilic (anthocyanins) bioactive compounds by Liu et al. (2022). Moreover, Melchior et al. (2021) and Marino et al. (2017) demonstrated the capacity of MG structured emulsions to protect probiotic bacteria during processing, cold storage, and *in vitro* digestion. This protective effect has been attributed to these systems' gel-network structure, which are able to effectively stabilize encapsulated compounds and bacterial cells against external stresses, while promoting their controlled release (Melchior et al. 2021; Liu et al. 2022). W/O gelled emulsions containing whey protein as a co-emulsifier were used to protect β-carotene during *in vitro* digestion trials (Lu et al. 2017). In this case, the *in vitro* bioavailability of β-carotene in the gelled emulsions was higher than that in the emulsion without MGs. MG emulsions also demonstrated a better cellular uptake of β-carotene by Caco-2 cells, which was attributed to the formation during digestion of micelles rich in MG-deriving hydroxyl groups on the surface leading to a higher hydrophilicity of these micelles and thus a better cellular uptake (Lu et al. 2017). The structure of MG-based oleogels was also shown to efficaciously protect curcuminoids during *in vitro* digestion, and to slow down oil digestibility, due to the steric impediment of the gel network to the lipolytic enzyme activity (Calligaris et al. 2020; Alongi et al. 2022).

In vivo studies also demonstrated that oil structuration in O/W MG-based emulsions promotes the controlled release of the lipids into the blood, which in turn attenuates the post-prandial increases in triglycerides, free fatty acids, and insulin levels induced by acute fat ingestion (Marangoni et al. 2007).

2.4.6 Conclusions

The wide number of possible structures formed by mono- and diglycerides in different environments, previously described have mainly been studied and designed to successfully serve as alternatives of saturated and trans fatty acids. However, most of the literature studies are limited to the characterization of the structure and stability of the materials to be used as fat mimetics without considering their final target application. Although necessary, the accurate knowledge of their composition and structure is not the only condition successfully introducing mono- and di-glyceride based fat alternatives into food products. This is due to the complex interactions potentially occurring among ingredients during formulation but also during processing when destructuring/structuring events occur, concomitantly with the development of chemical reactions forming novel species. In other words, the characteristics of the final reformulated food pass through the re-design of the relationship between the fat replacer and the food matrix.

References

Abdallah, D.J. and Weiss, R.G. (2000). Organogels and low molecular mass organic gelators. *Advanced Materials* 12: 1237–1247.

Alfutimie, A., Curtis, R., and Tiddy, G.J.T. (2014). Gel phase (Lβ) formation by mixed saturated and unsaturated monoglycerides. *Colloids and Surfaces A: Physicochemical and Engineering Aspects*. Elsevier B.V., 456: 286–295.

Alongi, M., Lucci, P., Clodoveo, M.L., Schena, F.P., and Calligaris, S. (2022). Oleogelation of extra virgin olive oil by different oleogelators affects the physical properties and the stability of bioactive compounds. *Food Chemistry* 368: 130779.

Barfod, N.M. and Sparsø, F.V. (2007). Structure and function of emulsifiers and their role in microstructure formation in complex foods. In: *Understanding and Controlling the Microstructure of Complex Foods* (ed. D.J. McClements), 113–152. Woodhead Publishing.

Barroso, N.G., Okuro, P.K., Ribeiro, A.P.B., and Cunha, R.L. (2020). Tailoring properties of mixed-component oleogels: Wax and monoglyceride interactions towards flaxseed oil structuring. *Gels* 6: 1–20.

Batte, H.D., Wright, A.J., Rush, J.W., and Idziak, S.H.J. (2007a). Effect of processing conditions on the structure of monostearin-oil-water gels. *Food Research International* 40: 982–988.

Batte, H.D., Wright, A.J., Rush, J.W., Idziak, S.H.J., and Marangoni, A.G. (2007b). Phase behavior, stability, and mesomorphism of monostearin-oil-water gels. *Food Biophysics* 2: 29–37.

Bin Sintang, M.D., Danthine, S., Brown, A., Van de Walle, D., Patel, A.R., Rimaux, T., and Dewettinck, K. (2017). Phytosterols-induced viscoelasticity of oleogels prepared by using monoglycerides. *Food Research International* 100: 832–840.

Binks, B.P. and Rocher, A. (2009). Effects of temperature on water-in-oil emulsions stabilised solely by wax microparticles. *Journal of Colloid and Interface Science* 335: 94–104.

Blake, A.I. and Marangoni, A.G. (2015). Factors affecting the rheological properties of a structured cellular solid used as a fat mimetic. *Food Research International* 74: 284–293.

Blázquez, C., Emond, E., Schneider, S., Dalmazzone, C., and Bergeron, V. (2014). Non-aqueous and crude oil foams. *Oil and Gas Science and Technology* 69 (3): 467–479. doi: 10.2516/ogst/2013186.

Brun, M., Delample, M., Harte, E., Lecomte, S., and Leal-Calderon, F. (2015). Stabilization of air bubbles in oil by surfactant crystals: A route to produce air-in-oil foams and air-in-oil-in-water emulsions. *Food Research International* 67: 366–375.

Calligaris, S., Manzocco, L., Valoppi, F., and Nicoli, M.C. (2013). Effect of palm oil replacement with monoglyceride organogel and hydrogel on sweet bread properties. *Food Research International* 51: 596–602.

Calligaris, S., Marino, M., Maifreni, M., and Innocente, N. (2018). Potential application of monoglyceride structured emulsions as delivery systems of probiotic bacteria in reduced saturated fat ice cream. *LWT - Food Science and Technology* 96: 329–334.

Calligaris, S., Alongi, M., Lucci, P., and Anese, M. (2020). Effect of different oleogelators on lipolysis and curcuminoid bioaccessibility upon *in vitro* digestion of sunflower oil oleogels. *Food Chemistry* 314: 126146.

Calligaris, S., Plazzotta, S., Barba, L., and Manzocco, L. (2021). Design of roll-in margarine analogous by partial drying of monoglyceride-structured emulsions. *European Journal of Lipid Science and Technology* 123: 2000206.

Cameron, N.R. and Sherrington, D.C. (1996). High internal phase emulsions (HIPEs)- structure, properties and use in polymer preparation. In: *Biopolymers Liquid Crystalline Polymers Phase Emulsion*, 163–214. Springer Berlin Heidelberg.

Cerqueira, M.A., Fasolin, L.H., Picone, C.S.F., Pastrana, L.M., Cunha, R.L., and Vicente, A.A. (2017). Structural and mechanical properties of organogels: Role of oil and gelator molecular structure. *Food Research International* 96: 161–170.

Chen, C.H. and Terentjev, E.M. (2009). Aging and metastability of monoglycerides in hydrophobic solutions. *Langmuir* 25: 6717–6724.

Chisholm, H., Gunes, Z.D., Gehin-Delval, C., Nouzille, C.A., Garvey, e., Destribats, M.J., Chandrasekaran, S.N., Vieira, J.B., German, J., and Binks, B.P. (2018). Aerated confectionery material. US20180064127A1.

Co, E.D. and Marangoni, A.G. (2012). Organogels: An alternative edible oil-structuring method. *JAOCS, Journal of the American Oil Chemists' Society* 89: 749–780.

Contreras-Ramírez, J.I., Gallegos-Infante, J.A., Pérez-Martínez, J.D., Dibildox-Alvarado, E., Rocha-Guzman, N.E., Moreno-Jiménez, M.R., González-Laredo, R.F., and Rosas-Flores, W. (2020). Influence of vegetable oil, monoglycerides and polyglycerol polyricinoleate into the physical stability of organogel-emulsion (w/o) systems. *SN Applied Sciences* 2: 1–11.

Da Pieve, S., Calligaris, S., Co, E., Nicoli, M.C., and Marangoni, A.G. (2010). Shear nanostructuring of monoglyceride organogels. *Food Biophysics* 5: 211–217.

Da Pieve, S., Calligaris, S., Panozzo, A., Arrighetti, G., and Nicoli, M.C. (2011). Effect of monoglyceride organogel structure on cod liver oil stability. *Food Research International* 44: 2978–2983.

da Silva, T.L.T., Arellano, D.B., and Martini, S. (2019a). Effect of water addition on physical properties of emulsion gels. *Food Biophysics* 14: 30–40.

da Silva, T.L.T., Arellano, D.B., and Martini, S. (2019b). Use of high-intensity ultrasound to change the physical properties of oleogels and emulsion gels. *JAOCS, Journal of the American Oil Chemists' Society* 96: 681–691.

da Silva, T.L.T. and Danthine, S. (2021). Effect of high-intensity ultrasound on the oleogelation and physical properties of high melting point monoglycerides and triglycerides oleogels. *Journal of Food Science* 86: 343–356.

Data Bridge Market Research (2020). *Global mono and diglycerides and derivatives market – industry trends and forecast to 2027*. https://www.databridgemarketresearch.com/reports/global-mono-and-diglycerides-and-derivatives-market (accessed 7 December 2021).

Drenckhan, W. and Saint-James, A. (2015). The science of foaming. *Advances in Colloid and Interface Science* 222: 228–259.

Du, L., Jiang, Q., Li, S., and Zhou, Q. (2021). Microstructure evolution and partial coalescence in the whipping process of oleofoams stabilized by monoglycerides. *Food Hydrocolloids* 112: 106245.

Du, L., Li, S., Jiang, Q., and Tan, Y. (2021). Interfacial interaction of small molecular emulsifiers tea saponin and monoglyceride: Relationship to the formation and stabilization of emulsion gels. *Food Hydrocolloids* 117: 106737.

EC (2008) *1333/2008. Regulation (EC) No 1333/2008 of the European Parliament and of the Council of 16 December 2008 on Food Additives*. OJ L 354, 31.12.2008, 16–33. Consolidated Version: 18/ 06/2019. http://data.europa.eu/eli/reg/2008/1333/oj (accessed 7 December 2021).

EFSA (2017). Re-evaluation of mono- and di-glycerides of fatty acids (E 471) as food additives. *EFSA Journal*, 15(11): 5045, 43 pp.

Fameau, A.L. and Saint-Jalmes, A. (2017). Non-aqueous foams: Current understanding on the formation and stability mechanisms. *Advances in Colloid and Interface Science* 247: 454–464.

Fasolin, L.H., Cerqueira, M.A., Pastrana, L.M., Vicente, A.A., and Cunha, R.L. (2018). Thermodynamic, rheological and structural properties of edible oils structured with LMOGs: influence if gelator and oil phase. *Food Structure* 16: 50–58.

Fayaz, G., Goli, S.A.H., Kadivar, M., Valoppi, F., Barba, L., Calligaris, S., and Nicoli, M.C. (2017). Potential application of pomegranate seed oil oleogels based on monoglycerides, beeswax and propolis wax as partial substitutes of palm oil in functional chocolate spread. *LWT – Food Science and Technology* 86: 523–529.

Fayaz, G., Polenghi, O., Giardina, A., Cerne, V., and Calligaris, S. (2021). Structural and rheological properties of medium-chain triacylglyceride oleogels. *International Journal of Food Science and Technology* 56: 1040–1047.

Fayaz, G., Calligaris, S., and Nicoli, M.C. (2020). Comparative study on the ability of different oleogelators to structure sunflower oil. *Food Biophysics* 15: 42–49.

Ferro, A.C., Okuro, P.K., Badan, A.P., and Cunha, R.L. (2019). Role of the oil on glyceryl monostearate based oleogels. *Food Research International* 120: 610–619.

Ghosh, S., Pradhan, M., Patel, T., Haj-Shafiei, S., and Rousseau, D. (2015). Long-term stability of crystal-stabilized water-in-oil emulsions. *Journal of Colloid and Interface Science* 460: 247–257.

Giacomozzi, A.S., Carrín, M.E., and Palla, C. (2018). Muffins elaborated with optimized monoglycerides oleogels: from solid fat replacer obtention to product quality evaluation. *Journal of Food Science* 83: 1505–1515.

Giacomozzi, A.S., Carrín, M.E., and Palla, C. (2021). Storage stability of oleogels made from monoglycerides and high oleic sunflower oil. *Food Biophysics* 16: 306–316.

Goldstein, A. and Seetharaman, K. (2011). Effect of a novel monoglyceride stabilized oil in water emulsion shortening on cookie properties. *Food Research International* 44: 1476–1481.

Gunes, D.Z., Murith, M., Godefroid, J., Pelloux, C., Deyber, H., Schafter, O., and Breton, O. (2017). Oleofoams: properties of crystal-coated bubbles from whipped oleogels-evidence for pickering stabilization. *Langmuir* 33: 1563–1575.

Gunes, Z.D., Schafer, O., Chisholm, H., Deyber, H., Pelloux, C., and Binks, B.P. (2016). Lipid based foam. US10765125B2.

Hasenhuettl, G.L. (2008). Synthesis and commercial preparation of food emulsifiers. In: *Food Emulsifiers and Their Applications*, 2e (ed. G.L. Hasenhuettl and R.W. Hartel), 11–37. New York, NY: Springer.

Heertje, I., Hendrickx H.A.C., Knoops, A.J., Royers, E.C., Turksma, H., and Wesdorp, L.H. (2002). Use of mesomorphic phases in food products. United States Patent US6368653B1.

Heertje, I., Roijers, E.C., and Hendrickx, H. (1998). Liquid crystalline phases in the structuring of food products. *LWT – Food Science and Technology* 31: 387–396.

Heymans, R., Tavernier, I., Danthine, S., Rimaux, T., Van der Meeren, P., and Dewettinck, K. (2018). Food-grade monoglyceride oil foams Tthe effect of tempering on foamability, foam stability and rheological properties. *Food and Function* 9: 3143–3154.

Hughes, N.E., Marangoni, A.G., Wright, A.J., Rogers, M.A., and Rush, J.W.E. (2009). Potential food applications of edible oil organogels. *Trends in Food Science and Technology* 20: 470–480.

Kesselman, E. and Shimoni, E. (2007). Imaging of oil/monoglyceride networks by polarizing near-field scanning optical microscopy. *Food Biophysics* 2: 117–123.

Krog, N. (2001). Crystallization properties and lyotropic phase behavior of food emulsifiers: relation to technical applications. In: *Crystallization Processes in Fats and Lipid Systems* (ed. N. Garti and S.K.M. Dekker), 505–526. CRC Press.

Krog, N. and Larsson, K. (1968). Phase behavior and rheological properties of aqueous systems of industrial distilled monoglycerides. *Chemistry and Physics of Lipids* 2: 129–143.

Krog, N. and Sparsø, F.V. (1997). Food emulsifiers. In: *Food Emulsifiers and Their Applications* (ed. S.E. Friberg, K. Larsson, and J. Sjoblom), 521–534. New York: Marcel Dekker Inc.

Krog, N. and Sparsø, F.V. (2004). Food emulsifiers: Their chemical and physical properties. In: *Food Emulsions*, 4e (ed. S.E. Friberg, K. Larsson, and J. Sjoblom), 45–91. New York: Marcel Dekker Inc.

Lei, M., Zhang, N., Lee, W.J., Tan, C.P., Lai, O.M., Wang, Y., and Qui, C. (2020). Non-aqueous foams formed by whipping diacylglycerol stabilized oleogel. *Food Chemistry*. Elsevier, 312: 126047.

Lim, J., Hwang, H.S., and Lee, S. (2017). Oil-structuring characterization of natural waxes in canola oil oleogels: rheological, thermal, and oxidative properties. *Applied Biological Chemistry* 60: 17–22.

Liu, Y., Lee, W.J., Tan, C.P., Lai, O.M., Wang, Y., and Qui, C. (2022). W/O high internal phase emulsion featuring by interfacial crystallization of diacylglycerol and different internal compositions. *Food Chemistry* 372: 131305.

Lo, S.K., Tan, C.P., Long, K., Yusoff, M.S.A., and Lai, O.M. (2008). Diacylglycerol oil-properties, processes and products: a review. *Food and Bioprocess Technology* 1: 223–233.

Lu, W., Kelly, A.L., and Miao, S. (2017). Improved bioavailability of encapsulated bioactive nutrients delivered through monoglyceride-structured O/W emulsions. *Journal of Agricultural and Food Chemistry* 65: 3048–3055.

Lutton, E.S. (1971). The phases of saturated 1-monoglycerides C14-C22. *Journal of the American Oil Chemists Society* 48: 778–781.

Manzocco, L., Calligaris, A., Pleve, S.D., and Marzona, S. (2012). Effect of monoglyceride-oil-water gels on white bread properties. *Food Research International* 49: 778–782.

Mao, L., Calligaris, S., Barba, L., and Miao, S. (2014). Monoglyceride self-assembled structure in O/W emulsion: Formation, characterization and its effect on emulsion properties. *Food Research International* 58: 81–88.

Marangoni, A.G., Idziak, S.H.J., Vega, C., Batte, H., Ollivon, M., Jantzi, P.S., and Rush, J.W.E. (2007). Encapsulation-stucturing of edible oil attenuates acute elevation of blood lipids and insulin in humans. *Soft Matter* 3: 183–187.

Marino, M., Innocente, N., Calligaris, S., Maifreni, M., Marangone, A., and Nicoli, M.C. (2017). Viability of probiotic *Lactobacillus rhamnosus* in structured emulsions containing saturated monoglycerides. *Journal of Functional Foods* 35: 51–59.

Melchior, S., Marino, M., D'Este, F., Innocente, N., Nicoli, M.C., and Calligaris, S. (2021). Effect of the formulation and structure of monoglyceride-based gels on the viability of probiotic *Lactobacillus rhamnosus* upon *in vitro* digestion. *Food and Function* 12: 351–361.

Miao, S. and Lin, D. (2018). Monoglycerides: Categories, structures, properties, preparations, and applications in the food industry. In: *Encyclopedia of Food Chemistry* (ed. P. Varelis, L. Melton, and F. Shahidi), 155–163. Elsevier.

Mishima, S., Suzuki, A., Sato, K., and Ueno, S. (2016). Formation and microstructures of whipped oils composed of vegetable oils and high-melting fat crystals. *JAOCS, Journal of the American Oil Chemists' Society* 93: 1453–1466.

Naik, M.K., Naik, S.N., and Mohanty, S. (2014). Enzymatic glycerolysis for conversion of sunflower oil to food based emulsifiers. *Catalysis Today* 237: 145–149.

Nicholson, R.A. and Marangoni, A.G. (2018). Diglycerides. In: *Encyclopedia of Food Chemistry* (ed. P. Varelis, L. Melton, and F. Shahidi), 70–73. Elsevier.

O'Sullivan, C.M., Barbut, S., and Marangoni, A.G. (2016). Edible oleogels for the oral delivery of lipid soluble molecules: composition and structural design considerations. *Trends in Food Science and Technology* 57: 59–73.

Ojijo, N.K.O., Neeman, I., Eger, S., and Shimoni, E. (2004). Effects of monoglyceride content, cooling rate and shear on the rheological properties of olive oil/monoglyceride gel networks. *Journal of the Science of Food and Agriculture* 84: 1585–1593.

Palla, C., Giacomozzi, A., Genovese, D., and Carrín, M. (2014). Multi–objective optimization of high oleic sunflower oil and monoglycerides Oleogels:Ssearching for rheological and textural properties similar to margarine. *International Journal of Food Science* 2014: 1–14.

Palla, C., de Vicente, J., Carrín, M.E., and Ruiz, M.J.G. (2019). Effects of cooling temperature profiles on the monoglycerides oleogel properties: a rheo-microscopy study. *Food Research International* 125: 108613.

Palla, C., Wasinger, M.F., and Carrín, M.E. (2021). Monoglyceride oleogels as fat replacers in filling creams for sandwich cookies. *Journal of the Science of Food and Agriculture* 101: 2398–2405.

Patel, A.R. (2017). Oil structuring: concepts, overview and future perspectives. In: *Edible oil Structuring: Concepts, Methods and Applications* (ed. A.R. Patel), 3–22, Royal Society of Chemistry.

Pouilloux, Y., Abro, S., Vanhove, C., and Barrault, J. (1999). Reaction of glycerol with fatty acids in the presence of ion-exchange resins: preparation of monoglycerides. *Journal of Molecular Catalysis A: Chemical* 149: 243–254.

Puşcaş, A., Mureşan, V., and Muste, S. (2021). Application of analytical methods for the comprehensive analysis of oleogels—a review. *Polymers* 13: 1–23.

Rocha-Amador, O.G., Gallegos-Infante, J.A., Huang, Q., Rocha-Guzman, N.E., Moreno-Jiménez, M.R., and González-Laredo, R.F. (2014). Influence of commercial saturated monoglyceride, mono-/diglycerides mixtures, vegetable oil, stirring speed, and temperature on the physical properties of organogels. *International Journal of Food Science* 2014: 1–8.

Rodríguez-Hernández, A.K., Pérez-Martínez, J.D., Gallegos-Infante, J.A., Toro-Vazquez, J.F., and Ornelas-Pax, J.J. (2021). Rheological properties of ethyl cellulose-monoglyceride-candelilla wax oleogel vis-a-vis edible shortenings. *Carbohydrate Polymers* 252: 117171.

Rosen-Kligvasser, J. and Davidovich-Pinhas, M. (2021). The role of hydrogen bonds in TAG derivative-based oleogel structure and properties. *Food Chemistry* 334: 127585.

Sagalowicz, L., Leser, M.E., and Michel, W.M. (2006). Monoglyceride self-assembly structures as delivery vehicles. *Trends in Food Science and Technology* 17: 204–214.

Sagiri, S.S., Behera, B., Rafanan, R.R., Bhattacharya, C., Pal, K., Banerjee, I., and Rousseau, D. (2014). Organogels as matrices for controlled drug delivery: A review on the current state. *Soft Materials* 12: 47–72.

Sato, K. (2001). Crystallization behavior of fats and lipids – a review. *Chemical Engineering Science* 56: 2255–2265.

Sein, A., Verheij, J.A., and Agterof, W.G.M. (2002). Rheological characterization, crystallization, and gelation behavior of monoglyceride gels. *Journal of Colloid and Interface Science* 249: 412–422.

Shrestha, L.K., Aramaki, K., Kato, H., Takase, Y., and Kunieda, H. (2006). Foaming properties of monoglycerol fatty acid esters in nonpolar oil systems. *Langmuir* 22: 8337–8345.

Shrestha, L.K., Shrestha, R.G., Sharma, S.C., and Aramaki, K. (2008). Stabilization of nonaqueous foam with lamellar liquid crystal particles in diglycerol monolaurate/olive oil system. *Journal of Colloid and Interface Science* 328: 172–179.

Shrestha, R.G., Shrestha, K., Solans, C., and Azon, C.G. (2010). Nonaqueous foam with outstanding stability in diglycerol monomyristate/olive oil system. *Colloids and Surfaces A: Physicochemical and Engineering Aspects* 353: 157–165.

Susi, H., Morris, S.G., and Scott, W.E. (1961). Differentiation of 1- and 2-Monoglycerides by near-infrared absorption spectroscopy. *Journal of the American Oil Chemists Society* 38: 199–201.

Toro-Vasquez, J.T. et al. (2001). Crystallization and solidification properties of lipids. In: *Crystallization and Solidification Properties of Lipids* (ed. N. Widlak, R. Hartel, and S. Narine), 53–78. AOCS Press.

Toro-Vazquez, J.F., Maurico-Pérez, R., González-Chávez, R., Marco, M., Sánchez-Becerril, M., Orenelas-Paz, J.J., and Pérez-Martínez, J.D. (2013). Physical properties of organogels and water in oil emulsions structured by mixtures of candelilla wax and monoglycerides. *Food Research International* 54: 1360–1368.

Truong, T., Prakash, S., and Bhandari, B. (2019). Effects of crystallisation of native phytosterols and monoacylglycerols on foaming properties of whipped oleogels. *Food Chemistry* 285: 86–93.

Valoppi, F., Calligaris, S., Barba, L., and Nicoli, M.C. (2015a). Compositional phase diagram, rheological and structural properties of systems containing uht skim milk, sunflower oil, saturated monoglycerides and co-surfactants. *Food Biophysics* 10: 94–102.

Valoppi, F., Calligaris, S, Barba, L., and Nicoli, M.C. (2015b). Structural and viscoelastic characterization of ternary mixtures of sunflower oil, saturated monoglycerides and aqueous phases containing different bases. *Food Research International* 74: 224–230.

Valoppi, F., Calligaris, S., Barba, L., Šegatin, N., Ulrih, N.P., and Nicoli, M.C. (2017). Influence of oil type on formation, structure, thermal, and physical properties of monoglyceride-based organogel. *European Journal of Lipid Science and Technology* 119: 1–10.

Walstra, P. and van Vilet, T. (2008). Dispersed systems: basic considerations. In: *Fennema's Food Chemistry*, 4e (ed. S. Damodaran, K.L. Parkin, and O.R. Fennema), Chapter 13, 784–845. Boca Raton, FL: CRC Press Taylor & Francis group.

Wang, F.C. and Marangoni, A.G. (2014). Nature and dynamics of monostearin phase transitions in water: stability and the sub-α-gel phase. *RSC Advances* 4: 50417–50425.

Wang, F.C. and Marangoni, A.G. (2015). Effect of intrinsic and extrinsic factors on the stability of the α-gel phase of a glyceryl monostearate-water system. *RSC Advances* 5: 43121–43129.

Wijarnprecha, K., de Vries, A., Santiwattana, P., and Sonwai, S. (2019). Microstructure and rheology of oleogel-stabilized water-in-oil emulsions containing crystal-stabilized droplets as active fillers. *LWT – Food Science and Technology* 115: 108058.

Wijarnprecha, K., de Vries, A., Sonwai, S., and Rousseau, D. (2021). Water-in-oleogel emulsions—from structure design to functionality. *Frontiers in Sustainable Food Systems* 4: 1–6.

Yılmaz, E. and Öğütcü, M. (2014). Properties and stability of hazelnut oil organogels with beeswax and monoglyceride. *JAOCS, Journal of the American Oil Chemists' Society* 91: 1007–1017. https://www.coasun.com Coasun – A smart ingredient alternative (accessed 7 December 2021).

2.5

Oleogels Based on Fatty Acids and Fatty Alcohols

Toward Oil Foams

Anne-Laure Fameau[1,*] *and Alejandro G. Marangoni*[2]

[1] *Unité Matériaux et Transformations (UMET) – UMR 8207, INRAE, Team PIHM, Villeneuve d'Ascq, France*
[2] *Department of Food Science, University of Guelph, Guelph, ON, Canada*
* *Corresponding author*

2.5.1 Introduction

When the first works about the gelation of organic liquids using low-molecular-weight structuring agents emerged in the late 1990s, the research topic gained a lot of attention, especially within the food industry, and more recently in the cosmetic and pharmaceutical industry (Terech and Weiss 1997). In the food industry, oil structuring was traditionally achieved by using solid fats in food products due to high melting point triglycerides (Marangoni et al. 2012, 2020). The triglycerides' crystals form a colloidal network capable of immobilizing the liquid oil. This crystalline network is responsible for both organoleptic properties such as the melt-in-mouth effects, but also physical and functional properties such as: texture, spreadability, hardness, and crispiness (Marangoni 2004; Marangoni and Wesdorp 2013). However, these high melting temperature triglycerides are often predominantly made of saturated and/or *trans* fatty acids. Recently, the food industry has been under pressure to find alternative solutions to structure edible oils using low amounts of saturated fats and to eliminate industrial sources of *trans* fatty acids. Therefore, using low molecular weight structuring agents seem to be a possible alternative for structuring edible oils (Marangoni and Garti 2018). The structuring agents are called oleogelators, and the resulting gel based on edible oils are called oleogels (Co and Marangoni 2012; Patel et al. 2020). Twenty years ago, Eini and Tamarkin first reported using long-chain saturated fatty acids as oleogelators with the formation of a crystalline network that immobilized the liquid oil (Eini and Tamarkin 2001). A few years later, Daniel and Rajasekharan described using long-chain saturated fatty alcohols in oil structuring (Daniel and Rajasekharan 2003). Then, Gandolfo et al. showed a synergistic effect for specific weight ratio (R) in oleogels based on the mixture of fatty alcohols and fatty acids with the same chain length (Gandolfo et al. 2004). Thus, mixtures of fatty acids and fatty alcohols were some of the first identified two-component gelator systems with potential for food applications. The alcohol and acid ratio strongly influences textural and structural properties of the resulting oleogels (Gandolfo et al. 2004). Combinations of stearyl alcohol and stearic acid, or behenyl alcohol and behenic acid, have shown to exhibit a high oil structuring capacity in edible oils as a function of this ratio (Gandolfo et al. 2004; Schaink et al. 2007; Blach et al. 2016; Callau et al. 2020). The advantages of fatty acids and fatty alcohols systems for oleogelation is their similarity with triglycerides (Bot and Flöter 2018). Fatty acids can be obtained from triglycerides by hydrolysis. Fatty alcohols can be derived from triglycerides by transesterification with methanol followed by high-pressure hydrogenation (Valoppi et al. 2018). Neither fatty acids nor fatty alcohols are regular food additives (Bot and Flöter 2018). However, fatty acids are used as additives in foods as

Fat Mimetics for Food Applications, First Edition. Edited by Miguel Ângelo Parente Ribeiro Cerqueira and Lorenzo Miguel Pastrana Castro.

lubricants, anti-foaming agents, or binders. For fatty alcohols, only stearyl alcohol has been considered generally recognized as a safe (GRAS) ingredient; this was established in 2001 by the Food and Drug Administration (FDA) in the USA. Therefore, it can be used "as an oil structuring and solidifying agent in margarine, shortening, and foods containing semi-solid and solid fats, at levels of up to 10% of the oil content of the food" (Bot and Flöter 2018). Oil foams or oleofoams based on the aeration of oleogels based on fatty acids and/or fatty alcohols appeared in the literature within the past five years (Binks et al. 2016; Callau et al. 2020). The interest in using oil foams for edible applications is high since they combine the advantages of oil structuring of the oleogels with the presence of gas bubbles, resulting in both a reduced fat content and new appealing textures and sensorial properties (Fameau and Saint-Jalmes 2020).

This chapter will discuss the potential for fatty alcohols and fatty acids alone and in combination to form oleogels and oil foams. We will first focus on the oleogel based on pure fatty acids or fatty alcohols as an oleogelator. Then, the effect of the ratio between fatty acids and fatty alcohols on the oleogel properties, microstructure, and crystalline structures will be discussed for two systems: stearyl alcohol/stearic acid and behenyl alcohol/behenic acid. The formation and stabilization of oil foams based on the oleogel previously described will be explained. We will also explain the potential of oil foams in food products described in the recent patents.

2.5.2 Structure and Properties of Oleogel Based on Fatty Acids or Fatty Alcohols

2.5.2.1 Definition and Properties of Fatty Alcohol

Fatty alcohols are amphiphilic molecules due to the polar hydroxyl head groups helping to form self-assembled structures in different solvents (Valoppi et al. 2018). Fatty alcohols are widely used in various industries such as the cosmetic and pharmaceutical industries and in lubricants. (Valoppi et al. 2018; de Oliveira et al. 2019). Fatty alcohols can have one or more hydroxyl groups localized on the aliphatic chain (Monick 1979). Most of the fatty alcohols used have a chain length between 6 and 22 carbon atoms (Table 2.5.1). When the chain length is larger than 22 carbons, they are called waxy alcohols. A large variety of fatty alcohols exist depending on: the carbon number of the chain, the nature of the aliphatic chain which can be linear or branched chain, the position and number of hydroxyl groups, and possible unsaturation in the aliphatic chain. Fatty alcohols are named based on the IUPAC names coming from the corresponding alkanes by adding the end suffix "-ol" (Table 2.5.1). However, they also have common names based on their extraction source. The best example is 1-hexadecanol commonly named cetyl alcohol because in 1818, Chevreul made the first 1-hexadecanol by using sperm oil from the sperm whale. In Greek, the term "ketos" is used for whale and "cetyl" comes from the Greek term "ketos". The sperm oil is rich in the cetyl ester of palmitic acid, which by saponification with alcoholic potassium hydroxide results in cetyl alcohol (Monick 1979).

Different methods for producing fatty alcohols are used industrially (Valoppi et al. 2018). There are processes based on natural sources starting from fats and oils. Here, the triglycerides in the natural sources are hydrolyzed first into glycerol and fatty acids, then glycerol is removed. Fatty acids or the methyl esters of fatty acids are fractioned and hydrogenated to fatty alcohols (Monick 1979). Naturally sourced fatty alcohol products depend on the fatty acid and wax ester composition of the starting material. There are petrochemical processes based on the controlled hydrogenation and oxidation processes in the presence of catalysts, and the starting materials are petrochemical feedstocks (Noweck and Grafahrend 2006). More recently, biotechnological methods are based on using yeasts and bacteria to synthetize fatty alcohols (Miklaszewska et al. 2017).

Table 2.5.1 Example of saturated fatty alcohols and fatty acids with even chain length.

Chain length	Corresponding alkanes	IUPAC fatty alcohol names	IUPAC fatty acid names	Common fatty alcohol names	Common fatty acid names
10	Decane	1-decanol	Decanoic acid	Capric alcohol	Capric acid
12	Dodecane	1-dodecanol	Dodecanoic acid	Lauryl alcohol	Lauric acid
14	Tetradecane	1-tetradecanol	Tetradecanoic acid	Myristil alcohol	Myristic acid
16	Hexadecane	1-hexadecanol	Hexadecanoic acid	Cetyl alcohol	Palmitic acid
18	Octadecane	1-octadecanol	Octadecanoic acid	Stearyl alcohol	Stearic acid
20	Cosane	1-eicosanol	Icosanoic acid	Arachidyl alcohol	Arachidic acid
22	Docosane	1-docosanol	Docosanoic acid	Behenyl alcohol	Behenic acid
24	Tetracosane	1-tetracosanol	Tetracosanoic acid	Lignoceryl alcohol	Lignoceric acid

Above 10 carbons, fatty alcohols are insoluble in water but are soluble in short-chain alcohols such as ethanol or in ethers. The melting temperatures of fatty alcohols increase by increasing the number of carbon atoms. From 12 carbon atoms, fatty alcohols are crystalline materials with a melting temperature around room temperature. Fatty alcohols show a peculiar behavior for the transition from a liquid to a solid state. There is first the transition from the liquid phase to the rotator phase and then to the solid phase. The rotator phase is commonly described for alkanes and correspond to a solid-state phase in which the molecules are packed like in a crystalline form but rotate around their long axis (Valoppi et al. 2018). In the rotator phase, fatty alcohols form lamellar structures. Three rotator phases have been determined in the literature for fatty alcohols with chain lengths ranging from 14 to 22 carbon atoms. In the solid state, different crystal polymorphs can be formed. Polymorphism is a solid-state phenomenon related to the different ways molecules can pack to form a crystalline structure (Marangoni 2004; Marangoni and Wesdorp 2013). The β-phase or γ-phase can be obtained with a monoclinic unit cell, but they are all characterized by lamellar structures (Valoppi et al. 2016).

2.5.2.2 Definition and Properties of Fatty Acids

Fatty acids are the fundamental building blocks of all lipids in living matter. They are biocompatible since lipids are the main components of the skin and cellular membranes. Fatty acids are amphiphilic molecules due to the polar carboxylic head groups helping to form self-assembled structures in different solvents (Fameau et al. 2014). The polar hydrophilic headgroup can be in its protonated (-COOH) or deprotonated ($-COO^-$) state. The aliphatic hydrophobic tail, which is saturated or unsaturated, can have additional groups such as hydroxyl groups (Fameau et al. 2014). So a large variety of fatty acids exists and some are classified in Table 2.5.1. As for fatty alcohols, fatty acids are named based on the IUPAC convention, corresponding to the equivalent chain-length alkanes by adding the end suffix "-oic" and the word acid (Table 2.5.1). However, they have also common names shown in Table 2.5.1. Fatty acids are also anionic surfactants under their deprotonated forms and are commonly called « soaps ». The term "soaps" consist of negatively charged long-chain carboxylates with positively charged counter-ions (inorganic or organic) (Wolfrum et al. 2016). They are formed by the reaction of either an inorganic or organic base with an organic fatty acid. Soaps are one of the oldest surfactants and are perhaps most natural surfactants since they were discovered as early as in 2800 BC. Babylonians boiled fats with wood ashes to obtain soap-like substances (Toedt et al. 2005; Wolfrum et al. 2016). Fatty acids are widely used in industry. For example, today, annually, over 5 million tons of metal soaps find their way into various cosmetic products such as bar soaps, facial cleansers, liquid washes, personal care creams, and toothpastes (Raut et al. 2008). In almost all of these products, the soap molecules serve as a surface-active agent that performs various functions such as lather and emulsion stabilization, detergency, foam stabilization, and surface wetting. Soap molecules' ability to form self-assembled structures even in hydrophobic liquids also makes them suitable for use as so-called low-molecular-mass organic gelators (LMOGs) in a diverse range of products. For example, lithium salts of 12-hydroxystearic acid are used to structure free-flowing lubricant liquids into viscous, gel like lubricant greases (Fameau and Rogers 2020). Soaps are also used in paper flotation de-inking or as paint dispersing agents.

Fatty acids are obtained by hydrolysis of oils from various oleochemical sources: animal, marine and vegetable (Johansson and Svensson 2001). The composition of fatty acids in the oil is determined by the source of oils. Animal sources, e.g. lard and tallow, are characterized by high concentrations of saturated fatty acids, whilst marine sources such as fish oils are characterized by long-chain and unsaturated acids. The fatty acid composition of oils from vegetable sources varies depending on plant origin and sort (Johansson and Svensson 2001). Commercially exploited seeds such as soya, rape, sunflower, and linseed have been the subject of many years of breeding programs to obtain oils with particular fatty acid compositions. Work is being done to domesticate alternative oil-rich plants that may yield new potentially useful fatty acids such as hydroxylated fatty acids. For example, it has been discovered recently that the seed oil of *Orychophragmus violaceus* has high concentrations of dihydroxy (at positions 7 and 18) C_{24} fatty acids (Li et al. 2018). Demand for specific fatty acids such as hydroxylated fatty acids in the oleo-chemical industry has increased as chemical feedstock has recently led to research on the production of these molecules using metabolic engineering approaches in transgenic plants (Lee et al. 2015).

For saturated fatty acids, the melting point increases by increasing the alkyl chain length. For example, the melting point of lauric acid (12 carbons) is ~45°C and ~70°C for stearic acid (18 carbon atoms) (Table 2.5.1). Unsaturation decreases the melting point. For example, the melting point of oleic acid (18 carbon atoms with an unsaturation) is ~16°C. Long-chain fatty acids above 10 carbons are insoluble in water but are soluble in different solvents such as ethanol. The transition of fatty acids from a liquid to a solid state during cooling shows different transitions from the liquid phase to the rotator phase to the solid phase, as for fatty alcohols. The types of rotator phases depend on the aliphatic chain length, composition, and solvent content. Crystal polymorphs are formed in the solid state, and fatty acids exhibit complex polymorphism (Bond 2004). The type of crystalline structure formed depends strongly on various factors such as the chain length and the degree of unsaturation in the alkyl chain. Another factor that influences the polymorphism of fatty acids is the solvent (and especially the solvent's polarity) used in the crystallization process (Silva et al. 2019). In the literature, different polymorphs have been observed for fatty acids. Fatty acids display forms A_1, A_2, A_3, A_{super}, C, $B_{o/m}$, and $E_{o/m}$ (Sato 2018). The B and E forms exhibit two polytypes: orthorhombic and monoclinic, respectively (Kaneko 2001).

2.5.2.3 Fatty Alcohols as Oleogelators

The first study of fatty alcohols as oleogelators was carried out by Daniel and Rajasekharan in 2003 (Daniel and Rajasekharan 2003). They studied the oil gelling properties of different long-chain fatty alcohols such as arachidyl, behenyl, and myricyl alcohol in different oils such as sunflower oil, lavender oil, and diesel oil (Table 2.5.1). Then in 2004, Gandolfo et al. studied the gelling properties of cetyl, stearylique, arachidyl, and behenyl alcohols in sunflower oil (Gandolfo et al. 2004). They showed that all fatty alcohols studied have the potential to structure sunflower oil at concentrations as low as 2 wt.%. They also showed that the observed hardness increased with increasing concentration due to an increase of solid particles in the liquid oil phase. The oil type strongly affected oleogel hardness. For example, stearyl alcohol at 5 wt.% showed a lower hardness in rapeseed oil than in sunflower oil. Almost ten years after these findings, other studies focused on understanding the structure of the oleogels made by fatty alcohols. A multi-scale approach was used to link the nano and microstructure with the resulting oleogel properties in terms of oil retention capacity, and for mechanical properties such as hardness, thermal properties, and stability over time (Schaink et al. 2007; Blach et al. 2016; Valoppi et al. 2017, 2018). It was showed using scanning electron microscopy (SEM) that fatty alcohols in vegetable oils formed platelet-like crystalline microstructures as seen by Differential Interference Contrast (DIC) microscopy (Figure 2.5.1.a). The interconnection between the crystalline particles is responsible for oleogel formation. The size of the platelet-like crystals depended on cooling and shearing rates. For example, for stearyl alcohol at 5 wt.% in peanut oil, a slow cooling rate of 5°C/min formed crystalline particles ten times larger than those formed at a fast cooling rate of 40°C/min (Valoppi et al. 2017). Moreover, the polymorphism and molecular arrangement of fatty alcohol-structured oleogels varied as a function of the cooling rate and the fatty alcohol chain length (Blach et al. 2016; Valoppi et al. 2016). The longer the chain length, the longer the time needed for adopting the most stable β form. A slow cooling rate supports formation of the most stable oleogels compared to fast cooling rates (Valoppi et al. 2018). The oil retention capacity of oleogels, which is also commonly called oil-binding capacity, was also directly linked to the polymorphic form of the fatty alcohol and to the cooling rate. A polymorphic transformation during oleogel storage does not have an oil release as high as an oleogel directly obtained with a stable polymorphic form. The oil release during storage was 100 times higher for a fast cooling at 40°C/min than a slow cooling rate at 5°C/min for arachidyl-based oleogels (Valoppi et al. 2018). Therefore, oleogel properties are directly linked to

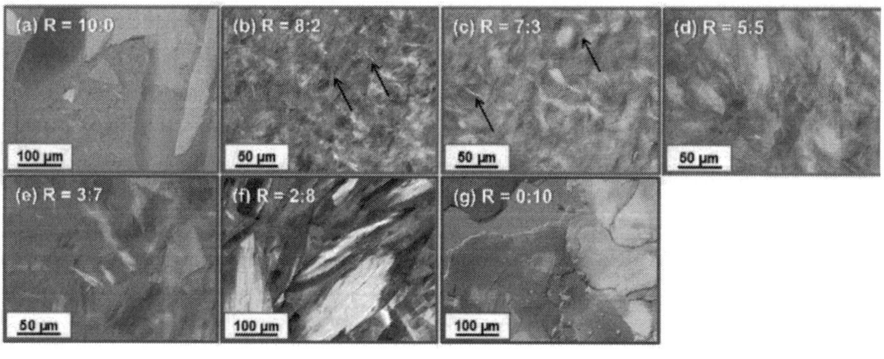

Figure 2.5.1 Differential Interference Contrast (DIC) Microscopy images of oleogels as a function of R between behenyl alcohol:behenic acid (BO:BA) in soya oil: 10:0 (a), 8:2 (b), 7:3 (c), 5:5 (d), 3:7 €, 2:8 (f), 0:10 (g). The total concentration of the structurant was kept constant at 10 wt.%. The scale is indicated on the images. (Callau et al. 2020, Reproduced with permission from Elsevier) with Permission from Elsevier.

the size of the platelet-like structures, their quantity, and their molecular arrangement, which are directly linked to the fatty alcohol chain length, their concentration, and the process parameters used to produce the oleogel (shearing and cooling rate).

2.5.2.4 Fatty Acids as Oleogelators

Fatty acids used to produce oleogels can be distinguished into two main categories (Bot and Flöter 2018). The first category is saturated, non-branched, long-chain fatty acids as described in Table 2.5.1. We focus on this category in this chapter. The other category includes all other long-chain fatty acids such as fatty acids with two carboxylic groups (called fatty di-acids), unsaturated fatty acids, and fatty acids with hydroxyl groups (Cui et al. 2016; Bot and Flöter 2018). The best known fatty acid oleogelator is the 12-hydroxystearic acid, which creates fibrillar networks leading to high gel firmness at low concentrations (Fameau and Rogers 2020). Eini and Tamarkin were the first ones to report using long-chain saturated fatty acids as oleogelators to form a crystalline network immobilizing the liquid oil (Eini and Tamarkin 2001). Then, Gandolfo et al. studied the potential of palmitic, stearic, arachidic, and behenic acids to gel sunflower oil in greater detail (Gandolfo et al. 2004). An interesting point to notice is that in the first study that compared the oil gelling properties of fatty acids and fatty alcohols, Gandolfo et al. already showed that at the equivalent concentration of fatty alcohols yield harder gels than fatty acids (Gandolfo et al. 2004). For example, stearyl alcohol at 5 wt.% showed a lower hardness of around 50 g-force at 5°C, compared to stearyl alcohol at the same concentration, which showed a hardness of 150 g-force in sunflower oil. Long-chain saturated fatty acids, such as behenic acid in sunflower oil, form platelet-like crystalline microstructures as shown in Figure 2.5.1.f. As for fatty alcohol oleogels, the fatty acid polymorph present depends on the process used to produce the oleogel. This is why different groups have different results in terms of polymorphs for the stearic acid based oleogels (Schaink et al. 2007; Blach et al. 2016). For example, Schaink et al. found that stearic acid in sunflower oil formed oleogels, which were made by stearic acid molecules with an orthorhombic structure, but Blach et al. found that stearic acid molecules in canola oil organized into a double layer structure corresponding to the C-form found in pure stearic acid (Schaink et al. 2007; Blach et al. 2016). Also, the nature of the vegetable oil phase (sunflower, canola, or sesame oils) had a slight effect on the fatty acid polymorphs, but less pronounced than the process effect (Bot and Flöter 2018). As described in the previous section for oleogels based on fatty alcohols, process parameters such as cooling and shear rate, are key for modulating the structure and functionality of fatty acid oleogels.

2.5.3 Mixture of Fatty Acids and Fatty Alcohol to Improve Oleogel Properties

Here, we describe two types of fatty acids and fatty alcohols systems: stearyl alcohol/stearic acid and behenyl alcohol/behenic acid systems. Several studies focused on mixtures of stearic acid and stearyl alcohol because stearyl alcohol is the only fatty alcohol registered as a GRAS ingredient for oleogelation (Valoppi et al. 2018). It is an edible material that is useful as an oleogelator in food applications. However, for other applications, such as cosmetic or pharmaceutical products, longer alkyl chain fatty acids and fatty alcohols can be used, which display better gelling properties (Callau et al. 2020). For example, behenyl alcohol and behenic acid are raw materials widely used in the cosmetic and pharmaceutical industries; their gelling properties are promising compared to stearyl alcohol and stearic acid due to their longer alkyl chain length. The weight ratio (R) between the alcohol and the acid strongly influences their texture and the structure of the resulting oleogels. We describe here how R changes the crystal structure at the nanometer scale, which leads to important changes at the meso and macroscopic scales of the oleogels.

2.5.3.1 Effect of R on the Crystal Structure

Small-Angle (SAXS) and Wide-Angle X-ray Scattering (WAXS) experiments are performed on oleogel systems to determine the molecular packing inside the crystals, the so-called "nanostructure" (Figure 2.5.2). In the SAXS region, we can obtain information about long spacing, which is related to chain length, chain tilt, number of chains per layer, the correlation length, and the thickness of the crystalline structures such as the thickness of crystalline nanoplatelets (Peyronel et al. 2014). In the WAXS region, we can obtain information about the short spacing corresponding to the chains' lateral packing. For oleogels with pure fatty alcohols or fatty acids, depending on the preparation process, different structures can be obtained. For example, for behenyl alcohol in sunflower oil, two *d*-spacings at around 60.4 Å and 50.2 Å have been detected in the SAXS region (Figure 2.5.2) (Callau et al. 2020). This was consistent with behenyl alcohol molecules crystallized in a double layer structure associated with $d = 60.4$ Å and the remaining molecules were organized into a triple layer structure with a *d*-spacing around 100 Å deduced from the position of the second order reflection peak at $d = 50.2$ Å. In the WAXS region, two possible polymorphs were also deduced from the peak positions, the β–form and the sub-α-form. For pure behenic acid, only one long-spacing at around 49.0 Å was obtained corresponding to a double layer structure associated with possibly the β-form (Blach et al. 2016; Callau et al. 2020). For R = 8:2, 7:3 and 5:5, with an excess of behenyl alcohol, only one long spacing at around 57.0 Å was observed. This *d*-spacing could not be associated to the *d*-spacings measured for oleogels containing only behenyl alcohol (R = 10:0) or only behenic acid (R = 0:10). Therefore, only one type of crystals in those systems was present: mixed crystals of behenyl alcohol and behenic acid. The two components co-crystallized. Moreover, these mixed crystals showed a different polymorphic form in the WAXS region as compared to the pure components. When the concentration of behenic acid was higher than the concentration of behenyl alcohol (R = 3:7 and 2:8), the long spacing corresponding to the mixed crystals that were present was combined with additional peaks corresponding to the pure behenic acid crystals. These oleogels were made of pure fatty acid crystals and of mixed alcohol/acid crystals. In the WAXS region, the same observations could be made. The two characteristic WAXS reflections correspond to the mixed crystals and the pure fatty acid crystals. The same results were obtained by Blach et al. for the stearyl alcohol/stearic acid system in canola oil that showed that mixed crystals of the same long chain fatty alcohol

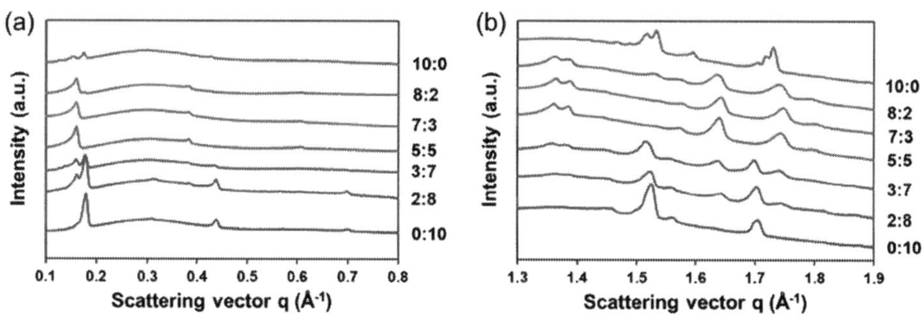

Figure 2.5.2 SAXS (a) and WAXS (b) spectra of oleogels with varying ratios of behenyl alcohol:behenic acid (BO:BA), with 10 wt.% structurant in sunflower oil. The measurements were carried out at 25 °C. The spectra were shifted in intensity for clarity. Reproduced from (Callau et al. 2020) with Permission from Elsevier.

and fatty acids were obtained for the same weight ratio whatever the chain length (18 carbons versus 22 carbons), as it was already supposed by Gandolfo et al. in the pioneering work (Gandolfo et al. 2004).

2.5.3.2 Effect of R on the Microstructure of the Oleogels

To obtain qualitative information about the microstructure of the oleogels as a function of R, Differential Interference Contrast Microscopy (DIC), polarized light microscopy (PLM), or SEM are commonly used (Figure 2.5.1). For both stearyl alcohol/stearic acid and behenyl alcohol/behenic acid systems in vegetable oils, for R = 10:0 and 0:10 (*i.e.*, containing only one fatty component), large platelet-like crystals were seen (Schaink et al. 2007; Blach et al. 2016; Callau et al. 2020). For example, in Figure 2.5.1, platelet structures were seen for pure behenyl alcohol (R = 10:0) and for pure behenic acid (R = 10:0). For two specific ratios, R = 8:2 and 7:3, small platelet-like crystals were seen in this mixed system (Figures 2.5.1.b-c). When the concentration of behenic acid increased, which corresponds to R = 5:5, 2:8 and 3:7, the size of the crystals increased again (Figure 2.5.1.d-f). The same observations have been made for stearyl alcohol/stearic acid systems showing that in all cases, oleogels have platelet-like crystals wherein the size varied depending on R. It is important to notice here that some studies describe needle-like structures at intermediate ratios, but it was only an artifact of sample preparation, since the platelet edge can align with the glass slide or coverslip with DIC experiments (Blake and Marangoni 2015a). The platelet-like structures for all R were shown by using cryo-SEM experiments (Blach et al. 2016). The smallest crystals' size is obtained for the ratio corresponding to mixed crystals of fatty alcohol/fatty acid.

2.5.3.3 Effect of R on the Thermal Behavior and Solid Fat Content of Oleogels

The thermal behavior of oleogels can be studied by Differential Scanning Calorimetry (DSC) experiments where the melting and crystallization profiles of oleogels as a function of R are obtained. Previous studies reported that for the behenyl alcohol/behenic acid oleogel system, only a dominant primary peak during the melting and the crystallization was seen for all oleogels (Callau et al. 2020). However, the two R (3:7 and 2:8) ratios containing mixed crystals and pure fatty acid crystals showed a secondary peak in their profiles. The secondary peak temperatures were lower than the primary ones. The presence of two endotherms and two exotherms for these R values showed mixed crystals and pure fatty acid crystals. The mixed crystals melt before the pure behenic acid crystals. For oleogels with mixed crystals (R = 8:2, 7:3, and 5:5), and the oleogel based on only behenyl alcohol (R = 10:0) or behenic acid (R = 0:10), the DSC spectra showed only one endothermic peak for the melting and one exothermic peak for the crystallization due to only one type of crystalline structure inside the oleogel. The pure behenyl alcohol oleogel (R = 10:0) had a melting temperature of 54.9 ± 0.3°C and a crystallization peak at 45.1 ± 0.9°C, whereas the pure behenic acid oleogel (R = 0:10) melted at 62.4 ± 0.8°C and crystallized oleogel at 51.5 ± 0.3°C. Thus, behenic acid had a higher melting and crystallization temperature than behenyl alcohol. The melting temperatures for the mixed crystals were lower than for pure behenyl alcohol and pure behenic acid (Callau et al. 2020). Therefore, the thermal stability of the oleogels with mixed crystals was lower than oleogels with pure components. The same results have also been described for stearyl alcohol/stearic acid systems (Schaink et al. 2007; Blach et al. 2016). From DSC heating curves and SAXS/WAXS experiments, a schematic phase diagram was proposed for fatty alcohol/fatty acid mixed systems (Bot and Flöter 2018; Callau et al. 2021). As shown in Figure 2.5.3 for the behenyl alcohol/behenic acid system, for

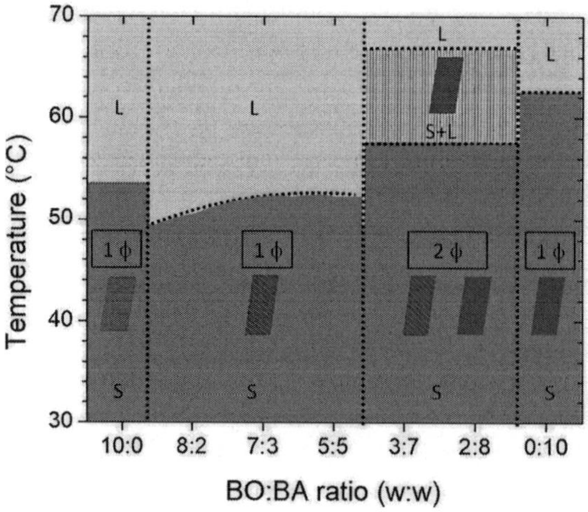

Figure 2.5.3 Schematic phase diagram for BO/BA oleogelator system in oil. One phase region is indicated by 1ϕ, and the two phase region is indicated by 2ϕ. S corresponds to solid and L to liquid. Platelets crystals are drawn: pure BO in red, mixed BO/BA crystals in red and blue dashed, and pure BA in blue. Reproduced from (Callau et al. 2021) with Permission from Society of Cosmetic Chemists.

R = 8:2, R = 7:3 and R = 5:5, mixed crystals of behenyl alcohol and behenic acid were observed corresponding with one crystalline phase. For R = 3:7 and R = 2:8, the mixed crystals were present with pure behenic acid crystals leading to two crystalline phases dispersed in the liquid oil phase. As shown by DSC, the mixed crystals melted before the pure behenic acid crystals for R = 3:7 and R = 2:8. Thus, by heating, there was a phase containing only behenic acid crystals since the mixed crystals already melted.

The solid fat content (SFC) is also an important parameter to consider to understand the resulting properties of these oleogels produced with 10 wt.% of gelator(s). For behenyl alcohol/ behenic acid systems, the SFC varied greatly as a function of R (Callau, Sow-Kébé, Nicolas-Morgantini et al. 2020). For example, at 20°C, the lowest SFC was around 2.7 % for pure behenic acid, indicating that only 2.7 % of behenic acid was crystallized, and 7.3 % of BA was solubilized in the sunflower oil. For pure behenyl alcohol, the SFC was much higher and around 6.9 %. For oleogels with mixed crystals, the SFC increased to reach values around 9.5 %, showing that the solubility of behenic alcohol and behenic acid decreased in the oil. When mixed crystals were present, most the behenic acid and behenic alcohol molecules were crystallized (SFC close to 10 %). Then, the SFC of oleogels decreased with an increasing concentration of behenic acid (R = 7:3, SFC = 9.07 %; R = 5:5, SFC = 8.59 %; and R = 3:7, SFC = 7.94 %). Similar results were obtained for mixed crystals based on stearyl alcohol and stearic acid in different oils (Schaink et al. 2007; Blach et al. 2016).

2.5.3.4 Effect of R on Oleogel Properties: Mechanical Strength and Stability

The mechanical strength and the stability of oleogels are critical factors in practical applications and depend on the microstructure and nanostructure of the oleogels. To obtain information on the mechanical strength of the oleogels as a function of R, the average hardness of the oleogels is commonly determined by the well-known Texture Profile Analysis (TPA) method (Blach et al. 2016). To obtain information on the stability oleogels' oil retention as a function of R, oil loss during centrifugation at high speed is also measured.

For both systems (stearyl alcohol/stearic acid and behenyl alcohol/behenic acid), when oleogels had mixed crystals, the lowest percentage of oil loss was comparatively measured against all the other R. In terms of oleogel hardness, when mixed crystals are present, the hardness is much higher than it is for pure components. This suggests that the amount of solids in oil determined by SFC is a key parameter that governs the mechanical strength and the stability of the oleogels. Mixed crystals have a strong effect on the oleogels' properties in terms of oil loss and hardness. The best Rs to obtain the best oleogel properties are the ones with mixed crystals corresponding to high SFC. The two ideal R were R = 8:2 and 7:3 for behenyl alcohol/behenic acid in sunflower oil and stearyl alcohol and stearic acid in canola oil (Blach et al. 2016; Callau et al. 2020). Fatty acid crystals have a greater solubility compared to the mixed crystals. Therefore, fatty acid crystals with mixed crystals decreases the stability and hardness of these oleogels compared to the ones containing only mixed crystals. The two ideal weight ratios (R = 8:2 and R = 7:3) correspond to molar ratios are close to 3:1. This specific 3:1 molecular ratio is already highlighted for other colloidal systems as a peculiar ratio in mixed surfactants. For example, the stability of foams or emulsions is enhanced for this specific ratio as shown in various systems (lecithin and cholesterol, decanoic acid and decanol, sodium laurate and lauryl alcohol, stearyl alcohol and stearic acid, etc.) (Shah 1971; Patist et al. 1999). Moreover, Langmuir monolayers experiments on the mixed systems stearyl alcohol and stearic acid have shown that monolayers exhibit a minimum area per molecule for a molar ratio equal to 3:1 (Shah 1971). That is why it is presumed based on the work from Gandolfo et al. that a minimum in area per molecule could also occur for a molar ratio of 3:1 corresponding to an ideal weight ratio comprised of the weight ratio R = 8:2 and R = 7:3 (Gandolfo et al. 2004). A minimum in area per molecule leads to a decrease in interfacial tension. In the literature, it is well described that the nucleation rate depends on the interfacial energy (Garside 1987). Thus, a decrease of the interfacial energy for a specific ratio leads to a higher nucleation rate which leads to a decrease in the average crystal size. Therefore, it is the effect of the 3:1 molecular ratio in mixed surfactant systems described by Shah et al. over 50 years ago for foams, emulsions, and Langmuir monolayers; these also occur on the crystallization of mixed fatty alcohol and fatty acid systems in oils leading to better oleogels properties compared to other R (Shah 1971). It is important to notice that not only the presence of mixed crystals, their specific morphology, and size, but also the SFC can better explain the oleogel properties obtained for R = 8:2 and R = 7:3 compared to all of the other oleogels tested (Blach et al. 2016; Gravelle et al. 2017). Blach et al. demonstrated that the differences in oleogels' properties were also linked to another parameter—the spatial distribution of the crystalline mass inside the oleogel. They used a modeling approach based on the cellular-solid model of Gibson and Ashby to show the importance of the spatial distribution of the crystalline particles inside the oil for stearyl alcohol/stearic acid oleogels (Miyazaki and Marangoni 2014; Blach et al. 2016).

An interesting and recent study has also shown that a polymer oleogelator (ethylcellulose) could further improve the mechanical gel strength of the oleogels obtained from stearyl alcohol/stearic acid in canola oil as a function of R (Gravelle et al. 2017). However, the ethylcellulose did not change neither the polymorphism described as a function of R, nor the thermal behavior of oleogels. The ethylcellulose affected the microstructure due to the production of clusters of the crystals (Gravelle et al. 2017). The increase of gel strength came from a reinforcement of the gel by the ethylcellulose network forming a secondary network inside the oil phase. A slight increase of the SFC was also seen in addition to ethylcellulose. So, the oleogel properties (mechanical strength and stability) could be further improved via a judicious combination with other polymeric oleogelators.

2.5.4 Oil Foams Based on Fatty Acids and Fatty Alcohols

2.5.4.1 Definition of Oil Foams Stabilized by Crystalline Particles

Foams are gas bubbles dispersed in a liquid phase containing a surface-active species. Two categories of liquid foams exist depending on the continuous phase: aqueous (water-based) and non-aqueous (oil-based) (Fameau and Saint-Jalmes 2017). Aqueous foams have been widely used for industrial applications for years, but oil foams were not discovered until the 1970s and were then essentially ignored for about 30 years (Binks and Vishal 2021). In the last five years, interest in using oil foams based on oleogels for edible applications has grown tremendously due to their potential to develop food products with both a reduced fat content and new appealing textures and sensorial properties (Heymans et al. 2017). Notice that the production of edible oil foams for molecular gastronomy started well before researchers becamae interested in these systems (Fameau and Saint-Jalmes 2020). Oil foams were used for molecular gastronomy to provide a different mouthfeel and appearance to oil products by incorporating air bubbles in the vegetable oils (Fameau 2021).

The main difference between aqueous and oil foams comes from the relatively large difference in the surface tension of the solvents (Fameau and Saint-Jalmes 2017). Water is a unique solvent due to extensive hydrogen bonding leading to a high surface tension of 71.9 mN m^{-1} at 25°C. For example, in the case of oils, the surface tension is relatively low between *ca.* 5 and 35 mN m^{-1} depending on the oil type. Hydrocarbon-based surfactants can efficiently adsorb and reduce the high air-water surface tension to lower values ranging from 30–40 mN m^{-1}. For oil systems, the low surface tension makes the adsorption of hydrocarbon-based surfactants energetically unfavourable. One way to produce and stabilize edible oil foams is to use surfactant crystalline particles based on fatty components, which can adsorb at the air-liquid surface (Fameau and Binks 2021). Fatty acids and fatty alcohols crystalline particles are efficient oil foam stabilizers (Fameau et al. 2015; Binks et al. 2016). The formation and stabilization mechanisms of oil foams are based on the adsorption of the crystalline particles at the air-liquid surface (Fameau and Binks 2021). Foam formation is associated with the presence of these crystals, which reduce the bare surface area by their presence rather than lowering the surface tension. The key parameter for the crystals to adsorb at the air-oil surface is to show a suitable three-phase contact angle below 90° (Binks and Vishal 2021). The presence of a dense layer of adsorbed crystalline particles at the bubble surface protects the gas bubbles and lead to textured surfaces and the formation of non-spherical bubbles (Figure 2.5.4).

These oil foams are based on the presence of crystalline particles dispersed in oil, meaning they are on the formation of oleogels as described. So the foam can be produced in two ways: directly from the oleogel or during cooling of the oil-containing oil soluble fatty components, which crystallized during foam formation (Fameau and Saint-Jalmes 2017). Whichever way the oil foam is produced, the crystalline particles are key parameters to produce and stabilize oil foams. Foam formation is only possible when crystals are present in dispersion (Fameau and Saint-Jalmes 2017). Different methods used

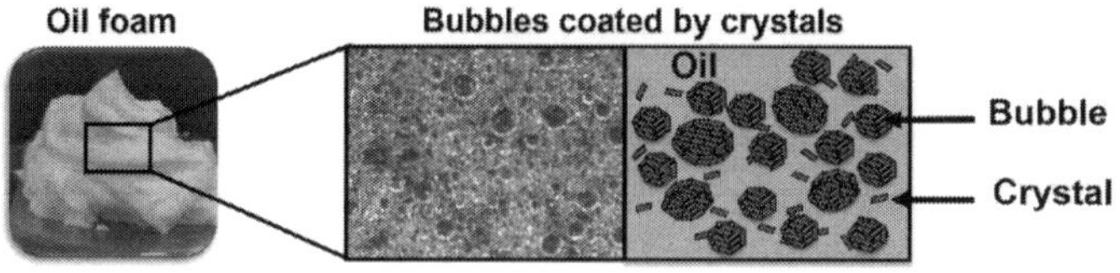

Figure 2.5.4 Illustration of oleofoam stabilized by crystals both in bulk and at the interface as observed by optical microscopy.

to fabricate oil foams such as whipping or depressurization are described in the literature. Whipping is the most popular method for producing edible oil foams at lab scale or for molecular gastronomy (Fameau and Saint-Jalmes 2020). For whipping, different devices are used: electric beater, rotor-stator homogenizer, or kitchen food mixer. Foam production by whipping is based on air entrainment and progressive bubble breakup under shear (Drenckhan and Saint-Jalmes 2015). During shearing, the gas fraction increases and the average bubble size decreases. Over time, at the macroscopic scale, this process leads to oil foam, which becomes more opaque and solid-like. When the equilibrium state is reached, the gas fraction and the bubble size remain fixed with further whipping. Both the rheological properties of the oil dispersion and the whipping speed control affect the characteristic gas fraction and bubble size of the foaming mixture. To manufacture edible oil foams on an industrial scale, high shear rotor-stator mixers such as Mondomix VL (by Bühler) are often used in the food industry and works well for oil foam production (Gunes et al. 2017). The gas incorporation process is a key parameter, since it controls the size of the bubbles, the gas introduced in the foam, the rheology of the oleogel forming the continuous phase, and the morphology of the crystals. All these intermediate-scale parameters then control the macroscopic stability and rheology of the produced oleofoam.

The majority of these foams are stable for at least several months (Fameau and Saint-Jalmes 2020). This high stability is due to the dense layer of adsorbed crystals at bubble surfaces that considerably reduce both disproportionation and coalescence (Figure 2.5.4). The fraction of non-adsorbed crystals, which increases with crystal concentration, strengthens the oleogel network in the continuous oil phase, reducing the buoyancy-driven creaming of bubbles within the foam. When the strength of the gel is high enough, oil foams can sustain their own weight and do not flow when inverted. The resulting rheological properties of oleofoams is still under study, but they are linked to both the contribution of the crystals coating the bubbles and to the rheology of the oleogel continuous phase in which the bubbles are dispersed (Gunes et al. 2017; Mishra et al. 2020; Saha et al. 2020). At low frequencies, some oleofoams behaves as viscous liquids (Fameau et al. 2015). So the oleofoams have a good flowability without collapsing due to their rheological properties (Gunes et al. 2017; Heymans et al. 2018). This is a key parameter for the food industry, which needs easy-to-handle products that could be pumped through pipes and easy mixed with other components (Heymans et al. 2018). However, the flowable character strongly depends on the amount of incorporated gas and the size of the gas pockets. Keep in mind that to produce oleofoams from oleogels, they need to behave as viscous dispersions once the whipping or mixing process starts to allow air incorporation, implying specific rheological properties (such as yielding) of these oleogels (Mishra et al. 2020).

2.5.4.2 Oil Foams Based on Fatty Acids and Fatty Alcohols

Different examples are available in the literature dealing with oil foam production based on pure fatty acids or pure fatty alcohols crystals (Table 2.5.2).

Fatty acids and fatty alcohols are described in the literature for the production of oil foams form platelet-like crystals in vegetable oils. Platelet-like is a suitable shape to stabilize air-liquid surfaces (Fameau and Binks 2021). These crystals are adsorbed at the air-oil surface due to the three-phase contact angle below 90°. For example, the contact angle of behenic acid and behenyl alcohol in sunflower oil was around 45° showing that the crystalline particles were partially hydrophobic (Callau et al. 2020). Similar values have also been obtained for myristic acid in different vegetable oils (Binks et al. 2016). Each gas bubble is coated by adsorbed crystalline particles leading to textured surfaces and non-spherical bubbles (Figure 2.5.4). The driving force for their adsorption is to reduce the surface energy of the air–oil surface. In the case of alkane waxes in oil solvent, crystals consist of flat plates of alkane with large, slow growing flat faces exposing terminal methyl groups

Table 2.5.2 Fatty alcohols and fatty acids used to produce oil foams in the literature.

Vegetable oil	Crystals Particle	Gelator Concentration(wt.%)	Foam production method	References
Sunflower oil	Myristyl alcohol, cetyl alcohol, stearyl alcohol	From 1 to 20	whipping	(Fameau et al. 2015)
9 vegetable oils (High oleic sunflower oil, rapeseed oil, coconut oil, soybean oil, corn oil, olive oil, peanut oil, cottonseed oil and sesame oil)	Myristic acid	From 1 to 12	whipping	(Binks et al. 2016)
5 vegetable oils (sunflower, olive, camelina, rapeseed, and apricot)	Stearic acid, behenic acid, stearyl alcohol, stearic acid	10	whipping	(Callau, Sow-Kébé, Jenkins et al. 2020; Callau et al. 2021)

and small, fast growing edge faces, which expose mainly methylene groups (Binks et al. 2016). The difference in surface energy between the low energy methyl-exposing surface and the higher energy methylene-exposing surface is approx. 8 mN m^{-1}. For fatty acids and fatty alcohols crystals, faces expose methyl groups, while edges expose mainly methylene and carboxylic acid or hydroxyl groups. Therefore, faces are likely to be in contact with air (and oil), while edges interact with each other through carboxylic acid or hydroxyl groups within the air–oil surface (Mishima et al. 2016).

In oleogel systems, the two main parameters for tuning the crystals' shape and size are cooling and shearing rates, as described in Section 2.5.2 (Marangoni et al. 2012, 2020). It is well described in the literature that a fast cooling process leads to smaller crystal sizes in oleogel systems than a slow cooling process (Blake and Marangoni 2015b; Valoppi et al. 2017). The effect of cooling rate on the oil foam properties has been studied for behenic acid and behenyl alcohol in sunflower oil (Callau et al. 2020). Two processes were applied to produce the oleogels before foam production by whipping: the slow (5°C/min) and the fast cooling (40°C/min) process. For the slow cooling process, big crystalline particles from 100–500 μm were obtained. The crystalline size was decreased with the fast cooling process to 50–200 μm, but the shape of the crystalline particles remained platelet-shaped between the two processes. In both cases (behenic acid and behenyl alcohol), because the fast cooling process gave rise to smallest crystalline particles in the oleogel, foam overrun increased, and bubble size decreased, which showed the effect of the crystalline particles size on the quantity of foam produced. In terms of foam stability, a decrease in particle size was correlated with half the amount of liquid drained after one month of room temperature aging. Foam firmness also increased by decreasing crystal size. By increasing the cooling rate, the platelet-shape remained, but the crystal size decreased, and the quantity of crystals increased, leading to enhanced foamability and foam stability (Callau et al. 2020). By modifying the cooling process, both oleogel and foam properties were enhanced by decreasing crystals size and by increasing their stability against oil loss and their firmness. There was a direct link between oil foam and oleogel properties.

One interesting property of these oil foams is their responsiveness to temperature (Fameau and Fujii 2020). Since the crystals form from molecular solutions upon cooling below their solubility limit, and melt upon heating above their solubility limit, which corresponds to the transition between oleogel and liquid, the oil foams stabilized by fatty acids or fatty alcohols crystalline particles can lead to thermoresponsive foams. When ultrastable oil foams are stored at temperatures at which crystals melt, they collapse due to a combination of oil drainage and of bubble coalescence in a few minutes. No foam remains and only a molecular solution is present. Both the particles at the bubble surface as well as those in the continuous oil phase melt leading to the rapid disappearance of air bubbles. As the melting of crystal particles is reversible, foam stability can be easily switched between high and low by tuning the temperature above or below the solubility boundary through multiple temperature cycles. For example, thermoresponsive oil foams produced from stearyl alcohol in sunflower oil have been described in the literature (Fameau et al. 2015). The temperature at which the crystals melted was around 48°C; they are formed at around 32°C, as determined by differential scanning calorimetry. At 20°C, the foam was stable for months since the temperature was below the solubility limit. However, when the temperature was increased to 50°C (above the solubility boundary), the crystals melted, and the continuous phase between the bubbles shifted from an oleogel to a liquid state, leading to the onset in foam destabilization. By cooling the foam back to 20°C, the destabilization process was halted, and the foam restabilized due to the recrystallization process in the continuous phase. By increasing the temperature to 50°C again, foam destabilization mechanisms were reactivated and the foam collapsed in a few minutes. Foam could reproduced with whipping once the crystals were reformed. The same results have been obtained for other oil foam systems based on fatty acids (Binks et al. 2016).

2.5.4.3 Controlling Oil Foam Properties by Tuning the Ratio between Fatty Alcohol and Fatty Acids

As described in Section 2.5.3, an easy way to tune the quantity of crystals and their size, while keeping the platelet shape, is to use a mixture of fatty alcohols and fatty acids with the same hydrocarbon chain length but at different weight ratios to induce co-crystallization. The effect of R has also been studied for oil foams for two systems in sunflower oil: stearyl alcohol/stearic acid and behenyl alcohol/behenic acid (Callau et al. 2020). The contact angle, bubble size, overrun, firmness and stability of the oil foams have been studied as a function of R (Figure 2.5.5).

For all oleogel systems studied, the contact angle varied between 45–49° regardless of the crystal size and nature of the crystals for the oleogels described in Section 2.5.3. These contact angles were below 90° showing that all the crystalline particles were partially oleophobic and could adsorb to the interface at any R. The bubbles' size of the oil foams was directly linked to the crystals with the initial crystalline particle size and the solid fat content in the oleogel used to produce the oil foam. For example, for R = 10:0 and R = 0:10 for behenyl alcohol/behenic acid oleogels containing the largest crystalline particles, the mean bubble diameter was the highest, varying from 100 to 200 μm. Whereas the smallest particle size in the oleogels (R = 8:2 and R = 7:3) led to the smallest bubble sizes of around 25–50 μm. The oleogels with intermediate particle size produced foams with air bubbles ranging from 50 to 100 μm. In the same way, the lowest overruns were obtained for the oleogels containing pure fatty acid or pure fatty alcohols with the largest crystalline particles and the lowest SFC. The highest overruns were obtained for the oleogels at R = 8:2 and R = 7:3, based only on mixed crystals with the smallest crystalline particles and the highest SFC. The oleogels with intermediate particle sizes and intermediate SFC gave intermediate overruns. The smaller the crystalline particle size and the higher the SFC, the higher the overrun of the oil foams. The firmness of the foam was also directly linked to R (Figure 2.5.5). The firmness of the oil foams can be determined using the TPA method as described for oleogels. The foam firmness corresponds to the maximum strength measured in compression. The lowest values for firmness were obtained for the foams made with

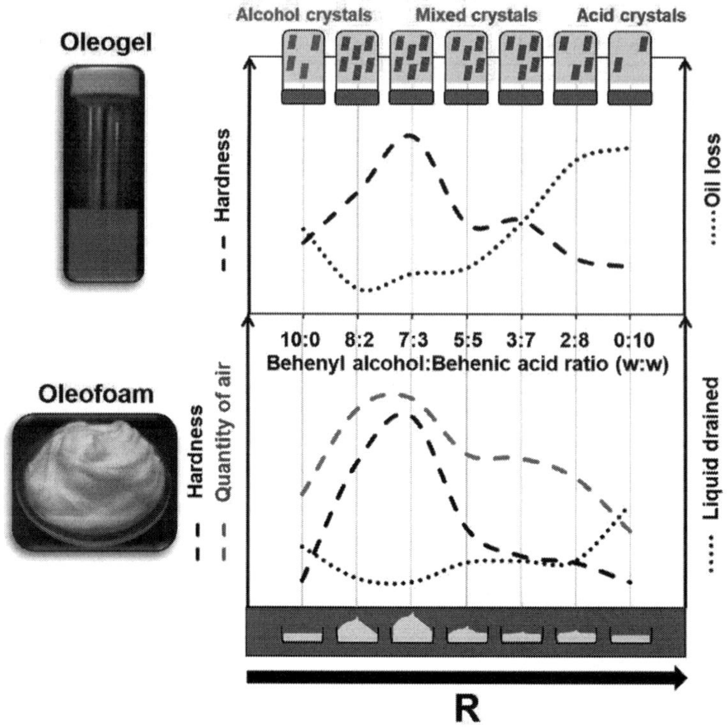

Figure 2.5.5 Illustration of the links between oleogel and oleofoam properties as a function of the weight ratio (R) between fatty alcohol and fatty acid in sunflower oil. When mixed crystals of behenyl alcohol and behenic acid are formed inside the oleogel, the hardness is the maximum, and the oil loss is the minimum, leading to the highest oleofoam hardness and quantity of air and to the lowest liquid drained inside the foam.

R = 10:0 and R = 0:10, which is to say, for the pure fatty alcohol and fatty acid. The firmness was highest for the foams obtained with R = 8:2 and R = 7:3. These foams could be inverted without any flow, which showed that they behave like a strong gel by sustaining their own weight. The oleogels with both intermediate particle size and SFC gave foams with intermediate firmness. R = 8:2 and R = 7:3 contained the highest amount of solid in oil and were associated with mixed crystals with the smallest initial crystalline particle size, and highest foam firmness. To explain the resulting foam overrun and firmness, have a look to the stability of the initial oleogels. The whipping process used to create the oil foams is known to rapidly create a heterogeneous material: liquid oil and gel (Gunes et al. 2017). If the oleogels are not stable against oil loss, they could release a high quantity of liquid oil when the whipping process starts. The presence of this free liquid oil could be at the origin of foams with low firmness and overrun. As described in section 2.5.3, for R = 10:0 and R = 0:10, the oleogels had the lowest stability against oil loss during centrifugation at high speed, but R = 8:2 and R = 7:3 exhibited the highest stability against oil loss for behenyl alcohol/behenic acid system (Figure 2.5.5).

The most stable foams were obtained for oleogels at R = 8:2 and R = 7:3 for both fatty alcohol/fatty acid systems, which had the smallest mixed crystal particles and the highest solid fat content (see Section 2.5.3). The crystals had a suitable size to efficiently stabilize small air bubbles during foam production by whipping and to completely over the bubbles' surface with a dense layer to protect them against coalescence and Oswald ripening. The small crystalline particles that remained in a high quantity (high level of SFC) in the oil continuous phase formed a gel and reduced the liquid drainage. The combination of the oil phase properties with a high quantity of small air bubbles helped form creamy foams with exceptional firmness. These two ratios also gave rise to oleogels with the highest hardness and stability against oil loss (see Section 2.5.3). On the contrary, the fewer stable foams were obtained for the oleogels containing pure fatty alcohol (R = 10:0) or pure fatty acid (R = 0:10). These oleogels showed the lowest hardness and stability against oil loss. They contained the biggest particles and the lowest SFC. The particles were too big to efficiently trap the air bubbles. Furthermore, the low quantity of crystals meant that an efficient gel network in the oil continuous phase could not be formed to help slow down the drainage. The bubbles' surface was not covered by the crystals. This enhanced foam destabilization and led to the instability of the resulting oil foams.

The highest overrun, foam firmness, and foam stability were obtained from oleogels with the highest SFC and lowest particles size, which resulted from mixed crystals (Callau, Sow-Kébé, Jenkins et al. 2020). Therefore, there was a direct link between the crystals, crystalline particle size, solid fat content, foamability, and foam firmness in this oleogel system (Figure 2.5.5). Oleogel properties such as hardness and stability against oil loss were correlated with their resulting foaming properties in terms of foamability, foam firmness, and stability. In other words, high foam stability was dictated by mixed crystals with a small particle size in high concentrations. Pure fatty alcohol or fatty acids were less efficient in terms of foam properties than mixed crystals obtained for the best weight ratio of the two fatty components between 7 (alcohol):3 (acid) and 8:2 (corresponding to molar ratios around 3), for which co-crystallisation occurs, forming mixed crystals of small size in high quantity. It is interesting to note here that these molar ratios are close to those reported 50 years ago by Shah et al. in mixed surfactant systems to obtain the most stable aqueous foams (Shah 1971). Co-crystallisation seems to be a promising route to enhance the properties and stability of oil foams.

2.5.4.4 Application of Oil Foams to Develop Food Products

The increased interest in oil foams based on oleogels can be explained by oil foams helping create food products with reduced fat content combined with new textures and sensorial properties (Gunes et al. 2017; Chisholm et al. 2018; Gehin-Delval et al. 2019). The industrial potential for oil foams in the food industry is also high due to other advantages such as long term stability even above room temperature and clean labelling (Heymans et al. 2018). Oil foams show very long term stability against the common foam destabilization mechanisms, which leads to an enhancement of food shelf-life helping to reduce food waste and keep food product quality during storage. To obtain oil foams, few or even no additives are necessary (Binks and Marinopoulos 2017). Few or no preservatives are needed in oil foams since no water is present inside the product, which drastically reduces microbial spoilage (Heymans et al. 2017).

Oil foams have been successfully used in sponge cakes, biscuits, and laminated pastry to replace part of the butter leading to less fat by volume in the final product as described in several recent patents (Chisholm et al. 2018; Gehin-Delval et al. 2019). Oil foams can also serve as a building block to design new types of edible emulsions such as air-in-oil-in-water (A/O/W) emulsions (Goibier et al. 2019). To obtain these A/O/W emulsions, an oil foam is produced with surfactant crystals based on monoglycerides in an aqueous continuous phase containing both an emulsifier to stabilize the oil droplets and a thickener to protect the droplets against creaming. The air bubbles become encapsulated within oil drops dispersed in water. Such emulsions may be useful in, for example, mayonnaise, wherein the oil phase is replaced by the oil foam thus reducing the overall fat content. In this example, monoglycerides crystalline particles are used, but we could expect that fatty acids or fatty alcohols crystalline particles could also lead easily to the formation of A/O/W emulsions. This example of A/O/W emulsions show new oil foam systems will emerge soon with a broader range of applications (Fameau and Binks 2021). However, all of these food products' examples have been made by using mono- or di- or triglycerides. No example of real food products based on oil foams made with fatty alcohol and fatty acid is available in the current literature. It is also important to notice that since their first use in molecular gastronomy ten years ago, oil foams are becoming more and more popular. Oil foams are easy to produce based on different vegetable oils with a kitchen whisk (Fameau 2021). Anyone can do it at home with no chemistry background.

Notice that the cosmetic and pharmaceutical industries are also interested in oil foams stabilized by crystalline particles based on fatty acids and fatty alcohols, since oil is one of the main components in many cosmetic and pharmaceutical products as well as the fatty acids and fatty alcohols (Fameau and Saint-Jalmes 2020; Callau et al. 2020). The application of oil foams is easier for the consumer than pure oil products for skin and hair, and provides simultaneously a new sensorial texture for the consumer with less greasy feel applications (Callau et al. 2020). Oil foams can also act as efficient carriers for drug delivery (Fameau and Saint-Jalmes 2020). Fatty acid crystalline particles can also stabilize another type of non-aqueous foams: alcohol-based foams (Fameau et al. 2021). Liquid foams based on ethanol or isopropanol are important nowadays since the global pandemic due to COVID-19 and the frequent use of alcohol-based hand sanitizers. Foam-type disinfectant formulations reduce dripping problems, runoff, and difficulty to cover the entire hand surface during their use (Fameau et al. 2021). Foams can be visually spread on the skin and are easier to handle and spread than traditional gels. Fatty acid crystalline particles can act as foam stabilizers in alcohol-based liquid foams in the same way than described for oil foams (Fameau et al. 2021). We can expect that crystalline particles based on fatty alcohols and fatty acids could also produce edible and drinking foams made with a liquid phase containing a high percentage of alcohol.

2.5.5 Conclusion and Perspectives

The mixture of fatty alcohol and fatty acid of the same chain length in an oil phase in a specific ratio drastically improves the oleogel properties compared to the corresponding pure fatty alcohol based oleogel or the pure fatty acid based oleogel. The link between the molecular scale to the macroscopic scale has been described in different studies. Using mixtures of fatty alcohols and fatty acids to produce oil foams lead also to an enhancement of the foaming properties. However, some scientific questions still need to be addressed to improve these oleogel and oil foam systems to better control their properties for potential industrial applications (Scharfe and Flöter 2020). The role of the oil phase polarity and viscosity still need to be studied carefully for oleogels based on the mixture of fatty alcohols and fatty acids, since they are parameters known to affect the gelation and crystallization behaviour of oleogel systems such as for monoglycerides based oleogels (Valoppi et al. 2017). Recent work on oleogels based γ-oryzanol and β-sitosterol have clearly shown the effect of minor polar compounds in oils by removing them carefully (Scharfe et al. 2019; Scharfe et al. 2021). The understanding how minor polar components effects the fatty alcohol/ fatty acid oleogelator system in terms of oleogel structures and properties and how they affect their foaming properties could be a topic of interest; it is one that still needs to be studied. Moreover, no description of oil foams based on fatty alcohol/fatty acid mixture has been published to produce real food products such as for monoglycerides, and no data about the sensorial and mouthfeel aspects of these systems is available. Data about the oxidative stability of the oil foams also needs to be obtained before going further with these mixed oleogelator systems in terms of potential applications in food, but also in other industries such as the cosmetic and pharmaceutical industries (Fameau and Saint-Jalmes 2020).

References

Binks, B.P., Garvey, E.J., and Vieira, J. (2016). Whipped oil stabilised by surfactant crystals. *Chemical Science* 7 (4): 2621–2632. doi: 10.1039/C6SC00046K.

Binks, B.P. and Marinopoulos, I. (2017). Ultra-stable self-foaming oils. *Food Research International* 95: 28–37.

Binks, B.P. and Vishal, B. (2021). Oil foams. *Advances in Colloid and Interface Science* 291: 102404. doi: 10.1016/j.cis.2021.102404.

Blach, C., Gravelle, A.J., Peyronel, F., Weiss, J., Barbut, S., and Marangoni, A.G. (2016). Revisiting the crystallization behavior of stearyl alcohol: Stearic acid (SO: SA) mixtures in edible oil. *Rsc Advances* 6 (84): 81151–81163. doi: 10.1039/C6RA15142F.

Blake, A.I. and Marangoni, A.G. (2015a). Plant wax crystals display platelet-like morphology. *Food Structure* 3: 30–34.

Blake, A.I. and Marangoni, A.G. (2015b). The use of cooling rate to engineer the microstructure and oil binding capacity of wax crystal networks. *Food Biophysics* 10 (4): 456–465.

Bond, A.D. (2004). On the crystal structures and melting point alternation of the n-alkyl carboxylic acids. *New Journal of Chemistry* 28 (1): 104–114. doi: 10.1039/B307208H.

Bot, A. and Flöter, E. (2018). Structuring edible oil phases with fatty acids and alcohols. In: *Edible Oil Structuring* (ed. A.G. Marangoni and N. Garti), 95–105. San Diego: AOC Press.

Callau, M., Jenkins, N., Sow-Kébé, K., Levivier, C., and Fameau, A.L. (2021). The effect of vegetable oil composition on the structural properties of oleogels based on behenyl alcohol/behenic acid oleogelator system. *Journal of Cosmetic Science* 72 (4): 399–405.

Callau, M., Sow-Kébé, K., Jenkins, N., and Fameau, A.L. (2020). Effect of the ratio between fatty alcohol and fatty acid on foaming properties of whipped oleogels. *Food Chemistry* 333: 127403. doi: 10.1016/j.foodchem.2020.127403.

Callau, M., Sow-Kébé, K., Nicolas-Morgantini, L., Fameau, A.L. (2020). Effect of the ratio between behenyl alcohol and behenic acid on the oleogel properties. *Journal of Colloid and Interface Science* 560: 874–884.

Chisholm, H., Gunes, Z.D., Gehin-Delval, C., Nouzille, C.A., Garvey, E., Destribats, M.J., Chandrasekaran, S.N., Vieira, J.B., German, J., and Binks, P.B. (2018) *Aerated confectionery material.*

Co, E.D. and Marangoni, A.G. (2012). Organogels: an alternative edible oil-structuring method. *Journal of the American Oil Chemists' Society* 749–780. doi: 10.1007/s11746-012-2049-3.

Cui, L., Jia, Y., Cheng, Z.W., Gao, Y., Zhang, G.L., and He, C.F. (2016). Advancements in the maintenance of skin barrier/skin lipid composition and the involvement of metabolic enzymes. *Journal of Cosmetic Dermatology* 15 (4): 549–558.

Daniel, J. and Rajasekharan, R. (2003). Organogelation of plant oils and hydrocarbons by long-chain saturated FA, fatty alcohols, wax esters, and dicarboxylic acids. *Journal of the American Oil Chemists' Society* 80 (5): 417–421.

de Oliveira, T.E., Leonforte, F., Nicolas-Morgantini, L., Fameau, A.L, Querleux, B., Thalmann, B., and Marques, C.M. (2019). *A fluid bilayer phase in aqueous mixtures of fatty alcohol and cationic surfactant. arXiv.*

Drenckhan, W. and Saint-Jalmes, A. (2015). The science of foaming. *Advances in Colloid and Interface Science* 222: 228–259.

Eini, M. and Tamarkin, D. (2001). Thickened oil compositions of edible oil. *World Patent* WO2002102169A3.

Fameau, A.-L., Lam, S., Arnould, A., Gaillard, C., Velev, O.D., and Saint-James, A. (2015). Smart nonaqueous foams from lipid-based oleogel. *Langmuir* 31 (50): 13501–13510.

Fameau, A.-L. (2021). Foams: pickering edible oil foam–toward new food products. In: *Handbook of Molecular Gastronomy* (ed. R. Burke et al.), 357–364. New York: CRC Press. doi: 10.1201/9780429168703.

Fameau, A.-L., Ma, Y., Siebenbuerger, M., and Bharti, B. (2021). Foamitizer: High ethanol content foams using fatty acid crystalline particles. *Journal of Colloid and Interface Science* 600: 882–886. doi: 10.1016/j.jcis.2021.05.076.

Fameau, A.-L., Arnould, A., and Saint-Jalmes, A. (2014). Responsive self-assemblies based on fatty acids. *Current Opinion in Colloid and Interface Science* 19: 5. doi: 10.1016/j.cocis.2014.08.005.

Fameau, A.-L. and Binks, B.P. (2021). Aqueous and oil foams stabilized by surfactant crystals: New concepts and perspectives. *Langmuir* 37 (15): 4411–4418. doi: 10.1021/acs.langmuir.1c00410.

Fameau, A.-L. and Fujii, S. (2020). Stimuli-Responsive liquid foams: From design to applications. *Current Opinion in Colloid & Interface Science* 50: 101380. doi: 10.1016/j.cocis.2020.08.005.

Fameau, A.-L. and Rogers, M.A. (2020). The curious case of 12-hydroxystearic acid–the Dr. Jekyll & Mr. Hyde of molecular gelators. *Current Opinion in Colloid & Interface Science* 45: 68–82.

Fameau, A.-L. and Saint-Jalmes, A. (2017). Non-aqueous foams: Current understanding on the formation and stability mechanisms. *Advances in Colloid and Interface Science* 247: 454–464. doi: 10.1016/j.cis.2017.02.007.

Fameau, A.-L. and Saint-Jalmes, A. (2020). Recent advances in understanding and use of oleofoams. *Frontiers in Sustainable Food Systems* 4: 110. doi: 10.3389/fsufs.2020.00110.

Gandolfo, F.G., Bot, A., and Flöter, E. (2004). Structuring of edible oils by long-chain FA, fatty alcohols, and their mixtures. *Journal of the American Oil Chemists' Society* 81 (1): 1–6.

Garside, J. (1987). General principles of crystallization. In: *Food Structure and Behaviour* (ed. J.M.V. Blanshard and P. Lillford). London: Academic.

Gehin-Delval, C., Chisholm, H., Chung, W., Deyber, H., Destribats, M.J., Gunes, Z.D., and Pelloux, C. (2019) *Method for forming a laminated pastry*.

Goibier, L., Pillement, C., Monteil, J., Faure, C., and Leal-Calderon, F. (2019). Emulsification of non-aqueous foams stabilized by fat crystals: Towards novel air-in-oil-in-water food colloids. *Food Chemistry* 293: 49–56.

Gravelle, A.J., Blach, C., Weiss, J., Barbut, A., and Marangoni, A.G. (2017). Structure and properties of an ethylcellulose and stearyl alcohol/stearic acid (EC/SO: SA) hybrid oleogelator system. *European Journal of Lipid Science and Technology* 119 (11): 1700069.

Gravelle, A.J., Davidovich-Pinhas, M., Barbus, S., and Marangoni, A.G. (2017). Influencing the crystallization behavior of binary mixtures of stearyl alcohol and stearic acid (SOSA) using ethylcellulose. *Food Research International* 91: 1–10.

Gunes, D.Z., Murith, M, Godefroid, J., Pelloux, C., Dyber, H., Schafer, O., and Breton, O. (2017). Oleofoams: properties of crystal-coated bubbles from whipped oleogels–evidence for pickering stabilization. *Langmuir* 33 (6): 1563–1575.

Heymans, R., Tavernier, I., Dewettinck, K., and Van der Meeren, P. (2017). Crystal stabilization of edible oil foams. *Trends in Food Science & Technology* 69: 13–24.

Heymans, R., Tavernier, I., Danthine, S., Rimaux, T., Van der Meeren, P., and Dewettinck, K. (2018). Food-grade monoglyceride oil foams: the effect of tempering on foamability, foam stability and rheological properties. *Food & Function* 9 (6): 3143–3154. doi: 10.1039/C8FO00536B.

Johansson, I. and Svensson, M. (2001). Surfactants based on fatty acids and other natural hydrophobes. *Current Opinion in Colloid & Interface Science* 6 (2): 178–188. doi: 10.1016/S1359-0294(01)00076-0.

Kaneko, F. (2001). Polymorphism and phase transitions of fatty acids and acylglycerols. In: *Crystallization Processes in Fats and Lipid Systems* (ed. N. Garti and K. Sato), 67–112. CRC Press.

Lee, K.-R., Chen, G.Q., and Kim, H.U. (2015). Current progress towards the metabolic engineering of plant seed oil for hydroxy fatty acids production. *Plant Cell Reports* 34 (4): 603–615.

Li, X., Teitgen, A.M., Shirani, A., Ling, J., Busta, L., Cahoon, R.E., Zhang, W., Li, Z., Chapman, K.D., Berman, D., Zhang, C., Minto, R.E., and Cahoon, E.B. (2018). Discontinuous fatty acid elongation yields hydroxylated seed oil with improved function. *Nature Plants* 4 (9): 711–720.

Marangoni, A.G. (2004). *Fat Crystal Networks*. CRC Press.

Marangoni, A.G., Acevedo, N., Maleky, F., Co, E., Peyronel, F., Mazzanti, G., Quinn, B., and Pink, D. (2012). Structure and functionality of edible fats. *Soft Matter* 8 (5): 1275–1300. doi: 10.1039/C1SM06234D.

Marangoni, A.G., Van Duynhoven, J.P.M., Acevedo, N.C., Nicholson, R.A., and Patel, A.R. (2020). Advances in our understanding of the structure and functionality of edible fats and fat mimetics. *Soft Matter* 16 (2): 289–306.

Marangoni, A.G. and Wesdorp, L.H. (2013). Viscoelastic properties of fats. In: *Structure and Properties of Fat Crystal Networks*, 2e (ed. A.G. Marangoni and L.H. Wesdorp). London: CRC press, Taylor and francis Group.

Marangoni, A.G. and Garti, N. (2018). *Edible Oleogels: Structure and Health Implications* (ed. A.G. Marangoni and N. Garti). San Diego: AOC Press.

Miklaszewska, M., Banaś, A., and Królicka, A. (2017). Metabolic engineering of fatty alcohol production in transgenic hairy roots of Crambe abyssinica. *Biotechnology and Bioengineering* 114 (6): 1275–1282.

Mishima, S., Suzuki, A., Sato, K., and Ueno, S. (2016). Formation and microstructures of whipped oils composed of vegetable oils and high-melting fat crystals. *Journal of the American Oil Chemists' Society* 93 (11): 1453–1466. doi: 10.1007/s11746-016-2888-4.

Mishra, K., Dufour, D., and Windhab, E.J. (2020). Yield stress dependent foaming of edible crystal-melt suspensions. *Crystal Growth & Design* 20 (2): 1292–1301.

Miyazaki, Y. and Marangoni, A.G. (2014). Structural-mechanical model of wax crystal networks—a mesoscale cellular solid approach. *Materials Research Express* 1 (2): 25101.

Monick, J.A. (1979). Fatty alcohols. *Journal of the American Oil Chemists' Society* 56 (11Part2): 853A–860A.

Noweck, K. and Grafahrend, W. (2006). Fatty alcohols. In: *Ullmann's Encyclopedia of Industrial Chemistry*, vol. 14, 117–141, Hoboken, NJ Book: Wiley-VCH Verlag GmbH & Co.

Patel, A.R., Nicholson, R.A., and Marangoni, A.G. (2020). Applications of fat mimetics for the replacement of saturated and hydrogenated fat in food products. *Current Opinion in Food Science* 33: 61–68.

Patist, A., Devi, S., and Shah, D.O. (1999). Importance of 1: 3 molecular ratio on the interfacial properties of mixed surfactant systems. *Langmuir* 15 (21): 7403–7405.

Peyronel, F., Pink, D.A., and Marangoni, A.G. (2014). Triglyceride nanocrystal aggregation into polycrystalline colloidal networks: Ultra-small angle X-ray scattering, models and computer simulation. *Current Opinion in Colloid & Interface Science* 19 (5): 459–470.

Raut, J.S., Naik, V.M., Singhal, S., and Juvekar, J.A. (2008). Soap: The polymorphic genie of hierarchically structured soft condensed-matter products. *Industrial & Engineering Chemistry Research* 47 (17): 6347–6353. doi: 10.1021/ie0714753.

Saha, S., Saint-Michel, B., Leynes, B., Binks, B.P., and Garbin, V. (2020). Stability of bubbles in wax-based oleofoams: decoupling the effects of bulk oleogel rheology and interfacial rheology. *Rheologica Acta* 59: 255–266.

Sato, K. (2018). Polymorphism of lipid crystals. In: *Crystallization of Lipids* (ed. K. Sato), 17–60. Chichester, UK: John Wiley & Sons, Ltd.

Schaink, H.M., van Malssen, K., Morgado-Alves, S., Kalnin, D.J.E., and van der Linden, E. (2007). Crystal network for edible oil organogels: possibilities and limitations of the fatty acid and fatty alcohol systems. *Food Research International* 40 (9): 1185–1193.

Scharfe, M., Ahmane, Y., Seilert, J., Keim, J., and Flöter, E. (2019). On the effect of minor oil components on β-sitosterol/γ-oryzanol oleogels. *European Journal of Lipid Science and Technology* 121 (8): 1800487.

Scharfe, M. and Flöter, E. (2020). Oleogelation: From scientific feasibility to applicability in food products. *European Journal of Lipid Science and Technology* 122 (12): 2000213.

Scharfe, M., Prange, D., and Flöter, E. (2021). The composition of edible oils modifies β-sitosterol/γ-oryzanol oleogels. Part II: Addition of selected minor oil components. *Journal of the American Oil Chemists' Society* 99 (1): 57–77.

Shah, D.O. (1971). Significance of the 1: 3 molecular ratio in mixed surfactant systems. *Journal of Colloid and Interface Science* 37 (4): 744–752.

Silva, L.F.L., Paschoal Jr., W., Pinheiro, G.S., da Silva Filho, J.G., Freire, P.T., de Sousa, F.F., and Moreira, S.G.C. (2019). Understanding the effect of solvent polarity on the polymorphism of octadecanoic acid through spectroscopic techniques and DFT calculations. *CrystEngComm* 21 (2): 297–309. doi: 10.1039/C8CE01402G.

Terech, P. and Weiss, R.G. (1997). Low molecular mass gelators of organic liquids and the properties of their gels. *Chemical Reviews* 97 (8): 3133–3160.

Toedt, J., Koza, D., and Van Cleef-Toedt, K. (2005). *Chemical Composition of Everyday Products*. Greenwood Publishing Group.

Valoppi, F., Calligaris, S., and Marangoni, A.G. (2016). Phase transition and polymorphic behavior of binary systems containing fatty alcohols and peanut oil. *Crystal Growth & Design* 16 (8): 4209–4215.

Valoppi, F., Calligaris, S., and Marangoni, A.G. (2017). Structure and physical properties of oleogels containing peanut oil and saturated fatty alcohols. *European Journal of Lipid Science and Technology* 119 (5): 1600252.

Valoppi, F., Calligaris, S., and Marangoni, A.G. (2018). Stearyl alcohol oleogels. In: *Edible Oleogels. Structure and Health Implications*, 2e (ed. A.G. Marangoni and N. Garti), 219–234. Elsevier.

Wolfrum, S., Marcus, J., Touraud, D., and Kunz, W. (2016). A renaissance of soaps?—How to make clear and stable solutions at neutral pH and room temperature. *Advances in Colloid and Interface Science* 236: 28–42. doi: 10.1016/j.cis.2016.07.002.

2.6

Proteins as Fat Replacers in the Food Industry

Davanam Srikanth, Dharani Gopi*, Sunil, C. K., Karunairaj Michael, and Ashish Rawson***

National Institute of Food Technology Entrepreneurship and Management, Thanjavur, Tamil Nadu, India
* *Equal Contribution*
** *Corresponding author*

2.6.1 Introduction

Carbohydrates, proteins, and fats are essential macronutrients in the human diet. Carbohydrates and fats are major energy sources for humans, but proteins, vitamins, and minerals are mainly responsible for the human body's growth and maintenance. Fats provide a high amount of energy contributing about 9.3 kcal/g compared to proteins and carbohydrates, which give around 4.1 kcal/g of energy (Akoh 1998; Sons 2005). Fats also help with transporting fat-soluble vitamins A, D, E, and K and are an important source of essential fatty acids like omega-3 and omega-6 fatty acids (Akoh 1998). These fatty acids are building blocks of lipids such as triacylglycerols (TAGs), phospholipids, glycolipids, etc. (Sons 2005). Fats can be derived from both animal and plant sources. Saturated fats are generally from animal sources, but most unsaturated fats are from plant sources (Sons 2005). Fats are responsible for smell, taste, texture, aroma, and mouthfeel, which are the primary sensory attributes of any food (Akoh 1998). Fats contribute to the texture, appearance, palatability, lubricity, creaminess, consistency (Grossklaus 1996) of foods and it increases the feeling of fullness while having food (Patel et al. 2020). Many attributes are associated with fat in food. Some are mentioned in Table 2.6.1.

Presence of excessive fat in the diet has been reported to be responsible for obesity, type-II diabetes, high blood cholesterol, cancer (like breast and colon cancer) and some cardiovascular-related diseases (Yashini et al. 2021) which is why reducing fat intake in the human diet is the major concern for many people. The guidelines for a healthy human diet to avoid excessive weight gain suggests limiting total intake of fat to less than 30%, saturated fats to less than 10%, and trans-fats to less than 1% of total energy intake (calories) with a shift in consumption of saturated fats and trans-fats to unsaturated fats (Brennan and Tudorica 2008). According to American Heart Association (AHA), those with elevated LDL cholesterol levels or cardiovascular diseases should restrict saturated fats to less than 7% of total calories (Ognean et al. 2006). Though health-conscious people are consuming a reduced-fat diet by changing their dietary habits, most are not ready to compromise on the flavor and taste given by fats to the food (Ognean et al. 2006). Fat contributes its own flavor (like creamy, nutty, buttery, oily) to the food and is also responsible for the intensity, duration, and balance of other flavors in the food. Fat can slow down the release of other flavors in the food (Lucca and Tepper 1994). When a full-fat food is chewed it results in the release of fat-soluble flavorants. Fat present in the food disturbs the time of the release of these flavorants. So, the perception of saltiness, bitterness, sweetness, acidity becomes more apparent by the reduction or removal of fat in the food. Since these fat-soluble flavorants are more stable in aqueous and

Fat Mimetics for Food Applications, First Edition. Edited by Miguel Ângelo Parente Ribeiro Cerqueira and Lorenzo Miguel Pastrana Castro.

Table 2.6.1 Attributes associated with fat for several categories of food, adapted from Lucca and Tepper (1994), p. 13.

Types of products	Attributes associated with fats	Examples
Baked foods	Flavor, Viscosity/body, Richness, Texture, Aeration, Shortness, Tenderness, Elasticity, Flakiness, Leavening, Lubricity, Dough handling, Moisture retention	Breads, cookies, cakes, muffins, pizza, snack cakes, doughnuts, Danish pastry, biscuits, buns, sweet rolls.
Dairy products	Flavor, Creaminess, Saltiness, Acidity, Viscosity/body, yield stress, Smoothness, Mouthfeel, Melt, Emulsion, Aeration, Color, Sheen, Opacity	Milk, butter, cheese, yogurt, cream, ice cream, whey, casein.
Salad dressings	Flavor, Viscosity/body, yield stress, Saltiness, Acidity, Smoothness, Aeration, Mouthfeel, Shear sensitivity, Color, Opacity, Cold storage, Stability	Vegetable salad dressings, Greek yogurt ranch salad dressing.
Frozen deserts	Flavor, Viscosity/body, Creaminess, Mouthfeel, Melt, Heat-shock stability, Overrun, Opacity	Frozen mojito cake-tails, pomegranate sorbet, avocado ice pops, coffee ice cream pie.
Cream soup, Sauces and Gravies	Mouthfeel, Opacity, Flavor, Saltiness, Satiety	Tomato sauce, peanut sauce, Schewan sauce, Asian black bean sauce.
Fried foods	Flavor, Heat transfer, Crispness, Aroma, Color	French fries, chicken strips, cheese sticks.
Meat	Flavor, Mouthfeel, Juiciness, Tenderness, Texture, Bite, Handling, Heat transfer, Satiety	Chicken, beef, pork, fish, lamb.
Frostings and fillings	Viscosity/body, Mouthfeel, Creaminess, Aeration, Spread, Emulsion, Smoothness, Melt	Buttercream frosting, Fudge frosting, chocolate filling.

hydrophilic environments, the creation of this flavor system in reduced-fat foods similar to that in high-fat foods is a great challenge (Lucca and Tepper 1994). Fat replacers are gaining importance and there is an increasing demand for them in the current market. Food companies are also aiming at producing no-or-low-fat foods by replacing the fat with fat substitutes and fat mimetics (Lucca and Tepper 1994). Substituting water or air in place of fat, using skim milk instead of whole milk in frozen desserts, using lean meats in frozen entrees, baking instead of frying in snack foods are some of the conventional methods used to replace fats (Ognean et al. 2006). Mainly, there are two groups of fat replacers: (i) Fat Substitutes (ii) Fat Mimetics (Akoh 1998). Fat substitutes are often called lipid or fat-based replacers, but fat mimetics are often called protein-based or carbohydrate-based fat replacers (Yashini et al. 2021). Fat substitutes are energy-free macromolecules with the physical structure and cooking properties like fats. They are chemically synthesized or are derived from conventional oils and fats by enzymatic change (Ognean et al. 2006). They provide no calories to the diet. These fat substitutes are used for foods involved in cooking and frying because these are stable at those temperatures. While Fat Mimetics are energy-reduced fat replacers, their chemical structure differs completely from fats and they cannot substitute fats on a one-to-one basis. These fat mimetics are used in baking and retorting (Yashini et al. 2021). This chapter mainly focuses on protein-based fat mimetics. Topics like sources of protein-based fat mimetics, their applications, factors affecting their acceptability, and the future of protein-based fat mimetics are also discussed in this chapter.

Table 2.6.2 Examples of fat replacers (Yashini et al. 2021, p. 4).

Source of milk based fat replacers	Name of the fat replacer	Extraction method
Milk protein	Complete Milk Protein (CMP-I) Lactalbumin 75 Lactalbumin 70 SGH Lactomil HF	Ultrafiltration & subsequent diafiltration Membrane filtration Membrane filtration Membrane filtration
Whey protein	Dairy-Lo Dairy light Simplesse 100 Simplesse 300 Simplesse D-100 Simplesse D-500 Simplesse Bakery Blend 730 Simplesse Bakery Blend 710 Simplesse Bakery Blend 720 Simplesse D-550 PALs (ALACO) K-Blazer Leancreme™ Optipep™ Nutrilac	Partial denaturation and & heat treatment of whey protein Heating and homogenizing of whey protein
Egg white	Trailblazer range	Biphasic method/Salting out method
Gelatin	Slimgel 100 Slimgel 100/i Slimgel 200 Supercreme C Supercreme SD	Conditioning followed by extraction, filtration and concentration and subsequent drying and grinding

2.6.2 Fat Mimetics

Fat mimetics help replace fat in terms of mouthfeel, viscosity, and appearance, but they cannot provide the same energy as fats. They do not have the non-polar chemical properties of fats, so they are not used for frying because of water concentration and heat liability (Akoh 1998). From a nutritional perspective, fat mimetics should mimic features of fat and it should give a similar mouthfeel, texture, and energy to the foods. Also, from an ecological viewpoint, they should be biodegradable (Grossklaus 1996) and should cause no deteriorating effects on consumer health. Considering all these factors, developing a fat mimetic can be challenging. Fat mimetics can be categorized into three types based on their source: (i) protein-based, (ii) carbohydrate/starch-based, and (iii) cellulose-based (Figure 2.6.1).

Protein-based fat mimetics are structurally modified natural proteins made from proteins like casein, egg white protein, wheat gluten protein, soy protein, etc. Simplesse D-100, Trialblazer, Slimgel, and Supercreme C are some of the product names of protein-based fat mimetics (Yashini

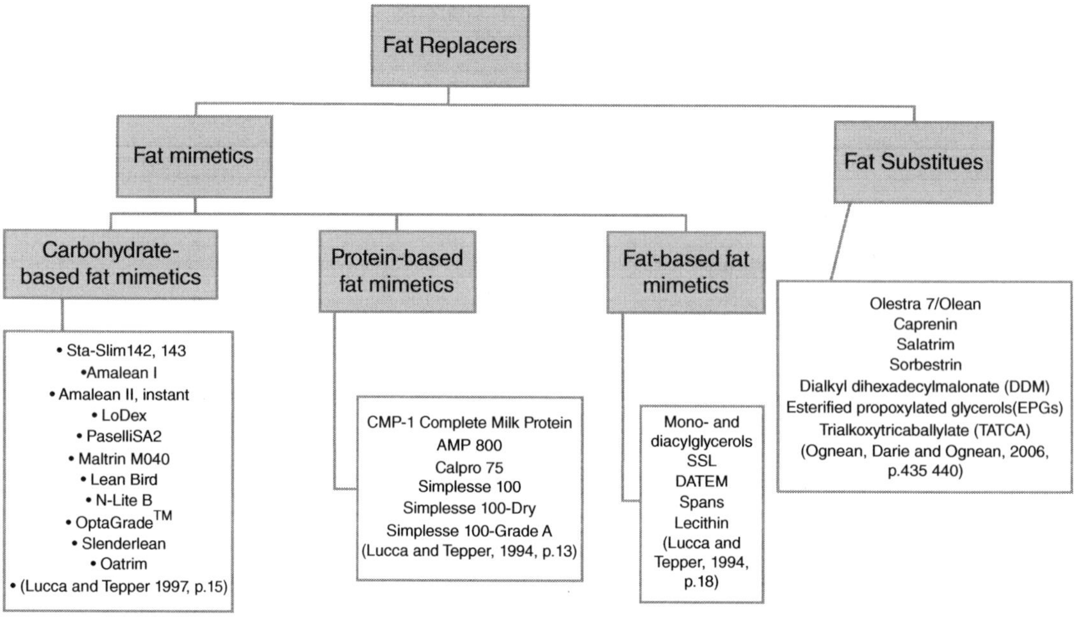

Figure 2.6.1 Classification of fat replacers.

Table 2.6.3 Commercial plant-protein-based fat replacers and their applications (Yashini et al. 2021, p. 14).

Source	Product name	Type/form	Applications
Soy protein	–	Soy protein concentrate	Meat and dairy products, sour cream, soup bases, fillings, cream cheese, toppings
	Supro range	Isolated soy proteins	Meat products, coffee creamers, pasta, beverages, food bars, non-dairy products
	Ultra-Freeze 500	Modified starch, soya Protein	Dairy products
Corn zein protein	Lita A Lita C Lita D Lita KC	Microparticulated Zein protein	Mayonnaise, salad dressing, sour cream, topping, frosting, soups, sauces, cream or cottage cheese, dips, meat products, confectionery
Pea protein	–	Pea protein, extracted using water and an ultrafiltration process	Meat products, sauces, baked products, dressings
Vegetable protein	UltraBake NF	Modified vegetable protein, Granular starch, xanthan Gum	Bakery products
	Ultra-Freeze 400	Vegetable protein, modified starch	Sherbets, cheese sauces, frozen products such as yogurt and waffles mixes

et al. 2021). Carbohydrate/starch-based fat mimetics are mainly extracted from starch derivatives like corn, potato, tapioca, maltodextrins, polydextrose, gums, and fibers (Peng and Yao 2017). These are formulated from GRAS oligo or polysaccharides. They taste bland and are digested and absorbed to some extent by giving about 4.5kcal/g of carbohydrate (Akoh 1998). These are more cost-effective than protein-based fat mimetics and are generally used in dairy products in the form of dietary fibers (DF). Charles et al. studied the carbohydrate-based fat replacers such as beta-glucan, partially hydrolyzed guar gum, inulin in low-fat yogurt production (Brennan and Tudorica 2008). They stabilize water in a gel-like matrix resulting in lubrication and flow properties like fats. The major drawback of fat replacers is that they are highly associated with a water content which increases in water activity thus reducing the shelf life of the product (Lucca and Tepper 1994). Cellulose-based fat mimetics contribute no calories to the diet. One such example is Avicel, which is a blend of microcrystalline cellulose (MCC) and carboxy-methyl cellulose (CMC). They generally give creamy, mouthful, and little taste to foods. These fat mimetics can be used uniquely or in different combinations to improve the flavor of the food (Akoh 1998). Gibis et al. (Gibis et al. 2015) studied the effect of CMC and MCC as fat replacers on the microstructure and sensory features of fried beef patties. Ana Salvador et al. (Martínez-Cervera et al. 2015) used cellulose-based emulsions as fat replacers in muffins and they studied the rheological, thermal, and textural properties of muffins with fat replacers. They reported that cellulose-based emulsions can reduce 49% of fat content in muffins, but the sensory acceptability of these muffins was slightly lower than normal muffins. Of all the fat replacers discussed, protein-based fat replacers had the highest success rate as a fat replacer and proved nutritionally-beneficial, which means it is good to include high protein food in our diet.

2.6.3 Protein-Based Fat Mimetics

Proteins comprise amino acids linked with peptide bonds. These proteins are responsible for functional properties like water absorption and retention, solubility, viscosity, texture, and gelation in foods. Because of these properties, proteins can be a replacement for fats in food. Protein-based fat mimetics have a positive impact on low-fat foods; however, they have limited applications in the food industry because of their denaturation and coagulation properties at high temperatures. So protein-based fat mimetic food processing is limited to baking and retorting (Ognean et al. 2006). Proteins are generally classified as animal proteins and plant proteins. Major sources of proteins (Fig 2.6.2) and their contributions to total protein production across the world are discussed.

Animal proteins have all essential amino acids so they are generally termed as complete proteins (table 2.6.2). They have good digestibility and higher biological value than plant proteins. Milk proteins like casein and whey protein, egg white protein, collagen protein, and plasma proteins are being used as animal protein-based fat replacers. Plant proteins are generally derived from sources like cereals, pulses, vegetables, and fruits and they have a major role in substituting for carbohydrates and fats. Plant proteins like soy protein, corn zein protein, pea protein, and other vegetable proteins find their applications as protein-based fat replacers in low-fat foods. Most plant proteins are considered incomplete proteins because they have lower amounts of amino acids groups such as methionine, lysine, threonine, and sulfur compounds compared to animal proteins. So, a mixture of plant protein and animal protein is a good source of protein-rich food for human consumption. Proteins require some special physical-chemical properties such as isoelectric precipitation

(Akbari et al. 2019), protein denaturation and coagulation, protein-carbohydrate complex formation, protein-protein complex formation, hydrophobic effects, protein-stabilized emulsions, and protein-carbohydrate stabilized emulsion to use as fat replacers. Usage of protein-based fat replacers increases the total amount of protein in food. Lowering cholesterol, fat content, and calorie density, and improving nutritional and functional properties are major advantages of protein-based fat replacers over other types of fat replacers (Yashini et al. 2021). Because of all these factors, there is an increasing demand for protein-based fat replacers in the current market.

2.6.3.1 Animal Protein-Based Fat Replacers (Table 2.6.2)

2.6.3.1.1 Casein

A major portion of milk protein (around 80%) has casein. It can be a fat replacer in milk products such as ice creams, dips, spreads, cheese, yogurts, etc. Casein can be extracted by bringing the pH of the milk to the isoelectric point of casein (pH = 4.6) and subsequent protein coagulation. When casein was a fat substitute, sensory qualities such as mouth-coating, coolness, smoothness, firmness, melting rate, and overall meal satisfaction increased. In addition, the product's viscosity is raised. Milk Protein Concentrate (MPC) has both casein and whey proteins in the same proportions as in milk. It is the most widely used milk protein-based fat replacer in dairy products such as ice creams, cheese and yogurts. Peter Eisner et al. (Schädle et al. 2020) studied the effect of a different combination of fat replacers (inulin, corn dextrin, polydextrose, and microparticulated whey protein) with rennet casein in low-fat processed cheese. They concluded that higher concentrations of inulin or corn dextrin can be used for flowability without affecting the properties of unheated low-fat cheese samples. The following is a list of fat replacers made from milk protein.

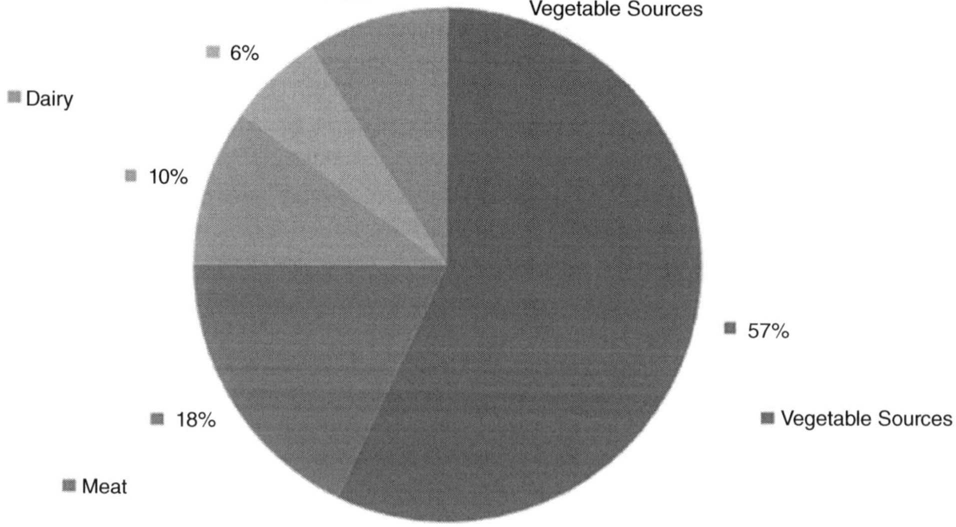

Figure 2.6.2 Major sources of protein and their contribution in total protein production. Data from Henchion, M. et al. 2017).

2.6.3.1.2 Whey Protein

Whey protein contributes about 20% of the total protein in milk. The dairy industry produces whey protein as a byproduct during cheese or yogurt production in liquid form. This liquid undergoes several processing steps to increase the protein content in it. When enough protein concentration is obtained, the excess liquid can be drained off to obtained Whey Protein Concentrate (WPC). WPC contains around 80% of protein and some amounts of carbohydrates and fats. When this WPC is further processed to increase protein concentration to around 90% by removing carbohydrates and fats in it we will get Whey Protein Isolate (WPI). WPC and WPI both can be used as fat replacers in the food industry. When heat is applied to whey protein, it begins to de-naturize. This happens in two steps. (i) unfolding (ii) aggregation (Yashini et al. 2021). Azizinia et al. (Aziznia et al. 2008) studied the effect of WPC and gum tragacanth (GT) as fat replacers on the chemical, physical, microstructural properties of non-fat yogurts. They concluded that (i) WPC yogurts had a more compact structure, better hardness, and less water drainage, and (ii) WPC as a fat replacer improves the physical and microstructural qualities of non-fat yogurts. Sandoval-Castilla et al. (Sandoval-Castilla et al. 2004) looked at the microstructure and texture of yogurt as a function of fat replacers in another study. They made reduced-fat yogurts by combining reconstituted milk (15 g of fat and 120 g of total solids per liter of milk) with fat replacers such Whey Protein Concentrate (WPC), Microparticulated Whey Protein (MWP), and Modified Tapioca Starch (MTS), both separately and together. They found that yogurt containing WPC and MWP as fat replacers had the same textural qualities as full-fat yogurt (FFY), but yogurt containing MWP had lower tension and stiffness, but higher cohesiveness. Kun Liu et al. studied the impact of MWP on the sensory qualities of liquid and semi-solid model foods (Liu et al. 2016). Razavi et al. (Poursani et al. 2021) investigated how fat replacers affected the rheological, physical, and sensory aspects of non-fat ice cream. Because of their ability to interact with water molecules and milk proteins, they determined that BSSG (Balangu Shirazi Seed Gum) and WPC might imitate the features of low-fat ice creams.

2.6.3.1.3 Microparticulated Whey Protein

Microparticulation is producing larger amounts of reduced protein particles of the size and shape to give a mouthfeel like fat (O'Brien et al. 2003). MWP can be produced by heat coagulation of whey proteins. These MWP proteins are of spherical shape with a diameter of around 0.1–0.3μm. Because of its superior emulsifying, heat stabilizing, and gelation qualities, MWP has a wider range of applications as a fat replacer in the dairy industry, particularly in manufacturing ice creams, cheese, and yogurt (Yashini et al. 2021). The principle behind MWP as a fat replacer is due to its lubrication effects. It reduces friction coefficient in water and oil-in-water emulsions due to the ball-bearing mechanism (Liu et al. 2016) *i.e.*, they act as ball bearings at the point of contact and reduce the friction coefficient. This friction reduction coefficient is not only due to the increase in viscosity but also due to ball-bearing lubrication, especially in liquids. Kun Liu et al. (Liu et al. 2016) also suggest that other food colloidal microparticles having similar ball-bearing lubrication effects can also be used as fat replacers. Micro fluidization and superfine grinding are two methods to produce food microparticles from MWP. The superfine grinding process reduces particles such that those particles enable oil extraction from particles to enhance solubility, oil-binding ability, surface hydrophobicity, and foam stability and capacity. The size of MWP particles is reduced from 0.062 mm and 0.001 mm to 0.015 mm and 0.009 mm in superfine grinding. Superfine MWP particles exhibit properties that almost are similar to fats, so they can be used as effective fat replacers because of improved creaminess, viscosity, mouthfeel, and lowered water holding capacity (Yashini et al. 2021). Sun et al. (Sun et al. 2018) looked at how fat mimetics based on MWP and high-methoxy pectin affect the caloric value, stability, and rheological aspects of mayonnaise. They found that adding fat mimetics to the mix reduced pseudoplastic behavior, viscosity index, relative thixotropic area, and a gradual increase in the elasticity index.

2.6.3.1.4 Simplesse & Dairy-Lo

Simplesse is a fat replacer that can be prepared by an egg protein and whey protein mix. It is available in many forms including Simplesse 100, Simplesse 300, Simplesse D-100, Simplesse D-500, Simplesse Bakery Blend 710, Simplesse Bakery Blend 720, Simplesse Bakery Blend 730 with whey as a main ingredient. Simplesse is made from whey protein heated and homogenized. When the procedure ended, an insoluble protein's microparticle structure is obtained. The radius of disulfide cross-linked protein spherical aggregates is roughly 0.065nm. Simplesse's ideal particle size ranges from 0.1 to 3 μm. Simplesse is the most widely used fat substitute in dairy products, meat products, bakery foods, and oil-based items such as salad dressings, mayonnaise, and margarine. (Stargel et al. 1991). However, Simplesse weakens the gluten network in wheat bread resulting in a more porous structure unable to keep gas under stress (O'Brien et al. 2003). Dairy-Lo is an animal protein-based fat replacer that is made from partially denaturated MWP. Since the protein is partially denatured, the water binding and emulsification properties of MWP will not be lost. Protein unfolding and self-aggregation are the first steps in partial denaturation. Whey protein concentrate is ultrafiltered at a pH of 6.1 and mildly heated (80°C for 17 sec) to achieve 60–80% denaturation in Dairy-Lo. Because it is not microparticulated, it keeps the majority of the features of whey protein. To increase moisture content and improve the functional qualities of Mozzarella cheese, McMahon et al. used two protein-based fat replacers (Dairy-Lo and Simplesse D100) and two carbohydrate-based fat replacers (Stellar 100X and Novagel RCN-15).(McMahon et al. 1996) and they saw that cheese with Novagel RCN-15 fat replacer showed the highest moisture content.

2.6.3.1.5 Egg White Protein

The translucent liquid that forms surrounding fertilized or unfertilized egg yolks is known as egg white. Egg whites are 90% water and 10% of dissolved proteins (including albumins, mucoproteins, and globulins). Egg whites have around 56% of the total protein in the egg. Ovalbumin (54%), Ovomucoid (11%), Ovatransferrin (12%), Ovamucin (3.5%), Lysozyme (3.5%) are dominant proteins in egg white. Egg white protein can be typically extracted by size exclusion liquid chromatography, ion-exchange chromatography, ultrafiltration, electrophoresis, adsorption techniques, etc. (Pereira et al. 2016). Heat treatment at varying temperatures causes protein denaturation, which affects functional qualities such as heat stability, solubility, foaming ability, emulsifying ability. Sala et al. (Rovers et al. 2016) studied the potentia of dispersed microbubbles to act as fat replacers and studied their effect on rheological, tribological, and sensorial properties of model food systems. It was proven to be an effective fat replacer in model food systems with and without thickeners. They found that microbubble dispersions with microbubbles up to a volume fraction of 5% and a size of about 2 m have lower friction and mouthfeel than emulsions with similar emulsion droplet volume fractions and sizes; they further found that the sensory properties of microbubble dispersions with microbubbles up to a volume fraction of 5% and a size of about 2 m are not comparable with those emulsions with a similar emulsion droplet volume. Trailblazer is a protein-based fat mimicker with micro fragments made from egg white protein, serum protein, and xanthan gum. Xanthan gum is a polysaccharide that electrostatically interacts with egg white protein to form Trailblazer. The functional features of combining protein and polysaccharide solutions were explained by Tolstoguzov (Tolstoguzov 2006). It generates a single homogeneous solution, a liquid two-phase system, and an insoluble electrostatic protein complex with anionic polysaccharides, according to the researchers. They mixed the substances and kept the pH below the proteins' isoelectric point to use electrostatic force to react protein with anionic polysaccharides. They discovered that the heating process results in the formation of stabilized fibers. To make micro fragmented protein-based fat replacer, the stabilized fibers are subjected to high shear processing. The shape of the microparticles is not regular, with the longest size exceeding 10 µm. They resemble an electrostatic complex and a liquid two-phase system. The Trailblazer particles are in their gel state in a non-acidic environment by trapping xanthan in the gel interstices, which is a two-phase system.

2.6.3.1.6 Plasma Protein

Plasma proteins are mainly of two types: (i) albumins; which provide amino acids to body tissues and prevent fluids leakage, (ii) globulins; which help support your immune system, blood clotting, and other vital functions. Many slaughterhouses dumping the blood of bovines and porcines after their slaughter but the plasma protein from their blood can be a potential food ingredient due to its properties like gelation, emulsification, and foamability. Albumin is the most abundant protein in animal plasma, accounting for nearly half of the total protein. When combined with soy fiber in low-fat bologna sausage, it can be used as fat replacers since it improves binding and textural qualities. This plasma protein from bovine blood can be extracted using the ion-exchange chromatography technique (Moure et al. 2004), cold ethanol precipitation, anion exchange chromatography (Ekwumemgbo et al. 1970). In laboratories, bovine blood is mixed with an anticoagulant and centrifuged for 15 min to separate bovine plasma. The separated bovine plasma can be stored at −18°C (Yashini et al. 2021). Padilla et al. (Rodriguez Furlán et al. 2014) looked into the production of low-fat minced meats using inulin and plasma protein as fat substitutes. Proteins are obtained using ultrafiltration and freeze-drying. They looked at chemical composition, sensorial analysis (color, flavor, taste, and consistency), emulsion stability, and instrumental texture analysis of samples, and found that a combination of plasma protein (2.5 percent w/w) and inulin (2 percent w/w) had the best consistency and fat drain from the emulsion. Viana et al. (2005) (Viana et al. 2005) assessed the quality of ham pâé incorporating globin and plasma as fat substitutes. They studied the effect of using globin (10%), plasma (10%), or both (5%) as fat replacers on the quality of ham pâté and found that fat replacement with globin and/or plasma improved the quality of ham pâté.

2.6.3.1.7 Collagen Protein

Collagen is a fibrous protein found in connective tissue and the extracellular matrix that is insoluble. It is the most abundant protein in the animal world. There are at least 16 forms of collagen, yet only three account for 90% of the collagen in the body. Collagen is a vital protein for skin elasticity. This protein can be obtained from flesh and connective tissues of vertebrates such as cattle, pork, fish, etc. Collagen can be a fat replacer in sausages, dark chocolate, beef burger, frankfurters, etc. due to its moisturizing, visco-elastic, emulsifying properties. The quality of reduced-fat Vienna sausages with a blend of Hydrolyzed Collagen and Man-Sao powder (HCMSM) was assessed by Olanwanit et al. (Olanwanit and Rojanakorn 2019). They discovered that including 40% HCMSM into reduced-fat Vienna sausages can result in great quality and acceptability. In another investigation on the effects of Fish Collagen Hydrolysate (FCH) as a fat replacer in the production of low-fat buffalo patties by Nur Ibrahim et al. (Ibrahim et al. 2018), it was reported that FCH can be a potential fat replacer in low-fat buffalo patties. Youn-Kyung Ham et al. (Ham et al. 2013) investigated the effects of pig collagen and wheat fiber on the qualitative attributes of reduced-fat sausages as a fat replacer. The chemical composition, calorie, cooking yield, emulsion stability, texture, and sensory properties of emulsion sausage were determined in these studies. They concluded that collagen along with wheat fiber is proved to be an excellent fat replacer in meat sausages.

2.6.3.1.8 Gelatin Protein

Gelatin is a mixture of peptides and proteins made by the partial hydrolysis of collagen collected from animal body parts such as bones, skins, and connective tissues from swine, fish, poultry, and insects, among other sources. Gelatin is generally considered a nutritionally incomplete protein since it is missing amino acids such as tryptophan, isoleucine, threonine, and methionine. Acidulation, washing, alkaline treatment, filtration, deionization, concentration, sterilization, chilling, drying, milling, sitting & blending are the processing steps involved in gelatin production. Gelatin is a fat replacer due to its emulsifying, water binding, foaming, gelation properties. The influence of duck-feet gelatin (DFG) on the physicochemical, textural, and sensory aspects of low-fat frankfurters was investigated by Eui-Joo Yeo et al. (Kim et al. 2014). DFG enhanced all textural metrics in low-fat frankfurters, including hardness, springiness, cohesiveness, chewiness, and gumminess. Gelatin along with Basil Seed Gum (BSG) was more effective fat replacers in low-fat comminuted sausages (LFS) than using them alone was reported by Chang Hoon Lee et al. (Lee and Chin 2017). They also reported that the use of BSG as a water binding agent and gelatin to improve gel formation in combination is a good fat replacer in LFSs. The effects of various quantities of pork skin gelatin on low-fat sausages have been studied (Lee and Chin 2016). Low-fat sausages with gelatin have reduced cooking loss with good texture and water holding capacity. Gelatin from chicken byproducts proved to be a good replacer for fat in the chocolate spread but the major drawback in this method is, it decreases the shelf life of the chocolate spread.

2.6.3.2 Plant Protein-Based Fat Replacers (Table 2.6.3)

2.6.3.2.1 Soy Protein

Soy protein is a protein obtained from defatted and de-hulled soybean meal. Since it provides the essential amino acids, soy protein is a complete protein. It comprises different proteins such as globulin, glycinin and conglycinin. Globulin acts as a storage protein. Different techniques like the pH shift method, cryoprecipitation, etc. can isolate main proteins like glycinin and conglycinin from soy protein. Glycinin is responsible for gelling, emulsifying and foaming properties in soy protein, as a corollary, soy protein can supplant fat in low-fat diets. Soybeans are finely ground to obtain soy flour which has around 50% of soy protein. To obtain soy protein concentrate (SPC), which has around 70% protein, this soy flour is defatted and a portion of the carbohydrates are removed. Soy Protein Isolate (SPI), which can be prepared from defatted soy flour slurry *i.e.*, SPC is a dried food additive that can be used in dairy, meat, baking, and confectionery applications. It has around 90% of protein on a moisture-free basis. Siva Kumar et al. (Kumar et al. 2011) investigated the efficacy of SPI as a replacer of fat on physico-chemical and sensory parameters of low-fat paneer. They concluded that SPI (0.2%) is a good fat replacer in low-fat (12.7%) paneer. SPI over 0.2% imparts beany flavor to the paneer. Soy protein Isolate combined with polysaccharides showed enhanced fat mimetic properties. In ice cream, SPI with cellulose nanofiber complex gels resulted in improved texture qualities. Carrageenan, SPI, and soybean oil have been effectively employed as a fat substitute in Panela-type cheeses (Rojas-Nery et al. 2015). Some fat replacers derived from soy protein include the Supro range and Ultra-Freeze500, which can be used in dairy and meat products, soup bases, fillings, cream cheese, sour cream, toppings, and more.

2.6.3.2.2 Corn Zein Protein

Zein is a key corn storage protein, contributing to 35–65% of total corn protein. Zein is mainly found in endosperm and is deficient in tryptophan and lysine. Zein alone determines the hardness of the endosperm. Corn kernels have 9–12 percent protein, but only half of that is used as zein in the industrial world. Corn zein is a protein that belongs to the prolamine family. Pure zein is transparent, odorless, and tasteless and is hard and insoluble in water, and edible. Glutamic acid (21–26%), leucine (20%), proline (10%), and alanine (10%) are all plentiful in zein. Alpha zein and beta zein are the two primary proteins of zein. A-zein is soluble in 95% ethanol and accounts for 80% of total prolamine in maize, but B-zein is only soluble in 60% ethanol and is insoluble in 95% ethanol (Shukla and Cheryan 2001). Zein can effectively be extracted by solvent extraction using different concentrations of alcohol since it is soluble in alcohol (Parris and Dickey 2001; Shukla and Cheryan 2001). In non-aqueous fluids, zein has a globular shape with conformational features like globular proteins. Hydrophobic proteins, such as zein protein, have been an excellent fat substitute since they have properties like lipids. Jiahui Gu et al. (Gu et al. 2016) developed a zein-based fat mimic for mayonnaise formulation using a reduced-pressure distillation process. They determined that a satisfactory mayonnaise can be made with a fat substitution ratio of no more than 40% based on appearance, stability, and total calorific value, as well as rheological, microstructure, and sensory evaluations. Microparticulation of zein protein results in a product called LITA which can be a fat replacer. LITA is a protein that can be dissolved in water. LITA particles range in size from 0.1 to 8 micrometers. LITA is said to be distinct from other fat mimics since it has roughly 87 percent protein and little water, so it does not act through its water-controlling capacity. Salad dressings, mayonnaise, sour cream, toppings, frostings, soups, sauces, dips, cream or cottage cheese, meat products, confectionery, and so forth can all employ LITA as fat substitutes (Yashini et al. 2021)

2.6.3.2.3 Wheat Gluten Protein

Gluten is a primary storage protein found in wheat, accounting for 70–75 % of the total protein content. Gluten is naturally found in grains like wheat, barley, oat, rye etc. Wheat gluten is typically water-insoluble. Gluten is mainly made up of two proteins—glutenin and gliadin. Glutenin folds into a more stretched wire form, but gliadin folds into a compact spherical form. Gliadin contributes to the extensibility of dough and glutenin is largely responsible for dough's elastic features. Gluten can be obtained by biological, chemical, and mechanical methods. However mechanical method is commonly used to extract gluten. When the dough is prepared with wheat flour to the required consistency, the dough is continuously washed in water to remove starch. Washing should be continued until no starch is left. The gluten is then placed on a plate (not on paper because of stickiness) and should be rested for 15 min. This gluten is called wet gluten and if it is fried until a constant weight is obtained then it is called dry gluten. SWPI (Solubilized Wheat Protein Isolate) is a wheat protein enhanced to strengthen its functional features. Ahmedna et al. (Ahmedna et al. 1999) assessed the functional qualities of solubilized wheat protein isolate (SWPI). They compared SWPI functional properties with other protein isolates. They analyzed SWPI's acceptance in a variety of items including ice cream, hamburger patties, cookies, and muffins. They suggested that SWPI is a good fat replacer in full-fat foods due to its emulsifying, foaming, fat, and water binding capacity, solubility, and other properties. In their investigation, Serdaroğlu & Özsümer et al. (Serdaroğlu and Özsümer 2003) found that wheat gluten, soy protein, and whey protein can be used as fat replacers in low-fat beef sausages since they have no influence on color or sensory attributes but increase binding properties. Wheat gluten can also be used in meat food products as a filler or extender. In comparison to soy protein, wheat gluten and whey protein give sausages a soft finish. Anam Khalid et al. (Khalid et al. 2021) developed low-fat beef patties using Wheat Germ Oil (WGO) and Wheat Bran Fiber (WBF) as fat replacers. Cooked beef patties treated with WGO (4.5%) and WBF (3%) had superior quality, stability, and lower cholesterol content, according to the researchers. Milićevć et al. employed wheat bran gel and oat gels to create low-fat cookies, and they investigated the physico-chemical properties of these low-fat cookies (Milićević et al. 2020). They observed the best results in sensory properties of these cookies were when fat reduction is 30%. They discovered that replacing 30% of the fat in cookies with bran gels resulted in fat-reduced, added-value cookies in terms of phenolics, minerals, and dietary fiber.

2.6.3.2.4 Pea Protein

Pea protein is extracted in powder form from yellow and green split pea beans. It is abundant in essential amino acids but is low in amino acids like cysteine, methionine, and sulfur-containing amino acids. Pea Protein Concentrate (PPC) and Pea Protein Isolate (PPI) can be obtained from dry peas. Because of their functional qualities, such as foaming, solubility, gelation, water binding, emulsifying, viscosity, and flavor, PPC and PPI are used as fat mimetics in food products. In a study by Gouveia et al. the fat mimetic capacity of Chlorella vulgaris biomass in oil-in-water food emulsions was strengthened by pea protein (Raymundo et al. 2005). Emulsions with C.vulgaris and pea protein were made by changing the oil and protein ratios. They studied the fat mimetic potential of emulsions by characterizing the textural and rheological parameters of emulsions. Tabarestani also investigated the optimization of physicochemical features of low-fat hamburger preparations using a blend of soy flour, split-free flour, and wheat starch as a fat replacer system (Shahiri Tabarestani and Mazaheri Tehrani 2014). They discovered that combining soy flour with starch increases cooking yield and that adding split-pea flour to a mixed formula reduces shrinkage while increasing textural attributes. Using a combination of protein and carbohydrate-based fat replacers resulted in a textural quality better when compared to that of a full-fat control hamburger. UltraBake NF, Ultra-Freeze 400 are a few of those vegetable proteins which can be used as fat replacers in bakery products, sherbets, cheese, sauces, etc.

2.6.3.2.5 Lupin Protein

Lupin protein is derived from lupin seeds and has high emulsifying qualities, suggesting that it could be a fat substitute. It has albumins and globulins. In food items, it has good foaming features, water binding capacity, emulsifying qualities, and textural features. It is a fat replacer in the form of Lupin Protein Isolate (LPI). Isoelectric Lupin Protein Isolate is obtained using isoelectric precipitation, but Micellar Lupin Protein Isolate is obtained through salt-induced extraction followed by dilutive precipitation. Salad dressings and mayonnaise both use protein-stabilized emulsions. The protein structure, concentration, and interactions with other components all influence the rheological and textural qualities of lupin protein-stabilized emulsions. The qualities of lupin protein-stabilized emulsions are also influenced by processing factors. These fat replacers can be mainly applied in mayonnaise and salad dressing due to their excellent emulsifying properties. To use as fat replacers, low-fat emulsions made from xanthan gum, lupin protein, and oil should have less than 60 percent oil, 6–7 percent lupin protein, and 0.3–0.4 percent xanthan (Raymundo et al. 2002).

2.6.4 Properties of Protein-Based Fat Mimetics

Fat mimetics should mimic fats in foods to get acceptance from potential consumers. So, the properties of fat mimetics should be similar to that of fats. Fat mimetics should provide similar mechanical and rheological properties as fats to foods. Some properties of fat mimetics they should include to use them as fat replacers in foods are mentioned below:

- Flavor
- Texture

- Lubricating property
- Emulsifying property
- Water holding capacity
- Appearance
- Nutrition
- Melting point
- Heat transfer
- Satiety

Of all the mechanical properties of edible fats, the spreadability of fat mimetics is the most desirable property. Spreadability is a broad word that includes hardness, elastic constant, and post-yield stress flow properties. When stress greater than the fat's yield stress is applied, the fat experiences plastic deformation instead of just brittle fracture. When compared to room temperature butter, this is useful when spreading cold butter onto a piece of bread. Fat mimetics make it difficult to produce this plastic flow. Most fat mimics are hard yet fragile, and while the hardness of the fat can be matched, the plasticity cannot. So, properties like spreadability, plasticity, and harness are important for fat mimetics.

2.6.5 Factors Affecting the Acceptability of Protein-Based Fat Mimetics

Fat is being over-consumed by many people. This consumption of high-fat results in obesity, type-II diabetes, cancer, and some chronic cardiovascular diseases. In 2016, the World Health Organization (WHO) estimated that 1.9 billion individuals were overweight and 0.65 Billion people were unfit. So, there is a need to reduce the fat from the diet. Fat provides more calories even when consumed in small quantities. It makes people feel surfeit. Given the widespread belief that the amount and type of fat consumed has a role in the development of a variety of chronic diseases, it is no surprise that consumers are more willing to follow fat intake standards. Consumers choose low-fat food not only to support medical condition but also to reduce cholesterol, to maintain weight, to maintain proper health and some also chooses it for its characteristic taste. Nowadays, there is an increasing demand in the consumer market for low-fat bakery products, no-or-reduced-fat meat products, etc which are prepared with protein-based fat replacers. However, some factors affect the acceptability of these fat replacers in the market. Some of the important factors which determine the acceptability of protein-based reduced-fat versions of food are sensory attributes, nutritional properties, hygienic aspects, cost, consumers health, marketing, convenience and heat stability.

2.6.5.1 Sensory Attributes

Sensory properties are the main deciding factors for the acceptance of protein-based fat replacers. Since fat contributes little to odor, the three main sensory properties are taste, texture, and appearance. Taste is the most important sensory attribute. Low-fat foods with improved taste reportedly gained more acceptance. Foods made with protein based fat substitutes do not have the same flavor as full-fat foods. Although health-conscious consumers are still looking for ways to enhance their dietary habits without losing psychological enjoyment, the majority of consumers choose flavorful foods. Full fat-food has characteristic gloss and gleam limited in protein-based fat-reduced versions of food. The next important property is the texture of the modified food. Products made from other sources of fat replacers such as cellulose, carbohydrates have shown increased textural properties and have better acceptability than products made using protein-based fat replacers. To obtain good protein-based replacers in terms of aspects it should be used combined with other nutrients such as cellulose, pectin, etc.

2.6.5.2 Nutritional Properties

Consumers are not ready to compromise even in the nutritional properties that fat replacers provide. They are expecting almost similar nutritional properties of fats from fat replacers. So, fat replacers are providing the same appetite, flavor, odor, calories after consuming them. Since proteins are nutritionally sound compared to fat, consumers are demanding protein-based fat replacers in foods.

2.6.5.3 Hygienic Aspects

Food safety is everyone's business. It is not only for food business operators, food handlers but also for consumers. Food safety is a critical part of today's modern food sector, as both manufacturers and customers are increasingly worried about food-related concerns. Maintaining good hygiene not only ensures food safety but also hasa good market due to consumer awareness. For example, Consumer demands for acceptable hygienic conditions have risen because of the usage of protein-based fat replacers in restructured meat, as these products are sensitive to spoiling if not handled properly.

2.6.5.4 Cost

The cost of production has a significant impact on the acceptance of protein-based fat substitutes. The increased cost of production reflects in the increased product price. Protein being an important nutrient, its cost of production is also high and makes the consumer choose affordable products with actual fat. Consumer acceptance is influenced byproduct price variances due to buying power, societal, and educational differences. Some consumers prefer fat products despite knowing it causes certain consequences because of their cultural influence. But customers being unaware of the fat replacers will choose fat products due to their educational difference. The product obtained via technological developments must not only be of top standard but also be more cost-effective and less time-consuming.

2.6.5.5 Health Aspects

The consumption of fat results in severe health consequences. Food companies have reacted to consumer needs for fat-free products, and the market for consumables with a healthy image has grown exponentially. Protein being the essential macronutrient in the human diet its over consumption does not result in any diseases compared to fat. So, consumers prefer their diet to be enriched with proteins. Substituting protein-based replacers for fat may improve customer acceptance.

2.6.5.6 Marketing

Industries using protein-based fat replacers should adopt wise marketing methods to achieve utmost reach to the customers. They should educate the consumers about the benefits of replacing the daily diet with protein-based fat replacers. But the use of other nutrients along with protein makes the products incapable to be claimed as made of pure protein-based fat replacers and it makes it less acceptable. For example, protein-based fat replacers that are used in the meat industry should have the property of Water Holding Capacity (WHC), which is only achieved when integrated with carbohydrates. In bakery products, using protein alone as a fat replacer makes it harder and unacceptable. It was reported that Simplesse, a protein-based fat replacer in bread, was shown to have disappointing outcomes, including a weak gluten network, low loaf volume, and high hardness (O'Brien et al. 2003).

2.6.5.7 Convenience

In this fast-paced world where everyone is working, the need for simple-to-prepare food items that also provide high-quality nutrients has skyrocketed. People are worried about their food habits. So, if a product is convenient to use and has protein supplements in it, its demand has increased among consumers.

2.6.5.8 Heat Stability

Speaking of the stability of protein molecules, they become unstable and denatures at high temperatures and they lose their functional properties. At high temperatures, proteins get coagulated, and the creamy-fat texture is lost in fried foods. Hence the application of protein-based fat replacers is limited to use at moderate temperatures like in baking and retorting but not for frying applications.

Using protein as a fat replacer has shown a better decrease in fat (lipid content) than any other substitute. Thus, protein contributes to food in terms of health features but does not meet taste requirements. A study revealed that the use of protein as a fat substitute in beef frankfurters did not even reach 50% acceptability in terms of taste, but it provided significant tenderness to the product (Pszczola 2006). On a whole, using proteins in reduced-fat versions can gain acceptability by improving the taste of the product.

2.6.6 Applications of Protein Based Fat Mimetics

2.6.6.1 Bakery Products

Bakery fats such as shortening or butter are required for bakery items such as bread, biscuits, bakery, cakes, and so on. To act as fat replacers in bakery foods, fat mimetics should impart properties such as water-holding (for keeping moisture over storage and preventing drying), long-term softness (for improving shelf life), aeration (for volume and uniform cell structure), emulsifying, and lubrication (for tender texture) in the final baked products. Although the rheology of intermediate goods using fat mimetics is less researched, fat mimetics such as wax-based oil gels and polymer gels can be used in bakery products since they have equivalent qualities in baked foods to foods with typical bakery fats. Natural waxes derived from oil gels such as sunflower wax, candelilla wax, beeswax, rice bran wax are proven effective as fat substitutes in cookies and other baked goods. Laguna et al. (Laguna et al. 2014) studied the sensory and instrumental evaluation of Hydroxy Propyl Methyl Cellulose (HPMC) and Inulin as fat substitutes in biscuits. They saw that biscuits with HPMC and Inulin are comparatively harder than normal biscuits when evaluated under instrumental texture measurements. They found that fat substitution of up to 15 g/100 g with Inulin or HPMC resulted in acceptable biscuits, but that higher fat replacement reduced overall acceptability. Anne Forker et al. researched the development of a fat replacer combination that allows for the production of low-fat short dough biscuits with high sensory quality (Hadnadev et al. 2014). They conducted this study to see how a single fat replacer (corn fiber, maltodextrin, lupin extract) or a binary combination of fat replacers (corn fiber, maltodextrin, lupin extract) about 30–40% of fat affected product features (moisture, volume, color, and texture) and sensory properties of biscuits. They concluded that a combination of corn fiber and lupin extract in 1:1 as fat replacers showed the best results.

2.6.6.2 Chocolate and Confectionery Products

The migration and leaking of oil during storage is the main issue in chocolate and confectionery items. Because solid fat works as an oil binder, fat mimetics must be carefully designed to prevent jeopardizing product stability. Paste-like dispersions or highly concentrated suspensions of non-fat particles (plant-cell fragments and small sugar particles) are distributed in an oil continuous phase in confectionery items such as chocolate paste and peanut butter. In terms of rheology, these paste-like dispersions act like an arrested "soft" solid at rest (to resist dispersed particle deposition and to reduce liquid oil segregation) and exhibit a more "liquid-like" spreading characteristic (or yielding) under an applied force. The crystalline fat network in the continuous oil phase and specialized stabilizers (*i.e.*, oil binders) that reduce surface interactions/friction among non-fat particles contribute to this intriguing rheological behavior. The most often used oil binders in the chocolate and confectionery sector are hydrogenated oils, which are normally added at 2% wt or even less. Miroslav investigated the physical and sensory properties of maltodextrin gel additions, which are used as fat replacements in confectionery filling systems (Hadnadev et al. 2014). They discovered that increasing the quantity of fat reduction resulted in the final product being harder and changing color. They concluded that confectionery filling with 5 g/100 g has the highest acceptability.

2.6.6.3 Dairy Products

Everyone consumes dairy products in their diet, such as butter, cheese, and so on. Dairy products are divided into three categories based on their colloids: foamed emulsions (ice creams and whipped creams), coagulated gels (different types of cheese), and structured emulsions (butter and butter spreads). The microstructure of these products differs, with a large quantity of milk fat dispersed in bulk or scattered phases. The underlying colloidal network of crystalline particles is explained by the high melting milk fat. Milk fat provides greater functionalities like spreadability, stabilization effect, emulsifying effect in oil-in-water emulsions (milk). Replacing high melting milk fat with liquid oils without compromising product qualities is a difficult task.In the aqueous phases, Wladimir Silva et al. used double emulsions as prospective fat replacers, with gallic acid and quercetin nano emulsions (Silva et al. 2018). Wax-based gels are said to be effective substitutes for milk fat in dairy products such as ice cream. Incorporating fat mimetics resulted in ice creams comparable to the standard samples made with milk fat. Some features of reformulated ice creams beat those of normal ice creams, including such melting starting point and overrun. Because phytosterols and phytostanols are present, such reformulated items might be labeled as having extra plant sterols and stanols for people who desire to reduce their cholesterol levels. Schmidt et al. looked into the rheological properties of ice cream mixes and frozen ice creams using fats and fat replacers (Adapa et al. 2000). By adding 6% protein-based fat replacer and 6% carbohydrate-based fat replacer to ice cream mixes and frozen ice creams with varying quantities of milk fat (12%, 8%,

6%), they assessed the qualities of ice cream mixes and frozen ice creams. They concluded that adding protein-and carbohydrate-based fat replacers to ice creams improved the viscous qualities but not the elastic properties. Emulsifying (fat-continuous) items such as tables and different types of margarine are made with varying fat content. For example, low fat and low fat spreads are used in bread and high fat spread is used in cooking/frying applications. Wax-based gels are said to be good fat substitutes in spreads and margarine products since they fulfil the functions that fat fulfils in them, including i)forms a crystalline fat network in the continuous oil phase to influence structure properties (such as firmness, texture, and spreadability); ii) forms fine fat crystals at water-oil interfaces to stabilise dispersed water droplets; and iii) contributes to sensorial properties such as mouthfeel and release of flavor (Akoh 1998).

2.6.6.4 Meat Products

Fat mimetics are being used in meat products such as sausages, burger patties, and frankfurters to reduce animal fats and cholesterol, improve the nutritional composition, and reduce overall calories. Oil gels composed of hydrophilic polymers were used to substitute animal fat in prepared meat patties. Fructooligosaccharides can be used as fat replacers in fermented cooked sausages (Dos Santos et al. 2012). Collage hydrolase is a good replacement for cocoa butter in dark chocolate because it increases viscosity, hardness, moisture, and protein while having no influence on sensory attributes and water activity. Emulsion gels (EG) containing soybean oil, functional ingredients, stabilizers,and/or gelling agents (chia flour and/or soy protein isolate, inulin, carrageenan, sodium caseinate, and sodium tripolyphosphate) were evaluated and used to replace 50% and 100% of pork back fat (PBF) in Bolognese sausage (de Souza Paglarini et al. 2019; Paglarini et al. 2020). Santos et al. looked at canola-oil-emulsion gels stabilized by pigskin and dietary fiber to replace the entire pork back fat content (Dos Santos et al. 2020). They found that emulsion gels (EGs) behaved similarly to pork back fat, suggesting that they could be employed as fat substitutes in meat food stuffs. Paz Robert et al. investigated double emulsions (DEs) with olive leaves extract (OLE) as fat replacers in meat systems with great oxidative stability (Robert et al. 2019). They concluded that encapsulating live leave extract in double emulsion could be a good way to prevent the lipid oxidation process in meat systems with better lipid profiles. In squid emulsion sausages and bologna-style sausages, pork skin collagen blended with green banana flour and coconut powder work well as a fat replacer (Zhou et al. 2018).

2.6.6.5 Other Applications

In addition to all the above applications, fat mimetics are also applied in salad dressing, frostings, and fillings, cream soups, sauces, and gravies, etc. Cheryl Chung et al. (Chung et al. 2014) studied the overall appearance and rheological properties using protein microparticles and dietary fibers as fat replacers to produce low-calorie salad dressings and emulsion-based sauces. They used microparticulated wheat protein with either starch or locust bean gum (LBG) as fat replacer. Fat mimetics have also been proposed for the physiological effects of oils and change digestion, which increases the delivery of hydrophobic bioactives and tastes by altering food-body interactions. Fat mimetics, such as oleogels, can provide multiple benefits as delivery vehicles including, i) stabilization against recrystallization (due to the high solvency of liquid oil) and prevention of oxidation of oily bioactive constituents; ii) controlling and slowing down the rate of bioactive release; and iii) increasing bioaccessibility/bioavailability of bioactives due to direct micellization. Sullivan et al. investigated the use of oleogels for bioactive delivery, using B-carotene as a model bioactive (O

Sullivan et al. 2017). Fat mimetics have been employed as delivery mechanisms to create microstructures with variable flavor release. This study's positive findings helps with the design of novel foodstuffs with improved flavor combinations.

2.6.7 Future of Protein-Based Fat Mimetics

Obesity, diabetes, and other non-communicable diseases (NCDs) are linked to the consumption of fats in foods. Easy availability and access to unhealthy foods are some reasons behind this. According to the latest WHO guidelines, dietary reference intake (DRI) of fats 25%-30% of total calories intake should be from fats. However, people consumed excess fats from their foods as oils, butter, cheese, etc. This problem raised the concern of developing no-or-reduced fat food by supplementing fat with other macronutrients like proteins, carbohydrates, dietary fibers, etc. People chose protein-based fat mimetics over carbohydrates-based due to their ability to increase muscle mass and strength, aiding in weight loss, boosting metabolism, lowering blood pressure, good for bones, etc. Protein-based fat mimetics thus gaining importance among the increasing population. According to reports, the world population will be over 9 billion by 2050. This surge in population not only pressures resources to produce more food but also to produce different foods. People also being aware of the importance of a nutritional diet *i.e.*; including proteinaceous food to maintain their healthy diet. So there is increasing demand for producing protein-rich food. According to a report, global demand for animal-derived protein is expected to increase two times by 2050. Even though animal-based diets emit more green house gases (GHG) than plant-based diets, and these gases contribute to climate change, producing this much animal-based protein is exceptionally hard. Due to the requirement to generate more animal feed, rising demand for animal-based protein pressures land, resulting in the transformation of forest lands, marshes, and pastures into agricultural lands. Socio-economic reasons such as increased income, an aging generation, modernization, and so on are driving the global increase in protein demand. WHO's protein recommended dietary allowance (RDA) for a healthy adult with moderate physical fitness is now 0.83 g protein per kg body weight (BW) per day (Wu 2016) which is estimated to cover the protein demands of 97.5 % of the world's healthy adult population. However, there are few concerns associated with the over-consumption of protein, especially meat-based protein. Over-consumption of meat-based proteins is linked to rising levels of global obesity, type-II diabetes, as well as cardiovascular diseases, which may also lead to death (Institute 2016). According to estimations for 7.3 billion people there is a protein demand of 202 million tonnes globally (Henchion et al. 2017). By 2050 this figure may increase by 50%. However, not all proteins are considered equal. Bioavailability, nutritional profile, digestibility, environmental implications, customer acceptance, and other aspects all differ. This suggests that future protein supply is not just about producing more of the same protein, but also about replacing it with proteins from various sources.

We also know that major sources of proteins are vegetables followed by animal meat. This also suggests using plant proteins as fat replacers in food because they reduce NCD complications due to the consumption of excess amounts of fats. However, plant-based proteins are lack of some essential amino acids, which is a major drawback, so they should always be used combined with other plant proteins to replace fat efficiently in food.

Consumers prefer high protein–low GI (Glycemic Index) foods these low GI – high protein foods are believed to boost the gut hormones response to food and help in avoiding overeating of the meal. There is a growing demand for these foods in the present market. People also look for the sensory and textural properties while using fat mimetics in foods. Consumer awareness is one of the major concerns to look at while using protein-based fat replacers. Most of the consumers take a backstep while buying food with protein-based fat mimetic due to its high cost. Some consumers may not be aware of what the fat replacer is due to their educational background. So effective marketing should be done before using the protein as fat mimetic in food.

References

Adapa, S., Dingeldein, H., Schmidt, K.A., and Herald, T.J. (2000). Rheological properties of ice cream mixes and frozen ice creams containing fat and fat replacers. *Journal of Dairy Science* 83 (10): 2224–2229. doi: 10.3168/jds.S0022-0302(00)75106-X.

Ahmedna, M., Prinyawiwatkul, W., and Rao, R.M. (1999). Solubilized wheat protein isolate: functional properties and potential food applications. *Journal of Agricultural and Food Chemistry* 47 (4): 1340–1345. doi: 10.1021/jf981098s.

Akbari, M., Eskandari, M.H., and Davoudi, Z. (2019). Application and functions of fat replacers in low-fat ice cream: a review. *Trends in Food Science and Technology* 86 (February): 34–40. doi: 10.1016/j.tifs.2019.02.036.

Akoh, C.C. (1998). Fat replacers. *Food Technology* 52 (3): 47–53. doi: 10.5650/jos1996.46.1299.

Aziznia, S., Madamlous, A., and Rahimi, J. (2008). Whey protein concentrate and gum tragacanth as fat replacers in nonfat yogurt: chemical, physical, and microstructural properties. *Journal of Dairy Science* 91 (7): 2545–2552. doi: 10.3168/jds.2007-0875.

Brennan, C.S. and Tudorica, C.M. (2008). Carbohydrate-based fat replacers in the modification of the rheological, textural and sensory quality of yogurt: comparative study of the utilisation of barley beta-glucan, guar gum and inulin. *International Journal of Food Science and Technology* 43 (5): 824–833. doi: 10.1111/j.1365-2621.2007.01522.x.

Chung, C., Degner, B., and McClements, D.J. (2014). Reduced calorie emulsion-based foods: protein microparticles and dietary fiber as fat replacers. *Food Research International* 64: 664–676. doi: 10.1016/j.foodres.2014.07.034.

de Souza Paglarini, C., de Figueriredo Furtado, G., Honório, A.R., Mokarzel, L., Vidal, V.A.S. Cunha, R.R.L., and Pollonio, M. (2019). Functional emulsion gels as pork back fat replacers in Bologna sausage. *Food Structure* 20 (January): 100105. doi: 10.1016/j.foostr.2019.100105.

Dos Santos, B.A., Campagnol, P.C.B., Pacheco, M.T.B., and Pollonio, M.A.R. (2012). Fructooligosaccharides as a fat replacer in fermented cooked sausages. *International Journal of Food Science and Technology* 47 (6): 1183–1192. doi: 10.1111/j.1365-2621.2012.02958.x.

Dos Santos, M., Ozaki, M.M., Ribeiro, W.S., de Souza Paglarini, C., Vidal, V.A.S., Campagnol, P.C.B., and Pollonio, M.A.R. (2020). Emulsion gels based on pork skin and dietary fibers as animal fat replacers in meat emulsions: An adding value strategy to byproducts. *Lwt* 120: 108895. doi: 10.1016/j.lwt.2019.108895.

Ekwumemgbo, P.A., Kagbu, J.A., Nok, A.J., Omoniyi, I.K., Ameh, P.O., and Oladunni, N. (1970). Isolation of plasma proteins from bovine blood by cold ethanol precipitation and anion exchange chromatography. *Journal of Nepal Chemical Society* 26: 2–12. doi: 10.3126/jncs.v26i0.3624.

Gibis, M., Schuh, V., and Weiss, J. (2015). Effects of carboxymethyl cellulose (CMC) and microcrystalline cellulose (MCC) as fat replacers on the microstructure and sensory characteristics of fried beef patties. *Food Hydrocolloids* 45: 236–246. doi: 10.1016/j.foodhyd.2014.11.021.

Grossklaus, R. (1996). Fat replacers — Requirements from a nutritional physiological point of view. *Lipid* 98 (4): 136–141.

Gu, J. et al. (2016). A "reduced-pressure distillation" method to prepare zein-based fat analogue for application in mayonnaise formulation. *Journal of Food Engineering* 182: 1–8. doi: 10.1016/j.jfoodeng.2016.01.026.

Hadnadev, M., Hadnadev, T.D., Dokić, L., Pajin, B., Torbica, A., Šarić, L., and Ikonić, P. (2014). Physical and sensory aspects of maltodextrin gel addition used as fat replacers in confectionery filling systems. *LWT – Food Science and Technology* 59 (1): 495–503. doi: 10.1016/j.lwt.2014.04.044.

Ham, Y., Kim, Y.J., Yeo, E.J., Lim, Y.B., and Kim, C.J. (2013). *Effects of pork collagen and wheat fiber on quality properties of reduced-fat sausage as a fat replacer.* (August), 3–6.

Henchion, M., Hayes, M., Mullen, A.M., Fenelon, M., and Tiwari, B. (2017). Future protein supply and demand: Strategies and factors influencing a sustainable equilibrium. *Foods* 6 (7): 1–21. doi: 10.3390/foods6070053.

Ibrahim, F.N., Yusoff, M.M., and Ismail-Fitry, M.R. (2018). Effects of fish collagen hydrolysate (FCH) as fat replacer in the production of buffalo patties. *Journal of Advanced Research in Applied Sciences and Engineering Technology* 11 (1): 108–117.

Institute, W.R. (2016). *The global protein challenge.* http://www.wri.org.

Khalid, A., Sohaib, M., Nadeem, M.T., Saeed, F., Imran, A., Imran, M., Afzal, M.I., Ramzan, S., Nadeem, M., Anjum, F.M., and Arshad, M.S. (2021). Utilization of wheat germ oil and wheat bran fiber as fat replacer for the development of low-fat beef patties. *Food Science and Nutrition* 9 (3): 1271–1281. doi: 10.1002/fsn3.1988.

Kim, H.W., Park, J.H., Yeo, E.J., Hwang, K.E., Song, D.H., Kim, Y.J., Ham, Y.K., Jeong, T.J., Choi, Y.S., and Kim, C.J. (2014). Effect of duck feet gelatin concentration on physicochemical, textural, and sensory properties of duck meat jellies. *Korean Journal for Food Science of Animal Resources* 34 (3): 387–394. doi: 10.5851/kosfa.2014.34.3.387.

Kumar, S.S., Balasubramanian, S., Biswas, A.K., Chatli, M.K., Devatkal, S.K., and Sahoo, J. (2011). Efficacy of soy protein isolate as a fat replacer on physico-chemical and sensory characteristics of low-fat paneer. *Journal of Food Science and Technology* 48 (4): 498–501. doi: 10.1007/s13197-010-0193-z.

Laguna, L., Primo-Martín, C., Varela, P., Salvado, A., and Sanz, T. (2014). HPMC and inulin as fat replacers in biscuits: Sensory and instrumental evaluation. *LWT - Food Science and Technology* 56 (2): 494–501. doi: 10.1016/j.lwt.2013.12.025.

Lee, C.H. and Chin, K.B. (2016). Effects of pork gelatin levels on the physicochemical and textural properties of model sausages at different fat levels. *LWT – Food Science and Technology* 74: 325–330. doi: 10.1016/j.lwt.2016.07.032.

Lee, C.H. and Chin, K.B. (2017). Development of low-fat sausages using basil seed gum (Ocimum bacilicum L.) and gelatin as a fat replacer. *International Journal of Food Science and Technology* 52 (3): 733–740. doi: 10.1111/ijfs.13328.

Liu, K., Stieger, M., van der Linden, E., and van de Velde, F. (2016). Effect of microparticulated whey protein on sensory properties of liquid and semi-solid model foods. *Food Hydrocolloids* 60: 186–198. doi: 10.1016/j.foodhyd.2016.03.036.

Liu, K., Tian, Y., Stieger, M., van der Linden, E., and van de Velde, F. (2016). Evidence for ball-bearing mechanism of microparticulated whey protein as fat replacer in liquid and semi-solid multi-component model foods. *Food Hydrocolloids* 52: 403–414. doi: 10.1016/j.foodhyd.2015.07.016.

Lucca, P.A. and Tepper, B.J. (1994). Fat replacers and the functionality of fat in foods. *Trends in Food Science and Technology* 5 (1): 12–19. doi: 10.1016/0924-2244(94)90043-4.

Martínez-Cervera, S., Salvador, A., and Sanz, T. (2015). Cellulose ether emulsions as fat replacers in muffins: Rheological, thermal and textural properties. *LWT – Food Science and Technology* 63 (2): 1083–1090. doi: 10.1016/j.lwt.2015.04.067.

McMahon, D.J., Alleyne, M.C., Fife, R.L., and Oberg, C.J. (1996). Use of fat replacers in low fat Mozzarella cheese. *Journal of Dairy Science* 79 (11): 1911–1921.

Milićević, N., Saka°, M., Hadnadev, M., Škrobot, D., Šarić, B., Hadnadev, T., Jovanov, P., and Pezo, L. (2020). Physico-chemical properties of low-fat cookies containing wheat and oat bran gels as fat replacers. *Journal of Cereal Science* 95 (July). doi: 10.1016/j.jcs.2020.103056.

Moure, F., Rendueles, M., and Díaz, M. (2004). Bovine plasma protein fractionation by ion exchange chromatography. *Bioprocess and Biosystems Engineering* 27 (1): 17–24. doi: 10.1007/s00449-004-0372-2.

O'Brien, C.M., Mueller, A., Scannell, A.G.M., and Arendt, E.K. (2003). Evaluation of the effects of fat replacers on the quality of wheat bread. *Journal of Food Engineering* 56 (2–3): 265–267. doi: 10.1016/S0260-8774(02)00266-2.

Ognean, C.F., Darie, N., and Ognean, M. (2006). Fat replacers – review. *Journal of Agroalimentary Processes and Technologies* XII (2): 433–442.

Olanwanit, W. and Rojanakorn, T. (2019). Effect of hydrolysed collagen and Man-sao powder mixture as a fat replacer on quality of Vienna sausages. *International Food Research Journal* 26 (5): 1525–1533.

O Sullivan, C.M., Davidovich-Pinhas, M., Wright, A.J., Barbut, S., and Marangoni, A.G. (2017). Ethylcellulose oleogels for lipophilic bioactive delivery-effect of oleogelation on: in vitro bioaccessibility and stability of beta-carotene. *Food and Function* 8 (4): 1438–1451. doi: 10.1039/c6fo01805j.

Paglarini, C. de S., Vidal, V.A.S., Martini, S., Cunha, R.L., and Pollonio, M.A.R. (2020). Protein-based hydrogelled emulsions and their application as fat replacers in meat products: a review. *Critical Reviews in Food Science and Nutrition* 62 (3): 640–655. doi: 10.1080/10408398.2020.1825322.

Parris, N. and Dickey, L.C. (2001). Extraction and solubility characteristics of zein proteins from dry-milled corn. *Journal of Agricultural and Food Chemistry* 49 (8): 3757–3760. doi: 10.1021/jf0011790.

Patel, A.R., Nicholson, R.A., and Marangoni, A.G. (2020). Applications of fat mimetics for the replacement of saturated and hydrogenated fat in food products. *Current Opinion in Food Science* 33: 61–68. doi: 10.1016/j.cofs.2019.12.008.

Peng, X. and Yao, Y. (2017). Carbohydrates as fat replacers. *Annual Review of Food Science and Technology* 8: 331–351. doi: 10.1146/annurev-food-030216-030034.

Pereira, M.M., Cruz, R.A.P., Almeida, M.R., Lima, A.S., Coutinho, J.A.P., and Freire, M.G. (2016). Single-step purification of ovalbumin from egg white using aqueous biphasic systems. *Process Biochemistry* 51 (6): 781–791. doi: 10.1016/j.procbio.2016.03.002.

Poursani, P., Razavi, S.M.A., Tehrani, M.M., and Javidi, F. (2021). Rheological, physical, and sensory properties of non-fat ice creams as affected by selected fat replacers. *Journal of Food Processing and Preservation* 45 (1): 0–3. doi: 10.1111/jfpp.15010.

Pszczola, D.E. (2006). Future strategies for fat replacement. *Food Technology* 60 (6).

Raymundo, A., Franco, J.M., Empis, J., and Sousa, I. (2002). Optimization of the composition of low-fat oil-in-water emulsions stabilized by white lupin protein. *JAOCS, Journal of the American Oil Chemists' Society* 79 (8): 783–790. doi: 10.1007/s11746-002-0559-6.

Raymundo, A., Gouveia, L., Batista, A.P., Empis, J., and Sousa, I. (2005). Fat mimetic capacity of Chlorella vulgaris biomass in oil-in-water food emulsions stabilized by pea protein. *Food Research International* 38 (8–9): 961–965. doi: 10.1016/j.foodres.2005.02.016.

Robert, P., Zamorano, M., González, E., Silva-Weiss, A., Cofrades, S., and Giménez, B. (2019). Double emulsions with olive leaves extract as fat replacers in meat systems with high oxidative stability. *Food Research International* 120 (November): 904–912. doi: 10.1016/j.foodres.2018.12.014.

Rodriguez Furlán, L.T., Padilla, A.P., and Campderrós, M.E. (2014). Development of reduced fat minced meats using inulin and bovine plasma proteins as fat replacers. *Meat Science* 96 (2): 762–768. doi: 10.1016/j.meatsci.2013.09.015.

Rojas-Nery, E. Güémes-Vera, N., Meza-Marquez, O.G., and Totosaus, A. (2015). Carrageenan type effect on soybean oil/soy protein isolate emulsion employed as fat replacer in panela-type cheese. *Grasas y Aceites* 66: 4. doi: 10.3989/gya.0240151.

Rovers, T.A.M., Sala, G., Van der Linden, E., and Meinders, M.B.J. (2016). Potential of microbubbles as fat replacer: Effect on rheological, tribological and sensorial properties of model food systems. *Journal of Texture Studies* 47 (3): 220–230. doi: 10.1111/jtxs.12175.

Sandoval-Castilla, O., Lobato-Calleros, C., Aguirre-Mandujano, E., and Vernon-Carter, E.J. (2004). Microstructure and texture of yogurt as influenced by fat replacers. *International Dairy Journal* 14 (2): 151–159. doi: 10.1016/S0958-6946(03)00166-3.

Schädle, C.N., Eisner, P., and Bader-Mittermaier, S. (2020). The combined effects of different fat replacers and rennet casein on the properties of reduced-fat processed cheese. *Journal of Dairy Science* 103 (5): 3980–3993. doi: 10.3168/jds.2019-17694.

Serdaroğlu, M. and Özsümer, M.S. (2003). Effects of soy protein, whey powder and wheat gluten on quality characteristics of cooked beef sausages formulated with 5, 10 and 20% fat. *Electronic Journal of Polish Agricultural Universities* 6 (2): 3.

Shahiri Tabarestani, H. and Mazaheri Tehrani, M. (2014). Optimization of physicochemical properties of low-fat hamburger formulation using blend of soy flour, split-pea flour and wheat starch as part of fat replacer system. *Journal of Food Processing and Preservation* 38 (1): 278–288. doi: 10.1111/j.1745-4549.2012.00774.x.

Shukla, R. and Cheryan, M. (2001). Zein: the industrial protein from corn. *Industrial Crops and Products* 13 (3): 171–192. doi: 10.1016/S0926-6690(00)00064-9.

Silva, W., Torres-Gatica, M.F., Oyarzun-Ampuero, F., Silva-Weiss, A., Robert, P., Cofrades, S., and Giménez, B. (2018). Double emulsions as potential fat replacers with gallic acid and quercetin nanoemulsions in the aqueous phases. *Food Chemistry* 253 (June 2017): 71–78. doi: 10.1016/j.foodchem.2018.01.128.

Sons, J.W. (2005). *Fat Replacers, Kirk-Othmer Food and Feed Technology*, 1, 393–426. United Kingdom: Wiley. (This is a part of Fifth Edition of Wiley's Kirk-Othmer Encyclopedia of Chemical Technology).

Stargel, W.W., Butchko, H.H., Hjelle, J.J., and Tschanz, C. (1991). SIMPLESSE®: a natural fat substitute. In: *Obesity: Dietary Factors and Control* (ed. D.R. Romsos, J. Himms-Hagen, M. Suzuki), 25–35. Basel: Karger. doi: 10.1159/000420913.

Sun, C., Liu, R., Liang, B., Wu, T., Sui, W., and Zhang, M. (2018). Microparticulated whey protein-pectin complex: a texture-controllable gel for low-fat mayonnaise. *Food Research International* 108 (2017): 151–160. doi: 10.1016/j.foodres.2018.01.036.

Tolstoguzov, V.B. (2006). Microstructural elements and their interactions-ingredient interactions: aggregation and phase separation. In: *Understanding and Controlling the Microstructure of Complex Foods* (ed. D.J. McClements), 185–206. Woodhead: Cambridge.

Viana, F.R., Silva, V.D.M., Delvivo, F.M., Bizzotto, C.S., and Silvestre, M.P.C. (2005). Quality of ham pâté containing bovine globin and plasma as fat replacers. *Meat Science* 70 (1): 153–160. doi: 10.1016/j.meatsci.2004.12.013.

Wu, G. (2016). Dietary protein intake and human health. *Food and Function* 7 (3): 1251–1265. doi: 10.1039/c5fo01530h.

Yashini, M., Sunil, C.K., Sahana, S., Hemanth, S.D., Chidanand, D.V., and Rawson, A. (2021). Protein-based fat replacers–a review of recent advances. *Food Reviews International* 37 (2): 197–223. doi: 10.1080/87559129.2019.1701007.

Zhou, T., Zhao, Y., Wu, X., Wang, W., and Liu, A. (2018). Production of squid emulsion sausages using pork skin and coconut powder mixture as fat replacers. *International Journal of Food Science and Technology* 53 (3): 747–754. doi: 10.1111/ijfs.13650.

2.7

Polysaccharide-Based Oleogels

Andrew J. Gravelle

Department of Food Science and Technology, University of California, Davis, One Shields Avenue, Davis, CA, United States

2.7.1 Introduction

Polysaccharides are a diverse class of macromolecules present throughout nature, and serve various biological roles including energy storage, structural support, and intercellular communication. They comprise individual monosaccharide building blocks bound via glycosidic linkages and may form either linear or branched segments. The structural and functional properties of polysaccharides are largely dictated by the composition and arrangement of monosaccharides, degree of branching, polymer molecular weight, and functional groups or charged species. The majority of polysaccharides used in food applications are largely hydrophilic and are often used for their strong viscosity enhancing, water stabilizing, and gelation abilities (X. Yang et al. 2020). Polysaccharides have been extensively used as bulking agents, extenders, and direct fat replacers across the food industry (e.g. for fat reduction strategies), and these applications continue to be actively investigated (Lu et al. 2019; X. Yang et al. 2020). These traits are also commonly used to improve stability in emulsions and foams, reduce ice crystal formation, and impart specific structural, textural, or sensory attributes (Lu et al. 2019; Cao and Mezzenga 2020; X. Yang et al. 2020). While polysaccharides have been widely used in fat reduction/replacement strategies, shifting consumer demands have also led to their use in lipid structuring through various novel and innovative approaches. In this regard, this chapter will focus on those approaches specifically targeted at structuring liquid oil.

Using polysaccharides as structuring agents for edible oils presents a significant technological challenge due to their inherent hydrophilic properties. The most widely investigated polysaccharides used for oil structuring have been the commercially produced, chemically modified cellulose derivatives ethyl cellulose (EC), methyl cellulose (MC), and hydroxypropyl methylcellulose (HPMC), as the modification processes introduce functional groups along the cellulose backbone which impart varying degrees of amphiphilic or hydrophobic character. Depending on the oil structuring strategy, these polymers may be used with other hydrophilic polysaccharides to improve or modulate the oil binding and functional properties of the resulting oleogel. Additional innovative chemical and physical approaches have also been reported using some naturally derived polysaccharides including chitin, k-carrageenan, and modified starch.

The most common strategies for generating polysaccharide-based oleogels include (1) direct dispersion in the oil phase, or multi-step indirect structuring methods using either (2) emulsification, (3) foaming, or (4) hydrogel formation as an initial step in the oleogelation process. The indirect approaches also require removal of the aqueous phase after the initial structuring procedure, and in the case of foaming and hydrogel formation, the oil must be subsequently introduced. Each

Fat Mimetics for Food Applications, First Edition. Edited by Miguel Ângelo Parente Ribeiro Cerqueira and Lorenzo Miguel Pastrana Castro.

distinct strategy will be discussed in greater detail below. While it is noteworthy that the indirect approaches have also been used to develop protein- and protein/polysaccharide-based oleogels (Bascuas et al. 2021), this chapter will focus on oleogels in which polysaccharides are the primary structuring agent.

2.7.2 Direct Polymeric Structuring

The general process for forming polymeric gels involves two major steps. First, the polymer must be introduced into the dispersing medium under suitable conditions which let it take on an extended conformation and form a space-filling, three-dimensional network. The second phase involves crosslinking the polymer chains through the formation of chemical or physical bonds. In food polymers, this is often achieved either via progressive heating, heating/cooling, enzymatic crosslinking, or altering the electrostatic interactions between polymers by adjusting the pH or addition of salts (Cao and Mezzenga 2020). This step produces the formation of a self-supporting matrix in which the solvent becomes physically entrapped. The polymer must be sufficiently compatible with the dispersing phase to achieve the requisite extended conformation and keep the liquid phase upon gelation. The functional properties of the resulting gel will also depend on both the nature and density of polymer-polymer junction zones, as well as polymer-solvent interactions. Incorporating compounds which act as plasticizing agents can also influence the degree of interaction and nature of the junction zones.

The necessity to balance polymer-polymer and polymer-solvent interactions greatly limits the applicability of most polysaccharides as direct oil structuring agents. While some relatively amphiphilic polysaccharides may be dispersed in the lipid phase before being used as emulsifiers, they cannot transition to an extended conformation until becoming hydrated at the oil/water interface. To date, ethylcellulose is the only polymer gelator identified that can be directly incorporated into edible oils and form a polymeric network with potential for food applications. As such, it has received considerable attention as an oleogelator, and has been evaluated both as a single-component gelator and used in conjunction with other oil structuring systems.

2.7.2.1 Ethylcellulose

Ethylcellulose (EC) is a chemically modified cellulose derivative commonly used in pharmaceutical, cosmetic, and food packaging for barrier coating and controlled release applications. Commercial varieties are generally produced to have a degree of substitution of ~ 2.5 (~48.5% ethoxy content), which corresponds to a minimum in the glass transition temperature (T_g) of ~ 140°C (Rekhi and Jambhekar 1995). The molecular weight of EC is reported according to the viscosity of a 5 wt% solution in 80:20 toluene/ethanol, and commercial variety generally ranges from ~ 4 to 300 cP (Davidovich-Pinhas et al. 2014). Although EC is soluble in a variety of organic solvents, it cannot be dissolved in triglyceride oils. However, when heated above the T_g, the amorphous regions of this semi-crystalline polymer become more flexible (Davidovich-Pinhas et al. 2014), letting it adopt an extended conformation and become dispersed in the oil phase. Upon cooling, the formation of intermolecular hydrogen bonds between the unsubstituted hydroxyl groups along the EC backbone produces a rigid, cross-linked network, physically entrapping the liquid oil (Laredo et al. 2011; Zetzl et al. 2012, 2014). This process is shown schematically in Figure 2.7.1A, with physical crosslinks highlighted in the final gel state. The macroscopic behavior of the resulting oleogel is highly dependent on the polymer molecular weight, oil composition, and gelation conditions

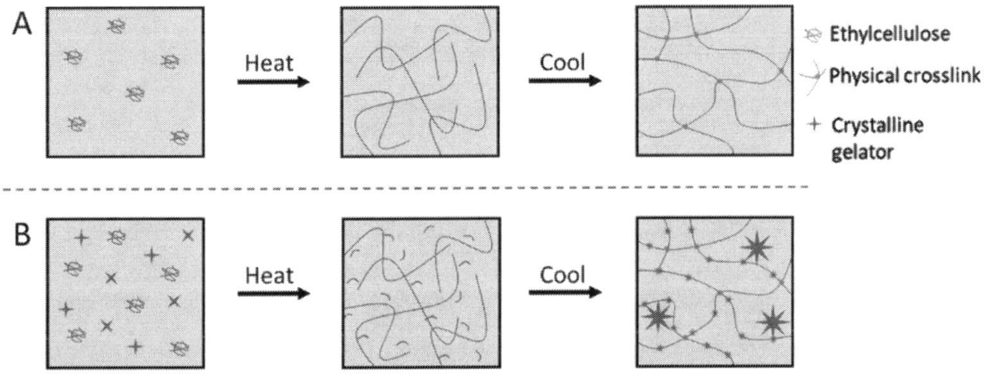

Figure 2.7.1 Schematic representation outlining the direct structuring methods used to produce EC-based polymeric oleogels (A) and EC/crystalline gelator multi-component oleogels (B).

(Davidovich-Pinhas et al. 2015a, 2015b; Gravelle et al. 2016). The positive correlation between gel strength and polymer molecular weight can be attributed to an increased number of intermolecular junction zones per polymer chain (Gravelle et al. 2014; Davidovich-Pinhas et al. 2015a). Cooling rate has also been shown to strongly affect gel strength, where a more rapid cooling results in inefficient network formation, and this effect becomes increasingly pronounced with increasing molecular weight (Gravelle et al. 2013; Davidovich-Pinhas et al. 2015b). In contrast, allowing EC-based oleogels to slowly progress through the T_g produces firmer gels, as the polymer chains remain in a more extended state during network formation.

The composition of the oil phase also has a pronounced impact on the mechanical properties of EC oleogels. It was first reported that oils with more unsaturation produced firmer gels (Laredo et al. 2011; Zetzl et al. 2012), and this effect was later confirmed to be primarily caused by polar compounds naturally present in the oil, or arising from thermally-induced lipid oxidation (Gravelle et al. 2012). To evaluate the effect of solvent polarity, Gravelle et al. supplemented canola oil with either castor oil or mineral oil, and demonstrated gel strength was linearly correlated to the content of polar compounds, as shown in Figure 2.7.2A, Figure 2.7.2B (Gravelle et al. 2016). The Hansen solubility parameter formalism was used to show these effects were primarily mediated by solvent/polymer hydrogen bonding interactions, while dispersive and intermolecular forces had little impact on EC solubility and gelation behavior (Stortz and Marangoni 2014; Gravelle et al. 2016).

The mechanical strength of EC oleogels has also been shown to drastically increase when surface-active small molecules are present in the oil. Adding oleic acid or oleyl alcohol to canola oil (Figure 2.7.2C) produced a dramatic increase in gel strength at low incorporation levels (<1 wt%). This reinforcing effect depended on the chemical structure of the additive, but approached a plateau for both molecules at ~ 5 wt%. This trend suggests a site-specific interaction between EC and the surface-active molecules, which is supported by the Langmuir binding curve fits depicted in in Figure 2.7.2C. Higher supplementation plasticized the polymer network, leading to a gradual reduction in gel strength as more additive was introduced; however, this was also noted to further reduce brittleness and oil release upon fracture, and promote gel elasticity (Figure 2.7.2C, insets) (Gravelle et al. 2016). Conversely, removing free fatty acids from both cold pressed flaxseed oil (Gravelle et al. 2016) and extra virgin olive oil (Giacintucci et al. 2018) has been shown to decrease gel strength. Using atomic force microscopy, the latter investigation showed that progressive refinement of olive oil led to a reduction in network connectivity and wall thickness in partially de-oiled gels, indicating an overall reduction in polymer/solvent interaction and network connectivity. The effect of common food grade, lipid-based small molecule surfactants has also been investigated, both

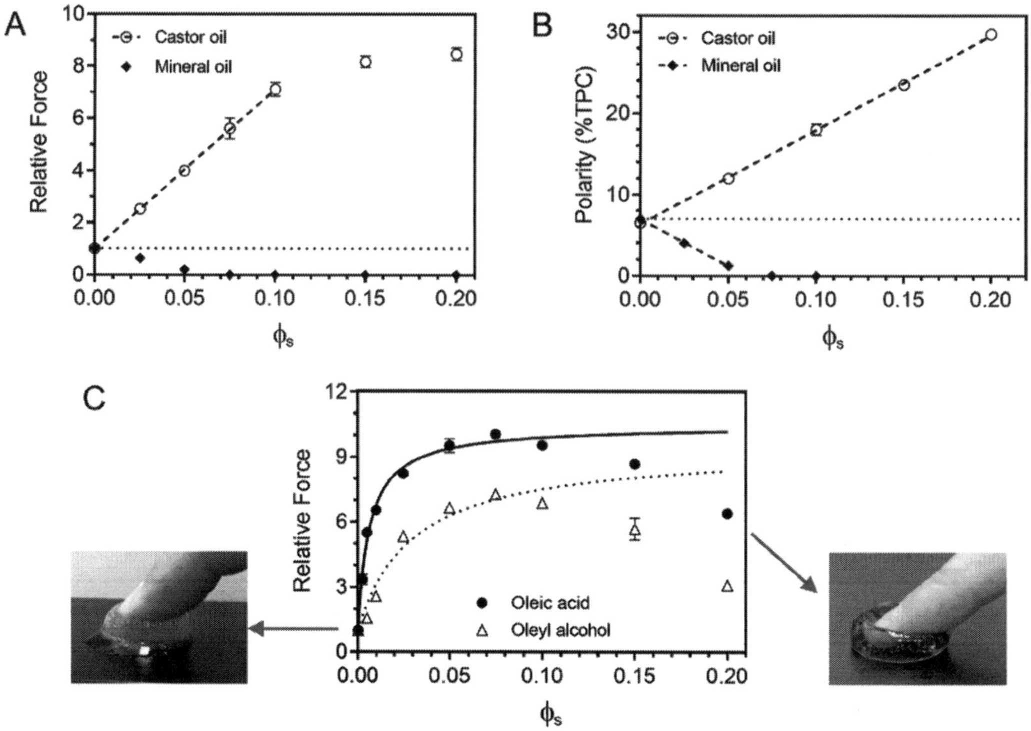

Figure 2.7.2 Normalized gel strength of ethylcellulose (EC) oleogels (A) resulting from changes in oil polarity (B) associated with supplementing canola oil with mineral oil or castor oil (wt% supplementation, φ_S). Effect of oleic acid and oleyl alcohol on normalized gel strength (C); Insets in (C) show the improved oil binding and elasticity with increasing plasticization. Adapted from Gravelle, A.J. et al. (2016), with permission from Elsevier.

to modulate the functional properties of EC oleogels and elucidate the physicochemical interactions responsible for these macroscopic behaviors (Gravelle et al. 2013, 2014; Davidovich-Pinhas et al. 2015). While the effect of surfactants depends largely on the properties of the polar headgroup, smaller groups and those with greater charge elicit the most pronounced reinforcement in EC oleogels. The observed enhancement occurs because these amphiphilic compounds can establish stronger interaction between the oil phase and the polymer via interactions with the unsubstituted hydroxyl groups (Davidovich-Pinhas et al. 2015; Gravelle et al. 2016). Strong plasticizers such as oleic or stearic acid also promote chain flexibility, which has been reported to increase the elastic behavior and reduce brittleness and oil leakage upon deformation (Gravelle et al. 2016). Due to these properties, adding various amphiphilic lipid-based small molecules has been explored as a menas of manipulating the properties of EC-based oleogels or forming hybrid gelator systems that could be used to design fat mimetics with appropriate techno-functional attributes.

2.7.2.2 EC-Based Hybrid Oleogelation Systems

It has been recognized that combining oleogelators as a general strategy may enhance or promote unique functional features in the resulting bulk structured oil (Shakeel et al. 2021). Many studies have focused on combining various small molecule gelator compounds, with the intent of identifying co-operative or co-crystallizing synergistic networks. Depending on the chemical properties and composition of the selected gelators, these systems may form independent, non-interacting networks (Doan et al. 2017; Harris et al. 2019), or they may promote a synergistic enhancement in gelation properties and impart unique physical features to the resulting oleogel (Okuro et al. 2018, 2020; Bin Sintang et al. 2020). Identifying complementary oil structuring agents has been recognized as a promising strategy for manipulating the bulk rheological, mechanical, and techno-functional attributes of fats, which may be tailored for the desired application (Pakseresht and Mazaheri Tehrani 2020; Shakeel et al. 2021).

The direct interaction between EC and surface-active small molecules noted above presents an alternative approach for developing multi-component oleogels. The unsubstituted hydroxyl groups along the polymer backbone provide a means of forming direct interactions with a secondary crystalline gelator phase (as depicted schematically in Figure 2.7.1B), so long as these structures possess enough polar character. Further, the physical nature of these bonds may be useful for plasticizing the polymer network, and could enhance critical techno-functional features in the combined system, such as plasticity and ductility (Gravelle et al. 2017b). Introducing a crystalline structuring network into an EC-based polymer matrix thus provides an opportunity to modulate the properties of directly dispersed oleogels.

2.7.2.2.1 Monoacylglycerol

The first study to use a direct hybrid polymer/crystalline gelator approach evaluated the effect of introducing EC into oleogels structured with monoacylglycerols (MAGs) (Lopez-Martínez et al. 2015). In this work, EC was incorporated below its critical gelation concentration, such that it could be suspended within the crystalline MAG network without providing a system-spanning polymer scaffold. This approach is notably distinct from incorporating a low concentration of amphiphilic molecules in an EC-based oleogel, as the crystalline network was the primary oil structuring compound. It was shown that adding 6 wt% of a low molecular weight EC (4 cP) increased the complex storage modulus (G^*) of the MAG-based oleogels, independent of the MAG concentration (2 wt% or 8 wt%). Oleogels augmented with EC also showed markedly improved oil binding capacity over 14 days of storage. X-ray diffraction (XRD) and differential scanning calorimetry (DSC) analyses indicated the MAG crystals underwent a polymorphic transition from the sub-α to the β phase during storage, which promoted oil syneresis. The addition of EC was shown to slow this transition, which directly correlated to the improved oil binding properties. This work thus demonstrated incorporating EC below its critical gelation concentration could provide a potential means of modulating the crystallization behavior and associated gelation properties of low molecular weight oleogelators.

Combining additional polymers with EC and MAGs was recently explored by Naeli et al. to develop a functional fat substitute with reduced saturates (Naeli et al. 2020). This work demonstrated a combination of the amphiphilic polymers hydroxypropyl methylcellulose (HPMC) and gum arabic could be directly incorporated into EC-based oleogels as part of the gel preparation process. The EC and MAG were first heated in oil above the T_g of EC, and the additional polymers could be directly dispersed in the mixture after cooling to ~ 90°C. Oleogels were formulated with either 1, 3, or 5 wt% 100 cP EC, 0 or 10 wt% palm stearin, and 0–1 wt% HPMC (with 0.2 wt% gum arabic when HPMC was employed). It is noteworthy that an EC content < ~ 4 wt% falls below the critical gelation concentration; however, it can be expected that synergistic interactions between MAG and EC contributed to the oil structuring capacity (Lopez-Martínez et al. 2015). Nonetheless, a centrifugal accelerated oil binding test showed that increasing HPMC content led to an improvement in oil binding for oleogels containing sunflower oil or sunflower oil supplemented with 10 wt% palm stearin.

Interestingly, while the gels containing palm stearin produced the same trends in oil retention, they showed greater oil loss relative to the equivalent formulations containing only liquid oil, despite the higher solid fat content. This can likely be attributed to a direct interaction between EC and palm stearin, which would be expected to enhance crystal nucleation and produce a network of smaller crystals which undergo plastic flow more easily (Gravelle et al. 2017b). Similar changes in the morphology of palm stearin crystals have been reported by Patel and coworkers when MC was successfully incorporated into the lipid phase through an indirect emulsion-based approach (Patel 2017). These interactions may also further plasticize EC, making it more susceptible to the accelerated oil loss test. The thermo-rheological profiles of all gels displayed similar overall behavior for both G′ and G″. A clear reduction in both moduli was reported between 40–60°C, which could be attributed to the melting of the MAG and MAG/palm stearin crystal networks. No crossover was observed up to 80°C, indicating EC caused the gels to maintain their elastic-dominated mechanical response. Interestingly, while the addition of palm stearin increased the firmness of EC-based gels (higher G′), the opposite trend was seen when HPMC and gum arabic were also present. This further suggests interactions between the various polymers and crystallizing agents have a strong impact on the structural and mechanical properties of multi-component polymer/crystalline oleogelator systems. It is therefore apparent that employing amphiphilic hydrocolloids may provide an additional strategy to optimize the thermal and mechanical behavior of structured edible oils.

2.7.2.2.2 Stearyl Alcohol and Stearic Acid

While incorporating EC as a crystal habit modifier may provide a promising approach to optimize the functional behavior of self-assembling small molecules, most published works have focused on using EC as the primary gelator. As EC may generally be viewed as providing a system-spanning polymeric scaffold to which the small molecule gelators can be introduced, the latter will be referred to as the secondary structuring agent. This approach to the hybrid polymer/crystalline gelator structuring strategy was first investigated using a combination of stearyl alcohol (SO) and stearic acid (SA) as the crystalline phase. Stearyl alcohol and stearic acid can each independently structure liquid oil; however, intermediate ratios of these small molecules will co-crystallize (Gandolfo et al. 2003, 2004; Schaink et al. 2007). The particular SO/SA ratio strongly affects the resulting crystal morphology and functional attributes, such as gel strength and oil binding capacity (Schaink et al. 2007; Blach et al. 2016; Yadav et al. 2017).

Gravelle et al. (Gravelle et al. 2017b) evaluated the potential of using a hybrid polymer/crystalline gelator approach to structure canola oil by combining 6 wt% 45 cP EC and 5 wt% SO/SA using a 7:3 alcohol/acid mass ratio. This ratio has been reported to produce the greatest gel strength and highest oil binding capacity (Schaink et al. 2007; Blach et al. 2016). Thermal analysis of the native SO/SA oleogels displayed a single crystallization event corresponding to the mixed SO/SA crystals. With the addition of EC, both the onset and peak crystallization temperatures increased, indicating the polymer served as a nucleation platform to promote crystal formation along the EC backbone. An additional minor shoulder which overlapped with the crystallization peak of the native SO/SA oleogels was also observed, suggesting free crystals also formed in the bulk oil phase. This independent thermal event indicated a secondary space-filling crystalline network persisted in the presence of the primary network-spanning EC polymer network.

Microstructural analysis further demonstrated the SO/SA crystals interact directly with the EC network. Absent of EC, brightfield micrographs of SO/SA oleogels displayed a clear platelet morphology when imaged in bulk. In stark contrast, the crystals in the hybrid system appeared as arching, fan-like clusters of elongated needles which could be attributed to the nucleation and growth of crystals along the polymer backbone. Images prepared directly on glass slides tend to show needle-like morphology due to the preferential edge-on growth of the platelets which align with the hydrophilic glass surface (Blake and Marangoni 2015). However, introducing EC led to a similar transition from a disordered network of needle-like crystals to a collection of smaller needles which assembled into arching, highly organized feather- or palm leaf-like structures (Figures 2.7.3A, 2.7.3B). This analysis provided direct evidence that EC served as a scaffold upon which the SO/SA crystals nucleated to form a hybrid gelator system, as opposed to maintaining two distinct domains which form a bicontinuous or interpenetrating network.

Thermo-rheological analysis further demonstrated the direct interaction between the two gelators. The crossover point during cooling (*i.e.*, where G′ = G″) is a commonly used indicator of the gel

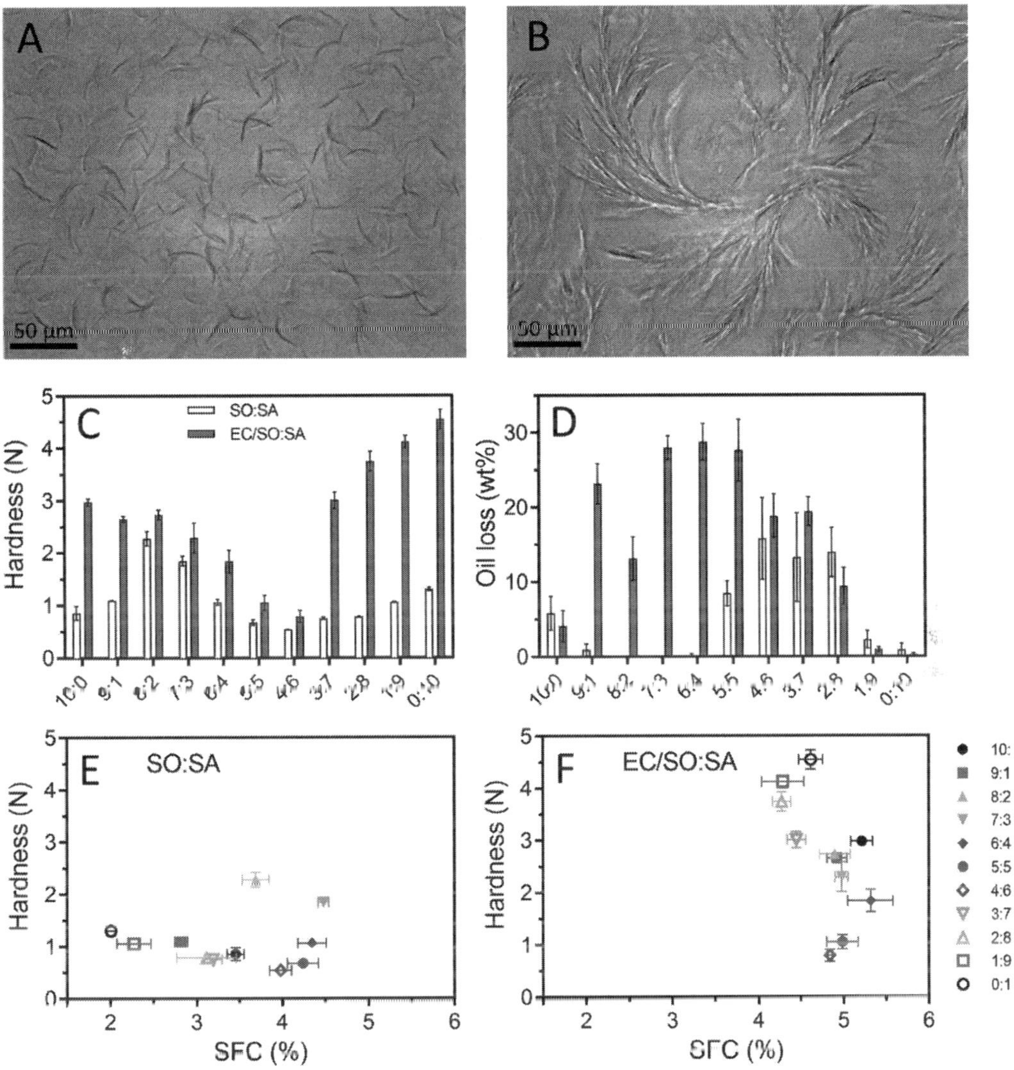

Figure 2.7.3 Microstructure (A,B) and physical properties (C-F) of oleogels prepared with mixtures of stearyl alcohol (SO) and stearic acid (SA) at varying SO:SA ratios (5 wt%), in the absence or presence of ethylcellulose (EC; 4 wt% 45cP). Micrographs were prepared with a 7:3 SO/SA mass ratio without (A) and with EC (B). Images in panels (C-F) reproduced from Gravelle, A.J. et al. (2017a) with permission of John Wiley & Sons.

point temperature. Gravelle et al. reported the crossover temperature decrease from ~ 110°C for EC oleogels to ~ 90°C for the EC-SO/SA hybrid system. It was additionally shown that absent oil, the glass transition temperature of neat EC progressively decreased, and the thermal event broadened with increasing SO/SA content. These observations provide evidence that the SO/SA crystals plasticize the polymer network. In support of this finding, the large deformation flow behavior of the native and hybrid gels were evaluated. While SO/SA oleogels displayed a flow profile indicative of brittle fracture, the hybrid system produced a synergistic increase in gel strength, while also nearly eliminating the brittle flow behavior. This reduction in brittle flow could be attributed to the increased nucleation and formation of smaller crystals, as well as the simultaneous plasticization of the polymer scaffold. Further, when EC was incorporated below the critical gelation concentration, the hybrid gel was weaker, but displayed highly plastic flow similar to commercial spreadable tub margarines (Gravelle et al. 2017b). In this formulation, EC did not form a system-spanning network and mainly promoted nucleation of smaller crystallites and altered crystal-crystal interactions. Both of these factors are known to influence the mechanical and functional traits of fat crystal networks, such as spreadability (Ramel et al. 2016).

Blach et al. explored the effect of the stoichiometric SO/SA ratio on the gelation properties and crystallization behavior of 5 wt% mixed SO/SA oleogels (Blach et al. 2016). These authors confirmed that ratios of 8:2 and 7:3 SO/SA produced oleogels with the highest gel strength and greatest oil binding (Figures 2.7.3C, 2.7.3D, white bars). In contrast, formulations with a higher SA content (5:5 and 4:6) exhibited much lower gel strength and poor oil binding, despite these ratios also being structured exclusively by mixed crystals (Blach et al. 2016). This phenomenon was attributed to the ratio of SO/SA molecules within the unit cell, as it has been reported that a 3:1 molecular ratio provides an optimal packing arrangement to minimize interfacial free energy in mixed SO/SA systems (Shah 1971). A combination of brightfield microscopy and cryo-SEM were used to show these optimal ratios produced larger, flatter platelet-like crystals, while those ratios with poor gelation properties were thinner and curved (Blach et al. 2016). It was further shown that the extreme ratios were predominantly structured by pure SO or pure SA crystals and exhibited intermediate oil structuring properties, despite having a lower solid fat content (*i.e.*, not all of the structurants crystallized).

In a follow-up study by Blach's group, they evaluated the impact of combining 4 wt% EC (45 cP) with 5 wt% mixed SO/SA oleogels at varying ratios to structure canola oil (Gravelle et al. 2017a). This EC concentration is below the critical gelation concentration and would not be expected to form a system-spanning polymer scaffold (Zetzl et al. 2012; Gravelle et al. 2014). However, all formulations displayed some increase in gel strength, with those containing pure SO (10:0, 9:1) or pure SA (8:2, 9:1, 10:0) crystals having a more pronounced effect (Figure 2.7.3C). Additionally, the oil binding capacity of the hybrid gels structured with mixed SO/SA crystals displayed higher oil losses than the corresponding gels without EC (Figure 2.7.3D). These effects were all correlated to the positive interaction between the polar headgroups of SO/SA molecules and the free hydroxyl groups of EC, which promoted crystal nucleation along the polymer backbone. These interactions, and a greater number of crystal nuclei led to an increase in total solids contributing to oil structuring. This effect was most pronounced in the formulations containing either pure SO or pure SA, which had more fatty alcohol or fatty acid molecules remaining in solution (Figure 2.7.3E vs 2.7.3F). This increase in pure SO and pure SA crystals was directly correlated to the observed increase in gel strength, and caused no detrimental impact on oil binding. In contrast, the gels predominantly structured by mixed SO/SA crystals had a comparatively small increase in solid fat content, and a smaller impact on gel strength. However, the large increase in accelerated oil loss for the gels structured by mixed crystals in the presence of EC was attributed to the change in crystal morphology and the ability of this network to enhance plasticization, resulting in greater flow when subjected to an externally applied load (centrifugation). These results further demonstrate that both the network structure and polymer/crystalline gelator interactions are critical parameters which can be manipulated to optimize the functional attributes of hybrid oil structuring systems.

2.7.2.2.3 Lauric Acid

Lauric acid, having a 12-carbon acyl chain, is generally classified as a medium-chain fatty acid, and has become a molecule of particular interest for plant-based meat and dairy alternatives. The latter is primarily due to the high content of lauric acid present in coconut oil, of which it accounts for roughly 50% of the total fatty acid composition (Bhatnagar et al. 2009). While coconut oil also contains longer chain saturated fatty acids (~90% total saturates), the high proportion of lauric acid imparts a relatively sharp melting profile slightly above room temperature, which can be useful for many food applications.

Adding lauric acid as a secondary structuring agent in EC oleogels has also been reported (Haj Eisa et al. 2020). It was shown that while the polymorphic form of lauric acid crystals was unaffected by the addition of 20 cP EC, these free fatty acids caused a decrease in the polymer's gelation temperature (*i.e.*, G′/G″ crossover). Both effects were consistent with that observed for EC-SO/SA hybrid gels outlined above (Gravelle et al. 2017b). However, EC also caused a depression in the crystallization temperature of lauric acid, which became more pronounced with increasing polymer concentration. It was suggested that interactions with the EC chains result in steric hindrance between the fatty acid molecules, thus impeding crystallization. This effect contrasted with that reported for SO/SA at a 7:3 mass ratio, resulting from crystal nucleation along the polymer backbone. This difference in crystallization behavior may be due to the mixed nature of the SO/SA crystals. The more polar nature of the carboxylic acid group will preferentially interact with EC, while the smaller hydroxyl group of stearyl alcohol may be less restricted in its ability to crystallize in the immediate proximity to the polymer backbone.

The EC-lauric acid hybrid gels displayed a synergistic enhancement in gel strength for all polymer concentrations above the CGC (6 wt% and 8 wt% EC), whereas 4 wt% and 2 wt% EC had a neutral impact and reduced gel strength, respectively. The latter is again consistent with that reported for EC-SO/SA hybrid gels, which was attributed to the formation of smaller crystallites. The neutral effect of 4% EC may have been due to a balance between partial network formation (close to the CGC) and a reduction in crystal size. Finally, the thixotropic recovery of the EC-lauric acid gels was evaluated for 8 wt% EC and 10 wt% lauric acid using a three-cycle oscillatory test. The hybrid system produced a greater structural recovery than the corresponding systems structured by EC or lauric acid alone, further demonstrating the potential for optimizing the techno-functional properties of oleogels using a hybrid polymer/crystalline gelator approach.

2.7.2.2.4 Behenic Acid

Behenic acid is a 22-carbon long-chain fatty acid, and has recently been shown to act as a crystal promoter in palm oil-based fats which can manipulate crystallization kinetics and post-crystallization hardening (Kim and Marangoni 2017; Cordova-Barragan et al. 2021). Combinations of behenic acid and behenyl alcohol have also been shown to structure oil, with analogous trends in the molecular ratio to that of SO/SA (Callau et al. 2020). EC-behenic acid hybrid gels were evaluated by Ahmadi et al. using a fixed 6 wt% total structurant, but altering the EC/behenic acid ratio (Ahmadi et al. 2020). While this experimental design makes it difficult to determine the direct impact of behenic acid on the polymer network, some interesting trends were reported. Although the rheological properties of the native systems were quite comparable, an enhancement in gel strength was seen for 5:1, 2:4, and 1:5 EC/behenic acid. In contrast, the intermediate ratios 4:2 and 3:3 produced much weaker gels (lower G′ and G″). These results indicate that behenic acid can potentially reinforce EC-based oleogels when the polymer forms a system-spanning network. Conversely a small amount of EC was shown to alter the crystallization behavior and morphology of behenic acid. Polarized light micrographs displayed a distinct transition from platelet-like crystals in native behenic acid gels to organized structures of fine needles, similar to that reported in EC-SO/SA hybrid systems (Gravelle et al. 2017a, 2017b). Solid fat content measurements also showed that EC promoted full crystallization of the behenic acid compared to native oleogels, which again follows that reported for stearic acid-based oleogels (Gravelle et al. 2017a).

The ability of EC to promote and alter the crystallization behavior of behenic acid resulted in higher oil binding for the 2:4 and 1:5 ratios relative to all other formulations evaluated. The weakest gels were those consisting of intermediate ratios with EC below its CGC (4:3 and 3:4 EC/behenic acid). In these formulations, the polymer primarily served to nucleate behenic acid crystals, and the lower gel strength can thus be attributed to the lower crystalline gelator concentration. In addition to gelation, it was shown that the 1:5 EC/behenic acid ratio produced stable emulsions containing 55 – 95 wt% oleogel and no added surfactant. It was suggested through this preliminary screening that EC may act as a stabilizing agent via a Pickering-type mechanism. During emulsification, a change in EC solubility when exposed to water may promote the formation of micro- or nanoparticles which could coat and stabilize the oil/water interface. Although recent studies have demonstrated the ability of EC nanoparticles to form Pickering emulsions, these are generally formed before emulsification through an anti-solvent precipitation procedure (Kibbelaar et al. 2022; Zhang et al. 2022). Thus, further analysis would be needed to confirm the proposed Pickering stabilization mechanism. Overall, the properties of EC-behenic acid hybrid oleogels were analogous to that observed for SO/SA; however, the higher melting point of behenic acid may provide an additional means of manipulating the temperature-dependent techno-functional properties of the hybrid oleogels.

2.7.2.2.5 Lecithin

Lecithin is a mixture of phospholipids commonly isolated from plant or animal cell membranes, and whose composition will depend on the biological source. Soy lecithin (Soy-Lec) is predominantly made up of phosphatidyl choline and phosphatidyl ethanolamine, and is widely used as a natural emulsifier in food applications. It has been shown that Soy-Lec can also structure liquid oils by introducing small amounts of water, inducing the formation of reverse micelles that grow uniaxially and assemble into elongated bundles (Shchipunov 2001; Bodennec et al. 2016). Water-induced Soy-Lec oleogels were recently evaluated in combination with EC to evaluate the thermo-rheological properties of the resulting hybrid networks (Aguilar-Zárate et al. 2019). This study compared the impact of two lecithin variants consisting predominantly of either unsaturated (C18:2) or saturated (C18:0) fatty acid moieties. The hybrid gels elicited up to a 10-fold increase in the G′ relative to native EC gels (10 wt% 20 cP EC), but this effect diminished with increasing EC content due to the increase in inter-polymer crosslink density. 1 wt% Soy-Lec was sufficient to elicit the increase in EC rheological properties, while higher concentrations diminished the visco-elastic response. The stress-strain response of EC/Soy-Lec hybrid gels (10% EC, 1% Soy-Lec) was also comparable to a commercial lard shortening, as these gels displayed a similar yield stress, but produced a higher yield strain. The authors suggested this discrepancy may be associated with the ability of EC polymer chains to deform to a higher extent before yielding relative to a polycrystalline triglyceride network. This "stress overshoot" behavior was also seen in native EC oleogels and was maintained in the high saturated variety of Soy-Lec, but was less pronounced for the hybrid gels structured with the more unsaturated surfactant.

Large amplitude oscillatory shear (LAOS) analysis was also used to evaluate the non-linear rheological behavior of the hybrid EC/Soy-Lec oleogels. It was shown that the transition from solid-like to fluid-like response was similar for the native EC and both EC/Soy-Lec varieties, suggesting the polymer network dominates this transitional behavior. The unsaturated Soy-Lec produced a less brittle mechanical response under LAOS, as indicated by a lower intra-cycle thickening/thinning. Further, the commercial lard underwent fluidization at much lower strain, and displayed little intracycle thickening/thinning. This may suggest that the polymer network would require a high level of plasticization to accurately mimic the large deformation rheological behavior of colloidal fat crystal networks. However, further work is needed to confirm the molecular interactions responsible for the dynamic macroscopic behavior seen in the EC/Soy-Lec hybrid oleogels.

2.7.2.2.6 EC/MAG Binary and Ternary Blends (Edible Shortenings)

EC/MAG binary and ternary blends have recently been explored as potential systems that could be optimized to mimic the performance of shortenings and margarines. The combination of surface activity and positive interaction with EC makes EC/MAG a promising hybrid system for this application. The work of Naeli et al. evaluating the potential of EC/MAG oleogels and EC/MAG used with a combination of HPMC and gum arabic as candidates for developing low saturated-fat shortenings has been further investigated using a response surface methodology (Naeli et al. 2022). Although 2% MAG was reportedly used in all formulations for its surfactant properties, its notable interaction with EC makes this worth highlighting, particularly as EC was incorporated below the CGC (≤ 3 wt%). Shortenings were prepared with 0–35 wt% palm stearin blended with sunflower oil, either without HPMC or 0.5–1.0 wt% HPMC and 0.2 wt% gum arabic. All formulations were prepared in a lab-scale scraped surface heat exchanger, and a response surface optimization methodology was used to separately evaluate the single and multi-polymer systems. Key optimization attributes were compared to two commercial margarines, including fatty acid composition, oxidative stability, and mechanical and rheological performance. The formulations structured with EC/MAG were evaluated over a polymer concentration range of 1–3 wt%, while those containing HPMC were conducted with fixed 3 wt% EC and varied the HPMC composition from 0.5–1.0 wt%. It was shown that mechanical parameters such as product firmness, G′ in the linear viscoelastic regime, yield strain, and strain at crossover (where G′ = G″) could be optimized to mimic that of commercial products with a substantial reduction in saturated fat content (15–17 % relative to ~ 50 % in commercial products). However, the intended application should also be considered when using a gold-standard optimization approach, as formulation and processing conditions can be used to alter critical functional properties of the final product (e.g., spreadability or ductility). These differences can manifest in relatively nuanced mechanical and rheological behaviors (Macias-Rodriguez and Marangoni 2016; Macias-Rodriguez et al. 2018). Further work into elucidating the mechanical signatures and underlying structural features which produce key performance attributes in fat crystal networks will therefore be needed to broaden this approach.

The mechanical, thermo-rheological and microstructural properties of EC/MAG-based oleogels were also recently evaluated with respect to the functional properties of various commercial margarines (Rodríguez-Hernández et al. 2021). This study further evaluated the impact of incorporating an additional high melting point natural wax as a second crystalline gelator. Oleogels were prepared by varying EC concentration (4, 8, or 12 wt% 45 cP), and incorporating either 5 wt% MAG or a combination of 5 wt% MAG and 3 wt% candelilla wax as the crystalline gelator phase. The low concentration of candelilla wax was selected due to its high melting temperature, which can produce an undesirable mouth-coating effect (Pérez-Martínez et al. 2019), and earlier reports indicating poor shear recovery of wax-based oleogels with > 3 wt% structurant. Low-magnification polarized light micrographs indicated the MAG crystals were visibly smaller when formed in the presence of candelilla wax but were drastically reduced with the introduction of EC (both in the binary and ternary mixtures). This is consistent with the effect of EC on the crystallization behavior of SO/SA mixtures (Gravelle et al. 2017a). During cooling, both binary and ternary gels displayed a sharp increase in gel stiffness starting at ~ 35°C, which could be attributed to the MAG crystal network formation. In the ternary system, the extent of stiffening was greater, and plateaued at slightly lower temperatures. This enhanced stiffness (measured as the complex modulus, G^* at 5°C) was observed independent of polymer concentration. Additionally, the oils structured with either MAG or candelilla wax alone showed very poor mechanical recovery after application of high shear (< 5%), whereas adding EC resulted in > 100% shear recovery in both systems. This recovery behavior was attributed to a rearrangement in the polymer chains facilitated by EC/MAG interactions but was also similar to that of various commercial margarines. This work demonstrated that incorporating complementary crystalline gelators could be an effective approach to tailor the temperature-dependent mechanical performance of EC-based oleogels. Further, polymer/crystalline gelator interactions may promote useful functional features to the crystalline phase, such as improving structural recovery in otherwise inherently brittle networks.

The production of EC-based water-in-oleogel (W/OG) emulsions has also been reported, which are of particular relevance to developing margarine-type products with low saturated fat content. While the high heating step to prepare EC oleogels is a challenge for incorporating water, García-Ortega et al. developed a protocol to prepare 20 wt% W/OG emulsions using high oleic safflower oil and either 0.5 or 1.0 wt% glycerol monostearate and 7 wt% 4 cP EC (García-Ortega et al. 2021). Briefly, a concentrated EC oleogel was first separately prepared, partially cooled, and combined with a concentrated GMS-based W/O emulsion under shear at 70°C. The mixture was then cooled to 20°C under mixing to form the final W/OG emulsion. This process also allowed GMS to assemble at the W/O interface before being exposed to EC, thus ensuring greater interfacial association which would promote maximal EC/emulsion droplet interactions. In the absence of EC, GMS-based emulsions were unstable, and phase separated after 24 hours. Reference EC/GMS-based gels were prepared with the same emulsification protocol (excluding water addition) and did not form self-supporting gels due to the shear-sensitivity of the EC network. In contrast, the hybrid emulsion gels showed no evidence of phase separation, which could be attributed to the cooperative nature of the structuring agents. While the gels prepared with 0.5 wt% GMS showed viscous flow, 1 wt% GMS was sufficient to impart a self-supporting structure and solid-like rheological response. Brightfield and confocal micrographs showed the latter formulation produced smaller water droplets, which likely contributed to their more solid-like character. The viscosity-enhancing effect of EC would also be a significant contributing factor to the gel stability. The gelled emulsions were also stable after freeze-thaw temperature cycling, which has been shown to promote surfactant desorption and droplet coalescence (Ghosh et al. 2011). No emulsion destabilization was observed, indicating the favorable EC/GMS interactions and high continuous phase viscosity allowed the surfactant molecules to remain at the interface.

The thermo-rheological behavior of the gelled emulsions was compared with two commercial low-fat margarine spreads. It was reported that the gelled emulsions had a lower G' both in the linear viscoelastic regime and at the crossover point ($G' = G''$). This difference was attributed to both the amount of structurant present in the gelled emulsions and the water droplet size, as it was noted that the commercial products had finer water droplets. Therefore, while EC/GMS hybrid gelled emulsions were shown to be a promising strategy for developing low saturated fat spreads, future work must optimize both formulation and processing conditions to fine-tune the functional properties.

2.7.2.3 EC-Based Oleogels in Food Applications and Nutrient Delivery

In addition to serving as shortening substitutes, EC and EC-based oleogels have been extensively explored in a variety of food applications, both as fat substitutes and as a crystal habit modifier (see Table 2.7.1). EC- and EC/hybrid oleogels have been incorporated into a variety of meat products, including frankfurters (Zetzl et al. 2012; Barbut et al. 2016a, 2016b; Barbut and Marangoni 2019), breakfast sausages (Barbut et al. 2016c), burgers (Gómez-Estaca et al. 2019b; Adili et al. 2020), cooked salami (Woern et al. 2021), and liver pâté (Barbut et al. 2019; Gómez-Estaca et al. 2019a; Barbut et al. 2021). In general, oleogels have been used as a means to improve the fatty acid composition of processed meat products, which are generally high in saturated fats. The high heat tolerance of EC oleogels also makes them desirable candidates for applications requiring cooking, as this can reduce oil migration and fluid loss; however, this can also lead to some undesirable sensory properties. The addition of secondary structuring agents has also been used in many cases to alter their mechanical strength and functional performance for the targeted application. In this regard, lipid-based crystalline gelators are a promising candidate, as they may provide both increased plasticity and a desirable melting behavior. However, the ability of EC to form a scaffolding-like network within the lipid phase suggests it may also be a promising candidate for developing adipose tissue analogs that could be used in whole-cut plant-based meat products (McClements and Grossmann 2021).

EC and EC-based oleogels have also been incorporated into chocolate by several groups. Using a solvent-substitution approach, Stortz et al. reported that directly adding EC to various types of chocolate can provide heat resistance (Stortz and Marangoni 2013; Stortz et al. 2014, 2015). This effect was associated with polymer-sucrose hydrogen bonding interactions, which promoted oil structuring when the crystalline fat network melted (Stortz et al. 2014). EC has also been evaluated as a potential surfactant replacer and viscosity modifier in dark chocolate (Ceballos et al. 2016). It was shown that EC could successfully modify the rheological properties of liquid chocolate, which is a critical aspect of processing and production. Further, the introduction of EC did not interfere with the formation of the form V cocoa butter polymorph. Corn oil-based EC oleogels have also been evaluated as a cocoa butter substitute for saturated fat reduction in dark chocolate (Li and Liu 2019; Li et al. 2021). This work demonstrated that 50 wt% replacement of cocoa butter with oleogel was possible and did not interfere with the tempering process; however, the reduction in solid fat content resulted in softer chocolates. It was further shown the cocoa butter remained in form V over a 60-day storage period, indicating EC did not promote fat bloom.

EC oleogels have also been used as a strategy to reduce the saturated fat or total fat content in cream cheese and various baked goods. Bemer et al. achieved 25% fat reduction in a cream cheese product via direct substitution with EC oleogels, although this led to a reduction in product firmness and adhesiveness (Bemer et al. 2016). To address this issue, introducing a secondary crystalline gelator may provide a means of improving the performance of EC oleogels for dairy applications. Ye and coworkers conducted a broad screening of EC-based oleogels intended to serve as shortening alternatives in wheat-based breads (Ye et al. 2019). The effect of varying EC molecular weight, concentration, content of hard stock fat (palm stearin), and addition of various small molecule surfactants were cataloged. Using this approach, optimal oleogel formulations were identified which provided excellent air incorporation and bread volume equivalent to a commercial shortening. Several patents have also been awarded outlining the production and use of EC oleogels in cookies to impede oil migration (Cattaruzza et al. 2014) and the development of a reduced fat laminating shortening substitute (Ergun et al. 2017). The diversity of food products in which EC and EC-based multicomponent oleogel systems have been successfully used demonstrates the great potential of this strategy for developing novel fat mimetics. Further studies investigating the molecular

Table 2.7.1 Summary of studies using EC-based oleogels in food systems.

Food	Product	EC content	Oil type	Additive	Reference	Notes
Chocolate	Compound chocolate	1.0 or 2.2 wt% of chocolate 10cP	Palm kernel oil, cocoa butter	–	(Stortz and Marangoni 2013)	EC used to promote heat resistance
		2.17 wt% of chocolate 10 cP	Palm kernel oil, cocoa butter	–	(Stortz et al. 2014)	EC used to promote heat resistance
	Chocolate (milk, dark, white)	2.17 wt% of chocolate 10 cP	Palm kernel oil	–	(Stortz et al. 2015)	EC used to promote heat resistance
	Dark chocolate	0–0.20 wt%, 10, 20, 45, 100 cP	Medium chain triglyceride	–	(Ceballos et al. 2016)	EC used as surfactant replacer, MCT replaced up to 2% cocoa butter
		10% 45cP	Corn oil	–	(Li and Liu 2019; Li et al. 2021)	50% cocoa butter replacement
Dairy	Cream cheese	10% 45 cP	Soybean or high oleic soybean	–	(Bemer et al. 2016)	100% milk fat replacement
Meats	Frankfurters	10% 10cP	Canola	–	(Zetzl et al. 2012)	100% beef fat replacement
		8, 10% 10 cP	Canola	0, 1.5, 3% SMS	(Barbut et al. 2016a)	100% beef fat replacement
		8, 10, 12, 14% 10 cP	Canola	0, 1.5, 3% SMS	(Barbut et al. 2016b)	100% beef fat replacement
		10% 10 cP	Canola	5% SMS	(Barbut and Marangoni 2019)	100% beef fat replacement
	Breakfast sausages	8, 10, 12, 14% 10 cP	Canola	0, 1.5, 3% SMS	(Barbut et al. 2016c)	100% pork back fat replacement
	Low-fat pork burger	11% 10 cP	Olive, flaxseed, and fish (blend)	–	(Gómez-Estaca et al. 2019b)	100% pork back fat replacement
	Beef burger	2% (*vns)	Soybean	4% adipic acid	(Adili et al. 2020)	50% beef fat replacement
	Cooked salami	6, 8, 9, 10, 11, 12, 14% 20 cP	Canola	3% GMS, 8:2 SO:SA, or soy lecithin	(Woern et al. 2021)	50% pork/beef fat replacement
	Pâté (pork liver)	12, 14% 20 cP	Canola	0, 1.5, 3% GMS or 1% commercial stabilizer	(Barbut et al. 2019)	100% pork fat replacement

		12% 20 cP	Canola	3% GMS	(Barbut et al. 2021)	0–100% pork fat replacement
		11% 10 cP	Olive, flaxseed, and fish blend	3.67% SMS	(Gómez-Estaca et al. 2019a)	60% or 100% pork fat replacement
Baked goods	Cake	2% (*vns)	Soybean	4% adipic acid	(Adili et al. 2020)	50% shortening replacement
	Bread	1–6% 7, 20, 50, 100, or 200 cP	Soybean with palm stearin (20–30%)	1% surfactant (various)	(Ye et al. 2019)	EC used to structure soybean/palm stearin blend
Edible shortening	–	0, 4, 8, 12% 45 cP	High oleic safflower	MAG, MAG + CLX	(Rodríguez-Hernández et al. 2021)	EC used below CGC; altered MAG, CLX crystallization
		1–5% 100 cP	Sunflower with palm stearin (0 or 10%)	HPMC + MAG + gum arabic	(Naeli et al. 2020)	EC used below CGC;
		1–3% 100 cP	Sunflower oil with palm stearin (0–35%)	HPMC + MAG + gum arabic	(Naeli et al. 2022)	cellulose derivatives altered crystallization behavior of MAG and palm stearin

Additive denotes an additional component when a multi-component oleogelator approach was used.
*vns–viscosity not specified; CGC–critical gelation concentration.

interactions which dictate the functional behavior of EC/crystalline gelator hybrid networks will undoubtedly aid in expanding this approach to more accurately mimic the rheological, mechanical, and thermal properties of traditional solid fats.

Structuring edible oils also provides an opportunity to introduce lipid-soluble nutraceuticals and could have additional implications for digestion behavior. O'Sullivan et al. investigated the potential of using EC-based oleogels as a delivery vehicle for β-carotene (O'Sullivan et al. 2017). Using a static *in vitro* digestion model, it was shown that oleogel hardness was negatively correlated to the extent of lipid hydrolysis, and the associated transfer of β-carotene. Thus, physical entrapment of the oil phase could be used as a potential controlled release approach for the delivery of lipid-soluble nutraceuticals. The *in vivo* physiological response associated with triglyceride- and oleogel-based lipid structuring has also been addressed in the context of consuming a carbohydrate rich meal (Tan et al. 2017a). Rice porridge containing either palm oil or rice bran oil produced a significant increase in plasma triglycerides, with palm-based meals producing a markedly higher response. In contrast, no significant change in plasma triglycerides was seen compared to baseline levels up to 6 hours post-consumption when either of these oils was structured with EC (11 wt% 45cP). While other biological markers such as plasma glucose, insulin, and non-esterified fatty acids were not affected by oleogelation, a similar effect on plasma triglycerides was observed when test meals were formulated with coconut oil and EC-based coconut oil oleogels (Tan et al. 2017b). While the scope of these investigations was limited to healthy Chinese males, this work suggests physical entrapment of lipids within a polymer-based oleogel network may provide a means of modulating the rate of lipid digestion and the associated physiological response. This may be of particular interest for addressing populations who regularly consume high saturated fat foods in excess.

2.7.2.4 Chitin

Chitin is a naturally occurring structural polysaccharide found in the outer cell walls of fungi and is also a ubiquitous component in the exoskeleton of arthropods, where it is integrated into composite biomaterials with varying rigidity. Chitin is a homopolymer made up of repeating N-acetylglucosamine units, with the acetylamino group at the C2 position facilitates extensive intermolecular hydrogen bonding. This results in the formation of ordered microcrystalline domains, from which its structural rigidity is derived. After ethylcellulose, chitin is the only polymer that has been used to directly structure edible oils for potential food applications. This was first demonstrated by Nikiforidis and Scholten (Nikiforidis and Scholten 2015), who evaluated the oleogelation properties of both chitin and chitin nanocrystals in sunflower oil. It was reported that while crude purified chitin could be dispersed in oil, extensive aggregation led to sedimentation and syneresis. This was consistent with previous studies using chitin as a potential biolubricant, in which "oil bleed" was noted upon deformation (Sánchez et al. 2011). However, stable oleogels could be formed by introducing various surfactants, including phosphatidyl choline, enzymatically hydrolyzed phosphatidyl choline, and sorbitan monostearate. It was proposed these amphiphilic small molecules interacted with the hydrophilic domains of chitin, thus impeding polymer-polymer hydrophilic interactions and improving dispersibility. The chemical structure of the surfactants also had a strong impact on the mechanical strength and temperature response of the resulting oleogels. It was reported that phosphatidyl choline and sorbitan monostearate promoted structural order in the chitin network during heating, while the rheological behavior of hydrolyzed phosphatidyl choline displayed no temperature dependence up to 90°C.

As a second approach to improve the oil structuring properties of chitin, nanocrystals were generated from the crude polymer using an acid hydrolysis procedure, which were then combined with surfactants to produce oleogels. The needle-like nanocrystals were ~ 200–500 nm and produced higher gel strength than the cruse chitin at lower structurant concentrations (≤10 wt%). This improved gelation efficiency was attributed to more extensive particle-particle interactions. Furthermore, adding water to both the crude and nanocrystalline chitin oleogels prepared with phosphatidyl choline substantially increased gel strength. It was shown that increasing water content from 5% to 25% produced a ~ 10-fold increase in the shear modulus for both variations of chitin. Confocal micrographs suggested this dramatic increase in mechanical strength resulted from swelling of the polymer, which was expected to increase particle jamming. This result may be of interest for producing oleogels with reduced caloric content.

An alternative approach to improving the gelation properties of chitin nanocrystals has been explored by Huang et al. in which saturated fatty acyl chains were incorporated onto the polymer backbone via chemical modification (Huang et al. 2015). By altering the reaction conditions, the degree of substitution, and thus the overall hydrophobicity of the purified chitin nanowhiskers could be successfully modulated. The modified polymers displayed a distinct X-ray diffraction pattern and a reversible melting peak between 40–50°C associated with the alkyl chains which positively correlated to the degree of substitution. The hydrophobic chitin could thus be directly dispersed in sunflower oil at 60°C, resulting in structured oleogels after subsequent refrigeration, analogous to the procedure for preparing oleogels with natural waxes and other lipid-based structurants (Zhao et al. 2021). These gels could also be formulated to exhibit melting behavior below or above room temperature, depending on the DS and nanocrystal concentration. While this approach shows promise as a novel method for structuring triglyceride oils, further work will be needed to evaluate the structural morphology and rheological properties of this system to optimize the techno-functional behavior.

2.7.3 Indirect Structuring

While the vast majority of food biopolymers cannot structure edible oils through the direct dispersion approach, several alternative methods have been identified which broaden the potential for producing oleogels using polysaccharides. These strategies take advantage of the intrinsic amphiphilic (or hydrophilic) character of select polysaccharides widely used in the food industry to develop aqueous "templates" from which water is removed to generate an oil-continuous structured network. The most common approaches involve initially forming either a concentrated emulsion (Figure 2.7.4A) or aqueous foam (Figure 2.7.4B) from which the water phase is subsequently removed. In the latter case, dehydration of the polymer matrix generates an aerogel, to which the oil phase may be incorporated via passive absorption, or this network may be chopped before combining with the liquid oil. For both strategies, a secondary hydrocolloid may also be added to enhance viscosity and interfacial stability to ensure the polymeric network maintains its integrity during the drying process. The resulting oil-filled structures are then subjected to high shear, producing an oil-continuous system in which the polymers remain suspended, thus effectively structuring the liquid phase. An alternative variation to the aerogel templating approach has also been demonstrated by forming cross-linked hydrogels and subsequently using a progressive solvent substitution to introduce CO_2, which may then be removed via supercritical drying (Figure 2.7.4C). While these strategies have also been used to develop protein-based and protein/polysaccharide templates (Feichtinger and Scholten 2020), this overview will highlight those systems in which polysaccharides are the primary gelling agent.

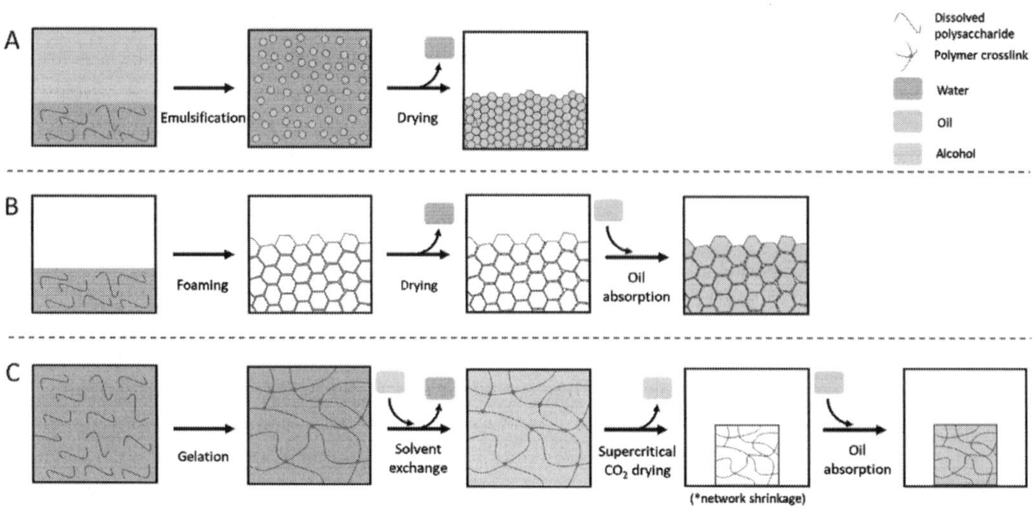

Figure 2.7.4 Schematic representation outlining the standard workflows for preparing polysaccharide-based oleogels using indirect methods; emulsion-templating (A), foam-templating (B), and hydrogel-templating via supercritical CO_2 drying after substitution with an intermediate solvent (C).

2.7.3.1 Emulsion-Templating

The emulsion-templating approach was first conceptualized and demonstrated by Romoscanu and Mezzenga, using β-lactoglobulin as the emulsifying agent (Romoscanu and Mezzenga 2006). Oil in water emulsions were prepared and freeze dried, effectively forming a viscoelastic protein-in-oil colloidal gel network. However, this procedure required the removal of excess unabsorbed protein and either thermal or chemical crosslinking to enhance the stability of the interfacial film. Patel and coworkers later expanded this approach using the surface-active polysaccharides methylcellulose (MC) and hydroxypropyl methylcellulose (HPMC) to form concentrated emulsions (~60 wt% oil), from which water was removed via passive oven drying (Patel et al. 2014a). Xanthan gum (XG) was also incorporated as a stabilizing agent to enhance the solution viscosity and increase interfacial elasticity of the emulsion droplets, allowing them to withstand the drying process without cross-linking (Patel 2018). All oleogels were formulated with a final concentration of 1.3 wt% polymer and ~ 1 wt% XG in sunflower oil. A schematic representation of this process and both photographs and scanning electron micrographs of a concentrated emulsion, and the resulting dried oleogel (before shearing) are shown in Figure 2.7.5A. After drying, the emulsion droplets became deformed as they were forced to pack into a highly dense network, reminiscent of a foam structure; however, these dried emulsions were sheared before evaluation to develop an oil-continuous network. This work contrasted low- and high-viscosity grades for each polymer, and showed for both cellulose derivatives the lower viscosity grades produced finer emulsion droplets. While these smaller droplets enhanced the rheological response of the concentrated emulsion (*i.e.*, increased G′), they produced weaker oleogels after drying and shearing. This effect was attributed to the relative stiffness of the interfacial film formed around the emulsion droplets, which was expected to be greater for the higher viscosity grades (Patel et al. 2014a). All formulations displayed substantial structural recovery after application of high shear (Figure 2.7.5B), and a modest increase in gel strength during heating, which occurred between ~ 45–65°C (Figure 2.7.5C). The latter effect can be attributed to the ability of these polymers to undergo a thermally-induced

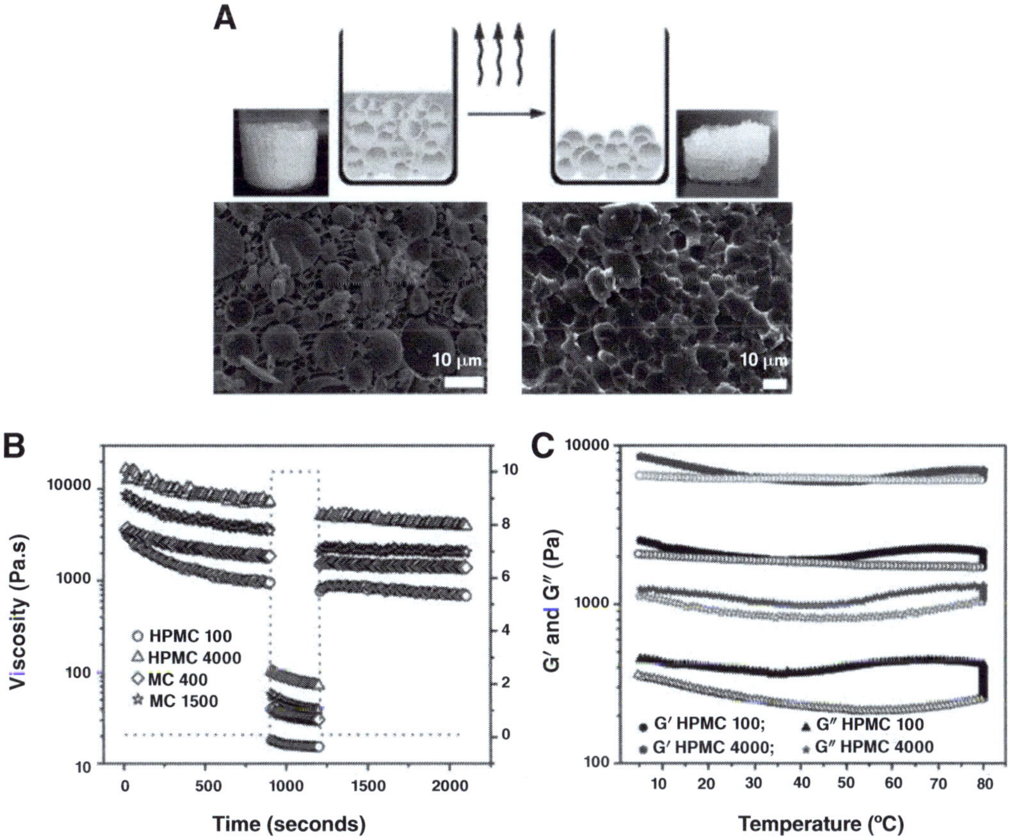

Figure 2.7.5 Structural and rheological properties of emulsion-templated oleogels prepared from methylcellulose and hydroxypropyl methylcellulose. (A) Schematic representation of emulsion-templated oleogel formation process, with corresponding photographs and scanning electron micrographs; (B) Shear recovery and (C) temperature-dependent elastic (G′) and viscous (G″) moduli of emulsion-templated oleogels. Reproduced from Patel, A.R. et al. (2014a) with permission from John Wiley & Sons.

sol-gel transition associated with changes in the relative contribution of hydrogen bonding and hydrophobic interactions (Coffey et al. 2006).

Following this initial work, MC/XG emulsion-templated oleogels with varying polymer content were evaluated as potential shortening alternatives in cake batters (Patel et al. 2014b). Interestingly, while the MC viscosity grade had a dramatic impact on the oleogel rheological properties, reducing the polymer concentration from 1.3 wt% to 0.65 wt% had no effect on the shear stress or frequency response. As a shortening alternative, the strongest oleogels were compared to liquid oil, margarine, and cake shortening. The rheology and microstructure of batters containing oleogels was most like those prepared with liquid oil, which were weaker and provided less aeration than margarine or shortening. This behavior was attributed to the absence of both solid fat crystals and emulsifiers which would be present in the commercial shortenings. The resulting cake textures containing oleogels were also intermediate between liquid oil and cake shortening, thus demonstrating the potential opportunity for optimizing emulsion-templated oleogels for food applications.

Meng et al. further evaluated the properties of HPMC/XG and MC/XG emulsion-templated oleogels, elaborating on the microstructural and physicochemical interactions supporting these systems (Meng et al. 2018a). These authors showed that HPMC produced concentrated emulsions with finer droplets than MC (60% oil, 0.6% HPMC or MC, and 0.3% XG). Microstructural analysis was conducted using polarized light and scanning electron microscopy, and confirmed this trend was consistent for both oven- and freeze-dried emulsions. As the difference in emulsion droplet size was not previously observed between these two cellulose derivatives (Patel et al. 2014a), it is likely that the emulsification properties depend strongly on polymerization, in addition to molecular weight and processing conditions. However, in agreement with earlier work, increasing polymer molecular weight was found to enhance the rheological response, and all gels displayed a high structural recovery and marginal temperature dependence, irrespective of the formulation. Fourier Transform Infrared (FTIR) spectroscopy analysis suggested that hydrogen bonding contributed to the oleogel structuring; however, interfacial polymer-polymer interactions may have also aided in stabilizing the individual droplets. Finally, it was shown that vacuum drying the emulsions decreased the drying time needed to achieve full water removal, resulting in improved oxidative stability. This group further showed increasing HPMC concentration reduced oil droplet size, which produced a corresponding increase in emulsion stability, oleogel firmness, and oil retention (Meng et al. 2018b).

The type of secondary thickening hydrocolloid has also been shown to influence the functional behavior of HPMC-based emulsion-templated oleogels. Meng and coworkers evaluated oleogels formulated with 0.6 wt% 1,500 cP HPMC and 0.3 wt% thickener, with the latter being either carboxymethyl cellulose, xanthan gum, sodium alginate, acacia gum, guar gum, fenugreek gum, or locust bean gum (Meng et al. 2018c). All the resulting concentrated emulsions (60 wt% oil) were visually similar, and only minor differences were apparent in the microstructure of vacuum dried emulsions visualized by polarized light. All oleogels displayed analogous rheological behavior, with a similar limit of linearity and crossover point, a modest shear frequency dependence, and pronounced shear-thinning and structural recovery. While some differences in microstructure and performance were noted across the thickeners, these effects could be attributed to the ability of the additives to interact with HPMC and their thickening behavior in the concentrated emulsions. In summary, emulsion-templated oleogel performance is predominantly dictated by the ability of the emulsion to remain stable during the drying phase, which can be strongly influenced by the secondary thickening hydrocolloid.

The need to remove water from emulsion-templated oleogels may be viewed as one of the major technological hurdles associated with this strategy, as it requires subjecting the oil phase to unfavorable conditions which can promote oxidation and rancidity. The drying procedure (*i.e.*, temperature and time) can therefore have a substantial impact on both oil stability and the physical properties of the resulting oleogels (Bascuas et al. 2020b). To address this, Bascus et al. evaluated the effect of conventional oven drying and vacuum drying on HPMC/XG emulsion-templated oleogels prepared with sunflower oil, olive oil, or flaxseed oil (Bascuas et al. 2020a). Each gel was prepared from a concentrated emulsion formulated with 60 wt% oil, 1 wt% 4,000 cP HPMC and 0.6 wt% XG, and either conventionally dried at 80°C for 10.5 h or under vacuum at 60°C for 14 h. The selected conditions were found to be the minimum times required to achieve a constant dry weight (Bascuas et al. 2020b). Cryo-SEM of the emulsions showed both a distinct surface coating stabilizing the oil droplets and a polymeric network extending throughout the aqueous phase. After drying, the olive oil emulsion droplets were densely packed and deformed, but similar in size to the initial emulsion, irrespective of the drying conditions. In contrast, the sunflower oil emulsions displayed evidence of coalescence, with fewer smaller droplets in the conventionally dried emulsions, and distinctly larger droplets resulting from the more aggressive vacuum drying conditions. This same trend was more evident in the flaxseed oil emulsions, with very large droplets present after conventional drying, while stable dried emulsions could not be produced under vacuum. These trends were correlated to the organization of the hydrocolloid network in the aqueous phase, which was thicker and more dispersed in the sunflower oil emulsions, and notably less prevalent between the flaxseed oil droplets. Oil type and drying conditions had only a marginal effect on the rheological behavior, except for flaxseed oil-based oleogels which were much weaker due to the high degree of coalescence. Finally, vacuum drying successfully reduced the formation of peroxides induced by the drying procedure. This approach could thus improve the shelf-life of emulsion-templated oleogels, particularly if used in conjunction with anti-oxidants and by employing more thermally stable oils. This work further highlights the importance of the interaction between the emulsifying polymer and the oil phase in dictating the structural features and associated functional behavior of emulsion-templated oleogels. Similar trends were also reported for both HPMC- and MC-based oleogels prepared without a thickening hydrocolloid (Espert et al. 2020). Specifically, both increasing the emulsifier content or reducing the oil content of the initial emulsion produced firmer gels with higher oil binding capacity. As previous studies have universally incorporated a thickening hydrocolloid, excess emulsifying hydrocolloid may serve a similar role upon drying by promoting increased rigidity of the interfacial film and enhanced polymer-polymer interactions in the final oil-continuous gel matrix.

In addition to cellulose ethers, several other polysaccharide-based oleogels have been developed using the emulsion-templated approach. While numerous polymer-based emulsifier systems combine proteins and polysaccharides (Abdolmaleki et al. 2020; Alizadeh et al. 2020; Urbánková et al. 2021), those which exclusively employ polysaccharides are highlighted here. Citrus pectin-based emulsions have been combined with tea polyphenol-palmitate as a secondary crystalline oleogelator to structure Camilla seed oil (Luo et al. 2019). Consistent with other stabilizing hydrocolloids, increasing pectin concentration stiffened the interfacial film of the emulsion droplets, leading to firmer, more viscoelastic oleogels and improved oil binding properties. The presence of tea polyphenols also improved the oil stability. Finally, when used as a butter substitute in cake batter, the oleogels performed well, but the performance deteriorated with increasing pectin content, indicating formulations could be optimized based on the intended application.

A combination of locust bean gum and XG has also been explored as a potential emulsifying system for producing polysaccharide-based oleogels (Espert et al. 2022). The initial emulsions (20 wt%) had substantially higher viscosity and viscoelastic properties compared to emulsions produced using other surface-active molecules (whey protein, polysorbate 80, or soy lecithin) in combination with XG. This effect could be attributed to the well characterized synergistic interaction between locust bean gum and XG (Sandolo et al. 2010; Grisel et al. 2015). Despite these differences, only soy lecithin produced oleogels with notably distinct rheological properties, but retained similar oil binding behavior. However, it is noteworthy that relatively high emulsifier concentrations were used in this study, which may have marginalized differences in the performances of the resulting oleogels. Thus, while using alternative surface-active hydrocolloids may show promise for further developing emulsion-templated oleogels, further research will be needed to catalog the functional performance of such systems.

Chitosan (*i.e.*, deacylated chitin) cross-linked with vanillin has also been explored as a novel approach to producing emulsion-templated oleogels (Brito et al. 2022). It was shown that 0.75 wt% chitosan alone produced unstable oleogels, but chemical crosslinking improved surface activity and enhanced the oil structuring properties. FTIR analysis indicated the formation of a Schiff base supported the chitosan/vanillin interaction upon solvent removal. Varying vanillin concentration (1 or 3 wt%) and addition of the non-ionic, lipid-based surfactant polysorbate 60 were used to modify the resulting oleogel structure and mechanical performance. Higher vanillin content led to extensive crystal formation, which appeared to interfere with the compact dried emulsion structure and was associated with a weaker gel network. Adding polysorbate 60 (0.4 wt%) did not produce a detectible change in the chemical interactions; however, this added emulsifier drastically altered the oleogel microstructure, which resulted in a marginal enhancement of the large-deformation mechanical properties. Performance as a shortening alternative in cookies was found to be most favorable when the oleogels were formulated with a lower vanillin content (0.75 wt% chitosan, 1 wt% vanillin) and without polysorbate 60. This work demonstrates the potential to expand the scope of polysaccharides which can be used as oil structuring systems with potential for food applications.

Chitin has also been used to produce solid particle suspensions capable of forming oil in water emulsions for the generation of emulsion-templated oleogels (Baraki et al. 2021). Particulate dispersions of chitin were used to form highly stable Pickering emulsions capable of withstanding the drying step necessary for oleogel formation. The emulsion droplet size decreased with increasing chitin content (0.6–1.4 wt% chitin), producing both emulsions and corresponding dried oleogels with higher rheological properties (G′ and G″). Pickering-stabilized emulsions may thus represent an alternative strategy for structuring liquid oils for food, pharmaceutical, and cosmetic applications. This approach has also been explored using a variety of biopolymeric systems, including proteins, protein/polysaccharide conjugates, and complexes of polyphenols with either protein or polysaccharide nanoparticles. The latter has been primarily investigated as newly developed emulsifier systems, but may also hold promise for emulsion templated oleogels (Hu et al. 2016; Yan et al. 2020; T. Yang et al. 2020). Starch-lipid complexes have also been explored as a means of modifying the functional behavior of lipid-based crystalline gelators such as plant-derived waxes (Gao and Wu 2019, 2020).

Another innovative approach to produce novel edible oleogels was established by Mert and Vilgis, in which plant-derived oleosomes were transformed into emulsion-templated oleogels using XG or pectin (Mert and Vilgis 2021). While this approach relies on the native protein and phospholipid membrane as the primary emulsifying agent, this matrix is naturally present, and only the addition of stabilizing hydrocolloids is needed to produce the emulsion-based oleogels. The anionic polysaccharides were incorporated into aqueous suspensions of oleosomes derived from raw hazelnuts at a pH 4 to facilitate electrostatic deposition. The resulting suspensions were freeze-dried and sheared to produce the templated oleogels. Differences in the rheological and large deformation behavior were dependent on the type of hydrocolloid used, with XG producing a notably higher storage modulus and large-deformation gel strength. Increasing polymer content was also positively correlated to gel strength for both polysaccharides. Further, the shear stress response, thixotropic flow behavior, and oil binding properties were similar across both polymer type and concentration, with ~ 90 wt% oil retention being achieved for all stable oleogels evaluated (1–2 wt% polysaccharide).

While many of the systems discussed above are produced using concentrated emulsions, long drying times are still often required to achieve complete water removal, which can have a detrimental impact on the oil quality and shelf life of the resulting oleogel. The production of high internal phase emulsions (HIPEs) may therefore reduce the initial water content and overall drying time. HIPEs are defined as having an internal phase greater than the maximum packing limit for monodisperse spheres (~0.74), resulting in a jammed network of deformed, polyhedral emulsion droplets, generally resulting in a highly viscous or soft solid rheological response (Bascuas et al. 2021). The interfacial film must be sufficiently rigid to prevent coalescence, but have enough elastic character to retain their structural integrity upon deformation. To achieve this balance, protein-polysaccharide complexes have been successfully used, including whey protein isolate or sodium caseinate complexed with low methoxyl pectin, sodium caseinate-alginate complexes (Wijaya et al. 2018), and pea protein combined with various stabilizing polysaccharides (Vélez-Erazo et al. 2020), among others. It is noted that HIPEs in which the aqueous phase is retained are often described as emulsion gels and should be distinguished from emulsion-templated oleogels in which the water phase has been actively removed. Despite this difference, HIPEs may also be suitable as fat replacers, as they often have a similar lipid and water composition to butter and margarine (Abdullah et al. 2020). An alternative approach to producing HIPEs is to use colloidal Pickering-type particles capable of adhering to the oil/water interface and serve as a physical barrier to coalescence due to their high detachment energy (Cui et al. 2021). An extensive array of polysaccharide-based HIPE gels have been developed in recent years, but these systems have most commonly been evaluated as water-continuous emulsion gels (Bascuas et al. 2021; Cui et al. 2021). However, the potential of using Pickering-type emulsions as oleogel templates has been demonstrated using octenyl succinic anhydride-modified starch nanocrystals (Liu et al. 2021) and both soluble and insoluble fiber particles (Phoon and Henry 2020). Further exploration and optimization of these systems to tailor the interfacial composition and functional performance shows great promise as an emerging strategy to develop fat mimetics for a broad range of food applications.

The potential of emulsion-templated polysaccharide-based oleogels as direct substitutes for traditional fats has also been evaluated in several food systems. The replacement of cocoa butter with HPMC-based oleogels has been explored as a means of improving the lipid profile of chocolate (Espert et al. 2021; Alvarez et al. 2021a). The introduction of emulsified liquid oil drastically reduced product firmness and the peak melting temperature arising from cocoa butter. These trends were caused by a reduction in the continuous fat crystal network which provides the characteristic firmness and snap associated with high quality chocolate. Bascus et al. evaluated the performance of a coconut oil-based chocolate spread in which 50 wt% or 100 wt% of the lipid phase was substituted with HPMC/XG-based emulsion-templated oleogels containing olive oil or sunflower oil. The spreads prepared with oleogels performed similar to the control, but complete substitution caused an increase in the shear modulus and imparted a gummy and "lumpy" texture in the spreads resulting from the intact emulsion droplets. Emulsion-templated oleogels have also been explored as shortening substitutes in baked goods. A HIPE-based oleogel stabilized with octenyl succinic anhydride-modified starch was shown to produce a more uniform crumb and improved textural attributes compared to liquid vegetable oil when used as a margarine replacement in a simple sponge cake (Liu et al. 2021). While only a select set of food systems have been explored for polysaccharide-based emulsion-templated oleogels, further developments using emulsion-templated oleogels could expand their applicability. Two promising approaches include the use of protein/polysaccharide complexes as emulsifier systems (Tavernier et al. 2017), and adding secondary crystalline gelators in the oil phase to form structured emulsions (Tavernier et al. 2018; Gao and Wu 2019).

2.7.3.2 Aerogel-Templating

Aerogel-templating provides a relatively passive means of introducing liquid oil into the polymer matrix, since most of the processing (*i.e.*, water removal) is completed prior to this step. As outlined in Figure 2.7.4, the standard approach involves first using a polymer suspension to form a stable polymer network (either via foaming or gelation), and subsequent removal of the dispersing liquid to form an open, porous aerogel structure. Liquid removal can be achieved by freeze drying under vacuum (Figure 2.7.4B) or through supercritical CO_2 drying (Figure 2.7.4C). These methods greatly reduce the capillary forces generated during conventional drying, which can lead to wall collapse and pore shrinkage (Mikkonen et al. 2013). Both strategies have been successfully used to generate polysaccharide aerogels through physical sorption of liquid oil to produce aerogel-templated oleogels. Polysaccharide aerogels have received substantial interest in various biomedical (Maleki et al. 2016; El-Naggar et al. 2020) and food-related applications (Mikkonen et al. 2013; Nešić et al. 2020; Zou and Budtova 2021), including controlled drug delivery, incorporation of lipid-soluble nutraceuticals, and food packaging and insulation.

2.7.3.2.1 Foam-Templating

Polysaccharide aerogels may be produced by first forming a stable aqueous foam, which is then dehydrated by freeze-drying, leaving an intact and highly porous aerogel polymer network. Foam-templating lends itself well to amphiphilic, surface-active polymers which can generate large foam volumes. The foaming properties of the polysaccharide solution are of critical importance, as this will greatly affect the resulting templated structure (Jiang et al. 2022a). Higher foaming capacity indicates a larger foam volume can be generated (*i.e.*, higher overrun), whereas foam stability is a measure of the rate of foam collapse arising from fluid drainage and disproportionation. While a balance between these two behaviors is critical for generating an effective foam, interfacial stability is of particular importance for producing aerogel templates capable of withstanding the freeze-drying process (Patel et al. 2013).

Patel and coworkers first demonstrated the concept of producing polysaccharide-based foam-templated oleogels with no crosslinking agent using HPMC, which was selected for its notable high surface activity (Patel et al. 2013). Preliminary screening to characterize the foaming behavior of various viscosity grades (10, 50, 100, and 4000 cP) confirmed that while foamability was higher for lower viscosity varieties, foam stability was positively correlated to polymer viscosity (*i.e.*, molecular weight). This trend was attributed to the larger polymers causing a higher degree of entanglement, producing a stiffer, more cohesive film at the air/water interface. Freeze-dried foam templates were produced with the high viscosity polymer and no thickening agent (such as XG). The resulting aerogel templates displayed rapid, passive oil absorption throughout the network, indicative of a reticulated open-celled solid foam. Due to the passive nature of this interaction, the oil was not strongly bound; however, subjecting this structure to high shear resulted in the formation of soft solid-like materials which did not show oil leakage. Microstructural analysis indicated the shearing processes produced a network of dispersed polymer sheets suspended in the liquid oil, giving rise to the observed soft solid-like behavior. Oleogels with varying gel strength were produced using 1–5 wt% HPMC, demonstrating the high gelator efficiency of these polymer networks. While the oleogels were stable over several months, they displayed strong shear-thinning and negligible structural recovery. The potential of these templated systems to serve as practical fat replacers in food products was also demonstrated using a simple sponge cake recipe (Patel and Dewettinck 2015). Oleogels structured with 2 wt% HPMC (4,000 cP) improved air entrapment (decreased batter density) relative to liquid oil but were less effective than commercial shortening. The viscoelastic response of these batters was comparable in magnitude to that of liquid oil, but the overall strain-dependence was more like that of a commercial shortening. In contrast, a high internal phase emulsion-templated oleogel provided a stronger initial gel structure but exhibited a rapid decrease in the viscoelastic response with increasing strain. The presence of amphiphilic polymer sheets suspended in the oil phase therefore seemed to maintain the expected rheological profile, suggesting this strategy could be further optimized to more appropriately mimic the full range of functional behavior expected of traditional plastic fats.

Following this initial proof-of-concept investigation, the applicability of foam-templated oleogels using cellulose derivatives has been explored in several other food systems, including confectionary creams, natural peanut butter, chocolate, muffins, and meat patties. Tanti and coworkers evaluated oleogels prepared from either HPMC or MC, while also using a modified preparation procedure in which the freeze-dried aerogels were ground before incorporating oil (Tanti et al. 2016a). The resulting powders were directly introduced into natural peanut butters as an alternative to high melting point triglyceride-based stabilizers. Complete stabilization of the oil was achieved with 2.0 wt% freeze-dried HPMC or MC powder, as no oil loss was observed for over 6 months. Textural and oil binding properties similar to commercial stabilized products could be achieved with 1.0–2.0 wt% polymer, with little differences seen between the two cellulose derivatives.

This same group also prepared HPMC- and MC-based oleogels to partially replace the fat in sandwich cookie creams, which typically contain ~ 25–40% solid fat such as palm-based shortening, butter, or hydrogenated oils (Tanti et al. 2016b). The optimized oleogels were formulated with 3 wt% polymer in canola oil (polymer viscosity not reported). Shortening-based cream fillings (40% fat) were produced with either 50, 75, or 100 wt% oleogel in the lipid phase. 24-hour oil loss profiles showed the oleogel-based fillings provided less oil retention as compared to a shortening-based control, but were similar to a commercial standard, with 100% HPMC providing the greatest oil retention. Increasing either HPMC or MC content also increased the product hardness but was negatively correlated to other large deformation textural parameters, including adhesiveness, cohesiveness, and gumminess. The latter trend could be attributed to the weak polymer/oil interactions, resulting in poor structural recovery (Patel et al. 2013). The small amplitude oscillatory strain profiles of the oleogel-based cream fillings were similar to both the commercial and full-fat controls, and product stiffness (magnitude of G′) was positively correlated to oleogel content. These results indicate the promise of using foam-templated oleogels in confectionary applications; however, further optimization of the physical performance may benefit from using a multi-component oleogelator strategy to improve the plastic deformation behavior characteristic of commercial shortenings (Macias-Rodriguez and Marangoni 2016).

Substitution of cocoa butter with HPMC foam-templated oleogels in chocolate has also been investigated as a saturated fat reduction strategy by Alvarez et al. (Alvarez et al. 2021b). Blending cocoa butter with oleogels had a negative impact on the melting behavior and fat firmness, which directly translated to chocolate performance. While mechanical and sensory evaluation indicated key functional properties were acceptable with up to 50% fat replacement, further work will be needed to evaluate the impact of blending on cocoa butter structure and the practical performance and sensory properties of the resulting chocolate products.

Oh and coworkers evaluated the HPMC-based foam-templated oleogels as a potential saturated fat reduction strategy in both muffin batters (Oh and Lee 2018) and beef patties (Oh et al. 2019). In muffins, acceptable functional performance was maintained when replacing up to 50% shortening with oleogels prepared with 4 wt% 4,000 cP HPMC in sunflower oil. This level of substitution maintained textural attributes and had a negligible impact on the specific volume of the baked product, despite producing larger, heterogeneously distributed air cells, as visualized by X-ray tomography. For incorporation in meat patties, canola oil oleogels structured with 4 wt% HPMC were firmer than beef tallow at room temperature and were found to reduce both total fluid loss during cooking and final product firmness. Sensory analysis indicated an increase in perceived juiciness and tenderness at 50% fat replacement, while also achieving a ~ 25% reduction in the saturated fat content.

While the studies outlined above used a single surface-active hydrocolloid, the effect of incorporating an additional stabilizing polysaccharide has also been addressed. A detailed characterization of foam-templated HPMC oleogels was conducted by Jiang et al. to evaluate the effect of xanthan gum (XG) as a secondary foam stabilizing agent (Jiang et al. 2021). In this work, 400 cP HPMC was varied from 0.8–3.0 wt%, while XG was fixed at 0.3 wt%. Although a higher overrun was produced in the aqueous foams prepared without XG, they underwent rapid creaming and collapse (within 1 hour) and could not be used to prepare aerogel templates. While this effect could be attributed to the low viscosity variety of HPMC employed (Patel et al. 2013), the addition of XG substantially increased the foam stability, which was further enhanced with increasing HPMC content. The much higher stability of these foams (8 to > 24 hours) was attributed to an increase in solution viscosity and reduction in air bubble diameter. However, at the highest HPMC content evaluated (3 wt%) a wider distribution of air bubbles was produced, due to excess viscosity enhancement which hindered size reduction during the foaming process. The passive absorption of oil into the aerogel templates was very rapid, plateauing after ~ 1 min, irrespective of the polymer concentration. Further, while total oil uptake was negatively correlated to HPMC content, the absorbed oil was more constrained (on a wt% basis) when subjected to an accelerated centrifugal oil loss test. The rheological behavior of the oleogels were evaluated after shearing to produce an oil-continuous polymer suspension. G′ was positively correlated to HPMC content and displayed little frequency or temperature dependence. However, the viscous modulus showed a moderate increase at ~ 55–60°C, independent of polymer concentration. This was likely associated with an increase in solvent-polymer hydrophobic interactions, as this temperature range aligns with the thermo-reversible physical gelation of HPMC. Polarized light microscopy, FTIR, and X-ray diffraction analysis confirmed that HPMC formed a semi-crystalline polymer network supported by hydrogen bonds. Further, the structural recovery of these oleogels after applying high shear was dramatically improved by the addition of XG, as a negligible recovery in G′ was previously reported for foam-templated oleogels structured by HPMC alone (Patel et al. 2013). This effect was attributed to the physical interactions (*i.e.*, hydrogen bonds) between the semi-crystalline HPMC network and XG, indicating a secondary stabilizing hydrocolloid may provide distinct benefits for both the formation and functional properties of polysaccharide-based foam-templated oleogels.

In a follow-up study, Jiang and coworkers evaluated various hydrocolloids as thickening agents in HPMC-based foam templates, including carboxymethylcellulose, κ-carrageenan, and flaxseed, guar, and acacia gums (Jiang et al. 2022a). Foams were prepared with 1.5 wt% HPMC (400 cP) and 0.3 wt% thickener, resulting in aqueous solutions with a broad range in viscosity. κ-carrageenan produced a far greater increase in viscosity than any other hydrocolloid evaluated (Table 2.7.2) and maintained stable foams for over 24 hours, while all other hydrocolloids showed substantial destabilization after 2–4 hours. However, both κ-carrageenan and flaxseed gum produced foam templates with similar performance, having the highest firmer (>10 N) and oil absorption capacity (54–55 g/g polymer), as shown in Table 2. G′ of the oleogels was also largest with the addition of κ-carrageenan (>10 kPa), followed by flaxseed gum (>7 kPa), while the remaining thickeners produced oleogels with a G′ less than 5 kPa. Further, all the added hydrocolloids provided similar structural recovery after application of high shear, while only acacia gum and guar gum displayed a notable temperature sensitivity. FTIR analysis again indicated that physical polymer-polymer interactions were established in the foam templates and persisted in the resulting sheared oleogels. From this work, it is apparent that secondary hydrocolloids can drastically alter the performance of HPMC-based foam-templated oleogels, but the functional behavior of the resulting gel will depend on a variety of factors, including solution viscosity, surface activity, and polymer-polymer interactions, among others.

Table 2.7.2 Impact of various stabilizing polysaccharides on the solution viscosity and physical properties of hydroxypropyl methylcellulose-based aerogels used as foam-templated oleogels. All solutions were formulated with 1.5 wt% 400 cP HPMC and 0.3 wt% stabilizer. Adapted from (Jiang et al. 2022a), with permission.

Stabilizer	Solution viscosity (cP)	Aerogel hardness (N)	Oil absorption capacity (wt/wt)
Carboxymethyl cellulose	568 ± 5.5^{a}	4.77 ± 0.10^{a}	38.84 ± 0.19^{a}
κ-carrageenan	1072 ± 12.1^{b}	12.10 ± 0.01^{b}	55.15 ± 0.15^{b}
Flaxseed gum	186.4 ± 1.4^{c}	10.79 ± 0.61^{b}	54.08 ± 0.25^{c}
Guar gum	328 ± 5.0^{d}	6.52 ± 0.20^{c}	40.22 ± 0.49^{d}
Acacia gum	108.8 ± 1.4^{e}	3.24 ± 0.04^{d}	32.42 ± 0.02^{e}

Different letters within each column indicate significant differences ($p < 0.05$).

Incorporating a secondary, crystalline gelator in a polymer network is an alternative approach to modulating oleogel functionality and may provide a means of both enhancing structuring capacity and improving the fat-like techno-functional properties of the combined system. This strategy has been extensively used in directly structured EC-based oleogels, as outlined in Section 2.7.2.2. Jiang et al. recently explored this method for modulating the properties of foam templates, in which HPMC/GMS hybrid oleogels were prepared as a function of GMS concentration (Jiang et al. 2022b). The two structuring agents were combined by heating the liquid oil and dissolving GMS before being passively absorbed by the HPMC aerogels. Upon cooling, the polymer sheets became suspended in a space-filling GMS crystalline network, and both hardness and oil binding capacity were positively correlated to the GMS content. Additionally, these gels were firmer than those containing secondary hydrocolloids, but showed a substantial strain-weakening and gradual decrease in G′ with increasing temperature. However, the HPMC/GMS hybrid system displayed a similar structural recovery to that observed in the HPMC-based oleogels with added hydrocolloids. These effects were attributed to the physical state of the GMS, as well as hydrogen bonds occurring between the HPMC and GMS networks. Finally, the hybrid gels were used as a partial replacement for butter in both cookies and cakes. In general, increasing GMS produced a corresponding increase in product hardness, springiness, and chewiness, but was negatively correlated with cohesiveness and resilience, as determined by texture profile analysis. These trends indicate the oleogels became more brittle as the GMS content increased, and may further suggest that factors such as the size and distribution of air cells and resulting batter density may affect the breakdown behavior. This work, and those outlined above, suggest that adopting a suitable polymer/crystalline hybrid gelator system using the foam-templating approach may provide a promising approach for developing functional fat mimetics for a diverse range of food applications.

2.7.3.2.2 Supercritical CO_2-Derived Templates

Supercritical CO_2 drying has been widely used to produce a broad range of biopolymer aerogel scaffolds for biomedical applications (Maleki et al. 2016; El-Naggar et al. 2020), but has also garnered interest for various food applications, including nutrient delivery, controlled release, and active packaging (Mikkonen et al. 2013). This approach has recently been explored as an alternative means of producing oil-sorbing biopolymer aerogel templates. The supercritical CO_2 drying process has the advantage of eliminating the surface tension arising from a liquid/gas phase transition, which can greatly reduce pore collapse during evacuation and maintain the high surface area and low density of the polymer scaffold. Producing biopolymer aerogels through this approach generally involves forming a stable hydrogel and substituting the aqueous phase with an intermediate solvent such as ethanol, which has a lower critical point and density closer to CO_2, and thus reduces pore constriction (Manzocco et al. 2017). The potential of using supercritical drying to produce oil-absorbing polysaccharide aerogels was first demonstrated using barley β-glucan as a potential carrier for oxidatively-prone flaxseed oil (Comin et al. 2012). Heat-induced β-glucan hydrogels were transitioned to alcohol-based gels (alcogels), and then subjected to supercritical CO_2 drying and impregnated with oil in the pressurized drying chamber. The resulting oleogels retained ~ 45–60% oil, depending on the processing parameters used. However, the oil binding and mechanical properties were not evaluated, as these materials were intended to serve as high-value oil carriers.

Using an approach analogous to foam-templating, Manzocco et al. generated polysaccharide aerogel scaffolds capable of passively absorbing oil using salt-induced κ-carrageenan hydrogels as the scaffolding template (Manzocco et al. 2017). Polymer concentration had a strong impact on network density and shrinkage during the hydrogel-alcogel-aerogel progression, which directly affected oil absorption behavior and oleogel performance. Increasing the polysaccharide concentration reduced network shrinkage, which was attributed to the higher polymer-polymer crosslink density in the initial hydrogel. While this resulted in greater oil uptake, the absorption kinetics remained on the order of hours, but slowed with increasing polymer concentration. Further, oil binding capacity and gel strength were negatively impacted by the more open, porous network present in the gels with a higher κ-carrageenan content. This structural arrangement provided less mechanical support and weaker capillary forces, which contributed to lower oil binding, as determined via a centrifugal accelerated oil loss test.

In an effort to expand the functional properties of κ-carrageenan-based oleogels, Plazzotta et al. substituted the aqueous phase with iceberg lettuce homogenate as a means of incorporating bio-based filler particles (Plazzotta et al. 2019). The lettuce homogenate dramatically decreased the hydrogel network density, but also sharply reduced shrinkage in the subsequent drying steps (Figure 2.7.6A vs Figure 2.7.6B). The net effect was a ~ 5-fold decrease in network density and a ~ 6-fold reduction in aerogel firmness. These templates had a much higher oil sorption rate and oil absorption capacity, the latter of which increased from ~ 3.4 to 13 g oil/g polymer. These changes also produced a decrease in oleogel hardness and oil binding (Figure 2.7.6E), which was attributed to the more open, porous structure and associated decrease in capillary forces. This study further explored the impact of the drying method by directly freeze-drying both the standard and homogenate-filled κ-carrageenan hydrogels. The freeze-dried aerogels showed far less network shrinkage (Figure 2.7.6), high oil absorption, and rapid sorption kinetics relative to the corresponding supercritical CO_2-dried templates. However, the lettuce filler particles increased the freeze-dried aerogel network density, resulting in a decrease in the maximum oil absorption capacity from ~ 30 to 15 g oil/g polymer. This change in network structure produced an associated increase in oleogel hardness and oil binding capacity; the latter (presented in Figure 2.7.6E) was determined by passive oil desorption on a filter paper. The freeze-dried aerogels also displayed oil sorption kinetics similar to that reported for freeze-dried HPMC aerogels (Jiang et al. 2021), reaching saturation within 1–2 min, which was roughly two orders of magnitude faster than the supercritical CO_2 dried gels.

While supercritical CO_2 drying is a relatively intensive process that may limit its potential for developing fat mimetics, this approach could be well suited for specialized applications, such as the delivery of bioactive compounds. This approach is also not limited to polysaccharides, as proteins

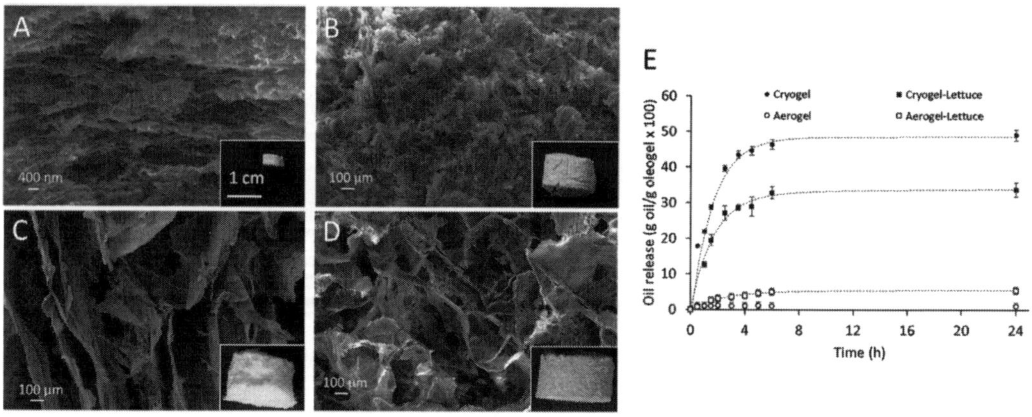

Figure 2.7.6 SEM micrographs and photographs (A-D and insets, respectively) and oil desorption (E) of supercritical CO_2 (aerogel) and freeze-dried (cryogel) templates produced from salt-induced κ-carrageenan hydrogels. Reproduced from Plazzotta, S. et al. (2019) with permission from Elsevier.

are also actively being explored as oil-sorbing biopolymer scaffolds (Plazzotta et al. 2020, 2021; Manzocco et al. 2021b). Finally, another innovative application using supercritical CO_2 extraction is the development of hydrophobic aerogel scaffolds using ethyl cellulose oleogels (Manzocco et al. 2021a), which may provide unique functionality for encapsulation and delivery applications.

2.7.4 Conclusion

With the rising demand for health-promoting, sustainable, and ethically produced food products, edible oleogels continue to be viewed as a promising solution for developing structured lipids with favorable techno-functional performance. While polysaccharides are ubiquitously employed across the food industry for their excellent structuring and stabilizing properties in aqueous environments, they have also recently garnered particular interest as oil structuring systems, despite the technical challenges arising from their inherent hydrophilic nature. Modified cellulose derivatives have played a large role in establishing both direct and indirect approaches for producing oleogels, and several strategies have been identified which provide a means of manipulating and expanding their functional properties. Synergistic interactions with surface-active small molecules or small molecular weight crystalline oleogelators represents a promising approach to tailor the techno-functional properties of polysaccharide-based fat mimetics. Combinations of amphiphilic and viscosity-enhancing polysaccharides have also been shown to improve the performance of emulsion- and aerogel-templated oleogels. These indirect structuring methods are also being combined with the hybrid polymer/crystalline gelator approach to impart desirable attributes, such as melting behavior and plasticity.

While not reviewed here, combining proteins and polysaccharides represents an additional broad area for further developments in structuring edible oils (Bascuas et al. 2021). These hybrid biopolymer systems are of particular relevance to the emulsion- and foam-templating approaches, as polysaccharides can be used to enhance the viscosity and stability of protein-based templates (Mohanan et al. 2020). Emulsions may also be formed by either sequential deposition of proteins and carbohydrates at the interface, or by generating a mixed viscoelastic biopolymer layer. Protein/polysaccharide complexes have also been used to produce particles capable of sterically stabilizing oil droplets, forming highly stable Pickering-type emulsions (T. Zhou et al. 2019; T. Yang et al. 2020). Finally, high internal phase emulsions, emulsion-filled gels, and bigels are all promising strategies in which polysaccharides could provide considerable utility, and show great promise for developing edible soft materials that mimic fatty tissue (Hu and McClements 2022). Thus, while practical solutions for producing oleogels which mimic the broad range of functional attributes provided by animal-derived fats and tropical oils remain relatively elusive, the continued development of polysaccharide-based hybrid gelator approaches shows great promise for closing this gap.

References

Abdolmaleki, K., Alizadeh, L., Nayebzadeh, K., Hosseini, S.M., and Shahin, R. (2020). Oleogel production based on binary and ternary mixtures of sodium caseinate, xanthan gum, and guar gum: Optimization of hydrocolloids concentration and drying method. *Journal of Texture Studies* 51: 290–299.

Abdullah, Weiss, J., Ahmad, T., Zhang, C., and Zhang, H. (2020). A review of recent progress on high internal-phase pickering emulsions in food science. *Trends in Food Science and Technology* 106: 91–103.

Adili, L., Roufegarinejad, L., Tabibiazar, M., Hamishehkar, H., and Alizadeh, A. (2020). Development and characterization of reinforced ethyl cellulose based oleogel with adipic acid: Its application in cake and beef burger. *LWT – Food Science & Technology* 126: 109277.

Aguilar-Zárate, M., Macias-Rodriguez, B.A., Toro-Vazquez, J.F., and Marangoni, A.G. (2019). Engineering rheological properties of edible oleogels with ethylcellulose and lecithin. *Carbohydrate Polymers* 205: 98–105.

Ahmadi, P., Tabibiazar, M., Roufegarinejad, L., and Babazadeh, A. (2020). Development of behenic acid-ethyl cellulose oleogel stabilized Pickering emulsions as low calorie fat replacer. *International Journal of Biological Macromolecules* 150: 974–981.

Alizadeh, L., Abdolmaleki, K., Nayebzadeh, K., and Hosseini, S.M. (2020). Oleogel fabrication based on sodium caseinate, hydroxypropyl methylcellulose, and beeswax: Effect of concentration, oleogelation method, and their optimization. *JAOCS, Journal of the American Oil Chemists' Society* 97: 485–496.

Alvarez, M.D., Cofrades, S., Espert, M., Salvador, A., and Sanz, T. (2021a). Thermorheological characterization of healthier reduced-fat cocoa butter formulated by substitution with a hydroxypropyl. *Foods* 10: 793.

Alvarez, M.D., Cofrades, S., Espert, M., Sanz, T., and Salvador, A. (2021b). Development of chocolates with improved lipid profile by replacing cocoa butter with an oleogel. *Gels* 7: 220.

Baraki, S.Y., Jiang, Y., Li, X., Debeli, D.K., Wang, B., Feng, X., Mao, Z., and Sui, X. (2021). Stable sunflower oil oleogel from oil/water pickering emulsion with regenerated chitin. *LWT* 146: 111483.

Barbut, S. and Marangoni, A.G. (2019). Organogels use in meat processing – effects of fat/oil type and heating rate. *Meat Science* 149: 9–13.

Barbut, S., Marangoni, A.G., Thode, U., and Tiensa, B.E. (2019). Using canola oil organogels as fat replacement in liver Pâté. *Journal of Food Science* 84: 2646–2651.

Barbut, S., Tiensa, B.E., and Marangoni, A.G. (2021). Partial fat replacement in liver pâté using canola oil organogel. *LWT – Food Science & Technology* 139: 110428.

Barbut, S., Wood, J., and Marangoni, A.G. (2016a). Potential use of organogels to replace animal fat in comminuted meat products. *Meat Science* 122: 155–162.

Barbut, S., Wood, J., and Marangoni, A.G. (2016b). Effects of organogel hardness and formulation on acceptance of frankfurters. *Journal of Food Science* 81: C2183–C2188.

Barbut, S., Wood, J., and Marangoni, A.G. (2016c). Quality effects of using organogels in breakfast sausage. *Meat Science* 122: 84–89.

Bascuas, S., Hernando, I., Moraga, G., and Quiles, A. (2020a). Structure and stability of edible oleogels prepared with different unsaturated oils and hydrocolloids. *International Journal of Food Science & Technology* 55: 1458–1467.

Bascuas, S., Morell, P., Hernando, I., and Quiles, A. (2021). Recent trends in oil structuring using hydrocolloids. *Food Hydrocoll* 118: 106612.

Bascuas, S., Salvador, A., Hernando, I., and Quiles, A. (2020b). Designing hydrocolloid-based oleogels with high physical, chemical, and structural stability. *Frontiers in Sustainable Food Systems* 4: 1–8.

Bemer H.L., Limbaugh, M., Cramer, E.D., Harper, W.J., and Maleky, F. (2016). *Food Research International* 85: 67–75.

Bhatnagar, A.S., Prasanth Kumar, P.K., Hemavathy, J., and Gopala Krishna, A.G. (2009). Fatty acid composition, oxidative stability, and radical scavenging activity of vegetable oil blends with coconut oil. *JAOCS, Journal of the American Oil Chemists' Society* 86: 991–999.

Bin Sintang, M.D., Danthine, S., Khalenkow, D., Tavernier, I., Sosa, D.A.T., Julmohammad, N.B., Van de Walle, D., Rimaux, T., Skirtach, A., and Dewettinck, K. (2020). Modulating the crystallization of phytosterols with monoglycerides in the binary mixture systems: mixing behavior and eutectic formation. *Chemistry and Physics of Lipids* 230: 104912.

Blach, C., Gravelle, A.J., Peyronel, F., Weiss, J., Barbus, S., and Marangoni, A.G. (2016). Revisiting the crystallization behavior of stearyl alcohol: Stearic acid (SO: SA) mixtures in edible oil. *RSC Advances* 6.

Blake, A.I. and Marangoni, A.G. (2015). Plant wax crystals display platelet-like morphology. *Food Structure* 3: 30–34.

Bodennec, M., Guo, Q., and Rousseau, D. (2016). Molecular and microstructural characterization of lecithin-based oleogels made with vegetable oil. *RSC Advances* 6: 47373–47381.

Brito, G.B., Peixoto, V.O.D.S., Martins, M.T., Rosário, D.K.A., Ract, J.N., Conte-Júnior, C.A., Torres, A.G., and Castelo-Branco, V.N. (2022). Development of chitosan-based oleogels via crosslinking with vanillin using an emulsion-templated approach: structural characterization and their application as fat-replacer. *Food Structure* 32: 100264.

Callau, M., Sow-Kébé, K., Nicolas-Morgantini, L., and Fameau, A.L. (2020). Effect of the ratio between behenyl alcohol and behenic acid on the oleogel properties. *Journal of Colloid and Interface Science* 560: 874–884.

Cao, Y. and Mezzenga, R. (2020). Design principles of food gels. *Nature Food* 1.

Cattaruzza, A., Radford, S., and Marangoni, A.G. (2014). United states patent *US 2014/0044839 A1 1*.

Ceballos, M.R., Bierbrauer, K.L., Faudone, S.N., Cuffini, S.L., Beltramo, D.M., and Bianco, I.D. (2016). Influence of EC-medium chain triglycerides blend on the flow behaviour and beta-V polymoprh retention of dark chocolate. *Food Structure* 10: 1–9.

Coffey, D.G., Bell, D.A., and Henderson, A. (2006). Cellulose and cellulose derivatives. In: *Food Polysaccharides and Their Applications* (ed. A.M. Stephen, G.O. Phillips, and P.A. Williams), 148–179. Boca Raton, FL: CRC Press.

Comin, L.M., Temelli, F., and Saldaña, M.D.A. (2012). Barley β-glucan aerogels as a carrier for flax oil via supercritical CO2. *Journal of Food Engineering* 111: 625–631.

Cordova-Barragan, M., Marangoni, A.G., Peyronel, F., and Dibildox-Alvarado, E. (2021). Crystallization enhancement by a high behenic acid stabilizer in a palm oil-based model fat blend and its corresponding fat-reduced water-in-oil emulsion. *JAOCS, Journal of the American Oil Chemists' Society* 98: 413–424.

Cui, F., Zhao, S., Guan, X. McClements, D.J., Xuebo, L., Liu, F., and Ngai, T. (2021). Polysaccharide-based pickering emulsions: Formation, stabilization and applications. *Food Hydrocoll* 119: 106812.

Davidovich-Pinhas, M., Barbut, S., and Marangoni, A.G. (2014). Physical structure and thermal behavior of ethylcellulose. *Cellulose* 21: 3243–3255.

Davidovich-Pinhas, M., Barbut, S., and Marangoni, A.G. (2015). The role of surfactants on ethylcellulose oleogel structure and mechanical properties. *Carbohydrate Polymers* 127: 355–362.

Davidovich-Pinhas, M., Barbut, S., and Marangoni, A.G. (2015a). The gelation of oil using ethyl cellulose. *Carbohydrate Polymers* 117: 869–878.

Davidovich-Pinhas, M., Gravelle, A.J., Barbut, S., and Marangoni, A.G. (2015b). Temperature effects on the gelation of ethylcellulose oleogels. *Food Hydrocoll* 46: 76–83.

Doan, C.D., Tavernier, I., Sintang, M.D., Danthine, S., Van de Walle, D., Rimaux, T., and Dewettinck, K. (2017). Crystallization and gelation behavior of low- and high melting waxes in rice bran oil: A case-study on berry wax and sunflower wax. *Food Biophysics* 12: 97–108.

El-Naggar, M.E., Othman, S.I., Allam, A.A., and Morsy, O.M. (2020). Synthesis, drying process and medical application of polysaccharide-based aerogels. *International Journal of Biological Macromolecules* 145: 1115–1128.

Ergun, R., Appell, R.B., and Malotky, D.L. (2017). *Process for preparing an oleogel.* US Pat. 9,655,376.

Espert, M., Hernández, M.J., Sanz, T., and Salvador, A. (2021). Reduction of saturated fat in chocolate by using sunflower oil-hydroxypropyl methylcellulose based oleogels. *Food Hydrocoll* 120: 106917.

Espert, M., Hernández, M.J., Sanz, T., and Salvador, A. (2022). Rheological properties of emulsion templated oleogels based on xanthan gum and different structuring agents. *Current Research in Food Science* 5: 564–570.

Espert, M., Salvador, A., and Sanz, T. (2020). Cellulose ether oleogels obtained by emulsion-templated approach without additional thickeners. *Food Hydrocoll* 109: 106085.

Feichtinger, A. and Scholten, E. (2020). Preparation of protein oleogels: Effect on structure and functionality. *Foods* 9: 1745.

Gandolfo, F.G., Bot, A., and Flöter, E. (2003). Phase diagram of mixtures of stearic acid and stearyl alcohol. *Thermochim Acta* 404: 9–17.

Gandolfo, F.G., Bot, A., and Flöter, E. (2004). Structuring of edible oils by long-chain FA, fatty alcohols, and their mixtures. *JAOCS, Journal of the American Oil Chemists' Society* 81: 1–6.

Gao, Y. and Wu, S. (2019). Thermal and oxidation stability of functional oleogels formed by edible wax/starch and Schisandra chinensis oil. *Food Function* 10: 8056–8068.

Gao, Y. and Wu, S. (2020). Development and evaluation of a novel oleogel system based on starch-water-wax-oil. *Food Function* 11: 7727–7735.

García-Ortega, M.L., Toro-Vazquez, J.F., and Ghosh, S. (2021). Development and characterization of structured water-in-oil emulsions with ethyl cellulose oleogels. *Food Research Internationa* 150: 110763.

Ghosh, S., Tran, T., and Rousseau, D. (2011). Comparison of pickering and network stabilization in water-in-oil emulsions. *Langmuir* 27: 6589–6597.

Giacintucci, V., Di Mattia, C.D., Sacchetti, G., Flamminii, F., Gravelle, A.J., Baylis, B., Dutcher, J.R., Marangoni, A.G., and Pittia, P. (2018). Ethylcellulose oleogels with extra virgin olive oil: the role of oil minor components on microstructure and mechanical strength. *Food Hydrocoll* 84: 508–514.

Gómez-Estaca, J., Herrero, A.M., Herranz, B., Álvarez, M.D., Jiménez-Colmenero, F., and Confrades, S. (2019a). Characterization of ethyl cellulose and beeswax oleogels and their suitability as fat replacers in healthier lipid pâtés development. *Food Hydrocoll* 87: 960–969.

Gómez-Estaca, J., Pintado, T., Jiménez-Colmenero, F., and Cofrades, S. (2019b). Assessment of a healthy oil combination structured in ethyl cellulose and beeswax oleogels as animal fat replacers in low-fat, PUFA-enriched pork burgers. *Food and Bioprocess Technology* 12: 1068–1081.

Gravelle, A.J., Barbut, S., and Marangoni, A.G. (2012). Ethylcellulose oleogels: manufacturing considerations and effects of oil oxidation. *Food Research International* 48: 578–583.

Gravelle, A.J., Barbut, S., and Marangoni, A.G. (2013). Fractionation of ethylcellulose oleogels during setting. *Food Function* 4: 153–161.

Gravelle, A.J., Barbut, S., Quinton, M., and Marangoni, A.G. (2014). Towards the development of a predictive model of the formulation-dependent mechanical behaviour of edible oil-based ethylcellulose oleogels. *Journal of Food Engineering* 143: 114–122.

Gravelle, A.J., Blach, C., Weiss, J., Barbus, S., and Marangoni, A.G. (2017a). Structure and properties of an ethylcellulose and stearyl alcohol/stearic acid (EC/SO:SA) hybrid oleogelator system. *The European Journal of Lipid Science and Technology* 119: 1700069.

Gravelle, A.J., Davidovich-Pinhas, M., Barbut, S., and Marangoni, A.G. (2017b). Influencing the crystallization behavior of binary mixtures of stearyl alcohol and stearic acid (SOSA) using ethylcellulose. *Food Research International* 91: 1–10.

Gravelle, A.J., Davidovich-Pinhas, M., Zetzl, A.K., Barbut, S., and Marangoni, A.G. (2016). Influence of solvent quality on the mechanical strength of ethylcellulose oleogels. *Carbohydrate Polymers* 135: 169–179.

Grisel, M., Aguni, Y., Renou, F., and Malhiac, C. (2015). Impact of fine structure of galactomannans on their interactions with xanthan: Two co-existing mechanisms to explain the synergy. *Food Hydrocoll* 51: 449–458.

Haj Eisa, A., Laufer, S., Rosen-Kligvasser, J., and Davidovich-Pinhas, M. (2020). Stabilization of ethyl-cellulose oleogel network using lauric acid. *European Journal of Lipid Science and Technology* 122: 1–10.

Harris, L., Rosen-Kligvasser, J., and Davidovich-Pinhas, M. (2019). Gelation of oil using combination of different free fatty acids. *Food Structure* 21: 100121.

Hu, X. and McClements, D.J. (2022). Construction of plant-based adipose tissue using high internal phase emulsions and emulsion gels. *Innovative Food Science and Emerging Technologies* 78: 103016.

Hu, Z., Marway, H.S., Kasem, H., Pelton, R., and Cranston, E.D. (2016). Dried and redispersible cellulose nanocrystal pickering emulsions. *ACS Macro Letters* 5: 185–189.

Huang, Y., He, M., Lu, A., Zhou, W., Stoyanov, S.D., Pelan, E.G., and Zhang, L. (2015). Hydrophobic modification of chitin whisker and its potential application in structuring oil. *Langmuir* 31: 1641–1648.

Jiang, Q., Du, L., Li, S., Liu, Y., and Meng, Z. (2021). Polysaccharide-stabilized aqueous foams to fabricate highly oil-absorbing cryogels: application and formation process for preparation of edible oleogels. *Food Hydrocoll* 120: 106901.

Jiang, Q., Li, P., Ji, M., Du, L., Li, S., Liu, Y., and Meng, Z. (2022a). Synergetic effects of water-soluble polysaccharides for intensifying performances of oleogels fabricated by oil-absorbing cryogels. *Food Chemistyr* 372: 131357.

Jiang, Q., Yu, Z., and Meng, Z. (2022b). Double network oleogels co-stabilized by hydroxypropyl methylcellulose and monoglyceride crystals: Baking applications. *International Journal of Biological Macromolecules* 209: 180–187.

Kibbelaar, H.V.M., Dekker, R.I., Morcy, A., Kegel, W.K., Velikov, K.P., and Bonn, D. (2022). Ethyl cellulose nanoparticles as stabilizers for pickering emulsions. *Colloids and Surfaces A: Physicochemical and Engineering Aspects* 641: 128512.

Kim, G.Y. and Marangoni, A.G. (2017). Engineering the nucleation of edible fats using a high behenic acid stabilizer. *European Journal of Lipid Science and Technology* 119: 1–13.

Laredo, T., Barbut, S., and Marangoni, A.G. (2011). Molecular interactions of polymer oleogelation. *Soft Matter* 7: 2734–2743.

Li, L. and Liu, G. (2019). Corn oil-based oleogels with different gelation mechanisms as novel cocoa butter alternatives in dark chocolate. *Journal of Food Engineering* 263: 114–122.

Li, L., Liu, G., and Lin, Y. (2021). Physical and bloom stability of low-saturation chocolates with oleogels based on different gelation mechanisms. *LWT Food Sciene Technology* 140: 110807.

Liu, X., Ding, S., Wu, J., Liu, G., Wei, J., Yang, F., and Liu, X. (2021). Molecular structures of octenyl succinic anhydride modified starches in relation to their ability to stabilize high internal phase emulsions and oleogels. *Food Hydrocoll* 120: 106953.

Lopez-Martínez, A., Charó-Alonso, M.A., Marangoni, A.G., and Toro-Vazquez, J.F. (2015). Monoglyceride organogels developed in vegetable oil with and without ethylcellulose. *Food Research International* 72: 37–46.

Lu, X., Chen, J., Guo, Z., Zheng, Y., Rea, M.C., Su, H., Zheng, X., Zheng, B., and Miao, S. (2019). Using polysaccharides for the enhancement of functionality of foods: a review. *Trends in Food Science and Technology* 86: 311–327.

Luo, S.Z., Hu, X.F., Jia, Y.J., Pan, L.H., Zheng, Z., Zhao, Y.Y., Mu, D.D., Zong, X.Y., and Jiang, S.T. (2019). Camellia oil-based oleogels structuring with tea polyphenol-palmitate particles and citrus pectin by emulsion-templated method: Preparation, characterization and potential application. *Food Hydrocoll* 95: 76–87.

Macias-Rodriguez, B. and Marangoni, A.G. (2016). Rheological characterization of triglyceride shortenings. *Rheologica Acta* 55: 767–779.

Macias-Rodriguez, B.A., Ewoldt, R.H., and Marangoni, A.G. (2018). Nonlinear viscoelasticity of fat crystal networks. *Rheologica Acta* 57: 251–266.

Maleki, H., Durães, L., García-González, C.A., del Gaudio, P., Portugal, A., and Mahmoudi, M. (2016). Synthesis and biomedical applications of aerogels: Possibilities and challenges. *Advances in Colloid and Interface Science* 236: 1–27.

Manzocco, L., Basso, F., Plazzotta, S., and Calligaris, S. (2021a). Study on the possibility of developing food-grade hydrophobic bio-aerogels by using an oleogel template approach. *Current Research in Nutrition and Food Science* 4: 115–120.

Manzocco, L., Plazzotta, S., Powell, J., de Vries, A., Rousseau, D., and Calligaris, S. (2021b). Structural characterisation and sorption capability of whey protein aerogels obtained by freeze-drying or supercritical drying. *Food Hydrocoll* 122.

Manzocco, L., Valoppi, F., Calligaris, S., Andreatta, F., Spillimbergo, S., and Nicoli, M. (2017). Exploitation of κ-carrageenan aerogels as template for edible oleogel preparation. *Food Hydrocoll* 71: 68–75.

McClements, D.J. and Grossmann, L. (2021). The science of plant-based foods: constructing next-generation meat, fish, milk, and egg analogs. *Comprehensive Reviews in Food Science and Food Safety* 20: 4049–4100.

Meng, Z., Qi, K., Guo, Y., Wang, Y., and Liu, Y. (2018a). Physical properties, microstructure, intermolecular forces, and oxidation stability of soybean oil oleogels structured by different cellulose ethers. *The European Journal of Lipid Science and Technology* 120.

Meng, Z., Qi, K., Guo, Y., Yong, W., and YuanFa, L. (2018b). Macro-micro structure characterization and molecular properties of emulsion-templated polysaccharide oleogels. *Food Hydrocoll* 77: 17–29.

Meng, Z., Qi, K., Guo, Y., Wang, Y., and Liu, Y. (2018c). Effects of thickening agents on the formation and properties of edible oleogels based on hydroxypropyl methyl cellulose. *Food Chemistry* 246: 137–149.

Mert, B. and Vilgis, T.A. (2021). Hydrocolloid coated oleosomes for development of oleogels. *Food Hydrocoll* 119: 106832.

Mikkonen, K.S., Parikka, K., Ghafar, A., and Tenkanen, M. (2013). Prospects of polysaccharide aerogels as modern advanced food materials. *Trends in Food Science and Technology* 34: 124–136.

Mohanan, A., Tang, Y.R., Nickerson, M.T., and Ghosh, S. (2020). Oleogelation using pulse protein-stabilized foams and their potential as a baking ingredient. *RSC Advances* 10: 14892–14905.

Naeli, M.H., Milani, J.M., Farmani, J., and Zargaraan, A. (2020). Development of innovative ethyl cellulose-hydroxypropyl methylcellulose biopolymer oleogels as low saturation fat replacers: Physical, rheological and microstructural characteristics. *International Journal of Biological Macromolecules* 156: 792–804.

Naeli, M.H., Milani, J.M., Farmani, J., and Zargaraan, A. (2022). Developing and optimizing low-saturated oleogel shortening based on ethyl cellulose and hydroxypropyl methyl cellulose biopolymers. *Food Chemistry* 369: 130963.

Nešić, A., Cabrera-Barjas, G., Dimitrijević-Branković, S., Davidović, S., Radovanović, N., and Delattre, C. (2020). Prospect of polysaccharide-based materials as advanced food packaging. *Molecules* 25.

Nikiforidis, C.V. and Scholten, E. (2015). Polymer organogelation with chitin and chitin nanocrystals. *RSC Advances* 5: 37789–37799.

O'Sullivan, C.M., Davidovich-Pinhas, M., Wright, A.J., Barbut, S., and Marangoni, A.G. (2017). Ethylcellulose oleogels for lipophilic bioactive delivery-effect of oleogelation on: in vitro bioaccessibility and stability of beta-carotene. *Food Function* 8: 1438–1451.

Oh, I., Lee, J.H., Lee, H.G., and Lee, S. (2019). Feasibility of hydroxypropyl methylcellulose oleogel as an animal fat replacer for meat patties. *Food Research International* 122: 566–572.

Oh, I.K. and Lee, S. (2018). Utilization of foam structured hydroxypropyl methylcellulose for oleogels and their application as a solid fat replacer in muffins. *Food Hydrocoll* 77: 796–802.

Okuro, P.K., Martins, A.J., Vicente, A.A., and Cunha, R.L. (2020). Perspective on oleogelator mixtures, structure design and behaviour towards digestibility of oleogels. *Current Opinion in Food Science* 35: 27–35.

Okuro, P.K., Tavernier, I., Bin Sintang, M.D., Skirtach, A.G., Vicente, A.A., Dewettinck, K., Cunha, R.L. (2018). Synergistic interactions between lecithin and fruit wax in oleogel formation. *Food Function* 9: 1755–1767.

Pakseresht, S. and Mazaheri Tehrani, M. (2020). Advances in multi-component supramolecular oleogels – A review. *Food Research International* 38 (4): 760–782.

Patel, A.R. (2017). Methylcellulose-coated microcapsules of palm stearine as structuring templates for creating hybrid oleogels. *Materials Chemistry and Physics* 195: 268–274.

Patel, A.R. (2018). Structuring edible oils with hydrocolloids: Where do we stand? *Food Biophysics* 13: 113–115.

Patel, A.R., Cludts, N., Bin Sintang, M.D., Lewille, B., Lesaffer, A., and Dewettinck, K. (2014a). Polysaccharide-based oleogels prepared with an emulsion-templated approach. *ChemPhysChem* 15: 3435–3439.

Patel, A.R., Cludts, N., Sintang, M.D., Lesaffer, A., and Dewettinck, K. (2014b). Edible oleogels based on water soluble food polymers: Preparation, characterization and potential application. *Food Function* 5: 2833–2841.

Patel, A.R. and Dewettinck, K. (2015). Comparative evaluation of structured oil systems: Shellac oleogel, HPMC oleogel, and HIPE gel. *The European Journal of Lipid Science and Technology* 117: 1772–1781.

Patel, A.R., Schatteman, D., Lesaffer, A., and Dewettinck, K. (2013). A foam-templated approach for fabricating organogels using a water-soluble polymer. *RSC Advances* 3: 22900–22903.

Pérez-Martínez, J.D., Sánchez-Becerril, M., Marangoni, A.G., Toro-Vazquez, J.F., Ornelaṣ-Pax, J.J., and Ibarra-Junquera, V. (2019). Structuration, elastic properties scaling, and mechanical reversibility of candelilla wax oleogels with and without emulsifiers. *Food Research International* 122: 471–478.

Phoon, P.Y. and Henry, C.J. (2020). Fibre-based oleogels: effect of the structure of insoluble fibre on its physical properties. *Food Function* 11: 1349–1361.

Plazzotta, S., Calligaris, S., and Manzocco, L. (2019). Structure of oleogels from κ-carrageenan templates as affected by supercritical-CO2-drying, freeze-drying and lettuce-filler addition. *Food Hydrocoll* 96: 1–10.

Plazzotta, S., Calligaris, S., and Manzocco, L. (2020). Structural characterization of oleogels from whey protein aerogel particles. *Food Research International* 132: 109099.

Plazzotta, S., Jung, I., Schroeter, B., Subrahmanyam, R.P., Smirnova, I., Calligaris, S., Gurikov, P., and Manzocco, L. (2021). Conversion of whey protein aerogel particles into oleogels. Effect of oil type on structural features. *Polymers* 13: 4063.

Ramel, P.R., Co, E.D., Acevedo, N.C., and Marangoni, A.G. (2016). Structure and functionality of nanostructured triacylglycerol crystal networks. *Progress in Lipid Research* 64: 231–242.

Rekhi, G.S. and Jambhekar, S.S. (1995). Ethylcellulose – a polymer review. *Drug Development and Industrial Pharmacy* 21: 61–77.

Rodríguez-Hernández, A.K., Pérez-Martínez, J.D., Gallegos-Infante, J.A., Toro-Vazquez, J.F., and Ornelas-Paz, J.J. (2021). Rheological properties of ethyl cellulose-monoglyceride-candelilla wax oleogel vis-a-vis edible shortenings. *Carbohydrate Polymers* 252: 117171.

Romoscanu, A.I. and Mezzenga, R. (2006). Emulsion-templated fully reversible protein-in-oil gels. *Langmuir* 22: 7812–7818.

Sánchez, R., Stringari, G.B., Franco, J.M., Valencia, C., and Gallegos, C. (2011). Use of chitin, chitosan and acylated derivatives as thickener agents of vegetable oils for bio-lubricant applications. *Carbohydrate Polymers* 85: 705–714.

Sandolo, C., Bulone, D., Mangione, M.R., Margheritelli, S., Meo, C., Alhaique, F., Matricardi, P., and Coviello, T. (2010). Synergistic interaction of locust bean gum and xanthan investigated by rheology and light scattering. *Carbohydrate Polymers* 82: 733–741.

Schaink, H.M., van Malssen, K.F., Morgado-Alves, S., Kalnin, D.J.E., and van der Linden, E. (2007). Crystal network for edible oil organogels: Possibilities and limitations of the fatty acid and fatty alcohol systems. *Food Research International* 40: 1185–1193.

Shah, D.O. (1971). Significance of the 1:3 molecular ratio in mixed surfactant systems. *Journal of Colloid and Interface Science* 37: 744–752.

Shakeel, A., Farooq, U., Gabriele, D., Marangoni, A.G., and Lupi, F.R. (2021). Bigels and multi-component organogels: An overview from rheological perspective. *Food Hydrocoll* 111: 106190.

Shchipunov, Y.A. (2001). Lecithin organogel: A micellar system with unique properties. *Colloids Surfaces A Physicochem. Eng. Asp.* 183–185: 541–554.

Stortz, T.A., De Moura, D.C., Laredo, T., and Marangoni, A.G. (2014). Molecular interactions of ethylcellulose with sucrose particles. *RSC Advances* 4: 55048–55061.

Stortz, T.A., Laredo, T., and Marangoni, A.G. (2015). The role of lecithin and solvent addition in ethylcellulose-stabilized heat resistant chocolate. *Food Biophysics* 10: 253–263.

Stortz, T.A. and Marangoni, A.G. (2013). Ethylcellulose solvent substitution method of preparing heat resistant chocolate. *Food Research International* 51: 797–803.

Stortz, T.A. and Marangoni, A.G. (2014). The replacement for petrolatum: Thixotropic ethylcellulose oleogels in triglyceride oils. *Green Chemistry* 16: 3064–3070.

Tan, S.Y., Peh, E., Lau, E., Marangoni, A.G., and Henry, C.J. (2017a). Physical form of dietary fat alters postprandial substrate utilization and glycemic response in healthy Chinese men. *Journal of Nutrition* 147: 1138–1144.

Tan, S.Y., Wan-Yi Peh, E., Marangoni, A.G., and Henry, C.J. (2017b). Effects of liquid oil vs. oleogel co-ingested with a carbohydrate-rich meal on human blood triglycerides, glucose, insulin and appetite. *Food Function* 8: 241–249.

Tanti, R., Barbut, S., and Marangoni, A.G. (2016a). Oil stabilization of natural peanut butter using food grade polymers. *Food Hydrocoll* 61: 399–408.

Tanti, R., Barbut, S., and Marangoni, A.G. (2016b). Hydroxypropyl methylcellulose and methylcellulose structured oil as a replacement for shortening in sandwich cookie creams. *Food Hydrocoll* 61: 329–337.

Tavernier, I., Doan, C.D., Van der Meeren, P., Heyman, B., and Dewettinck, K. (2018). The potential of waxes to alter the microstructural properties of emulsion-templated oleogels. *European Journal of Lipid Science and Technology* 120: 1700393.

Tavernier, I., Patel, A.R., Van der Meeren, P., and Dewettinck, K. (2017). Emulsion-templated liquid oil structuring with soy protein and soy protein: κ-carrageenan complexes. *Food Hydrocoll* 65: 107–120.

Urbánková, L., Sedlác, T., Kašpárková, V., and Bordes, R. (2021). Formation of oleogels based on emulsions stabilized with cellulose nanocrystals and sodium caseinate. *Journal of Colloid and Interface Science* 596: 245–256.

Vélez-Erazo, E.M., Bosqui, K., Rabelo, R.S., Kurozawa, L.E., and Hubinger, M.D. (2020). High internal phase emulsions (HIPE) using pea protein and different polysaccharides as stabilizers. *Food Hydrocoll* 105: 105775.

Wijaya, W., Van der Meeren, P., and Patel, A.R. (2018). Oleogels from emulsion (HIPE) templates stabilized by protein-polysaccharide complexes. In: *Edible Oil Structuring: Concepts, Methods and Applications* (ed. A.R. Patel), 175–197. The Royal Society of Chemistry.

Woern, C., Marangoni, A.G., Weiss, J., and Barbut, S. (2021). Effects of partially replacing animal fat by ethylcellulose based organogels in ground cooked salami. *Food Research International* 147: 110431.

Yadav, I., Kasiviswanathan, U., Soni, C., Paul, S.R., Nayak, S.K., Sagiri, S., Anis, A, and Pal, K. (2017). Stearic acid modified stearyl alcohol oleogel: analysis of the thermal, mechanical and drug release properties. *Journal of Surfactants and Detergents* 20: 851–861.

Yan, X., Ma, C., Cui, F., McClements, D.J., Liu, X., and Liu, F. (2020). Protein-stabilized pickering emulsions: formation, stability, properties, and applications in foods. *Trends in Food Science and Technology* 103: 293–303.

Yang, T., Li, X.T., and Tang, C.H. (2020). Novel edible pickering high-internal-phase-emulsion gels efficiently stabilized by unique polysaccharide-protein hybrid nanoparticles from Okara. *Food Hydrocoll* 98: 105285.

Yang, X., Li, A., Li, X.X., Sun, L., and Guo, Y. (2020). An overview of classifications, properties of food polysaccharides and their links to applications in improving food textures. *Trends in Food Science and Technology* 102: 1–15.

Ye, X., Li, P., Lo, Y.M., Fu, H., and Cao, Y. (2019). Development of novel shortenings structured by ethylcellulose oleogels. *Journal of Food Science* 84: 1456–1464.

Zetzl, A.K., Gravelle, A.J., Kurylowicz, M., Dutcher, J., Barbut, S., and Marangoni, A.G. (2014). Microstructure of ethylcellulose oleogels and its relationship to mechanical properties. *Food Structure* 2: 27–40.

Zetzl, A.K., Marangoni, A.G., and Barbut, S. (2012). Mechanical properties of ethylcellulose oleogels and their potential for saturated fat reduction in frankfurters. *Food Function* 3: 327–337.

Zhang, R., Yu, J., Liu, N., Gao, Y., and Mao, L. (2022). W/O emulsions featuring ethylcellulose structuring in the water phase, interface and oil phase for multiple delivery. *Carbohydrate Polymers* 283: 119158.

Zhao, W., Wei, Z., and Xue, C. (2021). Recent advances on food-grade oleogels: Fabrication, application and research trends. *Critical Reviews in Food Science and Nutrition* 0: 1–18.

Zhou, F.Z., Yu, X.H., Zeng, T., Yin, S.W., Tang, C.H., and Yang, X.Q. (2019). Fabrication and characterization of novel water-insoluble protein porous materials derived from pickering high internal-phase emulsions stabilized by gliadin-chitosan-complex particles. *Journal of Agricultural and Food Chemistry* 67: 3423–3431.

Zou, F. and Budtova, T. (2021). Polysaccharide-based aerogels for thermal insulation and superinsulation: an overview. *Carbohydrate Polymers* 266: 118130.

Section III

Methodologies for the Characterisation of Fat Mimetics

3.1

Rheology and Texture Analysis

Luiz Henrique Fasolin, Carolina Siqueira Franco Picone, Gabrielli Nunes Clímaco, and Felipe de Andrade Maia*

Department of Food Engineering and Technology, School of Food Engineering, University of Campinas – UNICAMP, Campinas, SP, Brazil
** Corresponding author*

3.1.1 Introduction

Rheology is based on the stress-strain relation, *i.e.*, this science describes the material physical response to the applied forces (e.g., deformation or flow). Rheology of fluids, solids (also called mechanical properties), and semi-solid foods can be related to material processing, their physical stability, and to textural and sensorial properties. For fat mimetics, several rheological properties are important in terms of their production application and need to be evaluated.

The sizing and choice of equipment, e.g., pumps, stirrers, temperature exchanges, and process conditions in fat mimetics processing depend on viscosity and temperature. Since the consumer has expectations regarding the sensory properties during handling, cooking and consumption, hardness, spreadability, and melting behavior of fat mimetics, the afore-mentioned aspects must be assessed. In addition, their behavior within the gastrointestinal tract will be affected by the oil type and the oleogelator used. Thus, the free fatty acids release and the ability of the fat mimetic to act as a vehicle for the controlled delivery of bioactive compounds will be affected by rheological and mechanical properties.

In this chapter, techniques for evaluating rheology and the texture of fat mimetics will be discussed. At first, a compilation of some basic principles will be presented so the reader can understand the solid, liquid, and viscoelastic behavior and the properties related to them. There are several textbooks that have a more in-depth discussion of the principles of rheology and mechanical properties (Steffe 1996; Barnes 2000; Hibbeler 2001; Beer et al. 2013; Mezger 2020). Finally, small and large deformation tests are listed, and some fat mimetics' results are presented as are some empirical tests for texture analyses.

3.1.2 Rheology Principles

Rheology is based on the relationship between deformation and the force needed to cause strain in the material. The relationship between the applied stress (force per area) and the strain (relative deformation) is the key to rheological evaluation. The loading process (stress application) can cause materials to store or dissipate energy. Ideal or Hookean solids can store all the energy and show a linear relationship with strain, obeying the Hooke's law. These materials will be explored better in subsequent sections. Likewise, ideal or Newtonian fluids do not resist the shear forces,

Fat Mimetics for Food Applications, First Edition. Edited by Miguel Ângelo Parente Ribeiro Cerqueira and Lorenzo Miguel Pastrana Castro.

dissipating all the energy and flowing. In this case, they are characterized by Newton's viscosity law, a linear relationship between shear stress and shear rate (Equation 3.1.1).

$$\tau = \eta \dot{\gamma} \tag{3.1.1}$$

where τ is the shear stress (Pa); η is the viscosity (Pa.s) and $\dot{\gamma}$ is the shear rate (s^{-1}).

Most food materials, including fat mimetics do not have only solid (elastic) or fluid (viscous) characteristic, being classified as viscoelastic materials, which combine the ability to store and dissipate part of the energy applied simultaneously. Viscoelastic fluids dissipate more energy than they store, while viscoelastic solids show the opposite behavior. Viscoelastic behavior can be classified as linear viscoelasticity and non-linear viscoelasticity. In the linear viscoelastic region (LVE region) the ratio of stress to strain does not depend on the stress or stress magnitude, being only a function of time. But in the non-linear viscoelastic region (non-LVE region) the amount of stress and strain bean to be relevant, increasing its theoretical complexity. The linear viscoelasticity is generally used for rheological characterization.

Viscoelastic behaviors can be determined using transient or oscillatory tests. Creep-recovery and stress relaxation are transient tests. They can be carried out under shear or uniaxial loading and are performed evaluating the structural changes over time with the application of a constant strain or stress. In the stress relaxation test, the material is submitted to a constant stress and force needed to maintain such strain is evaluated. The measured force is usually converted to stress or stress relaxation modulus (Equation 3.1.2).

$$G(t) = \frac{\tau(t)}{\gamma} \text{ or } E(t) = \frac{\sigma(t)}{\varepsilon} \tag{3.1.2}$$

where *G(t)* and *E(t)* are the stress relaxation modulus under shearing and axial loading, respectively; σ is the normal stress (Pa); τ is the shear stress (Pa), and γ and ε (dimensionless) are the strain under shearing and axial loading respectively.

On the other hand, constant stress and the correspondent strain are tracked in the creep-recovery test. Here, the measured strain is expressed as creep compliance *J(t)* that along with the stress relaxation modulus, can be a function of normal or of shear strain (Equation 3.1.3).

$$J(t) = \frac{\gamma(t)}{\sigma} \text{ or } J(t) = \frac{\varepsilon(t)}{\sigma} \tag{3.1.3}$$

The relationship between the stress relaxation modulus or creep-compliance and time are usually fitted to linear and non-linear viscoelastic equations characterized in terms of massless mechanical models, composed of different connections of ideal spring (obeying Hooke's law) and dashpot (obeying Newton's law) elements. Among them, the generalized Maxwell model is the most popular for the stress relaxation test and comprises Maxwell elements in parallel with an independent spring. The time-dependent behavior of a viscoelastic material in the stress relaxation process is described by Equation (3.1.4). For creep-recovery assays, the generalized Kelvin-Voigt model is the most used (Equation 3.1.5)

$$Y(t) = Y_0 + \sum_{i=1}^{n} a_i . e^{\frac{-t}{\tau_i}} \tag{3.1.4}$$

$$J(t)= J_0 + \sum_{i=1}^{n} J_i \left(1 - e^{\frac{-t}{\tau_i}}\right) + \frac{t}{\eta_0} \tag{3.1.5}$$

where *Y(t)* is the decaying parameter (force, stress, or stress relaxation modulus); t represents time; *Y*0 is the residual value of *Y(t)* when $t = \infty$; *ai*, and τi are constants characteristic of the material. *J(t)* is the variation of the creep compliance with time *t*; *J*0 is the instantaneous compliance; *J*i are the retardation compliances; τi *and* $\eta 0$ are characteristic parameters of the material.

For oscillatory tests, as the name suggests, an oscillatory strain input $\gamma(t)$ at determined frequency ω (rad/s) is applied (Equation 3.1.6). The result is a sinusoidal shear stress output $\tau(t)$ (Equation 3.1.7). Oscillatory measurements are usually carried out in shear; however, it is possible to take those measurements with normal force.

$$\gamma(t)= \gamma_0 \, sin(\omega t) \tag{3.1.6}$$

$$\tau(t)= \tau_0 \, sin(\omega t+\delta) \tag{3.1.7}$$

The phase angle δ (rad) describes the viscoelastic behavior. Solid materials instantly respond to the signal, while fluid materials show the opposite behavior. Phase angles close to 0°and 90° correspond to viscoelastic solids and viscoelastic fluids, respectively. Thus, Equation 3 can be decoupled into an in phase and out of phase response, generating two new moduli. $G'(\omega)$ (Pa) is defined as storage modulus and $G''(\omega)$ (Pa) is defined as loss modulus. (Equations 3.1.8 and 3.1.9). Other useful information can be obtained from G' and G'' relation, as damping factor ($tan\ \delta$), complex modulus (G^*), and complex viscosity (η^*).

$$G'(\omega)=\left(\frac{\sigma_0}{\gamma_0}\right)\cos(\delta) \tag{3.1.8}$$

$$G''(\omega)=\left(\frac{\sigma_0}{\gamma_0}\right)\sin(\delta) \tag{3.1.9}$$

There are many rheological tests that can be used to completely characterize rheological and viscoelastic behaviors. They are called large and small deformation tests. Small deformation tests are performed in the LVE region through dynamic oscillatory rheology. They can be related to the structure of the food materials and correlated to composition, molecular interactions, texture, consumer perception, etc. Large deformation tests are outside the LVE region, which breaks the structure of the material. These tests are important for characterizing the rheological behavior, time-dependence, and viscosity; these parameters are important for all food product steps (processing, storage, handling, consumption, etc.).

In addition, rheological and viscoelastic behavior are highly temperature dependent. At higher temperatures, the molecular movement is enhanced because of internal energy increase. Viscosity decreases exponentially when the temperature rises, making food structure softer. Since food materials are subject to a wide range of temperatures during processing, handling, and consumption, setting the proper temperature is important when performing rheological tests.

3.1.2.1 Large Deformation Tests

3.1.2.1.1 Rheological Behavior/Viscosity Measurements

The flow tests or flow curves should ideally be performed with a fully developed, laminar and steady flow. The temperature must also be kept constant and uniform throughout the sample. Different geometries can perform the test, however, for fat mimetics, parallel plate or cone-plate geometries are generally used, which allow a uniform forces distribution for this material. Sample handling is another priority of concern as structural damage must be kept to a minimum during its settle into the equipment. The proper amount of raw material also must be considered so the measurements will be not over or underestimated. After conditioning, a stress or shear rate is applied to the sample. The output is its corresponding variable (shear rate and stress, respectively). For ideal Newtonian fluids, measuring a single stress or shear rate is sufficient, since its viscosity is constant, and the stress and shear rate are proportional.

However, because of their complexity, most food materials are not ideal, being classified as non-Newtonian fluids. As a result, they show a non-linear relationship between shear stress and shear rate. Depending on their composition and structure, food materials such oleogels, bigels, and emulgels can show different rheological behavior. Shear-thinning behavior, *i.e.*, the viscosity decreases with the shear rate increase without or with the presence of yield stress (Herschel-Bulkley fluid) are the most common rheological behavior seen for fat mimetics.

For non-Newtonian fluids, such as fat mimetics, measurements must be carried out at a wide range of shear rate, which must include the characteristic shear rate of the process to which the material will be subjected during processing, handling, or consumption. For example, mixing can be characterized by 1–100 s^{-1}, 10–100 s^{-1} corresponds to swallowing, and 10–1000 s^{-1} to a pipeline flow. Since the viscosity of these fluids depends on the shear rate, it is called apparent viscosity, and it is represented by ηx, where x is the shear rate and where viscosity assumes the corresponding value.

From the stress versus shear rate data, the best rheological model that describes the material's behavior and its viscosity in different processes can be determined. . Structured materials, such as oleogels, bigels, and emulgels, generally exhibit a Herschel-Bulkley or shear-thinning behavior. When the increase in shear starts to break the structure, flow resistance and viscosity decrease. Differences in rheological behavior and in the viscosity of the material, can be attributed to differences in composition to molecular interactions that can change the structure of the gel.

Liu et al. (2022) evaluated the rheological behavior of linseed oil and rapeseed oleogels structured with stigmasterol. The authors observed a shear-thinning behavior, where the stigmasterol concentration increase led to an increase in the apparent viscosity of the oleogel. This result indicates that more gelling agents' molecules can increase the crosslinking and the system structure. The oleogels prepared with rapeseed oil had a more apparent viscosity, which promotes a more stable contract between the crystals. Bigels formulated with different proportions of 10% beeswax in soybean oil and 3% HPMC in water also showed pseudoplastic behavior. In this case, higher oleogel fractions increased the bigel viscosity (Z. Chen et al. 2023). On the other hand, oleogels formulated with candelilla wax and other structuring agents showed a shear thinning behavior and yield stress, fitting better to Herschel-Bulkley model. The authors observed that increasing the complexity of the gelling agent increased the yield stress value, the initial stress needed for the oleogel flows. The authors observed the existence of overshoot for some formulations (Figure 3.1.1). This phenomenon can be explained by the force necessary to promote the initial breakage of the elastic structure subjected to flow. After reaching such value there is a sudden drop in the stress with shear rate increase (da Silva et al. 2018). Cerqueira et al. (2017), observed similar

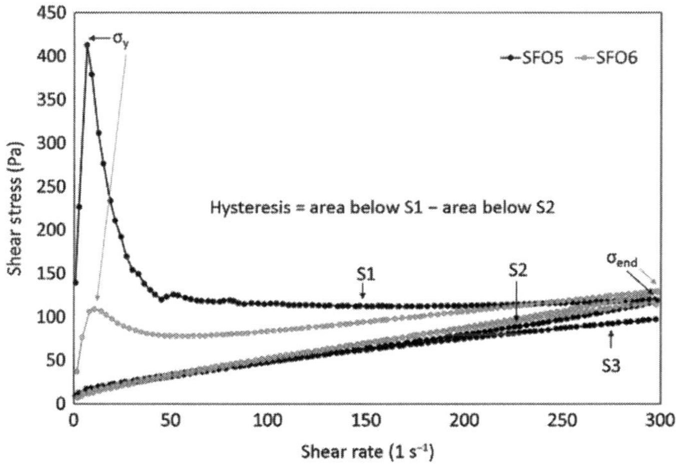

Figure 3.1.1 Flow curves of the oleogels produced with candelilla wax, monoacylglycerols, and hard fat using high oleic sunflower oil (systems SFO5 and SFO6) at 25°C (da Silva et al. 2018).

behavior in oleogels formulated with long and medium chain triacylglycerols with different low molecular oleogelators (LMOGs). The authors correlated this effect with a more structured network and with the time-dependent behavior characteristic of fat mimetics.

3.1.2.1.2 Time-Dependence

Time-dependence is not of great relevance during food processing, as they occur in the steady state. However, the recovery of the structure is important after processing and in food storage, handling, and consumption. If a material is time-dependent, it means that when sheared at a constant shear rate, its viscosity changes over time. For thixotropic materials, the viscosity decreases with shearing, which is a consequence of a gradual structure's breakdown. After a rest period, the sheared material can recover its structure totally, partially, or none. Materials that show the opposite behavior are classified as rheopectic or anti-thixotropic.

There are different ways to assess the time-dependence. A constant shear rate can be applied, and the viscosity tracked over time. Step or linear changes in shear rate can also be applied. The shear stress decay data (*i.e.*, the viscosity decay) can be fitted to different rheological models that describe the time-dependent rheological behavior. Weltman model (Equation 3.1.10) can characterize both thixotropic and rheopectic behavior. But Tiu-Boger model (Equation 3.1.11) describes the thixotropic behavior of Herschel-Bulkley fluids, *i.e.*, fluids that present yield stress.

$$\tau = A - B\log(t) \tag{3.1.10}$$

$$\tau = \lambda[\tau_0 + k_H(\dot{\gamma})^{n_H}] \tag{3.1.11}$$

where, τ is the shear stress (Pa); t is the time (s); A and B are constants; $\dot{\gamma}$ is the shear rate (s^{-1}); λ is a time-dependent structural parameter; $\tau0$ is the yield stress (Pa); kH is the consistency index (Pa s^n); and nH is the flow index behavior.

In addition, there is a qualitative evaluation of the time-dependence based on the flow curves. For this purpose, the shear rate is evaluated in a determined range (that must include the shear rate of interest) in a three-step program, *i.e.*, an up-down-up cycles. Here, if the material is thixotropic or rheopectic, there will be a hysteresis between the curves. The greater the hysteresis, the greater the time-dependence (Figure 3.1.1)

Besides these tests, in recent years, three interval thixotropy test (3ITT) were used to simulate the food structure deformation during processing. This test comprises three intervals in the linear viscoelasticity region (LVE region) and non-linear viscoelasticity region (non-LVE region). The first one, shear conditions are at rest, thus low shear is applied in the LVE region to evaluate the original structure of the material. In the second interval, high shear is applied (non-LVE region), simulating material processing or handling, which breaks its structure. The last interval is performed in the LVE region to evaluate the regeneration ability of the structure. Low shear rate is applied to simulate what happens after processing (Figure 3.1.2) (Mezger 2020).

Oleogel, bigels, emulgels, and other fat mimetics are generally classified as thixotropic and most studies used the qualitative or 3ITT methods to evaluate such behavior. Alizadeh et al. (2020) evaluated oleogels prepared with hydroxypropyl methylcellulose, sodium caseinate, beeswax, and their combinations by different methods (foam and emulsion-template). The authors used the 3ITT test to evaluate the structural recovery of the oleogels compared to a commercial shortening. Except for the oleogel prepared only with beeswax and sodium caseinate, the others showed structure recovery up to 26 %. This result was higher than with commercial shortening (0.24 %), suggesting these oleogels are less strain sensitive. Oleogels prepared with soybean oil or high oleic sunflower oil using candelilla wax, monoacylglycerols and hydrogenated and their combination were evaluated throughout the hysteresis between the flow curves. Oleogels with candelilla wax only were most influenced by shear rate compared to oleogelator blends (da Silva et al. 2018).

Thixotropy measurements were also used to evaluate fat mimetics application. Patel et al. (2014) presented edible applications of shellac oleogel, including chocolate paste. Compared to the reference, the shellac chocolate paste showed a higher thixotropy value, *i.e.*, the formulation was more

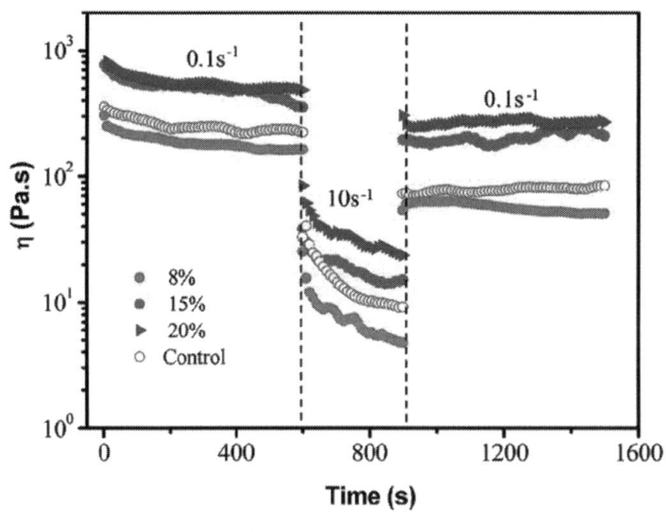

Figure 3.1.2 Thixotropic behavior (3ITT test) of emulsion-template oleogels stabilized with soy protein isolated (X. W. Chen et al. 2019).

shear sensitive. This result can be attributed to the rheological properties and partial structure recovery of shellac oleogels.

However, shear-induced structures, *i.e.*, rheopectic behavior have already been reported. Martins et al. (2022) produced sterol-based oleogels to assess the effect of environmental conditions on structural formation and stability. The authors observed that the structuring process can be prompted through shearing, which affects the rheological properties of the oleogel. The importance of a small mechanical force application was reported for similar systems (Bot and Agterof 2006) and for olive oil + monoglyceride oleogels (Ojijo et al. 2004).

3.1.2.2 Small Deformation Tests

3.1.2.2.1 Transient Tests

3.1.2.2.1.1 Stress Relaxation

To perform a stress relaxation test in shear, the same care described for large shear deformation tests must be taken (e.g., handling, amount of sample, temperature uniformity etc). From stress relaxation modulus data, ideal and viscoelastic behavior can be analyzed. For ideal Hookean solids, its value remains constant, since this material absorbs all the energy applied, deforming itself proportionally.

Conversely, ideal Newtonian materials present an instant stress output followed by rapid stress relaxation due to the energy dissipation during flow. Most part of food material, as fat mimetic, present viscoelastic behavior and after an instantaneous response to the input strain, they show stress decreasing during time due to the structure relaxation until reach a constant equilibrium stress (no longer time-dependent). The rate of decreasing stress and the equilibrium stress reflect the viscoelastic character of the material. The higher the elastic behavior, the less intense the stress relaxation is and the higher the equilibrium stress is due to the amount of energy stored during deformation. If the viscous character of material prevails, the stress shows a sharp decrease during time while the energy is dissipated. Data analyses can be done evaluating the relaxation percentage (Equation 3.1.12) or using the generalized Maxwell or other rheological models (e.g., Weichert model–an adaptation of Maxwell model) to fit the data. In this case characteristic parameters of the material are estimated.

$$relaxation(\%) = \left(\frac{\sigma_0 - \sigma_r}{\sigma_0}\right) \times 100\% \tag{3.1.12}$$

where $\sigma 0$ is the initial normal stress and σr is the residual normal stress after relaxation.

The effect of Span 60 on physicochemical properties of oleogels was evaluated by stress relaxation assays (Uvanesh et al. 2016). The probe compressed 1 mm inside the sample at 1 mm/s and was held for 60 s. Higher Span 60 concentrations increased the relaxation degree. This result was correlated with the more rigid network of these systems since they are more sensitive to stress application. The structure broke, allowing the residual stress reduction. The authors used the Wiechert model with three Maxwell elements, which was reported as sufficient to describe soft materials (Maria et al. 2014). The parameters were under the% relaxation. Relaxation times ($\tau 1$, $\tau 2$ and $\tau 3$) were determined and related to elastic behavior and with network destabilization at intermediate and longer times.

Sagiri et al. (2015) evaluated bigels produced with gelatin hydrogel. Oleogel phase was composed of stearic acid in sesame and soybean oils and mixture from 5% to 15% in gelatin solution. They also produced emulgels (gelatin solution with liquid oil). For the stress relaxation test, the probe compressed the bigels 5 mm at 1 mm/s. The probe was held in this position for 60 s. The authors observed that emulgels showed a sharp decrease in the stress values, since they are softer than bigels (Figure 3.1.3). They used the Maxwell model to calculate the relaxation time (a characteristic of the material), that was lower for bigels. This indicates that they took less time to dissipate the stress.

3.1.2.2.1.2 Creep-Recovery

On the creep-recovery tests, the step stress input will induce the instantaneous strain of the material. Hookean elastic materials present constant strain during the loading, because of the lack of flow and return to the original shape upon stress removal. Ideal viscous materials show steady flow and a linear stress-strain relationship and the inability to recover the imposed deformation. Conversely, the viscoelastic material presents an intermediate behavior with an instantaneous strain (related to its elastic character) followed by a nonlinear region of strain curve because of its

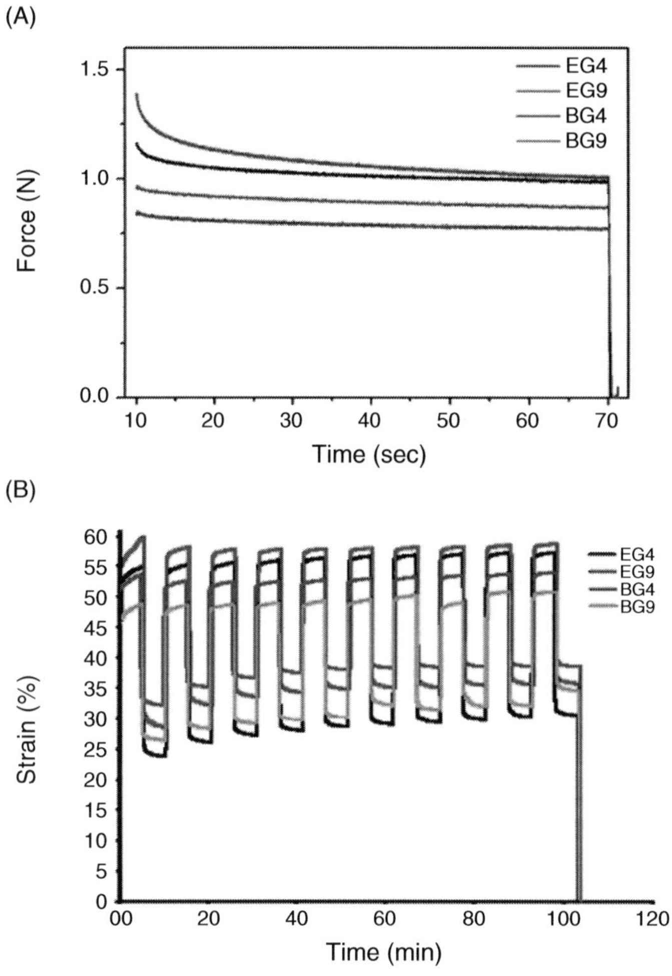

Figure 3.1.3 (A) Stress relaxation and B) Cyclic creep-recovery test for emulgels (EM) and bigels (BG). (Adapted from Sagiri et al. (2015)).

partial flow. After the stress removal (recovery region), viscoelastic materials present permanent deformation (less intense than the total deformation applied to the sample). The greater the creep-recovery, the closer the material is to an ideal elastic solid. Viscoelastic materials with predominant viscous behavior are less prone to storage energy and show intense permanent deformation.

Creep-recovery tests were performed in oleogels made of beeswax, canola oil, and β-carotene to evaluate the influence of the carotenoid in the gel structure (Barragán-Martínez et al. 2022). A constant shear stress (0.3535 mN m^{-1}) during 900 s and then the shear stress were removed to allow the oleogel's structure to recover during another 900 s. The data were fitted to Burger's model. A decrease in the elasticity of the gel with β-carotene addition increase was observed. However, the initial viscosity showed a slight and non-significant decrease. The authors conclude that β-carotene disrupted the molecular network and negatively influenced its recovery.

Sagiri et al. (2015) used the same gelatin/stearic acid bigels and emulgels and performed a cyclic creep-recovery test. A 30 mm diameter probe applied a constant shear force (200 g) for 5 min, followed by a 5 g force for the same time (recovery phase). This sequence was repeated 10 times (Figure 3.1.4B). The strain (%) was higher for emulgels compared to bigels, meaning that the presence of a gelled oil phase reduced the deformation (increase the elastic component). The authors fit the data to the Burger's model with four elements. They observed a decrease in the initial viscosity throughout the cycles, indicating that the systems lost the viscous element gradually. On the other hand, the elastic element increased after each cycle. However, no gel-sol transition was seen. The change was faster for emulgels than for bigels, which can be associated with the leaching of the liquid oil phase with the shear stress.

3.1.2.2.2 Oscillatory Tests

3.1.2.2.2.1 Amplitude Sweep

Strain or stress amplitude sweeps are used generally to determine LVE region, since as this region indicates the interval where the structure of the material is not destroyed. This test is carried out before any other oscillatory measurement. *G′* is evaluated as a function of strain/stress at a fixed frequency, showing a plateau in a determined range which characterize the LVE region. With the increase in the strain/stress, the structure of the material is disturbed, reflecting in deviation of the linearity. A tolerance of *G′* deviation around 5% or 10% is adopted to define the LVE region, according to international standards (ISO 6721–10 and EN/DIN EN 14770 or ASTM D7175 and DIN 51810–2, respectively).

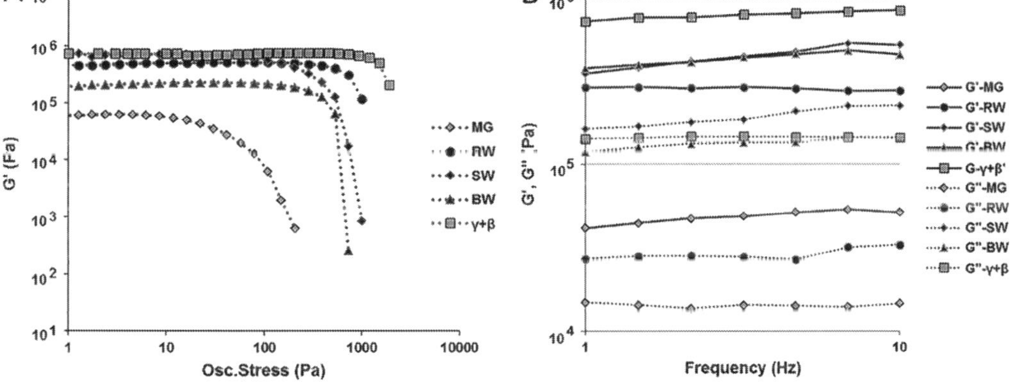

Figure 3.1.4 Oscillatory rheology curves of MCT oleogels containing 10% (w/w) monoglyceride (MG), rice bran wax (RW), sunflower wax (SW), beeswax (BW), and mixture of γ-oryzanol and β-sitosterol (γ − β) at 20°C. A) amplitude sweep and B) frequency sweep. (Adapted from Fayaz et al. (2021)).

A stress sweep was performed for medium chain triacylglycerol oleogel, structured with rapeseed wax, rice wax, sunflower wax, beeswax, saturated monoglyceride, and a mixture of γ-oryzanol and β-sitosterol $\gamma + \beta$ (Fayaz et al. 2021). The authors observed a sharp decrease in G' after the LVE region (Figure 3.1.4A). They adopted a 10% deviation from linearity as a critical stress to determine LVE region. The differing critical stresses observed were correlated with the resistance to deformation attributed to different intermolecular forces holding up the structure in the network of particles.

From amplitude sweeps is also possible to infer about the viscoelastic character of the material through G'/G'' relation. If G'' is higher than G' the material is a viscoelastic fluid and vice versa. This test can also evaluate flow point and yield stress. Aguilar-Zárate et al. (2019) prepared high oleic canola oil-based oleogels with ethylcellulose and lecithin and used a stain sweep to evaluate the yielding behavior of the gels. With the increase of the strain the intermolecular bonds of all oleogels break, and the gel gradually changes from solid-like to liquid-like behavior. The authors observed that adding lecithin increased the yield stress. They attributed this increase to the interaction between functional groups of ethylcellulose and lecithin.

Amplitude sweeps have also been performed for bigels to determine the LVE region (Zhang et al. 2021). A strain sweep was carried out from 0.01 % to 10% in bigels produced with glycerol monostearate and beeswax in soybean oil as oleogel phase and gellan gum as hydrogel phase. The authors fixed the strain at 1% to perform all oscillatory tests. Zhu et al. (2021) prepared surimi gels with beeswax + rapeseed and fish oil oleogels and performed a strain sweep to determine the LVE region. The authors found that surimi sausages were stable in the 0.1–5% strain region.

3.1.2.2.2.2 Frequency Sweep

Frequency sweeps describes the time-dependent behavior of the material since low frequencies emulate long timescale and high frequencies short timescales. This test can evaluate long-term stability of the fat mimetics. They must be performed within LVE region, oscillating usually between 0.01 and 100 radians/s. From the mechanical spectra generated it is possible to classify the material as dilute or concentrated solutions or gelled systems. For gels, including the fat mimetics, generally G' always prevail over G'', So the phase angle is between 0° and 45° and is lower than 1.

Oleogels prepared with ethylcellulose and lecithin in high oleic canola oil were submitted for a frequency sweep test from 0 to 100 Hz at 0.1% strain. The oleogels were independent of the frequency with G' prevailing. The authors reported that 1% of lecithin increased the storage modulus about an order of magnitude because of the interaction with the ethylcellulose (Aguilar-Zárate et al. 2019). In the same way, canola oil-based oleogels structured with diosgenin were evaluated between 0.01 and 100 Hz at fixed strain within LVE. These formulations also showed a strong gel behavior with no frequency dependence and sol-gel transition. The loss modulus (G'') increased with the frequency increase, which was attributed to non-covalent interactions present in the oleogel network (Zeng et al. 2020). Fayaz et al. (2021) produced oleogels with medium chain triacylglycerol oleogel and different oleogelators. They observed a viscoelastic behavior and $\tan\delta$ between 0.1 and 1, indicating that the oleogels behave as weak gels (Figure 3.1.4B).

For bigels, Ghiasi and Golmakani (2022), formulated a novel food-grade bigel with an oleogel produced with ethylcellulose + sunflower and wheat starch hydrogels to be used as partial fat replacer in beef burgers. The bigels were evaluated at 0.1 % strain between 0.1 and 100 Hz. They saw that both G' and G'' moduli presented a weak frequency-dependence, *i.e.*, slight positive slopes. However, storage modulus was higher for all frequency range. In comparison to pure oleogel and hydrogel, bigels showed a stronger elastic network.

3.1.2.2.2.3 Temperature Sweep

Thermal behavior is an important information regarding oleogel. Their thermal reversibility, melting and gelling (crystallization) behavior can be assessed throughout temperature sweeps They are performed in the LVE at constant shear and frequency. A temperature ramp must be defined (e.g., 5°C/min). The reversible behavior can be assessed by carrying out different cooling and heating ramp. If the G' and G'' coincide, the oleogel is thermoreversible. Oleogel produced with LMOGs (Fasolin et al., 2018) and with ethylcellulose and lecithin (Aguilar-Zárate et al. 2019) are examples of oleogels thermoreversible behavior.

The temperature dependence of oleogels produced with canola oil and diosgenin was reported by Zeng et al. (2020). Oleogels prepared at higher temperatures showed higher temperature stability, characterized by the crossover between G' and G'' moduli. The authors attributed this result to differences in the supramolecular structure caused by the different process conditions. On the other hand, different slopes in the G' curve was reported because of the more complex structure of bigels based on monoglyceride oleogels and κ-carrageenan hydrogel. The onset of destabilization of the structure started after a critical temperature, around 40°C. At temperatures ranging 50–60°C, the oleogel shows gel-sol transition, and when it reaches the 60°C, the bigels become a viscous liquid system (Zheng et al. 2020). Similar behavior was also reported in oleogel systems. However, similar behavior was also reported for oleogels produced with LMOGs (Fasolin et al. 2018). The authors compared rheological results with the thermal events seen in differential scanning calorimetry measurements (DSC). Despite G' curves presented more than one slope, at least one was seen in both analyses. The other thermal changes were attributed to physical rearrangements with temperature increase. Papadaki et al. (2019) also compared the rheological results with DSC for oil wax-based oleogel.

Several studies reported that for self-assembled systems, such as oleogel and bigels for example, the cooling rate is closely related to the network structure formation and physicochemical properties. The choice of the cooling rate has great importance. Chia seed oil was structured with glyceryl and sorbitan monostearate and the effect of the cooling rate on the oleogel were assessed (Trujillo-Ramírez et al. 2019). Higher rates formed more compact networks with smaller crystals, while bigger crystals were seen for small rates. Similar results were seen for shellac oleogels (Patel et al. 2013). Pérez-Monterroza et al. (2014) observed that for avocado oleogels the cooling rate influence the formation of the β or β′ crystals in the structure and the onset temperature was greater with smaller rates.

3.1.3 Texture Principles

Texture or mechanical properties tests are useful to evaluate the behavior of fat mimetics under large deformations, as during consumption and process conditions. The mechanical properties at large deformations can be evaluated under different stress conditions (torsion, tension, or compression), being the uniaxial compression tests the most usual on fat mimetics characterization. On uniaxial tests the direct output of the assays is the force of sample to the applied deformation. To compare the behavior of samples with different sizes and shapes, the force intensity must be evaluated regarding the area of the loading distribution. The intensity of the forces distributed over a section is defined as Stress (Equation 3.1.13). So, a force distributed in small surface areas will induce high stresses and structural demand if compared to the same force over applied to a larger surface.

$$\sigma = \frac{F(t)}{A} \tag{3.1.13}$$

where σ is the normal stress (Pa), F is the output force (N) of the material to the applied strain, and A (m^2) is the area of the loading distribution.

Conversely, the change on sample length can be related to the initial sample height to evaluating material strain, according to Equation 3.1.14:

$$\varepsilon = \frac{\delta}{L_0} \tag{3.1.14}$$

where ε is the strain (dimensionless), δ is the difference between the sample's length in a time t and the sample first length ($L0$).

On these tests, the isotropic sample is compressed at a constant rate until a certain deformation. As the test output, the force of the sample against the deformation is recorded. The stress *vs* strain behavior helps better define fat mimetics mechanical properties (Figure 3.1.5). At small deformation, fat mimetics behave as an ideal solid and show a linear stress-strain relationship described by the Hooke's equation (Equation 3.1.15):

$$\sigma = E.\varepsilon \tag{3.1.15}$$

where σ is the normal stress (Pa), E is the Young's modulus (Pa) and ε is the strain (-).

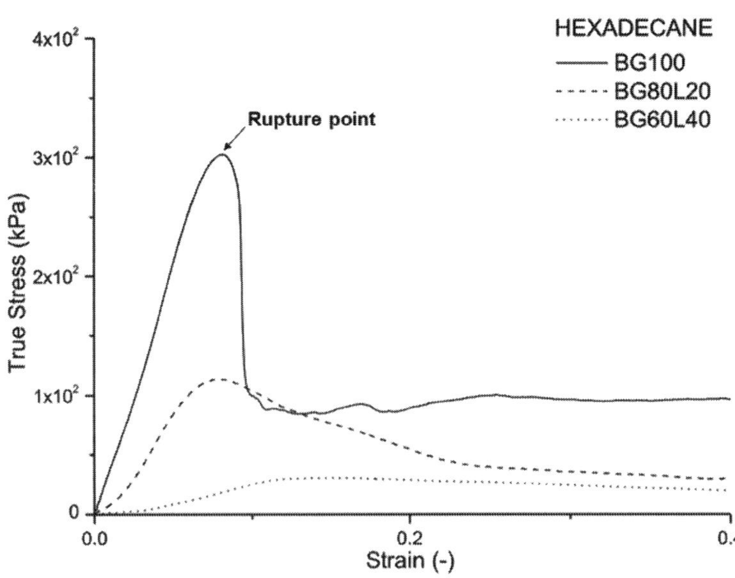

Figure 3.1.5 Stress vs strain curves for oleogels produced with lecithin and phytosterols using hexadecane or sunflower oil. (Adapted from Okuro et al. (2018)).

Within this region, all the energy involved on the strain process is stored by the structure that recovers its initial shape after the load is removed. The rate of stress increasing with strain depends on the Young's modulus (E) that reflects the stiffness of the sample. The linear region of stress-strain curve is seen until the elastic limit of the material. Further, increasing strain values will cause microstructural reorganization with energy dissipation, and the material will lose its total strain recovery after load removal. A partial recovery of its original shape is seen, accomplished by a permanent strain. This characterizes the plastic behavior of the material at large strain values where the rheological behavior is more complex to predict. The increase in strain within the plastic region will induce structural discontinuities that will cause the structure to collapse at the failure stress. Failure or rupture stress is often used to characterize the strength of fat mimetics.

During materials compression with constant density, as the sample height decreases, its cross-section area increases to keep the same sample volume. Thus, the longitudinal strain in length is usually associated with a lateral strain expressed by the Poisson coefficient (Equation 3.1.16):

$$\nu = -\frac{lateral\ strain}{longitudinal\ strain} \tag{3.1.16}$$

For example, oleogels prepared by a hydrogel-templated approach are reported to present a Poisson ration of 0.5 (Feichtinger and Scholten 2020). This implies considerable changes on the cross-section area of gels during compression and on stress and strain within the plastic region. As the stress depends on the area of the applied load, it changes during the compression. Thus, the stress and strain during uniaxial compression up to rupture are calculated by Hencky relations (Equations 3.1.17 and 3.1.18, respectively).

$$\sigma_H = \frac{F(t).h(t)}{A_0 h_0} \tag{3.1.17}$$

$$\varepsilon_H = -ln\left(\frac{h(t)}{h_0}\right) \tag{3.1.18}$$

where σ_H and ε_0 is the Hencky stress (Pa) and strain (-), respectively, h_0 and A_0 are the sample first height and area, respectively, *h(t)* is the sample height in time *t* and *F(t)* is the reaction force of the material to the imposed strain at a time *t*.

The temperature is of outcome importance during the measurements and the thermal history of the samples. The stress-strain behavior of fat mimetics depends straight on the temperature of the material, as well as the strain rate of the compression, which must be set to the mechanical analysis.

Fundamentals tests based on Hencky stress – strain curves are useful to characterize the material's mechanical properties under uniaxial compression, providing information about the stress and strain at rupture of fat mimctics and their Young's modulus. These parameters can be associated with the degree of structuration and intensity of network interactions. However, the behavior of fat mimetics in some practical situations as consumption conditions, mastication, or swallowing often involve complex compression and shearing forces of different magnitudes, which cannot be predicted by fundamental tests. In these circumstances the empiric tests provide information about the performance of fat mimetic systems compared to other fats.

Empiric tests evaluate the force response of the material under diverse mechanical solicitations. As the force is recorded regardless the area of the loading application, the data are dependent of the material shape, in addition to temperature, the velocity of the assay and the probe used on the measurements. Puncture and spreadability tests are the most common empiric tests to evaluate fat mimetics.

3.1.3.1 Fundamental Tests

3.1.3.1.1 Uniaxial Compression

On uniaxial tests the direct output of the assays is the force of the sample to the applied deformation. To compare the behavior of samples with different sizes and shapes, the force intensity must be evaluated regarding the area of the loading distribution. Besides the temperature, parameters such as compression speed and percentage of deformation must be chosen to obtain the desired response (e.g., 80% deformation from first height and compression speed of 1 mm/s). The choice of the compression probe and its adequate lubrication must assure the prevalence of normal forces during the test and absence of shearing forces. In addition, to evaluate the elasticity or Young's modulus is essential to make sure data are into the linear region, where Hooke's law is valid.

Hybrid oleogels produced with lecithin and phytosterols were produced with hexadecane or sunflower oil (Okuro et al. 2018). Uniaxial compression was performed at a speed of 1 mm/s until 80% of its original height. The force vs distance data was fitted to Hencky's equations and to evaluate the elasticity modulus, it was used the data in the linear interval ($0.01–0.02 < \varepsilon H < 0.06$). The authors observed a defined rupture point; however, the peak was less defined with lower fracture stress with lecithin addition. For hexadecane and medium-chain triacylglycerol oleogels, the structure was less stiff with a lower elasticity modulus compared to sunflower oil oleogels (around 10^3 Pa).

Dreher et al. (2020) proposed an innovative structure approach to mimic animal fat. The authors structured canola oil with plant-derived materials. For texture analysis they compressed the oleogels twice until 50% of the initial height at 1.5 mm/s. By compressing the material a second time, it is possible to obtain other information about the structure. Hardness represents the maximum force during the first compression cycle, springiness represents the difference in recoverable

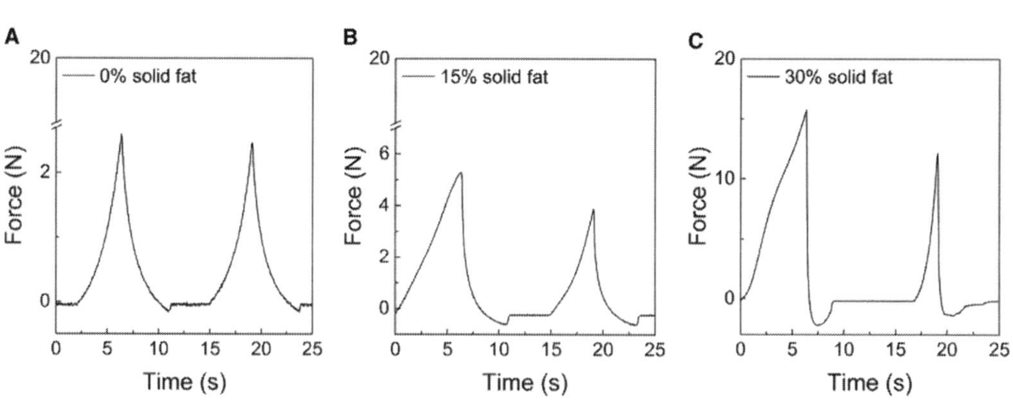

Figure 3.1.6 Force–time plots of two cyclic uniaxial compression tests of crosslinked emulsified fat crystal networks with 0% (A), 15% (B), and 30% (C) solid fat content (Dreher et al. 2020).

specimen height between first and second compression, and cohesiveness denotes the ratio of the work done during the first and the second compression (Figure 3.1.6).

3.1.3.2 Empiric Tests

3.1.3.2.1 Puncture

On the puncture tests, the force necessary to penetrate a probe onto the sample is measured as function of time. The higher the recorded force, the more resistant and firmer is the material. Some authors mistakenly identify the puncture test as compression. Only normal forces are involved in the fundamental compression test. During puncture tests both compression and shear forces are present being their balance dependent on the amount of perimeter and surface area of the probe used (BOURNE 1966). Cylindrical or needle probes are often used for evaluating fat mimetics firmness (Patel et al. 2015; Jadhav et al. 2022; Liu et al. 2022) and the collected data is plotted as force–displacement curves. Similar to compression tests, the velocity, penetration depth, and temperature must be carefully chosen.

For grapeseed oil, oleogels gelled with candelilla wax and glyceryl monostearate, a puncture test was performed using a 5 mm diameter rod probe (Choi et al. 2020). The probe penetrate 10 mm of the samples at 100 mm/min. Hardness was related to the maximum peak from the force-time plots. The hardness of the oleogels prepared with 100% candelilla wax, which was four times higher than for oleogels prepared only with glyceryl monostearate. When the oleogelators were blended with higher candelilla wax proportion, the hardness reached greater values. These results suggested that the synergistic interaction between the oleogelators reinforced the gel structure.

Puncture tests were also reported for food-grade bigels. Bigels were produced with gelatin hydrogel and glycerol monostearate oleogel to simultaneously act as hydrophilic and lipophilic compounds vehicles (Lu et al. 2022). A 10 mm diameter cylindrical probe penetrates 7 mm in the bigels at 60 mm/min. Bigels with lower oleogelator concentrations had weaker structures; the oil phase showed higher deformation with some oil release. The mechanical properties of the systems were improved with oleogelator concentration increase. These results can be related with the ability of oleogel particles to act as active fillers, agreeing with other studies (Lupi et al. 2016; Rousseau 2020).

This texture test has been successfully applied to food products made with oleogels. It is useful when the shape of the sample is not standardized. Li and Liu (2019) evaluated the influence of oleogels with different compositions and that were produced by different gelation mechanism on the dark chocolate preparation. They replaced 50% or 100% of cocoa butter. A 6 mm diameter cylindrical probe penetrated 6 mm in the chocolate samples at 2 mm/s to measure their hardness. The study showed that dark chocolate produced with oleogels was softer compared to cocoa butter chocolates. However, highly unsaturated oleogels are promising systems for reducing the use of cocoa butter. Frankfurt sausages (Wolfer et al. 2018), bread (Meng et al. 2019), meat patties (Oh et al. 2019), and other food products with oleogels were also evaluated with the puncture test.

3.1.3.2.2 Spreadability

Spreadability measurements are of particular interest to fat mimetics for products like butter, margarine, spreads, etc. It is a subjective property related to the consumer perception of the easiness of applying uniformly such food material on a surface. The most used test is composed of a male/female conical set probe. The negative cone (female) is fulfilled with the sample, and the positive cone (male) penetrates a fixed distance into the sample at pre-determined velocity and

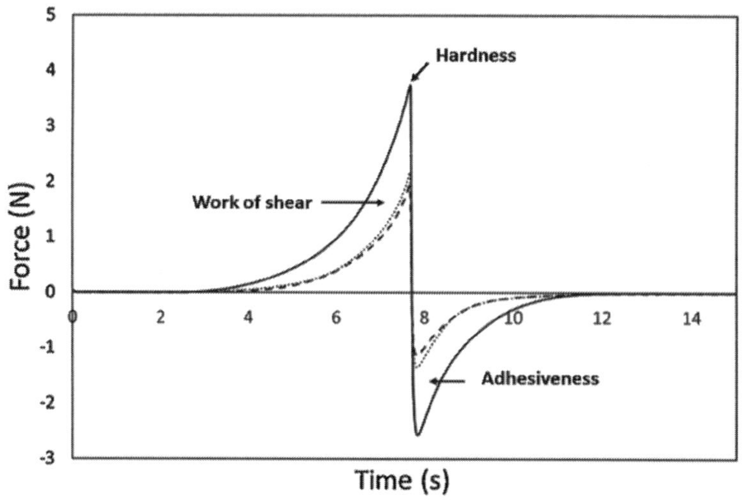

Figure 3.1.7 Characteristic curve of spreadability test obtained from bigels produced with gellan gum hydrogels and high oleic sunflower oil with glycerol monostearate oleogels. (Adapted from Fasolin et al. (2021)).

temperature. Hardness, work of shear, and adhesiveness can be obtained from force versus time plots. Spreadability and hardness (or firmness) are inversely proportional, since harder gels that demand more shear work are less spreadable (Wright et al. 2001). This must be taken into consideration since some authors positively and erroneously correlate firmness with spreadability.

The influence of process parameters on the production of bigels composed by gellan gum hydrogel and glycerol monostearate in high oleic sunflower oil oleogel was evaluated by Fasolin et al. (2021). Spreadability evaluation was performed with a male/female conical set probe at 1 mm/s velocity and 23 mm penetration. From force-time plots, hardness, work of shear, and adhesiveness (maximum force, positive and negative areas, respectively) were assessed (Figure 3.1.7). Spreadability was evaluated as a secondary parameter. The results followed rheological results (frequency sweep). All parameters were higher for bigels compared to control gels (pure oleogel and hydrogel). These results showed once again that the disperse solid phase (the gellan gum particles) can act as active fillers, increasing the mechanical properties. In addition, bigels with oleogel fraction prevailing showed lower spreadability (higher hardness).

Kupiec et al. (2020) structured rapeseed oil with different oleogelators (beeswax, candelilla wax, sunflower wax and monoglycerides). They used a conical set probe where the male one penetrated 25 min in the oleogels at 3 mm/s. The authors observed that candelilla and sunflower waxes oleogels were harder and that they presented higher shear work. These oleogels were less spreadable.

References

Aguilar-Zárate, M., Macias-Rodriguez, B.A., Toro-Vazquez, J.F., and Marangoni, A.G. (2019). Engineering rheological properties of edible oleogels with ethylcellulose and lecithin. *Carbohydrate Polymers* 205. doi: 10.1016/j.carbpol.2018.10.032.

Alizadeh, L., Abdolmaleki, K., Nayebzadeh, K., and Hosseini, S.M. (2020). Oleogel fabrication based on sodium caseinate, hydroxypropyl methylcellulose, and beeswax: Effect of concentration, oleogelation method, and their optimization. *JAOCS, Journal of the American Oil Chemists' Society* 97 (5). doi: 10.1002/aocs.12341.

Barnes, H.A. (2000). A handbook of elementary rheology. *Polymer Composites* 6.

Barragán-Martínez, L.P., Alvarez-Poblano, L., Vernon-Carter, W.J., and Alvarez-Ramirez, J. (2022). Effects of β-carotene on the color, textural, rheological, and structural properties of canola oil/ beeswax oleogel. *Journal of Food Measurement and Characterization* 16 (5). doi: 10.1007/ s11694-022-01449-4.

Beer, F.P., Johnston, R., Dewolf, J., and Mazurek, D. (2013). *Mechanics of Materials*. New York: McGraw-Hill.

Bot, A. and Agterof, W.G.M. (2006). Structuring of edible oils by mixtures of γ-oryzanol with β-sitosterol or related phytosterols. *JAOCS, Journal of the American Oil Chemists' Society* 83 (6). doi: 10.1007/s11746-006-1234-7.

BOURNE, M.C. (1966). Measure of shear and compression components of puncture tests. *Journal of Food Science* 31 (2). doi: 10.1111/j.1365-2621.1966.tb00494.x.

Cerqueira, M.A., Fasolin, L.H., Picone, C.S.F., Pastrana, L.M., Cunha, R.L., and Vicente, A.A. (2017). Structural and mechanical properties of organogels: Role of oil and gelator molecular structure. *Food Research International* 96. doi: 10.1016/j.foodres.2017.03.021.

Chen, X.W., Sun, S.D., Yang, G.L., and Ma, C.G. (2019). Engineering phytosterol-based oleogels for potential application as sustainable petrolatum replacement. *RSC Advances* 10 (1). doi: 10.1039/ c9ra06950j.

Chen, Z., Bian, F., Cao, X., Shi, Z., and Meng, Z. (2023). Novel bigels constructed from oleogels and hydrogels with contrary thermal characteristics: Phase inversion and 3D printing applications. *Food Hydrocolloids* 134: 108063.

Choi, K.O., Hwang, H.S., Jeong, S., Kim, S., and Lee, S. (2020). The thermal, rheological, and structural characterization of grapeseed oil oleogels structured with binary blends of oleogelator. *Journal of Food Science* 85 (10). doi: 10.1111/1750-3841.15442.

da Silva, T.L.T., Arellano, D.B., and Martini, S. (2018). Physical properties of candelilla wax, monoacylglycerols, and fully hydrogenated oil oleogels. *JAOCS, Journal of the American Oil Chemists' Society* 95 (7). doi: 10.1002/aocs.12096.

Dreher, J., Blach, C., Terjung, N., Gibis, M., and Weiss, J. (2020). Formation and characterization of plant-based emulsified and crosslinked fat crystal networks to mimic animal fat tissue. *Journal of Food Science* 85 (2). doi: 10.1111/1750-3841.14993.

Fasolin, L.H., Cerqueira, M.A., Pastrana, L.M., Vicente, A., and Cunha, L. (2018). Thermodynamic, rheological, and structural properties of edible oils structured with LMOGs: Influence of gelator and oil phase. *Food Structure* 16: 50–58. doi: 10.1016/j.foostr.2018.03.003.

Fasolin, L.H., Martins, A.J., Cerqueira, M.A., and Vicente, A.A. (2021). Modulating process parameters to change physical properties of bigels for food applications. *Food Structure* 28. doi: 10.1016/j. foostr.2020.100173.

Fayaz, G., Polenghi, O., Giardina, A., Cerne, V., and Calligaris, S. (2021). Structural and rheological properties of medium-chain triacylglyceride oleogels. *International Journal of Food Science and Technology* 56 (2). doi: 10.1111/ijfs.14757.

Feichtinger, A. and Scholten, E. (2020). Preparation of protein oleogels: Effect on structure and functionality. *Foods*. doi: 10.3390/foods9121745.

Ghiasi, F. and Golmakani, M.T. (2022). No TiFabrication and characterization of a novel biphasic system based on starch and ethylcellulose as an alternative fat replacer in a model food systemtle. *Innovative Food Science & Emerging Technologies* 78: 103028.

Hibbeler, R.C. (2001). *Mechanics of Materials Eigth Edition*. Pearson Prentice Hall.

Jadhav, H.B., Pratap, A.P., Gogate, P.R., and Annapure, U.S. (2022). Ultrasound-assisted synthesis of highly stable MCT based oleogel and evaluation of its baking performance. *Applied Food Research* 2 (2): 100156. doi: 10.1016/j.afres.2022.100156.

Kupiec, M., Zbikowska, A., Marciniak-Lukasiak, K., and Kowalska, M. (2020). Rapeseed oil in new application: Assessment of structure of oleogels based on their physicochemical properties and microscopic observations. *Agriculture (Switzerland)* 10 (6). doi: 10.3390/agriculture10060211.

Li, L. and Liu, G. (2019). Corn oil-based oleogels with different gelation mechanisms as novel cocoa butter alternatives in dark chocolate. *Journal of Food Engineering* 263. doi: 10.1016/j.jfoodeng.2019.06.001.

Liu, Y., Ma, S., Xia, H., Guo, S., and Zeng, C. (2022). Edible oleogels stabilized solely by stigmasterol: effect of oil type and gelator concentration. *Journal of the Science of Food and Agriculture* 102 (11). doi: 10.1002/jsfa.11841.

Lu, Y., Zong, Y., Guo, X., Zhang, J., Gao, Y., and Mao, L. (2022). Structural modification of O/W bigels by glycerol monostearate for improved co-delivery of curcumin and epigallocatechin gallate. *Food Science and Technology* 2 (6): 975–983.

Lupi, F.R., Shakeel, A., Greco, V., Rossi, C.O., Baldino, N., and Gabriele, D. (2016). A rheological and microstructural characterisation of bigels for cosmetic and pharmaceutical uses. *Materials Science and Engineering: C* 69: 358–365. doi: 10.1016/j.msec.2016.06.098.

Maria, H.J., Lyczko, N., Nzihou, A., Joseph, K., Mthew, C., and Thomas, S. (2014). Stress relaxation behavior of organically modified montmorillonite filled natural rubber/nitrile rubber nanocomposites. *Applied Clay Science* 87. doi: 10.1016/j.clay.2013.10.019.

Martins, A.J., Cerqueira, F., Vicente, A.A., Cunha, R.L., Pastrana, L.M., and Cerqueira, M.A. (2022). Gelation behavior and stability of multicomponent sterol-based oleogels. *Gels* 8 (1). doi: 10.3390/gels8010037.

Meng, Z., Guo, Y., Wang, Y., and Liu, Y. (2019). Oleogels from sodium stearoyl lactylate-based lamellar crystals: structural characterization and bread application. *Food Chemistry* 292. doi: 10.1016/j.foodchem.2018.11.042.

Mezger, T. (2020). *The Rheology Handbook*. The Rheology Handbook. doi: 10.1515/9783748603702.

Oh, I., Lee, J.H., Lee, H.G., and Lee, S. (2019). Feasibility of hydroxypropyl methylcellulose oleogel as an animal fat replacer for meat patties. *Food Research International* 122. doi: 10.1016/j.foodres.2019.01.012.

Ojijo, N.K.O., Neeman, I., Eger, S., and Shimoni, E. (2004). Effects of monoglyceride content, cooling rate and shear on the rheological properties of olive oil/monoglyceride gel networks. *Journal of the Science of Food and Agriculture* 84 (12). doi: 10.1002/jsfa.1831.

Okuro, P.K., Malfatti-Gasperini, A.A., Vicente, A.A., and Cunha, R.L. (2018). Lecithin and phytosterols-based mixtures as hybrid structuring agents in different organic phases. *Food Research International* 111. doi: 10.1016/j.foodres.2018.05.022.

Papadaki, A., Cipolatti, E.P., Aguieiras, E.C.G., Pinto, M.C.C., Kopsahelis, N., Freire, D.M.G., Mandala, I., and Koutinas, A.A. (2019). Development of microbial oil wax-based oleogel with potential application in food formulations. *Food and Bioprocess Technology* 12 (6). doi: 10.1007/s11947-019-02257-3.

Patel, A.R., Rajarethinem, P.S., Cludts, N., Lewille, B., De Vos, W.H., Lesaffer, A., and Dewettinck, K. (2015). Biopolymer-based structuring of liquid oil into soft solids and oleogels using water-continuous emulsions as templates. *Langmuir* 31 (7). doi: 10.1021/la502829u.

Patel, A.R., Rajarethinem, P.S., Grędowska, A., Turhan, O., Lesaffer, A., De Vos, W.H., Van de Walle, D., and Dewettinck, K. (2014). Edible applications of shellac oleogels: spreads, chocolate paste and cakes. *Food and Function* 5 (4). doi: 10.1039/c4fo00034j.

Patel, A.R., Schatteman, D., De Vos, W.H., Lesaffer, A., and Dewettinck, K. (2013). Preparation and rheological characterization of shellac oleogels and oleogel-based emulsions. *Journal of Colloid and Interface Science* 411. doi: 10.1016/j.jcis.2013.08.039.

Pérez-Monterroza, E.J., Márquez-Cardozo, C.J., and Ciro-Velásquez, H.J. (2014). Rheological behavior of avocado (Persea americana Mill, cv. Hass) oleogels considering the combined effect of structuring agents. *LWT* 59 (2P1). doi: 10.1016/j.lwt.2014.07.020.

Rousseau, D. (2020). Aqueous droplets as active fillers in oil-continuous emulsions. *Current Opinion in Food Science* 33: 173–186. doi: 10.1016/j.cofs.2020.06.004.

Sagiri, S.S., Singh, V.K., Kulanthaivel, S., Banerjee, I., Basak, P., Battachyra, M.K., and Pal, K. (2015). Stearate organogel–gelatin hydrogel based bigels: Physicochemical, thermal, mechanical characterizations and in vitro drug delivery applications. *Journal of the Mechanical Behavior of Biomedical Materials* 43: 1–17. doi: 10.1016/j.jmbbm.2014.11.026.

Steffe, J.F. (1996). *Rheological Methods in Food Process Engineering*, 2e. East Lansing: Freeman Press.

Trujillo-Ramírez, D., Lobato-Calleros, C., Jaime Vernon-Carter, E., and Alvarez-Ramirez, J. (2019). Cooling rate, sorbitan and glyceryl monostearate gelators elicit different microstructural, viscoelastic and textural properties in chia seed oleogels. *Food Research International* 119: 829–838. doi: 10.1016/j.foodres.2018.10.066.

Uvanesh, K., Sagiri, S.S., Senthilguru, K., Anis, B.A., Al-Zahrani, S.M., and Pal, K. (2016). Effect of span 60 on the microstructure, crystallization kinetics, and mechanical properties of stearic acid oleogels: An in-depth analysis. *Journal of Food Science* 81 (2). doi: 10.1111/1750-3841.13170.

Wolfer, T.L., Acevedo, N.C., Prusa, K.J., Sebranek, J.G., and Tarté, R. (2018). Replacement of pork fat in frankfurter-type sausages by soybean oil oleogels structured with rice bran wax. *Meat Science* 145. doi: 10.1016/j.meatsci.2018.07.012.

Wright, A.J., Scanlon, M.G., Hartel, R.W., and Marangoni, A.G. (2001). Rheological properties of milkfat and butter. *Journal of Food Science*. doi: 10.1111/j.1365-2621.2001.tb16082.x.

Zeng, C., Wan, Z., Xia, H., Zhao, H., and Guo, S. (2020). Structure and properties of organogels developed by diosgenin in canola oil. *Food Biophysics* 15 (4). doi: 10.1007/s11483-020-09643-x.

Zhang, R., Zhang, T., Hu, M., Xue, Y., and Xue, C. (2021). Effects of oleogels prepared with fish oil and beeswax on the gelation behaviors of protein recovered from Alaska Pollock. *LWT* 137. doi: 10.1016/j.lwt.2020.110423.

Zheng, H., Mao, L., Cui, M., Liu, J., and Gao, Y. (2020). Development of food-grade bigels based on κ-carrageenan hydrogel and monoglyceride oleogels as carriers for β-carotene: roles of oleogel fraction. *Food Hydrocolloids*. doi: 10.1016/j.foodhyd.2020.105855.

Zhu, Q., Gao, J., Han, L., Han, K., Wei, W., Wu, T., Li, J., and Zhang, M. (2021). Development and characterization of novel bigels based on monoglyceride-beeswax oleogel and high acyl gellan gum hydrogel for lycopene delivery. *Food Chemistry* 365. doi: 10.1016/j.foodchem.2021.130419.

3.2

Application of Small-Angle X-Ray Scattering and Small-Angle Neutron Scattering to Fat Mimetics

*Elliot Paul Gilbert**

Australian Centre for Neutron Scattering, Australian Nuclear Science and Technology Organisation, Lucas Heights, NSW, Australia
Australian Institute for Bioengineering and Nanotechnology and Centre for Nutrition and Food Sciences, The University of Queensland, St. Lucia, Brisbane, QLD, Australia
** Corresponding author*

3.2.1 Introduction

The term fat mimetics was originally used to describe protein and carbohydrate-based materials physically or chemically changed to imitate the organoleptic or functional properties of natural fats and oils. More recently, they have taken on the sense of the non-chemical transformation of liquid oils into solid or semi-solid structures able to mimic the physical properties and functionality of solid fats (Marangoni et al. 2020). The motivation here is to replace lipids rich in saturated fatty acids, and whose consumption raises serum LDL cholesterol levels, with monounsaturated and polyunsaturated fatty acids to help in reducing the risk of cardiovascular disease. However, plant-based oils, rich in unsaturated fatty acids, do not provide a proper macroscopic structure essential for many food products because they are liquid at ambient temperatures. In addition, saturated fats with their higher melting temperature impart texture and pleasurable mouthfeel, and possibly influence aroma and flavor perception, to products (Davidovich-Pinhas et al. 2016). Hydrogenation of unsaturated fats has been one approach explored. However, this approach yields trans fats having two negative effects, both raising low-density lipoprotein while simultaneously lowering high-density lipoprotein. Driven by consumer demands for regulation, the food industry has emphasized finding alternative methods to replace such trans and saturated fats with "healthier" alternatives while simultaneously accommodating consumer perception and acceptance.

There are three fat mimetic approaches: (i) oleogelation, (ii) indirect oil structuring, and (iii) structured biphasic systems (Marangoni et al. 2020). The indirect method is typically based on colloidal templates, such as foams and emulsions and uses solvent exchange; it involves the creation of a conformational framework of hydrophilic polymers, either in the bulk or by first promoting their adsorption at air–water or oil–water interfaces, where the water is removed by drying or exchanged with oil solvent. For example, β-lactoglobulin can be a stabiliser for water-continuous emulsions subsequently used as templates to obtain protein-in-oil gels via spray or freeze drying. Similarly, K-carrageenan has been used to produce porous aerogels. A slightly different example involves aerating oloegels to form oleofoams (Fameau and Saint-Jalmes 2020). A recent example that is directly applicable to confectionary products described the crystalline arrangement of cocoa butter at the air/oil interfaces around air bubbles that support a Pickering stabilization mechanism, providing stabilization from coalescence and coarsening (Metilli et al. 2021). Structured biphasic systems encompass structured emulsions, oil bulking systems (and emulsion gels), and gelled emulsions including high internal phase oil-continuous emulsions (Marangoni et al. 2020).

Fat Mimetics for Food Applications, First Edition. Edited by Miguel Ângelo Parente Ribeiro Cerqueira and Lorenzo Miguel Pastrana Castro.

Oleogelation involves incorporating gelator molecules into liquid oil systems to form gels. Of the three fat mimetic approaches, only oleogels could be described as relying on a regular gelation mechanism; they are this chapter's focus. Oleogelation can be achieved using natural waxes, phytosterols and other sterols, phospholipids, including lecithin, mono- and diglycerides, fatty alcohols and fatty acids, proteins, and functionalised polysaccharides with the molecules arranging themselves into a 3D network that can entrap oil (Scharfe and Flöter 2020). The gelling ability naturally relates to the balance between the soluble and insoluble fractions of the gelator in the solvent; the molecule must be sufficiently insoluble to organise within the oil phase but with a sufficiently soluble fraction to interact with the oil moieties. The composition of the oleogel will thus differ depending on the means of gelation, and by changing the gelator concentration, it is possible to tune properties, such as the hardness and melting point of the gel, to suit the desired application. Liquid oils are structured and solidified/gelled by non-triglyceride networks to take on solid-like properties. Oleogels can also be used in a variety of applications beyond food, from cosmetics, where it is used to prevent oil leakage, to pharmaceutical and biotechnology applications, where they can be used for the encapsulation and controlled release of hydrophobic bioactive molecules. Oleogel technology can also be used in the coating, paint, and plastic industries, where it can be used to entrap organic solvents (Davidovich-Pinhas et al. 2016) and for lubrication. In food, they have been explored to reduce oil migration in confectionery products, to immobilise water droplets in water-in-oil emulsions, and to control the release of functional ingredients such as flavours or bioactive compounds (Bodennec et al. 2016). There are several recent reviews that discuss oleogelation approaches and opportunities (Pernetti et al. 2007; Daniel and Alejandro 2012; Idziak 2012; Rogers et al. 2014; Zetzl and Marangoni 2014; Patel and Dewettinck 2015; Davidovich-Pinhas et al. 2016; Singh et al. 2017; Davidovich-Pinhas 2019; Cao and Mezzenga 2020; Feichtinger and Scholten 2020; Marangoni et al. 2020; Puscas et al. 2020; Scharfe and Flöter 2020; Cerqueira et al. 2021; Flöter et al. 2021; Shakeel et al. 2021).

Oleogels are conventionally studied with a range of characterization techniques. Many of these are ostensibly concerned with gel properties including storage and loss moduli (via rheological studies), and texture. Melting is a relevant property that can be studied via differential scanning calorimetry (DSC). Useful information can be extracted from spectroscopies, e.g., infra-red, to yield details on local structural organization via vibrational data. Naturally, larger-scale structural information is critical and this can be obtained from optical (often polarized) and electron microscopies (SEM, TEM). Some techniques using electron microscopy require special preparation and many require the samples being exposed to vacuum and intense beams, potentially generating artefacts. Many methods are also used independently *i.e.*, one at a time.

Small-angle scattering (SAS) methods have several unique advantages. There are no special sample preparation requirements; the samples can be presented to the beam at atmospheric pressure and without staining. In addition, a range of so-called sample environments exist, which enables scattering and other measurements to be conducted simultaneously (e.g., DSC, Couette shear, and rheology); this will be discussed subsequently. Finally, the spatial dimensions extend from the sub-nanometre to multiple micron range. The latter is attractive to enable a structural connection to be drawn between the gelling behaviour, the underlying structure necessary for the gel formation to occur, and over a range of length scales that directly overlap with microscopy. In addition, due to the beam size dimensions used, particularly for neutron techniques, the sample volume probed is typically many times larger than that achieved with electron microscopy methods, providing an opportunity to extract statistically averaged details representative of the bulk structure. While the avoidance of sample artefacts itself is a significant advantage of scattering methods, there are inevitably, disadvantages. Unlike microscopy techniques, small-angle scattering does not generate images but provides structural information in "reciprocal space" with dimensions of inverse length. This means that scattering data must be inverted to real space or reciprocal space data fitted with real space models that, themselves, need to be inverted. Unquestionably, this has been the most significant barrier to establishing these methods within the food materials science community. The corollary of this is that it is imperative for complementary characterisation tools (e.g., microscopy) and some knowledge of the system to be available (e.g., concentration, composition) to develop appropriate, chemically- and physically-sensible, mathematical models to the data. It is not an exaggeration to state that without such details, one should not embark on SAS techniques. In addition, the availability of instrumentation is a challenge, particularly for neutron beams where access is obtained following competitive peer-review allocation. Nonetheless, even though there are difficulties and there is effort required to surmount these challenges, there is a wealth of rich and detailed information available from the data.

Although not exhaustive, this chapter summarizes the latest findings in the application of SAS to fat mimetics with a focus on oleogels. The abbreviations SANS, SAXS, or SAS, will be used where neutron, X-ray, or both radiation sources have been used in a study; for ultra-SAS techniques, USANS, USAXS, or USAS will be used. The chapter will start with a description of the fundamentals of small-angle scattering, followed by SAS instrumentation, and experimental design and data collection. The chapter progresses into data analysis and interpretation. The chapter will then report on specific examples of SAS applied to oleogels, followed by an outlook on future perspectives.

3.2.2 Fundamentals of Small-Angle Scattering

Due to an inverse relationship, the scattering of X-ray and neutrons (SAXS and SANS) at "small" angles provides information on the structure of materials on "long" length scales; here, "long" means larger than atomic length scales from about one to several hundred nanometres, and extending to microns in the case of ultra-small angle scattering (USAS = USAXS and USANS) methods. On this size range, the technique is sensitive to the arrangement of assemblies of atoms, enabling molecular structure and superstructure as well as their spatial distribution to be determined. While the detailed theory of SAS may be found in a wide range of texts, here the focus is given to the fundamental parts of the technique and particularly the approaches necessary for interpreting the data from oleogels. For further information, the reader is encouraged to consult the SAS Portal at http://smallangle.org, which has a useful list of recommended books, journal articles, web-based tools, blogs, and primers. The freely available book by Hammouda (Hammouda 2008) is also an excellent resource, although, focusing on SANS. Gilbert has recently reviewed the application of SAS and USAS for food materials (Gilbert 2019). Figure 3.2.1 illustrates the size range accessible with SAS and USAS methods alongside complementary structural characterization techniques.

The independent variable in a SAS experiment is the scattering vector, q; it defines the relationship between the incident and scattered beams in Figure 3.2.2a. q has dimensions of reciprocal length. Despite not being an SI unit, Ångstrom is commonly used as a unit of dimension, which corresponds to 0.1 nm; q is quoted in units of $Å^{-1}$. As an alternative, it is also common to report q in units of nm^{-1} and real-space dimensions correspondingly in nm. Here, the former convention is used, but both approaches are equally acceptable; it need only be recalled that the latter has a numerical value ten times greater than the former *i.e.*, since 1 nm = 10 Å, 0.1 $Å^{-1}$ = 1 nm^{-1}.

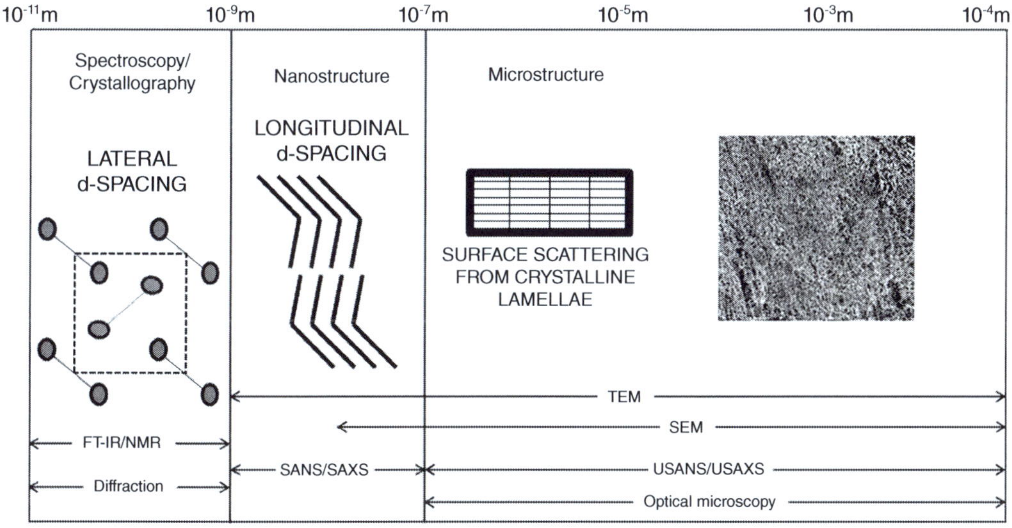

Figure 3.2.1 Small-angle scattering and ultra-small angle scattering techniques and associated size range with complementary structural characterisation methods relevant for oleogels; Cornet et al. 2021 / with permission from Elsevier.

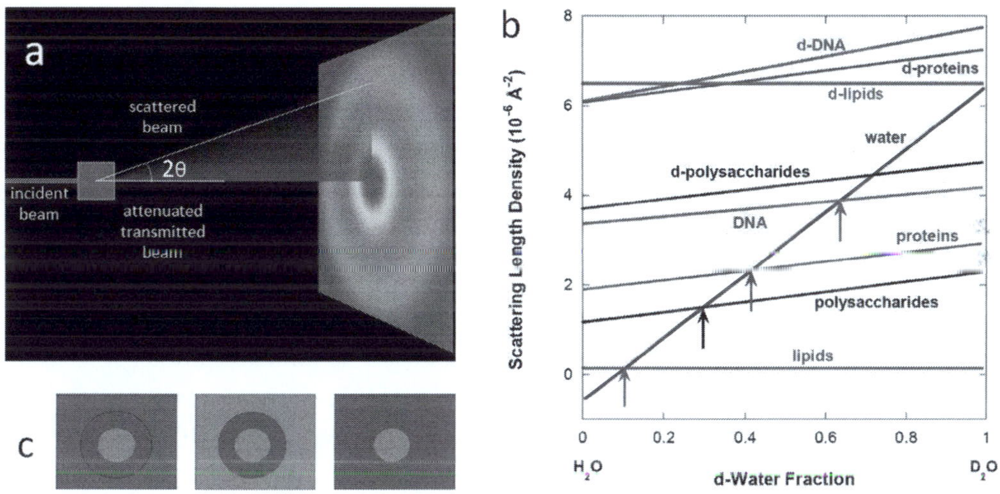

Figure 3.2.2 (a) Scattering geometry for a SAXS or SANS measurement; (b) Neutron scattering length densities for common food-based materials (Hammouda 2008); (c) Core-shell nanostructure in solution showing the possibility of selectively contrast matching either phase through changing solvent $C_{16}H_{34}$/$C_{16}D_{34}$ composition: by matching the SLD of the solvent to core, only the shell is visible, while matching the solvent to the shell yields information only on the core. Reproduced from Gilbert 2019 / with permission of Elsevier.

q is related to the scattering angle as follows (equation 1):

$$q = \frac{4\pi}{\lambda} sin\theta$$

where 2θ is the scattering angle and λ is the wavelength of the incident radiation.

A small-angle scattering experiment conventionally measures the number of photons (SAXS) or neutrons (SANS) of a given wavelength, scattered with respect to the direct beam through a particular angle, that arrive in a small area of a detector within a unit time. Most of the incident radiation is transmitted (*i.e.*, unscattered) through the sample and some proportion, depending on the radiation source and elemental composition of the sample, may also be absorbed. A small proportion of the beam will undergo such a change in direction, *i.e.*, $\theta \neq 0$, as a result of interacting with the sample and it is this scattering discussed here that provides valuable structural information.

If one introduces the latter equation into Bragg's law (equation 2):

$$\lambda = 2d sin\theta$$

one obtains the relationship: $q = 2\pi/d$ where d represents a typical dimension. Thus, the lower the value of q, the greater the dimension that can be studied. Many oleogel systems show peaks in the small-angle scattering whereby a corresponding dimension or d-spacing is reported. Just as peaks observed in X-ray (or neutron) diffraction or wide angle scattering (WAXS, WANS) yield structural information on the sub-nm length scale, these SAS measurements are a special case of diffraction. They are at a small angle to reveal characteristic, but longer, structural distances in the sample; one may see the abbreviation SAXD used representing small-angle X-ray diffraction. Figure 3.2.1 shows the range of dimensions accessed with SAS and USAS. At angles greater than those accessed in SAS—a rather vague and imprecise statement to make—smaller length scales covered by wide angle scattering or powder X-ray diffraction are probed; note that these are subtly different methods due to the relative geometry of the incident and diffracted beams regarding the sample (Gilbert et al. 2012). A typical SAS geometry is shown in Figure 3.2.2a; USAS instruments are employed to reach lower q and to access information on larger length scales; a range of microscopies overlap with USAS and a combination of such methods is effective in revealing oleogel structure.

To understand what is measured in a SAS experiment, consider a sample of thickness, t (units: cm, which for oleogels is typically 0.2 cm or less) that is illuminated by a source of radiation of wavelength, λ, which has an incident flux (of X-ray photons or neutrons), I_0 (units: $cm^{-2}s^{-1}$). The beam size is defined by an aperture with an area, A, (units: cm^2), which is located in front of the sample. At a scattering angle, 2θ, with respect to the incident beam of radiation, and far away, L (units: cm), the scattered intensity, I, is detected within a solid angle element, ΔΩ, defined by the area of an individual detector pixel divided by L^2 and with an efficiency, η (dimensionless). The flux of radiation undergoing scattering, conventionally described as the intensity, is expressed (equation 3):

$$I(q) = I_0(q)\Delta\Omega\eta(\lambda)T(\lambda)At\frac{d\Sigma}{d\Omega}(q)$$

where T is the transmission of the sample, defined as the ratio of outgoing to incoming intensity through the sample at zero angle. Both η and T are wavelength dependent. $d\Sigma/d\Omega$ is the differential macroscopic scattering cross-section and is the probability of a photon or neutron being scattered out of the sample volume into the solid angle $\Delta\Omega$; it is the dependent variable in a SAS experiment and has units of inverse length, typically cm^{-1}. Depending on the instrument used, however, and whether the data have been collected on an absolute scale, the units may still be only be on a relative scale. Data can still be interpreted even in the absence of absolute intensity. Indeed, sometimes it is difficult to achieve absolute scaling (e.g., uncertainty in knowledge of the precise packing density of a powder sample); however, it is always preferable to design the experiment so that absolute data are obtained and this is discussed below.

Most terms in equation 3 are specific to the instrument; however, transmission, thickness, and $d\Sigma/d\Omega$ depend on the sample. The aim, through the procedure of data reduction, is to relate the intensity measured in an experiment to $d\Sigma/d\Omega(q)$ and to the structure of the sample, giving rise to the scattering. As will be seen below, it is necessary to consider all contributions to the scattering signal since some will not be of relevance to the oleogel e.g., scattering from the sample holder, continuous oil phase scattering and specific instrumental background and cosmic radiation.

Critically, the differential macroscopic scattering cross-section is quantitative, as opposed to merely qualitative; a representation of the interaction between the radiation and the sample and has both structural and density-dependent information. It may be written (equation 4):

$$\frac{d\Sigma}{d\Omega}(q) = (\Delta\rho)^2 V_p^2 N_p P(q) S(q)$$

where $\Delta\rho$ is the scattering contrast between the species of interest, *i.e.*, gelator system, and its surroundings (*i.e.*, continuous oil phase), V_p is the particle volume and N_p the number of scattering particles per unit volume. $d\Sigma/d\Omega(q)$ contains all the information regarding the size, shape and interactions between the scattering particles. The form factor, P(q), describes the scattering from an individual scattering species, detailing their size and shape while the structure factor, S(q), describes the scattering arising from interparticle interactions [e.g., hard sphere (infinite potential at zero contact and zero elsewhere), attractive, repulsive etc]. For isotropic scatterers, *i.e.*, with no preferred orientation, data are radially averaged, and it is the magnitude of the scattering vector that is relevant. This is the most common case for oleogel systems; however, in systems with preferred orientation, e.g., prepared under shear or deformation, the directional information should be conserved during the reduction process; the vector notation is relevant.

For small angle scattering to occur, there must be a "contrast" between the objects—for which the experiment is designed and intended to reveal structural information–and its surroundings. The scattering contrast, $\Delta\rho$, is equal to $\Delta\rho = \rho_p—\rho_m$, where ρ_p and ρ_m are the scattering length densities (SLD) of the species of interest (p), and the surrounding material (m), respectively. The SLD, ρ, of a molecule with i atoms is defined as $\delta N_A \Sigma b_i/M$, where δ is the bulk physical density of the molecule (units: $g\,cm^{-3}$), N_A is Avogadro's number (mol^{-1}), b_i is the scattering length of the ith atom in the molecule (cm) and M is the molecular weight ($g\,mol^{-1}$). Thus, the SLD can easily be calculated from the molecule's physical density, its atomic composition, and the contrast from the differences between them.

The atomic scattering length is radiation source-dependent since X-rays are sensitive to atomic electron density, while neutrons are sensitive to nuclear density (*i.e.*, the physical details of the atomic nucleus). As a consequence SANS and SAXS provide complementary information. It follows then that SLD, and contrast, depend also on radiation source. For example, the X-ray SLD for H_2O is 9.47×10^{10} cm^{-2} whereas the neutron SLD is -0.56×10^{10} cm^{-2}. Note that the units of SLD are an inverse area (*i.e.*, length per unit volume). Much like the choice in the dimensional units used for q, SLD is often quoted in units of $Å^{-2}$, which differs from cm^{-2} by a factor of 10^{-16}, *i.e.*, the neutron SLD for H_2O is also -0.56×10^{-6} $Å^{-2}$; again, both conventions are equally acceptable. The latter is used here. Table 3.2.1 summarises the neutron and X-ray SLDs for some typical components in oleogels. While it is informative to calculate SLDs based on tables of atomic scattering lengths, molecular weights, and physical densities, online calculators are available (e.g., https://www.ncnr.nist.gov/resources/activation)

Contrast can be manipulated through "contrast variation", which, in SANS, often relies on the different scattering lengths of hydrogen and deuterium. For example, if one considers that the apolar oil, hexadecane, the SLD for $C_{16}H_{34}$ is negative (for nuclear physics reasons, the atomic scattering length of hydrogen is negative) but its heavier, isotopically-substituted analogue for $C_{16}D_{34}$ is positive. Figure 3.2.2b illustrates the neutron SLDs for common molecular species found in food. With this knowledge, components may be strategically contrast-matched using mixtures of deuterated and hydrogenated hexadecane so they effectively become transparent to neutrons (Table 3.2.1). This greatly simplifies the scattering contributions in multi-component oleogel systems. Figure 3.2.2c shows the approach for obtaining structural information from a model core-shell particle with regions of different SLD; by matching the SLD of the solvent to the core, only the shell is visible, while matching the solvent to the shell only yields information about the core. This approach is widely exploited in small-angle neutron scattering, most commonly for aqueous systems since D_2O is readily available in bulk quantities. Again, the scattering of one component in a mixed system can be contrast-matched to enable structural information to be extracted on the remaining non-matched component. A non-aqueous example is the formation of inverse micelles of oleic acid and sodium oleate gelator molecules in apolar oil (Cornet et al. 2021). Cornet et al. measured the SANS from a 2:1 oleic acid–sodium oleate mixture in hexadecane in which 1 wt% water was present, where either the hexadecane, water, or both were deuterated. By modeling the three data sets simultaneously, they extracted information on the individual dimensions of the core (water) and shell (gelator) regions of the inverse micelle.

Neutron contrast may already be intrinsic to the system through differences in chemical composition and physical density between species of interest; however, if this is not the case or is insufficient, deliberate isotopic labeling strategies can also be pursued at a dedicated establishment, e.g., the National Deuteration Facility at the Australian Nuclear Science and Technology Organisation (www.ansto.gov.au/facilities/national-deuteration-facility). While X-ray solvent contrast can be manipulated, the

Table 3.2.1 Chemical formulae, physical densities and X-ray and neutron scattering length densities (SLD) for typical oleogel parts. SLDs were determined using https://www.ncnr.nist.gov/resources/activation. Olive oil values based on triolein, sunflower oil based on equimolar mixture of triolein and trilinolein ($C_{57}H_{104}O_6$) and medium chain triglyceride based on equimolar mixture of tricaprylin ($C_{27}H_{50}O_6$) and tricaprin ($C_{33}H_{62}O_6$). Oleic acid and sodium oleate are included here as they are discussed in an example.

	Chemical formula	Physical density(g/cm^3)	X-ray SLD(10^{-6} Å^{-2})	Neutron SLD(10^{-6} Å^{-2})
Sunflower oil	$C_{57}H_{101}O_6$	0.919	8.70	0.226
Olive oil	$C_{57}H_{104}O_6$	0.919	8.73	0.155
Medium chain triglyceride oil	$C_{30}H_{56}O_6$	0.945	8.91	0.275
Hexadecane H_{34} (hydrogenated)	$C_{16}H_{34}$	0.770	7.52	−0.426
Hexadecane D_{34} (deuterated)	$C_{16}D_{34}$	0.887	7.53	6.83
Water (hydrogenated)	H_2O	1.00	9.47	−0.561
Water (deuterated)	D_2O	1.11	9.46	6.39
Oleic acid	$C_{18}H_{34}O_2$	0.895	8.52	0.0784
Sodium oleate	$C_{18}H_{33}O_2Na$	0.90	8.46	0.204

options are somewhat limited for oleogel systems. The utility of deuteration will be shown below for two examples. However, deuteration is not necessarily benign as far as phase behaviour and the replacement of H with D in a molecule can lead to subtle changes in behaviour (e.g., phase transition temperature, strength of hydrogen bonds). For example, Nikiforidis et al. reported differences in flow behaviour when oil, water, or both were deuterium-labelled in oleic acid–sodium oleate organogels (Nikiforidis et al. 2015); although, such changes could inform an interaction mechanism, rather than being an impediment. Without question, the potential for detailed structural information to be revealed through deuteration has proven to be a crucial method for small-angle scattering.

3.2.2.1 SAS Instrumentation

Conducting a SAS experiment requires access to SAS instrumentation. Many universities or publicly funded research organizations have laboratory-based SAXS equipment. Their ease of operation is also being continuously enhanced by manufacturers around the world; this extends to data analysis and interpretation discussed below. However, as with many techniques, the latter has sometimes resulted in a "black box" understanding of the analytical process leading to incorrect interpretations and conclusions. SAXS instrumentation has also been constructed at synchrotron-based X-ray sources; a list of synchrotrons can be found at en.wikipedia.org/wiki/List_of_synchrotron_radiation_facilities. To access synchrotron-based SAXS, as for SANS instrumentation, preparation of an experimental beamtime proposal submitted for competitive review is required; the latter is necessary as, like neutron sources, the demand for synchrotron radiation outweighs its availability. SANS is also available only at specialized, large-scale facilities, namely steady-state reactors and spallation sources. At reactors, neutrons are continuously produced by a fission process; whereas, at a spallation facility, a pulsed neutron beam (with a typical frequency of between 25 and 60 Hz) is generated by the collision of high-energy protons with a target that also produces heavy atom fragments. In the future, facilities may also become available at accelerator-based compact sources. An example of a monochromatic SANS instrument at a reactor-based facility is QUOKKA at the OPAL reactor in Australia (Gilbert et al. 2006; Wood et al. 2018) which has been widely used in the study of food materials including oleogels (Blazek and Gilbert 2011; Doutch et al. 2012; Doutch and Gilbert 2013; Lopez-Sanchez et al. 2015; Martínez-Sanz, Gidley et al. 2015; Martínez-Sanz, Lopez-Sanchez et al. 2015; Yang et al. 2016; Yu et al. 2018, 2019; Martínez-Sanz et al. 2020 Yang et al. 2023, Gilbert 2022, Cheng et al. 2022, Yang et al. 2022). SANS2d at the ISIS Neutron and Muon Source is an example of a spallation-based SANS instrument (Heenan et al. 2011). The specific details of each instrument and the subtle differences as to how data are collected are outside the scope of this chapter; however, the fundamental difference is that a reactor provides a continuous source of neutrons from which a narrow wavelength band is selected; however, spallation-based instruments use the inherent pulse structure where a range of neutron wavelengths are used and whose wavelengths are determined by time of arrival at the detector (or time-of-flight). It is also possible to run time-of-flight SANS instruments at a reactor source, as is the case e.g., BILBY (Sokolova et al. 2019). By combining SAS measurements with ultra-SAS, the q range can be substantially extended to reveal information on larger length scales. For example, USANS can be measured on the KOOKABURRA (Rehm et al. 2018) instrument at OPAL and USAXS can be measured on ID2 at the European Synchrotron Radiation Facility (Narayanan et al. 2018). The 9ID beamline at the Advanced Photon Source at Argonne National Laboratory enables USAXS, SAXS, and WAXS to be measured on the same instrument (Ilavsky et al. 2018). In all cases, it is advised to discuss the proposed experiments and associated design with the scientist operating the beamline.

3.2.2.2 SAS Experiment and Data Collection

There are several measurements required to convert the data into a format that can be interpreted, irrespective as to whether one uses X-rays or neutrons; the instrument scientist will typically guide the experimenter through this largely automated process. Such measurements include detector sensitivity (to account for variations in response between detector pixels), beam center determination, and an electronic noise measurement (in the case of neutron scattering by using a neutron absorber such as cadmium). In addition, a method of placing the data on an absolute scale is highly desirable—e.g., measuring a calibration standard of known absolute intensity or an attenuated direct beam transmission measurement with known attenuation factor. It is also necessary to consider the scattering from what would represent a proper background, *i.e.*, what should be subtracted from the sample scattering to only leave what is relevant. Background—a term used rather loosely—can be separated into two categories: external and internal. External backgrounds include scattering from the sample container and the solvent, *i.e.*, oil in an oleogel system. The internal background is defined as a background arising from the sample itself; this is a common feature in SANS where a background associated with incoherent scattering from hydrogenated components in the material occurs.

As well as scattering measurements, the transmission through all samples also needs to be measured; the latter is the ratio between the flux through the sample and the incident flux at $q = 0$. In summary, for a sample, these measurements need to be taken to generate the data on an absolute intensity scale, most of which need to be repeated for each instrument configuration used (e.g., sample-to-detector distance):

- Sample (which includes contributions from the sample container): both transmission and scattering;
- Background (e.g., solvent, which also includes contributions from the sample container) both transmission and scattering;
- Attenuated direct beam—transmission only (or with a secondary standard, both transmission and scattering).

After collecting the above data, the next stage is to "reduce" the data to generate $d\Sigma/d\Omega(q)$ on an absolute scale. All facilities have their own software to carry out the various stages of data reduction, and the steps necessary can be found in standard texts, but the steps include:

- Calculating transmissions;
- Calculating the beam centers for the different configurations used;
- Performing radial averaging (or alternative if orientational order in the sample is present) to yield $d\Sigma/d\Omega(q)$.

3.2.2.3 Data Analysis and Interpretation

Having collected and reduced the data, the differential macroscopic scattering cross-section, $d\Sigma/d\Omega(q)$, also colloquially known as the scattering intensity, is obtained. Before fitting the data, it is valuable to assess the fundamental features of the scattering; this also helps to consider the possible models and starting parameters before beginning with fitting. As far as data fitting is concerned, this is often achieved using dedicated software packages. There is a range of software for fitting, some free and open-source, as well as commercial platforms, onto which specific programs have been written; many are accessible from the "SAS Portal" at http://smallangle.org. A freely available open source software package is SASView available from www.sasview.org.

It is good practice to plot SAS data on a log-log scale; this not only highlights the features in the scattering, but it also readily shows how appropriate a model is in describing the SAS data. Except in simple fitting of Bragg peaks, publications in which data and fits are shown on a linear scale should be viewed with extreme caution as the latter makes even very poor fits appear acceptable. Though this occurs often, there are opportunities to substantially extend information that may be derived from oleogel systems, even from existing scattering data.

There are three regions to consider in the SAS: the forward scattering in the limit of $q = 0$, the Guinier regime (if there is one), and the typical power-law decay if it is present in the data, including a Porod region (Figure 3.2.3). With forward scattering, if the intensity in the limit of low q is flat, analysis via a Guinier approximation should be considered and information on the size of the scattering objects should be directly accessible. If the intensity is increasing at low q, the maximum dimension of the scattering objects in the system is larger than q_{min}^{-1}. This is expected in a space-filling oleogel system that shows structures at the micron length scale and above, but it is unlikely to be the case for inverse spherical micelles. A reason such scattering arises is aggregation or an attraction between the individual particles. Scattering that decreases in the limit of low q indicates a repulsive interparticle interaction. Both attractive and repulsive interactions are intrinsic to the structure factor, S(q). A flat intensity in the limit of low q indicates either diluted particles or a weak interparticle interaction. In this case, it may be reasonable to assume that $S(q) = 1$ with the form factor, P(q), providing information on the scattering particles.

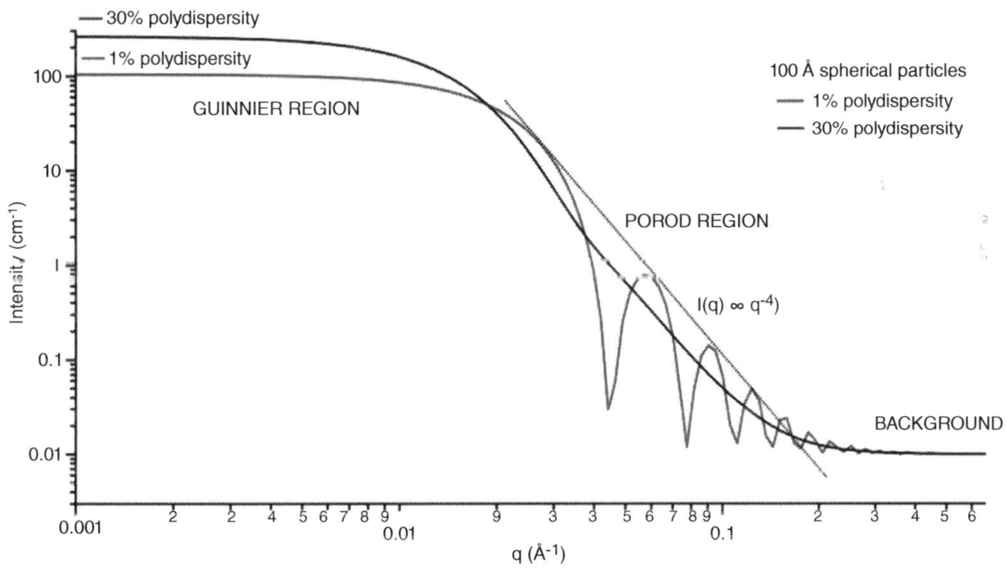

Figure 3.2.3 SAS from spherical particles with radius 100 Å in the dilute regime (here, 1% volume fraction) and SLD contrast of 5×10^{-6} Å^{-2} and background of 0.01 cm^{-1}. The scattering is described by the spherical form factor, P(q). It is characterized by a flat Guinier region at low q and a Porod region at high q. The decay from the plateau occurs at $q \sim R_g^{-1}$ where R_g is the radius of gyration of the sphere. The Porod Region shows oscillations because of the monodispersity of the model system. The crests of the oscillations give rise to q^{-4} scattering associated with a sharp interface between the particle and solvent. More realistic polydispersity results in smearing of the oscillations as is also shown for the same average particle radius with 30% size distribution.

In flat scattering, Guinier analysis enables a parameter known as the radius of gyration, R_g, to be extracted from a plot of the natural logarithm of $d\Sigma/d\Omega(q)$ against q^2. This approximation is broadly valid for $qR_g \leq \sim 1$ via equation 5:

$$\frac{d\Sigma}{d\Omega}(q) = \frac{d\Sigma}{d\Omega}(q=0)\exp\left(\frac{-q^2R_g^2}{3}\right)$$

R_g is independent of particle shape, but, if the shape is known, then R_g can be related to characteristic dimensions. For example, for a spherical object of uniform SLD, the particle radius $= \sqrt{\frac{5}{3}}R_g$. The relationship between R_g and other particle shapes can be found in standard texts. Note this approach is only valid for dilute systems where no interparticle effects are expected nor observed. Note that $\frac{d\Sigma}{d\Omega}(q=0) = (\Delta\rho)^2 V_p^2 N_p$ and, if the particle concentration is known (e.g., from details of the sample preparation), scattering contrast information can be extracted or vice versa. Care should be taken when $qR_g > 1$ as the Guinier approximation increasingly breaks down and to an extent dependent on the particle's anisometry.

The Porod regime describes scattering indicative of the interface between the particle and its surroundings. It is valid for "large" q where q is much greater than the inverse of the maximum particle dimension, ${d_{max}}^{-1}$. For a sharp transition in SLD across an interface, the scattering decays with q^{-4} behavior. The Porod equation (6) is:

$$\lim_{q\to\infty}\frac{d\Sigma}{d\Omega}(q) = \frac{2\pi(\Delta\rho)^2}{q^4}S/V$$

where S/V represents the interfacial surface area per unit volume. Deviations from Porod's law can arise from situations where the interface is rough (e.g., $q^{-3.8}$) or diffuse (e.g., $q^{-4.2}$). If power-law behaviour, whether Porod-like or not, is observable over an extended range in q then the structure of the system may be fractal *i.e.*, self-similar on multiple length scales. Slopes seen over an extended q range between −4 and −3 are described as surface fractal, but those between −3 and −1 are described as mass fractal. Diffusion-limited reactions show slopes of $q^{-1.8}$ and reaction-limited diffusion have slopes of $q^{-2.5}$. However, such description needs to be made with care; power-law behaviour over only one order of magnitude or less cannot describe such structure as fractal. It should also be apparent now that what is meant by "low" q and high "q" is a descriptor that is only relevant in the dimensions of the particles being studied. If one imagines a smooth scattering object of micron radius, no Guinier region would be seen in the SAS and, the "low q" region in the data set would be characterised by Porod scattering. q should therefore be considered as a scattering particle-centric parameter.

There are two commonly used models that consider both the Guinier and Porod regimes, and these are called the unified model (Beaucage 1996) and the more mathematically constrained Guinier-Porod model (Hammouda 2010) (equation 7i and 7ii respectively):

$$\frac{d\Sigma}{d\Omega}(q) = \frac{G}{q^s}exp\left(\frac{q^2R_g^2}{3-s}\right), for\ q \leq q_1$$

$$\frac{d\Sigma}{d\Omega}(q) = \frac{D}{q^m}, for\ q \geq q_1$$

The empirical model can be applied to intermediate shapes between spheres and rods or between rods and plates. G and D represent scale factors for the Guinier and Porod regions and are related by mathematical continuity functions. M is the Porod exponent and s is a dimension variable of 0 for spheres (and where the standard Guinier formula is recovered), 2 for plates, and 1 for rods. Such approaches are likely to have broad application to oleogels. A sample Guinier-Porod model is shown in Figure 3.2.4 for a scattering particle with R_g of 100 Å exhibiting a sharp interface with its surroundings (*i.e.*, Porod exponent of q^{-4}) and with a dimension variable of 1, indicative of a rod-like scatterer. For hierarchical systems, multiple Guinier-Porod layers can be summed.

Between the Guinier and Porod regimes, other typical slopes may be apparent. One can define a distance below which a polymer will show rod-like characteristics known as the persistence length, l_p; the region above $q \sim l_p^{-1}$ will exhibit q^{-1} scattering. At larger distances (lower q), the overall conformation of the polymer may be apparent: Gaussian features in a theta solvent yields q^{-2} scattering, in a good solvent gives $q^{-5/3}$ whereas a randomly branched Gaussian chain exhibits $q^{16/7}$ scattering. These concepts lend themselves to the description of flexible wormlike micelles.

Typically data are fitted to models describing structures in reciprocal space where mathematical functions are required whose parameters must be based on existing knowledge of the system; it is therefore essential that information obtained from complementary techniques is available. The models considered, and the associated fit parameters used in a model, should be based on realistic expectations of the structure of the system. An acceptable "fit" to the data should never only be assessed visually. One should question whether the returned parameters from the structural model obtained have low uncertainties, whether they are chemically reasonable (e.g., based on concentration, contrast) and physically reasonable (e.g., are the dimensions consistent with molecular dimensions in the system?). There are no unique fits to scattering data as scattering intensity only is measured in an experiment; critical information is lost which prevents the full retrieval of the original structure

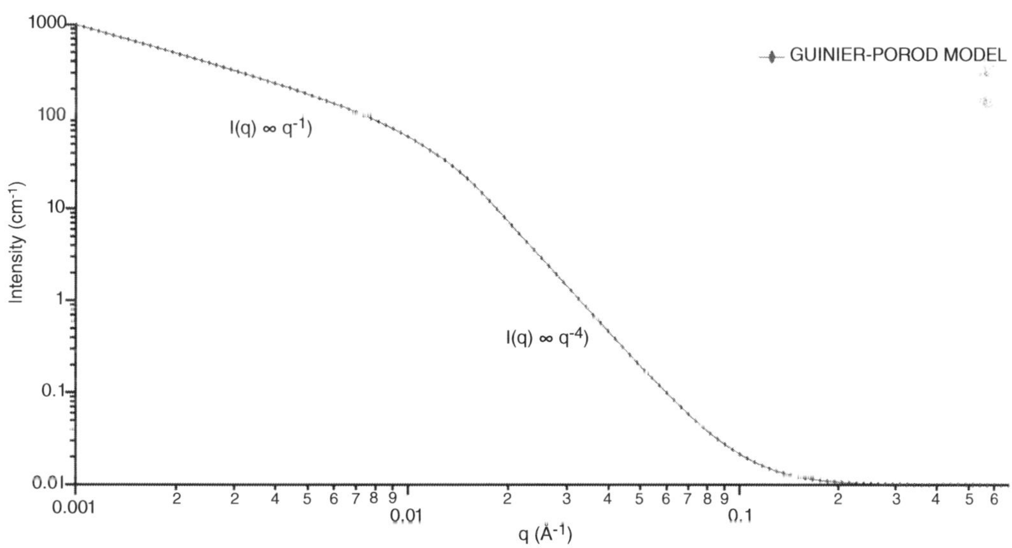

Figure 3.2.4 SAS from a Guinier-Porod model of a rod-like scattering particle with R_g of 100 Å exhibiting a sharp interface with its surroundings and a background of 0.01 cm^{-1}. Note that this background limits the upper value at which Porod behaviour is seen.

(the "phase problem"). Indeed, a multitude of solutions (that) may be equally valid for a particular set of collected intensities, which may be resolved only by obtaining structural information from other techniques. Conducting SAS measurements with multiple contrasts somewhat serves to provide additional constraints to reduce the number of possible models.

Mathematical form factors have been developed for many scattering particles of varying symmetry and some are described below. For example, for form factor for a homogeneous spherical particle is (equation 8):

$$P(q) = \left[\frac{3\sin(qR) - qR\cos(qR)}{(qR)^3} \right]^2$$

and yields the scattering in Figure 3.2.3 for a dilute (*i.e.*, S(q) = 1) array of monodisperse 100 Å particles. It can be seen that both a Guinier region and Porod region are seen, although the latter is only apparent as a gradient formed by the tops of the oscillations in the scattering and whose q range is truncated because of a finite scattering background. Such oscillations, unless a system can be produced with such substantial size control, are rarely seen. If they are, it makes fitting the data more straightforward. In practice, particle size polydispersity (resulting in the superposition of scattering from different particle sizes) combined with instrumental resolution smears out such oscillations. Such scattering is also shown for the same number average particle size but now with 30% size polydispersity; note how the intensity in the limit of low q is also greater because although the number average particle size is constant, the average particle volume, V_p has increased (equation 4). The form factors for other uniform shapes result in typical slopes between the Guinier and Porod regimes. For example, disk-like or lamellar structures give rise to q^{-2} regions of scattering; similarly, rod-like objects give rise to q^{-1} scattering. Modified Guinier approximations where $\ln(q\frac{d\Sigma}{d\Omega}(q))$ against q^2 for rods, or $\ln(q^2\frac{d\Sigma}{d\Omega}(q))$ against q^2 for disks, will yield the radius of gyration of the cross-section and thickness, respectively. While reporting of R_g values seems to be common for oleogel systems, there is often a lack of clarity as to how the values were obtained (e.g., q range and the extent to which the Guinier approximation is valid) or even to what structures the values specifically refer.

Many form factors have been coded into software packages. These form factors can be combined with a range of structure factors to describe interparticle interactions and separation, the latter giving rise to diffraction peaks in the small-angle region. Example structure factors include the hard sphere (which considers an infinite potential for interparticle distances less than particle diameter, otherwise zero), sticky hard sphere (which includes an attractive potential well) and screened Coulombic (for charged systems). Some assessment of the structure factors operating in a scattering system can be apparent from the intensity in the forward direction. An example of a spherical form factor and hard-sphere structure factor is shown in Figure 3.2.5 for 30% volume fraction, 100 Å particles with 30% size polydispersity with a contrast of 5×10^{-6} A^{-2} and background of 0.01 cm^{-1}.

3.2.3 Recent Experimental SAS and USAS Examples to Oleogels

In the following section, recent examples where SAS and USAS have been employed across the spectrum of oleogel systems are presented. However, to start, oleogels formed via the gelation of mixtures of oleic acid (OA) and sodium oleate (SO) are presented; this example is presented first as it is the most recent example in the literature that employs USAS and SAS and contrast variation, and illustrates the broad range of structural features that the oleogel formulation scientist is likely

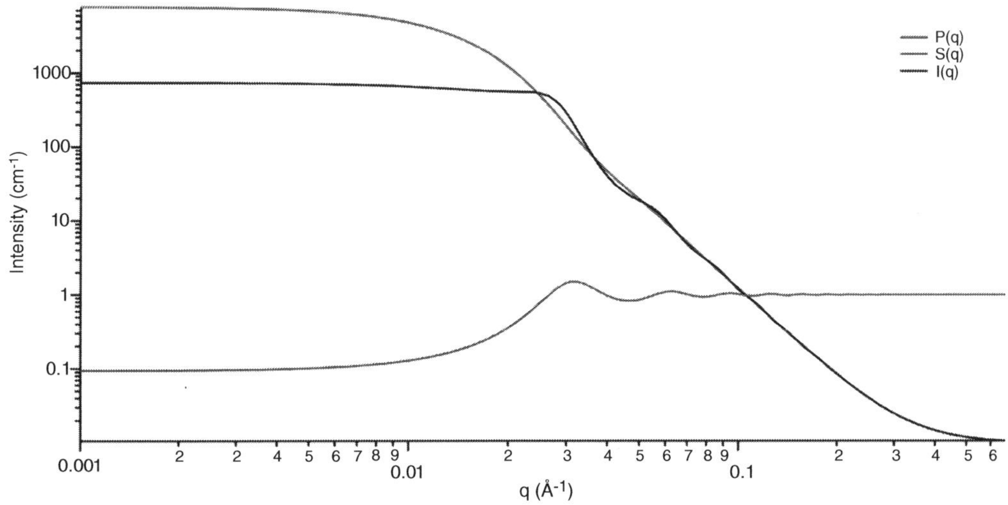

Figure 3.2.5 SAS from a 30% volume fraction, 100 Å particles with 30% size polydispersity with a contrast of 5.0×10^{-6} Å^{-2} and background of 0.01 cm^{-1}. The resultant intensity is the convolution of P(q) and S(q); note the effect of the structure factor, S(q) on reducing the intensity at low q due to the repulsive hard sphere interaction and generates a broad peak-like feature between q = 0.02 and 0.03 Å^{-1}.

to encounter. This system, depending on the OA:SO ratio, includes inverse micelles, power-law slopes and small-angle diffraction, often coexisting. The application of contrast variation and deuteration (both oil solvent and water) with neutron scattering reveals details of the role played by individual components in the gel and shows the rich array of information that can be extracted.

3.2.3.1 Oleic Acid–Sodium Oleate

A class of molecules that can structure edible oils is based on oleate-complexes which comprise an unsaturated fatty acid (oleic acid) and its associated sodium soap (sodium oleate). Previous research indicated that oleic acid cannot function alone as an oleogelator but, combined with sodium oleate, a stable gel can be formed (Nikiforidis et al. 2015). Preliminary SAXS data collected from oleogels exhibited power-law slopes and multiple diffraction peaks depending on the ratio of oleic acid (OA) to sodium oleate (SO). To gain further information on the nanoscale structure, small-angle neutron scattering (SANS) of related samples using the apolar (and non-edible) solvent, decane, was measured; the latter solvent makes contrast variation and manipulation of the scattering length density straightforward. While these more detailed investigations revealed that either inverse micellar phases or lamellar crystals were formed, depending on the OA:SO molar ratio (C. V Nikiforidis et al. 2015), there still remained unresolved questions as to how the structures formed on the micro- and nanoscale relate to the macroscopic gel behavior. More recently, Cornet et al. (Cornet 2016; Cornet et al. 2021) substantially extended these studies and investigated behaviour in different oils (medium-chain triglycerides (MCT), olive oil and sunflower oil (SFO)), at more gelator concentrations (4, 8 and 16 wt%) and OA:SO ratio.

Figure 3.2.6 shows the SANS from oleogels formed in sunflower oil, MCT and olive oil (OO) using 16 wt% of gelator and 1 wt% of water, the latter in its deuterated form to enhance scattering contrast. All patterns exhibit pseudo power-law scattering at low q and multiple diffraction peaks at higher q at

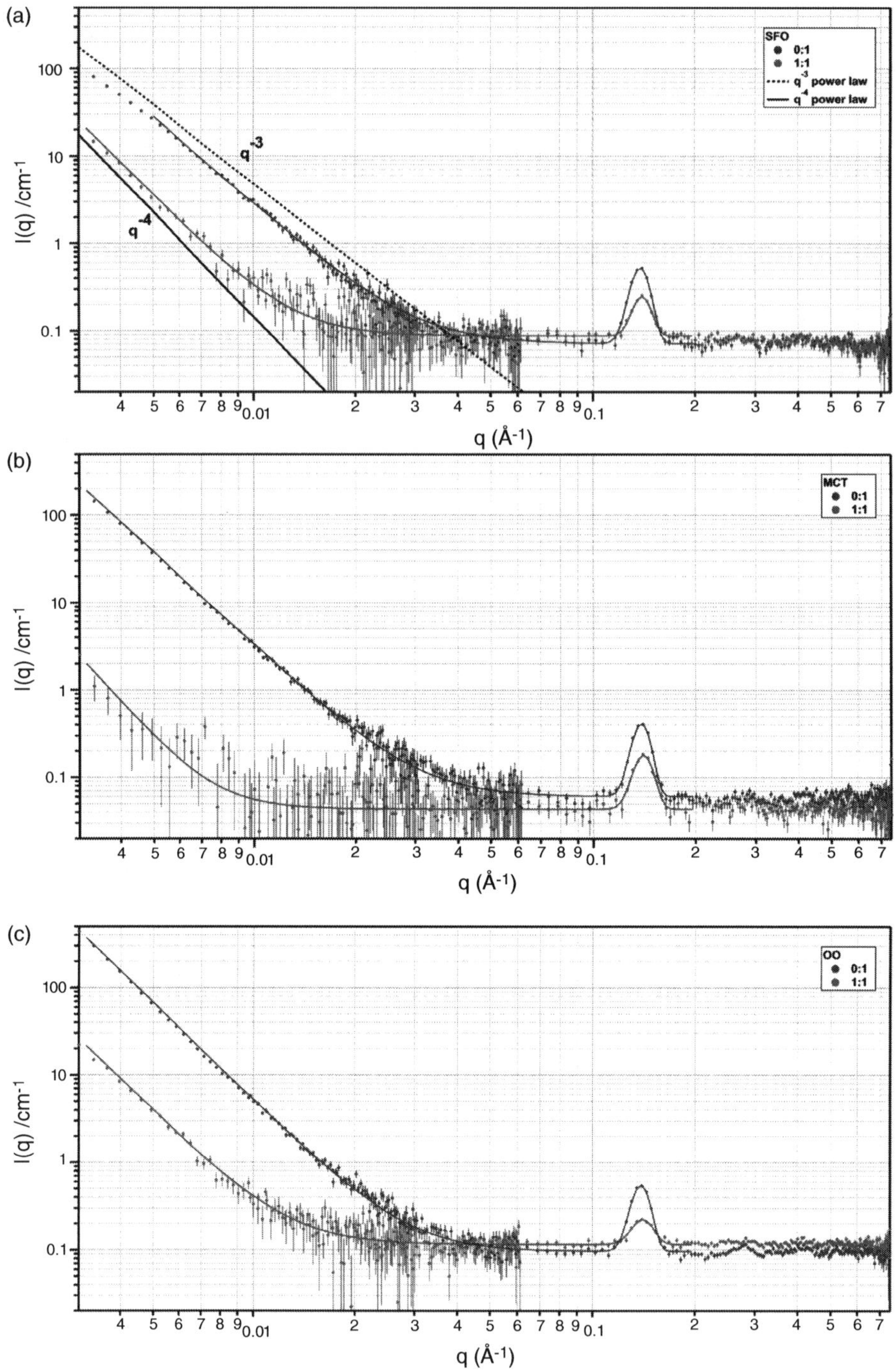

Figure 3.2.6 SANS from 1:1 and 0:1 oleogels: (a) SFO (with q^{-4} and q^{-3} power-law scattering superimposed); (b) MCT and (c) OO. Gels have been prepared with TAG oils (hydrogenated) and D_2O (deuterated). Fits using power law plus Gaussian peak model are shown as solid lines. Reproduced from Cornet et al. 2021 / with permission of Elsevier.

positions, q, 2q, 3q and 4q, indicative of a regular lamellar repeat. While the power-law scattering at low q is reminiscent of surface fractal-like structures (with exponent between ≈ –3 and –3.5), there is notable curvature with decreasing q indicating this interpretation is far too simplistic, particularly for the samples prepared in sunflower oil, pointing to the need to access even lower q. The lack of a Guinier region indicates that the structural entities giving rise to this scattering have a size, L, greater than the inverse of the minimum q measured *i.e.*, $L > \pi / q_{min}$ (≈ 1000 Å). In comparing the 0:1 samples, curvature in the SFO scattering at low q and its absence in the OO and MCT samples implies that the average size of the structures giving rise to the low q scattering are smaller for SFO. This is consistent with microscopy images, albeit microscopy provides structural information on an even larger length scale. For all 0:1 compositions, peaks are at q ≈ 0.138 $Å^{-1}$, which corresponds to a d-spacing of 45.5 Å; this value is in reasonable agreement with the d-spacing for pure SO of 45.1 Å. For the 1:1 samples, the peaks occur at a higher q of ≈ 0.140 $Å^{-1}$ (d ≈ 44.9 Å), *i.e.*, OA addition leads to mixed crystals with an overall shorter d-spacing; this is reasonable given the reported d-spacing for OA is less than SO at 41.4 Å. The effect of OA addition (cf. 0:1 with 1:1 here although intermediate ratios have been studied) consistently decreased peak intensities and decreased the intensity of the power-law scattering, indicating these two scattering features are directly related. Note that the intensity in the limit of low q can be considered a measure of the total material in the system that generates the space-filling network responsible for the macroscopic gel properties.

The solvents and structurants have similar values for the neutron scattering length density (Table 3.2.1) with deuterated water (D_2O) providing the principal contrast within the samples. To exploit the capabilities of SANS, deuterated solvents and/or structurants should be used. The majority component in the gels is oil (comprising 83–95 wt% of the total for 16, 8 and 4 wt% structurant respectively) and the use of a deuterated oil is an obvious strategy. Deuterated edible oils are not readily available but the deuterated alkanes, e.g., C_{16}-alkane hexadecane ($C_{16}H_{34}$), can be readily gained in deuterated form, $C_{16}D_{34}$. While alkanes are purely apolar solvents unlike edible oils and the substitution is not ideal, the additional insights obtained through the use of hexadecane proves to be invaluable in better understanding the TAG-based oleogels and particularly the influence of the OA:SO ratio.

For the hexadecane-based organogels, the SANS intensity in the limit of low q as a function of concentration of structurant for 4, 8, and 16 wt% organogels is shown in Figure 3.2.7a. For a fixed OA:SO ratio, there is a linear relationship between intensity and the wt% of structurant in the gel *i.e.*, the greater the structurant, the greater the network structure formation. Figure 3.2.7b shows the variation in SANS intensity in the limit of low q as a function of SO wt% where OA:SO ratios are stated in the labels. In the edible oil systems, there is a trend of increasing low q intensity with increasing SO content. However, OA has a clear negative effect as samples with the same SO exhibit reduced intensity. Similar negative effects are seen in gel strength where OA addition is observed to produce weaker gels for the same concentration of SO. Here the direct link between structure on the nanoscale and the macroscopic scale, *i.e.*, gel strength, is clear.

Hexadecane-based organogel SANS patterns are presented in Figure 3.2.8a where both the solvent and water are deuterated. Using this approach, the distribution of gelator molecules is revealed since these hydrogenated molecules, with low neutron scattering length density, are highlighted with respect to all other deuterated components in the formulation (Table 3.2.1). Naturally, more features are present than observed in the edible oil oleogel scattering. Given the lack of a Guinier region in the SANS, the benefits of more study on the USANS region is an obvious approach; the combination of SANS and USANS provides information on length scales between about 20 μm and 1 nm. The scattering data thus overlaps, at the upper end, with the length scales accessed in the microscopy

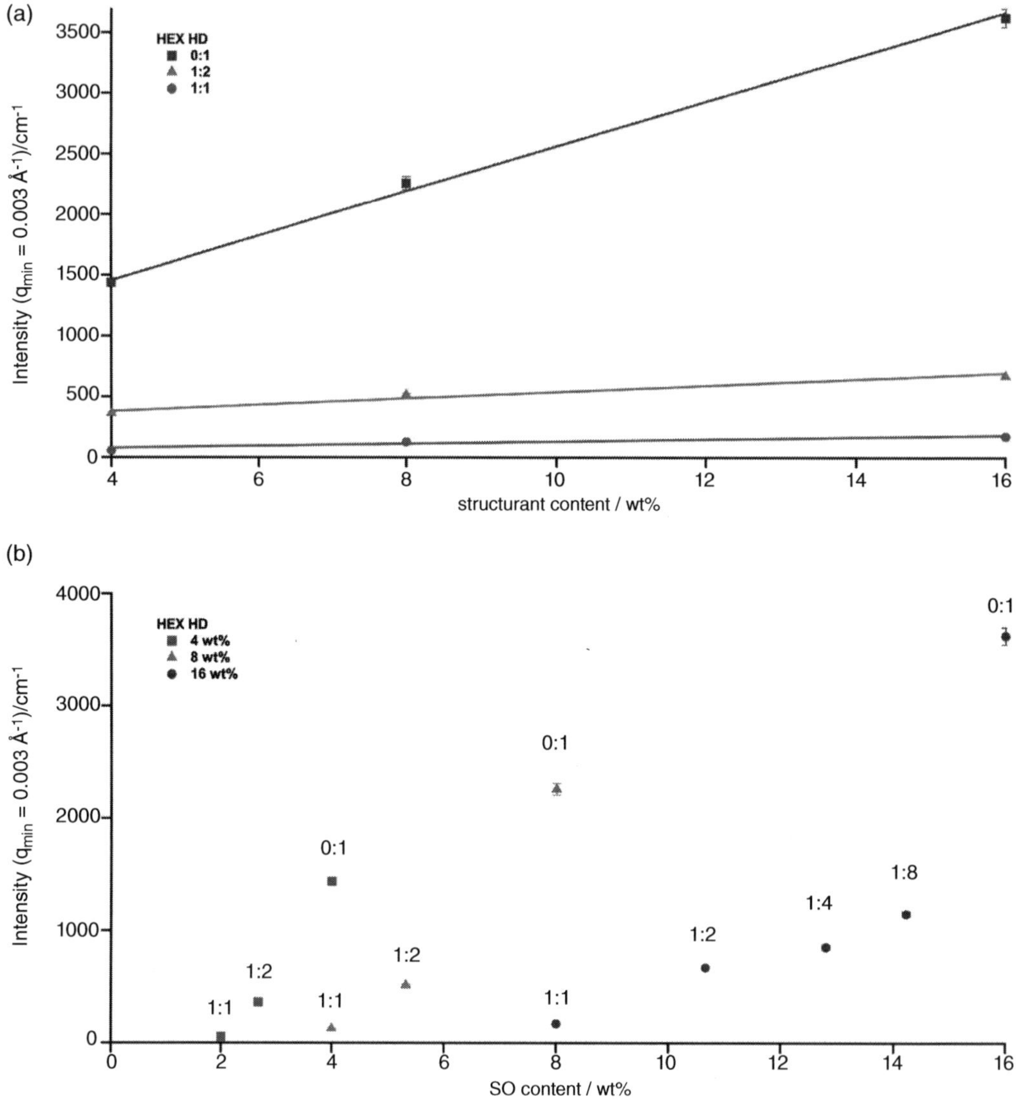

Figure 3.2.7 SANS intensity in the limit of low q = 0.003 Å^{-1} for hexadecane-based organogels with HD contrast (where the letters indicate the deuteration of the oil and water phases respectively thus H = hydrogenated oil and D = deuterated water, in this case) for 4, 8 and 16 wt% systems (a) as a function of the weight% structurant; and (b) as a function of the weight% SO; labels indicate associated OA:SO ratios. Reproduced from Cornet et al. 2021 / with permission of Elsevier.

and, at the lower end, extend down to the length scale of a structurant molecule. Combined, SANS and USANS provide information over five orders of magnitude in q and spatial dimension.

A fit over the full q range for 0:1 covering about 10 orders of magnitude in intensity and 5 orders of magnitude in q is shown in Figure 3.2.8a. The fit is based on a Guinier-Porod model combined with 5 Gaussian peaks (4 orders of reflections from a set of crystalline lamellar peaks at $q \approx 0.14$ Å^{-1} and the lower q peak at ≈ 0.055 Å^{-1}, the latter likely to be associated with a minor coexisting, possibly thermodynamically unstable lamellar liquid crystalline structure); the other fit parameters are $R_g = 121$ Å, dimension variable of 2.42 and Porod exponent of 3.67. The dimension variable for a

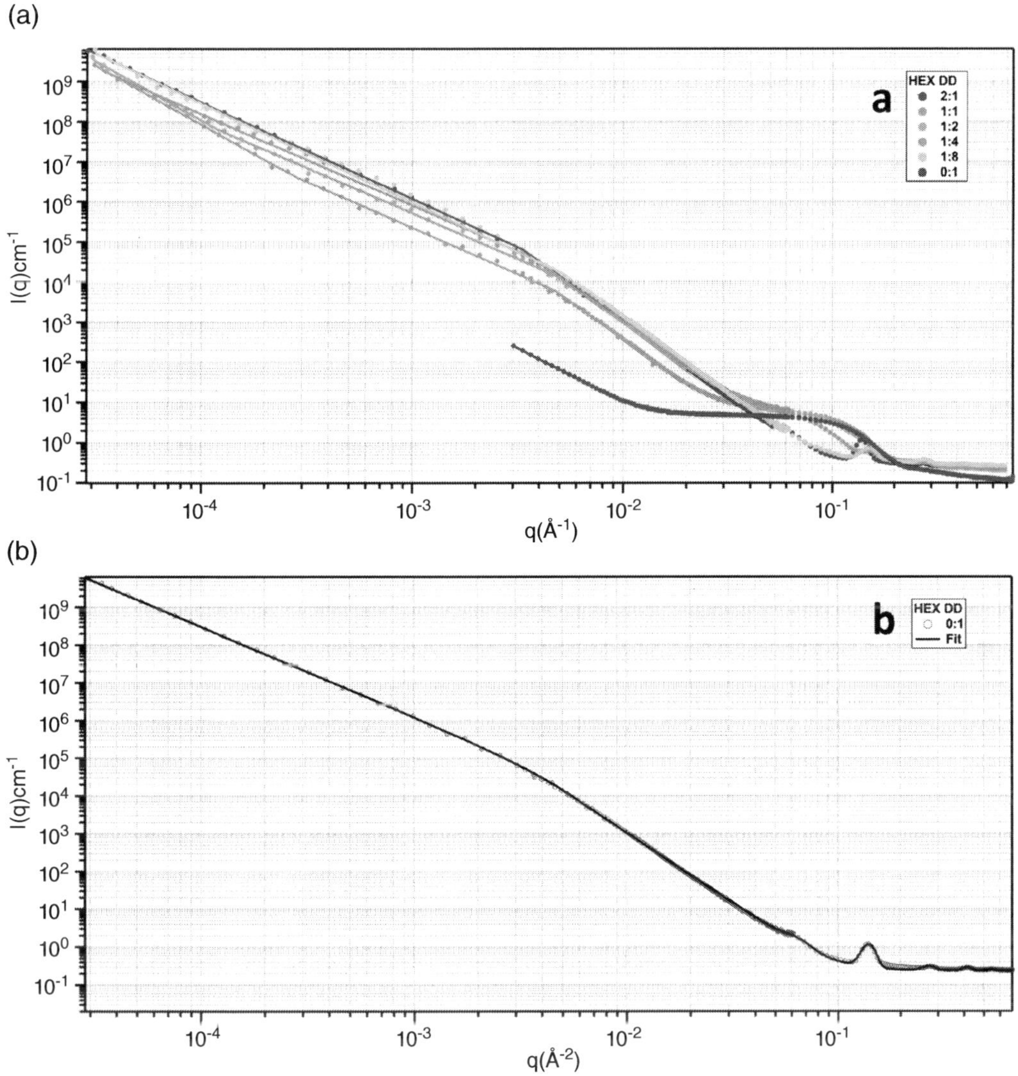

Figure 3.2.8 (a) Combined USANS and SANS from 16 wt% samples prepared with deuterated hexadecane and deuterated water (*i.e.*, DD contrast) including fits to single, double or triple power law; (b) 0:1 data with fit to Guinier-Porod model plus five Gaussian peak function. Reproduced from Cornet et al. 2021 / with permission of Elsevier.

plate-like scatterer would be 2, and the associated R_g for such an object would enable the thickness, T, to be determined based on the relationship $T = \sqrt{(12)}R_g$. It is not inconceivable that the observed low q power slope could be associated with plate-like scattering particles. This is typical of lamellar crystals with exponent = 2, which is contaminated by the coexistence of higher exponent power laws associated with the Porod region of larger particles. However, despite the acceptable Guinier-Porod model fit to the data, in the current case, the assembly of partially rough (Porod exponent < 4) crystalline particles undergoing mass fractal aggregation on a larger length scale is more structurally realistic and closer to microscopy observations; this is another reason complementary characterisation methods are invaluable. The USANS can therefore be equally well described with a double power-law function, which yields a low q power-law slope of ≈ 2.4, indicating mass fractal-like behavior, followed by a surface-fractal like slope of $\approx$ 3.8. The crossover between these two regions occurs at a q $\approx$ 0.0029 $Å^{-1}$ and can be broadly associated with a dimension of $\approx \pi/q_{crossover} \approx$ 1080 Å which is equivalent to approximately 24 longitudinal unit cell repeats of the structurant(s). This value is somewhat larger than observed in decane-based systems (Nikiforidis et al. 2015) where a crossover to Porod law scattering occurred at higher q. The average grain size, L, can also be estimated from the width of the Bragg peak, Δq, using the Scherrer equation where $L = 2\pi K/\Delta q$ and the Scherrer constant, K, is taken as 0.94. Taking into account the influence of instrumental broadening, Δq for the first Bragg peak is 0.0046 $Å^{-1}$; this gives a size of 1280 Å. This value is comparable to that obtained from the crossover value in q and indicates the typical lateral dimension of an SO crystal responsible for the gel behaviour.

As for the fractal-like behaviour in the USANS, compared to the pure SO system, the OA-containing samples exhibit broadly similar USANS in terms of the q value for the crossover region and a slightly greater Porod exponent but, in addition, exhibit a further crossover region at a lower q of $\approx 3–6 \times 10^{-4}$ $Å^{-1}$ where the power-law slope changes from $\approx$ 2.4 and $\approx$ 2.9 (in the limit of the USANS region); the latter exponent may suggest a transition towards denser mass fractal packing on the micron length scale. Returning to the SANS, the sample with 2:1 OA:SO ratio exhibited inverse micellar scattering with no evidence of Bragg peaks but some evidence of large-scale structure formation at low q although whose intensity was about two orders of magnitude lower than that of the other compositions; notably this sample does not form gels. Comparing the samples with 2:1, 1:1 and 1:2 OA:SO ratio, the low q intensity increased with SO content with inverse micelle scattering persisting. There is also evidence of the coexistence of Bragg peaks and crystalline material at these compositions although they are somewhat swamped by the micelle scattering. At an OA:SO ratio of 1:4, the micellar feature moves to lower q indicating larger particles which comes with the emergence of multiple Bragg reflections, the first at q $\approx$ 0.15 $Å^{-1}$. With this increasing amount of SO, such behavior may be understood as a greater proportion of structurant forming a separate crystalline phase leaving less structurant available to form inverse micelles. As a consequence, the same number of water molecules must be stabilized by fewer structurant molecules, resulting in a larger inverse micelle core. The micelle scattering is absent for 1:8 and 0:1 OA:SO ratios and the patterns are almost identical except that the Bragg peaks shift to lower q in 1:8 to 0:1 from 0.151 to 0.139 $Å^{-1}$, indicating a larger d-spacing with increasing SO content, as is also seen for equivalent edible oil systems. The gradual replacement of micellar with lamellar scattering may be explained simplistically based on the critical packing parameter, or cpp (Israelachvili et al. 1976). The cpp is given as $cpp = v_o/(al_o)$, where v_o and l_o are respectively the volume and the length of the average gelator molecule and a is the surface area of the structurant tail (which is the same for OA and SO). For 16 wt% 2:1 OA:SO ratio, the system is dominated by hydrophobic interactions between the OA tails and the oil. The OA headgroup is small compared to the oleic tail and $v_o/(al_o) > 1$. The 1 wt% water is located around the acidic headgroups and the formation of inverse spherical micelles is favored; such micellar structures are incapable of forming a space-filling network and gel therefore. The presence of sodium in the SO headgroup provides SO with greater polarity, hydrophilicity, and results in a reduction in the cpp. Hence, with increasing addition of SO, the average cpp of the structurant is reduced and structures with reduced curvature are favored, helping to form lamellar liquid crystals.

It is hopefully self-evident that the combination of USANS, SANS, microscopy and rheology can provide a detailed understanding of the OA-SO system and the basis for gel formation, particularly how changes in gel behaviour can be related to compositional changes and micro- and nanoscale structure. Since SO is largely insoluble in apolar solvents such as hexadecane and, to a lesser extent, edible oils, the 0:1 samples are likely to principally be dispersions of SO crystals. As OA is added to the formulation, a proportion of the SO is incorporated into the mixed crystal. The SO can also be partially incorporated in inverse micellar structures found in the OA-rich samples (*i.e.*, 2:1, 1:1) with the effect of decreasing the extent of large-scale structure formation. This also explains the accompanied reduction in low-q scattering, and the consequent reduction seen in gel strength. Overall, incorporating OA provides tunability in structure and resultant gel properties.

3.2.3.2 Natural Waxes

The SAS from oleogels formed using natural waxes has been principally interpreted based on the position of diffraction peaks in the small-angle region related to a crystalline lamellar structure. An example is beeswax which is a complex mixture of mostly straight-chain monohydric alcohol compounds and straight-chain acids; as well as being used as a gelator, it finds application as a glazing and coating agent, a texturizer for chewing gum base and a carrier for additives. Martins et al. (Martins et al. 2016) investigated the addition of beeswax to both medium chain (MCT) and long chain triglyceride (LCT) oils at a concentration of 1, 2, 4 and 8 wt%. Increasing concentration leads to stronger gels. Except for the solvent peak, discussed below, LCT comprises 3 peaks and MCT has 2 peaks. It is difficult to assess the extent to which peak intensities change with concentration, except for LCT, where the first peak, corresponding to a spacing of $\approx$ 93 Å increases with more gelator. The other peaks in LCT appear at $\approx$ 0.09 Å^{-1} ($\approx$ 67 Å) and 0.18 Å^{-1} (a weak second order reflection). In MCT, there is a shift in the position of the lowest angle peak from $\approx$ 0.05 Å^{-1} (106 Å) to $\approx$ 0.08 Å^{-1} between 4 and 8%, indicative of a change in structure. While there are differences in the peak positions between the MCT- and LCT-based gels at fixed gelator concentration, the most striking feature, although not noted by the author, are the broad peaks at $\approx$ 0.25 Å^{-1} and 0.35 Å^{-1} in LCT and MCT respectively. These peaks are associated with the oils themselves where the longer average chain length in LCT gives rise to a peak that appears at lower q. Indeed, Figure 3.2.9 shows unpublished SAXS data from the OA:SO oleogel study

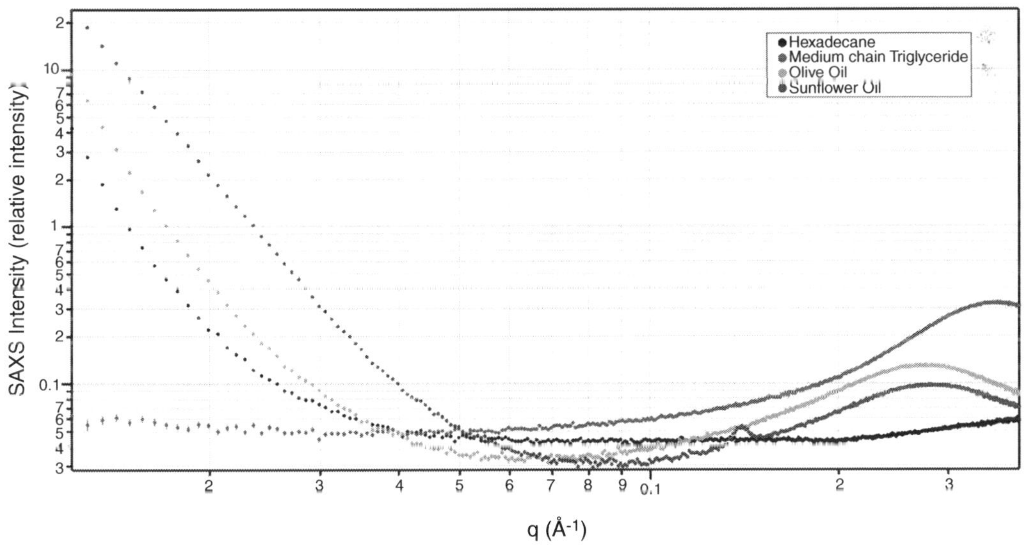

Figure 3.2.9 SAXS from typical edible oils along with apolar solvent, hexadecane. The weak peak in sunflower oil at q ≈ 0.14 Å^{-1} arises from a crystalline impurity.

discussed above from typical oleogel solvents and hexadecane. While reference is made to radii of gyration in the beeswax study, it is unclear how such values were determined or to what these values relate from a structural perspective. A similar comment can be made concerning a related study by the same authors where the influence of β-carotene on the properties of 2–8 wt% beeswax in LCT-based gels was investigated (Martins et al. 2017). Increasing beeswax concentration is increasingly effective in enhancing oxidative stability of β-carotene. The low q peak associated with a ≈ 93 Å repeat disappears in the presence of β-carotene, which is claimed to be related to the affinity for this part to the oil. Doan et al. (Doan et al. 2017) investigated beeswax (BW) and six additional natural waxes [rice bran wax (RBW), sunflower wax (SW), candelilla wax (CLW), carnauba Brazilian wax (CRBW), carnauba wild wax (CRWW) and berry wax (BEW)] at their minimum gelling concentrations ranging from 1.0 wt% (sunflower, candelilla and berry) to 5.0 wt% (rice bran) to transform liquid rice bran oil into oleogels. Using a conventional X-ray diffractometer to access the SAXS region, peaks were seen for all gels. RBW showed 4 reflections based on a repeat of 74.8 Å. SW and BEW had three reflections based on 66.9 Å and 50.7 Å repeats, respectively. The other waxes, BW, CRBW and CRWW and CLW, only exhibited one peak associated with 23.6 Å, 22.0 Å, 24.0 Å and 23.1 Å, respectively. The chain length breakdown in fatty hydrocarbons, free fatty acids and free fatty alcohols, as well as fatty acid and fatty alcohol moieties of wax ester components help with understanding differences in diffraction patterns. The authors reasonably state those gels that exhibit peaks with much lower intensity, indicate a less ordered layer structure; the latter is a consistent with mixtures of compounds with varying chain lengths. Carnauba wax was also studied at 3, 7 and 10 wt% in virgin olive oil, and compared to palm oil-derived monoglyceride and"breakfast margarine" (Öğütcü and Yilmaz 2014). Diffraction peaks are broadly consistent with those reported above, corresponding to 20.0 Å and 23.0 Å respectively for 7 and 10 wt% although absent at 3 wt%; peak positions are slightly lower in q than the corresponding gelator.

3.2.3.3 β-sitosterol (and Other Phytosterols) with γ-oryzanol

Without question, the most studied oleogelator with SAS is the plant sterol and corresponding plant sterol ester system, β-sitosterol and γ-oryzanol (BG); there are many publications with the first dating to 2006 (Bot 2018). While neither of the two molecules individually form oleogels, when the molecules are combined, oleogels are readily formed via the formation of an equimolar complex. SAXS from such gels shows a power law slope of q^{-2} at low q with broad oscillations at 0.099, 0.196, and 0.275 $Å^{-1}$. Homogeneous tubules would be expected to give rise to q^{-1} scattering; however, a hollow tube with finite wall thickness can theoretically reproduce q^{-2} scattering. It was proposed that such hollow tubules arise due to the establishment of a BG complex by intermolecular hydrogen bonding and, over longer length scales, can form a continuous three-dimensional network. The molecules position their sterane cores over each other to reduce the energy of the complex although a hydrogen bond prevents parallel stacking of the molecules, leading to a curvature in the stack to form a "helical ribbon". Evidence for these tubules is also seen in AFM where the dimensions of the finest tubules are 98 Å (Matheson et al. 2017). None of the characteristic reflections for the pure compounds are present in the oleogel scattering *i.e.*, 25 – 26 Å diffraction peak for γ-oryzanol; 36 – 38 Å bilayer spacing for sitosterol, with monolayer spacings of 17.6 Å, 18.8 Å and 17.6 Å monohydrate, hemihydrates and anhydrous respectively.

SAXS data can be reasonably fitted based on a hollow tubule of diameter 72 Å and wall thickness of 8 Å; the latter corresponds well with the dimension of the sterane moiety of the molecules. The question remains as to the position of the additional ferulic acid moiety on the γ-oryzanol and whether it points in or out of the tubule. To address this, more detailed studies were conducted using neutron scattering and solvent contrast variation using mixtures of sunflower oil and deuterated decane (Bot et al. 2012). SANS showed that the sterane cores of sterol and oryzanol form the wall of the tubules, with the hydrogen bond at the outside of the tubules; this conformation causes the ferulic acid moieties in the oryzanol to stick out of the exterior wall of the tubules. The corresponding models give rise to a diameter of 95 ± 9 Å and wall thickness of 23 ± 8 Å. However, it is possible that this is far from the complete story since tubule scattering appears even in the oil-free system at a sufficient temperature above the melting point of the gelators. In addition, some of the peak positions have a potential association to a liquid crystalline cubic structure although further study is needed (Sawalha et al. 2015). Related SAXS studies with phytosterols, other than β-sitosterol, combined with γ-oryzanol reveals that the tubule diameter increases with a decreasing number of double bonds in the sterane core (e.g., ergostanol – two double bonds – 67 Å, sitosterol – one double bond – 72 – 78 Å, and cholestenol – no double bonds – and 80 Å) although the wall thickness does not seem to be related to sterol structure. Sawalha et al. have explored the influence of oil type on oleogel structure (Sawalha et al. 2012, 2021) where increasing the dielectric constant of the oils results in tubules becoming more stretched, elongated and less curly (cf. sunflower oil and castor oil) and with further increase (e.g., eugenol), the tubules become less aggregated and bundled. Similarly, at low water content, curly tubules are formed, straighten with increasing water content and, at sufficiently high water content, become unstable and fall apart due to the formation of sterol monohydrates.

The scattering from phytosterol – monoglyceride oleogels is dominated by diffraction peaks, rather than the scattering from tubules (Bin Sintang, Danthine, Brown et al. 2017; D. X. Yang et al. 2018; S. Yang et al. 2018). Using a mixture of non-esterified phytosterols (including 50.4% β-sitosterol, 25.0% campesterol, and 15.3% stigmasterol) and monoglycerides from hydrogenated palm oil, changes in DSC after various aging times, highlights the role of kinetics in these mixed systems. Ratios of phytosterol: monoglyceride of 10:0, 8:2, 7:3, 6:4, and 0:10 at 10 wt% were investigated with ideal gel properties being obtained at a 2:8 ratio. Observations were interpreted based on incorporating the phytosterol into the alkyl tail bilayer structure of the monoglyceride. β-sitosterol with glyceryl monostearate in sunflower oil at 10 wt% in ratios of 10:0, 6:4, 4:6 and 0:10 were studied as a tunable delivery system for volatiles and flavor components. As expected, the monoglyceride system shows peaks associated with a 47.6 Å bilayer but a peak corresponding to a 45.1 Å sitosterol double layer is seen. In the mixed oleogels, a 35.4 Å peak is seen that is ascribed to a mixed double-layer ordered structure. However, in a similar system reported by the same authors (Yang et al. 2018), using stearic acid at a 2:8 sitosterol: stearic acid ratio, at 20 wt% in different edible oils, many diffraction peaks are reported; however, these peaks are not indexed even for the pure materials, except primary reflections associated with a 37.7 Å peak for sitosterol and 40.3 Å for stearic acid. Indeed, even a reflection for sitosterol as high as 120 Å is apparent but neither identified nor explained.

3.2.3.4 Lecithin

Lecithin is a complex mixture of glycerophospholipids with the major components being phosphatidylcholine, phosphatidylethanolamine and phosphatidylinositol. These individual molecules themselves comprise a variety of fatty acid moieties, differing in chain length, saturation, and with either a polar or charged headgroup esterified to a glycerol backbone; the detailed chemical composition depends on the origin of the lecithin (e.g., soy, egg etc.) and is far from representing a single molecular entity. Lecithin self-assembles into reverse micelles in organic solvents, and readily so, with the addition of water. With more water, reorganization occurs in which spheroidal micelles and replaced by cylindrical reverse micelles; at a critical water concentration, these tubular reverse micelles can overlap and entangle. This structural change manifests in the evolution of non-viscous solutions into gels; however, beyond a critical water concentration, the gel point is passed and viscosity decreases until a point where phase separation occurs. Bodennec et al. (Bodennec et al. 2016) investigated oleogels formed in canola oil with 5–30 wt% lecithin and 0.1 to 10 wt% water. The resultant networks critically depend on composition. A reverse hexagonal mesophase of densely-packed water-filled cylinders is formed in which each cylinder layer of phospholipid molecules is oriented perpendicular to the interface with their hydrophobic moieties pointing outward from the water cylinders and their fatty acid chains filling the voids between the hexagonally packed cylinders. On a larger length scale, these cylinders cross and form junction zones that trap oil, and are responsible for network formation. SAXS reveals a peak at $\approx$ 52 Å ascribed to the separation of cylinders within the micron sized bundles; with increasing temperature, this peak decreased in intensity and shifted to a larger dimension. The Scherrer equation is also used to assess intra-bundle order as well as its temperature dependence. The application of scattering to access much lower q_{min} would be valuable here to reveal the influence of different lecithin compositions on the resultant structure formation and consequent gel properties.

Due to their sensitivity to water, lecithin oleogels have limited potential for food application on their own; however, lecithin has been used as a co-oleogelator to tune the self-assembly of host structuring agents. Mixtures of lecithin, here of soy origin, with stearic acid (20 wt% and 30 wt% total gelator, with 1 wt% water) were investigated for oleogels and oleogel-based emulsions prepared in canola oil; ratios of lecithin:stearic acid 0:10, 3:7, 5:5, 7:3, and 10:0 were studied (Gaudino et al. 2019). With an increasing ratio of stearic acid to lecithin, both oleogels and oleogel-based emulsions showed an increase in physical stability via increased firmness and decreased oil loss. Two sets of peaks were seen in the oleogels with d-spacings between 50.1 and 57.6 Å – proposed to be due to the separation of inverse cylindrical micelles within bundles or fibres – and between 39.2 and 39.5 Å, proposed to arise from stearic acid arranged in double layers. However, it is noted that there is at least one more peak seen at lower q unrelated to the individual components and indicative of a larger length scale structure ($\approx$ 70 Å). As the concentration of stearic acid increased, the d-spacing of the lecithin reverse micelle increased suggesting that stearic acid probably interacts with the non-polar tails of the lecithin micelle, widening the micellar diameter. Of course, it is also possible that the polar portion of the stearic acid molecule interacts with the polar head of the lecithin within the micelle. It was also speculated that the stearic acid molecules are also distributed within the oleogel matrix as crystals and spherulites that contribute to overall gel stability. Bigels have been studied in which an oleogel emulsion composed of soy lecithin, stearic acid, soybean oil, and water (at 10 wt% or 20 wt% oleogel) has been homogenized at high shear with a hydrogel composed of whey protein concentrate (15 wt% or 25 wt%) and water (Bollom et al. 2020). The ratio of lecithin to stearic acid was 7:3 and the influence of oleogel-based emulsion to hydrogel was investigated. The results can be broadly interpreted based on the 7:3 lecithin to stearic acid oleogel system in the previous study since no Bragg peaks are seen in the hydrogel phase. However, the ability to reach lower q shows a greater array of peaks with the d-spacing of the largest corresponding to 110 Å. The authors speculate these peaks arise from hexagonal packing symmetry although peaks can be indexed perfectly to a simple cubic lattice coexisting with a stearic acid bilayer.

Nikiforidis and Scholten (Nikiforidis and Scholten 2014) studied purified phosphatidylcholine with α-tocopherol (*i.e.*, vitamin E) in ratios of 1:0, 1:1, 1:2, 2:1 and 0:1 at 25 wt% in sunflower oil in the absence and presence of 1 wt% water. While phosphatidylcholine has a large hydrophobic area (tails) and a small hydrophilic area (head), α-tocopherol has the opposite features. In the absence of water, neither phosphatidylcholine nor α-tocopherol form gels, but do so when mixed. With addition of water, only 1:1 and 1:0 were observed to form a gel. It Seems that the polarity of α-tocopherol has a similar effect to water in promoting the uniaxial growth of cylindrical, worm-like micelles. Limited SAXS is presented in this study where a single peak is seen that reflects the structure factor from inverse cylindrical micelles or lamellae. The addition of α-tocopherol leads to a larger but less ordered structure while adding water to lecithin:α-tocopherol mixtures reduces gel strength and melting temperature providing tunability in properties. Sucrose esters (Ses) are non-ionic surface-active surfactants consisting of sucrose as the hydrophilic group and fatty acids as lipophilic groups and have been investigated when combined with lecithin in sunflower oil to form oleogels; however, while SAXS was performed, no analysis of the peaks was reported (Bin Sintang, Danthine, Patel et al. 2017).

Lecithin has also been combined with mixtures of β-sitosterol and γ-oryzanol (BG), the latter at fixed equimolar ratio, equivalent to 2:3 on weight basis, in sunflower oil, MCT and hexadecane with ratios of BG:L (100:0; 80:20; 60:40; 50:50; 40:60; 20:80; 0:100) (Okuro et al. 2018). As discussed above, tubules are formed at the BG-rich end of the composition spectrum. With increasing lecithin, tubule scattering is gradually replaced by inverse micelles in hexadecane, which is similar to MCT; however, it is far more complex in SFO where multiple Bragg peaks, where three coexisting phases may be required and where SAXS from the 20:80 composition is most complicated. This weakening of tubule scattering on lecithin addition is proposed to be related to the polar head groups of lecithin disturbing the intermolecular interactions and hydrogen bonding necessary for BG complexes to form. At the other extreme, it is proposed that BG is incorporated into inverse micelles at higher lecithin content. Solvent polarity is critically important since the pure lecithin composition formed a gel only in the case of hexadecane. The authors recognise that, particularly in this system, access to even lower q scattering would be useful.

3.2.3.5 Mono-, Di- and Triglycerides

Oleogels formed using glyceryl monostearate (GM) or sorbitan monostearate (SM – an ester of the non-ionic surfactant, sorbitan, and stearic acid, also known as Span 60) at 9.1 wt% in chia seed oil have been investigated with SAXS including the influence of cooling rate (Trujillo-Ramírez et al. 2019). The study enabled the direct comparison between gelators of similar melting point differing only in headgroup structure. The influence of cooling rate on the gel properties has opposite effects depending on the gelator used and is thought to be related to differences in the hydrophile-lipophile balance between the two structurants (3.8 for GM and 4.7 for SM). Unfortunately, it is difficult to understand the conclusions drawn from the SAXS. The units of q are given in $Å^{-1}$ yet the q range shown contradicts the technique, the associated reported d-spacings from observed peaks seem unrelated to their q position and values for fractal dimension and R_g values are merely stated. A somewhat larger, more detailed (and earlier) study that also included the tri-fatty acid versions of the above gelators, namely glyceryl tristearate (GT) and sorbitan tristearate (ST, also known as Span 65), at 5 – 25 wt% in MCT or LCT was studied by Cerqueira et al. (Cerqueira et al. 2017). Diffraction peaks arising from a lamellar structure are seen in the SAXS in all cases with power law exponents between −2 and −3. As expected, gel properties increase with gelator concentration but not linearly and reach a threshold where the gelation mechanism changes and steric effects dominate. Glyceryl-based gelators produce the most structured organogels; monostearate helps to form stronger gels with greater lamellar spacing. Gels produced with LCT are stronger and gel at lower concentration than MCT. The peaks in the GT- and ST-based gels, irrespective of oil type and concentration, produce consistent peaks corresponding to d-spacings of

44–45 Å and 52–53 Å respectively. However, the GM- and SM-based gels do not show consistent values and vary between 48–68 Å and 48–62 Å in MCT and 48–52 Å and 57–65 Å in LCT respectively although it is noted that an additional peak at lower q of 0.06 $Å^{-1}$ is apparent in the 15 wt% SM in LCT system indicating a more complex, larger-scale structure. The authors note that sorbitol-based gelators lead to a higher d-spacing proposed to be related to the size of their hydrophilic head. Temperature dependence in the scattering has been investigated in the same system also (Fasolin et al. 2018).

Rosen-Kligvasser and Davidovich-Pinhas (Rosen-Kligvasser and Davidovich-Pinhas 2021) have conducted a detailed multi-technique (including FTIR, TGA, TEM and light microscopy) exploration of canola oil based oleogels formed from either GM or the corresponding fatty acid, stearic acid (SA), with a focus on understanding the role of headgroup (carboxylic acid versus glycerol plus two free hydroxyl groups) on hydrogen bonding and its corresponding influence on oleogel structure. Three concentrations of structurant were used for both GM and SA to form soft, medium and hard gels while ensuring their same molar composition enabling direct comparison *i.e.*, 6, 12 and 24 wt% for SA and 7.5, 15 and 30 wt% for GM, corresponding to 0.2, 0.4 and 0.8 M concentration respectively. SAXS data are reported only for the 0.4 M oleogels where peaks and higher order reflections are seen that are typical of the lamellar crystal structure of the gelators. For SA, these correspond to a 38 Å d-spacing while for GM peaks are seen at 48 Å and 39 Å that are associated with two forms of GM, namely glycerol-1-monostearate and glycerol-2-monostearate. Microscopy revealed ordered, anisotropically oriented needle-like crystals in SA and isotopically ordered braid-like crystals in GM oleogels. Stronger gels were formed with GM and it was proposed that gel strength is enhanced when crystal structure is isotropically oriented, as seen in composite materials, where the structurant crystals behave like a reinforcing agent within the oil matrix.

In more recent developments, Nicholson and Marangoni have explored lipase-catalyzed glycerolysis to generate fat-like structuring. In this enzymatic process liquid oils are converted into solid fat through the conversion of triacylglycerols (TAGs) in liquid oils into monoacylglycerols (MAGs) and diacylglycerols (DAGs). Since MAGs and DAGs crystallize at temperatures higher than TAGs containing the same fatty acids, glycerolysis can effectively increase the crystallization temperature of the oil without changing the overall fatty acid composition (Nicholson and Marangoni 2020). Glycerolysis has been investigated using *Candida antarctica* lipase B with cottonseed oil and which seems to be ideal at an equimolar glycerol:TAG ratio; at this composition microscopy reveals a crystal network containing smaller, more numerous fat crystals with a higher fractal dimension, and increased structural homogeneity. The observation between structural homogeneity and gel strength follows the findings of Rosen-Kligvasser and Davidovich-Pinhas (Rosen-Kligvasser and Davidovich-Pinhas 2021) above. Catalysis with the 1,3-specific lipase from *Rhizomucor miehei* as opposed to the non-specific *Candida antarctica* lipase B leads to different behavior presumably due to changes in the relative presence of 1,2- and 1,3-DAGs formed. Structural changes seem to be dominated by MAG (monopalmitin and monoolein), rather than DAG. SAXS shows multiple order reflections from a 46.5 Å repeat similar to that for monopalmitin. Related studies have also been performed on tigernut, peanut, rice bran, olive, high oleic canola, soybean, sesame and canola oils (Nicholson and Marangoni 2021) with SAXS displaying parent d-spacings of between 46 Å and 50 Å for each system (cf. monopalmitin, monostearin and monoolein d-spacings of 45.8 Å, 50.0 Å at 49.5 Å) and is consistent with the varying fatty acid compositions of the original oils. An additional peak was seen in the tigernut, olive, high oleic canola, sesame and canola oil systems believed to be associated with 1,3-diolein crystals with a long-spacing of 39.3 Å indicating co-crystallisation with the MAGs.

3.2.3.6 Carbohydrate and Proteins

Ethylcellulose (EC) is a linear polysaccharide derived from the ethylation of cellulose, an important structural component of the cell wall of many plants. The latter has been widely studied recently with SAS techniques (Lopez-Sanchez et al. 2015, 2020; Martínez-Sanz et al. 2017; Martínez-Sanz, Mikkelsen, B. Flanagan et al. 2016; Martínez-Sanz, Mikkelsen, B. M. Flanagan et al. 2016; Martínez-Sanz, Gidley et al. 2015). The final product is characterized the degree of substitution, DS, or ethoxy content; solubility in organic solvents is achieved with DS values in the range of 2.4–2.5. EC is usually produced and marketed in several viscosity grades where the viscosity increases as the length of the polymer chains increases. EC can structure edible oil primarily via interactions of unsubstituted hydroxyl groups which form inter- and intramolecular hydrogen bonds between the backbones of the polymer strands; the resultant gel hardness not only increases with EC concentration but also EC molecular weight and degree of unsaturation of the edible oil (Scharfe and Flöter 2020). There are no reported SAS studies of pure EC-based oleogels; however, the scattering from EC itself has been reported (Davidovich-Pinhas et al. 2014, 2016). SAXS shows a peak at 0.5 $Å^{-1}$ ascribed to the EC interlayer distance while USAXS shows two regions that can be described with a unified model with power-law exponents of −3.8 and −2.8 and radii of gyration of ≈ 1–2 and 5–6 μm, possibly dependent on EC viscosity. These slopes are suggested to arise from relatively smooth EC particles that aggregate by a particle-cluster diffusion-limited process in three dimensions, followed by some densification. The shape of the USAXS changed and intensity decreased above the glass transition temperature and decreased further beyond the melting transition while the SAXS peak changed only above the melting temperature. Gravelle et al. (Gravelle et al., 2017) investigated 4 wt% EC with 5 wt% mixtures of stearyl alcohol (*i.e.*, 1-octadecanol) and stearic acid in canola oil; 11 stearyl alcohol (SO):stearic acid (SA) ratios were studied. The SO:SA mixtures in the absence of EC show peaks associated with spacings varying from 50.0 Å (SO only) to 40.2 Å (SA only), but the 3:7–8:2 ratios exhibit peaks from a mixed phase although d-spacing is determined based on the third order reflection since lower angle data are unavailable. The addition of EC not only reduces the overall peak intensity but also the number of peaks.

Not strictly an oleogel in its own right, Tsung et al. report on gels formed from mixtures of zein – a major storage protein of corn – with ethanol and oleic acid (Tsung et al., 2020). The time-dependent gel formation was explored using the tube inversion method to reveal five qualitative regions of the ternary phase diagram. The rod-shaped structure of the protein with dimensions ≈ 130 x 12 × 30 Å is proposed to act as a building block for the resultant two- and three-dimensional structures responsible for gel formation. USAXS, SAXS and WAXS reveal three hierarchical structural levels. A unified model is applied to describe the data where the first level (highest q) corresponds to zein molecules (power-law exponent = 1–1.8, $R_g \approx 40 - 130$ Å) with higher structural levels interpreted as arising from ribbon-like structures with $R_g \approx 2500 - 12000$ Å) with rough interfaces (3.2 – 3.7) that ultimately entrap oleic acid.

3.2.4 Conclusions and Outlook

This chapter has described SAS methods with a focus on oleogel application. That the field of fat mimetics is far greater than these examples immediately indicates great opportunities for their use. The extent to which these methods are being exploited and the difference between what can be extracted from the data compared to what should be, varies enormously; unfortunately, this includes under-, over- and no interpretation whatsoever. As for many scientific fields, some concerning examples in the literature have supposedly passed expert review where data interpretation is wrong. There are opportunities therefore for SAS specialists to become more intimately involved in this important area, to help with understanding and to make sure data interpretation is more robust.

The examples presented are structurally hierarchical and complex, requiring the need for multiple characterization techniques spanning multiple length scales. This is critical to draw structure-function-property relationships and enable behaviour to be predicted so that new, more effective oleogel systems can be designed. Many examples presented here show aspects of behaviour that depend on one or more parameters beyond merely gelator composition combined with the added complication of metastability and time-dependence. More fundamental studies are needed to understand and predict oleogel behaviour. A coherent approach is also needed to elucidate the influence of minor dissolved polar components in edible oils (e.g., Scharfe and Flöter 2020) on the nanoscale, microscale and macroscopic properties of oleogels. More advanced characterisation will inevitably have a consequence in unravelling why different gelling behaviour is seen at varying critical gelling concentrations of gelator, depending on the gelator and solvent, the gelator-gelator, gelator-solvent interaction and the influence of cooling rate, shear rate, temperature and time of cooling. This is essentially an untapped opportunity for SAS where combined techniques are routine.

Where a sample is changing as a function of time, there is the inevitable question as to whether the material studied at a point-in-time with a first technique is the same system, studied some time later, with a second technique. Combined methods enable the possibility for properties to be correlated without a significant error in time. For example, SAXS and SANS can be measured simultaneously with DSC to provide insight into structural changes and phase transitions (Pullen et al. 2014) and the influence of cooling rate (Pullen et al. 2008). In principle, *in-situ* mixing and subsequent quenching of oleogel systems could be conducted so that the evolution of structure could be tracked with time to follow kinetics and its influence on gel behaviour. The extent of metastability on resultant structures and gel behavior could also be intimately studied as a function of processing conditions. Simultaneous SAS and rheological studies are also possible so that oleogel formation, in particular in orientational flow fields, could be probed. In addition, SAS combined with a range of spectroscopies is possible. The opportunities for such approaches with SAS for fat mimetics are virtually untapped and ripe to be exploited. It is hoped this chapter provides researchers in fat mimetics a greater understanding of the potential for SAS techniques in their research activities, stimulates an interest in exploiting these methods and helps to develop future fat mimetics.

References

Beaucage, G. (1996). Small-angle scattering from polymeric mass fractals of arbitrary mass-fractal dimension. *Journal of Applied Crystallography* 29 (2): 134–146.

Bin Sintang, M.D., Danthine, S., Brown, A., Van de Walle, D., Patel, A.R., Rimaux, T., Dewettinck, K., and Van de Walle, D. (2017). Phytosterols-induced viscoelasticity of oleogels prepared by using monoglycerides. *Food Research International* 100: 832–840.

Bin Sintang, M.D., Danthine, S., Patel, A.R., Rimaux, T., Van de Walle, D., Dewettinck, K.l. (2017). Mixed surfactant systems of sucrose esters and lecithin as a synergistic approach for oil structuring. *Journal of Colloid and Interface Science* 504: 387–396.

Blazek, J. and Gilbert, E.P. (2011). Application of small-angle X-ray and neutron scattering techniques to the characterisation of starch structure: A review. *Carbohydrate Polymers* 85 (2): 281–293.

Bodennec, M., Guo, Q., and Rousseau, D. (2016). A molecular and microstructural characterization of lecithin-based oleogels made with vegetable oil. *RSC Advances* 6 (53): 47373–47381.

Bollom, M.A., Clark, S., and Acevedo, N.C. (2020). Development and characterization of a novel soy lecithin-stearic acid and whey protein concentrate bigel system for potential edible applications. *Food Hydrocolloids* 101: 105570.

Bot, A. (2018). Phytosterols. *Encyclopedia of Food Chemistry* 66 (12): 225–228.

Bot, A., Gilbert, E.P., Bouwman, W.G., Sawalha, H., den Adel, R., Garamus, V.M., Venema, P., van der Linden, E., and Flöter, E. (2012). Elucidation of density profile of self-assembled sitosterol + oryzanol tubules with small-angle neutron scattering. *Faraday Discussions* 158: 223–238.

Cao, Y. and Mezzenga, R. (2020). Design principles of food gels. *Nature Food* 1 (2): 106–118.

Cerqueira, M.A., Fasolin, L.H., and Euston, S.R. (2021). Editorial: Structured edible oil: Towards a new generation of fat mimetics. *Frontiers in Sustainable Food Systems* 5: 652644.

Cerqueira, M.A., Fasolin, L.H., Picone, C.S.F., Pastrana, L.M., Cunha, R.L., and Vicente, A.A. (2017). Structural and mechanical properties of organogels: Role of oil and gelator molecular structure. *Food Research International* 96: 161–170.

Cheng, L., Ye, A., Yang, Z., Gilbert, E.P., Knott, R., de Campo, R., Storer, B., Hemar, Y., and Singh, H. (2022). *Food Structure* 32: 100276.

Cornet, S., de Campo, L., Martínez-Sanz, M., Scholten, E., and Gilbert, E.P. (2021). Small-angle neutron scattering reveals basis for composition dependence of gel behaviour in oleic acid – sodium oleate oleogels. *Innovative Food Science and Emerging Technologies* 73: 102763.

Cornet, S.H.V. (2016). *Structure formation in oleic acid – sodium oleate based organogels.* Wageningen University.

Daniel, E. and Alejandro, C. (2012). *Organogels : An alternative edible oil-structuring method.* 749–780.

Davidovich-Pinhas, M., Barbut, S., and Marangoni, A.G. (2014). Physical structure and thermal behavior of ethylcellulose. *Cellulose* 21 (5): 3243–3255.

Davidovich-Pinhas, M., Barbut, S., and Marangoni, A.G. (2016). Development, characterization, and utilization of food-grade polymer oleogels. *Annual Review of Food Science and Technology* 7: 65–91.

Davidovich-Pinhas, M. (2019). Oil structuring using polysaccharides. *Current Opinion in Food Science* 27: 29–35.

Doan, C.D., To, C.M., De Vrieze, M., Lynen, F., Danthine, S., Brown, A., Dewettinck, K., and Patel, A.R. (2017). Chemical profiling of the major components in natural waxes to elucidate their role in liquid oil structuring. *Food Chemistry* 214: 717–725.

Doutch, J., Bason, M., Franceschini, F., James, K., Clowes, D., and Gilbert, E. (2012). Structural changes during starch pasting using simultaneous rapid visco analysis and small-angle neutron scattering. *Carbohydrate Polymers* 88 (3): 1061–1071.

Doutch, J. and Gilbert, E.P. (2013). Characterisation of large scale structures in starch granules via small-angle neutron and X-Ray scattering. *Carbohydrate Polymers* 91 (1): 444–451.

Fameau, A.L. and Saint-Jalmes, A. (2020). Recent advances in understanding and use of oleofoams. *Frontiers in Sustainable Food Systems* 4: 110.

Fasolin, L.H., Cerqueira, M.A., Pastrana, L.M., Vicente, A.A., and Cunha, R.L. (2018). Thermodynamic, rheological and structural properties of edible oils structured with L. MOGs : Influence of gelator and oil phase. *Food Structure* 16 (January): 50–58.

Feichtinger, A. and Scholten, E. (2020). Preparation of protein oleogels: effect on structure and functionality. *Foods* 9 (12): 19–23.

Flöter, E., Wettlaufer, T., Conty, V., and Scharfe, M. (2021). Oleogels—their applicability and methods of characterization. *Molecules* 26 (6): 1673.

Gaudino, N., Ghazani, S.M., Clark, S., Marangoni, A.G., and Acevedo, N.C. (2019). Development of lecithin and stearic acid based oleogels and oleogel emulsions for edible semisolid applications. *Food Research International* 116: 79–89.

Gilbert, E.P. (2019). Small-angle X-ray and neutron scattering in food colloids. *Current Opinion in Colloid and Interface Science* 42: 55–72.

Gilbert, E.P. (2022). Small-angle X-ray scattering (SAXS) and small-angle neutron scattering (SANS) study on the structure of sodium caseinate in dispersions and at the oil-water interface: Effect of calcium ions. *Food Structure* 32: 100276.

Gilbert, E.P., Lopez-Rubio, A., and Gidley, M.J. (2012). *Characterisation techniques in food materials science.* Wiley-Blackwell, p. 52–93.

Gilbert, E.P., Schulz, J.C., and Noakes, T.J. (2006). 'Quokka'-the small-angle neutron scattering instrument at OPAL. *Physica B: Condensed Matter* 385–386: 1180–1182.

Gravelle, A.J., Blach, C., Weiss, J., Barbus, S., and Marangoni, A.G. (2017). Structure and properties of an ethylcellulose and stearyl alcohol/stearic acid (EC/SO:SA) hybrid oleogelator system. *European Journal of Lipid Science and Technology* 119 (11): 1700069.

Hammouda, B. (2008). *The SANS Toolbox.* NIST Center for Neutron Research, Gaithersburg, MD. [Online]. https://www.ncnr.nist.gov/staff/hammouda/the_SANS_toolbox.pdf (accessed March 24, 2020).

Hammouda, B. (2010). A New Guinier-Porod Model. *Journal of Applied Crystallography* 43 (4): 716–719.

Heenan, R.K., Rogers, S.E., Turner, D., Terry, A.E., Treadgold, J., and King, S.M. (2011). Small angle neutron scattering using Sans2d. *Neutron News* 22 (2): 19–21.

Idziak, S.H.J. (2012). *Structure-Function Analysis of Edible Fats* (ed. A.G. Marangoni). Urbana, IL 61802: AOCS Press.

Ilavsky, J., Zhang, F., Andrews, R.N., Kuzmenko, I., Jemian, P.R., Levine, L.E., and Allen, A.J. (2018). Development of combined microstructure and structure characterization facility for in situ and operando studies at the advanced photon source. *Journal of Applied Crystallography* 51: 867–882.

Israelachvili, J.N., John Mitchell, D., and Ninham, B.W. (1976). Theory of self-assembly of hydrocarbon amphiphiles into micelles and bilayers. *Journal of the Chemical Society, Faraday Transactions 2: Molecular and Chemical Physics* 72: 1525–1568.

Lopez-Sanchez, P., Martinez-Sanz, M., Bonilla, M.R., Sonni, F., Gilbert, E.P., and Gidley, M.J. (2020). Nanostructure and poroviscoelasticity in cell wall materials from onion, carrot and apple: roles of pectin. *Food Hydrocolloids* 98: 105253.

Lopez-Sanchez, P., Martı, M., Gidley, M.J., and Gilbert, E.P. (2015). Evidence for differential interaction mechanism of plant cell wall matrix polysaccharides in hierarchically-structured bacterial cellulose 22 (3): 1541–1563.

Marangoni, A.G., Van Duynhoven, J.P.M., Acevedo, N.C., Nicholson, R.A., and Patel, A.R. (2020). Advances in our understanding of the structure and functionality of edible fats and fat mimetics. *Soft Matter* 16 (2): 289–306.

Martínez-Sanz, M., Gidley, M.J., and Gilbert, E.P. (2015). Application of X-ray and neutron small angle scattering techniques to study the hierarchical structure of plant cell walls: a review. *Carbohydrate Polymers* 125: 120–134.

Martínez-Sanz, M., Lopez-Sanchez, P., Gidley, M.J., and Gilbert, E.P. (2015). Evidence for differential interaction mechanism of plant cell wall matrix polysaccharides in hierarchically-structured bacterial cellulose. *Cellulose* 22 (3): 1541–1563.

Martínez-Sanz, M., Mikkelsen, D., Flanagan, B., Gidley, M., and Gilbert, E.P. (2016). Multi-scale model for the hierarchical architecture of native cellulose hydrogels. *Carbohydrate Polymers* 147: 542–555.

Martínez-Sanz, M., Mikkelsen, D., Flanagan, B.M., Rehm, C., de Campo, L., Gidley, M.J., and Gilbert, E.P. (2016). Investigation of the micro- and nano-scale architecture of cellulose hydrogels with plant cell wall polysaccharides: A combined USANS/SANS study. *Polymer* 105: 449–460.

Martínez-Sanz, M., Pettolino, F., Flanagan, B., Gidley, M.J., and Gilbert, E.P. (2017). Structure of cellulose microfibrils in mature cotton fibres. *Carbohydrate Polymers* 175: 450–463.

Martínez-Sanz, M., Ström, A., Lopez-Sanchez, P., Knutsen, S.H., Balance, S., Zobel, H.K., Sokolova, A., Gilbert, E.P., and López-Rubio, A. (2020). Advanced structural characterisation of agar-based hydrogels: rheological and small angle scattering studies. *Carbohydrate Polymers* 236: 115655.

Martins, A.J., Cerqueira, M.A., Cunha, R.L., and Vicente, A.A. (2017). Fortified beeswax oleogels: Effect of β-carotene on the gel structure and oxidative stability. *Food and Function* 8: 4241–4250.

Martins, A.J., Cerqueira, M.A., Fasolin, L.H., Cunha, R.L., and Vicente, A.A. (2016). Beeswax organogels: Influence of gelator concentration and oil type in the gelation process. *Food Research International* 84: 170–179.

Matheson, A.B., Koutsos, V., Dalkas, G., Euston, S., and Clegg, P. (2017). Microstructure of β-sitosterol:γ-oryzanol edible organogels. *Langmuir* 33 (18): 4537–4542.

Metilli, L., Lazidis, A., Francis, M., Marty-Terrade, S., Ray, J., and Simone, E. (2021). The effect of crystallization conditions on the structural properties of oleofoams made of cocoa butter crystals and high oleic sunflower oil. *Crystal Growth and Design* 21 (3): 1562–1575.

Narayanan, T., Sztucki, M., Van Vaerenbergh, P., Léonardon, J., Gorini, J., Clauste, L., Sever, F., Morse, J., and Boesecke, P. (2018). A multipurpose instrument for time-resolved ultra-small-angle and coherent X-ray scattering. *Journal of Applied Crystallography* 51 (6): 1511–1524.

Nicholson, R.A. and Marangoni, A.G. (2020). Enzymatic glycerolysis converts vegetable oils into structural fats with the potential to replace palm oil in food products. *Nature Food* 1 (11): 684–692.

Nicholson, R.A. and Marangoni, A.G. (2021). Lipase-catalyzed glycerolysis extended to the conversion of a variety of edible oils into structural fats. *Current Research in Food Science* 4: 163–174.

Nikiforidis, C.V. and Scholten, E. (2014). Self-assemblies of lecithin and α-tocopherol as gelators of lipid material. *RSC Advances* 4 (5): 2466–2473.

Nikiforidis, C.V., Gilbert, E.P., and Scholten, E. (2015). Organogel formation via supramolecular assembly of oleic acid and sodium oleate. *RSC Advances* 5 (59): 47466–47475.

Öğütcü, M. and Yilmaz, E. (2014). Oleogels of virgin olive oil with carnauba wax and monoglyceride as spreadable products. *Grasas y Aceites* 65 (3): e040.

Okuro, P.K., Malfatti-Gasperini, A.A., Vicente, A.A., and Cunha, R.L. (2018). Lecithin and phytosterols-based mixtures as hybrid structuring agents in different organic phases. *Food Research International* 111: 168–177.

Patel, A.R. and Dewettinck, K. (2015). Function oleogelation : background. *Food & Function* 7: 20–29.

Pernetti, M., Van Malssen, K.F., Flöter, E., and Bot, A. (2007). Structuring of edible oils by alternatives to crystalline fat. *Current Opinion in Colloid and Interface Science* 12: 221–231.

Pullen, S.A., Booth, N., Olsen, S.R., Day, B., Franceschini, F., Mannicke, D., and Gilbert, E.P. (2014). Design and implementation of a differential scanning calorimeter for the simultaneous measurement of small angle neutron scattering. *Measurement Science and Technology* 25 (5): 055606.

Pullen, S.A., Gilbert, E.P., Olsen, S.R., Lang, E.A., and Doolan, K.R. (2008). An in situ rapid heat-quench cell for small-angle neutron scattering. *Measurement Science and Technology* 19 (6): 8.

Puscas, A., Muresan, V., Socaciu, C., and Muste, S. (2020). Oleogels in food: A review of current and potential applications. *Foods* 9 (1): 1–28.

Rehm, C., De Campo, L., Brûlé, A., Darmann, F., Bartsch, F., and Berry, A.. (2018). Design and performance of the variable-wavelength Bonse-Hart ultra-small-angle neutron scattering diffractometer KOOKABURRA at ANSTO. *Journal of Applied Crystallography* 51 (1): 1–8.

Rogers, M.A., Strober, T., Bot, A., Toro-Vazquez, J.F., Stortz, T., and Marangoni, A.G. (2014). Edible oleogels in molecular gastronomy. *International Journal of Gastronomy and Food Science* 2 (1): 22–31.

Rosen-Kligvasser, J. and Davidovich-Pinhas, M. (2021). The role of hydrogen bonds in TAG derivative-based oleogel structure and properties. *Food Chemistry* 334: 127585.

Sawalha, H., den Adel, R., Venema, P., Bot, A., Flöter, E., and van der Linden, E. (2012). Organogel-emulsions with mixtures of β-sitosterol and γ-oryzanol: Influence of water activity and type of oil phase on gelling capability. *Journal of Agricultural and Food Chemistry* 60 (13): 3462–3470.

Sawalha, H., Venema, P., Bot, A., Flöter, E., den Adel, R., and van der Linden, E . (2015). The phase behavior of γ-oryzanol and β-sitosterol in edible oil. *JAOCS, Journal of the American Oil Chemists' Society* 92 (11–12): 1651–1659.

Sawalha, H., Venema, P., Bot, A., Flöter, E., Lan, Y., and van der Linden, E.. (2021). Effects of oil type on sterol-based organogels and emulsions. *Food Biophysics* 16 (1): 109–118.

Scharfe, M. and Flöter, E. (2020). Oleogelation: from scientific feasibility to applicability in food products. *European Journal of Lipid Science and Technology* 122 (12): 2000213.

Shakeel, A., Farooq, U., Gabriele, D., Marangoni, A.G., and Lupi, F.R. (2021). Bigels and multi-component organogels: an overview from rheological perspective. *Food Hydrocolloids* 111: 106190.

Singh, A., Auzanneau, F.I., and Rogers, M.A. (2017). Advances in edible oleogel technologies – a decade in review. *Food Research International* (March): 97307–317.

Sokolova, A., Whitten, A.E., De Campo, L., Christoforidis, J., Eltobaji, A., Barnes, J., Darmanna, F., and Berry, A. (2019). Performance and characteristics of the BILBY timeof- flight small-angle neutron scattering instrument. *Journal of Applied Crystallography* 52: 1–12.

Trujillo-Ramírez, D., Lobato-Calleros, C., Jaime Vernon-Carter, E., and Alvarez-Ramirez, J. (2019). Cooling rate, sorbitan and glyceryl monostearate gelators elicit different microstructural, viscoelastic and textural properties in chia seed oleogels. *Food Research International* 119: 829–838.

Tsung, K.L., Ilavsky, J., and Padua, G.W. (2020). Formation and characterization of zein-based oleogels. *Journal of Agricultural and Food Chemistry* 68 (46): 13276–13281.

Wood, K., Mata, J.P., Garvey, C.J., Wu, C-M., Hamilton, W.A., Abbeywick, P., Bartlett, D.B., Friedl, B., Baxter, P., Booth, N., Brown, W., Christoforidis, J., Clowes, D., d'Adam, T., Darmann, F., Deura, M., Harrison, S., Hauser, N., Horton, G., Gilbert, E.P. (2018). QUOKKA, the pinhole small-angle neutron scattering instrument at the OPAL Research Reactor, Australia: Design, performance, operation and scientific highlights. *Journal of Applied Crystallography* 51 (2): 294–314.

Yang, D.X., Chen, X.W., and Yang, X.Q. (2018). Phytosterol-based oleogels self-assembled with monoglyceride for controlled volatile release. *Journal of the Science of Food and Agriculture* 98 (2): 582–589.

Yang, S., Zhu, M., Wang, N., Cui, X., Xu, Q., Saleh, A.S.M., Duan, Y., and Xiao, Z. (2018). Influence of oil type on characteristics of β-sitosterol and stearic acid based oleogel. *Food Biophysics* 13 (4): 362–373.

Yang, Z., de Campo, L., Cheng, L., Gilbert, E.P., Mittelbach, R., and Hemar, Y. (2023). Microstructural evolution during acid induced gelation of cow, goat, and sheep milk probed by time-resolved (ultra)-small angle neutron scattering. *Food Hydrocolloids* 137: 108381.

Yang, Z., de Campo, L., Gilberet, E.P., Knott, R., Cheng, L., Storer, B., Lin, X., Luo, L., Patole, S., and Heman, Y. (2022). Effect of NaCl and CaCl2 concentration on the rheological and structural characteristics of thermally-induced quinoa protein gels. *Food Hydrocolloids* 124: 107350.

Yang, Z., Hemar, Y., Hilliou, L., Gilbert, E.P., McGillivray, D.J., Williams, M.A.K., and Chaieb, S. (2016). Nonlinear behavior of gelatin networks reveals a hierarchical structure. *Biomacromolecules* 17 (2): 590–600.

Yu, L., Yakubov, G.E., Gilbert, E.P., Sewell, K., van de Meene, A.M.L., and Stokes, J.R. (2019). Multi-scale assembly of hydrogels formed by highly branched arabinoxylans from plantago ovata seed mucilage studied by USANS/SANS and rheology. *Carbohydrate Polymers* 207: 333–342.

Yu, L., Yakubov, G.E., Martínez-Sanz, M., Gilbert, E.P., and Stokes, J.R. (2018). Rheological and structural properties of complex arabinoxylans from plantago ovata seed mucilage under non-gelled conditions. *Carbohydrate Polymers* (March): 193: 179–188.

Zetzl, A.K. and Marangoni, A.G. (2014). Structured emulsions and edible oleogels as solutions to trans fat. Kodali, D.R. (Ed.), *Trans Fats Replacement Solutions* 215–243. AOCS Press.

3.3

Sensory Evaluation of Fat Reduction in Foods

Patricia Severiano-Pérez[1,], Aurora Pintor-Jardines[2], Mariel Calderón-Oliver[3], and Hector Escalona-Buendía[2]*

[1] *Departamento de Alimentos y Biotecnología, Facultad de Quimica, UNAM, Mexico*
[2] *Departamento de Biotecnología, UAM-Iztapalapa, Mexico*
[3] *Tecnologico de Monterrey, Escuela de Ingeniería y Ciencias, Mexico*
[*] *Corresponding author*

3.3.1 Introduction

Sensory evaluation is a scientific discipline involving developing, using principles, and measuring human responses to products and ingredients from various areas such as food, medicine, household, and personal care items. This evaluation focuses on identifying and/or measuring the features of a product perceived by the senses; notable among them are appearance, color, smell, taste, texture, mouthfeel (such as irritation sensations), and sound. These sensations also involve secondary sensory systems like pain and thermal receptors in lips, and in internal parts of the mouth and nose.

There are several types of sensory tests; the ideal test for a product depends on the goal and the scope desired, since there are tests that allow to determine if the samples differ significantly from each other (discriminative tests) while others allow describing the sensory properties and intensities of them (descriptive tests). There is a third group of tests (affective) that are the most common and that measure subjective attitudes such as preference, acceptability, or qualification. Most of the established sensory methods have been compiled per the American Society for Testing and Materials (ASTM) standards (Wolf 2020) and the International Standard Organization (ISO 2022). Likewise, commercially there are already specialized software with the most common methodologies and that collecting and/or analyzing the data obtained, or even run a statistical software useful to analyze the answers from a test. Several alternatives of software can be used as tools for design and collecting data in sensory tests, such as Fizz (www.biosystemes.com), Compusense (www.compusense.com), EyeQuestion (www.eyequestion.nl) and RedJade (www.redjade.net) and within the statistical software XLSTAT (www.xlstat.com) and the Sensominer Package of R (http://sensominer.free.fr).

In addition to the goals of the diverse methodologies, they also depend on the type and number of people who will perform the test. For the discriminative ones, around 12 to 20 people familiar with the product are needed; the descriptive ones involve 6 to 15 selected and trained people (also called judges); affective ones may require anywhere from 75 to up to over 100 consumers (depending on the size of the targeted population). Thus, another factor inherent in the above is time, since selecting and training judges may take longer than six months, while tests involving consumers could be faster but could also require pre-recruiting enough consumers for a representative sample.

As a response to this timing issue and the demand for innovative sensory tools, rapid tests have been developed that allow quick qualitative or semi-quantitative evaluations that do not involve such a high training time and that describe important parameters or attributes. An example of them is the CATA test, flash profile, among others (Valentin et al. 2012). Using consumers is also feasible for some of these tests, so the evaluation could be complemented with hedonic aspects. A

Fat Mimetics for Food Applications, First Edition. Edited by Miguel Ângelo Parente Ribeiro Cerqueira and Lorenzo Miguel Pastrana Castro.

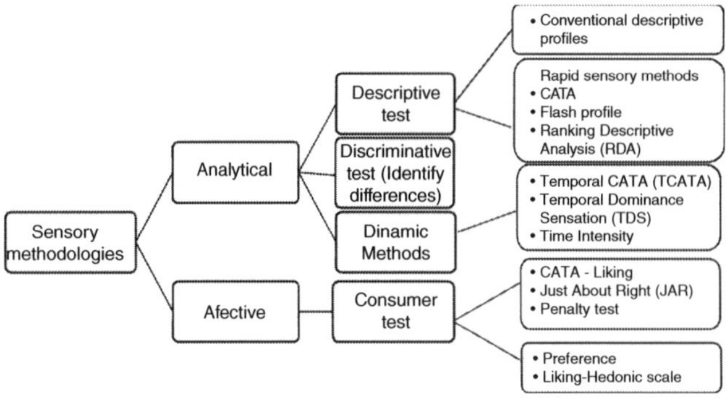

Figure 3.3.1 Sensory analysis methods.

deeper discussion of novel techniques in sensory evaluation and consumer studies is presented by Varela and Ares (2014).

This chapter has been developed as a review of the main information published on the sensory impact of fat reduction on food products, mainly about those studies where there was a strategy for fat replacement by substitutes and mimetics. The main sensory methodologies used to evaluate the sensory effect of fat replacement are described and discussed in this review. Figure 3.3.1 shows a general summary of the sensory methodologies that will be discussed in this review.

The chapter is organized in these sections: First, Generalities on Fat Replacement and Sensory Evaluation where those aspects are presented including their relevance and main fundamentals. The second section deals with the "Effect of Fat Replacement in Food Products", where some impacts of diverse food products are presented and research carried out over the last 5 years is discussed; there is also a table of selected publications from 2010–2016 on the subject of sensory methodologies that formed sensory attributes affected by developing fat reduced products through diverse strategies. Then, the next sections are focused on "Cereal and Baking Products", "Meat Products", and "Dairy Products". It is noteworthy that there is a higher number of publications focused on meat and dairy product alternatives, as they are important sources of fat consumption. Finally, there is a section of "Conclusions and Perspectives" where the main outputs and final remarks of the chapter are discussed.

3.3.2 Generalities on Fat Replacement and Sensory Evaluation

Sensory evaluation and consumer studies are powerful tools that help develop new and reformulated products based on sensory features such as flavor, taste, and aroma, textural appearance, and rheological properties. Acceptability of these products also depends on functionality, the convenience of use, price, and nutritional value. There are plenty of studies where the sensory impact of fat reduction in foods has been explored by hedonic tests, such as the one reported by Santiago-García et al. (2021).

In the last four decades, the role of fat in the diet has changed. Some studies maintain that fat intake is related to increased risk for some cancers and that saturated fat intake is associated with high blood cholesterol and coronary heart diseases. However, the new results indicate that it is the opposite, and the consumption of some fatty acids have health benefits, protecting against cardiovascular disease and type-II diabetes. Also, fat intake has been related to being overweight and to obesity; excessive energy consumption and a sedentary lifestyle are also related to being overweight and to obesity (Astrup et al. 2019; Hashim et al. 2021). Fat is an important energy source; fats have twice the number of calories compared to carbohydrates and proteins. Many investigations have focused on developing low-fat products that contribute to a proper energy balance.

There is a global trend to create new food products that are more nutritious and that contribute to reducing the prevalence of non-transmissible diseases such as diabetes, obesity, and others that belong to metabolic syndrome and cancer. That is why studies are carried out on the impact on purchase intention, sensory evaluation, impact on claims and nutritional information on packaging, among others, to discover how successful this product would be (Hill et al. 2002; Bolha et al. 2021). The evaluation of the sensory properties provides highly relevant information for substituting sugar, fat, and salt; these reformulations commonly affect appearance, flavor, texture, and even shelf-life stability. Sensory perception of fat is multifactorial and is not focused only on the property of either texture or flavor, but on them, oil and fat substitutions can affect one or more sensory parameters (Glumac and Chen 2020).

The most commonly used fat substitutes and mimetics are polysaccharides such as inulin, changed starches, combinations of gums such as guar and xanthan gum, emulsions involving the combination of vegetable oils, water and polysaccharides, pre-emulsions, seed flours, maltodextrins, and fibers (Hadnadev et al. 2014; Zeng et al. 2016; Carcelli et al. 2020; Li et al. 2020). Carbohydrates and proteins are molecules that can to bind great amounts of water and can change the rheology and overcome sensory problems related to fat reduction (Elkhidier and Hamid 2017; Aydinol and Ozcan 2018). A new fat substitutes trend includes oleogels, which are semi-solid systems where the hydrophobic liquid phase, usually an oil or fatty acid, is immobilized in a gel-like, three-dimensional network of solids that shows similar characteristics to solid fats. This process is also known as organogelation or oleogelation (Gutiérrez-Luna et al. 2020; Domínguez et al. 2021).

3.3.3 Effect of Fat Replacement in Food Products

Fat provides multiple functionalities relevant to the sensory properties in many types of foods. For example, in meat products the reduction of fat content or fat substitution implies changes in color, flavor, juiciness, hardness, and stability, while in dairy products, it can destabilize emulsions, present syneresis, and changes in color and nutritional value. In other products, the fat contributes the properties of elasticity, plasticity, spreadability and lubricant as in margarine and butter; and in others, fat has the function of being a carrier of aroma and vitamins, a satiating agent, among other mainly sensory changes that could be a reason for rejection or negative comments towards this product. It is also known that the fat should not be 100% replaced, since it considerably changes sensory features (Zahn et al. 2010); likewise, there is no rule or physicochemical parameter such as viscosity or any other aspect that directly relates to the sensation of fat (taste, smell, and aroma) to make it easier to reduce the fat content to a level of sensory acceptability (Le Calvé et al. 2015).

Even the same fat substitute in a product can affect sensory quality, such as in the case of oleogels. It has been reported that the type of oleogel and the materials used to make it can alter how acceptable it is from a sensory standpoint. Incorporating oils, such as sesame oil, in oleogels gives products a peculiar taste and smell and that can favor or not favor the product's sensory acceptance (Moghtadaei et al. 2018); even incorporating colorful compounds, such as curcumin, in oleogels can affect the acceptability and color score of the final product, likewise with pork burgers (with oleogels of ethyl cellulose and beeswax replacing 100% pork backfat) (Gómez-Estaca et al. 2020).

Table 3.3.1 summarizes selected relevant publications that enabled to understand the sensory impact of either fat substitution or reduction in diverse foods.

Table 3.3.1 Main sensory techniques used in the study of reduced-fat foods.

Food product	Fat modification	Sensory technique	Main findings and characteristics	Reference
Chocolate	Substitution of cocoa butter (0–50%) with β-glucan concentrate	Acceptance test and flash profile	– The higher substitution of fat decreases the acceptance of the product – Increase of undesirable features in chocolate such as bitter taste, non-characteristic chocolate flavor, oat/flour flavor and aftertaste	(Rezende et al. 2015)
Cheddar cheese	Commercial full-fat and 75% reduced-fat cheddar cheese	Descriptive analysis and consumer test	Sensory identification of compounds that contribute to the flavor of Cheddar cheese of complete formulation and 75% reduction in fat. Critical flavors for full-fat cheese are butyric acids, furaneol, methional and 2-methyl butanal and for reduced-fat butyric acid, 1-octene-3-one, dimethyl trisulfide and homofuraneol were most critical for low-fat flavor By adding 12 important compounds and present in the full-fat to the low-fat, it presents better acceptability scores	(Kim et al. 2011)
Lyon-style sausages and liver sausages	Substitution of pork fat with inulin and citrus fiber	Triangle tests according to ISO standards 2004 and Descriptive analysis according to ISO standards 2003, Nine-point hedonic scale	Lyon-style sausage (32–88% of fat substitution) – Significant differences in sensory profile between formulations – Decreased meat flavor, aftertaste meat flavor, greasiness, and juiciness intensities – Enhanced color intensity, spiciness, spicy aftertaste, raspy throat, coarseness, and firmness scores – Preferred sausages featured medium to high intensities in juiciness, greasiness, and meat flavor. – Too dark red, spicy, coarse, and firm, causing a raspy throat were disliked Liver sausages (33–90% of fat substitution) – Significant differences between the pure liver sausage samples in all the attributes (26) except for porous, off-odor, meat flavor, spiciness (taste), saltiness, sourness, and graininess. – Meat flavor, saltiness and spiciness intensities were perceived more intensive with fat reduction – Decreased greasiness, creaminess, lumpiness, and foamy scores – The fat reduced liver sausages (33% fat substitution) reached the high liking scores in all groups.	(Tomaschunas et al. 2013)

Fresh sausage	Substitution of pork fat (0–14%) with corn oil	Panelists were trained according to guidelines of American Meat Science Association (1995).	– Texture was influenced by both treatment types (oil inclusion and fat source). – Oil inclusion reduced texture scores. – Lean color of fresh sausage was decreased with oil inclusion.	(Baer and Dilger 2014)
Frankfurters (raw and cooked)	Substitution of pork back fat (50%) with glasswort hydrate and other non-meat ingredients: carboxyl methyl cellulose; carrageenan; isolated soy protein; and sodium caseinate	Descriptive análisis and overall acceptability	– In color scores, all frankfurters with glasswort hydrate had scores lower than reduced-fat treatment. – Tenderness and juiciness scores with glasswort hydrates were improved when compared to reduced-fat treatment. – Formulations reduced in fat, even with the incorporation of glasswort hydrate, had acceptability lower than the full fat control.	(Lim et al. 2015)
Low-fat beef burger	Substitution of beef fat (12.5–50%) with tomato pomace fiber pectin (TFP)	Triangle test Acceptability	– There were differences in acceptance between the control sample and the beef burgers containing TFP. – The more TFP, the lower the acceptance of samples.	(Namir et al. 2015)
Non-acid fermented sausages (fuet type)	Substitution of fat (35 and 50%) with sunflower oil. In addition, the salt content of the formulations was reduced from 2.5 to 1.5%.	Quantitative descriptive analysis according to the American Society for Testing and Materials [ASTM], 1981; International Organization for Standardization [ISO] 8586-1, 1993; International Organization for Standardization [ISO] 8586-2, 1994	– The simultaneous reduction significantly increased the darkness, hardness, and elasticity and decreased the crumbliness and fat mouthfeel. – Resulted in a in a significant increase in acceptability – The addition of sunflower oil decreased darkness, hardness, elasticity, and chewiness – Sunflower oil increased the brown color and crumbliness to values closer to those of the control batch.	(Mora-Gallego et al. 2016)

(Continued)

Table 3.3.1 (Continued)

Food product	Fat modification	Sensory technique	Main findings and characteristics	Reference
White puddings (oatmeal pudding)	Substitution of added pork fat (50–100%) with one of the next lists: – wheat bran, – sodium citrate, – carrageen, – pectin, – potassium chloride, – potassium chloride and glycine (KClG), – carboxymethylcellulose, – PuraQ Arome NA4, – seaweed wakame (seaweed), – mixture of potassium citrate, potassium phosphate and potassium chloride (KCPCl) – waxy maize starch In addition, the salt content of the formulations was reduced around 32.89%	Ranking descriptive analysis (RDA) using the consensus list of sensory descriptors including grain quantity, fatness, spiciness, saltiness, juiciness, toughness and off-flavor (intensity), which was also measured on a 10-cm line scale. Hedonic scale on a continuous line from 1 to 10 cm	– The white pudding samples formulated with seaweed and KCPCl had lower sensory acceptance for color. – No ultimate replacer combination was found which increased all the three sensory attributes: saltiness, fatness, and spiciness -Most of the added replacers showed an enhancement in at least one taste or flavor attribute. – Formulations with sodium citrate and potassium chloride with glycine have overall acceptance in some levels of addition	(Fellendorf et al. 2016)

Small caliber fermented sausages	Substitution of pork back fat (25, 50 and 75%) with a mixture of collagen, dietary fiber and ice (1:1:8)	Descriptive analysis conducted according to Mora-Gallego et al. (2013)	– As storage time progressed, sensory analysis scores decreased in most treatments despite fat content. – Regarding color, control and reduced-fat sample showed significantly decreased scores after the third and fourth weeks of storage, respectively. – Significant decrease in scores was seen in flavour and off-flavour of control from the third week of storage, not so in the samples with fat substitution. – Control and reduced-fat sample had significantly decreasing scores in overall acceptability as storage time increased.	(Ham et al. 2016)
Low-fat beef burger	Substitution of beef fat (50%) with pineapple byproducts (peel and pomace) and/or canola oil	Quantitative Descriptive Analysis	– Fat reduction decrease the intensity of brown color – Fat reduction increase cohesiveness, however, adding pineapple byproducts positively affected the products, resulting in burgers with similar scores to full fat sample. – Fat reduction decrease juiciness except with the addition of canola oil and pineapple byproducts	(Selani et al. 2016)

3.3.3.1 Cereal and Baking Products

The substitution of fat in products derived from cereals is mainly focused on bread, cookies, and biscuits, where the main substitution ingredients involve mixtures of vegetable oils with polysaccharides such as guar gum, changed starches, and vegetable flours (Curti et al. 2018; Li et al. 2020). The reason for including emulsions or mixtures of these ingredients is to match the functionality of the fat during the lubrication and handling of the dough and to match the final texture of the product; although, most studies substitute fat (mainly butter) to reduce calories, cholesterol content, and trans fats incorporation (Inglett et al. 2004; Bordi et al. 2007; Scheuer et al. 2016). In other cases, replacing fat is focused on partially or totally replacing shortening in these products and reducing saturated fats and trans fats to make a healthier product. Using oleogels in cookies can help with this. Results are good since there are no significant differences in acceptability, smell, flavor, texture, and appearance ($p<0.05$) when incorporating beeswax and sunflower wax oleogels (Yilmaz and Öǫütcü 2015). In this same study, trained panelists found nuanced differences in texture with compactness, bite hardness, fracturability, and meltdown parameters being more intense than in biscuits without oleogels. Another example is the comparison of cakes and their overall acceptability when made with various oleogel bases. Findings reveal that oleogels made with an oil-only base (50% cotton seed oil and 50% high oleic sunflower oil) presented better ratings compared to those with water (18% water with 82% of the 1:1 mixture of cotton seed oil and sunflower oil) (Pehlivanoglu et al. 2018).

Sensory evaluation in most studies where fat substitution is carried out uses overall acceptability in hedonic scales of 9 points (1 dislike extremely and 9 like extremely), it is seen that the increase in fat substitution, reduces its acceptability (Bordi et al. 2007; Curti et al. 2018; Li et al. 2020). Sensory tests, such as liking, serve as an evaluation method in stimulating and exposing the low-salt or low-sugar products as a function of time and seeing if this parameter changes for modified formulations or not, such as in biscuits (Biguzzi et al. 2015).

In addition to the hedonic scale, products such as cookies have been evaluated through the CATA methodology that presents a list of descriptive and/or consumer perception terms that the assessors (mainly consumers) must select all that apply in each product. The study on cookies (Curti et al. 2018) also included selecting those terms expected in an ideal product, to compare with fat replacement formulas. The attributes considered were good, mediocre, excellent, crumbly, hard, soft, rubbery, crunchy, buttery, oily, dark, golden, and pale. The ideal cookie was golden, excellent, and buttery, close to the original product, while the cookie made with only vegetable oil presented oppositional features such as rubbery, hard, and soft.

3.3.3.2 Meat Products

Reducing fat in meat products began in the 1980s as a technological and academic exercise; the focus was mainly on changes produced by the substitution in composition, physicochemical makeup, and texture features. However, in recent years, changes in eating habits have led to increased incidence and prevalence of diseases caused by consuming foods with a high caloric intake, such as diabetes and cardiovascular problems. Because of health problems, academia and industry have evaluated different fat substitutes to reduce caloric content and nutrient intake without affecting the sensory quality of the products with which they are already familiar by the consumers.

The main product fat substitutes were used for in studies were different sausages made with pork, chicken, or mixed meat (Alaei et al. 2018; de Souza Paglarini et al. 2021), and fried chicken meat products (Alaei et al. 2018), mortadella (Auriema et al. 2022), and black pudding (Fellendorf et al. 2017).

Among the studies that used the 9-point hedonic scale, Guzek et al. (2017) evaluated the sensory quality of pork-ostrich sausages ("Krakowska" style), which were reduced in fat content (50% lower than in the typical product without adding other fat substituting ingredients), and contained salt and allergen control during storage, using a 10 cm unstructured hedonic scale anchored at the end with the option "dislike very much" and "like very much". As a result, correlations were found between overall quality and (b+) component of color as well as color acceptability and NaCl content of pork-ostrich allergen-controlled sausages. Correlations between consumer color acceptability and moisture content, consumer sensory saltiness acceptability, protein content, and overall quality and fat content of pork-ostrich allergen-controlled sausages were observed. The taste was rated at over 80% of the maximum rating. Results revealed a higher overall quality and higher color acceptance correlated with lower fat and lower NaCl levels.

Alaei et al. (2018) used inulin as a fat substitute in chicken sausages and found mean sensory scores given to different sausage treatments. The increased level of inulin enhanced the scores for color, appearance, and texture, but decreased scores for smell. In conventional products, fat caused a good smell and mouthfeel to the consumer, which indicated that the reduced-fat was the main reason for the decline in the level of smell and mouthfeel in treatments with high levels of inulin. The sensory evaluation results for the texture of the sausage samples were like those obtained from the texture profile analysis tests. Both tests indicated lower values of textural hardness in treatments with higher inulin contents

Fellendorf et al. (2017) studied the effect of different salt (0.51–2.54%) and fat levels (12.76–46.90%) without other ingredients to replace fat with the physicochemical properties and sensory quality of black pudding without using more additives to produce a highly accepted product, using sensory acceptance and ranking descriptive analysis (RDA). The untrained assessors (n = 25–28, in the age range of 21–60 years) were asked to assess the following attributes on a continuous line scale from 1 to 10 cm: appearance, flavor, texture, color, and overall acceptability (Hedonic). The assessors then participated in an RDA (Richter et al. 2010). They were asked to assess sensory attributes, such as grain quantity, fatness, spiciness, saltiness, juiciness, toughness, and off-flavor (intensity), which were also measured on a continuous 10 cm line scale. The results showed that fat levels did not affect color acceptance. However, assessors preferred black pudding samples with had less yellowness. Therefore, liking product's appearance is strongly correlated to the fat and salt content. The perception of fat, salt, and spices was not affected by fat levels but was affected by salt levels. Varying the salt level in black pudding products affects the sensory acceptance of texture stronger than varying the fat content. The presence of blood in black puddings may constrict the visual appearance, texture, and taste of the grains used in these products. In black pudding products, salt influences the attributes of juiciness and toughness, varying salt, and fat levels in black puddings produced no off-flavors.

De Araujo et al. (2021) evaluated the effects of partial and total substitution by inulin on the quality parameters of chicken sausages. The inulin was used at levels of 25% and 5% and control. A panel of 100 untrained tasters participated using three tests, the 9-point hedonic scale and the purchase intention test using a 5-point scale (1 = certainly would not buy it to 5 = certainly would buy it), and finally, a preference test ranking the samples from the most to the least preferred. In the three tests, the attributes evaluated were color, appearance, texture, and flavor. The three sausage formulations did not present a significant difference ($p<0.05$) for the sensory attributes evaluated. Adding inulin to the formulation let to higher values in appearance, texture, and flavor, which was reflected in the fact that these were more likely to be purchased. Thus, with 50% fat reduction and 100% of fat with 2.5% and 5% of inulin added without reducing the sausage's quality.

In evaluating the chicken mortadella, in which the fat was replaced (25–100%) by green banana biomass (GBB) (Auriema et al. 2022), sensory perceptions during oral processing were evaluated by the TCATA method. Nine sensory attributes were evaluated: typical chicken mortadella flavor, condiment flavor, fat flavor, salty, smoky flavor, succulent, soft, fibrous, and firm. The procedures followed the recommendations of Esmerino et al. (2017) and Silva et al. (2020). One hundred regular consumers of chicken mortadella participated. Participants were asked to simultaneously click the start button as they placed the sample in their mouths. Then they were instructed to select all the attributes they considered applicable to describe the sensations in each sample during the 60-second evaluation time. This method allows for simultaneously choosing several sensory descriptors and completely describes the sensory features of the product. The time was determined in a preliminary test by the research group to make sure all the perceptions were seen during the trial. It was unnecessary to deselect the terms due to the use of an 8.0 s fade time (Esmerino et al. 2017). After the TCATA test, participants rated the samples for overall impression using a 9-point unstructured hedonic scale. A penalty analysis was performed on TCATA data and mean values of the acceptance test to identify the sensory descriptors that drive the samples' acceptance. At the beginning of evaluating TCATA, the texture was the main parameter perceived by the evaluators in each sample, with firm being the most cited attribute in the samples F0 (control treatment formulated with 100% chicken skin), F1, and F3, while in samples F2 and F4 (25, 50, 75, and 100% of chicken skin replacement by GBB, respectively), the attribute of "soft" was the mentioned most. Another attribute that stood out in F0, compared to the other samples was a fibrous texture. Then, the typical flavor of chicken mortadella was the most noticed attribute in all samples being mentioned throughout the evaluation period. More than 50% of GBB in the mortadella contributed to more indications of succulence. This is because of the increased ($p<0.05$) capacity to keep water in mortadella. In this study, the increasing substitution of fat for GBB produced a juicier mortadella. Regarding the flavor attributes: smoked, spicy, and salty, the citation curves were intermediate for all samples. The fat flavor, which is a typical characteristic of this product due to the high-fat content in its composition, was the attribute least noticed by participants in all samples during the evaluation. Therefore, replacing fat with GBB in chicken mortadella was an excellent strategy without affecting the mortadella flavor. According to the way obtained from the acceptance test, no significant differences ($p>0.05$) were seen between mortadella added with GBB and the control. In this context, this suggests that replacing up to 100% chicken skin with GBB in mortadella did not affect consumer acceptance. According to the results, increasingly replacing chicken skin by GBB in chicken mortadella did not affect did not affect how much consumers liked the attributes of fat and flavor in chicken mortadella. However, chicken mortadella that had 50% and 100% substitution were the least liked by the panelists in the areas of juiciness and softness. Through the association of TCATA with the acceptance test, it was possible to determine what made the product likable and to reformulate the mortadella to partially replace or fully replace the fat content. From TCATA, some observed that the main chicken attributes were firmness and flavor. The attributes that most positively stood out were fat flavor and fibrous texture, while the absence of juiciness and softness were considered drivers of disliking attributes. Although minimal juiciness and softness were attributed to mortadella with 100% chicken replacement by GBB, acceptance was not affected. Figure 3.3.2 shows an example of a TCATA procedure at three times of testing.

de Souza Paglarini et al. (2021) applied CATA and acceptance test to Bologna sausage with reduced sodium and fat-containing functional emulsion gels (4% soy protein, 50% soybean oil, 16.5% inulin and 29.5water). Sensorial properties (day 10 of storage at 4°C) were evaluated for PF20/NS (pork back fat 20%/normal salt), EG20/NS (emulsion gel 20%/normal salt 2%), EG20/RS (emulsion gel 20%/reduced-salt 1%), PF10-EG10/NS (pork back fat 10%/ emulsion gel 10%/ normal salt 2%) and PF10-EG10/RS (pork back fat 10/emulsion gel 10%/reduced-salt 1%) treatments. The sensory attributes of appearance (pale color, and ideal color), flavor (spicy, oil, little spicy, juicy,

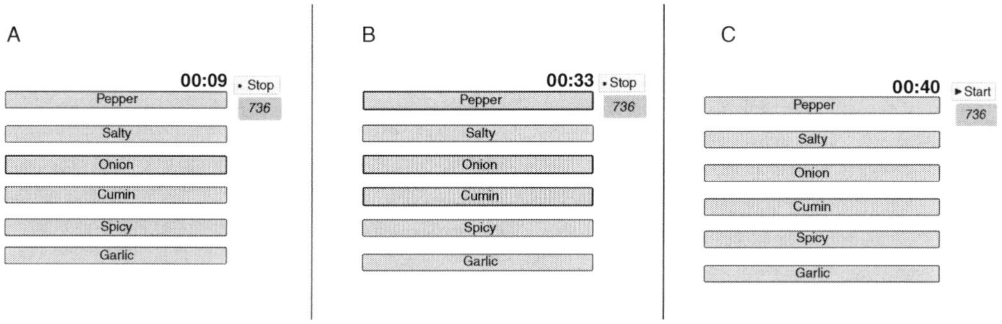

Figure 3.3.2 Example of TCATA evaluation screen. The highlighted button indicates the choice status and the terms regarded as non- applicable to describe the focal sample at this time are not selected.

mild, greasy, Bologna, salty, low-salt, and salt in the right amount), texture (spongy, gritty, firm, rubbery, and soft), and aroma (Bologna, and mild), were used to describe the sensory profile of the Bolognas. The results of the acceptance test showed that sensorial properties of reformulated Bologna sausages were judged acceptable (scores higher than 5). Except for color, no statistical difference ($p>0.05$) was found between control (PF20/NS) and Bolognas with emulsion gel EG + pork back fat PF with no salt reduction (PF10-EG10/NS) for texture flavor, aroma, and overall liking. Samples with simultaneously salt and fat reduction and adding EG (EG20/RS and PF10-EG10/RS) scored lowest when compared to other samples. Regarding buy intention, samples presented 3.7, 3.0, 2.5, 3.6, and 2.90 scores for PF20/NS, EG20/NS, EG20/RS, PF10-EG10/NS, and PF10-EG10/RS, respectively. For CATA results, the non-parametric Cochran's Q test was performed, and it shows there were significant differences for 16 of 19 descriptors used to characterize the samples. Descriptors with less than 10% of frequency mention were also removed. By correspondence analysis, three groups of samples with different sensory features were perceived. The first group PF20/NS and PF10-EG10/NS were close to the descriptors of the firm, juicy, ideal color, the right salt, Bologna flavor, Bologna aroma, and spicy flavor. The second group PF10-EG10/RS and EG20/RS were linked to the mild aroma, low-salt, little spicy, mild flavor, and pale color descriptors.

Paglarini et al. (2022) used inulin as a fat replacer and fiber carrier in healthier Bologna sausage and evaluated the influence of a gelled emulsion made from carrageenan, inulin, soy protein isolated, and soybean oil (10% fat reduction). The sensory evaluation tests carried out were acceptance tests and the attributes evaluated were aroma, color, appearance, flavor, texture, and overall liking, and CATA questions with 24 total descriptors. The descriptors evaluated were appearance (bright, opaque, dry, pale color, and ideal color), flavor (residual taste, seasoning, oil, little spicy, juicy, greasy, dry, salty, Bologna, and salt in the right amount), texture (spongy, firm, rubbery, and soft), and aroma (Bologna, mild, seasoning, meat, oil); they were used to characterize the sensory profile of the Bolognas. The consumer test results show that overall and excluding color grades, based on a hedonic scale of 9 points, assessors scored the samples from 6–7, indicating that the Bolognas were judged as liked slightly to moderately. Regarding flavor, texture, aroma, and overall liking, the control samples were more accepted than the reformulated samples ($p<0.05$). In general, all samples were considered acceptable by the consumers (scores higher than 5). On the other hand, the results of the CATA test show that color attributes were not affected by the reformulation. In general, these samples were associated with these descriptors: juciness, salt quantity, soft, spicy aroma, Bologna aroma, and Bologna flavor. The second group was formed by the reformulated samples with the addition of high-fiber emulsion gel (EG) that is associated with these descriptors: rubbery, little spicy, dry appearance, opaque appearance, firm, and aftertaste. Samples with EG were more associated with the rubbery descriptor. The reduction in the perception of flavor and aroma compounds in the reformulated samples can be attributed to the EG addition. In general, the reformulated Bologna sausages presented good sensorial stability during storage with no perception of oxidative attributes by the consumers.

The studies using oleogels in meat products have mostly used sensory analysis based on the consumers' general acceptance and perception parameters. Oleogels can change a product's sensory parameters and acceptability, as has been reported in pork burgers where ethyl cellulose and beeswax oleogels significantly change flavor and acceptability ($p<0.05$) (Gómez-Estaca et al. 2019) and the presence of ethyl cellulose oleogels with or without adipic acid in beef burgers reduced by 50% fat, reduce acceptability, texture, taste, and odor (Adili et al. 2020). Although in other products, such as fermented sausages, oleogels based on sterols and beeswax increase acceptability when they replace 40% of fat (Franco et al. 2020).

Using oleogels based on monoglycerides and phytosterols (15:5) to replace 50% of animal fat in frankfurters sausages were sensorily rated as juicier without affecting other parameters such as appearance, taste, and mouthfeel, in addition to overall acceptance (Kouzounis et al. 2017). A similar situation is seen in 50% and 100% beef tallow substitutions with canola oil oleogels in hydroxypropyl methylcellulose in meat patties where the samples were jucier and more tender when compared to the control; they presented even better sensory acceptability ($p<0.05$) (Oh et al. 2019). But it has also been reported that including canola oil oleogels in frankfurters with various levels of fat substitution (20–80%) did not present changes in sensory hardness, but there was more intensely the oiliness parameter (Barbut et al. 2016).

Additionally, using oleogels (10% of beeswax with sesame oil) to replace 50% of animal fat in beef burgers does not present sensory differences in texture and color; trained panelists rated the flavor of beef burgers significantly better (scores = 1.69 oleogel vs 0.19 control on a 7-point scale from −3 to 3) ($p<0.05$) (Moghtadaei et al. 2018).

In the case of incorporating canola oil oleogels in pâté, the materials and their matrix curiously do not influence the main sensory properties between the products with oleogel and the control when evaluated with consumers, since it does not change the juiciness, cohesiveness, hardness and oiliness significantly ($p<0.05$), however, if they change the perceived off-flavors such as rancid, earthy, chemical and grassy, being the formulation of 14% ethyl cellulose with 1.5% glycerol monostearate that substitutes the 100% added fat (Barbut et al. 2019). However, when evaluated with a trained panel, there is a difference in the parameters of cohesiveness, hardness, juiciness, and oiliness in various percentages of fat substitution and oleogels (0–100%) (Barbut et al. 2021).

3.3.3.3 Dairy Products

In dairy products, fat has an important role in texture, appearance, flavor, and color. Reducing or substituting fat leads to sensory and texture defects such as lack of flavor, weak body, and undesirable textures. Fat mimetic or replacers have been used to simulate the functionality of fat and to reduce the caloric content.

Labneh cheese was formulated with low-fat (6%) and fat reduced (12%) inulin and β-glucan. Ten panelists familiar with the sample evaluated attributes of appearance (syneresis, stickiness, shininess); texture (spreadability, lupiness, compactness); color, odor, and flavor (milky, fermented/acetaldehyde); taste (sweet, bitter, sour/fermented, milky, salty), and overall acceptance on a 5-point hedonic scale (from 1 = strongly disliked to 5 = strongly liked). The panelists found significant differences ($p<0.01$) in attributes such as appearance, texture, flavor, taste, and overall acceptability among cheese with inulin and β-glucan. The attributes odor and color were not significantly different ($p<0.01$). The panelists gave the highest sensory scores, on average, to the treatment of full-fat (18% fat), reduced-fat (12% fat), cheese with inulin and 12% fat and the control with 18% fat (Aydinol and Ozcan 2018).

Oligosaccharides have been used to mimic fat in dairy products such as yogurt. Santiago-García et al. (2021) used agavins from *Agave angustifolia* and *Agave potatorum* to replace fat in yogurt. Sensory evaluation was carried out with 42 untrained consumers, which were frequent consumers of fermented milk. The consumers were asked to rate appearance, odor, texture, flavor, and overall acceptability using a 9-points hedonic scale. The consumer found no significant difference ($p<0.05$) in global acceptance between low-fat yogurt with *Agave angustifolia* and *potatorum*. Yogurt with both agavins had higher liking values than full-fat. Color, smell, and taste did not impact agavins, but their use improved mouthfeel and texture.

Rodriguez Furlán and Campderrós (2017) evaluated the effect of stevia and sucralose as a sugar substitute and inulin as a fat mimetic on the physicochemical properties of starch-based model dairy dessert. Different combinations of sucrose, sucralose, stevia, and inulin were developed. Fifty-two regular consumers were recruited to evaluate the overall acceptability and attributes such as flavor, color, and texture using a 5-point unipolar scale of liking. Thirteen semi-trained panelists were asked to choose the best sample based on sensory and physicochemical analysis against a commercial sample. They found the blends of sucrose and stevia (50 + 50%) and stevia and sucralose (50 + 50%) showed the highest scores for acceptability ($p<0.05$), which even scored above the control (sucrose 100%). The formulation selected to be evaluated against commercial samples was 50% stevia + 50% sucralose + 2.5% inulin. Panelists did not find significant difference in flavor and color ($p<0.05$). This last sample was more acceptable than a commercial sample.

Chetachukwu et al. (2018) reduced fat with a short-chain inulin, coconut milk, yogurt set. About 20 yogurt consumers with knowledge in dairy generated the attributes list, and they were trained in three sessions. Then, panelists assessed the overall liking using a 9-point hedonic scale (9 = like extremely and 1 = dislike extremely). Per the results of the principal components analysis, using 10 to 15% of inulin in low fat yogurts provoked astringent, coconut, and soapy odor in addition to a slow melt down rate. Adding 15% of short-chain inulin had a significant positive effect ($p<0.05$) on attributes such as viscosity, flavor, and taste but did not improve the odor and color. The sweet attribute related to short-chain inulin level. The overall acceptance was higher for samples with 20% of short-chain inulin.

A non-fat yogurt was produced using whey protein isolate (WPI) to replace fat, and the final quality was studied using the sense-perception. Twelve trained panelists were recruited and attributes such as appearance (30), flavor (10), and texture (60) were evaluated by a sensory panel to give an overall score of 100 points. Based on hedonics results, significant differences were seen between samples that have WPI and full-fat control and non-fat control. Adding 3 and 5% of WPI improved the body, texture, and flavor of yogurts compared to control. Samples with full fat show an organoleptic acceptance higher than the rest of the samples ($p<0.05$). Low-fat yogurt had the highest score of appearance and taste but panelists considered those samples with protein isolate to have the best texture (Hashim et al. 2021).

Descriptive analysis is a traditional method that allows different foods to be adequately characterized, and its information is helpful during the development of new or reformulated products. Integrating descriptive methodologies and liking-hedonic tests is a powerful tool to detect those attributes highly relevant to increase or maintain consumer preference.

Gomez-Betancur et al. (2020) studied different sensory attributes in a low-fat yogurt mousse using long-chain inulin as fat replacer. The samples were codified and presented at 10º C to be evaluated by some trained panelists. They started a descriptive analysis using a 5-point scale to evaluate aerated appearance, sweet, dairy, and dairy ferment odors. Creaminess, smoothness, firmness, aerated, fatty, and chalky texture were studied. In this sensory study, the fat level affected the texture (except for firmness and chalky texture) of low-fat-mousse yogurt ($p<0.05$). The interaction of fat-inulin had a negative effect ($p<0.05$) on creaminess and smoothness while fat-calcium interaction increases the effect of these attributes.

Incorporating fiber in yogurt has also been studied. Different low-fat, full-fat, and fat-free yogurts fortified with fiber-rich persimmon and apple powders were developed and stored at $4 \pm 1^{o}C$ for 14 days. About seven expert panelists carried out a descriptive analysis to evaluate appearance, flavor, and consistency using a 5-point scale. The panelists found that samples with apple fiber and some with any fiber had better appearance value compared to those with persimmon fiber ($p<0.05$). Neither fat level, fiber type, or the interaction affected consistency ($p<0.05$). Adding apple fiber caused the lowest flavor values. Most of the sensory changes were seen in one to seven days (Karaca et al. 2019).

Low-fat Mozzarella cheeses were formulated using three samples at 6% of fat mimetics with different commercial protein levels (35% Dairy-Lo, 53% Simplesse, and 90% LSI907). The panelists were asked to analyze color using a 9-point ordinal scale (from dark to light). Chewiness was evaluated with a 3-point ordinal scale. Overall flavor preference was evaluated on a 7-point hedonic scale (from 1 = dislike extremely to 7 = like extremely). As a result, the highest values of color and flavor was reported for full-fat samples and low-fat cheese with 35% of protein (respectively). Chewiness unfavorable to Mozzarella cheese was significantly different ($p<0.05$) in low-fat samples with the highest values. Overall acceptance from consumers was better for full-fat cheese than in fat reduced, probably because reduced-fat cheese lacks texture and flavor (Tahereh et al. 2017).

Agave fructans (0–3.0%) was used to investigate the feasibility of replacing fat and sugar to produce both low-fat, and low-fat and low sugar ice cream. A CATA questionnaire was to better understand which sensory descriptors defined the samples and which were responsible for the hedonic response of consumers. The attribute list was selected from earlier studies, but the participants were new. About 30 attributes were generated based on appearance, flavor, odor, and texture. The evaluation was carried out in two sessions. A CATA questionnaire was evaluated by panelists selecting the attributes that describe each sample. Also, 70 consumers were asked to rate the overall liking on a 9-point hedonic scale. In low-fat set ice cream with lower *Agave fructans* concentration, the samples were correlated with a metallic and bitter flavor, fresh milk odor, crystallized (appearance and texture) grainy, sandy rough surface, cold, and hard texture. The low-fat samples with the highest *Agave fructans* concentration were close to the control (full-fat ice cream). Those were characterized by fluid, long melting time, frothy texture, and ease to scoop. The samples with a concentration of *Agave fructans* above 1.8% had the highest overall linking values. In low-fat and low sugar ice cream, the sample with the highest *Agave fructans* content were described as metallic, sour, bitter, powder milk flavors, mouth coating, dense, grainy, sandy, gummy, fatty, and creamy appearance. The acceptance was higher for samples with all the ingredients, so *Agave fructans* cannot compensate simultaneously for both (Pintor-Jardines et al. 2020). Figure 3.3.3 shows an example of a CATA questionnaire with an overall liking question.

RDA is an alternative fast descriptive method used in Mozzarella style cheeses. Both unheated ($4^{o}C$) and heated ($95^{o}C$), Mozzarella-style cheeses were developed by varying fat (23 to 11 w/w), salt (1.8 to 1.0 w/w) and calcium levels. Twelve Mozzarella cheeses at 15 and 35 days of aging were evaluated. Ten panelists evaluated the color of cheeses with a 9-ordinal type rating scale from dark to light; for color, they expressed their overall liking on a 7-point hedonic scale and for chewiness and creaminess an ordinal scale was used. The overall acceptance was assessed by 30 untreated consumers using a 7-point hedonic scale. Then, it was carried out RDA and samples were ordered by increasing

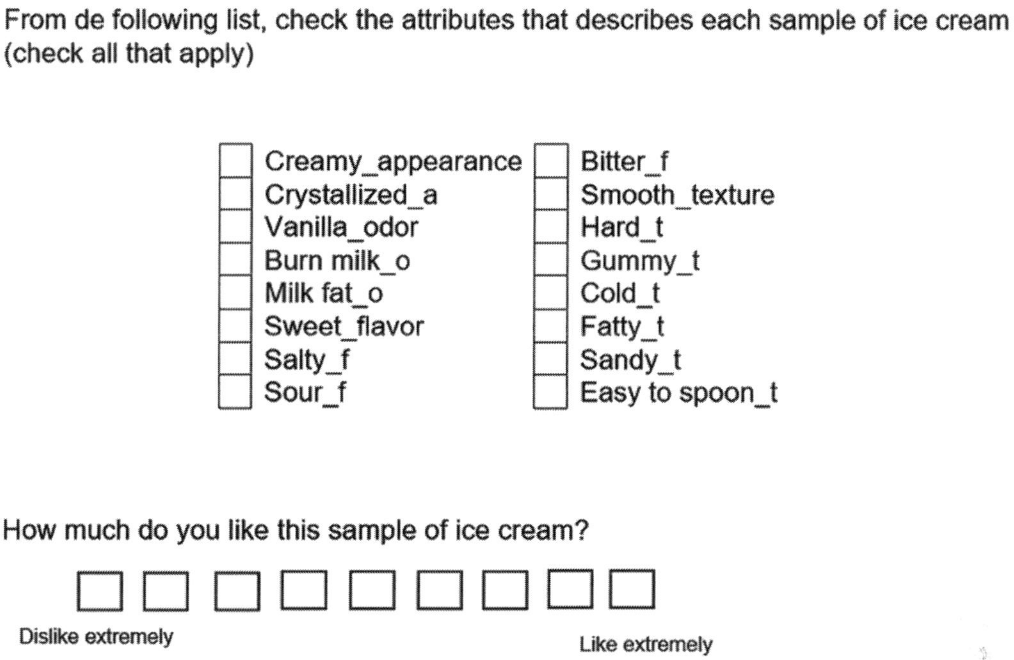

Figure 3.3.3 Example of CATA (Check-All-That-Apply) questionnaire with liking test.

intensity for each attribute. Unheated full-fat (FF), full-fat and full-salt (FFFS), full-fat, reduced-salt at 15 days of aging (FFRS15), and FFFS and FFRS at 35 days (FFRS35) of aging were positively ($p<0.05$) correlated to hedonic attributes, overall acceptance, and liking of texture and flavor. FFRS35 cheese was characterized by firmness, rubberiness, and chewiness. Reduced-fat (RF), reduced-fat and reduced-salt (RFRS15), reduced-fat and full salt at 15 days of aging, and reduced fat and full salt at 35 days of aging were negatively correlated with liking of texture and flavor and overall acceptability. In general, all reduce-fat cheeses were positively correlated with firmness, rubberiness, and chewiness. The addition of calcium showed no effect on hedonic sensory attributes. On the other hand, the different formulations at 95°C (FFRS15, FFR35, FFFS15, and FFFS35) were positive ($p<0.05$) correlated with overall acceptance, liking of texture, liking of flavor, color, oiling off, fluidity, succulence, cheese aroma intensity, fat flavor, cheese flavor intensity, and cooked cheese flavor. The reduction of calcium in reduced-fat cheese improved the sensory acceptability (Henneberry et al. 2016).

3.3.4 Conclusion and Final Considerations

It is well known that fat provides relevant flavors and textures in foods. Its reduction using substitutes and mimetics must compensate structural changes and aroma compounds lost as they come along with the fat. Liking tests, including hedonic scaling, has been widely used to evaluate the effect of fat reduction being the most used methods reported in the investigations discussed in this review. Even though it delivers very important information, it does not provide feedback of which were either affected or improved by using substitutes and mimetics when it is the only strategy to evaluate the sensory impact. Conventional descriptive profiling has been a powerful tool to quantitatively describe the attributes affected and need to be compensated using diverse alternatives for fat replacement. Recently, many studies were focused on dairy and meat products, and selected studies have incorporated alternative profiling methods such as RDA or Flash Profile, methods that are less time consuming for the judges training. A CATA test has also been successfully carried out with consumers and, when combined with overall liking, is also an excellent alternative as it informs estimate those attributes relevant for either liking or disliking, detecting direction to move toward product optimization.

It is also known that the sensory attributes of solid and semi-solid foods are evolving during mastication, and fat also impacts these dynamics because of its significant participation in the food's structure. Therefore, as structure is changed and broken, fat is released like aroma compounds. In this regard, dynamic methods such as TCATA have been used to understand how the fat reductions and incorporating substitutes are affecting perception during ingestion and mastication. Again, TCATA methodology has been used with consumers, also enabling estimates of sensory attributes that are drivers of liking.

For future research, recent nutrition and food product development trends have moved toward the consumption of healthier fat for healthier food. Research on sensory impact of fat reduction stays relevant, but now the main goal we expect to be the sensory and consumer studies to develop these healthier foods with consumer acceptance. For that, consider an integral sensory impact study considering profiling, dynamic evaluation, and consumer perception, always including the liking-hedonic tests. This integrated approach will provide better feedback for both ingredient development and food product innovation.

References

Adili, L., Roufegarinejad, L., Tabibiazar, M., Hamishehkar, H., and Alizadeh, A. (2020). Development and characterization of reinforced ethyl cellulose based oleogel with adipic acid: Its application in cake and beef burger. *LWT* 126. doi: 10.1016/j.lwt.2020.109277.

Alaei, F., Hojjatoleslamy, M., and Hashemi Dehkordi, S.M. (2018). The effect of inulin as a fat substitute on the physicochemical and sensory properties of chicken sausages. *Food Science & Nutrition* 6: 512–519. doi: 10.1002/fsn3.585.

American Meat Science Association (1995). *American Meat Science Association Research Guidelines for Cookery, Sensory Evaluation, and Instrumental Tenderness Measurements on Fresh Meat.* Chicago, IL: National Livestock Meat Board.

American Society for Testing and Materials (ASTM) (1981). *Guidelines for the selection and trainingfor sensory panel members. Special Technical Publication 758.* Philadelphia: American Society for Testing and Materials.

Astrup, A., Geiker, N.R.W., and Magkos, F. (2019). Effects of full-fat and fermented dairy products on cardiometabolic disease: Food is more than the sum of its parts. *Advances in Nutrition* 10: 924S–930. doi: 10.1093/advances/nmz069.

Auriema, B.E., Correa, F.J., Silva, R., Soares, P.T.S., Lima, A.L., Vidal, V.A.S., Raices, R.S.L., Pollonio, M.A.R., Luchese, R.H., Esmerino, E.A., and Mathias, S.P. (2022). Fat replacement by green banana biomass: Impact on the technological, nutritional and dynamic sensory profiling of chicken mortadella. *Food Research International* 152. doi: 10.1016/j.foodres.2021.110890.

Aydinol, P. and Ozcan, T. (2018). Production of reduced-fat Labneh cheese with inulin and β-glucan fibre-based fat replacer. *International Journal of Dairy Technology* 71: 362–371. doi: 10.1111/1471-0307.12456.

Baer, A.A. and Dilger, A.C. (2014). Effect of fat quality on sausage processing, texture, and sensory characteristics. *Meat Science* 96: 1242–1249. doi: 10.1016/j.meatsci.2013.11.001.

Barbut, S., Marangoni, A.G., Thode, U., and Tiensa, B.E. (2019). Using canola oil organogels as fat replacement in liver pâté. *Journal of Food Science* 84: 2646–2651. doi: 10.1111/1750-3841.14753.

Barbut, S., Tiensa, B.E., and Marangoni, A.G. (2021). Partial fat replacement in liver pâté using canola oil organogel. *LWT* 139. doi: 10.1016/j.lwt.2020.110428.

Barbut, S., Wood, J., and Marangoni, A.G. (2016). Effects of organogel hardness and formulation on acceptance of frankfurters. *Journal of Food Science* 81: C2183–C2188. doi: 10.1111/1750-3841.13409.

Biguzzi, C., Lange, C., and Schlich, P. (2015). Effect of sensory exposure on liking for fat- or sugar-reduced biscuits. *Appetite* 95: 317–323. doi: 10.1016/j.appet.2015.07.001.

Bolha, A., Blaznik, U., and Korošec, M. (2021). Influence of intrinsic and extrinsic food attributes on consumers' acceptance of reformulated food products: a systematic review. *Slovenian Journal of Public Health* 60: 72–78. doi: 10.2478/sjph-2021-0011.

Bordi, P.L., Jr., Rager, M.D., Hack, D.M., and Hessert, S.W., Jr. (2007). Development and sensory evaluation of a trans fat-free sugar cookie. *Journal of Foodservice* 18: 167–175. doi: 10.1111/j.1745-4506.2007.00063.x.

Carcelli, A., Crisafulli, G., Carini, E., and Vittadini, E. (2020). Can a physically modified corn flour be used as fat replacer in a mayonnaise? *European Food Research and Technology* 246: 2493–2503. doi: 10.1007/s00217-020-03592-y.

Chetachukwu, A.S., Thongraung, C., and Yupanqui, C.T. (2018). Effect of short chain inulin on the rheological and sensory characteristics of reduced fat set coconut milk yoghurt. *Journal of Texture Studies* 49: 434–447. doi: 10.1111/JTXS.12351.

Curti, E., Federici, E., Diantom, A., Carini, E., Pizzigalli, E., Symon, V.W., Pellegrini, N., and Vittadini, E. (2018). Structured emulsions as butter substitutes: effects on physicochemical and sensory attributes of shortbread cookies. *Journal of the Science of Food and Agriculture* 98: 3836–3842. doi: 10.1002/jsfa.8899.

De Araujo, C.D.L., Da Costa, G.F., De Oliveira, F.L.N., and Azerêdo, G.A. (2021). Elaboration of chicken sausages with fat reduction and inulin addition [Elaboração de salsichas de frango com redução de gordura e adição de inulina]. *Brazilian Journal of Food Technology* 24. doi: 10.1590/1981-6723.33419.

de Souza Paglarini, C., Vidal, V.A.S., Ribeiro, W., Ribeiro, A.P.B., Bernardinelli, O.D., Herrero, A.M., Ruiz-Capillas, C., Sabadini, E., and Pollonio, M.A.R. (2021). Using inulin-based emulsion gels as fat substitute in salt reduced Bologna sausage. *Journal of the Science of Food and Agriculture* 101: 505–517. doi: 10.1002/jsfa.10659.

Domínguez, R., Bohrer, B., Munekata, P.E.S., Pateiro, M., and Lorenzo, J.M. (2021). Recent discoveries in the field of lipid bio-based ingredients for meat processing. *Molecules* 26. doi: 10.3390/MOLECULES26010190.

Elkhidier, I. and Hamid, O. (2017). Effect of different levels of fat on the chemical composition and sensory characteristics of Sudanese low-fat cheese during storage. *IOSR Journal of Agriculture and Veterinary Science* 10: 11–15. doi: 10.9790/2380-1002011115.

Esmerino, E.A., Castura, J.C., Ferraz, J.P., Filho, E.R.T., Silva, R., Cruz, A.G., Freitas, M.Q., and Bolini, H.M.A. (2017). Dynamic profiling of different ready-to-drink fermented dairy products: a comparative study using Temporal Check-All-That-Apply (TCATA), Temporal Dominance of Sensations (TDS) and Progressive Profile (PP). *Food Research International* 101: 249–258. doi: 10.1016/j.foodres.2017.09.012.

Fellendorf, S., O'Sullivan, M.G., and Kerry, J.P. (2016). Effect of using ingredient replacers on the physicochemical properties and sensory quality of low-salt and low-fat white puddings. *European Food Research and Technology* 242: 2105–2118. doi: 10.1007/s00217-016-2707-z.

Fellendorf, S., O'Sullivan, M.G., and Kerry, J.P. (2017). Effect of different salt and fat levels on the physicochemical properties and sensory quality of black pudding. *Food Science & Nutrition* 5: 273–284. doi: 10.1002/fsn3.390.

Franco, D., Martins, A.J., López-Pedrouso, M., Cerqueira, M.A., Purriños, L., Pastrana, L.M., Vicente, A.A., Zapata, C., and Lorenzo, J.M. (2020). Evaluation of linseed oil oleogels to partially replace pork backfat in fermented sausages. *Journal of the Science of Food and Agriculture* 100: 218–224. doi: 10.1002/jsfa.10025.

Glumac, M. and Chen, J. (2020). Contribution analysis of sensory cues to oil/fat perception. *Journal of Sensory Studies* 35. doi: 10.1111/joss.12566.

Gomez-Betancur, A.M., Carmona-Tamayo, R., Martínez-Álvarez, O.L., Casanova-Yepes, H., and Torres-Oquendo, J.D. (2020). Effect of fat substitution using long-chain inulin and fortification with microencapsulated calcium in the rheological and sensory properties of yogurt mousse. *Journal of Food Process Engineering* e13433. doi: 10.1111/JFPE.13433.

Gómez-Estaca, J., Pintado, T., Jiménez-Colmenero, F., and Cofrades, S. (2019). Assessment of a healthy oil combination structured in ethyl cellulose and beeswax oleogels as animal fat replacers in low-fat, PUFA-enriched pork burgers. *Food and Bioprocess Technology* 12: 1068–1081. doi: 10.1007/s11947-019-02281-3.

Gómez-Estaca, J., Pintado, T., Jiménez-Colmenero, F., and Cofrades, S. (2020). The effect of household storage and cooking practices on quality attributes of pork burgers formulated with PUFA- and curcumin-loaded oleogels as healthy fat substitutes. *LWT* 119. doi: 10.1016/j.lwt.2019.108909.

Gutiérrez-Luna, K., Astiasarán, I., and Ansorena, D. (2020). Gels as fat replacers in bakery products: a review. *Critical Reviews in Food Science and Nutrition*. doi: 10.1080/10408398.2020.1869693.

Guzek, D., Głąbska, D., Brodowska, M.,, Godziszewska, J., Górska-Horczyczak, E., Pogorzelska, E., Wojtasik-Kalinowska, I., and Wierzbicka, A. (2017). The sensory quality of allergen-controlled, fat-reduced, salt-reduced pork-ostrich sausages during storage. *Journal of the Science of Food and Agriculture* 97: 5327–5334. doi: 10.1002/jsfa.8421.

Hadnadev, M., Dapčević Hadnadev, T., Dokić, L.,, Pajin, B., Torbica, A., Šarić, L., and Ikonić, P. (2014). Physical and sensory aspects of maltodextrin gel addition used as fat replacers in confectionery filling systems. *LWT - Food Science and Technology* 59: 495–503. doi: 10.1016/j.lwt.2014.04.044.

Ham, Y.K., Hwang, K.E., Kim, H.W.,, Song, D.H., Kim, Y.J., Choi, Y.S., and Kim, C.J. (2016). Effects of fat replacement with a mixture of collagen and dietary fibre on small calibre fermented sausages. *International Journal of Food Science & Technology* 51: 96–104. doi: 10.1111/ijfs.12960.

Hashim, M.A., Nadtochii, L.A., Muradova, M.B.,, Proskura, A.V., Alsaleem, K.A., and Hammam, A.R.A. (2021). Non-fat yogurt fortified with whey protein isolate: Physicochemical, rheological, and microstructural properties. *Foods* 10. doi: 10.3390/foods10081762.

Henneberry, S., O'Sullivan, M.G., Kilcawley, K.N.,, Kelly, P.M., Wilkinson, M.G., and Guinee, T.P. (2016). Sensory quality of unheated and heated Mozzarella-style cheeses with different fat, salt and calcium levels. *International Journal of Dairy Technology* 69: 38–50. doi: 10.1111/1471-0307.12300.

Hill, D.S., Knox, B., Hamilton, J., Parr, H., and Stringer, M. (2002). Reduced-fat foods: The shopper's viewpoint. *International Journal of Consumer Studies* 26: 44–57. doi: 10.1046/j.1470-6431.2002.00198.x.

Inglett, G.E., Carriere, C.J., Maneepun, S., and Tungtrakul, P. (2004). A soluble fibre gel produced from rice bran and barley flour as a fat replacer in Asian foods. *International Journal of Food Science & Technology* 39: 1–10. doi: 10.1111/j.1365-2621.2004.00739.x.

ISO (2022). *Sensory analysis*. 67.240 [WWW Document]. iso.org/ics/67.240/x.

Karaca, O.B., Saydam, İ.B., and Güven, M. (2019). Physical, chemical, and sensory attributes of low-fat, full-fat, and fat-free probiotic set yogurts fortified with fiber-rich persimmon and apple powders. *Journal of Food Processing and Preservation* 43: e13926. doi: 10.1111/JFPP.13926.

Kim, M.K., Drake, S.L., and Drake, M.A. (2011). Evaluation of key flavor compounds in reduced- and full-fat cheddar cheeses using sensory studies on model systems. *Journal of Sensory Studies* 26: 278–290. doi: 10.1111/J.1745-459X.2011.00343.X.

Kouzounis, D., Lazaridou, A., and Katsanidis, E. (2017). Partial replacement of animal fat by oleogels structured with monoglycerides and phytosterols in frankfurter sausages. *Meat Science* 130: 38–46. doi: 10.1016/j.meatsci.2017.04.004.

Le Calvé, B., Saint-Léger, C., Babas, R.,, Gelin, J.L., Erni, P., and Cayeux, L. (2015). Fat perception: How sensitive are we? *Journal of Texture Studies* 46: 200–211. doi: 10.1111/jtxs.12120.

Li, K., Zhang, M., Bhandari, B., Li, L., and Yang, C. (2020). Effect of pre-emulsified soybean oil as a fat replacer on the physical and sensory attributes of reduced-fat filling in steamed buns. *Journal of Food Process Engineering* 43. doi: 10.1111/jfpe.13306.

Lim, Y.B., Kim, H.W., Hwang, K.E.,, Song, D.H., Kim, Y.J., Youn-Kyung, H., Sung-Jin, J., Choong-Hee, L., Fu-Yi, H., Yung-Sang, C., and Cheon-Jei, K. (2015). Effects of glasswort (Salicornia herbacea L.) hydrates on quality characteristics of reduced-salt, reduced-fat Frankfurters. *Korean Journal for Food Science of Animal Resources* 35: 783–792. doi: 10.5851/kosfa.2015.35.6.783.

Moghtadaei, M., Soltanizadeh, N., and Goli, S.A.H. (2018). Production of sesame oil oleogels based on beeswax and application as partial substitutes of animal fat in beef burger. *Food Research International* 108: 368–377. doi: 10.1016/j.foodres.2018.03.051.

Mora-Gallego, H., Serra, X., Guardia, M.D., Miklos, R., Lametsch, R., Arnau, J. (2013). Effect of the type of fat on the physicochemical, instrumental and sensory characteristics of reduced fat nonacid fermented sausages. *Meat Science* 93: 668–674.

Mora-Gallego, H., Guàrdia, M.D., Serra, X., Gou, P., and Arnau, J. (2016). Sensory characterisation and consumer acceptability of potassium chloride and sunflower oil addition in small-caliber non-acid fermented sausages with a reduced content of sodium chloride and fat. *Meat Science* 112: 9–15. doi: 10.1016/j.meatsci.2015.10.008.

Namir, M., Siliha, H., and Ramadan, M.F. (2015). Fiber pectin from tomato pomace: characteristics, functional properties and application in low-fat beef burger. *Journal of Food Measurement and Characterization* 9: 305–312. doi: 10.1007/s11694-015-9236-5.

Oh, I., Lee, J., Lee, H.G., and Lee, S. (2019). Feasibility of hydroxypropyl methylcellulose oleogel as an animal fat replacer for meat patties. *Food Research International* 122: 566–572. doi: 10.1016/j.foodres.2019.01.012.

Paglarini, C.D.S., Vidal, V.A.S., Ozaki, M.M.,, Ribeiro, A.P.B., Bernardinelli, O.D., Câmara, A.K.F.I. (2022). Inulin gelled emulsion as a fat replacer and fiber carrier in healthier Bologna sausage. *Food Science and Technology International* 28: 3–14. doi: 10.1177/1082013220980586.

Pehlivanoglu, H., Ozulku, G., Yildirim, R.M.,, Demirci, O.S.T., and Sagdic, O. (2018). Investigating the usage of unsaturated fatty acid-rich and low-calorie oleogels as a shortening mimetics in cake. *Journal of Food Processing and Preservation* 42. doi: 10.1111/jfpp.13621.

Pintor-Jardines, A., Arjona-Román, J.L., Severiano-Pérez, P., Fiszman, A.T.S.S.,' Escalanocia, Buendia (2020). Agave fructans as fat and sugar replacers in ice cream: sensory, thermal and texture properties. *Food Hydrocolloids* 108: 106032. doi: 10.1016/j.foodhyd.2020.106032.

Rezende, N.V., Benassi, M.T., Vissotto, F.Z., Augusto, P.P.C., Grossmaann, M.V.E. (2015). Mixture design applied for the partial replacement of fat with fibre in sucrose-free chocolates. *LWT - Food Science and Technology* 62: 598–604. doi: 10.1016/j.lwt.2014.08.047.

Richter, V.B., de Almeida, T.C.A., Prudencio, S.H., and de Toledo Benassi, M. (2010). Proposing a ranking descriptive sensory method. *Food Quality and Preference* 21: 611–620. doi: 10.1016/j.foodqual.2010.03.011.

Rodriguez Furlán, L.T. and Campderrós, M.E. (2017). The combined effects of Stevia and sucralose as sugar substitute and inulin as fat mimetic on the physicochemical properties of sugar-free reduced-fat dairy dessert. *International Journal of Gastronomy and Food Science* 10: 16–23. doi: 10.1016/J.IJGFS.2017.09.002.

Rovers, T.A.M., Sala, G., der Linden, E., and Meinders, M.B.J. (2016). Potential of microbubbles as fat replacer: Effect on rheological, tribological and sensorial properties of model food systems. *Journal of Texture Studies* 47: 220–230. doi: 10.1111/jtxs.12175.

Santiago-García, P.A., Mellado-Mojica, E., León-Martínez, F.M.,, Dzul-Cauich, J.G., Lopez, M. G., and Garcia-Vieyra (2021). Fructans (agavins) from *Agave angustifolia* and *Agave potatorum* as fat replacement in yogurt: Effects on physicochemical, rheological, and sensory properties. *LWT* 140: 110846. doi: 10.1016/J.LWT.2020.110846.

Scheuer, P.M., Di Luccio, M., Zibetti, A.W., de Miranda, M.Z., and de Francisco, A. (2016). Relationship between instrumental and sensory texture profile of bread loaves made with whole-wheat flour and fat replacer. *Journal of Texture Studies* 47: 14–23. doi: 10.1111/jtxs.12155.

Selani, M.M., Shirado, G.A.N., Margiotta, G.B., Saldaña, E., Spada, F.B., Piedade, M.S., Contreras-Castillo, C.J., and Canniatti-Brazaca, S. (2016). Effects of pineapple byproduct and canola oil as fat replacers on physicochemical and sensory qualities of low-fat beef burger. *Meat Science* 112: 69–76. doi: 10.1016/j.meatsci.2015.10.020.

Silva, R., Rocha, R.S., Guimarães, J.T.,, Balthazar, C.S., Scludino, H., Ramos, G.L.G.P.A., Pimentel, T., Silva, M.C., Silva, P.H.F., Duarte, M.C.k.H., Freitas, M.Q., Cruz, A.G., and Esmerino, E. (2020). Dulce de leche submitted to ohmic heating treatment: consumer sensory profile using preferred attribute elicitation (PAE) and temporal check-all-that-apply (TCATA). *Food Research International* 134. doi: 10.1016/j.foodres.2020.109217.

Tahereh, J., Mahin, A., and Mojtaba, R. (2017). Determinative effects of fat replacement and homogenization on the sensory and rheological properties of low-fat Mozzarella. *Journal of Food Processing and Preservation* 41: e12753. doi: 10.1111/JFPP.12753.

Tomaschunas, M., Zörb, R., Fischer, J., Köhn, E., Hinrichs, J., and Busch-Stockfisch, M. (2013). Changes in sensory properties and consumer acceptance of reduced fat pork Lyon-style and liver sausages containing inulin and citrus fiber as fat replacers. *Meat Science* 95: 629–640. doi: 10.1016/j.meatsci.2013.06.002.

Valentin, D., Chollet, S., Lelièvre, M., and Abdi, H. (2012). Quick and dirty but still pretty good: a review of new descriptive methods in food science. *International Journal of Food Science & Technology* 47: 1563–1578. doi: 10.1111/j.1365-2621.2012.03022.x.

Varela, P. and Ares, G. (2014). *Novel Techniques in Sensory Characterization and Consumer Profiling*, 1e. Boca Ratón: CRC Press/Taylor & Francis. doi: 10.1201/b16853.

Wolf, M.B. (2020). *Sensory Testing Methods*, 3e. ASTM International. doi: 10.1520/MNL26-3RD-EB.

Yilmaz, E. and Öןütcü, M. (2015). The texture, sensory properties and stability of cookies prepared with wax oleogels. *Food & Function* 6: 1194–1204. doi: 10.1039/c5fo00019j.

Zahn, S., Pepke, F., and Rohm, H. (2010). Effect of inulin as a fat replacer on texture and sensory properties of muffins. *International Journal of Food Science & Technology* 45: 2531–2537. doi: 10.1111/j.1365-2621.2010.02444.x.

Zeng, H., Chen, J., Zhai, J., Wang, H., Xia, W., and Xiong, Y.L. (2016). Reduction of the fat content of battered and breaded fish balls during deep-fat frying using fermented bamboo shoot dietary fiber. *LWT - Food Science and Technology* 73: 425–431. doi: 10.1016/j.lwt.2016.06.052.

3.4

Gastrointestinal Fate of Lipid-Based Formulations as Fat Mimetics

Maria A. Azevedo[1] *and Catarina Gonçalves*[2,*]

[1] *Centre of Biological Engineering, University of Minho, Campus de Gualtar, Braga, Portugal*
[2] *International Iberian Nanotechnology Laboratory, Av. Mestre José Veiga, Braga, Portugal*
* *Corresponding author*

3.4.1 Introduction

This section is focused on the fate of lipid-based formulations as fat mimetics, under the gastrointestinal conditions. Ingestion of lipids is an important source of energy and lipid-soluble compounds, such as vitamins (A, D, E, and K), phytosterols, carotenoids, and ω-3 fatty acids to the human body (McClements and Öztürk, 2021). However, over-consumption of specific lipids (e.g., saturated fats, trans fatty acids, and cholesterol) is often linked to cardiovascular disease and obesity (Simopoulos 1999). There is a considerable research effort to develop new food products with health benefits, such as the controlled release of lipids, by manipulating the microstructure and/or physic-chemical properties of the foods (McClements et al. 2009).

Mimicking the human digestive process is crucial for the understanding of the complex processes that occur along the gastrointestinal tract and relationships that occur between the composition and structure of food and bioaccessibility, digestibility, or lipolysis. Different *in vitro* models are available in the literature to mimic human digestion, varying from simple mono-compartmental static methodologies up to complex multi-compartmental and dynamic systems. The choice of the digestion model depends on the goals of the study considering technical and economic factors (Lucas-González et al. 2018).

Human digestion is a complex biological process that provides nutrients to the human body from ingested food. Along the gastrointestinal tract, food is exposed to different conditions that induce physical and chemical (including enzymatic) transformations and that make nutrients available for absorption in the stomach or in the intestine to be used by different organs (*i.e.*, bioavailability).

Digestion starts in the mouth where food is broken down mechanically during the chewing process and mixed up with saliva. The bolus forms, starting the chemical digestion of starch. Then, in the stomach, gastric fluids, acidic pH, gastric lipase, and enzymes continue breaking down the bolus, along with peristaltic movements that promote mechanical mixing. In the gastric phase, proteins are primarily digested. The partly digested food, known as chyme, move on to the small intestine, where bile and pancreatin have the primordial actions. The full digestion of chyme ends

Fat Mimetics for Food Applications, First Edition. Edited by Miguel Ângelo Parente Ribeiro Cerqueira and Lorenzo Miguel Pastrana Castro.

up with nutrient absorption into the bloodstream. Later in the colon, some vitamins are produced and are absorbed into the bloodstream along with water and minerals.

3.4.2 Lipid Digestion

In the bolus, lipids are usually present as oil droplets, which have different sizes varying from less than a micrometer to more than a millimeter, for emulsions and bulk fats, respectively (McClements et al. 2009). Lipid digestion starts in the mouth and stomach where lingual and gastric lipases bind to the surface of the lipid droplets to hydrolyse triacylglycerols (TAGs), producing di-glycerides (DAGs), monoglycerides (MAGs), and free fatty acids (FFAs). In adults, 10–30% of ingested TAGs are digested in the stomach (Armand et al. 1994).

In the small intestine, lipases and co-lipases secreted by the pancreas follow up the lipid digestion at the oil-water interface. Bile made up of phospholipids, cholesterol, and bile salts plays an essential role in the formation of mixed micelles of digestion products. Mixed micelles comprise lipophilic cores surrounded by the phospholipids and lipolysis products with both acting as emulsifiers. In the lipolysis process, bile salts promote the displacement of lipid digestion products from the oil-water interface to the mixed micelles letting digestion continue (Bauer et al. 2005). Some short and medium-chain fatty acids can be absorbed directly in the stomach or in the small intestine and enter the portal vein without need of incorporation into mixed micelles (Hyun et al. 1967). MAGs and long-chain fatty acids incorporated into mixed micelles are absorbed by absorptive intestinal cells, enterocytes, and moved into lymphatic circulation. This pathway is a complex, multistep process that includes forming mixed micelles, intracellular de-esterification and re-esterification, then conversion to lipoproteins, and exocytosis through basolateral membrane of enterocytes, into the lymph (Mu and Høy 2004; Iqbal and Hussain 2009; Bernier-Latmani and Petrova 2017).

3.4.2.1 *In Vitro* Models

A huge portfolio of *in vitro* models has been proposed to mimic the human digestion, ranging from simple static methods to complex and dynamic but physiologically more-relevant models.

All models differ from one another in various parameters, such as number and type of steps included, chemical composition of the simulant fluids used in each phase (type and concentration of salts and enzymes), mechanical stresses (magnitude and direction), fluid flow, etc. (Lucas-González et al. 2018). These differences hamper the comparison of the results obtained between studies.

Static *in vitro* models recreate the oral, gastric, and intestinal phases in closed vessels incubating the sample at 37°C for 2 min, 2 h plus 2 h, respectively. Salivary, gastric, and intestinal simulant fluids are used at constant pH in each phase: pH 7.0, 3.0 and 7.0, respectively, and mixed up with enzymes. The most often used enzymes are pepsin and gastric lipase (commercialized by a French company–Lipolytech), chymotrypsin, trypsin, lipase, and amylase. Bile salts are also added in the intestinal phase. In the intestinal phase, enzymes can be added in a purified form or as an extract of pancreatic secretions containing all the digestive enzymes.

Aiming to harmonize methodologies, the COST INFOGEST network, an international consortium of scientists, proposed a standardized *in vitro* static digestion protocol based on physiological conditions, which has been extensively used (Minekus et al. 2014; Brodkorb et al. 2019). The protocol defines the composition of the simulant fluids and the enzymes' activities in each phase: amylase 75 $U.mL^{-1}$ in the oral phase, pepsin 2000 $U.mL^{-1}$ in the gastric phase and trypsin 100 $U.mL^{-1}$ in the intestinal phase. Despite the simplicity and low cost of static protocols that make them useful as pre-screening methodologies, before moving to more complex systems, they have several limitations. Static protocols cannot mimic the kinetics of digestion as they lack the gradual acidification and addition of gastric secretions and the absence of gastric emptying and nutrients absorption. The enzymatic activity is not adjusted to the caloric content or to the amount of food. The intestinal phase is not considering the sequential duodenal, jejunal, and ileal phases that show different conditions.

Recently, a semi-dynamic method built on the INFOGEST standardized protocol (Minekus et al. 2014) was proposed to bridge the gap between the convoluted dynamic models and the simplistic, static models, including digestion kinetics at the gastric phase through gradual acidification, continuous enzyme addition, and gastric emptying, while the intestinal phase remains static (Mulet-Cabero et al. 2020).

Dynamic *in vitro* models are more complex and expensive but are closer to human physiology considering the kinetics of secretions and enzymes addition and transit time in the gastric and intestinal phases. Several models are available, such as dynamic gastric model (DGM) developed by the Institute of Food Research in the UK, DIDGI at INRA in France, TNO-Intestinal Models (TIMs) by the Netherlands Organization for Applied Scientific Research, and the Human Gastric Simulator (HGS) by the University of California-Davis (Dupont et al. 2019). The application of different *in vitro* digestion models, the parameters assessed, and the physiological relevance of the data obtained when compared to *in vivo* data have recently been reviewed by experts from the INFOGEST network (Bohn et al. 2018).

3.4.2.2 Factors Affecting Lipid Digestion

Physicochemical properties of lipids, such as molecular/chemical structure, physical state, and lipid droplet size, can influence the rate and extent of lipid digestion. Besides that, the structure and interfacial composition of the food matrix where the lipids are inserted may also affect their digestion and final absorption (McClements et al. 2009; O'Sullivan et al. 2016). Understanding the relationship between lipids´ features and lipolysis is crucial on the design of new functional formulations. Some studies aiming to explore those effects are summarized in Table 3.4.1, and some are discussed below.

Lipolysis can be evaluated assessing the FFAs released during digestion. After the lipase´s addition, TAGs that make part of lipid-based formulations generate two FFAs and one MAG per TAG molecule. The FFAs released can be measured using the pH-stat method which is based on the assessment of NaOH needed to neutralize the FFAs generated during digestion, setting the pH to 7.0 (Figure 3.4.1) (Li and McClements 2010). *In vivo* models can also study lipid digestion despite associated ethical and economic issues. Limpimwong and co-workers (2017) studied the digestion of rice bran wax/rice bran oil oleogels using male Sprague–Dawley rats. To evaluate lipid digestion, they did blood analysis assessing the lipid profile, triacylglycerol, total cholesterol, free fatty acids, high-density lipoprotein cholesterol, and low-density lipoprotein cholesterol (Limpimwong et al. 2017).

- *Molecular/chemical structure* – The lipids can have different fatty acids compositions and TAGs structures. Their source (plants, oceanic algae, or animal) makes different the length of acyl chains, the degree of unsaturation and position of the fatty acids (FAs) in TAGs molecules, influencing their physical, chemical, and nutritional properties and, also lipolysis rate and extent

Table 3.4.1 Summary of studies exploring the *in vitro* or in vivo digestion of different formulations and their effect on lipid digestion.

Formulation Composition	Physical state of lipid phase	Loading	Food application	*In vitro* or *in vivo* digestion method	Assessment	Reference
Candelilla wax and canola oil	Oleogel	–	Butter replacement in sponge cake	Static method – Cake sample was crumbled and incubated with porcine pancreatic α-amylase and amyl glucosidase in sodium maleate buffer in a shaking water bath at 37°C for 120 minutes.	Starch digestibility–The influence of oleogel amount in starch digestion rate.	Alvarez-Ramirez et al. (2020)
Ethyl cellulose or mono- and di-glycerides or a mixture of β-sitosterol and γ-oryzanol with canola oil	Oleogel	–	–	*In vitro* digestive lipolysis of oleogels–The samples were placed in the reaction vessel with simulated small intestinal fluid.	Lipolysis–Free fatty acids release (%) using the pH-stat method.	Ashkar et al. (2019)
Phytosterols (γ-oryzanol and β-sitosterol) and sunflower oil	Small sizes tubular crystals with about 10 nm of diameter	–	–	Static gastrointestinal tract model that included mouth, stomach, and small intestine phases being used mucin in mouth, pepsin in gastric and bile salt solution and lipase in small intestine.	Free fatty acids release (%) using the pH-stat method; Interfacial binding of lipase measuring the lipase activity through standard potentiometric titration.	Dong et al. (2020)
Soybean oil dispersed within an oil-in-water emulsion (whey protein and oil) with rice bran wax as an oleogelator	Between 0.1 and 10 μm, depending on rice bran wax concentration	–	–	Static digestion–INFOGEST protocol	Free fatty acids release (%) using the pH-stat method.	Guo et al. (2019)
Rice bran wax and rice bran oil	Oleogel	–	Margarine replacement	*In vivo* experiment with male Sprague–Dawley rats	Measurement of blood parameters: lipid profile, triacylglycerol, total cholesterol, free fatty acids, high-density lipoprotein cholesterol and low-density lipoprotein cholesterol.	Limpimwong et al. (2017)
Emulsion gels of sunflower oil stabilized by whey protein isolate	301–328 nm at pH 4 and 260–275 nm at pH 7	–	–	*In vitro* mechanical gastric system (included a human stomach model and a mechanical system with realistic peristalsis); *In vitro* gastric digestion in a stirred beaker.	Intestinal Proteolysis (DH%) and Free fatty acids release (%) using pH-stat method.	Mella et al. (2021)

Canola oil-based monoacylglycerol gel	Structured emulsions		Sugar-free cookies and cakes	In vivo human study	Postprandial TAG, FFA, glucose and insulin responses through blood analysis.	Wright et al. (2014)
Sunflower oil gelled by adding 5% (w/w) saturated monoglycerides, rice bran waxes or a mixture of β-sitosterol and γ-oryzanol	Oleogel	Curcuminoids	–	Static digestion–INFOGEST protocol	Free fatty acids release (%) using pH-stat; Curcuminoid measurement and bioaccessibility by UHPLC analysis.	Calligaris et al. (2020)
Candelilla wax (3 wt%) and nut oils (peanut, pine nut and walnut oil)	Oleogel	β-carotene	–	Static digestion – INFOGEST protocol	Effect of high-intensity ultrasound on the β-carotene bioaccessibility in the gastric and intestinal phases, after extraction with hexane–quantification by absorbance at 450 nm.	Li et al. (2021)
Canola oil and ethylcellulose (EC)	Oleogel	β-carotene	–	Static model that simulates the digestive processes under fasted conditions (model described by Oomen et al. (2003)) being used α-amylase in mouth, pepsin and pyrogallol in stomach and bile salts solution and pancreatin for intestinal phase.	Moles of free fatty acids and Lipid Hydrolysis (%) using the non-esterified free fatty acid (NEFA) kit; β-carotene measurement by diode array spectrophotometer at 450 nm.	O'Sullivan et al. (2017)
Oleogels structured by β-sitosterol and lecithin	Oleogel	Curcumin	Delivery of drugs and nutrients	Digestion was conducted based on Yu et al. (2012) being only used simulated intestinal fluid with pancreatic lipase in lipolysis buffers.	Bioaccessibility of curcumin (fasted and fed state) by UV spectrophotometer at 425 nm; Lipolysis extent using the pH-stat.	Li et al. (2019)
Whey protein isolate and soybean oil containing capsaicinoids	Emulsion gels	Capsaicinoids	–	Dynamic gastric digestion model–Human Gastric Simulator (HGS)	FFAs release using the pH-stat method; Bioaccessibility of capsaicinoids by HPLC; Protein hydrolysis evaluated by SDS-PAGE.	Luo et al. (2021)

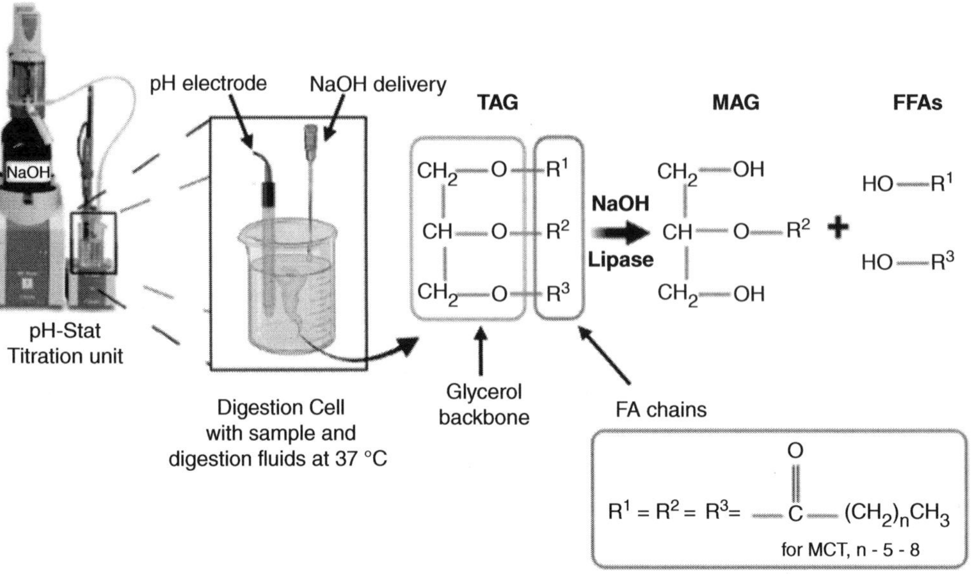

Figure 3.4.1 General schematic representation of the pH-stat technique to measure lipolysis during *in vitro* digestion. Each triacylglycerol (TAG) present in the sample is composed by three fatty acid (FA) chains. The lipase added catalyzes lipid digestion and each TAG molecule generates two FFAs and one monoacylglycerol (MAG). Created with BioRender.com.

(Ye et al. 2019). Some studies have shown that the rate of lipolysis increases with decreasing the length of FAs chain: long-chain triacylglycerol (LCT) < medium-chain triacylglycerol (MCT) < short-chain triacylglycerol (SCT) (O'Sullivan et al. 2016; Guo et al. 2017; Ye et al. 2019). The FFAs from LCTs accumulate at the surface of oil droplets, making it difficult for the lipase to access other TAGs. Alternatively, FFAs from MCT or SCT have higher water dispersibility, migrating rapidly to the surrounding aqueous phase, which makes access to lipase to hydrolyze the other TAGs easier (Li and McClements 2010; Guo et al. 2017). The position of FAs in TAGs also influences the lipolysis. The TAGs digestion mostly occurs by pancreatic lipases in the duodenum with higher specificity for FAs at sn-1 and sn-3 positions than to sn-2 position. This way, during lipid digestion, the lipase targets are in the sn-1 and sn-3 positions, and all the FAs in the sn-1 and sn-3 position are hydrolysed; however, sn-2 monoacylglycerols are absorbed more rapidly than FFAs. The degree of unsaturation of FFAs also interferes with their absorption. The long chain saturated FFAs form insoluble soaps. Due to their high melting points and strong electrostatic attraction with cationic mineral ions (e.g., calcium and magnesium), they are poorly absorbed (Karupaiah and Sundram 2007; McClements et al. 2009). Ashkar et al. (2019) have reported the impact of different gelators, such as ethylcellulose, mono- and di-glycerides, and β-sitosterol and γ-oryzanol mixture, on digestive lipolysis of canola oil oleogels. They have shown that gelators affect lipolysis patterns. Independently of the concentration of the ethylcellulose used, the ethylcellulose/canola oil oleogels presented susceptibility to lipolysis. Also, mono- and di-glycerides based oleogels showed higher lipolysis levels for higher concentrations of gelators. On the other side, the concentration of β-sitosterol + γ-oryzanol mixture reduced the rate and extent of free fatty acid released. Calligaris et al. (2020) have also evaluated the effect of different oleogelators (saturated monoglycerides, rice bran waxes, or a mixture of β-sitosterol and γ-oryzanol) on lipolysis and curcuminoid bioaccessibility upon *in vitro* digestion

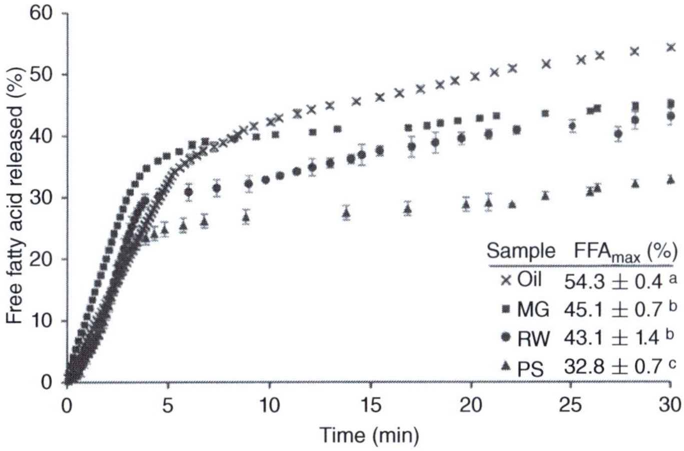

Figure 3.4.2 Free fatty acids released during *in vitro* digestion of oil and oleogels containing 5% (w/w) of monoglycerides (MG), rice wax (RW) or phytosterols (PS). Different letters (a-c) mean significant differences (p<0.05). Calligaris, S. et al. Food Chemistry, 314, 2020) / with permission of Elsevier.

of sunflower oil oleogels. In this work, the authors have shown the rate of fatty acids released was greatly affected by the oleogel structure. The extent of lipolysis decreased as oleogel strength increased in the following order: mixture of β-sitosterol and γ-oryzanol < rice bran waxes < saturated monoglycerides (Figure 3.4.2). On the other hand, the nature of the gelator affected the curcuminoid bioaccessibility, which was lower in oleogels containing crystalline particles (rice bran waxes and saturated monoglycerides).

- *Droplet size* – Lipids can be present in different forms in food, such as emulsion droplets or bulk phases. However, during digestion a complex variety of chemical, enzymatic, and physical processes occur leading to droplet coalescence, disruption, and/or dissolution of lipids (McClements et al. 2009). In the mouth and stomach, the lipids undergo changes and their emulsification (oil droplets) occurs before they reach the duodenum. *In vitro* studies have shown that the sizes of those oil droplets influence their digestion rate. Emulsions with small sizes have higher lipolysis rates than emulsions with larger sizes. This can be explained by the surface area of lipid exposed to lipase action, which is higher for small sizes (McClements et al. 2009; Li and McClements 2010; O'Sullivan et al. 2016).
- *Physical state of lipid phase* – Lipid-based fat mimetics can be formed by fat and/or oils. Both are complex mixtures of TAGs; the major difference between fat and oils are their physical state. Fats are solid at room temperature, but oils are liquid. The melting points of each TAG that makes part of the lipid and crystallinity are the main factors affecting the physical state (Marangoni et al. 2020). The physical state of lipids can significantly influence their digestibility/lipolysis. The *in vitro* and *in vivo* studies have shown that solid lipids are less susceptible to lipolysis. The limited access of pancreatic lipase to TAGs in the solid phase can be explained by the difficulty lipases have absorbing into the solid droplet surface and by the dense molecular packing of lipid molecules (McClements et al. 2009; O'Sullivan et al. 2016; Guo et al. 2017).
- *Interfacial composition* – Oil structuring based on physical processes consists of oil entrapment into a gelled continuous phase through polymer gelation or through the formation of structured emulsions with the support of a proper emulsifying agent (emulsifiers) (Curti et al. 2018). A big range of emulsifiers can be used, such as phospholipids, proteins, polysaccharides, and synthetic/natural small molecules known as surfactants (Tan et al. 2020). They are used to prevent some phenomenon like coalescence and phase separation. However, some can interfere with the kinetics of oleogelator solidification in the bulk and with the interactions

between oleogelator and dispersed droplets (Wijarnprecha et al. 2021). Additionally, the composition and properties of the emulsifier may also influence the lipid digestion and absorption (McClements et al. 2009; Li and McClements 2010). Emulsifiers can attach to the droplets' surface influencing the ability of bile salts and/or lipases to interact with the emulsions' surface. Also, the emulsifiers can change the surface area of lipids available to attach the bile salts and lipases. This way, depending on the emulsifier used, different rates and extension of lipolysis can be obtained (Tan et al. 2020). Mun et al. (2007) have shown that the access of pancreatic lipase to emulsified fats decreases in the following order: proteins (caseinate and WPI) > phospholipids (lecithin) > non-ionic surfactants (Tween 20). Tan and co-workers (2020) have also studied the effect of different classes of emulsifiers (Tween 20, *quillaja* saponin, caseinate, gum Arabic, and soy lysolecithin) during *in vitro* digestion (Figure 3.4.3). They showed the digestion of the lipid phase in emulsions stabilized with soy lysolecithin or caseinate was lower than those stabilized by gum Arabic, *quillaja* saponin, or Tween 20; this because of the lipids' lower surface area for lipase to attach to the lysolecithin and caseinate emulsions (Tan et al. 2020).

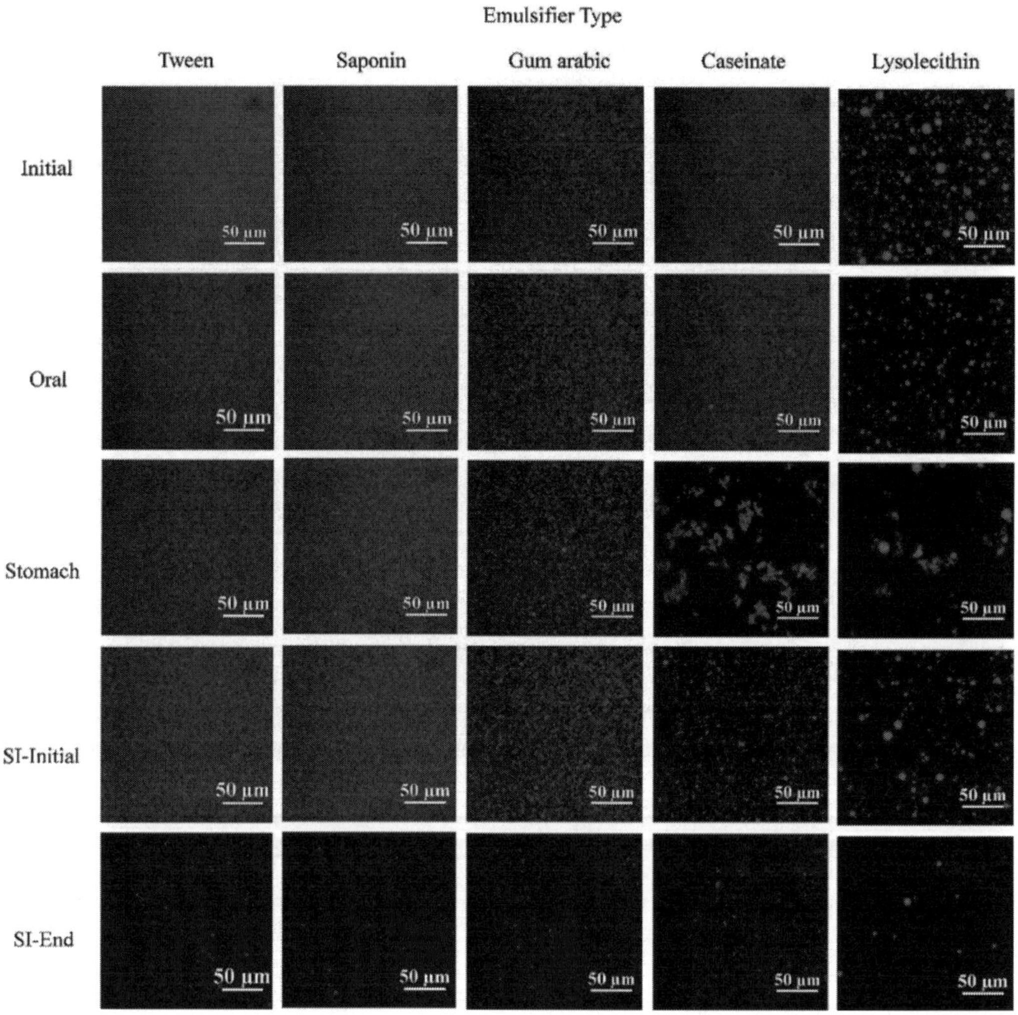

Figure 3.4.3 The effect of different emulsifier type on the confocal photos of corn oil-in-water emulsions during *in vitro* gastrointestinal digestion Tan et al. (2020), from Elsevier.

- *Food matrix* – Research on fat mimetics aims to find healthier alternatives to replace mainly saturated fatty acids in foods. Besides the factors affecting lipid digestion described before, incorporating fat mimetics into a food matrix makes the lipolysis much more complicated. Due to the food matrix, the TAGs are more difficult to access by lipases, and this way the release of fatty acids is impeded or retarded (Scharfe and Flöter 2020). Fat mimetics can be wrapped in a structure or surrounded by an interfacial layer and, due to that, the lipase access can be slowed or impeded. This way, the digestion time increases since it is necessary to break down the matrix surrounding the lipids. The bonds and the strength of the food matrix determine the rate of breakdown, erosion, and/or digestion (O'Sullivan et al. 2016). Wright et al. (2014) have developed canola oil-based monoacylglycerol gels (structured emulsions) as an alternative to saturated and trans fatty acids. They cooked sugar-free cookies and cakes with monoacylglycerol gels and did a human study with healthy adult men. The authors have evaluated the postprandial TAG, FFA, glucose, and insulin responses to cookies and cakes and have observed that although monoacylglycerol gel structure did not impact postprandial response, all cookies had higher TAG responses compared with cakes even when matched for fat content.

3.4.3 Conclusion

This section shows that a considerable effort has been made in the field of food research to understand the complex process of lipid digestion and the effect of lipids' physical and chemical properties in the final digestion products. This understanding supports developing new and healthier foods, such as controlling the release of lipids. The importance of following standardised protocols (static or dynamic) to mimic human digestion is also highlighted, allowing the comparison of results between research groups.

Acknowledgments

This research was funded by the SbDtoolBox – Nanotechnology-based tools and tests for Safer-by-Design nanomaterials, with the reference NORTE-01–0145-FEDER-000047, funded by Norte 2020 – North-Regional Operational Programme under the PORTUGAL 2020 Partnership Agreement, through the European Regional Development Fund (ERDF).

References

Alvarez-Ramirez, J., Vernon-Carter, E.J., Carrera-Tarela, Y., Garcia, A, and Roldan-Cruz, C. (2020). Effects of candelilla wax/canola oil oleogel on the rheology, texture, thermal properties and in vitro starch digestibility of wheat sponge cake bread. *LWT* 130: 109701. doi: 10.1016/J.LWT.2020.109701.

Armand, M., Borel, P., Dubois, C., Senft, M., Peyrot, J., Salducci, J., Lafont, J., and Lairon, D. (1994). Characterization of emulsions and lipolysis of dietary lipids in the human stomach. *American Journal of Physiology Gastrointestinal and Liver Physiology* 266: G372 G381. doi: 10.1152/ajpgi.1994.266.3.G372.

Ashkar, A., Laufer, S., Rosen-Kligvasser, J., Lesmes, U., and Davidovich-Pinhas, M. (2019). Impact of different oil gelators and oleogelation mechanisms on digestive lipolysis of canola oil oleogels. *Food Hydrocolloids* 97: 105218. doi: 10.1016/J.FOODHYD.2019.105218.

Bauer, E., Jakob, S., and Mosenthin, R. (2005). Principles of physiology of lipid digestion. *Asian-Australasian Journal of Animal Sciences* 18: 282–295. doi: 10.5713/ajas.2005.282.

Bernier-Latmani, J. and Petrova, T.V. (2017). Intestinal lymphatic vasculature: structure, mechanisms and functions. *Nature Reviews Gastroenterology & Hepatology* 14: 510–526. doi: 10.1038/nrgastro.2017.79.

Bohn, T., Carriere, F., Day, L., Deglaire, A., Egger, L., Freitas, D., Golding, M., Le Feunteun, S., Macierzanka, A., Menard, O., Miralles, B., Moscovici, A., Portmann, R., Recio, I., Rémond, D., Santé-Lhoutelier, V., Wooster, T.J., Lesmes, U., Mackie, A.R., and Dupont, D. (2018). Correlation between in vitro and in vivo data on food digestion. What can we predict with static in vitro digestion models? *Critical Reviews in Food Science and Nutrition* 58: 2239–2261. doi: 10.1080/10408398.2017.1315362.

Brodkorb, A., Egger, L., Alminger, M., Alvito, P., Assunção, R., Ballance, S., Bohn, T., Bourlieu-Lancal, C., Boutrou, R., Carrière, F., Clemente, A., Corredig, M., Dupont, D., Dufour, C., Edwards, C., Golding, M., Karakaya, S., Kirkhus, B., Le Feunteun, S., and...Recio, I. (2019). INFOGEST static in vitro simulation of gastrointestinal food digestion. *Nature Protocols* 14: 991–1014. doi: 10.1038/s41596-018-0119-1.

Calligaris, S., Alongi, M., Lucci, P., and Anese, M. (2020). Effect of different oleogelators on lipolysis and curcuminoid bioaccessibility upon in vitro digestion of sunflower oil oleogels. *Food Chemistry* 314: 126146. doi: 10.1016/J.FOODCHEM.2019.126146.

Curti, E., Federici, E., Diantom, A., Carini, E., Pizzigalli, E., Symon, V.U., Pellegrini, N., and Vittadini, E. (2018). Structured emulsions as butter substitutes: effects on physicochemical and sensory attributes of shortbread cookies. *Journal of the Science of Food and Agriculture* 98: 3836–3842. doi: 10.1002/jsfa.8899.

Dong, L., Lv, M., Gao, X., Zhang, L., Rogers, M., Cao, Y., and Lan, Y. (2020). In vitro gastrointestinal digestibility of phytosterol oleogels: influence of self-assembled microstructures on emulsification efficiency and lipase activity. *Food & Function* 11: 9503–9513. doi: 10.1039/D0FO01642J.

Dupont, D., Alric, M., Blanquet-Diot, S., Bornhorst, G., Cueva, C., Deglaire, A., Denis, S., Ferrua, M., Havenaar, R., Levieveld, J., Mackie, A.R., Marzorati, M., Menard, O., Minekus, M., Miralles, B., Recio, I., and Van den Abbeele, P. (2019). Can dynamic in vitro digestion systems mimic the physiological reality? *Critical Reviews in Food Science and Nutrition* 59: 1546–1562. doi: 10.1080/10408398.2017.1421900.

Guo, Q., Wijarnprecha, K., Sonwai, S., and Rousseau, D. (2019). Oleogelation of emulsified oil delays in vitro intestinal lipid digestion. *Food Research International* 119: 805–812. doi: 10.1016/J.FOODRES.2018.10.063.

Guo, Q., Ye, A., Bellissimo, N., Singh, H., and Rousseau, D. (2017). Modulating fat digestion through food structure design. *Progress in Lipid Research* 68: 109–118. doi: 10.1016/j.plipres.2017.10.001.

Hyun, S.A., Vahouny, G.V., and Treadwell, C.R. (1967). Portal absorption of fatty acids in lymph- and portal vein-cannulated rats. *Biochimica et Biophysica Acta (BBA) - Lipids and Lipid Metabolism* 137: 296–305. doi: 10.1016/0005-2760(67)90105-1.

Iqbal, J. and Hussain, M.M. (2009). Intestinal lipid absorption. *American Journal of Physiology-Endocrinology and Metabolism* 296: E1183–E1194. doi: 10.1152/ajpendo.90899.2008.

Karupaiah, T. and Sundram, K. (2007). Effects of stereospecific positioning of fatty acids in triacylglycerol structures in native and randomized fats: a review of their nutritional implications. *Nutrition & Metabolism* 4: 16. doi: 10.1186/1743-7075-4-16.

Li, L., Taha, A., Geng, M., Zhang, Z., Su, H., Xu, X., Pan, S., and Hu, H. (2021). Ultrasound-assisted gelation of β-carotene enriched oleogels based on candelilla wax-nut oils: physical properties and in-vitro digestion analysis. *Ultrasonics Sonochemistry* 79: 105762. doi: 10.1016/j.ultsonch.2021.105762.

Li, L., Wan, W., Cheng, W., Liu, G., and Han, L. (2019). Oxidatively stable curcumin-loaded oleogels structured by β-sitosterol and lecithin: Physical characteristics and release behaviour *in vitro*. *International Journal of Food Science & Technology* 54: 2502–2510. doi: 10.1111/ijfs.14208.

Li, Y. and McClements, D.J. (2010). New mathematical model for interpreting pH-stat digestion profiles: Impact of lipid droplet characteristics on in vitro digestibility. *Journal of Agricultural and Food Chemistry* 58: 8085–8092. doi: 10.1021/jf101325m.

Limpimwong, W., Kumrungsee, T., Kato, N., Yanka, N., and Thongngam, M. (2017). Rice bran wax oleogel: aA potential margarine replacement and its digestibility effect in rats fed a high-fat diet. *Journal of Functional Foods* 39: 250–256. doi: 10.1016/J.JFF.2017.10.035.

Lucas-González, R., Viuda-Martos, M., Pérez-Alvarez, J.A., and Fernández-López, J. (2018). In vitro digestion models suitable for foods: opportunities for new fields of application and challenges. *Food Research International* 107: 423–436. doi: 10.1016/J.FOODRES.2018.02.055.

Luo, N., Ye, A., Wolber, F.M., and Singh, H. (2021). Effect of gel structure on the in vitro gastrointestinal digestion behaviour of whey protein emulsion gels and the bioaccessibility of capsaicinoids. *Molecules* 26: 1379. doi: 10.3390/molecules26051379.

Marangoni, A.G., van Duynhoven, J.P.M., Acevedo, N.C., Nicholson, R.A., and Patel, A.R. (2020). Advances in our understanding of the structure and functionality of edible fats and fat mimetics. *Soft Matter* 16: 289–306. doi: 10.1039/C9SM01704F.

McClements, D. and Öztürk, B. (2021). Utilization of nanotechnology to improve the handling, storage and biocompatibility of bioactive lipids in food applications. *Foods* 10: 365. doi: 10.3390/foods10020365.

McClements, D.J., Decker, E.A., and Park, Y. (2009). Controlling lipid bioavailability through physicochemical and structural approaches. *Critical Reviews in Food Science and Nutrition* 49: 48–67. doi: 10.1080/10408390701764245.

Mella, C., Quilaqueo, M., Zúñiga, R.N., and Troncoso, E. (2021). Impact of the simulated gastric digestion methodology on the in vitro intestinal proteolysis and lipolysis of emulsion gels. *Foods* 10. doi: 10.3390/foods10020321.

Minekus, M., Alminger, M., Alvito, P., Ballance, S., Bohn, T., Bourlieu, C., Carrière, F., Boutrou R., Corredig, M., Dupont, D., Dufour, C., Egger, L., Golding, M., Karakaya, S., Kirkhus, B., Le Feunteun, S., Lesmes, U., Macierzanka, A., Mackie, A., and...Brodkorb, A. (2014). A standardised static in vitro digestion method suitable for food – an international consensus. *Food & Function* 5: 1113–1124. doi: 10.1039/C3FO60702J.

Mu, H. and Høy, C.E. (2004). The digestion of dietary triacylglycerols. *Progress in Lipid Research* 43: 105–133. doi: 10.1016/S0163-7827(03)00050-X.

Mulet-Cabero, A.-I., Egger, L., Portmann, R., Menard, O., Marze, S., Minekus, M., Le Feunteun, S., Sarkar, A., Grundy, M.M.L., Carrière, F., Golding, M., Dupont, D., Recio, I., Brodkorb, A., and Mackie, A. (2020). A standardised semi-dynamic in vitro digestion method suitable for food – an international consensus. *Food & Function* 11: 1702–1720. doi: 10.1039/C9FO01293A.

Mun, S., Decker, E.A., and McClements, D.J. (2007). Influence of emulsifier type on in vitro digestibility of lipid droplets by pancreatic lipase. *Food Research International* 40: 770–781. doi: 10.1016/J.FOODRES.2007.01.007.

Oomen, A. G., Rompelberg, C. J. M., Bruil, M. A., Dobbe, C.J. G., Pereboom, D. P. K. H., Sips A. J. A. M. 2003. Development of an In VitroDigestion Model for Estimating the Bioaccessibility of Soil Contaminants. *Arch. Environ. Contam. Toxicol.* 44: 281–287. doi: 10.1007/s00244-002-1278-0.

O'Sullivan, C.M., Barbut, S., and Marangoni, A.G. (2016). Edible oleogels for the oral delivery of lipid soluble molecules: Composition and structural design considerations. *Trends in Food Science and Technology* 57: 59–73. doi: 10.1016/J.TIFS.2016.08.018.

O'Sullivan, C.M., Davidovich-Pinhas, M., Wright, A.J., Barbut, S., and Marangoni, A.G. (2017). Ethylcellulose oleogels for lipophilic bioactive delivery – effect of oleogelation on in vitro bioaccessibility and stability of beta-carotene. *Food & Function* 8: 1438–1451. doi: 10.1039/C6FO01805J.

Scharfe, M. and Flöter, E. (2020). Oleogelation: From scientific feasibility to applicability in food products. *European Journal of Lipid Science and Technology* 122: 2000213. doi: 10.1002/ejlt.202000213.

Simopoulos, A.P. (1999). Essential fatty acids in health and chronic disease. *The American Journal of Clinical Nutrition* 70: 560s–569s. doi: 10.1093/ajcn/70.3.560s.

Tan, Y., Zhang, Z., Muriel Mundo, J., and McClements, D.J. (2020). Factors impacting lipid digestion and nutraceutical bioaccessibility assessed by standardized gastrointestinal model (INFOGEST): Emulsifier type. *Food Research International* 137: 109739. doi: 10.1016/j.foodres.2020.109739.

Wijarnprecha, K., de Vries, A., Sonwai, S., and Rousseau, D. (2021). Water-in-oleogel emulsions—from structure design to functionality. *Frontiers in Sustainable Food Systems* 4: 279. doi: 10.3389/FSUFS.2020.566445/BIBTEX.

Wright, A., Pinto, C., Tulk, H., McCluskey, J., Goldstein, A., Huschka, B., Marangoni, A., and Seetharaman, K. (2014). Monoacylglycerol gel offers improved lipid profiles in high and low moisture baked products but does not influence postprandial lipid and glucose responses. *Food Funct* 5: 882–893. doi: 10.1039/C3FO60596E.

Ye, Z., Li, R., Cao, C., Xu, Y.J., Cao, P., Li, Q., and Liu, Y. (2019). Fatty acid profiles of typical dietary lipids after gastrointestinal digestion and absorbtion: A combination study between in-vitro and in-vivo. *Food Chemistry* 280: 34–44. doi: 10.1016/j.foodchem.2018.12.032.

Yu, H., Shi, K., Liu, D. & Huang, Q. (2012). Development of a food-grade organogel with high bioaccessibility and loading ofcurcuminoids. *Food Chemistry* 131: 48–54. https://doi.org/10.1016/j.foodchem.2011.08.027.

3.5

Nutritional and Functional Properties of Fat Mimetics

Xiao-Wei Chen[1] *and Xiao-Quan Yang*[2,*]

[1] *Lipid Technology and Engineering, College of Food Science and Engineering, Henan University of Technology, Zhengzhou, People's Republic of China*
[2] *Laboratory of Food Proteins and Colloids, Department of Food Science and Engineering, South China University of Technology, Guangzhou, People's Republic of China*
[*] *Corresponding author*

3.5.1 Introduction

Driven by the need for trans and saturated fat replacement strategies, the field of fat mimetics has advanced tremendously over the past decade. Consuming a diet high in saturated fats and in trans fats as a well-accepted biomarker for increased risk of cardiovascular and coronary heart diseases has been highlighted in many published reports (DiNicolantonio et al., 2016). After the demise of partially hydrogenated oils (PHOs), industrially produced foods still have high levels of saturated fats because of their similar sensory impact and the suitability of existing processing technologies. So saturated fats have become the new target. Even though the metabolic effects of saturated fatty acids on human health acids remains controversial, a recent scientific review on this topic by Nettleton et al. (2017) and Briggs et al. (2017) have concluded there is strong evidence in favor of partially replacing saturated fatty acid-rich foods with those rich in polyunsaturated fatty acids to reduce cardiovascular health disease risk. To complicate matters, concerns over sustainability, veganism, genetically modified organisms, animal welfare, and religious beliefs severely limit our saturated fat sources. Thus, manufacturers and scientists have focused on seeking alternative ways to structure liquid oil with low in saturated and zero trans fatty acids and hence to provide the functionality of fats to food products without changing the nutritional profile of liquid oil. An exciting and novel strategy to eliminate trans fats from our diets and avoid excessive saturated fat lies in structuring edible oils to develop fat mimetics.

In addition to health issues, the functionality required in many food products must be considered in the development of fat mimetics. As hardstock fats, PHOs and saturated fats contribute to the structure, taste, texture, and oxidative stability of processed foods. Saturated fats are often added to foods to provide solid fat content, melting, and textural properties that consumers associate with certain products, including shortenings, butter, margarines, vegetable creams, spreads, and confectionary fats. Given the contribution of these fats to food's stability and sensory properties, these replacements must mimic the functionality of hardstock fats so the integrity of the food product is not compromised. Otherwise, consumers will not accept a particular food if the product does not meet their sensory and quality expectations.

Research into fat mimetics began almost two decades ago and many commercial carbohydrate-and-protein-based fat mimetics (e.g., Novagel™ NC200, Dairy Lo™, and ALACO PALS™ 6082) were developed and promoted for reduced-fat and low-fat foods. Carbohydrate-based fat mimetics are used to replace fat in foods because of their textural and organoleptic properties. Carbohydrates generally mimic fat by binding water, thus providing lubrication, slipperiness, body, and

Fat Mimetics for Food Applications, First Edition. Edited by Miguel Ângelo Parente Ribeiro Cerqueira and Lorenzo Miguel Pastrana Castro.

mouthfeel (Peng and Yao 2017). For example, maltodextrin and starch-based ingredients form a smooth, viscous solution or soft gel on hydration and create a finished product that has the textural and functional features of their full-fat counterparts. Microparticulated intact starch granules can also mimic the mouthfeel and structural contributions of fat crystals. Despite this, carbohydrate-based fat replacers are used most successfully to reduce, not replace, fat in a low moisture food system. Several protein-based fat mimetics are derived from a variety of protein sources, including egg, milk, whey, gelatin, soy, and wheat gluten (Kew et al., 2020). They are microparticulated to form microscopic, coagulated, round, deformable particles that mimic the texture and mouthfeel of natural fats and oils. Often, protein-based fat mimetics are used in margarines, butter, cheese, dairy products, sour cream, salad dressings, mayonnaise-containing products, soups, sauces, baked goods, and frozen desserts. These substances generally give a better mouthfeel than their carbohydrate-based counterparts. However, like carbohydrate-based substances, protein-based fat mimetics cannot be used for frying. To meet the diversified consumer's demands, some emerging fat mimetics developed over several years include oleogels from oleogelation by molecular and polymeric gelators networks, emulsion-templated networks of proteins and polysaccharides, colloidal and self-assembled fibrillar networks of polar lipid crystals and solid emulsion gels, and convert highly unsaturated liquid oil to semisolid trans fat-free, low-saturate functional fat mimetics (Patel 2017a; Scharfe and Flöter 2020). These fat mimetics have better nutritive value and adaptability than conventional fats and carbohydrate- and protein-based fat mimetics. Given this, the quest for novel fat mimetics is growing. Edible oil structuring may be an option for achieving fat-like consistency in food products.

3.5.2 Emerging Fat Mimetics

To replace saturated fats (solid fat) with unsaturated fats (liquid oil) without compromising the sensory and quality features of the food product, the properties of the oil must first be able to mimic those of the solid fat. In fats and oils, conventional crystallization involves forming fat crystals characterized by partial or total movement restriction, which is caused by chemical or physical bonds between the triacylglycerol molecules. As well known, edible fats and oils are complex mixtures of triacylglycerols (TAGs). The solid vs liquid nature of a TAGs mixture will thus depend on the melting points of TAGs and TAGs crystalline complexes within the mixture. From a structural perspective, edible fats are a special class of soft matter, where the higher melting point TAGs crystallize from the melt into different polymorphic states forming a crystal network that traps the liquid TAGs within (Marangoni et al., 2020). Crystallization into a particular polymorphic state is a function of external shear and temperature fields, supersaturation or undercooling, and molecular geometry. These are not always good predictors of mechanical strength (e.g., elastic modulus, yield stress) or of plasticity from a molecular composition. Solid-state structures define the melting behavior of a fat; thus, Marangoni et al. considered fats should be described as colloidal gels of polycrystalline particles (Figure 3.5.1), rather than as crystalline substances. In this seminal work, during TAGs crystallization, growing a fat crystal network starts with laterally packing triacylglycerol molecules into lamellae, which stack upright to form crystalline nanoplatelets (CNPs) with lamellae arrangement. CNPs aggregate with each other into single crystallites first and then clusters (crystalline aggregates) with sizes reaching 20–50 micrometers. Further collections of clusters help form a 3D crystal network. In this case, the macroscopic properties and functionality of the resulting fat is determined by the molecular composition and organization of all these levels of structure. The existence of similar lipid polymorphisms in fats is crucial as they influence the taste,

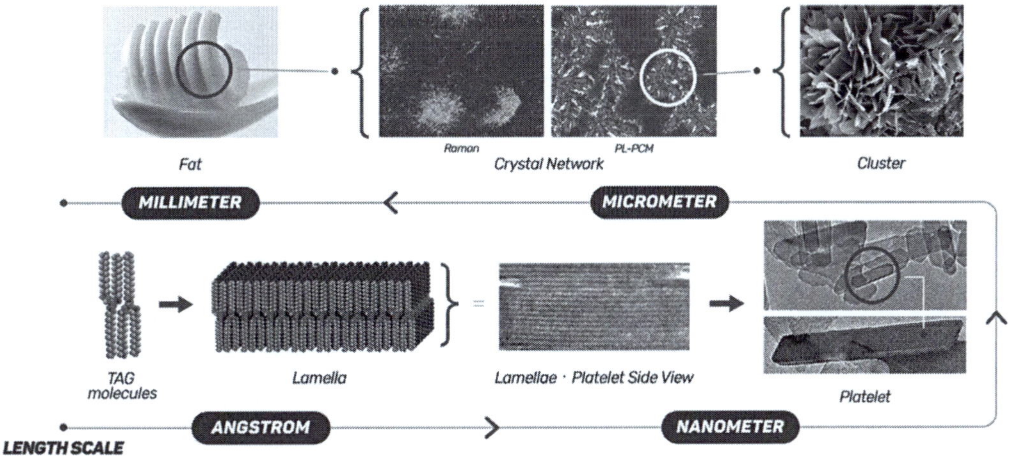

Figure 3.5.1 The hierarchy of fat crystal networks, where the model of the self-assembled supramolecular organization of a fat crystal network highlights the different structural levels and their corresponding length scale (Marangoni et al., 2020, from Royal Society of Chemistry).

texture, and mouthfeel of the foods. Therefore, it is key to successfully develop fat mimetics with a similar crystalline nature and/or polymorphism of lipidic amphiphiles as that of natural fats for desirable textural properties and mouthfeel (Patel et al., 2020). In this case, such information on structural levels of fat crystalline networks is useful for designing fat mimetics, which are based on crystalline particles as the building blocks.

Oil structuring (e.g., oleogelation, structured emulsification) offers an alternative route to entrap liquid oils in a macroscopically semi-solid lipid structure. Depending on the structuring mechanism, this route could offer independence from saturated fat with improved nutritional profiles. More recently, several novel fat mimetics have brought about many ways through which edible oils can be structured to produce a much more solid-like material (Figure 3.5.2A). Several of these fat mimetics techniques have achieved success when used in various food applications (Figure 3.5.2B), such as margarines, cakes, chocolate, cream cheese, meat products, cookies, spreads, and frozen foods.

3.5.2.1 Oleogels

Oleogel systems have emerged as alternatives to conventional fats with high levels of trans and saturated fatty acids. In oleogel systems, unsaturated liquid oil is immobilized into a three-dimensional network of lipophilic solids oleogelators forming a semi-solid system (Marangoni 2012; Puşcaş et al., 2020). Such soft matter can be formulated by a direct heating-cooling induced assembly of structuring agent into an oil phase or indirect routes, such as emulsion-templated, foam templated, and solvent replacement. For the direct method, oil structuring agents can be generally divided into two major groups: low and high molecular weight oil oleogelators (LMOG and HMOG) (Figure 3.5.2). LMOG refers to small molecules (<3000 Da) that can self-assemble and form a stable three-dimensional crystal network that entraps the surrounding oil phase. Oil-based gels have been developed using a direct mix of LMOGs with various oils at concentrations of 6–35%. In these systems, the powder of LMOGs was directly dispersed in the oil phase using a high-speed blender and was then heated to 85–90°C for 30–60 min, followed by cooling to room temperature. This process is driven by forming physical interactions such as Van

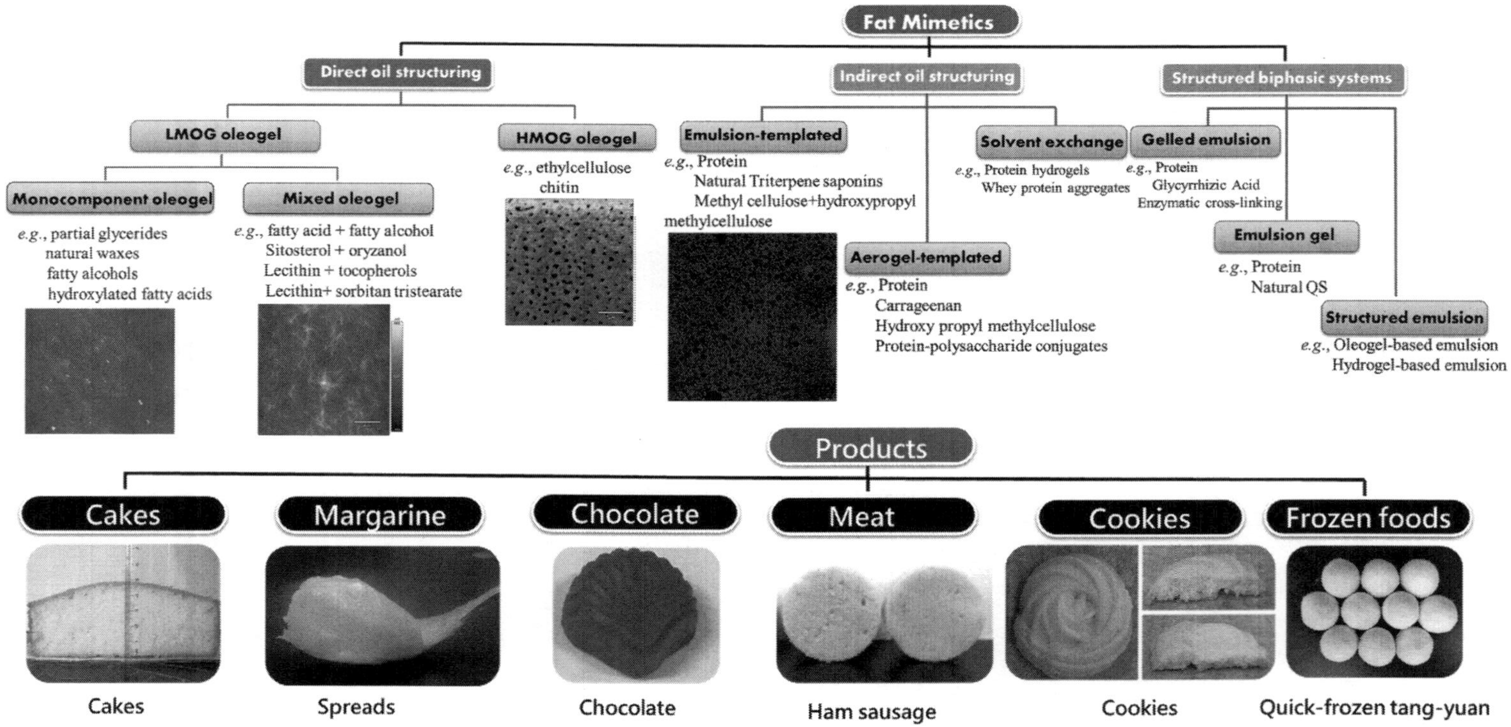

Figure 3.5.2 An overview diagram of approaches for oil structuring and their use in food products.

der Waals, hydrophobic, and hydrogen bonds between the small, temperature-controlled molecules. Among this group are waxes, dodecanoic acid, monoacylglycerols, fatty acids and alcohol, β-sitosterol/γ-oryzanol, and ceramides as gelators. Considering the role of physical interactions in the structure formation and the small molecular building blocks of this network, LMOGs are highly sensitive to temperature and shear forces. HMOGs are high molecular architectures such as ethylcellulose, which are able to form three-dimensional networks through physical interactions like hydrogen bonds. In this case, a polymer-based oleogel was also obtained by directly mixing ethylcellulose (EC) in oil while increasing the solution temperature above the EC glass transition, ~140°C, and cooling below its gelation temperature (Davidovich-Pinhas et al., 2016). This gelation process is driven by polymer dissolution in the oil and occurs above its glass transition temperature followed by hydrogen bonds' formation between the polymer chains upon cooling (Davidovich-Pinhas et al., 2015). The role of hydrogen bonds in gel formation leads to a direct relation between network stability and the temperatures used during preparation and storage. Because of the polymeric nature of this network, such oleogels have viscoelastic properties strongly dependent on the polymer molecular weight, conformation, and concentration (Davidovich-Pinhas 2017). Contrary to traditional triglyceride crystallization, these oleogels annealed at higher temperatures and formed harder gels compared to samples annealed at room temperature and below. This behavior was related to the ability to form a more efficient polymer–polymer hydrogen bonding network at higher temperatures (Davidovich-Pinhas et al., 2015). These oleogels are thermoreversible and behave similarly to solid fats, even when they have high unsaturated fatty acid content. Crucially, oleogel formation does not require chemical or structural changes in the triacylglycerol molecules, thus maintaining the nutritional features of the oil used, content of unsaturated fatty acids, and natural regiospecific distribution.

Among the various oleogelators available, phytosterols, a family of plant-produced molecules, stand out because they are natural and healthy. More importantly, phytosterols have a chemical structure similar to cholesterol, which has positive health implications for the phospholipid membrane in cell structures and the cholesterol-lowering functional features. Phytosterols can form plate-like crystals at high concentration (>10%) and result in turbid oleogel in edible vegetable oil (Figure 3.5.3A). This phenomenon was mainly due to relatively high total sterol concentrations because the microcrystalline suspension in oil solution—rather than the crystal network—was seen at low concentrations (Chen et al., 2016a, from Elsevier). Here, the oleogel-in-water structured emulsions improved colloidal stability during long-term storage and delayed volatiles release under real time dynamic conditions. It was confirmed that a novel phytosterols crystal-octenylsuccinate starch complex interface is generated in the structured emulsions by a gel trapping technology combined with polarized light microscopy and confocal laser scanning microscope. This unique interfacial microstructure might contribute to the strong retention of volatile compounds because of steric barrier and enhanced affinity to those lipophilic volatiles. So, a phytosterols self-assembled structure could be useful in formulating low-fat emulsions with designable volatiles release profile for better flavor perception. Bot et al. prepared the oleogels by adding natural oryzanol and found that the synergistic interactions between sitosterol and oryzanol reduced the sitosterol concentration, which resulted in the formation of highly stable oleogels (Bot and Agterof 2006). Although the crystals of sitosterol in edible oil form needle-like crystals (microcrystalline suspension) at 5% loading, the mixtures of β-sitosterol and γ-oryzanol can co-operative self-assembly through intermolecular hydrogen by intermolecular hydrogen bonds, helping to form oleogel with the hollow tubular microstructures and network (Figure 3.5.3B). Studies showed that in edible oils β-sitosterol and γ-oryzanol formed hollow, helical, fibrillar tubules that gel the edible oil. The gels were translucent, indicative of the relatively small size of the organogelating structures. These

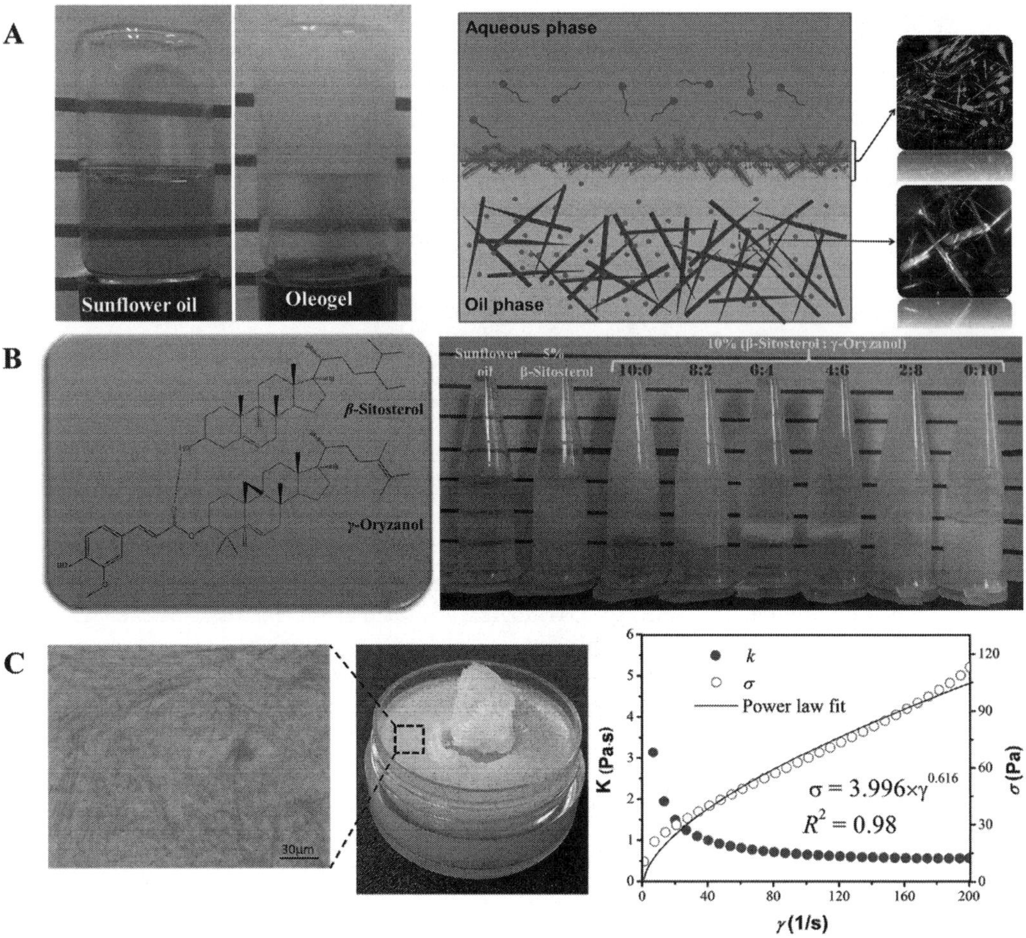

Figure 3.5.3 (A) Appearance of the sitosterol crystals in the oleogel, right is the schematic diagrams showing the formation of the interfacial network and controlling mechanism for volatile release of structured phytosterols oleogel-in-water emulsion, adapted from Chen et al. (2016a); (B) Interaction schematic and photographs of the different sitosterol concentration and ratios of sitosterol and oryzanol quenched in sunflower oil, adapted from Chen et al. (2017a) from Elsevier); (C) appearance, optical microscopy, viscosity and shear stress with power law fit of a sample made with 10 wt% phytosterol and monoglyceride in olive oil at the ratio of 6 : 4 (w/w) (Chen et al., 2020b, from American Chemical Society).

tubules have nanometer dimensions and outer diameter ranging between 6.0 and 8.0 nm and a wall thickness between 0.8 and 1.2 nm (Bot et al., 2008; Sawalha et al., 2011). In this case, structured oleogel-in-water emulsion can be controllable design based on the interface behavior of sitosterol and co-operative self-assembly from sitosterol and oryzanol and clearly showed the capability of delaying volatile releases (Chen et al., 2017a). Specifically, the sitosterol crystalline network interface shell was more effective in preventing volatile releases from directly contacting the capsule shell than the networking the internal phase and can further suppress the release of volatile compounds. The oleogel-in-water emulsion gives new insights into tunable volatile releases over conventional emulsions, promising a system to deliver fat-reduced and sterol-richen functional products with an improved flavor profile. On the other hand, inspired by the controllability of oleogelation and volatiles, an algae oleogel-in-water nanoemulsion was prepared using a one-pot ultrasound emulsification strategy (alternative heating-homogenization) that resulted in algae oil powders after spray drying. This has excellent oxidative stability and reconstitution ability, even after 30 days of storage (Chen et al., 2016b). Following the results of headspace volatiles using dynamic headspace-gas chromatography-mass spectrometry, the structured algae oleogel-loaded nanoemulsion and powder had lower levels of fishy off-flavor (e.g., (Z)-heptenal, decanal, ethanone, and hexadecenoic acid). The structure from oleogel-in-water emulsion was noted as an effective strategy to reduce the fishy off-flavor and to maximize the oxidative stability of both algae oil nanoemulsions and spray dried powders. With this finding, there's a possibility of an oleogel and oleogel-based emulsion enriched in polyunsaturated oils that can be applied to functional foods and beverages.

Generally, molecular and crystal engineering concepts could design and identify new oleogels (Patel 2017b). It is now generally accepted that oleogels are changed by self-assembly with other molecules through physical or chemical interactions into an extensive mesh network that enhances physical stability and functional features (Han et al., 2014). Recently, bicomponent oleogels based on phytosterol and monoglyceride were developed by Yang et al. (2018) and Sintang et al. (2017). Microscopy images of the bicomponent oleogels of sitosterol and monoglyceride showed the formation of a new three-dimensional network of entangled crystals and a controllable microstructure. From differential scanning calorimetry and small angle X-ray scattering, results suggest the self-assembly of sitosterol and monoglyceride via intermolecular hydrogen bonds into spherulitic microstructures. The combination at certain ratios helps form mixed component gels better than either of the mono component gels. On the one hand, the release rate, the maximum headspace concentrations, and the partition coefficients for oleogels had a very controlled release and were tunable via the microstructure of bicomponent oleogels under both dynamic and static conditions. But the solid-like oleogels had interesting thixotropic and thermoresponsive behavior because of intermolecular hydrogen bonding. Here, the viscosity of highly thixotropic oleogels could be tuned by manipulating the concentration of phytosterols and monoglycerides, and the surface-active small molecules and bulk vegetable oils (Chen et al., 2020b). After optimization, the developed viscous soft oleogels could be tuned to have favorable water vapour occlusive and wettability properties with a good texture and an outstanding thixotropic behavior, which can be a novel sustainable source and a green replacer of petroleum waxes in fruit preservation (Figure 3.5.3C). Sintang et al. (2016) also examined a mixture of monoglyceride and phytosterols in structuring sunflower oil. The authors noted a synergistic effect in that the elastic modulus of the multicomponent oleogel was higher than gels formed with the individual components. Han et al. (2014) also point out that adding lecithin can change the thermophology of the phytosterol crystals from a rosette-like morphology to fibrillar morphology.

3.5.2.2 Templated Oleogels

Using hydrophilic polymer as a structurant for oil structuring is challenging, as they do not readily form a network in oil or other hydrophobic environments. Recently, one of the success stories for oil structuring through an emulsion-template approach was shown with packed oil droplets to an interfacial film of emulsifiers followed by a drying process (Adelmann et al,. 2012; Patel et al., 2015). The interfacial film governs the success of oil structuring during dehydration from emulsion precursors. However, most emulsifiers are hydrophilic but do not have enough surface activity for the physical entrapment of liquid oil upon the stripping off of water (Patel 2018; Davidovich-Pinhas 2019; Scholten 2019; Bascuas et al., 2021b). Commonly, dehydration by using spray drying, freeze drying, and air drying often helps disrupt emulsion droplets collapse and phase separation, resulting in oil leakage (Chen and Yang 2019; Moradabbasi et al., 2021). Romoscanu and Mezzenga (2006) proposed to first use an emulsion-templated approach with the emulsifying agent. After the emulsification, with more assistance from thermal cross-linking of the interfacial protein layer, the protein-stabilized oil-in-water emulsions can be converted into protein-gelled oleogels and remove the water. The strengthening of the interfacial network is necessary to prevent the coalescence of the dispersed oil droplets. When the interfaces are strong enough, no oil leakage is seen during dehydration. Here, another strategy to enhance the interfacial properties is by using polysaccharides as an additional ingredient to strengthen the interface. Upon having successful interactions with the emulsifying proteins, the polysaccharides can co-adsorb at the interface and at the network properties of the dried emulsions. In recent years, several polysaccharides have been used to create emulsion-templated oleogels e.g., xanthan gum (Pan et al., 2021a), carrageenan (Tavernier et al., 2017), pectin (Luo et al. 2019), carboxymethylcellulose (Jiang et al,. 2018), and hydroxypropyl methyl cellulose (Meng et al., 2018), among others.

Innovatively, Gao and colleagues (2014) employed a novel strategy to prepare protein-based oleogels with a Pickering emulsion-templated approach. The zein-base colloidal particles were changed by complex with sodium stearate (SS). Here, these particles were used as stabilizers; the resulting Pickering type emulsions showed superior stability against both coalescence and creaming. After removing the water by freeze drying, this liquid emulsion turned into solid-like oil gels (Figure 3.5.4). The stability of these emulsions and oil gels depended on the critical complexation concentration (CCC) of SS (about 5 mM) (Figure 3.5.4A). When the concentration of sodium stearate was below the CCC (2.5 mM), the natural conformation of zein might not be disturbed completely. Here, the intact and inherently hydrophilic zein colloidal particles migrated to the oil-water interface, resulting in an unsaturated particle monolayer at the interface owing to the inefficient accumulation of particles (low surface protein loading) and weak adsorption at the interface. So, the particles did not provide a packed steric barrier for the oil droplets (Figure 3.5.4B). The oil gel obtained after freeze-drying had a little oil leakage (Figure 3.5.4C). However, when the SS concentration was above the CCC (10 mM), the C-16 alkyl chain of SS might be incorporated into the hydrophobic structure of zein, forming the zein-SS complexes with exposed hydrophobic microdomains. This not only improved the diffusive mobility of water-insoluble zein particles, but

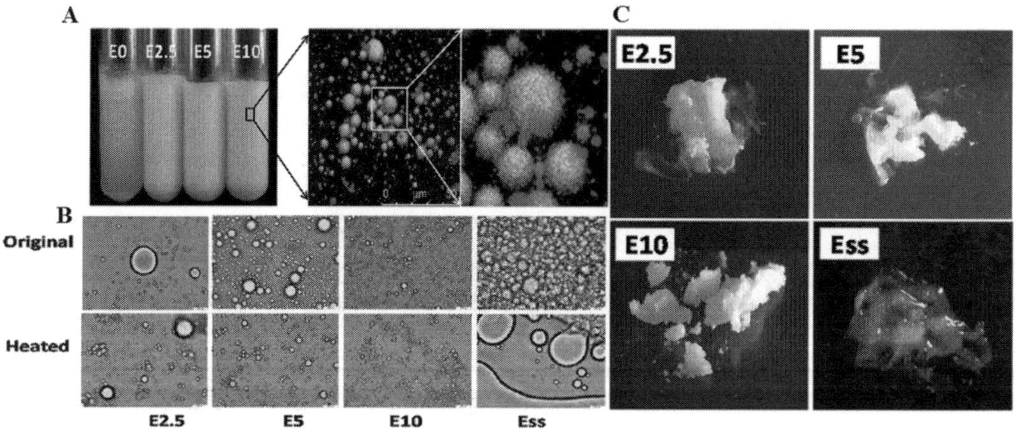

Figure 3.5.4 (A) Visual aspect of the freeze-dried emulsions stabilized by zein-SS and sole SS; (B) Optical microscope images of emulsions before and after heating-salting treatment (90°C/30 min, 200 mM NaCl); (C) Visual aspect of the freeze-dried emulsions stabilized by zein – SS and sole SS: E2.5, emulsions stabilized with 0.5 wt% zein and 2.5 mM SS; E5, emulsions stabilized with 0.5 wt % zein and 5 mM SS; E10, emulsions stabilized with 0.5 wt % zein and 10 mM SS; Ess, emulsions stabilized with 10 mM SS (Gao et al., 2014, from American Chemical Society)

it also endowed particles with equilibrium interfacial wetting properties, leading to a significant increase of surface particle coverage and to efficient packing of zein colloidal particles at the oil-water interface. Finally, a homogeneous and translucent gel without oil leakage was obtained after freeze-drying, meaning that the oil content entrapped in this matrix was over 92% wt. The emulsion only stabilized by 10 mM SS yielded a turbid flowing liquid, indicating that freeze drying did not produce an encapsulated oil gel. It was developed as a possible strategy for structuring liquid oils into semisolid fats structured hydrophobic cereal protein.

From the field's principle of economics and sustainability, it is still essential to identify natural and food-grade materials for liquid oil structuring. *Quillaja* saponin (QS), as natural triterpene glycosides, is mainly extracted from the soapbark tree (*Quillaja Saponaria Molina*) and has been approved for human consumption in many countries and organizations (e.g., China, Japan, US Food and Drug Administration, European Community, and Food Standards Australia New Zealand), owning to a generally recognized as safe (GRAS) status. As a triterpenoid bidesmosidic saponin, the sugar chain of QS is normally attached at C-3 and C-28 of the aglycone. In the molecular framework, the amphiphilic structures would adsorb at the interface to produce an overall free-energy reduction and to form viscoelastic films, which determines their rich, physicochemical properties and their biological activities (Böttcher and Drusch 2017; Tsibranska et al., 2020). Recently, edible oil powders oleogels were fabricated from emulsion templates stabilized by the natural triterpene QS; their rheological and texture properties can be tuned oil by the oil fraction and drying mode (Chen and Yang 2019). QS was found spontaneously, oil-water, interface-driven, self-assembly (SIDSA) into a micrometer-scale network in helical fibers by combined intermolecular π-π stacking and hydrogen bonding among saponins at the liquid–liquid interface. From SIDSA, multilayer films on the surfaces of dispersed droplets formed and emulsion stability was enhanced. Interfacial QS-based films on droplet surfaces limited interfacial diffusion processes by serving as transport barriers (Chen et al., 2020a). The fibrosis network interfacial film from self-assembly of QS at the oil-water interface protected oil droplets in the emulsion system, resulting in the fabrication of stable emulsion precursors and subsequent oil powder and oleogels (Figure 3.5.5). In fact, the transformation process of liquid oil into solid materials can be a function of water evaporation and droplet diffusion, owing to the heat transfer and mass transfer during drying. In comparison to conventional hydrophilic polymer-based emulsion templates, the role played by the interface assembly of QS provides a robust interfacial film enabling high mechanics to resist rupturing during dehydration. In this case, the initial templates stabilized only by natural triterpene QS have enough suitable precursor without a further processing step before dehydration, simplifying the procedure. Moreover, following the fast evaporation of water (e.g., spray drying), the interfacial films was likely to be sintered around oil drops from the QS self-assembly, resulting in a mechanical strength for an oil powder containing as high as 93 wt% oil. The high drying rate would drive the emulsion droplets to prefer diffusion and concentration at the air-liquid surface during drying and usually lead to a hierarchical shell configuration, showing excellent stability, flowability, and reconstitution ability. By using freeze drying to remove water, porous structured solid products followed by oleogels by simple shearing formed. Upon oven drying, the high internal phase emulsion template yielded a translucent oleogel with high oil loading of 98.2 wt%. The mechanical and rheological properties of the oleogels were well-tuned by both the oil fraction and the water evaporation. The fabricated oleogels with high gel strength had encouraging thixotropic recovery and reversibility to reconstituted emulsions, having a significant interest in the food area.

Although QS could supply a powerful precursor without a further processing step before dehydration to prepare oleogels at a relatively low concentrations (15 mg/g), the dose needs to be

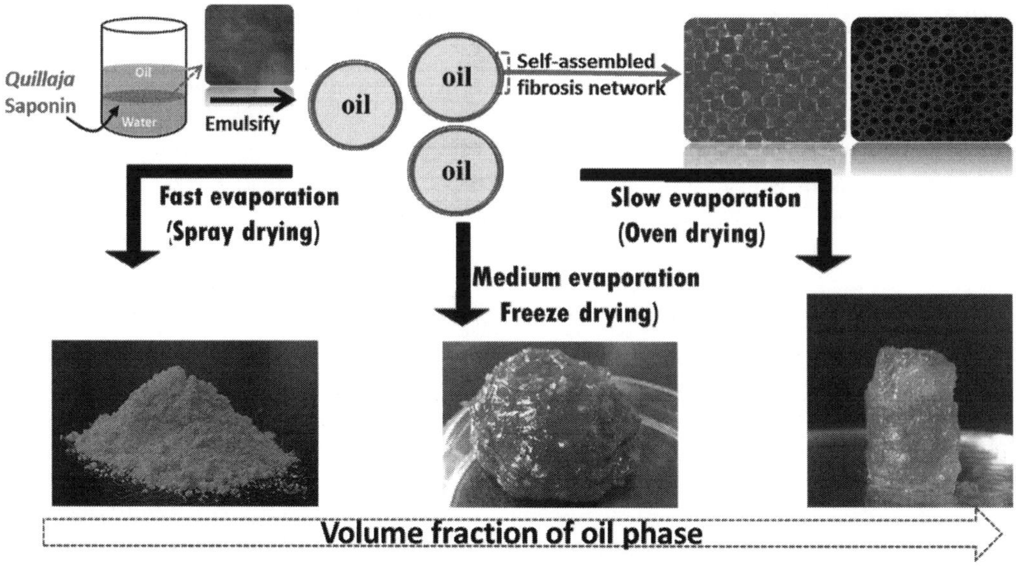

Figure 3.5.5 Structuring of liquid oil into powders and oleogels was achieved using emulsion templates stabilized only by natural triterpene *quillaja* saponin and can be tunable by modulating the oil fraction and water evaporation rate, based on the fibrosis self-assembly at oil-water interface (Chen et al., (2019), from American Chemical Society)

reduced for safety reasons. Inspired by the Pickering principle, relatively monodispersed QS-stabilized nanoemulsion droplets were prepared by ultrasonic emulsification and were used as block stabilizers for the fabrication of hierarchical high internal phase emulsions (HIPEs), where QS-coated nanodroplets were used as shell particles. The HIPEs were then easily converted into transparent oleogels upon water evaporation (Figure 3.5.6A). Generally, oleogels, especially transparent organogels, are only created based on the supramolecular assembly of oleogelators in the bulk oil phase (Rogers et al., 2010). Using hierarchical HIPEs, we firstly created transparent oleogels with an ultra-high oil content of about 99.7 wt% (Figure 3.5.6B). Characterizing by confocal laser scanning microscope, the nanoscaled droplets were stabilized by the QS fibrillar networks, and they were tightly packed to constitute the walls of the microscaled oil phase in the oleogels (Chen et al., 2017b). The nanodroplets were converted into a percolating 3D network that confined the microscale oil within the polyhedral packed nanodroplets upon water evaporation, which formed a self-supported 3D gel. A similar microstructure obtained from the emulsion-templated reversible protein-in-oil gels in the zein stabilized oil-in-glycerol emulsion gels was also found (Romoscanu and Mezzenga 2006; Chen et al., 2016c). Here, the oleogels exhibit unique hierarchical structure features based on the QS self-assembled fibrillar network, which could be explained by a significant amount of liquid oil being trapped by small amount of QS (2.4 mg/g), and the oleogel was transparent. The jelly-like oleogels exhibit weak elastic, shear thinning behavior, good thixotropic recovery, and thermostabilization properties, which are provided by the percolating 3D network of QS fibrils in the oil phase (Figure 3.5.6C). It spatially tuned the color performance of the HIPEs and subsequent oleogels by locating the compositions of food colorants in different sections of their hierarchal architecture (Figure 3.5.6D).

Besides the emulsion-templated approach, another strategy has been used to disperse the proteins in oil and to provide a protein network in hydrophobic environments. de Vries and co-workers used a solvent exchange approach to prepare oleogels using nanoparticles comprising heat-set

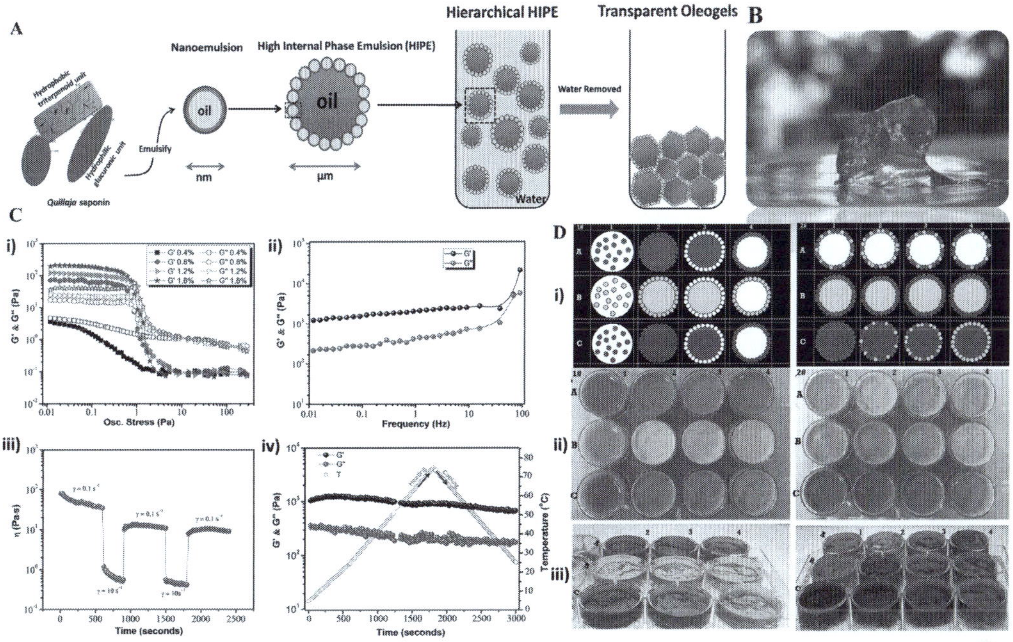

Figure 3.5.6 (A) Schematic for the formation of hierarchical high internal phase emulsions and oleogels from *quillaja* saponin; (B) Appearance of the oleogels obtained by drying the high internal phase emulsions (ϕ = 0.75) formed at 1.2% nanodroplets concentration; (C) Amplitude sweeps, frequency sweeps, flow measurements followed in time against alternating low (0.1 s^{-1}) and high (10 s^{-1}) shear rates and G' and G" versus time at varying temperatures; (D) Schematic of the position of the colorants in the hierarchical architecture (i) and representative image of the brightly colored emulsions (ii) and oleogels (iii) (Chen et al., 2017b, from the Royal Society of Chemistry)

whey proteins (de Vries et al., 2015). In this method, whey proteins were heated to form aggregates roughly 200 nm in diameter. The aggregates were washed with acetone to remove residual water to prevent agglomeration of the aggregates (de Vries et al., 2017). The particles were then dispersed throughout the edible oil solvent to prepare the oleogel. Similarly, Manzocco et al. (2017) generated edible oleogels by using k-carrageenan aerogels as template and affirmed that k-carrageenan oleogels with a maximum oil absorption capacity of 80% (w/w). A facile method to fabricate oleogels by oil absorption in the aerogel templates of protein-polysaccharide conjugates was reported by Chen and Zhang (2020). The alginate/soy protein conjugates characterized by excellent emulsifying activity and emulsion stability were prepared via a Maillard reaction and were freeze-dried to form the aerogel templates, which were then immersed in corn oil within 6 h to induce the oleogels. The resulting oleogels loaded with thymol showed excellent antimicrobial activities against *Staphylococcus aureus* and *Escherichia coli*. This stepwise methodology effectively benefits from the thermal and mechanical properties of many oleogelators in one system. It considerably expands the oleogelators useful for preparing oleogels, which affects the final product.

3.5.2.3 Emulsion Gels

Besides oleogels, emulsion gels are also recognized as a major strategy for structuring low-viscosity liquid oils into plastic fats for the replacement of saturated and hydrogenated fat in food products (Gutiérrez-Luna et al., 2020). Among them, protein-based emulsion gels are of particular interest because of their potential use on texture change, calorie and fat reduction, and amino acid nutrition. Zein, universally accepted as a natural all-grain protein from a byproduct extracted from corn or maize (*Zea mays L.*), has an amphiphilic nature, due to its hydrophobic (>50%, rich in glutamine, leucine, proline, and alanine) and hydrophilic amino acids. It has a highly robust structure made up of nine homologous repeat units arranged in an antiparallel distorted cylindrical form and stabilized by hydrogen bonds, which displays interesting features and offers enormous potential in the self-assembly into various microstructures, including spheres, rods, and films (Wang and Padua 2012). Osborne (1897) reported a method for dissolving zein by using glycerol as the solvent at the temperature of 150°C, and noted that such solution was heat stable, which could be heated to 200°C without apparent denaturation of the protein. The distinct conformation of zein helps form "rectangular tri-blocks" with a hydrophobic core, accompanied by two parallel hydrophilic patches at the surface. This unique geometry of zein resembles a tri-block, which provides an amphiphilic character. In this case, de Vries and co-workers (2014) reported a one-step process of homogenizing to obtain gel based on the orientation-controlled assembly behavior of the "Janus particle" zein (Figure 3.5.7A). Compared with the above preparation of conventional zein-based emulsion and oil gels (Gao et al., 2014; Zou et al., 2015), this strategy had not involved organic solvents and other surfactants, which food industry application possible. More importantly, the resulted gel samples could be obtained even after several cycles of heating and cooling, confirming the self-standing and thermo-reversibility of emulsion gel. Although viable strategies for structuring liquid oils into semi-solid fats that took advantage of the zein gelation behavior had been reported, the edible-related applications for the bioactive compounds delivery and the margarine or shortening alternatives has received little attention. Based on the research of de Vries et al., β-carotene was first used as a nutritional fortifier as well as texture modifier, which strengthened the gel-like network and improved spreadability in the emulsion gels stabilized by zein, which was

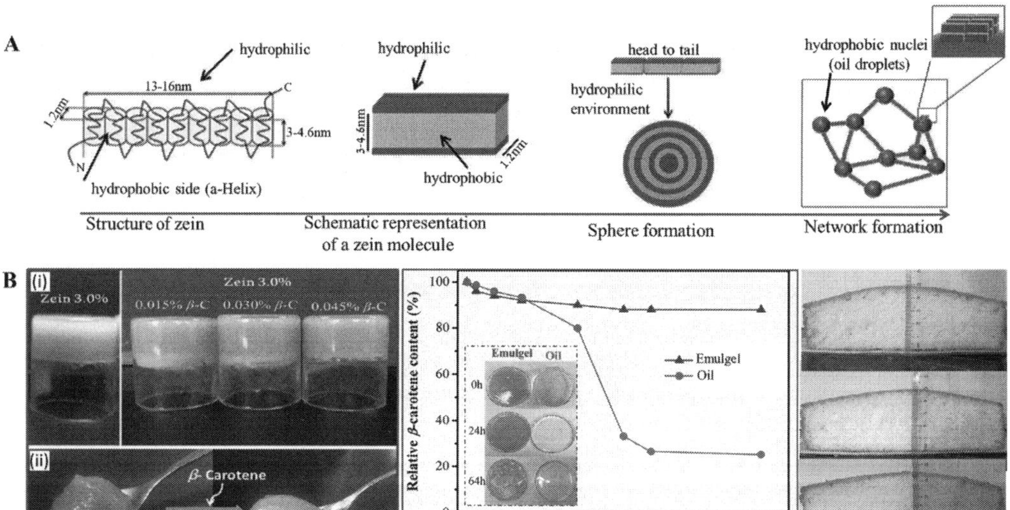

Figure 3.5.7 (A) Structure and schematic representation of a zein molecule, spherical particles, and emulsion gels formation by zein self-assembly in hydrophilic and hydrophobic environments (de Vries et al. (2014) / EPL Association). (B) Physical appearance (i) of zein-based emulsion gels incorporated varying concentrations of β-carotene; Panel (ii) presents the photograph of zein-based emulsion gels enriched with β-carotene (3.0% zein, 0.030% β-carotene) (Chen et al., (2016c), from Elsevier)

closely related to the hydrophobic interaction of zein and β-carotene (Figure 3.5.7B). Conversely, zein-based emulsion gels significantly enhanced the UV photo-stability of β-carotene and retarded oil oxidation (Chen et al., 2016c). Compared to a standard cake prepared by commercial bakery margarine, the low-fat cake prepared using zein-based emulgels showed comparable textural properties (e.g., firmness, cohesiveness, adhesiveness, and springiness), without statistical difference. The gel strength and crossover strain increased with an increase in oil polarity, indicating the formation of a stronger gel structure and higher stability against structure breakdown. the addition of water to the continuous phase can enhance the gel strength and the resistance against structure breakdown for more a polar oil, while the opposite results were seen for the more polar castor oil. The zein-based emulsion gels showed a thermo-responsive behavior, and different responses in structure collapse and rearrangements during heating and cooling were found for different oils (Zou et al., 2019). Based on these results, it is likely that zein-based emulgels could be a fat mimetic for highly saturated and trans fatty acid rich margarine in baked products.

Like zein, wheat gluten has a lot of proline, and they are insoluble in water. However, compared with zein, gluten proteins exhibit remarkable elastomeric properties, which are ascribed to the ability of gluten proteins to form long polymers. Meanwhile, the repetitive domains of gluten proteins are rich in glutamine, which is prone to hydrogen bonding. Liu and colleagues discovered hot glycerol (130°C) as a suitable solvent to dissolve the gluten, and then the wheat gluten-based emulsion gels with high corn oil loading (60%) were successfully fabricated *via* one-step homogenization (Liu et al., 2016). The self-supported emulsion gels with only wheat gluten concentration (0.5 wt%) were formed (Figure 3.5.8). In this system, hot edible glycerol as a green solvent could destroy the important hydrogen bonding interactions of wheat gluten protein chains, de-aggregate the wheat gluten networks into monomers and oligomers of gliadin and glutenin subunits, and form stable wheat gluten-glycerol suspensions. Gliadin as an amphiphilic protein played a dominant role in the formation of interfacial architecture through the assembly behavior at the oil-glycerol interface, while glutenin had synergistic effects on the interfacial networks' strength. Finally, both contributed to forming stable Pickering-type emulsion gels. Unlike typical, high internal phase emulsion gels stabilized by hydrophilic biopolymers, the wheat gluten-based emulsion gels showed higher elastic (G'~10^3–10^4 Pa), shear thinning behavior, good thixotropic recovery, and heat stability.

One of the most recent and exciting innovations in traditional plastic fats are high internal phase emulsion gels (HIPEGs). These gels are not strictly oleogels since they consist of water; however, it would be an encouraging fat mimetic. As the name suggests, HIPEGs are gelled emulsion that have a high-volume fraction of internal phase (oil is above 75% of the mass or volume) resulting in a soft gel. Recently, Liu et al. (2018) reported that oil-in-water HIPEGs for mayonnaise replacers

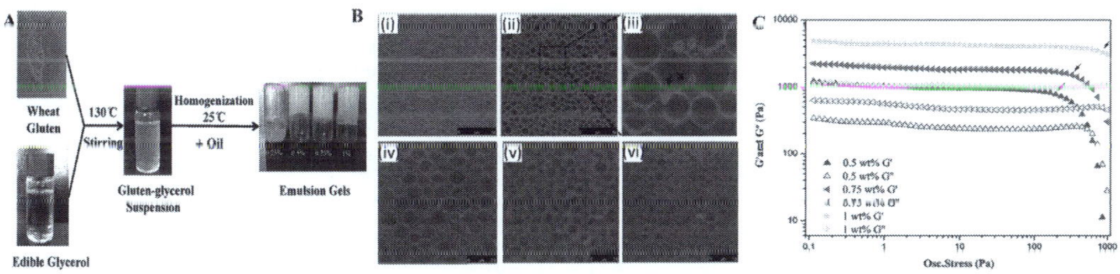

Figure 3.5.8 Schematic representation (A) of the fabrication process of the wheat gluten-based emulsion gels and their microstructure characterized by confocal laser scanning microscopy (B) and rheology and stress rheology (C) of emulsion gels Liu et al. (2016) / with permission from Elsevier).

can also be developed using wheat gluten as an available and low-cost plant protein ingredient via an emulsification-evaporation strategy (Figure 3.5.9A). First, wheat gluten was dissolved into aqueous acetic acid/ethanol (55/45, pH 3.0) solution to form phase-separated wheat gluten particles, and then HIPEGs were prepared through homogenization of high sunflower oil fraction (75 wt%) with wheat gluten particles suspension at 80C° after evaporating ethanol completely.

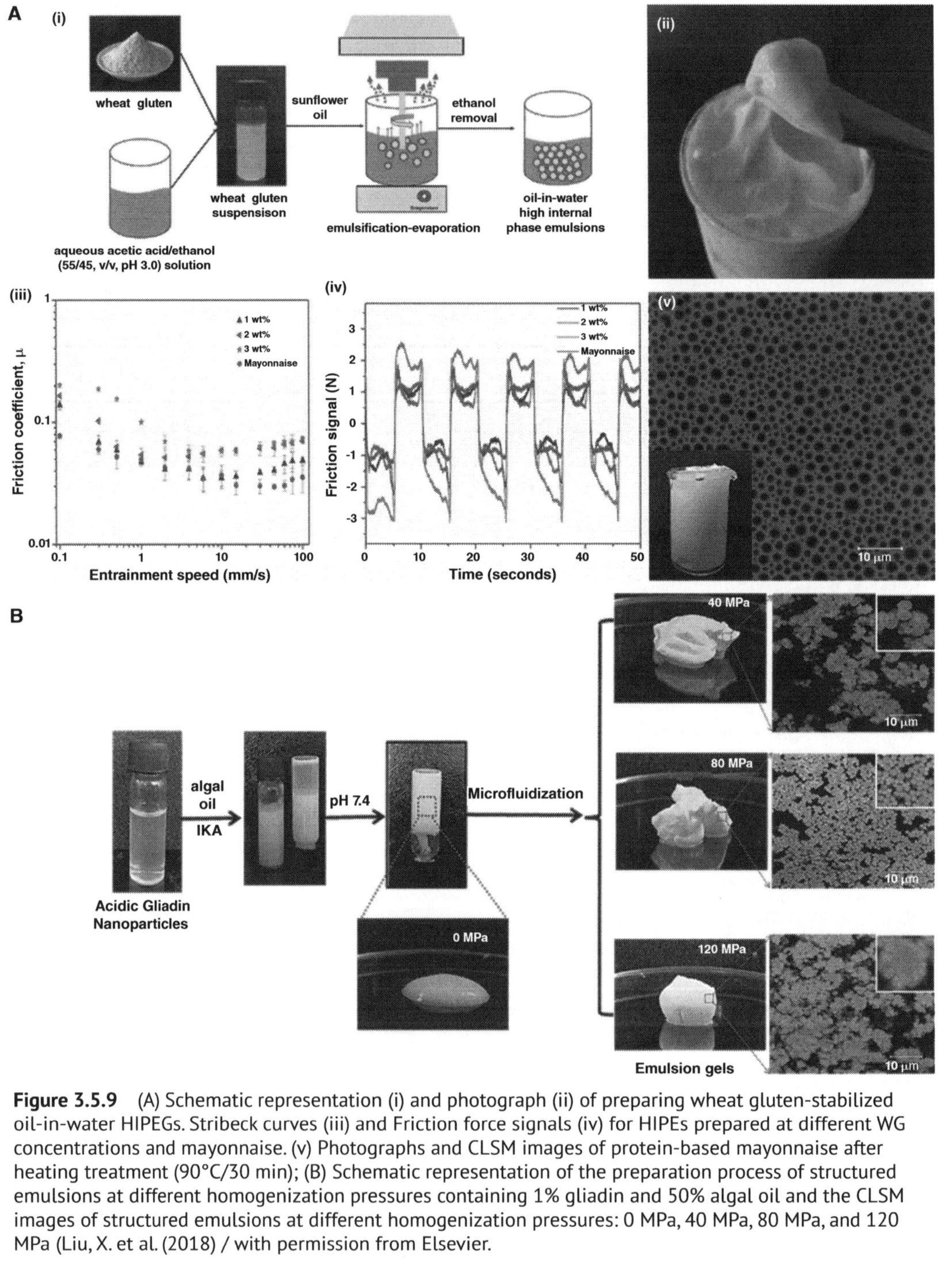

Figure 3.5.9 (A) Schematic representation (i) and photograph (ii) of preparing wheat gluten-stabilized oil-in-water HIPEGs. Stribeck curves (iii) and Friction force signals (iv) for HIPEs prepared at different WG concentrations and mayonnaise. (v) Photographs and CLSM images of protein-based mayonnaise after heating treatment (90°C/30 min); (B) Schematic representation of the preparation process of structured emulsions at different homogenization pressures containing 1% gliadin and 50% algal oil and the CLSM images of structured emulsions at different homogenization pressures: 0 MPa, 40 MPa, 80 MPa, and 120 MPa (Liu, X. et al. (2018) / with permission from Elsevier.

Compared with egg-based mayonnaise, HIPEGs (especially for 1.0 wt% wheat gluten) exhibited very similar droplet size distribution, rheological behavior, near-perfect thixotropic recovery, and tribological property. In addition, thermal stability of HIPEGs was much better than that of the mayonnaise. On the other hand, use of cysteine-rich protein to prepare emulsion gels via intermolecular disulfide bonds is a promising strategy for avoiding the use of chemical cross-linkers. As a result, the authors also reported a facile way to prepare emulsion gels with high polyunsaturated oil stabilized by gliadin particles via microfluidization interfacial covalent cross-linking of gliadin particles with no thermal or chemical treatment (Figure 3.5.9A). During microfluidization, intensive mechanical agitation can accelerate droplet-droplet hydrodynamic interactions and trigger cross-linking of gliadin through intermolecular disulfide bonds. The storage modulus of the emulsions significantly increased to above 10^3 Pa after microfluidization, which is about 10 times higher when compared with the conventional homogenization (Liu et al., 2017). Microfluidization treatment also improved the response of the force curve, and the hardness value of emulsions increased from 0.19 to 1.0^4 N with increasing homogenization pressure. Intermolecular disulfide cross-linking at the interface acted as a physical barrier, which reduced both the primary and secondary oxidation products of emulsions. Accordingly, unlike conventional cross-linking through thermal or chemical means, this facile approach for cross-linking at the interface by microfluidization provided a novel technology platform designing functional edible oil structuring. Monodispersed gliadin/chitosan hybrid particles used as a particulate emulsifier developed Pickering high internal phase emulsion gels as a potential alternative for partially hydrogenated and saturated fats (Zeng et al., 2017). The hybrid latexes are effectively adsorbed and anchored at the oil-water interface to exert steric hindrance against the coalescence. Concomitantly, the compressed droplets in Pickering HIPEGs to form a percolating 3D-network framework endowed the emulsions viscoelastic and self-supporting features. This strategy transformed liquid oils into viscoelastic solid-like gels without artificial trans. Additionally, the Pickering HIPEGs have a protective effect upon the chemical stability of the encapsulated curcumin after UV exposure as compared to bulk oil, and the primary lipid oxidation (the formation of lipid hydroperoxide) in HIPEGs was slightly retarded compared with that in bulk oil. Actually, fat mimetics available on the market are those of the continuous aqueous phase type. Coasun™ is a commercial example of a fat produced with monoacylglycerols (3%-5%). It consists of a structured O/W emulsion containing 35%-40% water; it is low in saturated fatty acids, has zero trans fats, and is highly stable on heating (Marangoni and Idziak 2010).

3.5.2.4 Structured Emulsions

Structured emulsions are other systems shown to be promising alternatives to trans and saturated fats, as they can be formulated to have a solid-like structure like that of hard-stock fats. Microscopically, structured emulsions are a type of complex soft-solid emulsion materials with both the properties of traditional multiphase emulsions and the behavior of physical gels, in which the oil droplets or water droplets are entrapped within a gel matrix. In comparison with oleogels, more biopolymers like proteins and dietary fiber can generate structured emulsions (Curti et al., 2018; Paglarini et al., 2020), but monoglycerides are commonly used as small-molecule structuring agents for oil-in-water (O/W) emulsions (Ferro et al., 2021). Given the enormous potential of structured oils in different fields, it is still essential to identify natural, sustainable, and food-grade materials for liquid oil structuring. Recently, increasing efforts have been made to use triterpenoid saponins like glycyrrhizic acid (GA) as structural building blocks to form supramolecular assemblies because of their multiple functional groups, rigid skeletons, and unique stacking behaviors

(Saha et al., 2015). GA, a natural triterpenoid saponin, is considered the main ingredient of licorice root extract and is widely used in candies and sweets due to its intense sweetness (50 times sweeter than sucrose). Chemically, GA is a monodesmosidic saponin comprised of a hydrophobic triterpenoid aglycon moiety (18β-glycyrrhetinic acid) attached to a hydrophilic diglucuronic unit, resulting in an expected hierarchical self-assembly behavior in aqueous solution forming supramolecular hydrogels (Figure 3.5.10A). The formation of GA nanofibrils (2.5 nm in diameter) results from lateral interactions between the hydrophobic triterpenoid moieties of GA molecules, yielding a head-to-head configuration and leaving the hydrophilic sugar groups exposed to water. Recently, Wan et al. found that the GA nanofibrils exhibit significantly remarkable behaviors at the liquid interfaces (Wan et al., 2017a). The GA nanofibrils can rapidly adsorb at the interfaces and can reduce the interfacial tension, which endow them with a high emulsifying capacity, providing a GA fibril-stabilized emulsion gels. In this system, the liquid oil was covered by semiflexible GA nanofibrils and embedded in supramolecular hydrogels constructed by fibril network. Here, the increase in GA fibril concentration positively affected the enhancement of gel strength and mechanical properties of emulsion gels (Figure 3.5.10C). The mechanical properties of the emulsion gels could be well-tuned by oil phase polarity, where the more polar oils and the higher mechanical strength (Wan et al., 2017b). Based on the gel strength, the emulsion gels emerged the shape of alphabets after cooling the extruded warm emulsion gels at room temperature (Figure 3.5.10B). They successfully prepared stable gel emulsions with many interesting rheological behaviors too, such as high gel strength and good ability of thixotropic recovery. Owing to the thermal sensitivity of the nanofibril self-assembly from the hydrogen bond interaction, the emulsion gels also have interesting temperature responsive behavior due to the thermo-reversible gel–sol transitions of the fibrillar

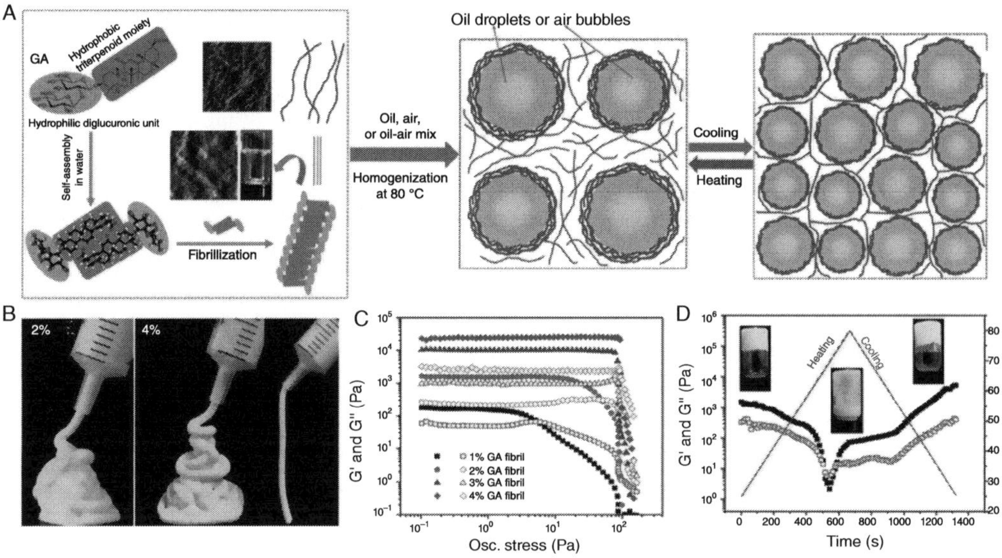

Figure 3.5.10 (A) Schematic illustration of the fibrillar self-assembly of glycyrrhizic acid (GA) molecules in water, forming GA nanofibrils and hydrogels, and of the formation of thermo-responsive gel emulsions by the multilayer adsorption of GA nanofibrils at the oil–water interface and the assembly of fibrillar hydrogel network in the continuous phase after cooling; (B) Photographs of the emulsion gels with the shape of alphabets after cooling the extruded warm emulsion gels (2 and 4% GA fibrils) using a plastic syringe at room temperature (25 °C) for 1 min; (C) Amplitude sweeps for emulsion gels containing 60 wt% olive oil prepared at different GA fibril concentrations (1–4 wt%); (D) Storage modulus (G') and loss modulus (G'') of emulsion gels containing 60 wt% olive oil prepared using 2 wt% GA fibrils, and photographs (Insets) of respective emulsion gels during heating and cooling (Wan et al,. (2017a), from Royal Society of Chemistry).

network in the continuous phase (Figure 3.5.10D). Thus, by simply changing the temperature, the emulsion gels can be switched reversibly between a gel and liquid dispersion (sol), promoting the application as fat mimetic in frozen food and enhancing the protection of functional ingredients (Wan et al., 2017a). Furthermore, such unique multilayer fibril shells of droplets with high electrostatic force could provide superior stability for the GA fibril-stabilized emulsion droplets and the whole emulsion gel during storage and heating. In the case of multilayer fibril shells protection, emulsion gels can be a template for preparing soft oleogels containing nearly 91 wt% vegetable oil (Wan et al., 2017b). Another commercial fat is CUBIQ FOODs' "Smart Fat," an O/W emulsion containing 40%-50% water stabilized with vegan ingredients and has a melting point between 60 and 80°C. Smart Fat is mainly used as a substitute for animal fat and coconut oil and reduces the number of ingredients required for processed products, enabling the so-called "clean label".

Replacing solid fat in water-in-oil emulsions for oleogel is a new approach to these structuring systems (Silva et al., 2021a). Low-molecular weight amphiphilic structures may stabilize and structure water-in-oil (W/O) emulsions simultaneously. Here, the aqueous phase remains immobilized in the continuous gelled oil phase. Recently, structuring edible liquid oil into oleogel-based emulsions was developed by trapping aqueous droplets inside the crystal network structure of phytosterols resulting in a bicomponent oleogels with monoglycerides. First, 10 wt% oleogelators (including 6 wt% phytosterols and 4 wt% monoglycerides) were dispersed into 90 wt% liquid olive oil and formed oleogel by self-assembly at a heat-cool strategy (Figure 3.5.11A and 3.5.11B). Then, soft oleogel-based emulsions were prepared by adding 20 wt% water or glycerol to oleogels followed by homogenizing at 50°C (Figure 3.5.11C). Visually, the resulting oleogels and emulsions were used to obtain the self-supporting soft solids with predetermined shapes by squeezing out the oleogel/emulsion using a plastic injector (Figure 3.5.11D-F). Both the oleogels and emulsions could be shaped into different geometry and had the characteristics of being malleable and smearing, which attributed to the crystal-network. Characterizing the microstructure by polarized light microscope (PLM), the emulsion has shown a water-in-oil type and glycerol-in-oil type, respectively. The emulsion's appearance differed with the oleogel's opacity given a dispersed aqueous phase. The dispersed aqueous phase was stabled by the crystals (arrows), suggesting Pickering-type stabilized emulsions consisted of crystal shells around the dispersed aqueous phase (Zembyla et al., 2020). Comparatively, glycerol within the oleogel enhanced storage modulus more so than expected by increasing water content, particularly at higher aqueous loads. Rheological measurements show that the gel strength of the oleogel-based emulsions depends on the fraction of the aqueous phase and is greater than that of corresponding oleogels. Also, these soft matters had a thermoreversible nature, which results in a higher flexibility for controlled release of different bioactives upon exposure to a proper thermal triggered switchable. Based on the health benefits of phytosterols, the soft oleogel-based emulsions provide opportunities to be applied in delivering functional active ingredients and in fat mimetics in various foods. Crucially, the emulsification of phytosterols could be an effective method to increase phytosterol solubility in the oil phase, decrease crystallization, modulate the rheological properties, and increase the effectiveness of these bioactive compounds (Moschakis et al., 2017). Silva et al. (2021b) developed low- and high-lipid margarines with oleogel-based emulsion. Although structured emulsions have not been used commercially yet, adding water could result in even more versatile systems wide for applications. For example, structured emulsions having a continuous oily phase could prepare margarines with low saturated fatty acids and healthier meat products (Kangchai et al., 2018). These food products face questions related to the high content of saturated and trans-fats, and structured emulsions are an alternative, healthier lipid source with tunable viscosity properties. Besides, these heat-sensitive emulsion gels could use as solid fat substitutes (such as cocoa butter) by 3D printing in the food industries (Shahbazi et al., 2021).

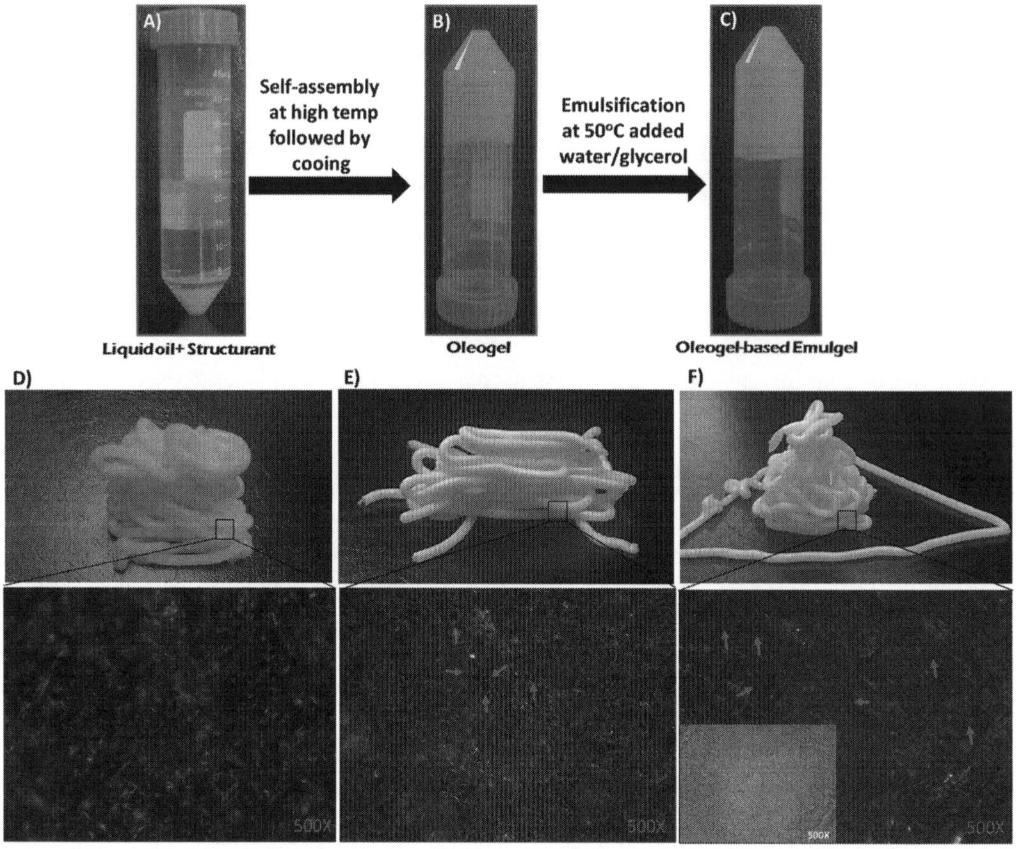

Figure 3.5.11 (A and B) The preparation process involved a high temperature (~85 °C) dissolving of a mixture of liquid oil in the presence of 10 wt% structurant (6 wt% phytosterols and 4 wt% monoglycerides) followed by cooling to room temperature helping to immediately form an oil continuous soft oleogels; (C) Soft emulsions prepared by emulsification at 50°C added 20 wt% water or glycerol to oleogels. (D-F) Spread-like appearance (upon) and polarized microscopy images (below) of oleogel (D), water-in-oil emulsion (E, W/O) and glycerol-in-oil emulsion (F, G/O) stabilized by crystals. The presence of both interfacial crystals (shown by arrows) and continuous phase crystals can be seen (Chen et al., (2022), from American Chemaical Society).

3.5.3 Multifunctionality of Fat Mimetics in Food Applications

Current nutritional guidelines recommend decreasing trans and saturated fats intake to improve blood lipids and reduce cardiovascular risk. Although the functional properties of fat mimetics differ greatly from those of conventional fats, they yield similar and, often, better results when applied in foods (Patel et al., 2020). Studies involving food-grade fat mimetics are applicable for in various food types (Figure 3.5.2B), such as sausages, cream cheese, margarine, cookies, ice cream, and spreads for replacing trans and saturated fats, reducing fat intake, improving *in vitro* and *in vivo* digestion, controlling delivery carriers and release of bioactive molecules, texture design and changes, reducing lipid oxidation and environmental applications (Figure 3.5.12) (Demirkesen and Mert 2020; Patel et al., 2020; Silva et al., 2021a). The detailed functionality of fat mimetics in food applications is presented as follows (Table 3.5.1).

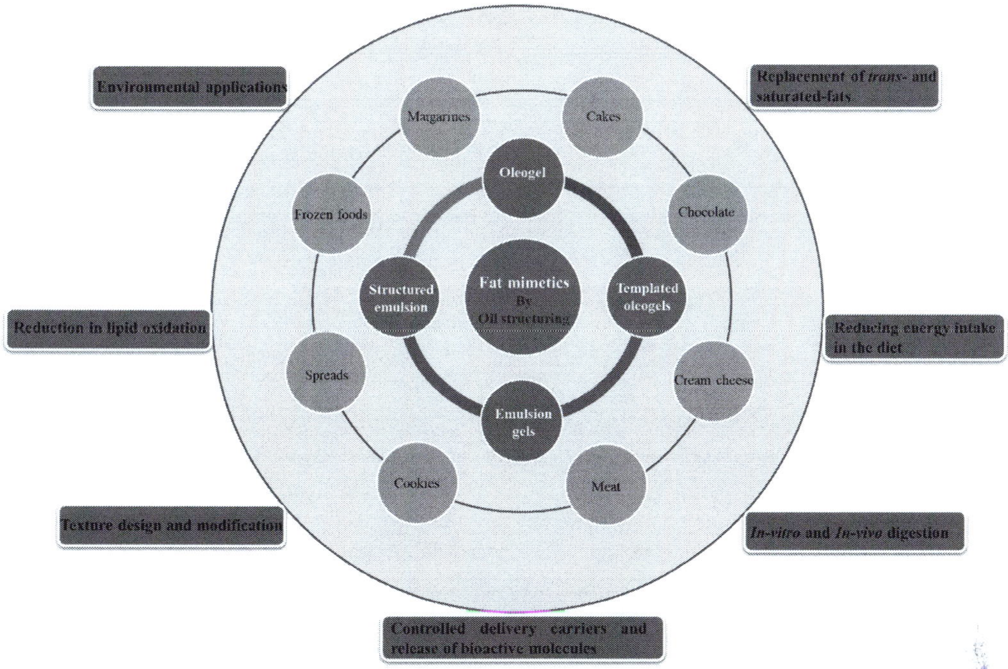

Figure 3.5.12 Nutritional and functional properties of fat mimetics in food systems.

3.5.3.1 Replacement of Saturated Fats

Driven by the need for saturated fat replacement strategies, the field of structuring of liquid oils rich in unsaturated fatty acids has made tremendous advances over the past decade; this is largely owing to human health benefits and to a richer constitution than typical of solid fats. Edible applications of fat mimetics have replaced the need for trans unsaturated and saturated fatty acids to structure processed foods. For example, Hwang et al. studied the feasibility of cookies in which the conventional margarine is replaced with an oleogel of vegetable oil and natural wax. It was confirmed that sunflower wax and rice bran wax provided the highest hardness to cookie dough. Several cookies rich in unsaturated fats made with wax-olive oil oleogels showed similar properties to cookies made with a commercial margarine (Hwang et al., 2016). When using protein oleogels obtained by the solvent-exchange approach, the consistency, hardness, and the stickiness of the dough was similar to the consistency of the dough with commercial margarine (Scholten 2019). Here, the dough resisted the rheological changes during baking. Resultantly, the appearance and internal structure of the oleogel cookie was like that of the reference cookie with margarine. Unfortunately, mechanical tests show that the cookies with protein oleogels break easier, which was also seen during sensory tests because the protein oleogel cookies scored lower on the sensory attribute of crunchiness. However, some panelists also preferred the crumblier structure of the cookies prepared with the protein oleogel. Meanwhile, the protein oleogel was successfully used to replace pork fat in Frankfurter sausages (de Vries 2017). Similarly, sausages with the protein oleogels showed the same breaking/fracture force as the reference sausages with pork fat. Oh and Lee (2018) found that up to 50% of shortening could be replaced with fat mimetics in muffins without negatively affecting the finished baked products even though the viscoelasticity of the batter was reduced. Similarly, for gluten-free aerated products, a partial replacement of shortening (up to

Table 3.5.1 Examples of applications and functionalities of different types of fat mimetics in the food systems.

Food systems	Types of fat mimetics	Functional	Reference
Chocolate	Ethylcellulose	Heat-resistant	Stortz, Laredo and Marangoni et al. (2015)
Chocolate paste	Shellac oleogels	Replace saturated fat; heat-resistant	Patel et al. (2014a)
	Hydroxypropyl methylcellulose	Give a more homogeneous structure	Bascuas et al. (2021a)
Cakes	Polymer emulsion-templated oleogel	As a shortening alternative	Patel et al. (2014b)
	Rice bran, candelilla wax, beeswax oleogels	Solid fat replacer	Oh et al. (2017)
	Zein emulsion gel	100% Replacement of margarine	Chen et al. (2017b)
	10% Carnauba wax oleogel	25–50% of shortening replacement	Kim et al. (2017)
Muffins	Monoglycerides oleogels	Solid fat replacer	Giacomozzi et al. (2018)
	Hydroxypropyl methylcellulose foam-structured oleogel	Shortening replacer	Oh and Lee (2018)
Ice Cream	Rice bran wax	Replace saturated fat	Moriano and Alamprese (2017)
Cheese	Rice bran wax oleogel	To replace solid fat content and create ideal fat structure in ice cream	Park et al. (2018)
Bread	Ethylcellulose oleogels	Shortening replacer	Ye et al. (2019)
Cookies	Beeswax/sunflower wax	More desirable spreadable	Yılmaz and Ogütcü (2015)
	Candelilla wax	30% replacement of shortening	Mert and Demirkesen (2016a)
	Sunflower wax, rice bran wax, beeswax, and candelilla wax	Replacement of shortening	Hwang, Singh, Oh and Loo (2016).
	Foam templated oleogel	50% and 75% Replacement level of shortening	Tanti et al. (2016)
	Carnauba wax oleogel emulsion	Rich in unsaturated fatty acid and low in saturated fatty acid	Pehlivanoglu et al. (2018)
Frankfurter	20% Mixture of monoglycerides and phytosterols oleogel	50% Replacement of pork back fat	Kouzounis et al. (2017)
	10% or 20% Oryzanol and sitosterol oleogel	50% Replacement of pork back fat	Panagiotopoulou et al. (2016)
	8% Beeswax oleogel	25% and 50% of pork back fat	Franco et al. (2019)
	Ethylcellulose	Replace saturated fat	Zetzl et al. (2012)
Meat patties	Hydroxypropylmethylcellulose Foam templated oleogel	50% replacement of beef	Oh et al. (2019)
Burger	Gelled emulsion	High unsaturation contribution	Poyato et al. (2015)

Table 3.5.1 (Continued)

Food systems	Types of fat mimetics	Functional	Reference
	Ethylcellulose oleogel	Replace animal fat	Moghtadaei et al. (2021)
	Oleogel-based emulsion	Fat-reduced beef burgers	Özer and Çelegen (2021)
	10% Beeswax oleogel	50% Replacement of beef fat	Moghtadaei et al. (2018)
Meat Sauce	Monoglycerides of fatty acids and fatty alcohols		Lupi et al. (2014)
Bologna sausages	High oleic sunflower oil gelled by pork skin	Replacement of 50% pork back fat	da Silva et al. (2019)
Fermented sausages	Oryzanol and sitosterol or beeswax oleogel	20% and 40% Replacement of pork back fat	Franco et al. (2020)
Tahini halva	Sunflower, shellac, and beeswax	Prevent oil migration	Öğütcü et al. (2017)
Peanut butter	Sunflower wax oleogel	Replace hydrogenated or tropical fat	Winkler-Moser et al. 2019
	Rice bran wax	Replace hydrogenated or tropical fat	Huang et al. (2020)
Margarine	Oleogel-based emulsion	Replace saturated fat	Silva et al. (2021b)

45%) with fat mimetics resulted in final products with comparable properties to the control (Demirkesen and Mert 2019). To replace the hydrogenated or tropical oils, natural waxes were evaluated as stabilizers in peanut butter, reducing saturated fats and satisfying clean label requirements. At levels ≥0.5%, all waxes improved oil-binding capacity (OBC). Sunflower wax and hydrogenated cottonseed oil had the highest OBC, followed by rice bran wax, candelilla wax and beeswax. All waxes reduced oil separation after six months at 22 ± 2°C (Winkler-Moser et al., 2019). Huang et al. (2020) further optimized the formula and process by adding rice bran wax at 4.0 wt% with a heating temperature at 85°C and cooling temperature at 20°C, where the oil loss of peanut butter decreased from 12.19% to 4.04%, and the adhesiveness of peanut butter increased from 23.5 to 165.9 g·sec. The best stabilization effect was obtained when wax formed uniformly distributed needle-like crystals and peanut particles. Similarly, Blake and Marangoni (2015) successfully showed the use of structured oil in water emulsions as a zero trans and reduced saturated fat laminating shortening substitute for puff pastry products. These results suggest the promising potential of highly concentrated emulsions without trans fats as a safe alternative to partially hydrogenated oils to mitigate the risks of coronary heart diseases.

3.5.3.2 Reducing Energy Intake in Diets

The global prevalence of being overweight and obese and associated chronic health conditions continues to increase. Consumers demand reducing energy intake due to weight and chronic disease concerns. The amount of cake margarine used for baking contributes significantly to the total calories and the fat content (sometimes called "hidden fats") of the end products. Ferro and coworkers pointed out that monostearate-based oleogels can be a possible candidate for developing meat products with reduced fat content, or, more important, to provide a product richer in unsaturated fatty acids that retains the desired features of the traditional product (Ferro et al., 2021). Oleogel-based emulsion where structured liquid oil is a continuous phase was evaluated for bakery applications to decrease the total saturated fat content in bakery products such as cakes. Patel and co-workers (2015) reported that the absence of solid fat crystals in oleogel-based emulsions led to a runnier consistency of shellac oleogel emulsion batter as compared to the standard sample (prepared using cake margarine). The decrease in the batter's firmness also influences the air incorporation. However, despite the difference in the batter's consistency and density, baked cakes showed mostly comparable texture and sensorial attributes. Although replacing animal fat with oleogel-based emulsion in beef burger resulted in lower hardness, chewiness, and springiness values during storage, the oleogel-based emulsion incorporation group with 5% of oleogel-based emulsion, 10% of beef fat, and 1% of peapod powder had higher acceptability scores at the end of the storage (Özer and Çelegen 2021). In this case, innovative fat mimetics were developed based on low energy by-products with health benefits.

3.5.3.3 *In Vitro* and *In Vivo* Digestion

In addition to saturated fat reduction, fat mimetics have been proposed as a matrix to alter food-body interactions such as altering digestion and to influence the physiological effects of oils. Feeding experiments with oleogels have recently emerged, which aim to modulate lipid digestion, achieve controlled release of lipid-soluble nutraceuticals, reduce postprandial plasma triglycerides, reduce appetite, and decrease adipose tissue accumulation (Tan et al., 2017; Calligaris et al., 2020). For example, O'Sullivan, Davidovich-Pinhas et al. (2017) used an *in vitro* digestion model to show that oleogels with a higher concentration of ethylcellulose exhibited a significantly lower amount of both lipolysis and β-carotene transportation. A study by Duffy et al. on oleogel digestibility through *in vitro* models found that lipolysis occurring that during the digestion of phytosterol gelled oleogels decreased compared to un-gelled edible oils (Duffy et al., 2009). Similarly, the results reported by Tan et al. (2017) indicated that fats trapped in the three-dimensional network of digestive delays revealed meals containing postprandial levels of glucose and insulin of oleogel were more like meals containing no fat than meals containing un-gelled coconut oil. Postprandial triglyceride levels decreases the consumption of ethylcellulose oleogels when compared to un-gelled oil. Thus, oleogel structure in some way (e.g., lipase activity hindered by either gelation limiting migration of the oil to the surface or enzyme amount restriction caused by adsorption of the lipase to the hydrophobic surfaces within the oleogel structure) limited interactions between lipases and oil (Dong et al., 2020; Guo et al., 2019). However, contradictory results were reported by Hughes et al. (2018) that the post-prandial increase in serum lipids resultant from the digestion of the 12-HSA oleogel was not significantly different from that of liquid oil, but it differed significantly from consuming solid butter or margarine. These studies indicate that the digestion and lipolysis of an oleogel could depend on the gelator used. Ashka et al. pointed out that the confrontation between different LMOGs and HMOGs under simulated intestinal conditions showed different levels of breakdown and FFA release as the increase of the molecular weight of the gelators led to a lower lipid hydrolysis (Ashkar et al., 2019). Upon *in vitro* digestion, fatty acid release as a function of digestion time was greatly affected by oleogel structure: the extent of lipolysis decreased as oleogel strength increased (mixture of β-sitosterol and γ-oryzanol < rice bran waxes < monoglycerides). On the other hand, the nature of the oleogelator affected curcuminoid compounds bioaccessibility, which was lower in oleogels containing crystalline particles (Calligaris et al., 2020).

3.5.3.4 Controlled Delivery Carriers and Release of Bioactive Molecules

The health benefits of nutraceutical compounds that protect a person from lifestyle-related disorders have rapidly gained attention from the scientific community, industries, and consumers.

Unfortunately, the bioactivity and bioavailability of functional nutraceuticals at targeted sites within the gastrointestinal tract is largely compromised due to environmental stresses. Of all the applications, fat mimetics have the highest capacity to become controlled delivery carriers (CDCs) and to enhance the delivery of bioactive bioactives and flavors. Active ingredient delivery can easily be modulated as the structurant concentration directly influences the microarchitecture of the structurant network and, subsequently, the strength of the fat mimetics (Pinto et al., 2021). One of the main advantages of oleogels is the possibility of carrying lipophilic bioactive compounds, which is of great utility in food applications (Zhao et al., 2021). The potential of oleogels to deliver bioactives has been shown in a few studies involving β-carotene as the model bioactive (Martins et al., 2017; O'Sullivan, Barbut et al., 2017). Hughes et al. (2018) studies show that only a negligible amount of β-carotene was transferred during the gastric phase (evidencing structural maintenance of the gelator network and consequent bioactive protection); the major changes (bioactive transfer) occurred during the intestinal stage evidencing the delivery of highly aqueous-insoluble molecules through digestion of oleogels, representing a significant achievement. Li et al. (2019) also used a mixed oleogel structured by β-sitosterol and lecithin as a carrier for poorly water-soluble curcumin and found that the developed structure protected the curcumin from being oxidized and enhanced the bioaccessibility of curcumin in fasted state (up to 67.66%). Their incorporation in oleogel systems could be interesting in the way that such environmental conditions could favour polyphenols absorption in the gut since their interaction with proteins, fibers, and polysaccharides will determine their absorption (Luo et al., 2019). Moreover, the addition of a surfactant (e.g., sorbitan monopalmitate) could modify the structures of ethylcellulose oleogels for higher solubility and stability of curcumin (Liu et al., 2020). In fact, the gelled continuous phase in an emulsion may also work as multiple cavities for nutraceuticals, e.g. riboflavin (vitamin B_2), calcium, probiotics, and anthocyanin (Ojeda-Serna et al., 2019; Mao et al., 2020). Zhuang et al. (2021) reported that a semisolid soy lecithin-based oleogel (Hwang and Winkler-Moser, 2020) emulsion can entrap probiotics and increase the survival of *L. acidophilus* and *B. lactis* grown in de Man, Rogosa, and Sharpe broth for at least 42 days. In addition, the microstructure regulation and temperature sensitivities of many oleogels and structured emulsions make them useful in controlled release scenarios. For example, phytosterol-based oleogel was modifiable physicochemical properties by monoglyceride for achieving a desired controllable release (Yang et al., 2018). Davidovich-Pinhas, Barbut, and Marangoni (2016) confirmed that only sufficiently ethylcellulose hard oleogels can provide protection to β-carotene and have the potential for slow-release delivery application. This same observation was made in the context of increasing the strength of 12-HSA in soybean oleogel used for the delivery of ibuprofen and policosanol in olive oleogels containing ferulic acid (Iwanaga et al., 2010; Lupi et al., 2013). Tan et al. (2014) also conducted *in vitro* experiments to study the potential of these highly concentrated emulsion gels to deliver β-carotene. The results showed that at a low gelatin particle of 0.5 wt%, the initial release time was 36 h, and 35% β-carotene was released. In comparison, at a gelatin particle of 1.5 wt%, the initial fast release time was 14 h, and 15% β-carotene was released, followed by a sustained release pattern. These delivery systems are ideal for controlling the release of bioactive molecules.

3.4.3.5 Texture Design and Modification

Hardstock fats contribute to several food products' properties such as sensory, textural, physical structure, and aroma. Using liquid oil, the direct replacement of saturated and trans fat in foods always leads to compromised food hedonics due to changes in food flavor, texture, and mouthfeel. In this case, the properties of the oil must first be made to mimic those of the solid fat to replace saturated fats (solid fat) with unsaturated fats (liquid oil) without compromising the sensory and quality features of the food product. Some researchers tested the incorporation of oleogels (e.g., plant wax, monoglyceride) into baked products (e.g., cookies) as a replacement for traditional fat, which was unsuccessful owing to increased denser or harder in the final product (Blake and Marangoni 2015; Mert and Demirkesen 2016b). On comparing the dough properties (such as density, microstructure, and rheology) and finished product characteristics (texture, stability, and sensorial aspects), it is often found that although the rheology of intermediate products prepared using fat mimetics is less developed, the properties of baked products were comparable to the products prepared with bakery fats (Patel et al., 2014; Mert and Demirkesen 2016b; Curti et al., 2018). For instance, Hwang et al. (2016) studied replacing bakery fat with oil gels made with four natural waxes (sunflower wax, beeswax, rice bran wax, and candelilla wax) and three vegetable oils (olive, soybean, and flaxseed). They found that while the oleogel hardness and melting behavior were significantly affected by the wax and oil, the cookie properties (hardness, spread factor, and fracturability) were not significantly affected by different waxes and oils. Jang et al. (2015) reported that replacing shortening with wax oleogels increased the spread factor and reduced the snapping force, thus, producing cookies with soft eating features. Yılmaz and Öğütcü (2015) also showed that consumers would use these oleogels as alternative spreads to butter and that more than half of the consumers who tried or bought the oleogel purchased repeat. Bologna sausages were produced with 25, 50, 75, and 100% of their pork fat content replaced by monoglyceride-based oleogels (Ferro et al., 2021). Adding oleogels as a fat substitute made the sausages lighter and there was an observable small hardness increase in the sausages that had all of the fat replaced with oleogels. However, all samples were favorably accepted by consumers and no significant difference was seen between treatments. The results showed that monostearate-based oleogels can replace fats with with higher amount of unsaturated fatty acids to be used in meat products while keeping the desired features of the traditional products. Hwang and co-workers formulated a spread and a margarine with sunflower wax oleogels of a variety of oils (Hwang et al., 2016; Hwang and Winkler-Moser, 2020). Margarine formulations with 2% sunflower wax oleogels exhibited similar firmnesses as the commercial spreads while formulations with 10% sunflower wax oleogels had firmnesses like commercial margarines. These margarines were also used to make cookies. Texture Profile Analysis of the cookies showed that, texturally, they were indistinguishable from cookies made with commercial margarine. Zulim Botega et al. (2013) examined the effects on ice cream structure, overrun, and meltdown after replacing solid fat in ice cream with 10% (w/w) wax oleogels. They noted that the oleogels improved the foam structure of ice cream, for which reason ice cream containing rice bran wax oleogels displayed higher overrun, better texture, and an improved appearance. For replacing animal fat in comminuted meats to be considered realistic, sensory analysis must also be performed. Adding oleogels had similar characteristics in terms of texture and color when the products were compared to a formulation elaborated with animal fat (Alejandre et al., 2019), but contradictory results were observed about oxidative reactions. When soy protein-based emulsion gels were used as an animal fat alternative in Frankfurters, it was found that the sample was less red and lighter but presented the same hardness. Although a lower acceptability than control in the sensory score, using emulsion gels to replace fats in frankfurters is a viable strategy to make them healthier considering the improvement of the nutritional profile (de Souza Paglarini et al., 2019). Moreover, given the sensory and texture comparable with the standard cake, there is a potential for using zein-based oil-in-glycerol emulgels as a margarine alternative (Chen et al., 2016c). Incorporating fat into plant-based meat products is crucial for creating high-fidelity flavors, textures, and cooking experiences. However, bridging the sensory gap between unsaturated plant-based oils and saturated animal fats is challenging. Structuring plant-based oils can achieve the juiciness of legacy animal meat. Another challenge relates to the sensory experience of current products lacking enough fat or incorporate unsaturated oils as these do not keep well during cooking and mastication. As with all plant-based products, customer acceptability is the decisive factor for adopting the novel solution. Healthwise, functional and environmental considerations, structured emulsion, and oleogel from quillaja saponin may be considerable fat mimetics owing to their resistant to shear stress and oxidation.

Oil migration, the movement of liquid oil by diffusion or capillary action, is a concern in products such as creams, chocolate confectionaries, baked goods, and other oil-containing products. Oleogels have been used to replace solid fat in in chocolate to prevent oil migration and to reduce melting at room temperature (Li et al. 2021). Filled chocolates, and some chocolate covered confectionaries are susceptible to fat bloom due to oil migration from the inner filling to the outer surface, where visible white defects occur. In our research, phytosterol-based oleogel can partially replace cocoa butter in chocolate preparations, and more inhibit the bloom (data not published). When phytosterol-based oleogels were incorporated when making chocolate, the melting point significantly increased, and the resultant chocolate was termed heat-resistant chocolate. This chocolate is known for its ability to maintain its shape at temperatures above 40°C. Unfortunately, the fat mimetic decreased the resultant chocolate product's hardness. The same effect was also reported by Stortz and Marangoni et al. (2013), where heat-resistant chocolate was prepared by directly adding EC solution (20–25% EC in ethanol), melted, tempered, and evaporating the ethanol resulting in chocolate. Generally, the viscoelastic behavior of chocolate paste was achieved by incorporating significant amounts (>20%wt) of solid fats (hydrogenated oils or natural oils with high levels of saturated fatty acids such as palm oil). Patel and co-workers reported that shellac (at 1.5%wt of total formulation) could replace the oil binder and could partially replace (~27%) palm oil with liquid rapeseed oil, leading to a 30% reduction in the chocolate paste's saturated fat. Due to the replacement of palm oil, the oleogel paste had a much softer texture (Patel and Dewettinck 2015). However, even at low amounts of solid fat and the absence of an oil binder, shellac was effective in binding the oil in the paste formulation, which was confirmed by accelerated stability results where the oleogel paste showed no sign of "oiling out" even after four weeks of storage at an elevated temperature (30°C). Similarly, Wendt et al. (2017) incorporated a sitosterol-oryzanol oleogel into a praline confection filling and found that oil migration was retarded by adding an oleogel.

3.5.3.5 Reduction in Lipid Oxidation

Lipid oxidation produces hydroperoxides through interactions of unsaturated lipids with oxygen, resulting in potentially carcinogenic and inflammatory substances in foods. Pan et al. found that a lower oxidation rate was observed by the POV and TBARS values as a function of storage time in either the glycerol monolaurate-based oleogels or O/W oleogel emulsions (Pan et al. 2021). Ogütcü et al. (2015) prepared the oleogel emulsions with virgin olive oil gelled by 5 wt% beeswax and found that the developed emulsions were stable in terms of the texture, color, and oxidation during 90 days of storage. Previous research with oleogels using increasing beeswax concentrations, showed a follow-up improvement in regard to the oxidative stability of these oleogels when loaded with β-carotene (Martins et al., 2017). Meanwhile, Wang et al. (2018) prepared the camellia oil gelled emulsions to partly replace pork fat (10%-50%) in a cooked style Harbin sausage, and found the overall quality of Harbin sausage was improved by reducing lipid oxidation during storage at 4°C. The oxidative reactions could be delayed by the oleogel-based emulsion's (e.g., lecithin/stearic acid) physical and microstructural characteristics (Zhuang et al., 2021).

3.5.4 Conclusion and Outlook

Driven by increasing awareness of nutritional and functional properties, various novel fat mimetics have been studied and several applications have been tested based on these systems. Structuring oil into soft matters including oleogels, emulsion gels, and structured emulsions have been increasingly used to reduce or even abandon hydrogenated and saturated fats. These fat mimetics have provided the structure, functionality, and nutritional aspects necessary in many food products, such as sausages, cream cheese, margarine, cookies, ice cream, and spreads for replacing of trans and saturated fats. They can reduce fat intake, improve *in vitro* and *in vivo* digestion, control delivery carriers and release of bioactive molecules, texture design and changes, and reduce lipid oxidation and environmental applications.

Although these fat mimetics play an important role in health as food technology has shown an outstanding contribution of replacing convention trans and saturated fat made with unsaturated oils without compromising the taste and texture of the products, more work must exploit the full potential of structuring edible oils for specific foods. Extensive research is needed to seek natural structurants or directional design structurants with green chemistry principles from biocompatible, and nontoxic renewable and cost-competitive advantages resources. One crucial consideration when choosing the natural precursor materials is inexpensive or by-products or wastes from other processes to meet the demand for "natural" and "clean label" products. Another area of future work is exploring the structure-function relationships of fat mimic for the applications in different foods based on a scientific adage that structure determines function. As we all know, to mimic animal meat, plant-based meat needs to mimic the taste and texture of fat and create a "juicy" eating experience. But figuring out how to create and incorporate fat mimetics (replacing coconut fat) into plant-based meat has been the biggest challenge. Apart from the fat mimetics from natural plant ingredients (*e.g.*, *quillaja* saponin), plant protein-based fat mimetics may be a better option for the "clean label" and health-promoting food products. Plant protein-based fat mimetics would show strong molecular interactions with plant-based matrices. They are important to plant-based meats because their compatibility allows their retention during the extrusion, cooking, and mastication (chewing) process. Particularly, non-thermally sensitive protein-based fat mimetics (e.g., oleogels, emulsion gels, and structured emulsion) by design created a greasy liquid from forming in the pan during cooking while also encapsulating the volatile flavor compounds in the matrix. Besides this, how to scale-up structurants and fat mimetics by devising efficient routes is also paramount importance when industries are seeking for effective functional features and economical.

Though exemplary research has been carried out in designing plentiful fat mimetics, their industrial usage is limited due to meagre technologies or has at least been slow until now, and there has ben no product introduction into the food market to date. So, note the nutritional benefits, processing considerations, and the potential for application in more complex composite food products. In industrial applications, lipid oxidation should be widely concerned due to a lot of unsaturated fatty acids. Finally, despite the profound interest in various fat mimickers for replacing conventional trans and saturated fat, limited work has been done to evaluate the nutritional properties in *in vivo* digestion biochemical transformations, and excretion using bioevaluation trials. So, developing scientific knowledge to address fat mimetics' nutrition, function, and safety issues through *in vitro* and *in vivo* studies is needed.

Acknowledgments

The authors gratefully acknowledge the support of the National Natural Science Foundation of China (31801461, 31771923) and the Young Key Teacher Project of Henan University of Technology.

References

Adelmann, H., Binks, B.P., and Mezzenga, R. (2012). Oil powders and gels from particle-stabilized emulsions. *Langmuir* 28 (3): 1694–1697.

Alejandre, M., Astiasarán, I., Ansorena, D., and Barbut, S. (2019). Using canola oil hydrogels and organogels to reduce saturated animal fat in meat batters. *Food Research International* 122: 129–136.

Ashkar, A., Laufer, S., Rosen-Kligvasser, J., Lesmes, U., and Davidovich-Pinhas, M. (2019). Impact of different oil gelators and oleogelation mechanisms on digestive lipolysis of canola oil oleogels. *Food Hydrocolloids* 97: 105218.

Bascuas, S., Espert, M., Llorca, E., Quiles, A., Salvador, A., and Hernando, I. (2021a). Structural and sensory studies on chocolate spreads with hydrocolloid-based oleogels as a fat alternative. *LWT - Food Science and Technology* 135: 110228.

Bascuas, S., Morell, P., Hernando, I., and Quiles, A. (2021b). Recent trends in oil structuring using hydrocolloids. *Food Hydrocolloids* 118: 106612.

Blake, A.I. and Marangoni, A.G. (2015). Factors affecting the rheological properties of a structured cellular solid used as a fat mimetic. *Food Research International* 74: 284–293.

Bot, A. and Agterof, W.G.M. (2006). Structuring of edible oils by mixtures of γ-oryzanol with β-sitosterol or related phytosterols. *Journal of the American Oil Chemists' Society* 83: 513–521.

Bot, A., den Adel, R., and Roijers, E.C. (2008). Fibrils of γ-oryzanol and β-sitosterol in edible oil organogels. *Journal of the American Oil Chemists' Society* 85: 1127–1134.

Böttcher, S. and Drusch, S. (2017). Saponins-Self-assembly and behavior at aqueous interfaces. *Advances in Colloid and Interface Science* 243: 105–113.

Briggs, M., Petersen, K., and Kris-Etherton, P. (2017). Saturated fatty acids and cardiovascular disease: replacements for saturated fat to reduce cardiovascular risk. *Healthcare* 5 (2): 1–29.

Calligaris, S., Alongi, M., Lucci, P., and Anese, M. (2020). Effect of different oleogelators on lipolysis and curcuminoid bioaccessibility upon in vitro digestion of sunflower oil oleogels. *Food Chemistry* 314: 126146.

Chen, X.W., Chen, Y.J., Wang, J.M., Guo, J., Yin, S.W., and Yang, X.Q. (2017a). Tunable volatile release from organogel-emulsions based on the self-assembly of β-sitosterol and γ-oryzanol. *Food Chemistry* 221: 1491–1498.

Chen, X.W., Chen, Y.J., Wang, J.M., Guo, J., Yin, S., and Yang, X.Q. (2016b). Phytosterol structured algae oil nanoemulsions and powders: Improving antioxidant and flavor properties. *Food & Function* 7 (9): 3694–3702.

Chen, X.W., Fu, S.Y., Hou, J.J., Guo, J., Wang, J.M., and Yang, X.Q. (2016c). Zein based oil-in-glycerol emulgels enriched with β-carotene as margarine alternatives. *Food Chemistry* 211: 836–844.

Chen, X.W., Guo, J., Wang, J.M., Yin, S.W., and Yang, X.Q. (2016a). Controlled volatile release of structured emulsions based on phytosterols crystallization. *Food Hydrocolloids* 56: 170–179.

Chen, X.W., Li, X.X., Ma, C.G., Wan, Z.L., and Sun, S.D. (2022). Structuring of edible liquid oil into smart thermo-triggered soft matters for controlled bioactive delivery. *Journal of Agricultural and Food Chemistry* 70 (1): 309–318.

Chen, X.W., Sun, S.D., Ma, C.G., and Yang, X.Q. (2020a). Oil-water interfacial-directed spontaneous self-assembly of natural quillaja saponin for controlling interface permeability in colloidal emulsions. *Journal of Agricultural and Food Chemistry* 68 (47): 13854–13862.

Chen, X.W., Sun, S.D., Yang, G.L., and Ma, C.G. (2020b). Engineering phytosterol-based oleogels for potential application as sustainable petrolatum replacement. *RSC Advances* 10 (1): 244–252.

Chen, X.W., Wang, J.M., Guo, J., Wan, Z.L., Yin, S.W., and Yang, X.Q. (2017b). Hierarchical high internal phase emulsions and transparent oleogels stabilized by quillaja saponin-coated nanodroplets for color performance. *Food & Function* 8 (2): 823–831.

Chen, X.W. and Yang, X.Q. (2019). Characterization of orange oil powders and oleogels fabricated from emulsion templates stabilized solely by a natural triterpene saponin. *Journal of Agricultural and Food Chemistry* 67 (9): 2637–2646.

Chen, K. and Zhang, H. (2020). Fabrication of oleogels via a facile method by oil absorption in the aerogel templates of protein–polysaccharide conjugates. *ACS Applied Materials & Interfaces* 12 (6): 7795–7804.

Curti, E., Federici, E., Diantom, A., Carini, E., Pizzigalli, E., Symon, V.W., Pellegrini, N, and Vittadini, E. (2018). Structured emulsions as butter substitutes: effects on physicochemical and sensory attributes of shortbread cookies. *Journal of the Science of Food and Agriculture* 98 (10): 3836–3842.

da Silva, S.L., Amaral, J.T., Ribeiro, M., Sebastião, E.E., Vargas, C., de Lima Franzen, F., Schneider, G., Lorenzo, J.M., Fries, L.L.M., Cichoski, A.J., and Campagnol, P.C.B. (2019). Fat replacement by oleogel rich in oleic acid and its impact on the technological, nutritional, oxidative, and sensory properties of Bologna-type sausages. *Meat Science* 149: 141–148.

Davidovich-Pinhas, M. (2017). CHAPTER 7 Thermo-gelation of Ethyl-cellulose Oleogels. In: *Edible Oil Structuring: Concepts, Methods and Applications*, 133–149. London, UK: The Royal Society of Chemistry.

Davidovich-Pinhas, M. (2019). Oil structuring using polysaccharides. *Current Opinion in Food Science* 27: 29–35.

Davidovich-Pinhas, M., Barbut, S., and Marangoni, A.G. (2015). The gelation of oil using ethyl cellulose. *Carbohydrate Polymers* 117: 869–878.

Davidovich-Pinhas, M., Barbut, S., and Marangoni, A.G. (2016). Development, characterization, and utilization of food-grade polymer oleogels. *Annual Review of Food Science and Technology* 7: 65–91.

de Souza Paglarini, C., Martini, S., and Pollonio, M.A.R. (2019). Using emulsion gels made with sonicated soy protein isolate dispersions to replace fat in frankfurters. *LWT - Food Science and Technology* 99: 453–459.

de Vries, A. (2017). *Structuring oil by protein building blocks (Doctoral dissertation).* Wageningen University.

de Vries, A., Hendriks, J., van der Linden, E., and Scholten, E. (2015). Protein oleogels from protein hydrogels via a stepwise solvent exchange route. *Langmuir* 31: 13850–13859.

de Vries, A., Nikiforidis, C.V., and Scholten, E. (2014). Natural amphiphilic proteins as tri-block Janus particles: Self-sorting into thermo-responsive gels. *Europhysics Letters* 107 (5): 58003.

de Vries, A., Wesseling, A., van der Linden, E., and Scholten, E. (2017). Protein oleogels from heat-set whey protein aggregates. *Journal of Colloid and Interface Science* 486: 75–83.

Demirkesen, I. and Mert, B. (2019). Utilization of beeswax oleogelshortening mixtures in gluten-free bakery products. *Journal of the American Oil Chemists' Society* 96: 545–554.

Demirkesen, I. and Mert, B. (2020). Recent developments of oleogel utilizations in bakery products. *Critical Reviews in Food Science and Nutrition* 60 (14): 2460–2479.

DiNicolantonio, J.J., Lucan, S.C., and O'Keefe, J.H. (2016). The evidence for saturated fat and for sugar related to coronary heart disease. *Progress in Cardiovascular Diseases* 58 (5): 464–472.

Dong, L., Lv, M., Gao, X., Zhang, L., Rogers, M., Cao, Y., and Lan, Y. (2020). In vitro gastrointestinal digestibility of phytosterol oleogels: influence of self-assembled microstructures on emulsification efficiency and lipase activity. *Food & Function* 11 (11): 9503–9513.

Duffy, N., Blonk, H.C., Beindorff, C.M., Cazade, M., Bot, A., and Duchateau, G.S.M.J.E. (2009). Organogel-based emulsion systems, micro-structural features and impact on in vitro digestion. *Journal of the American Oil Chemists' Society* 86 (8): 733–741.

Ferro, A.C., de Souza Paglarini, C., Pollonio, M.A.R., and Cunha, R.L. (2021). Glyceryl monostearate-based oleogels as a new fat substitute in meat emulsion. *Meat Science* 174: 108424.

Franco, D., Martins, A.J., López-Pedrouso, M., Cerqueira, M.A., Purriños, L., Pastrana, L.M., Vicente, A.A., Zapata, C, and Lorenzo, J.M. (2020). Evaluation of linseed oil oleogels to partially replace pork backfat in fermented sausages. *Journal of the Science of Food and Agriculture* 100 (1): 218–224.

Franco, D., Martins, A.J., López-Pedrouso, M., Purriños, L, Cerqueira, M.A., Vicente, A.A., Pastrana, L.M., Zapata, C., and Lorenzo, J.M. (2019). Strategy towards replacing pork backfat with a linseed oleogel in frankfurter sausages and its evaluation on physicochemical, nutritional, and sensory characteristics. *Foods* 8 (9): 366.

Gao, Z.M., Yang, X.Q., Wu, N.N., Wang, L.J., Wang, J.M., Guo, J., and Yin, S.W. (2014). Protein-based pickering emulsion and oil gel prepared by complexes of zein colloidal particles and stearate. *Journal of Agricultural and Food Chemistry* 62 (12): 2672–2678.

Giacomozzi, A.S., Carrín, M.E., and Palla, C.A. (2018). Muffins elaborated with optimized monoglycerides oleogels: From solid fat replacer obtention to product quality evaluation. *Journal of Food Science* 83 (6): 1505–1515.

Guo, Q., Wijarnprecha, K., Sonwai, S., and Rousseau, D. (2019). Oleogelation of emulsified oil delays in vitro intestinal lipid digestion. *Food Research International* 119: 805–812.

Gutiérrez-Luna, K., Astiasarán, I., and Ansorena, D. (2020). Gels as fat replacers in bakery products: a review. *Critical Reviews in Food Science and Nutrition* 62 (14): 1–14.

Han, L., Li, L., Li, B., Zhao, L., Liu, G., Liu, X., and Wang, X. (2014). Structure and physical properties of organogels developed by sitosterol and lecithin with sunflower oil. *Journal of the American Oil Chemists' Society* 91: 1783–1792.

Huang, Z., Guo, B., Deng, C., Luo, S., Liu, C., and Hu, X. (2020). Stabilization of peanut butter by rice bran wax. *Journal of Food Science* 85 (6): 1793–1798.

Hughes, N., Rush, J.W., and Marangoni, A.G. (2018). CHAPTER 16 Feeding study of 12-hydroxystearic acid oleogels. In: *Edible Oleogels*, 2e. 381–399. Elsevier: Amsterdam, The Netherlands.

Hwang, H.S., Singh, M., Bakota, E.L., Winkler-Moser, J.K., Kim, S., and Liu, S.X. (2013). Margarine from organogels of plant wax and soybean oil. *Journal of the American Oil Chemists' Society* 90 (11): 1705–1712.

Hwang, H.S., Singh, M., and Lee, S. (2016). Properties of cookies made with natural wax-vegetable oil organogels. *Journal of Food Science* 81: 1045–1054.

Hwang, H.S. and Winkler-Moser, J.K. (2020). Properties of margarines prepared from soybean oil oleogels with mixtures of candelilla wax and beeswax. *Journal of Food Science* 85 (10): 3293–3302.

Iwanaga, K., Sumizawa, T., Miyazaki, M., and Kakemi, M. (2010). Characterization of organogel as a novel oral controlled release formulation for lipophilic compounds. *International Journal of Pharmaceutics* 388 (1–2): 123–128.

Jang, A., Bae, W., Hwang, H.S., Lee, H.G., and Lee, S. (2015). Evaluation of canola oil oleogels with candelilla wax as an alternative to shortening in baked goods. *Food Chemistry* 187: 525–529.

Jiang, Y., Liu, L., Wang, B., Sui, X, Zhong, Y., Zhang, L., Mao, Z., and Xu, H. (2018). Cellulose-rich oleogels prepared with an emulsion-templated approach. *Food Hydrocolloids* 77: 460–464.

Kangchai, W., Sangsirimongkolying, R., and Methacanon, P. (2018). Feasibility study of margarine substitute based on gelatin-oil emulsion gel. *Chiang Mai Journal of Science* 45 (1): 505–514.

Kew, B., Holmes, M., Stieger, M., and Sarkar, A. (2020). Review on fat replacement using protein-based microparticulated powders or microgels: A textural perspective. *Trends in Food Science and Technology* 106: 457–468.

Kim, J.Y., Lim, J., Lee, J., Hwang, H.S., and Lee, S. (2017). Utilization of oleogels as a replacement for solid fat in aerated baked goods: Physicochemical, rheological, and tomographic characterization. *Journal of Food Science* 82 (2): 445–452.

Kouzounis, D., Lazaridou, A., and Katsanidis, E. (2017). Partial replacement of animal fat by oleogels structured with monoglycerides and phytosterols in frankfurter sausages. *Meat Science* 130: 38–46.

Li, L., Liu, G., and Lin, Y. (2021). Physical and bloom stability of low-saturation chocolates with oleogels based on different gelation mechanisms. *LWT - Food Science and Technology* 140: 110807.

Li, L., Wan, W., Cheng, W., Liu, G., and Han, L. (2019). Oxidatively stable curcumin-loaded oleogels structured by sitosterol and lecithin: Physical characteristics and release behaviour in vitro. *International Journal of Food Science & Technology* 54: 2502–2510.

Liu, N., Lu, Y., Zhang, Y., Gao, Y., and Mao, L. (2020). Surfactant addition to modify the structures of ethylcellulose oleogels for higher solubility and stability of curcumin. *International Journal of Biological Macromolecules* 165: 2286–2294.

Liu, X., Chen, X.W., Guo, J., Yin, S., and Yang, X. (2016). Wheat gluten based percolating emulsion gels as simple strategy for structuring liquid oil. *Food Hydrocolloids* 61: 747–755.

Liu, X., Guo, J., Wan, Z.L., Liu, Y.Y., Ruan, Q.J., and Yang, X.Q. (2018). Wheat gluten-stabilized high internal phase emulsions as mayonnaise replacers. *Food Hydrocolloids* 77: 168–175.

Liu, X., Liu, Y.Y., Guo, J., Yin, S.W., and Yang, X.Q. (2017). Microfluidization initiated cross-linking of gliadin particles for structured algal oil emulsions. *Food Hydrocolloids* 73: 153–161.

Luo, S.Z., Hu, X.F., Jia, Y.J., Pan, L.H., Zheng, Z., Zhao, Y.Y., Mu, D.D., Zhong, X.Y., and Jiang, S.T. (2019). Camellia oil-based oleogels structuring with tea polyphenol-palmitate particles and citrus pectin by emulsion-templated method: Preparation, characterization and potential application. *Food Hydrocolloids* 95: 76–87.

Lupi, F.R., Gabriele, D., Baldino, N., Mijovic, P., Parisi, O.I., and Puoci, F. (2013). Olive oil/policosanol organogels for nutraceutical and drug delivery purposes. *Food & Function* 4 (10): 1512–1520.

Lupi, F.R., Gabriele, D., Seta, L., Baldino, N., and de Cindio, B. (2014). Rheological design of stabilized meat sauces for industrial uses. *European Journal of Lipid Science and Technology* 116 (12): 1734–1744.

Manzocco, L., Valoppi, F., Calligaris, S., Andreatta, F., Spilimbergo, S., and Nicoli, M.C. (2017). Exploitation of κ-carrageenan aerogels as template for edible Oleogel preparation. *Food Hydrocolloids* 71: 68–75.

Mao, L., Lu, Y., Cui, M., Miao, S., and Gao, Y. (2020). Design of gel structures in water and oil phases for improved delivery of bioactive food ingredients. *Critical Reviews in Food Science and Nutrition* 60 (10): 1651–1666.

Marangoni, A.G. (2012). Organogels: An alternative edible oil-structuring method. *Journal of the American Oil Chemists' Society* 89 (5): 749–780.

Marangoni, A.G. and Idziak, S.H.J. (2010). *Food Product*. USA: U.S. Patent 7, 718, 210 B2.

Marangoni, A.G., Van Duynhoven, J.P., Acevedo, N.C., Nicholson, R., and Patel, A.R. (2020). Advances in our understanding of the structure and functionality of edible fats and fat mimetics. *Soft Matter* 16 (2): 289–306.

Martins, A.J., Cerqueira, M.A., Cunha, R.L., and Vicente, A.A. (2017). Fortified beeswax oleogels: Effect of b-carotene on the gel structure and oxidative stability. *Food & Function* 8: 4241–4250.

Meng, Z., Qi, K., Guo, Y., Yong, W., and YuanFa, L. (2018). Macro-micro structure characterization and molecular properties of emulsion-templated polysaccharide oleogels. *Food Hydrocolloids* 77: 17–29.

Mert, B. and Demirkesen, I. (2016a). Reducing saturated fat with oleogel/shortening blends in a baked product. *Food Chemistry* 199: 809–816.

Mert, B. and Demirkesen, I. (2016b). Evaluation of highly unsaturated oleogels as shortening replacer in a short dough product. *LWT - Food Science and Technology* 68: 477–484.

Moghtadaei, M., Soltanizadeh, N., and Goli, S.A.H. (2018). Production of sesame oil oleogels based on beeswax and application as partial substitutes of animal fat in beef burger. *Food Research International* 108: 368–377.

Moghtadaei, M., Soltanizadeh, N., Goli, S.A.H., and Sharifimehr, S. (2021). Physicochemical properties of beef burger after partial incorporation of ethylcellulose oleogel instead of animal fat. *Journal of Food Science and Technology* 58: 4775–4784.

Moradabbasi, M., Goli, S.A.H., and Fayaz, G. (2021). Effect of biopolymers concentration and drying methods on physicochemical properties of emulsion-templated oleogel. *Journal of Food Science and Technology* 59: 1994–2003.

Moriano, M.E. and Alamprese, C. (2017). Organogels as novel ingredients for low saturated fat ice creams. *LWT - Food Science and Technology* 86: 371–376.

Moschakis, T., Dergiade, I., Lazaridou, A., Biliaderis, C.G., and Katsanidis, E. (2017). Modulating the physical state and functionality of phytosterols by emulsification and organogel formation: Application in a model yogurt system. *Journal of Functional Foods* 33: 386–395.

Nettleton, J.A., Brouwer, I.A., Geleijnse, J.M., and Hornstra, G. (2017). Saturated fat consumption and risk of coronary heart disease and ischemic stroke: a science update. *Annals of Nutrition & Metabolism* 70: 26–33.

Ogütcü, M., Arifoğlu, N., and Yılmaz, E. (2015). Preparation and characterization of virgin olive oil-beeswax oleogel emulsion products. *Journal of the American Oil Chemists' Society* 92 (4): 459–471.

Ogütcü, M., Arifoğlu, N., and Yılmaz, E. (2017). Restriction of oil migration in tahini halva via organogelation. *European Journal of Lipid Science and Technology* 119 (9): 1600189.

Oh, I., Lee, J., Lee, H.G., and Lee, S. (2019). Feasibility of hydroxypropyl methylcellulose oleogel as an animal fat replacer for meat patties. *Food Research International* 122: 566–572.

Oh, I.K., Amoah, C., Lim, J., Jeong, S., and Lee, S. (2017). Assessing the effectiveness of wax-based sunflower oil oleogels in cakes as a shortening replacer. *LWT - Food Science and Technology* 86: 430–437.

Oh, I.K. and Lee, S. (2018). Utilization of foam structured hydroxypropyl methylcellulose for oleogels and their application as a solid fat replacer in muffins. *Food Hydrocolloids* 77: 796–802.

Ojeda-Serna, I., Rocha-Guzmán, N., Gallegos-Infante, J., Cháirez-Ramírez, M.H., Rosas-Flores, W., Pérez-Martínez, J.D., Moreno-Jimenez, M.R., and González-Laredo, R.F. (2019). Water-in-oil organogel based emulsions as a tool for increasing bioaccessibility and cell permeability of poorly water-soluble nutraceuticals. *Food Research International* 120: 415–424.

Osborne, T.B. (1897). The amount and properties of the proteids of the maize kernel. *Journal of the American Chemical Society* 19 (7): 525–532.

O'Sullivan, C., Barbut, S., and Marangoni, A.G. (2017). Ethylcellulose oleogels for lipophilic bioactive delivery- effect of oleogelation on in vitro bioaccessibility and stability of beta-carotene. *Food & Function* 8: 1438–1451.

O'Sullivan, C.M., Davidovich-Pinhas, M., Wright, A.J., Barbut, S., and Marangoni, A.G. (2017). Ethylcellulose oleogels for lipophilic bioactive delivery–Effect of oleogelation on in vitro bioaccessibility and stability of beta-carotene. *Food & Function* 8: 1438–1451.

Özer, C.O. and Çelegen, Ş. (2021). Evaluation of quality and emulsion stability of a fat-reduced beef burger prepared with an olive oil oleogel-based emulsion. *Journal of Food Processing and Preservation* 45 (8): 14547.

Paglarini, C.D.S., Vidal, V.A.S., Martini, S., Cunha, R.L., and Pollonio, M.A.R. (2020). Protein-based hydrogelled emulsions and their application as fat replacers in meat products: a review. *Critical Reviews in Food Science* 62 (3): 640–655.

Pan, H., Xu, X., Qian, Z., Cheng, H., Shen, X., Chen, S., and Ye, X. (2021a). Xanthan gum-assisted fabrication of stable emulsion-based oleogel structured with gelatin and proanthocyanidins. *Food Hydrocolloids* 115: 106596.

Pan, J., Tang, L., Dong, Q., Li, Y., and Zhang, H. (2021). Effect of oleogelation on physical properties and oxidative stability of camellia oil-based oleogels and oleogel emulsions. *Food Research International* 140: 110057.

Panagiotopoulou, E., Moschakis, T., and Katsanidis, E. (2016). Sunflower oil organogels and organogel-in-water emulsions (part II): Implementation in frankfurter sausages. *LWT - Food Science and Technology* 73: 351–356.

Park, C., Bemer, H.L., and Maleky, F. (2018). Oxidative stability of rice bran wax oleogels and an oleogel cream cheese product. *Journal of the American Oil Chemists' Society* 95 (10): 1267–1275.

Patel, A.R. (ed.) (2017a). *Edible Oil Structuring*. Royal Society of Chemistry.

Patel, A.R. (2017b). A colloidal gel perspective for understanding oleogelation. *Current Opinion in Food Science* 15: 1–7

Patel, A.R. (2018). Structuring edible oils with hydrocolloids: Where do we stand? *Food Biophysics* 13: 113–115.

Patel, A.R., Cludts, N., Sintang, M.D.B. et al. (2014b). Edible oleogels based on water soluble food polymers: Preparation, characterization and potential application. *Food & Function* 5 (11): 2833–2841.

Patel, A.R. and Dewettinck, K. (2015). Comparative evaluation of structured oil systems: shellac oleogel, HPMC oleogel and HIPE gel. *European Journal of Lipid Science and Technology* 117: 1772–1781.

Patel, A.R., Nicholson, R.A., and Marangoni, A.G. (2020). Applications of fat mimetics for the replacement of saturated and hydrogenated fat in food products. *Current Opinion in Food Science* 33: 61–68.

Patel, A.R., Rajarethinem, P.S., Cludts, N., Lewille, B., De Vos, W.H., Lesaffer, A., and Dewettinck, K. (2015). Biopolymer-based structuring of liquid oil into soft solids and oleogels using water-continuous emulsions as templates. *Langmuir* 31: 2065–2073.

Patel, A.R., Rajarethinem, P.S., Grędowska, A., Turhan, O., Lesaffer, A., De Vos, W.H., Van de Walle, D., and Dewettinck, K. (2014a). Edible applications of shellac oleogels: Spreads, chocolate paste and cakes. *Food & Function* 5 (4): 645–652.

Pehlivanoglu, H., Ozulku, G., Yildirim, R.M., Demirci, M., Toker, O.S., and Sagdic, O. (2018). Investigating the usage of unsaturated fatty acid-rich and low-calorie oleogels as a shortening mimetics in cake. *Journal of Food Processing and Preservation* 42 (6): 13621.

Peng, X. and Yao, Y. (2017). Carbohydrates as fat replacers. *Annual Review of Food Science and Technology* 8: 331–351.

Pinto, T.C., Martins, A.J., Pastrana, L., Pereira, M.C., and Cerqueira, M.A. (2021). Oleogel-based systems for the delivery of bioactive compounds in foods. *Gels* 7 (3): 86.

Poyato, C., Astiasarán, I., Barriuso, B., and Ansorena, D. (2015). A new polyunsaturated gelled emulsion as replacer of pork back-fat in burger patties: Effect on lipid composition, oxidative stability and sensory acceptability. *LWT - Food Science and Technology* 62 (2): 1069–1075.

Puşcaş, A., Mureşan, V., Socaciu, C., and Muste, S. (2020). Oleogels in food: A review of current and potential applications. *Foods* 9 (1): 70–97.

Rogers, M.A., Bot, A., Lam, R.S.H., Pedersen, T., and May, T. (2010). Multicomponent hollow tubules formed using phytosterol and γ-oryzanol-based compounds: an understanding of their molecular embrace. *Journal of Physical Chemistry A* 114 (32): 8278–8285.

Romoscanu, A.I. and Mezzenga, R. (2006). Emulsion-templated fully reversible protein-in-oil gels. *Langmuir* 22 (18): 7812–7818.

Saha, A., Adamcik, J., Bolisetty, S., Handschin, S., and Mezzenga, R. (2015). Fibrillar networks of glycyrrhizic acid for hybrid nanomaterials with catalytic features. *Angewandte Chemie International Edition* 127: 5498–5502.

Sawalha, H., Venema, P., Bot, A., Flöter, E., and van der Linden, E. (2011). The influence of concentration and temperature on the formation of γ-oryzanol+ β-sitosterol tubules in edible oil organogels. *Food Biophysics* 6 (1): 20–25.

Scharfe, M. and Flöter, E. (2020). Oleogelation: from scientific feasibility to applicability in food products. *European Journal of Lipid Science and Technology* 122 (12): 2000213.

Scholten, E. (2019). Edible oleogels: how suitable are proteins as a structurant? *Current Opinion in Food Science* 27: 36–42.

Shahbazi, M., Jäger, H., Chen, J., and Ettelaie, R. (2021). Construction of 3D printed reduced-fat meat analog by emulsion gels. Part II: Printing performance, thermal, tribological, and dynamic sensory characterization of printed objects. *Food Hydrocolloids* 121: 107054.

Silva, T.J., Barrera-Arellano, D., and Ribeiro, A.P.B. (2021a). Oleogel-based emulsions: Concepts, structuring agents, and applications in food. *Journal of Food Science* 86 (7): 2785–2801.

Silva, T.J., Fernandes, G.D., Bernardinelli, O.D., da Rosa Silva, E.C., Barrera-Arellano, D., and Ribeiro, A.P.B. (2021b). Organogels in low-fat and high-fat margarine: a study of physical properties and shelf life. *Food Research International* 140: 110036.

Sintang, M.D.B., Danthine, S., Brown, A., Van de Walle, D., Patel, A.R., Tavernier, I., Rimaux, T., and Dewettinck, K. (2017). Phytosterols-induced viscoelasticity of oleogels prepared by using monoglycerides. *Food Research International* 100: 832–840.

Sintang, M.D.B., Rimaux, T., van de Walle, D., Dewettinck, K., and Patel, A.R. (2016). Studying the oil structuring properties of monoglycerides and phystosterols mixtures. *European Journal of Lipid Science and Technology* 119: 1500517.

Stortz, T.A., Laredo, T., and Marangoni, A.G. (2015). The role of lecithin and solvent addition in ethylcellulose-stabilized heat resistant chocolate. *Food Biophysics* 10 (3): 253–263.

Stortz, T.A. and Marangoni, A.G. (2013). Ethylcellulose solvent substitution method of preparing heat resistant chocolate. *Food Research International* 51 (2): 797–803.

Tan, H., Sun, G., Lin, W., Mu, C., and Ngai, T. (2014). Gelatin particle-stabilized high internal phase emulsions as nutraceutical containers. *ACS Applied Materials & Interfaces* 6 (16): 13977–13984.

Tan, S.Y., Peh, E.W.Y., Marangoni, A.G., and Henry, C.J. (2017). Effects of liquid oil vs. oleogel co-ingested with a carbohydrate-rich meal on human blood triglycerides, glucose, insulin and appetite. *Food & Function* 8: 241–249.

Tanti, R., Barbut, S., and Marangoni, A.G. (2016). Hydroxypropyl methylcellulose and methylcellulose structured oil as a replacement for shortening in sandwich cookie creams. *Food Hydrocolloids* 61: 329–337.

Tavernier, I., Patel, A.R., Van der Meeren, P., and Dewettinck, K. (2017). Emulsion-templated liquid oil structuring with soy protein and soy protein: κ-carrageenan complexes. *Food Hydrocolloids* 65: 107–120.

Tsibranska, S., Tcholakova, S., Golemanov, K., Denkov, N., Pelan, E., and Stoyanov, S.D. (2020). Role of interfacial elasticity for the rheological properties of saponin-stabilized emulsions. *Journal of Colloid and Interface Science* 564: 264–275.

Wan, Z.L., Sun, Y.E., Ma, L.L., Guo, J., Wang, J., Yin, S., and Yang, X. (2017a). Thermoresponsive structured emulsions based on the fibrillar self-assembly of natural saponin glycyrrhizic acid. *Food & Function* 8: 75–85.

Wan, Z., Sun, Y., Ma, L., Yang, X., Guo, J., and Yin, S. (2017b). Responsive emulsion gels with tunable properties formed by self-assembled nanofibrils of natural saponin glycyrrhizic acid for oil structuring. *Journal of Agricultural and Food Chemistry* 65 (11): 2394–2405.

Wang, X., Xie, Y., Li, X., Liu, Y., and Yan, W. (2018). Effects of partial replacement of pork back fat by a camellia oil gel on certain quality characteristics of a cooked style Harbin sausage. *Meat Science* 146: 154–159.

Wang, Y. and Padua, G.W. (2012). Nanoscale characterization of zein self-assembly. *Langmuir* 28 (5): 2429–2435.

Wendt, A., Abraham, K., Wernecke, C., Pfeiffer, J, and Flöter, E. (2017). Application of β-sitosterol+ γ-oryzanol-structured organogel as migration barrier in filled chocolate products. *Journal of the American Oil Chemists' Society* 94 (9): 1131–1140.

Winkler-Moser, J.K., Anderson, J., Byars, J.A., Singh, M., and Hwang, H.S. (2019). Evaluation of beeswax, candelilla wax, rice bran wax, and sunflower wax as alternative stabilizers for peanut butter. *Journal of the American Oil Chemists' Society* 96 (11): 1235–1248.

Yang, D.X., Chen, X.W., and Yang, X.Q. (2018). Phytosterol-based oleogels self-assembled with monoglyceride for controlled volatile release. *Journal of the Science of Food and Agriculture* 98 (2): 582–589.

Ye, X., Li, P., Lo, Y.M., Fu, H., and Cao, Y. (2019). Development of novel shortenings structured by ethylcellulose oleogels. *Journal of Food Science* 84 (6): 1456–1464.

Yılmaz, E. and Öğütcü, M. (2015). The texture, sensory properties and stability of cookies prepared with wax oleogels. *Food & Function* 6 (4): 1194–1204.

Zembyla, M., Murray, B.S., and Sarkar, A. (2020). Water-in-oil emulsions stabilized by surfactants, biopolymers and/or particles: a review. *Trends in Food Science and Technology* 104: 49–59.

Zeng, T., Wu, Z.L., Zhu, J.Y., Yin, S.W., Tang, C.H., Wu, L.Y., and Yang, K.Q. (2017). Development of antioxidant Pickering high internal phase emulsions (HIPEs) stabilized by protein/polysaccharide hybrid particles as potential alternative for PHOs. *Food Chemistry* 231: 122–130.

Zetzl, A.K., Marangoni, A.G., and Barbut, S. (2012). Mechanical properties of ethylcellulose oleogels and their potential for saturated fat reduction in frankfurters. *Food & Function* 3 (3): 327–337.

Zhao, W., Wei, Z., and Xue, C. (2021). Recent advances on food-grade oleogels: fFbrication, application and research trends. *Critical Reviews in Food Science and Nutrition* 62 (27): 7659–7676.

Zhuang, X., Gaudino, N., Clark, S., and Acevedo, N.C. (2021). Novel lecithin-based oleogels and oleogel emulsions delay lipid oxidation and extend probiotic bacteria survival. *LWT - Food Science and Technology* 136: 110353.

Zou, Y., Guo, J., Yin, S.W., Wang, J.M., and Yang, X.Q. (2015). Pickering emulsion gels prepared by hydrogen-bonded zein/tannic acid complex colloidal particles. *Journal of Agricultural and Food Chemistry* 63 (33): 7405–7414.

Zou, Y., Thijssen, P.P., Yang, X., and Scholten, E. (2019). The effect of oil type and solvent quality on the rheological behavior of zein stabilized oil-in-glycerol emulsion gels. *Food Hydrocolloids* 91: 57–65.

Zulim Botega, D.C., Marangoni, A.G., Smith, A.K., and Goff, H.D. (2013). The potential application of rice bran wax oleogel to replace solid fat and enhance unsaturated fat content in ice cream. *Journal of Food Science* 78 (9): 1334–1339.

Section IV

Food Applications

4.1

Processed Meat Products

Yogesh Kumar, Akhoon Asrar Bashir, and Poonam Choudhary*

ICAR-Central Institute of Post-Harvest Engineering and Technology (CIPHET), Ludhiana, India
Corresponding author

4.1.1 Introduction

Fat is an important dietary source, needed for growth and development, and provides essential fatty acids required for keeping cell membrane structure, and also, it helps in the absorption of fat-soluble vitamins and other phytochemicals. Due to its heat transfer proprieties, fat helps in the cooking process by enabling rapid heating and achievement of high temperatures. Fats are useful for texture and aroma, affecting the overall perception of food. Fat affects the melting point, viscosity, crystallinity, and spreadability in many food products (Sandrou and Arvanitoyannis 2000). In meat products, fat improves flavor and taste by adding creaminess, juiciness, and smoothness. It also improves the mouth-feel of processed meat products (Kumar 2021). However, the consumption of high-fat processed meat products is associated with certain health disorders viz. obesity, cancer, type-2 diabetes, coronary heart disease, stroke, high blood pressure, etc. ((ADA) 2005; Kumar et al. 2021). Consumers are turning to reduced or low-fat meat products because of the current drift in nutrition and health consciousness. The health organizations usually suggested that total dietary fat consumption should not be higher than 30% of the total daily calories in a diet. The saturated fat content should not exceed 10 percent of the total energy intake level while the level of monounsaturated and polyunsaturated should be 20 percent of the total energy Intake. The health organizations also have recommended reducing portion sizes and limiting the intake of highly processed high-fat meat products as a preventive measure and to lessen the worldwide burden of metabolic risk factors (Corpet 2011). The French National Cancer Institute, INCa and NACRe (2009) have advised to reduce the intake of cured meats, particularly high-fat items, and to eat them less often and in smaller proportions. These recommendations, which should reduce the use of several processed meat products, will undoubtedly affect the meat industry's long-term viability and profitability. Thus, an alternate positive approach must address this problem that does not compromise the meat industry's economics. The development of low- or reduced-fat processed meat products provides essential nutrients without impairing the consumers' health (Poyato et al. 2015; Kumar et al. 2016). The demand for these health meat products is predicted to skyrocket in the coming decade, owing to rising health concerns. Moreover, now consumers are favoring meat products with ingredients with health benefits over the ingredients that cause health problems (Guàrdia et al. 2006). This clearly shows low-fat processed meat products are promising health-benefiting ingredient which have the potential market opportunities.

Because fat adds significant sensory and physiological properties to processed meat products, replacing fat in a meat system to develop low-fat processed meat products is a difficult task (Kumar

Fat Mimetics for Food Applications, First Edition. Edited by Miguel Ângelo Parente Ribeiro Cerqueira and Lorenzo Miguel Pastrana Castro.

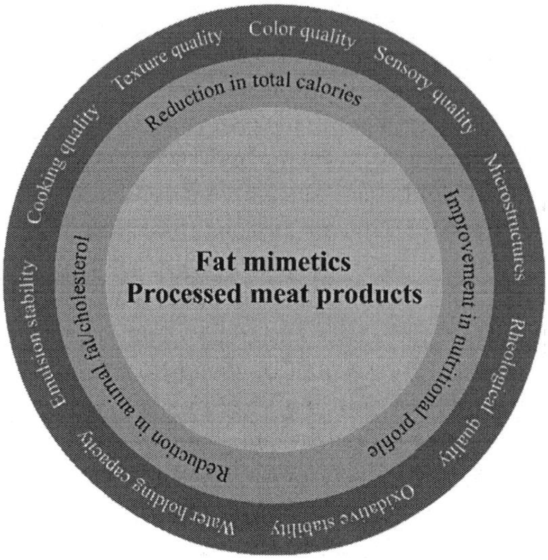

Figure 4.1.1 Application of fat mimetics in processed meat products: Required nutritional benefits and associated technological/quality parameters.

2021). Some processed beef products have up to 30% fat and are successful in various parts of the world. Reduced fat content results in lower product quality in terms of technological (water holding capacity, cooking losses, emulsion stability, and texture) and sensory (taste, aroma, mouth feel, juiciness, and overall acceptability) characteristics, affecting the marketing value and customer approval of newly developed low-fat meat products (Alves et al. 2016). The lower or reduced-fat in processed meat products makes them hard, rubbery, and negatively affects overall acceptance (Youssef and Barbut 2010; Brewer 2012). The substitution of saturated fats for unsaturated fats causes technological issues with food texture, spreadability, flavor, and rancidity (Yogesh et al. 2014; Kumar and Langoo 2016; Heck et al. 2017, 2018; Kumar et al. 2019, 2020; Kumar and Kumar 2020). As a result, the meat Industry is struggling with the challenge of producing low-fat processed meat products with desirable quality features. Therefore, it is necessary to reformulate and optimize the product so the final product may have good quality and better acceptance among the customers.

4.1.2 Definition and Classification

Fat mimetics are used for partial replacement of fat because they mimic certain specific functional properties of fat in a food (Figure 4.1.1). The phrase "fat mimetics" initially came into use in the late 1980s, when the food industry was trying to find different ways to reduce fats from its products. These differ from chemically-produced lipid-based fat replacers (or fat substitutes) like olestra. These include carbohydrate- and protein-based materials modified by physical and chemical means to mimic the functional, textural, and sensorial properties of natural fats and oils (Figure 4.1.2) (Owusu-Apenten 2004). Contrary to certain fat substitutes which have been shown to have side effects, fat mimetics have no major side effects and contributed to calorie reduction. Though, they have certain limitations *e.g.* not suitable for frying, not able to provide flavor, probabilities of allergenic reactions, and have different rheological properties than fat. The application of fat mimetics in the meat industry is extensively studied because it is helpful in reducing animal fat and cholesterol content, thereby lowering the total calories as well as improving the nutritional profile of meat products by increasing the levels of unsaturated fatty acids (Jimenez-Colmenero et al. 2015). In the last decade, various fat mimetics have been for overcoming the drawbacks of the above-mentioned fat replacers.

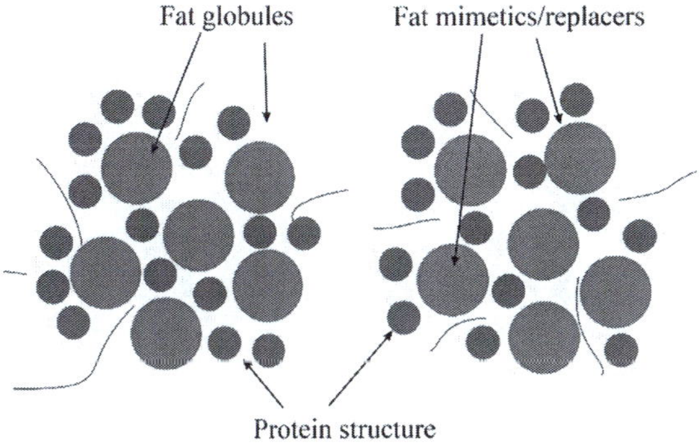

Figure 4.1.2 An ideal fat mimetic/replacer acts like fat globules or simulates the function of fat globules to provide a similar matrix and texture without affecting the technological qualities of meat emulsion/products.

4.1.3 Type of Fat Mimetics

4.1.3.1 Carbohydrate-Based Fat Mimetics

Carbohydrates, the prevalent macronutrient in the human diet, have distinct physicochemical properties and provide less energy than fat, can substitute fat. Carbohydrate-based fat replacers comprise both digestible and indigestible complexes that come from cereals, grains, and other plants. Maltodextrin and changed starches are digestible, but microcrystalline cellulose methylcellulose, gums (guar gum, gum arabic, and xanthan gum), pectin, hydrocolloid gums (alginates, carrageenan), and polydextrose are indigestible sources. Carbohydrate is used in the majority of fat mimetics nowadays that plays a key role in the development of low-fat processed meat products (Table 4.1.1).

Table 4.1.1 Application of fat mimetics/fat replacers in processed meat products.

Fat mimetic/replacer	Low- or reduced-fat processed meat products	Key findings	Reference
Polyunsaturated gelled (kappa-carrageenan emulsion)	Burger patties	↓ TBARS ↓ Cholesterol oxidation products	(Poyato et al. 2015)
Konjac flour, κ-carrageenan, and barley β-glucan	Frankfurters	↓ Total lipid and fat calorie value ↓ Cooking loss ↓ Emulsification stability ↑ Brightness	(Jingxin et al. 2021)
Bacterial nanocellulose	Sausages	↑ Water-binding properties ↑ Water holding capacity → shelf-life	(Marchetti et al. 2017)
Ethyl cellulose and beeswax oleogels	Liver pâtés	↑ Lipid oxidation → Emulsion stability → Texture → Color	(Gómez-Estaca et al. 2019a)

(Continued)

Table 4.1.1 (Continued)

Fat mimetic/replacer	Low- or reduced-fat processed meat products	Key findings	Reference
Chia and oat gel emulsion	Longaniza (fresh sausages)	↓ Cooking loss → Sensory quality	(Pintado et al. 2018)
Amorphous cellulose fiber (Z-trim®)	Emulsified and cooked meat model system	↑ Emulsion stability ↑ Consistency ↑ Hardness ↑ Cooking loss	(Schmiele et al. 2015)
Double emulsions with olive leave extract	Meat system	↓ Peroxide and TBARS ↑ Physical stability	(Robert et al. 2019)
Chitosan and whole and defatted golden flaxseed flour	Hamburgers	↑ Physical-chemical quality	(Hautrive et al. 2019)
Olive oil organogel with soy protein and water	Venison sausages	↓ Total saturated fatty acid → Lipolysis and lipid oxidation	(Utrilla et al. 2014)
Prebiotic fibers (inulin, FOS, polydextrose, and resistant starch)	Meat emulsions, bologna sausage	↓ Emulsion stability ↑ Tenderness *Showed porous matrix	(Felisberto et al. 2015)
Malva nut gum (0.6%)	Salami	↓ Titratable acidity and water activity ↓ Hardness and chewiness ↑ TBARS	(Pramualkijja et al. 2016)
Pineapple byproduct	Burgers	↓ Cooking loss and diameter reduction ↑ Water and fat retention ↑ Hardness, chewiness, and cohesiveness → Color	(Selani et al. 2016)
Pork skin gelatin powder and basil seed gum	Sausage	↓ Expressible moisture ↓ Cooking loss ↑ Gel formation ↑ Water binding	(Lee and Chin 2017)
Sugarcane dietary fiber	Meat batter	↑ Water and fat binding properties ↑ Hardness, gumminess and chewiness ↑ Dynamic rheology	(Zhuang et al. 2016)
Double emulsions (W1/O/W2)	Meat emulsion model system	↓ Jelly and fat separation ↑ Water holding capacity ↑ Emulsion stability	(Serdaroğlu et al. 2016b)
Pork skin and dietaryfibers (inulin, polydextrose, α-cyclodextrin, and bamboo fiber)	Meat emulsion model system	↑ Hardness ↑ Emulsion stability	(dos Santos et al. 2020)

Table 4.1.1 (Continued)

Fat mimetic/replacer	Low- or reduced-fat processed meat products	Key findings	Reference
Whey protein isolate and sodium dodecyl sulfate	Sausage	↑ Gelation ↑ Cooking loss ↑ Hardness and chewiness ↑ Water holding capacity ↑ Emulsion stability	(Kwon et al. 2021)
Hydroxypropyl methylcellulose oleogels	Meat patties	↓ Cooking loss ↑ Firmness and work of shear ↑ Resistance against oxidation	(Oh et al. 2019)
Hemicellulose B from defatted rice bran and rice bran insoluble dietary fibre	Meat balls	↑ Water binding and swelling capacity ↑ Fat binding capacity ↑ Higher unsaturation index	(Hu and Yu 2015)
Double emulsion of beetroot juice and whey protein isolate	Meat system	↓ Hardness ↑ Color ↑ Water and fat binding capacity	(Eisinaite et al. 2017)
Chia flour and/or soy protein isolate, inulin, carrageenan, sodium caseinate, and sodium tripolyphosphate	Bologna sausage	↓ Redness ↑ Brightness ↑ Hardness and chewiness ↑ Shear force	(de Souza Paglarini et al. 2019)
Monoglyceride-based oleogels with conventional or high oleic sunflower oils	Bologna sausage	↑ Brightness ↑ Hardness ↑ Sliceability ↑ Compactness	(Ferro et al. 2021)
Konjac gel matrix with olive, linseed and fish oils	Frankfurter	→ Water and fat binding properties → Color → Sensory quality	(Salcedo-Sandoval et al. 2013)
Whey protein isolate (WPI) and sugar beet pectin	Spreadable raw-fermented sausage and sliceable emulsion-type sausage	↓ Hardness → pH and water activity values *meat protein network became loose and disrupted	(Zeeb et al. 2018)
Soy protein isolates	Salami-type sausage	↑ Hardness ↑ Emulsion mean droplet size	(Dreher et al. 2021)
κ carrageenan gelled emulsion	Dryfermented sausage	→ Peroxide → TBRAS → Color → Sensory quality	(Alejandre et al. 2016)
Pigeon pea flour, corn flour, walnut paste and sesame paste	Sausage	↓ Cooking loss ↑ Emulsion stability ↑ Textural quality *Compact structures with low sponginess	(Tahmasebi et al. 2016)

(Continued)

Table 4.1.1 (Continued)

Fat mimetic/replacer	Low- or reduced-fat processed meat products	Key findings	Reference
Carrageenan and fish gelatin	Meatball	↑ Hardness and chewiness ↑ Cooking yield ↑ Water holding capacity	(Yasin et al. 2016)
Bovine skin gelatin powder and egg white powder	Emulsified sausage	↓ Color ↓ Texture ↑ Lipid oxidation ↑ Cooking yield ↑ Emulsion stability → Sensory quality	(Nacak et al. 2021)
Soy protein isolate, alginate, calcium sulphate 2-hydrate, tetra-sodium pyrophosphate 10-hydrate with phenolic compounds (grape seed and olive extract)	Frankfurter	↑ Oxidative stability → Sensory quality → Physico-chemical properties → Lipid structure	(Pintado et al. 2021)
Aloe gel	Meat emulsion	↓ Cooking loss ↑ Emulsion stability *More homogenous regular emulsion matrix with regular shaped oil droplets	(Kumar et al. 2016)
Pineapple dietary fibres	Beef sausage	↓ Chewiness, gumminess and springiness ↑ Purge ↑ Lightness, hue and chroma ↑ Total expressible fluid	(Henning et al. 2016)
Pork skin, bamboo fiber and inulin	Hot-dogsausage	↓ Redness ↑ Lightness → Water activity → Texture quality	(dos Santos et al. 2020)
Pork skin and green banana flour	Bologna sausage	↓ Cooking loss ↑ Emulsion stability → Microbiological quality → Color → Texture quality → Sensory quality	(Alves et al. 2016)
Aloe vera	Burger	↓ Cooking loss and diameter reduction ↑ Water absorption ↑ Texture quality ↑ Lipid stability	(Soltanizadeh and Ghiasi-Esfahani 2015)

Table 4.1.1 (Continued)

Fat mimetic/replacer	Low- or reduced-fat processed meat products	Key findings	Reference
Albedo-fiberpowder obtained from yellow passion fruit	Burger	↓ TBARS ↓ Aerobic, mesophilic and enterobacteria ↑ Cooking yield ↑ Moisture retention and fat retention ↑ Thickness	(López-Vargas et al. 2014)
Organogels prepared from ethyl cellulose and sorbitan monostearate	Breakfast sausage	↓ Objective hardness and springiness ↓ Sensory juiciness and oiliness ↓ Lightness ↓ Springiness	(Barbut et al. 2016)
Hydrolyzed collagen	Frankfurter-type sausage	↓ Lightness ↑ Texture ↑ Taste ↑ Water holding capacity ↑ Stability after cooking → Appearance, aroma and overall acceptability	(Sousa et al. 2017)
Adzuki beans flour	Meatball	↓ Lightness and yellowness ↑ Cooking yield ↑ Moisture content ↑ Hardness and chewiness	(Aslinah et al. 2018)
Oleogel of soybean oil and pumpkin seed paste	Meat batter	↓ Lightness and redness ↑ Oxidative stability ↑ Hue angle and saturation index *Hard and brittle texture	(Ferrer-González et al. 2019)
Polysaccharide-protein complex of xanthan gum and succinylated chicken liver protein with pre-emulsified sunflower oil	Emulsified sausage	↑ Emulsion stability ↑ WHC ↑ Heating stability ↑ Oxidative stability	(Lu et al. 2021)
Porcine plasmaprotein hydrolysates and oxidized tannic acid	Frankfurter	↑ Water and fat binding capacities ↑ Hardness, gumminess, and chewiness ↑ Oxidative stability	(Chen et al. 2020a)
Washed cashew apple fiber	Patties	↑ Cooking yield → Sensory quality	(Guedes-Oliveira et al. 2016)

↓ Lower than control; ↑ higher than control; → not changed or similar to control.

4.1.3.1.1 Starch and Starch Derivatives-Based Fat Mimetics

Native and modified starch granules that have comparable sizes and shapes to fat granules are ideal fat replacers (Malinski et al. 2003; Lindeboom et al. 2004). Their unique physicochemical properties such as thickening, viscous flow, water holding capacity, and gel formation ability make them a potential fat replacer in food systems (Table 4.1.2) (Chavan et al. 2016). The native individual starch granule can disperse similarly to emulsion droplets, imparting distinct textural and sensory properties (Peng and Yao 2017). However, changed starch is more extensively used in the food industry because native starch may not be suitable for all processing conditions, such as low pH, high temperature, and freezing.

Cross-linked starches are also attempted by several workers as an alternative to native starches in fat replacement applications (Radi et al. 2009). Cross-linked starch has a higher water holding capacity, forms gels, and creates viscous flow. It has digestibility lower than native starches generates fewer calories than native starches (Peng and Yao 2017). Modified starches, such as sodium octenyl succinate, can have emulsifying qualities and the ability to transport hydrophobic nutrients, thus enhancing the product's quality (Tesch et al. 2002; Peng and Yao 2017). As a result, the modified starches not only provide lipophilic nutrients but also interact with other food components (such as proteins) to improve the quality of low-fat foods (Lobato-Calleros et al. 2014). Corn starch (6%) improved the cooking yield, moisture content, pH, protein content, and sensory score In buffalo calf meat sausages while the level of TBARS and fat content decreased (Jairath et al. 2018). (Verma et al. 2015) developed pork low-fat patties by replacing the added fat with a combination of sweet potato power and chilled water. They observed lower texture profile attributes and hardness while a higher cohesiveness in the low-fat patties. They also found that sensory attributes (juiciness, texture, and overall acceptability) were higher in low-fat samples than control patties. Similarly, adzuki beans (*Vigna angularis*) flour at 50% (w/w) was used as a replacement of fat In beef meatballs and low-fat meatball samples exhibited a higher cooking yield, moisture content, hardness, chewiness, and overall acceptability while showing a lower lightless and yellowness values (Aslinah et al. 2018). (Fasuan et al. 2018) showed that the modification of amaranth starch by oxidation, osmotic pressure and cross-linking increased the presence of alkenyl functional group, which enhanced their characteristics of fat replacement. Low-fat buffalo meat patties

Table 4.1.2 Examples of existing starch-based fat mimetics (Chen et al. 2020b).

Product name	Producer	Characteristic
Maltrin M040	Grain Processing	Hydrolyzed from corn starch, DE = 5
Paselli SA-2	Avebe	Enzyme-modified potato starch, DE < 3
N-Oil	National Starch & Chemical	Acidification hydrolysis from tapioca starch
Oatrim	USDA laboratory in Peoria	Good heat stability, Gel properties, lower serum cholesterol levels
Z-Trim	USDA laboratory in Peoria	Digestibility, Zero-calorie, tasteless
Tapioca starch		DE 2.0–2.5, taste creamy, translucent
Wheat bran starch		DE 6–10
Potato starch		DE = 2.5

containing 3% tapioca starch had a higher cooking yield, better dimensional parameters, and higher scores for sensory qualities (Nisar et al. 2009).

Maltodextrins are generated by hydrolyzing natural starch with enzymes or chemicals. Maltodextrins with a low dextrose equivalency (DE < 20) are excellent fat mimetics (Chavan et al. 2016). Corn starch, wheat starch, tapioca starch, and rice starch are some of the raw ingredients used to generate maltodextrins. Maltodextrins have a slightly different functionality as fat replacers than native or changed starch since they are hydrolyzed products and lose the physical properties of starch granules. Maltodextrin gels made from potato starch have a shortening-like texture and are plastic-like, spreadable, and are non-elastic. These gels are thermoreversible, which means they melt when heated and that they return to a similar gel strength when cooled. They also show shear-thinning behavior and reform when shearing stops. Low-DE maltodextrins derived from potato starches add creaminess and a fatty coating to the food product. Maltodextrins have an affinity for water molecules, which lets them form gels with water in food systems with some similar features to fat. Maltodextrin gel is close In size to fat globules (1–3 μm), which makes them an excellent fat replacer that delivers a fat-like feel to dishes (Peng and Yao 2017). Low DE maltodextrins can improve physicochemical aspects of food such as viscosity, lubricity, and softness (Grzelak et al. 2016). Maltodextrin in meat products enhances juiciness and tenderness (Crehan et al. 2000; Telis and Nicoleti 2009). For example, maltodextrin has been a fat replacer for the production of low-fat chicken burger patties and it was reported that 90% chicken meat with 0.25% maltodextrin was most suitable with the lowest fat content and better sensory acceptance (Phiaw and Yaaku 2017). The maltodextrin showed a positive influence in reducing water loss in beef burgers and also improves the cooking yield, shrinkage, texture and sensory acceptance of the final product (Carvalho et al. 2017). Paselli SA2 from potato starch, Maltrin M040 from corn starch, N-Oil from tapioca starch, and Oatrim-1, Otrim-5, and Otrim-10 from oat starch are some of the commercial maltodextrin-based fat replacers on the market (Chavan et al. 2016). The Food and Drug Administration (FDA) has classified Oatrim as GRAS. Baked foods, fillings and frostings, frozen sweets, dairy beverages, cheese, salad dressings, processed meats, and confections can all benefit from this ingredient (Owusu-Apenten 2004). Oatrim-5, a unique combination of oat β-glucan and oat low DE-maltodextrins, has also been approved by the United States Department of Agriculture (USDA). However, there are several drawbacks associated with using maltodextrin as a fat replacer such as low freeze-thaw stability, acid stability, and heat stability (Chavan et al. 2016).

4.1.3.1.2 Cellulose-Based Fat Mimetics

Cellulose is a natural polymer with the molecular formula $(C_6H_{10}O_5)_n$ comprises β- (1→4) D-glucose units (Güzel and Akpınar 2020; Dong et al. 2021). The most broadly used cellulose-based fat replacers are microcrystalline cellulose, carboxymethylcellulose, and methylcellulose gums.

4.1.3.1.2.1 Microcrystalline Cellulose (MCC)

Microcrystalline cellulose is a refined wood pulp or colloidal substance which is widely used as a texturizer, anti-caking agent, emulsifier, extender, and a bulking agent in various food applications (Nsor Atindana et al. 2017). MCC is a porous, aggregate, white, odorless, impurity-free crystalline powder. It is characterized by lower polymerization when compared to the starting material (Leppänen et al. 2009). MCC is made from α-cellulose, which comprises paracrystalline and crystalline microfibrils, and is made through acid depolymerization and agglomeration, followed by drying to generate porous MCC particles (Chavan et al. 2016). The cellulose crystallites form a three-dimensional network highly thixotropic, temperature stable, and provides a clean mouthfeel with correct dispersion. It is widely available and now used as fat replacer in a variety of foods. It can be used at concentrations ranging from 0.1 to 10%, however, standard use levels are from 0.4 to 3.0%. Salad dressings, bakery products, dairy products, ice cream/frozen desserts, cheese spreads, and processed meats are among the foods that employ MCC as a fat replacer. MCC is used in most food systems as part of a fat mimetic system that also has soluble hydrocolloid starch, fat flavors, and antimicrobial agents. Gibis et al. (2015) investigated the effect of MCC as a fat replacer on the microstructure and sensory qualities of fried beef patties and observed that it could efficiently replace up to 50% fat when compared to the standard product. The authors indicated that because MCC is mostly crystalline and has no net charge, it forms a particle gel network as an inert molecule that fills the gaps in the tight meat fiber network, and MCC particles may not disrupt the protein network during heating. The addition of MCC with resistant starch and oat fiber influenced several technological parameters, including water activity, lactic acid bacteria, hardness, chewiness, and TBARS values of fermented sausage (Santos et al. 2021). The authors stated that adding these substances do not affect sensory attributes, but simultaneously reduce the fat content up to 25%. The mechanical qualities of emulsified sausage were significantly influenced by cellulose microcrystals, which improved the final product's firmness due to their high compactibility in the meat matrix (Schuh et al. 2013). *Methyl cellulose gums based fat mimetics*

Cellulose gums are an abundant food ingredient on earth and are also known as methyl cellulose (MC), hydroxypropylmethyl cellulose (HPMC) sodium carboxy methyl cellulose, etc. These gums are used in food and beverages to add viscosity, change texture, improve mouth-feel, binds water and for stabilizing air and carbon dioxide bubbles to reduce volume loss and enhance moisture retention in final products. Cellulose gums are produced through the chemical modification of cellulose mainly from wheat, maize, potato and rice. The properties of cellulose gums such as water retention, sensitivity to electrolytes, dissolution temperatures and gelation characteristics are dependent on molecular weight, degree of substitution and substitution groups (Naeli et al. 2020). The myofibrillar proteins (myosin and actin) present in meat are responsible for the three-dimensional gels during the production of the desired structure of meat products. The CMC interacts with its negatively charged carboxyl groups to modify functional properties of meat proteins like gel formation, solubility/emulsifying capacity and stability of the products (Ayadi et al. 2009). Gibis et al. (2015) showed that when CMC was used at a level of 2 to 3%, a significant decrease in the weight loss occurred when compared to the control group in fried beef patties. Similarly, Yadegari et al. (2021) reported that adding methyl cellulose at a concentration of 0.5% in low-fat frankfurters were most acceptable to the panelists. However, the cooking loss was greater in methyl cellulose added frankfurters compared to others. The meat patties developed using oleogels prepared by canola oil and HPMC as a replacement for animal fat (beef tallow) were used to reduce the level of saturated fat content (Oh et al. 2019). The developed oleogel at 50% replacement showed highest sensory acceptability, a firmer texture with reduced cooking loss and had gel-like behavior and higher oxidative stability of meat product during storage. The authors reported that addition of HPMC oleogels reduced saturated fat level from 42% to 15% in meat patties. Using ethylcellulose (2%) based oleogels with adipic acid (4%) showed improvement in fatty acid prolife with no adverse effect on texture, color as well as organoleptic attributes of beef burgers (Adili et al. 2020). The breakfast sausage prepared using oleogels made with canola oil, ethyl cellulose and sodium monostearate showed improvement in nutritional profile by replacing the saturated fat (pork fat) with mono- and polyunsaturated fatty acids (Barbut et al. 2016). The authors observed that springiness and objective hardness of developed sausages were lower than the control. Also, sensory juiciness, oiliness, lightness were lower than control while the redness and yellowness values were not affected. (Gómez-Estaca et al. 2019b) developed low-fat pork burgers using oleogels consisting of ethyl cellulose, beeswax and PUFA enriched oils (olive, linseed and fish oils) as fat replacers and reported that reformulated pork burgers were softer with a 3.6 fold increase of the PUFA/SFA ratio and a 23 fold decrease of the n-6/n-3 ratio as compared to the control. Barbut and Marangoni (2019) used ethyl cellulose (11%) and sodium monostearate (5%) with soybean/canola/linseed/rendered beef (100%) as fat replacers in meat batter and observed enhanced nutritional profile and reduced springiness in the formulated meat batter. Similarly, Alejandre et al. (2019b) and Ferrer-González et al. (2019) also developed meat batters using ethyl cellulose and sodium monostearate and mixture of ethyl cellulose, MCC (Avicel®) and α-cellulose with soybean oil, respectively. The reformulated batters showed increment in PUFA content and reduced lipid oxidation with no negative effect on texture and sensory acceptance.

4.1.3.1.3 Dietary Fiber-Based Fat Mimetics

Dietary fibers are present in cereals, fruits, nuts, and vegetables and are not digested by the human digestive system. Thus, a dietary fiber has two main components: soluble fibers and insoluble fibers. Based on their chemical composition, solubility, viscosity, and fermentability, dietary fibers are categorized into different groups. Compared to cellulosic fibers, soluble dietary fibers such as pectin and gums can hold higher water (up to 20 times of their weight). The capacity of a fiber to bind fat depends on the porosity, which is useful for avoiding excessive absorption of frying fat in batters. Additionally, the fiber networks enhance the stability of gels by holding the water within pores (Bengtsson et al. 2011). Dietary fibers are also essential nutrients as consuming fiber decreases diseases such as gastrointestinal disorders, obesity, diabetes, and hypertension (Anderson et al. 2009). Epidemiological, clinical, and biochemical studies showed that fiber has a strong positive impact on human health by lowering cholesterol, improving glucose tolerance and insulin response, lowering hyperlipidemia and hypertension, contributing to gastrointestinal health, and preventing certain cancers like colon cancer (Kaczmarczyk et al. 2012; Zhang et al. 2013). Fibers reduce the calorie amount of meat products while increasing the carbohydrate content, making them healthier (Mehta et al. 2019). Various fibers have been used as a fat replacer, a fat reduction agent during frying, and a volume enhancer, binder, bulking agent, and stabilizer in food items (Kumar 2021; Kumar et al. 2021). Many studies suggested that plant dietary fibers used as partial fat mimetics can improve the binding qualities, textural characteristics, and cooking yield of meat products and reduce formulation costs (Mehta et al. 2019). Adding albedo fiber powder obtained from yellow passion fruit improved textural properties and the cooking attributes such as yield, shrinkage and kept the fat and moisture content of pork burgers (López-Vargas et al. 2014).

4.1.3.1.3.1 Pectin

Pectin is a complex polysaccharide with a galacturonan backbone that may be methyl-esterified in various quantities and is commonly obtained from citrus peel and apple pomace. Structurally, α-(1→4)-glucosidic links connect D-galacturonic acid units. Pectins are available commercially as low methoxyl (LM) and high methoxyl (HM) pectins, based on their degree of esterification. Chemical amidation of HM pectins can result in low methyl-esterified and amidated (LMA) pectins. In the United States, pectin is given a GRAS status, but in the European Union it is designated E440 as a food additive. It acts as an emulsifier, stabilizer, gelling and thickening agent in food products (Sharefiabadi and Serdaroğlu 2020). Pectins from various plant sources are widely used in meat industry. Apple pomace pectin was used to formulate different meat products like chicken sausages (Yadav et al. 2016), buffalo meat sausages (Younis and Ahmad 2015), and reduced-fat chicken sausages (Choi et al. 2016). These studies revealed an increase in cooking yield, water holding capacity, and emulsion stability. (Kim et al. 2016) used 1% soy husk pectin to both fresh and frozen/thawed beef patties. The results showed reduced weight loss and increased cooking yield, but TBARS values were statistically unaffected and sensory acceptability were comparable to control. In another study, (Zeeb et al. 2018) formulated raw-fermented sausages and emulsion-type sausages by incorporating pectin at 3% and 1.88%, respectively. The developed raw-fermented sausages showed similar rheological properties as the full-fat control, but the emulsion-type sausages have more softening value than control. In another study, different fat replacers were tested in low-fat frankfurters. According to the findings, adding different hydrocolloids influenced the batter's emulsion stability in different ways. Cooking yield was best in samples with 1% pectin, and water holding capacity (WHC) was like control. (Thanakiatkrai and Kitpipit 2017) investigated the effect of pectin on a reduced-fat pig meat model system. It was reported that pectin-added samples had similar attributes to the control samples. The hamburger patty prepared using alginate-pectin with whey protein concentrate showed higher oxidative stability and tenderness (Barybina et al. 2019).

4.1.3.1.3.2 Inulin

Inulin is a naturally occurring polysaccharide present in plants like chicory and Jerusalem artichoke. It has a sweetening capability of 30–65% to that of sucrose and polymerization (DP) of 2–60. It belongs to a class of dietary fibers known as fructans. It is mainly present in roots and rhizomes of plants. Commercially, inulin is extracted from chicory roots by diffusing the roots in hot water, refining the concentration, and finally spray-drying the concentrate. The Food and Drug Administration in United States approved inulin as a dietary fiber ingredient in 2018 for improving the nutritional value of food products. It is a white, water soluble powder and has a natural taste and smell. At the concentration of 40–45%, it creates gel of cream which has similar fatty creamy-feel and has strong water binding and freeze thaw stability. Inulin with a lower DP of 25 is widely used for high-performance fat replacement. The fat content was reduced to 6–18% in emulsion type sausages manufactured by replacing fat with inulin (Berizi et al. 2017). The results showed that with 6% inulin and 12% water, reduced the energy content by 20–64% while the color, sensory and TPA were like the control samples. The addition of 6.58% fiber from inulin (5%), oat fiber (1%) and wheat fiber (0.58%) for partial replacement of pork backfat showed good sensory acceptability and reduced the pork fat level by 25–42.75% in bologna sausages (Barretto et al. 2015). Similarly, Silva-Vazquez et al. (2018) reported that the pork backfat replaced with inulin and pectin at 15% concentration maintained color and water holding capacity, cooking loss and total expressible fluid in formulated meat batter. Méndez-Zamora et al. (2015) studied the effect of substitution of animal fat with different formulations of pectin and inulin on chemical composition, textural, and sensory properties of emulsion-type sausages and concluded that 15% pectin and 15% inulin could be used as animal fat replacer.

4.1.3.1.3.3 β-glucan

B-glucan is a polysaccharide found in the cell walls of oat, barley, and other grains and in yeast. B-glucan (β-1–3),(1–4)-D-glucan) is a soluble dietary fiber fraction derived from oat with unique nutritional, functional, and technological attributes. Oat β-glucan has a good impact on human health in addition to its technological features. For diabetics and people with excessive cholesterol levels, fibre consumption, predominantly β-glucans, has become a good choice (Bozbulut and Sanlier 2019). In the meat product system, β-glucan forms a porous network that holds the protein-gel matrix and fat globules together. This mechanism explains the absorption of huge volumes of water and the creation of dense matrices, which are important for textural and rheological properties (Morin et al. 2004). As a result of its stability, gelation, and water-binding capacity, β-glucan

can have a role in altering the textural and rheological characteristics of low-fat foods (Piñero et al. 2008; Oliveira et al. 2012). Oat β-glucan contains hydrocolloid properties, which are useful in food processing. Water-binding properties of hydrocolloids increase viscosity, stabilize emulsions, and alter the gelation process (Lee et al. 2015; Pintado et al. 2016a). β-glucans at 30% concentration have been successfully used as fat replacer to improve water holding capacity and texture profile of low-fat beef burgers with similar sensory properties (Szpicer et al. 2020). It reduces the cholesterol content while increases pH and color value of final product. Similarly, several other workers have been used β-glucans in formulation of meat batters, sausages and chopped poultry products (Piñero et al. 2008; Álvarez and Barbut 2013; Ktari et al. 2016).

4.1.3.1.3.4 Z-trim

It is insoluble fiber derived from the hulls of oat, soybean, pea, and rice, as well as from corn and wheat bran. Z-trim refers to zero calories and is approved by the FDA for use as fat replacer. It provides an aqueous gel fiber structure, smoothness, moistness, density, and has a mouthfeel like fat providing no taste. Z-trim also has the added benefit of enhancing food fiber content. However, Z-trim is not recommended for frying (Owusu-Apenten 2004).

4.1.3.1.4 Gelling and Bulking Agent-Based Fat Mimetics

Polysaccharides can make a variety of gel structures that can immobilize oil droplets and build gelled emulsions that can include lipophilic substances with health benefits into food products (Figure 4.1.3). Using emulsion-based systems has enabled the production of nutritionally better

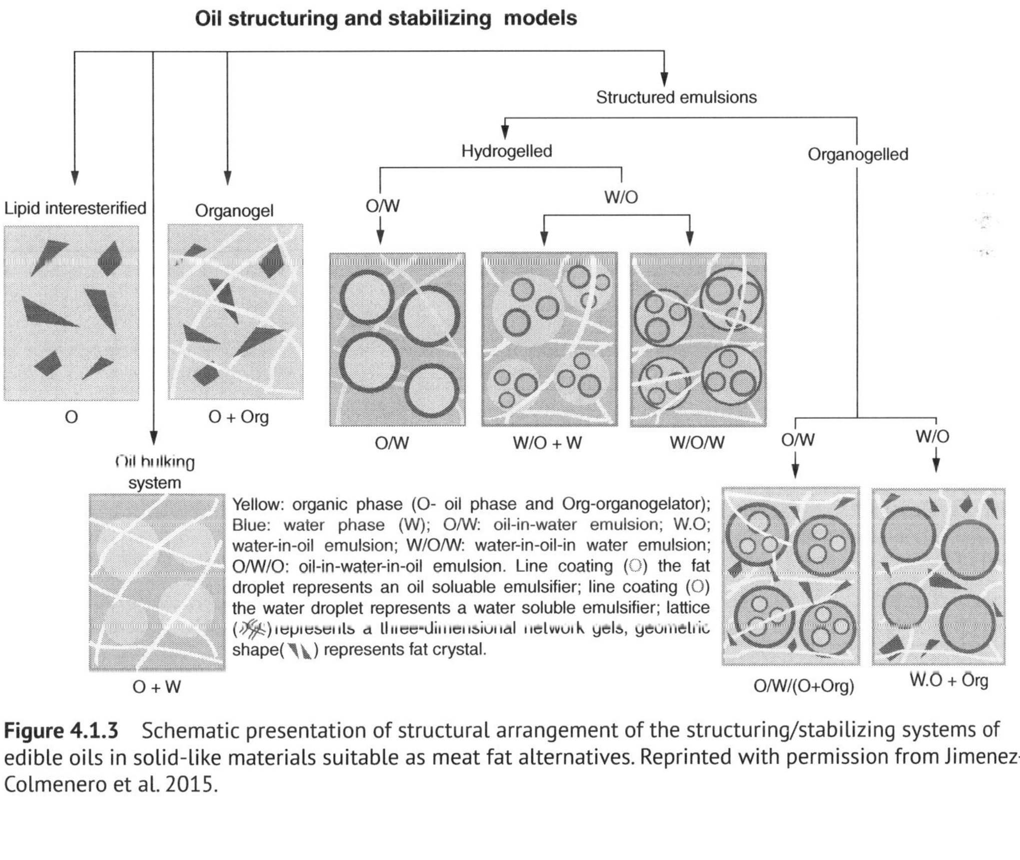

Figure 4.1.3 Schematic presentation of structural arrangement of the structuring/stabilizing systems of edible oils in solid-like materials suitable as meat fat alternatives. Reprinted with permission from Jimenez-Colmenero et al. 2015.

meat products using reformulation methodologies, with improvement in fat content and quality (Table 4.1.3). Substituting pork backfat by PUFA emulsified oils has been a good technique for achieving healthier lipid profiles in meat products (Serdaroğlu et al. 2016b; Robert et al. 2019). Some hydrocolloids are commonly used in food products to improve appearance (creaminess,

Table 4.1.3 Use of gelled emulsions to improve/replace animal fat content in meat products (Paglarini, C.D.S., et al. (2022) / Taylor & Francis).

Meat system	Oil	Protein/Polysaccharide/as a functional ingredient and gelling agents	Storage conditions	References
Fresh sausage (*longaniza*)	Olive	Chia flour or oat bran	18 days at 2°C	(Pintado et al. 2018)
Meat emulsion	Soybean	Soy protein isolate, sodium caseinate, carrageenan mix of kappa and iota, inulin, pectin, sodiumtripolyphosphate, and soy lecithin.	–	(Paglarini et al. 2018)
Frankfurter	Olive	Chia, transglutaminase, Sodium alginate, or gelatin	85 days at 2°C	(Pintado et al. 2016b)
Burger patties	Sunflower	Carrageenan, Polysorbate 80	Raw and cooked burger patties	(Poyato et al. 2015)
Burger patties	Algae	Carrageenan, Polysorbate 80	4°C/vacuum, 4°C/no vacuum, 25°C/vacum and 25°C/no vacuum for 31 days	(Alejandre et al. 2017)
Pork patties	Perilla	Gelatin and transglutaminase	Raw patties (2°C for 14 days) and cooked patties	(Freire et al. 2017)
Dry fermented sausage	Linseed	Carrageenan, Polysorbate 80	–	(Alejandre et al. 2016)
Bologna-type sausage	Linseed	Carrageenan, Polysorbate 80	–	(Poyato et al. 2014)
Frankfurter	Olive	Chia flour, sodium alginate, $CaSO_4$ and pyrophosphate	–	(Pintado et al. 2016b)
Turkey breast emulsion, chicken patties and meat emulsion	Olive	polyglycerol polyricinoleate, gelatin and inulin	–	(Serdaroğlu et al. 2017, Serdaroğlu et al. 2016a, Serdaroğlu and Öztürk 2017)
Burger patties	Linseed and chia	Carrageenan, Polysorbate 80	–	(Heck et al. 2019)
Meat emulsion	Canola	Carrageenan	–	(Alejandre et al. 2019b)
Burger patties	Microalgae	Blackthorn branch extract, carrageenan	–	(Alejandre et al. 2019a)
Fermented sausage	Linseed oil	Inulin, pork gelatin	30 days at 4°C	(Glisic et al. 2019)

homogeneity) and organoleptic qualities (mouthfeel, juiciness); to promote ease-of-use application (pourability, spreadability); to impart processing benefits such as freeze-thaw stability, sliceability; to keep freshness and improvement in shelf-life; and to increase yields. Carrageenan, konjac, alginate, and agar-agar are polysaccharide biopolymers that have been tested as potential fat alternatives to reduce or improve the lipid fraction of processed meat products, with promising results (Salcedo-Sandoval et al. 2013; Poyato et al. 2014; Felisberto et al. 2015; Alejandre et al. 2016; Yasin et al. 2016; dos Santos et al. 2020; Pintado et al. 2021).

4.1.3.1.4.1 Alginates

Alginates are the principal structural polysaccharides found in brown seaweeds (Phaeophyceae). *Macrocystis pyrifera* is the most common species of seaweed that produces alginates. *Laminaria hyperborea*, *Laminaria digitata*, and *Laminaria japonica* are other good sources of alginates. Alginates are brown seaweed extracts that are utilized in the production of animal feeds and certain processed meat products. Alginate is a water soluble polymer and forms viscous solution. Alginate was employed as a thickening and binding agent in meat products (Qin et al. 2020). It can used as fat replacers as it forms thermally stable gel matrix that exhibits a similar characteristics to that of fat substances hence it is used in formulation of salami and other similar meat products to produce products with relatively low calorie while maintaining the traditional texture and appearances (Qin et al. 2020). The cold gelling agents (alginates, transglutaminase and gelatin) were used with olive oil, as emulsion gels to develop low-fat frankfurter (Pintado et al. 2016b). The study showed that emulsion gel improved the fatty acid profile and fat/water-binding properties. Meat product with emulsion gel showed a lower lightness and redness, while yellowness was higher in the low-fat product. Similarly, alginate-based chia and oat emulsion gels were used as fat replacers in the formulation of low-fat fresh sausages (*longaniza* type) (Pintado et al. 2018).The study indicated that SFAs level decreased while the MUFAs level increased. Additionally, formulated sausages had higher mineral and amino acid composition. Cooking losses were lower while shear forces was higher only in the case of oat emulsion gels.

4.1.3.1.4.2 Carrageenan

Carrageenan denotes a family of water soluble polysaccharides extracted from marine algae/seaweed. It is a high-molecular-weight linear polysaccharide comprising repeating galactose units and 3,6-anhydrogalactose, which are both sulfated and non-sulfated and joined by alternating α-(1,3) and β-(1,4) glycosidic links (BeMiller 2019). There are three types of carrageenans: iota-, kappa- and lambda-carrageenans. Iota- and kappa-carrageenans act as gelling agents. The lambda-type is non-gelling and is more used for its thickening properties because it imparts a creamy texture rather than a typical gel texture. It is mainly used in desserts to provide creamy texture and mouthfeel. Carrageenan is widely used in processed meat products because it can improve texture, yield and appearance of the final product. Carrageenan's ability to form a gel matrix in meat products has been shown to improve uniformity, slice-ability, spreadability, and cohesion while lowering fat content and slicing loss. Nayak and Pathak (2016) formulated low-fat chevon patties by incorporating carrageenan as a fat replacer. They found that the emulsion stability and cooking yield increased with increasing levels of carrageenan. The carrageenan incorporated patties had lower fat and cholesterol contents and higher moisture and fat retention capacity. Texture properties indicated that the fat reduction in sausages showed higher firmness but the addition of κ-carrageenan, konjac, and tragacanth partially compensated the deficits in rheological properties (Atashkar et al. 2018). The incorporation of alginate improved texture more in comparison to carrageenan, whereas the later tended to release higher water after cooking and reheating in low-fat beef patties (Lin and Keeton 2006). Yasin et al. (2016) reported that carrageenan underwent several changes during cooking, freezing and showed significant changes in cooking yield, texture profile, water holding capacity and rheological properties of chicken meatballs. The study showed that κ-carrageenan at a concentration of 1.28 g/100 ml with fish gelatin (4.62 g/100 ml) and meat (60.52 g/100 g) was optimum for chicken ball formulation. Additionally, it has also been observed that the addition of calcium chloride at 0.04 M concentration improved the water holding capacity and gelling properties in porcine myosin-κ-carrageenan mixtures by enhancing the hydrophobic interactions and promoting the β-sheet structure formulation (Pan et al. 2017). In another study, the addition of 0.6–0.8% semi-refined carrageenan in formulation of chicken sausages, improved textural properties and inhibit lipid oxidation (Ganesan et al. 2019).

4.1.3.1.4.3 Agar-agar

Agar agar, a well-known gelling substance, is produced from seaweed (red algae: *Rhodophyta*). It is commonly used as a substitute for animal-derived gelatin and is used in meat processing applications (Ścieszka and Klewicka 2019). It is an easily available and cost-effective ingredient with a good gel forming ability upon cooling. The incorporation of pea protein and agar-agar as beef fat replacer in the formulation of meat emulsion showed higher emulsion stability, oxidative stability and cooking yield whereas unavoidable changes were recorded in color and texture profile (Öztürk-Kerimoğlu 2021). Similarly, Fitrianingsih et al. (2020) reported higher water holding capacity, cooking yield, color, aroma, texture, flavor and elasticity of chicken meatball formulated with agar-agar.

4.1.3.1.5 Gum-based Fat Mimetics

Guar gum, a galactomannan-rich substance derived from the seeds of the leguminous plant *Cyamopsis tetragonoloba* (L.) Taub, is widely utilized in a variety of industrial applications as a thickening and stabilizing agent. Guar gum is mostly made of polysaccharides with a high molecular weight, ranging from 50 kDa to 8,000 kDa. Guar gum is typically added to foods at a concentration < 1 g/100 g for gelling, thickening, firming, and emulsification applications. Guar gum can also be used as a source of soluble dietary fiber in food items, and it has been found to be safe at doses up to 20 g/day (Grabitske and Slavin 2009). Guar gum gives meat products a slick, creamy mouthfeel that resembles fat and is utilized for development of low-fat processed meat products. Ulu (2006) discovered that low-fat meatballs (10%) made with 0.5–1% carrageenan or guar gum had better cooking yield and fat retention than high-fat meatballs (25%). Rather et al. (2016a) reported that the addition of guar gum powder (1%) reduced the fat level from 20% to 10% in Indian traditional mutton product "goshtaba". Additionally, the final product had higher protein content and significantly lowered TBARS values. The author indicated that guar gum powder had higher water holding capacity because of its gel forming ability that resulted in higher juiciness of the product. Similarly, Goswami et al. (2019) found significant increase in the cooking yield while the diameter and spread ratio of cookies decreased significantly by incorporating guar gum as a fat replacer in carabeef cookies.

Another example of gum-based fat mimetics is xanthan gum, which is obtained by aerobic fermentation of sugars by *Xanthomonas campestris*. The back bone of xanthan gum consists of linear (1→4) β-D-glucopyranosyl unit that is substituted at C-3 on every other glucose residue with a charged trisaccharide side chain. The trisaccharide chain consists of a D-glucuronic acid unit between two D-mannose units. Xanthan gum is soluble in both hot and cold water, and it creates a pseudoplastic and viscous solution even at low concentrations. When salt is added to the formulation, the viscosity rises, and the solution remains stable despite changes in the pH and temperature. Pseudoplasticity affects the taste, texture, and appearance of food products (Lurueña-Martínez et al. 2004). Rather et al. (2015) restructured goshtaba using xanthan gum as a fat replacer and observed significantly lower fat content with lower values of fat and protein oxidation. In another study, the mixture of guar-xanthan gum (1:1) was used as a partial fat replacer in the formulation of meat emulsions. The results showed a higher emulsion stability, water holding capacity and cooking yield while lower TBARS, protein carbonyls and protein sulfhydryl groups were observed in the formulated meat product (Rather et al. 2016b).

Lee and Chin (2017) developed low-fat sausages using basil seed gum (*Ocimum bacilicum* L.) and gelatin as fat replacer and observed that the combination of basil seed gum and gelatin reduced the expressible moisture content, cooking loss and improved gumminess and cohesiveness in low-fat sausages. Similarly, water holding capacity, antioxidant activity and tissue hardness of low-fat meat hamburger were improved while cooking loss, lipid peroxidation were reduced with incorporation of quince seed gum (Yousefi et al. 2018). (Sayed et al. 2020) reported higher antioxidant activity, water holding capacity, and cooking yield in developed low-fat beef burgers using Gum Arabic as fat replacer. In another study, Kumar et al. (2016) used aloe gel with vegetable oil for the development of low-fat meat emulsion. The developed emulsion had 50% lowered fat content as compared to full-fat control without affecting sensory attributes. The developed emulsion showed a homogenous regular emulsion matrix with regular shaped oil droplets and decreased cooking loss. The addition of aloe gel with vegetable oil as a fat replacer in the emulsion matrix improved emulsion stability and textural attributes (Kumar et al. 2017)

4.1.3.2 Protein-Based Fat Mimetics

Protein-based fat mimetics such as microparticulated whey protein provide structure, viscosity, creaminess and opacity, a clean flavor base with only one-third of the calories to that of fat. These are made by modifying protein, using egg white or whey from milk and zein protein. Commercial brands of protein-based mimetics include simplesse and trailblazer. Simplesse is obtained from egg white and milk protein. Trailblazer is obtained from egg white and serum protein mixed with xanthan gum. Several protein-based fat mimetics have been used in processed meat products to increase the yield and functional properties while reducing the fat content in final meat products. Examples of these protein additives are: wheat flour/gluten, soy flour, soy protein concentrate/isolate, textured soy protein, cottonseed flour, oat flour, corn germ meal, non-fat dry milk, calcium-reduced non-fat dry milk, caseinates, whey proteins, surimi, blood plasma and egg proteins. A double emulsion was prepared containing beetroot juice in the inner water phase, sunflower oil as te oil phase and 0.5% whey protein as the outer water phase for use in meat systems as fat replacers. The addition of double emulsion significantly reduced pork backfat by 11% and also showed high water and fat binding properties in meat emulsion system (Eisinaite et al. 2017). Similarly, bologna sausages were developed using emulsion gels prepared with soybean oil, functional ingredients, stabilizers and gelling agents such as chia flour, soy protein isolate, inulin, carrageenan, sodium caseinate and sodium tripolyphosphate to replace the pork backfat (de Souza Paglarini et al. 2019). In another study, Freire et al. (2018) formulated emulsion gels with n-fatty acids, tannin and different emulsifiers such as sodium caseinate, whey protein isolates and isolated soy protein. The gelled emulsions showed higher emulsion stability and oxidative stability. Hence, this gel can be used as a fat replacer for the development of processed meat products. Recently, the emulsion gels containing soy protein isolate with solid phenolic extracts from grape seed and olive was used as fat replacers for the development of frankfurters with higher product quality and oxidative stability (Pintado et al. 2021).

4.1.3.3 Fat-Based Fat Mimetics

Oleogels are structured or tailored vegetable oils developed by means of partial hydrogenation, fractionation techniques that converts liquid vegetable oil into a solid-like gels with the use of organogelators. Oleogels are not only used to replace saturated fat and trans fat but also play an important role as carriers for water-insoluble bioactive compounds and imparting heat resistance to the developed product. Oleogels prepared from canola, soybean and flaxseed oil with 10% ethyl cellulose were used to develop low-fat frankfurters (Zetzl et al. 2012). The developed product had similar textural properties as the reference and showed improvement in nutritional attributes as well as fatty acid profile. Similarly, oleogels from sunflower oil mixed with phytosterols γ-oryzanol and β-sitosterol were used to reduce the total fat amount in the frankfurters (Panagiotopoulou et al. 2016). No differences were recorded in level of lipid oxidation, pH and textural properties in developed products and control. However, frankfurters containing oleogel emulsions had lower chewiness, hardness and gumminess. In another study, Kouzounis et al. (2017) replaced porkback fat with oleogels developed with monoglycerides and phytosterols (3:1) and 50% of porkback fat was replaced with sunflower oil in frankfurter sausage. The interaction between monoglycerides and phytosterols modify the crystallization behavior and forms a strong network with high gel strength and hardness in comparsion with the oleogels developed solely with monoglycerides. Similarly, Franco et al. (2019) replaced pork backfat with linseed oleogel gelled with beeswax and reported an improved nutritional profile of frankfurter sausages by decreasing saturated fatty acid content and cholesterol contents.

Meat patties were developed with oleogels prepared from canola oil and hydroxymethyl cellulose to reduce the level of saturated fat by replacing the beef tallow (Oh et al. 2019). The cooking loss of patties was reduced at 50% and 100% hydroxypropyl methylcellulose oleogels and the developed product showed highest overall acceptability at 50% level replacement level. Similarly, Moghtadaei et al. (2018) developed low fat beef burgers using sesame oil oleogel with beeswax (5–10%) and 25–50% of the animal fat was replaced with the developed oleogel. The authors observed a 1.6% decrease in fat absorption and 11% reduction in cooking loss. However, the textural properties viz., hardness, gumminess and chewiness were reduced in the developed beef burgers. Likewise, Martins et al. (2019) used linseed oil with γ-oryzanol and β-sitosterol based oleogels in the formulation of hamburger and suggested that hamburger developed with 25% oleogel had similar textural parameters to the control. In another study, Gómez-Estaca et al. (2020) developed a pork burger with mixture of vegetable oils (44.39% olive oil, 37.87% linseed oil, and 17.74% fish oils), with polyunsaturated fatty acids. The oleogels were developed with 11% ethyl cellulose, 11% beeswax with or without the addition of 2% curcumin for the development of five different samples of burgers. The developed low-fat burgers had soft texture and a higher antioxidant activity than the control. Among the five burgers, beeswax oleogel containing burgers showed the highest overall acceptability score.

4.1.4 Conclusion

Fat mimetics play an important role in the development of low-fat or reduced-fat processed meat products. Researchers have used various fat mimetics and their combinations for this purpose. However, the replacement of animal fat with fat mimetics and oils for the development of low-fat meat products with improved fatty acid profiles many times results in inferior product quality. Certain fat mimetics showed good results in terms of their fat mimicking effect. However, oxidative stability and consumer acceptability of low-fat or reduced-fat processed meat products is still the problem. Thus, there is still a requirement of identification, development, and assessment of fat mimetics for the development of low-fat processed meat products with higher consumer acceptability.

References

(ADA), A. D. A. (2005). Position of the American Dietetic Association: fat replacers. *Journal of the American Dietetic Association* 105: 266–275.

Adili, L., Roufegarinejad, L., Tabibiazar, M., Hamishehkar, H., and Alizadeh, A. (2020). Development and characterization of reinforced ethyl cellulose based oleogel with adipic acid: its application in cake and beef burger. *LWT* 126: 109277.

Alejandre, M., Ansorena, D., Calvo, M.I., Cavero, R.Y., and Astiasarán, I. (2019a). Influence of a gel emulsion containing microalgal oil and a blackthorn (Prunus spinosa L.) branch extract on the antioxidant capacity and acceptability of reduced-fat beef patties. *Meat Science* 148: 219–222.

Alejandre, M., Astiasarán, I., Ansorena, D., and Barbut, S. (2019b). Using canola oil hydrogels and organogels to reduce saturated animal fat in meat batters. *Food Research International* 122: 129–136.

Alejandre, M., Passarini, D., Astiasarán, I., and Ansorena, D. (2017). The effect of low-fat beef patties formulated with a low-energy fat analogue enriched in long-chain polyunsaturated fatty acids on lipid oxidation and sensory attributes. *Meat Science* 134: 7–13.

Alejandre, M., Poyato, C., Ansorena, D., and Astiasarán, I. (2016). Linseed oil gelled emulsion: a successful fat replacer in dry fermented sausages. *Meat Science* 121: 107–113.

Álvarez, D. and Barbut, S. (2013). Effect of inulin, β-Glucan and their mixtures on emulsion stability, color and textural parameters of cooked meat batters. *Meat Science* 94: 320–327.

Alves, L.A.A.D.S., Lorenzo, J.M., Gonçalves, C.A.A., Dos Santos, B.A., Heck, R.T., Cichoski, A.J., Campagnol, P.C.B. (2016). Production of healthier bologna type sausages using pork skin and green banana flour as a fat replacers. *Meat Science* 121: 73–78.

Anderson, J.W., Baird, P., Davis, R.H., Ferreri, S., Knudtson, M., Koraym, A., Waters, V., and Williams, C.L. (2009). Health benefits of dietary fiber. *Nutrition Reviews* 67: 188–205.

Aslinah, L.N.F., Mat Yusoff, M., and Ismail-Fitry, M.R. (2018). Simultaneous use of adzuki beans (Vigna angularis) flour as meat extender and fat replacer in reduced-fat beef meatballs (bebola daging). *Journal of Food Science and Technology* 55: 3241–3248.

Atashkar, M., Hojjatoleslamy, M., and Sedaghat Boroujeni, L. (2018). The influence of fat substitution with κ-carrageenan, konjac, and tragacanth on the textural properties of low-fat sausage. *Food Science & Nutrition* 6: 1015–1022.

Ayadi, M.A., Kechaou, A., Makni, I., and Attia, H. (2009). Influence of carrageenan addition on turkey meat sausages properties. *Journal of Food Engineering* 93: 278–283.

Barbut, S. and Marangoni, A. (2019). Organogels use in meat processing–effects of fat/oil type and heating rate. *Meat Science* 149: 9–13.

Barbut, S., Wood, J., and Marangoni, A. (2016). Quality effects of using organogels in breakfast sausage. *Meat Science* 122: 84–89.

Barretto, A.C.D.S., Pacheco, M.T.B., and Pollonio, M.A.R. (2015). Effect of the addition of wheat fiber and partial pork back fat on the chemical composition, texture and sensory property of low-fat bologna sausage containing inulin and oat fiber. *Food Science and Technology* 35: 100–107.

Barybina, L.I., Oboturova, N.P., Datsko, V.A., Nagdalian, A.A., Kopchekchi, M.E., Ozheredova, N.A., Verevkina, M.N., and Simonov, A.N. (2019). Hamburger patty development with alginate-pectin meat emulsion. *Journal of Hygienic Engineering and Design* 29: 111–118.

BeMiller, J.N. (2019). 13 – carrageenans. In: *Carbohydrate Chemistry for Food Scientists*, 3e (ed. J.N. BeMiller). AACC International Press.

Bengtsson, H., Montelius, C., and Tornberg, E. (2011). Heat-treated and homogenised potato pulp suspensions as additives in low-fat sausages. *Meat Science* 88: 75–81.

Berizi, E., Shekarforoush, S.S., Mohammadinezhad, S., Hosseinzadeh, S., Farahnaki, A. (2017). The use of inulin as fat replacer and its effect on texture and sensory properties of emulsion type sausages. *Iranian Journal of Veterinary Research* 18: 253–257.

Bozbulut, R. and Sanlier, N. (2019). Promising effects of β-glucans on glyceamic control in diabetes. *Trends in Food Science & Technology* 83: 159–166.

Brewer, M.S. (2012). Reducing the fat content in ground beef without sacrificing quality: a review. *Meat Science* 91: 385–395.

Carvalho, G.R.D., Milani, T.M.G., Trinca, N.R.R., Nagai, L.Y., da Silva Barretto, A.C. (2017). Textured soy protein, collagen and maltodextrin as extenders to improve the physicochemical and sensory properties of beef burger. *Food Science and Technology* 37: 10–16.

Chavan, R.S., Khedkar, C.D., and Bhatt, S. (2016). Fat replacer. In: *Encyclopedia of Food and Health* (ed. B. Caballero, P.M. Finglas, and F. Toldrá). Oxford: Academic Press.

Chen, Y., Jia, X., Sun, F., Jiang, S., Liu, H., Liu, Q., and Kong, B. (2020a). Using a stable pre-emulsified canola oil system that includes porcine plasma protein hydrolysates and oxidized tannic acid to partially replace pork fat in frankfurters. *Meat Science* 160: 107968.

Chen, Y., She, Y., Zhang, R., Wang, J., Zhang, X., and Gou, X. (2020b). Use of starch-based fat replacers in foods as a strategy to reduce dietary intake of fat and risk of metabolic diseases. *Food Science & Nutrition* 8: 16–22.

Choi, Y.-S., Kim, Y.-B., Hwang, K.-E.,Song, D-H., Ham, Y-K., Kim, H-W., Sung, J-M., and Kim, C-J. (2016). Effect of apple pomace fiber and pork fat levels on quality characteristics of uncured, reduced-fat chicken sausages. *Poultry Science* 95: 1465–1471.

Corpet, D.E. (2011). Red meat and colon cancer: should we become vegetarians, or can we make meat safer? *Meat Science* 89: 310–316.

Crehan, C.M., Hughes, E., Troy, D.J., and Buckley, D.J. (2000). Effects of fat level and maltodextrin on the functional properties of frankfurters formulated with 5, 12 and 30% fat. *Meat Science* 55: 463–469.

de Souza Paglarini, C., de Figueiredo Furtado, G., Honório, A.R., Mokarzel, L., da Silva Vidal, V.A., Ribeiro, A., Cunha, R.L., and Pollonio, M. (2019). Functional emulsion gels as pork back fat replacers in Bologna sausage. *Food Structure* 20: 100105.

Dong, Y.-D., Zhang, H., Zhong, G.-J., and Lai, B. (2021). Cellulose/carbon composites and their applications in water treatment – a review. *Chemical Engineering Journal* 405: 126980.

dos Santos, M., Munekata, P.E.S., Pateiro, M., Magalhães, G.C., Barretto, A., Lorenzo, J., and Pollonio, M. (2020). Pork skin-based emulsion gels as animal fat replacers in hot-dog style sausages. *LWT* 132: 109845.

Dreher, J., Weißmüller, M., Herrmann, K., Terjung, N., Gibis, M., and Weiss, J. (2021). Influence of protein and solid fat content on mechanical properties and comminution behavior of structured plant-based lipids. *Food Research International* 145: 110416.

Eisinaite, V., Juraite, D., Schroën, K., and Leskauskaite, D. (2017). Food-grade double emulsions as effective fat replacers in meat systems. *Journal of Food Engineering* 213: 54–59.

Fasuan, T.O., Gbadamosi, S.O., and Akanbi, C.T. (2018). Modification of amaranth (Amaranthus viridis) starch, identification of functional groups, and its potentials as fat replacer. *Journal of Food Biochemistry* 42: e12537.

Felisberto, M.H.F., Galvão, M.T.E.L., Picone, C.S.F., Cunha, R.L., and Pollonio, M.A.R. (2015). Effect of prebiotic ingredients on the rheological properties and microstructure of reduced-sodium and low-fat meat emulsions. *LWT – Food Science and Technology* 60: 148–155.

Ferrer-González, B.M., García-Martínez, I., and Totosaus, A. (2019). Textural properties, sensory acceptance and fatty acid profile of cooked meat batters employing pumpkin seed paste or soybean oil oleogel as fat replacers. *Grasas y Aceites* 70: e320.

Ferro, A.C., de Souza Paglarini, C., Rodrigues Pollonio, M.A., and Lopes Cunha, R. (2021). Glyceryl monostearate-based oleogels as a new fat substitute in meat emulsion. *Meat Science* 174: 108424.

Fitrianingsih, Isnaeni, P., Yaddi, Y., Libriani, R., Auza, F.A., and Prasanjaya, P.N.K. (2020). Physical and organoleptic properties of chicken meatball prepared with varied gelling agents. *IOP Conference Series: Earth and Environmental Science* 465: 012015.

Franco, D., Martins, A.J., López-Pedrouso, M., Purriños, L., Cerqueira, M.A., Vicente, A.A., Pastrana, L.M., Zapata, C., and Lorenzo, J.M. (2019). Strategy towards replacing pork backfat with a linseed oleogel in frankfurter sausages and its evaluation on physicochemical, nutritional, and sensory characteristics. *Foods* 8: 366.

Freire, M., Cofrades, S., Pérez-Jiménez, J., Gómez-Estaca, J., Jiménez-Colmenero, F., and Bou, R. (2018). Emulsion gels containing n-3 fatty acids and condensed tannins designed as functional fat replacers. *Food Research International* 113: 465–473.

Freire, M., Cofrades, S., Serrano-Casas, V., Pintado, T., Jimenez, M.J., and Jiménez-Colmenero, F. (2017). Gelled double emulsions as delivery systems for hydroxytyrosol and n-3 fatty acids in healthy pork patties. *Journal of Food Science and Technology* 54: 3959–3968.

Ganesan, A.R., Shanmugam, M., and Bhat, R. (2019). Quality enhancement of chicken sausage by semi-refined carrageenan. *Journal of Food Processing and Preservation* 43: e13988.

Gibis, M., Schuh, V., and Weiss, J. (2015). Effects of carboxymethyl cellulose (CMC) and microcrystalline cellulose (MCC) as fat replacers on the microstructure and sensory characteristics of fried beef patties. *Food Hydrocolloids* 45: 236–246.

Glisic, M., Baltic, M., Glisic, M., Trbovic, D., Jokanovic, M., Parunovic, N., Dimitrijevic, M., Suvajdzic, B., Boskovic, M., and Vasilev, D. (2019). Inulin-based emulsion-filled gel as a fat replacer in prebiotic- and PUFA-enriched dry fermented sausages. *International Journal of Food Science & Technology* 54: 787–797.

Gómez-Estaca, J., Herrero, A.M., Herranz, B., Álvarez, M.D., Jiménez-Colmenero, F., and Cofrades, S. (2019a). Characterization of ethyl cellulose and beeswax oleogels and their suitability as fat replacers in healthier lipid pâtés development. *Food Hydrocolloids* 87: 960–969.

Gómez-Estaca, J., Pintado, T., Jiménez-Colmenero, F., and Cofrades, S. (2019b). Assessment of a healthy oil combination structured in ethyl cellulose and beeswax oleogels as animal fat replacers in low-fat, PUFA-enriched pork burgers. *Food and Bioprocess Technology* 12: 1068–1081.

Gómez-Estaca, J., Pintado, T., Jiménez-Colmenero, F., and Cofrades, S. (2020). The effect of household storage and cooking practices on quality attributes of pork burgers formulated with PUFA- and curcumin-loaded oleogels as healthy fat substitutes. *LWT* 119: 108909.

Goswami, M., Sharma, B.D., Mendiratta, S.K., and Pathak, V. (2019). Quality evaluation of functional carabeef cookies incorporated with guar gum () as fat replacer. *Nutrition & Food Science* 49: 432–440.

Grabitske, H.A. and Slavin, J.L. (2009). Gastrointestinal effects of low-digestible carbohydrates. *Critical Reviews in Food Science and Nutrition* 49: 327–360.

Grzelak, T., Grupinska, J., Pelczynska, M., Sperling, M., and Czyzewska, K. (2016). The merits of fat replacers in low-calorie food. *Acta Scientiarum Polonorum Hortorum Cultus* 16: 87–94.

Guàrdia, M.D., Guerrero, L., Gelabert, J., Gou, P., and Arnau, J. (2006). Consumer attitude towards sodium reduction in meat products and acceptability of fermented sausages with reduced sodium content. *Meat Science* 73: 484–490.

Guedes-Oliveira, J.M., Salgado, R.L., Costa-Lima, B.R.C., Guedes-Oliveira, J., and Conte-Junior, C.A. (2016). Washed cashew apple fiber (Anacardium occidentale L.) as fat replacer in chicken patties. *LWT – Food Science and Technology* 71: 268–273.

Güzel, M. and Akpınar, Ö. (2020). Preparation and characterization of bacterial cellulose produced from fruit and vegetable peels by Komagataeibacter hansenii GA2016. *International Journal of Biological Macromolecules* 162: 1597–1604.

Hautrive, T.P., Piccolo, J., Rodrigues, A.S., Campagnol, P.C.B., and Kubota, E.H. (2019). Effect of fat replacement by chitosan and golden flaxseed flour (wholemeal and defatted) on the quality of hamburgers. *LWT* 102: 403–410.

Heck, R.T., Lucas, B.N., Santos, D.J.P.D., Pinton, M.B., Fagundes, M.B., Etchepare, M.A., Cichoski, A.J., de Menezes, C.R., Barin, J.S., Wagner, R., and Campagnol, P.C.B. (2018). Oxidative stability of burgers containing chia oil microparticles enriched with rosemary by green-extraction techniques. *Meat Science* 146: 147–153.

Heck, R.T., Saldaña, E., Lorenzo, J.M., Correa, L.P., Fagundes, M.B., Cichoski, A.J., de Menezes, C.R., Wagner, R., and Campagnol, P.C.B. (2019). Hydrogelled emulsion from chia and linseed oils: a promising strategy to produce low-fat burgers with a healthier lipid profile. *Meat Science* 156: 174–182.

Heck, R.T., Vendruscolo, R.G., de Araújo Etchepare, M., Cichoski, A.J., de Menezes, C.R., Barin, J.S., Lorenzo, J.M., Wagner, R., and Campagnol, P.C.B. (2017). Is it possible to produce a low-fat burger with a healthy n−6/n−3 PUFA ratio without affecting the technological and sensory properties? *Meat Science* 130: 16–25.

Henning, S.S.C., Tshalibe, P., and Hoffman, L.C. (2016). Physico-chemical properties of reduced-fat beef species sausage with pork back fat replaced by pineapple dietary fibres and water. *LWT* 74: 92–98.

Hu, G. and Yu, W. (2015). Effect of hemicellulose from rice bran on low fat meatballs chemical and functional properties. *Food Chemistry* 186: 239–243.

INCA, I.D.C.-F. and NACRE, R. (2009). Nutrition et Prévention des Cancers: des connaissances scientifiques aux recommandations. *Lessynthèses du Programme National Nutition Santé, PNNS, France,* 7: 1–50.

Jairath, G., Sharma, D., Dabur, R., Singh, P.K., and Bishnoi, S. (2018). Standardization of corn starch as a fat replacer in buffalo calf meat sausages and its effect on the quality attributes. *Indian Journal of Animal Research* 52: 1521–1525.

Jimenez-Colmenero, F., Salcedo-Sandoval, L., Bou, R., Cofrades, S., Herrero, A.M., and Ruiz-Capillas, C. (2015). Novel applications of oil-structuring methods as a strategy to improve the fat content of meat products. *Trends in Food Science & Technology* 44: 177–188.

Jingxin, X.U., Shuai, Z., and Jingyao, C. (2021). Application of polysaccharide-based fat mimetics to replace animal fat in frankfurter. *Science and Technology of Food Industry* 42: 85–93.

Kaczmarczyk, M.M., Miller, M.J., and Freund, G.G. (2012). The health benefits of dietary fiber: beyond the usual suspects of type 2 diabetes mellitus, cardiovascular disease and colon cancer. *Metabolism* 61: 1058–1066.

Kim, H.-W., Miller, D.K., Lee, Y.J., and Kim, Y.H.B. (2016). Effects of soy hull pectin and insoluble fiber on physicochemical and oxidative characteristics of fresh and frozen/thawed beef patties. *Meat Science* 117: 63–67.

Kouzounis, D., Lazaridou, A., and Katsanidis, E. (2017). Partial replacement of animal fat by oleogels structured with monoglycerides and phytosterols in frankfurter sausages. *Meat Science* 130: 38–46.

Ktari, N., Trabelsi, I., Bkhairia, I., Triki, M, Taktak, M.A., Moussa, H., Nasri, M., and Salah, R.B. (2016). Using Barley Beta Glucan, citrus, and carrot fibers as a meat substitute in Turkey meat sausages and their effects on sensory characteristics and properties. *Journal of Food Processing and Technology* 7: 1–7.

Kumar, Y. (2021). Development of low-fat/reduced-fat processed meat products using fat replacers and analogues. *Food Reviews International* 37: 296–312.

Kumar, Y., Kairam, N., Ahmad, T., and Yadav, D.N. (2016). Physico chemical, microstructural and sensory characteristics of low-fat meat emulsion containing aloe gel as potential fat replacer. *International Journal of Food Science & Technology* 51: 309–316.

Kumar, Y., Kumar, A., Vishwakarma, R.K., and Kumar Singh, R. (2021). The combined effects of multiple emulsion, plant gel, and fibers from pea pods on the characteristics of low-fat meat batter. *Journal of Food Processing and Preservation* 45: e15298.

Kumar, Y. and Kumar, V. (2020). Effects of double emulsion (W1/O/W2) containing encapsulated Murraya koenigii berries extract on quality characteristics of reduced-fat meat batter with high oxidative stability. *LWT* 127: 109365.

Kumar, Y., Kumar, V., and Sangeeta. (2020). Comparative antioxidant capacity of plant leaves and herbs with their antioxidative potential in meat system under accelerated oxidation conditions. *Journal of Food Measurement and Characterization* 14: 3250–3262.

Kumar, Y. and Langoo, B.A. (2016). Effects of aloe, green tea, and amla extracts on microbiological and oxidative parameters of refrigerated raw meat batter. *Agricultural Research* 5: 81–88.

Kumar, Y., Mehta, N., Anurag, R.K., Bashir, A.A., Sethi, S., Kumar, V., and Narsaiah, K. (2019). Improving meat safety through reformulation strategies: natural antioxidants and antimicrobials. In: *Reformulation as a Strategy for Developing Healthier Food Products* (ed. V. Raikos and V. Ranawana). Springer.

Kumar, Y., Tyagi, S.K., Vishwakarma, R.K., and Kalia, A. (2017). Textural, microstructural, and dynamic rheological properties of low-fat meat emulsion containing aloe gel as potential fat replacer. *International Journal of Food Properties* 20: S1132–S1144.

Kwon, H.C., Shin, D.-M., Yune, J.H., Jeong, C.H., and Han, S.G. (2021). Evaluation of gels formulated with whey proteins and sodium dodecyl sulfate as a fat replacer in low-fat sausage. *Food Chemistry* 337: 127682.

Lee, C.H. and Chin, K.B. (2017). Development of low-fat sausages using basil seed gum (Ocimum bacilicum L.) and gelatin as a fat replacer. *International Journal of Food Science & Technology* 52: 733–740.

Lee, S.H., Jang, G.Y., Hwang, I.G., Kim, H.Y., Woo, K.S., Kim, K.J., Lee, M.J., Kim, T.J., Lee, J., and Jeong, H.S. (2015). Physicochemical properties of β-glucan from acid hydrolyzed barley. *Preventive Nutrition and Food Science* 20: 110–118.

Leppänen, K., Andersson, S., Torkkeli, M., Knaapila, M., Kotelnikova, N., Serimaa, R. (2009). Structure of cellulose and microcrystalline cellulose from various wood species, cotton and flax studied by X-ray scattering. *Cellulose* 16: 999–1015.

Lin, K.-W. and Keeton, J.T. (2006). Textural and physicochemical properties of low-fat, precooked ground beef patties containing carrageenan and sodium alginate. *Journal of Food Science* 63: 571–574.

Lindeboom, N., Chang, P.R., and Tyler, R.T. (2004). Analytical, biochemical and physicochemical aspects of starch granule size, with emphasis on small granule starches: a review. *Starch – Stärke* 56: 89–99.

Lobato-Calleros, C., Ramírez-Santiago, C., Vernon-Carter, E.J., and Alvarez-Ramirez, J. (2014). Impact of native and chemically modified starches addition as fat replacers in the viscoelasticity of reduced-fat stirred yogurt. *Journal of Food Engineering* 131: 110–115.

López-Vargas, J.H., Fernández-López, J., Pérez-Álvarez, J.Á., and Viuda-Martos, M. (2014). Quality characteristics of pork burger added with albedo-fiber powder obtained from yellow passion fruit (Passiflora edulis var. flavicarpa) co-products. *Meat Science* 97: 270–276.

Lu, Y., Cao, J., Zhou, C., He, J., Sun, Y., Xia, Q., and Pan, D. (2021). The technological and nutritional advantages of emulsified sausages with partial back-fat replacement by succinylated chicken liver protein and pre-emulsified sunflower oil. *LWT* 149: 111824.

Lurueña-Martínez, M.A., Vivar-Quintana, A.M., and Revilla, I. (2004). Effect of locust bean/xanthan gum addition and replacement of pork fat with olive oil on the quality characteristics of low-fat frankfurters. *Meat Science* 68: 383–389.

Malinski, E., Daniel, J.R., Zhang, X.X., and Whistler, R.L. (2003). Isolation of small starch granules and determination of their fat mimic characteristics. *Cereal Chemistry* 80: 1–4.

Marchetti, L., Muzzio, B., Cerrutti, P., Andrés, S.C., and Califano, A.N. (2017). Bacterial nanocellulose as novel additive in low-lipid low-sodium meat sausages. Effect on quality and stability. *Food Structure* 14: 52–59.

Martins, A.J., Lorenzo, J.M., Franco, D., Vicente, A.A., Cunha, R.L., Pastrana, L.M., Quiñones, J., and Cerqueira, M.A. (2019). Omega-3 and polyunsaturated fatty acids-enriched hamburgers using sterol-based oleogels. *European Journal of Lipid Science and Technology* 121: 1900111.

Mehta, N., Chatli, M., Kumar, P., Malav, O.P., Verma, A.K., Kumar, Y., and Kumar, D. (2019). Development of dietary fiber-rich meat products: Technological advancements and functional significance. In: *Bioactive Molecules in Food* (ed. J.M. Merillon and K. Ramawat). Reference Series in Phytochemistry, Cham: Springer. 763–795.

Méndez-Zamora, G., García-Macías, J.A., Santellano-Estrada, E., Chavez-Martinez, A., Durán-Meléndez, L.A., Silva-Vázquez, R., and Quintero-Ramos, A. (2015). Fat reduction in the formulation of frankfurter sausages using inulin and pectin. *Food Science and Technology* 35: 25–31.

Moghtadaei, M., Soltanizadeh, N., and Goli, S.A.H. (2018). Production of sesame oil oleogels based on beeswax and application as partial substitutes of animal fat in beef burger. *Food Research International* 108: 368–377.

Morin, L.A., Temelli, F., and McMullen, L. (2004). Interactions between meat proteins and barley (Hordeum spp.) β-glucan within a reduced-fat breakfast sausage system. *Meat Science* 68: 419–430.

Nacak, B., Öztürk-Kerimoğlu, B., Yıldız, D., Çaginidi, O., and Serdaroglu, M. (2021). Peanut and linseed oil emulsion gels as potential fat replacer in emulsified sausages. *Meat Science* 176: 108464.

Naeli, M.H., Milani, J.M., Farmani, J., and Zargaraan, A. (2020). Development of innovative ethyl cellulose-hydroxypropyl methylcellulose biopolymer oleogels as low saturation fat replacers: physical, rheological and microstructural characteristics. *International Journal of Biological Macromolecules* 156: 792–804.

Nayak, N.K. and Pathak, V. (2016). Development and quality assessment of carrageenan incorporated low fat chevon patties. *Journal of Food Science and Technology* 53: 3477–3484.

Nisar, P., Chatli, M.K., and Sharma, D.K. (2009). Efficacy of tapioca starch as a fat replacer in low-fat buffalo meat patties. *Buffalo Bulletin* 28: 18–25.

Nsor-Atindana, J., Chen, M., Goff, H.D., Zhong, F., Sharif, H., and Li, YR. (2017). Functionality and nutritional aspects of microcrystalline cellulose in food. *Carbohydrate Polymers* 172: 159–174.

Oh, I., Lee, J., Lee, H.G., and Lee, S. (2019). Feasibility of hydroxypropyl methylcellulose oleogel as an animal fat replacer for meat patties. *Food Research International* 122: 566–572.

Oliveira, L.D.C., Oliveira, M., Meneghetti, V.L., Mazzutti, S., Colla, L.M., Elias, M.C., and Gutkoski, L.Z. (2012). Effect of drying temperature on quality of β-glucan in white oat grains. *Food Science and Technology* 32: 775–783.

Owusu-Apenten, R. (2004). *Introduction to Food Chemistry*, 1e. CRC Press.

Öztürk-Kerimoğlu, B. (2021). A promising strategy for designing reduced-fat model meat emulsions by utilization of pea protein-agar agar gel complex. *Food Structure* 29: 100205.

Paglarini, C.D.S., Furtado, G.D.F., Biachi, J.P., Vidal, V.A.S., Martini, S., Forte, M.B.S., Cunha, R.L., and Pollonio, M.A.R. (2018). Functional emulsion gels with potential application in meat products. *Journal of Food Engineering* 222: 29–37.

Paglarini, C.D.S., Vidal, V.A.S., Martini, S., Cunha, R.L., and Pollonio, M.A.R. (2022). Protein-based hydrogelled emulsions and their application as fat replacers in meat products: A review. *Critical Reviews in Food Science and Nutrition* 62: 640–655.

Pan, T., Guo, H., Li, Y., Song, J., Ren, F. (2017). The effects of calcium chloride on the gel properties of porcine myosin–κ-carrageenan mixtures. *Food Hydrocolloids* 63: 467–477.

Panagiotopoulou, E., Moschakis, T., and Katsanidis, E. (2016). Sunflower oil organogels and organogel-in-water emulsions (part II): Implementation in frankfurter sausages. *LWT* 73: 351–356.

Peng, X. and Yao, Y. (2017). Carbohydrates as fat replacers. *Annual Review of Food Science and Technology* 8: 331–351.

Phiaw, C.T. and Yaakub, N.M. (2017). Effect of maltodextrin as fat replacer on proximate composition and sensory characteristics of low fat chicken burger patty. *Proceedings of the International Conference on Food Science and Nutrition*, 115–118.

Piñero, M.P., Parra, K., Huerta-Leidenz, N., de Moreno, L.A., Ferrer, M., and Barboz, Y. (2008). Effect of oat's soluble fibre (β-glucan) as a fat replacer on physical, chemical, microbiological and sensory properties of low-fat beef patties. *Meat Science* 80: 675–680.

Pintado, T., Herrero, A.M., Jiménez-Colmenero, F., Cavalheiro, C.P., and Ruiz-Capillas, C. (2018). Chia and oat emulsion gels as new animal fat replacers and healthy bioactive sources in fresh sausage formulation. *Meat Science* 135: 6–13.

Pintado, T., Herrero, A.M., Jiménez-Colmenero, F., and Ruiz-Capillas, C. (2016a). Emulsion gels as potential fat replacers delivering β-glucan and healthy lipid content for food applications. *Journal of Food Science and Technology* 53: 4336–4347.

Pintado, T., Herrero, A.M., Jiménez-Colmenero, F., Cavalheiro, C.P., and Ruiz-Capillas, C . (2016b). Effects of emulsion gels containing bioactive compounds on sensorial, technological, and structural properties of frankfurters. *Food Science and Technology International* 22: 132–145.

Pintado, T., Muñoz-González, I., Salvador, M., Ruiz-Capillas, C., and Herrero, A.M. (2021). Phenolic compounds in emulsion gel-based delivery systems applied as animal fat replacers in frankfurters: Physico-chemical, structural and microbiological approach. *Food Chemistry* 340: 128095.

Poyato, C., Ansorena, D., Berasategi, I., Navarro-Blasco, I., and Astiasarán, I. (2014). Optimization of a gelled emulsion intended to supply ω-3 fatty acids into meat products by means of response surface methodology. *Meat Science* 98: 615–621.

Poyato, C., Astiasarán, I., Barriuso, B., and Ansorena, D. (2015). A new polyunsaturated gelled emulsion as replacer of pork back-fat in burger patties: Effect on lipid composition, oxidative stability and sensory acceptability. *LWT – Food Science and Technology* 62: 1069–1075.

Pramualkijja, T., Pirak, T., and Kerdsup, P. (2016). Effect of salt, rice bran oil and malva nut gum on chemical, physical and physico – chemical properties of beef salt – soluble protein and its application in low fat salami. *Food Hydrocolloids* 53: 303–310.

Qin, Y., Zhang, G., and Chen, H. (2020). The applications of alginate in functional food products. *Journal of Nutrition & Food Sciences* 3: 100013.

Radi, M., Niakousari, M., and Amiri, S. (2009). Physicochemical, textural and sensory properties of low-fat yogurt produced by using modified wheat starch as a fat replacer. *Journal of Applied Sciences* 9: 2194–2197.

Rather, S.A., Masoodi, F.A., Akhter, R., Gani, A., Wani, S.M., and Malik, A.H. (2015). Xanthan gum as a fat replacer in goshtaba-a traditional meat product of India: effects on quality and oxidative stability. *Journal of Food Science and Technology* 52: 8104–8112.

Rather, S.A., Masoodi, F.A., Akhter, R., Gani, A., Wani, S.M., and Malik, A.H. (2016a). Effects of guar gum as fat replacer on some quality parameters of mutton goshtaba, a traditional Indian meat product. *Small Ruminant Research* 137: 169–176.

Rather, S.A., Masoodi, F.A., Akhter, R., Rather, J.A., Gani, A., Wani, S.M., and Malik, A.H. (2016b). Application of guar–xanthan gum mixture as a partial fat replacer in meat emulsions. *Journal of Food Science and Technology* 53: 2876–2886.

Robert, P., Zamorano, M., González, E., Silva-Weiss, A., Cofrades, S., and Giménez, B. (2019). Double emulsions with olive leaves extract as fat replacers in meat systems with high oxidative stability. *Food Research International* 120: 904–912.

Salcedo-Sandoval, L., Cofrades, S., Ruiz-Capillas Pérez, C., Solas, M.T., and Jiménez-Colmenero, F (2013). Healthier oils stabilized in konjac matrix as fat replacers in n−3 PUFA enriched frankfurters. *Meat Science* 93: 757–766.

Sandrou, D.K. and Arvanitoyannis, I.S. (2000). Low-fat/calorie foods: Current state and perspectives. *Critical Reviews in Food Science and Nutrition* 40: 427–447.

Santos, J.M.D., Ignácio, E.O., Bis-Souza, C.V., and Silva-Barretto, A.C.D. (2021). Performance of reduced fat-reduced salt fermented sausage with added microcrystalline cellulose, resistant starch and oat fiber using the simplex design. *Meat Science* 175: 108433.

Sayed, M.E., Bakr, A., Gaafar, A., Ismaiel, A.I., and Salem, M.M. (2020). Production low-fat beef burger by using gum arabic as fat replacer. *Menoufia Journal of Food and Dairy Sciences* 5: 17–33.

Schmiele, M., Nucci Mascarenhas, M.C.C., da Silva Barretto, A.C., and Rodrigues Pollonio, M.A. (2015). Dietary fiber as fat substitute in emulsified and cooked meat model system. *LWT – Food Science and Technology* 61: 105–111.

Schuh, V., Allard, K., Herrmann, K., Gibis, M., Kohlus, R., and Weiss, J. (2013). Impact of carboxymethyl cellulose (CMC) and microcrystalline cellulose (MCC) on functional characteristics of emulsified sausages. *Meat Science* 93: 240–247.

Ścieszka, S. and Klewicka, E. (2019). Algae in food: A general review. *Critical Reviews in Food Science and Nutrition* 59: 3538–3547.

Selani, M.M., Shirado, G.A.N., Margiotta, G.B., Saldaña, E., Spada, F.P., Piedade, S.M.S., Contreras-Castillo, C.J., and Canniatti-Brazaca, S.G. (2016). Effects of pineapple byproduct and canola oil as fat replacers on physicochemical and sensory qualities of low-fat beef burger. *Meat Science* 112: 69–76.

Serdaroğlu, M., Nacak, B., and Karabıyıkoğlu, M. (2017). Effects of beef fat replacement with gelled emulsion prepared with olive oil on quality parameters of chicken patties. *Korean Journal for Food Science of Animal Resources* 37: 376.

Serdaroğlu, M., Nacak, B., Karabıyıkoğlu, M., and Keser, G. (2016a). Effects of partial beef fat replacement with gelled emulsion on functional and quality properties of model system meat emulsions. *Korean Journal for Food Science of Animal Resources* 36: 744.

Serdaroğlu, M. and Öztürk, B. (2017). Use of olive oil-in-water gelled emulsions in model turkey breast emulsions. *IOP Conference Series: Earth and Environmental Science*, IOP Publishing, 012071.

Serdaroğlu, M., Öztürk, B., and Urgu, M. (2016b). Emulsion characteristics, chemical and textural properties of meat systems produced with double emulsions as beef fat replacers. *Meat Science* 117: 187–195.

Sharefiabadi, E. and Serdaroğlu, M. (2020). Pectin: properties and utilization in meat products. *Food and Health* 7: 64–74.

Silva-Vazquez, R., Flores-Giron, E., Quintero-Ramos, A., Hume, M.E., and Mendez-Zamora, G. (2018). Effect of inulin and pectin on physicochemical characteristics and emulsion stability of meat batters. *CyTA – Journal of Food* 16: 306–310.

Soltanizadeh, N. and Ghiasi-Esfahani, H. (2015). Qualitative improvement of low meat beef burger using Aloe vera. *Meat Science* 99: 75–80.

Sousa, S.C., Fragoso, S.P., Penna, C.R.A., Arcanjo, N.M.O., Silva, F.A.P., Ferreira, V.C.S., Barreto, M.D.S., and Araújo, I.B.S. (2017). Quality parameters of frankfurter-type sausages with partial replacement of fat by hydrolyzed collagen. *LWT – Food Science and Technology* 76: 320–325.

Szpicer, A., Onopiuk, A., Półtorak, A., and Wierzbicka, A. (2020). The influence of oat β-glucan content on the physicochemical and sensory properties of low-fat beef burgers. *CyTA – Journal of Food* 18: 315–327.

Tahmasebi, M., Labbafi, M., Emam-Djomeh, Z., and Yarmand, M.S. (2016). Manufacturing the novel sausages with reduced quantity of meat and fat: the product development, formulation optimization, emulsion stability and textural characterization. *LWT – Food Science and Technology* 68: 76–84.

Telis, V. and Nicoleti, J. (2009). Reologia de suspensões de colágeno na presença de maltodextrina e lactose. *Brazilian Journal of Food Technology* 111: 119.

Tesch, S., Gerhards, C., and Schubert, H. (2002). Stabilization of emulsions by OSA starches. *Journal of Food Engineering* 54: 167–174.

Thanakiatkrai, P. and Kitpipit, T. (2017). Meat species identification by two direct-triplex real-time PCR assays using low resolution melting. *Food Chemistry* 233: 144–150.

Ulu, H. (2006). Effects of carrageenam and guar gum on the cooking and textual properties of low fat meatballs. *Food Chemistry* 95: 600–605.

Utrilla, M.C., García Ruiz, A., and Soriano, A. (2014). Effect of partial replacement of pork meat with an olive oil organogel on the physicochemical and sensory quality of dry-ripened venison sausages. *Meat Science* 97: 575–582.

Verma, A.K., Chatli, M.K., Kumar, D., and Mehta, N. (2015). Efficacy of sweet potato powder and added water as fat replacer on the quality attributes of low-fat pork patties. *Asian-Australasian Journal of Animal Sciences* 28: 252–259.

Yadav, S., Malik, A., Pathera, A., Islam, R.U., and Sharma, D. (2016). Development of dietary fibre enriched chicken sausages by incorporating corn bran, dried apple pomace and dried tomato pomace. *Nutrition & Food Science* 46: 16–29.

Yadegari, R.J., Kaya, E.E., and Vural, H. (2021). Thermal and microscopic properties and quality characteristics of low-fat frankfurters and emulsions produced with carboxymethyl cellulose, methyl cellulose and pectin. *Carpathian Journal of Food Science & Technology* 13.

Yasin, H., Babji, A.S., and Ismail, H. (2016). Optimization and rheological properties of chicken ball as affected by κ-carrageenan, fish gelatin and chicken meat. *LWT – Food Science and Technology* 66: 79–85.

Yogesh, K., Jha, S.N., and Ahmad, T. (2014). Antioxidant potential of aqueous extract of some food grain powder in meat model system. *Journal of Food Science and Technology* 51: 3446–3451.

Younis, K. and Ahmad, S. (2015). Waste utilization of apple pomace as a source of functional ingredient in buffalo meat sausage. *Cogent Food & Agriculture* 1: 1119397.

Yousefi, N., Zeynali, F., and Alizadeh, M. (2018). Optimization of low-fat meat hamburger formulation containing quince seed gum using response surface methodology. *Journal of Food Science and Technology* 55: 598–604.

Youssef, M.K. and Barbut, S. (2010). Physicochemical effects of the lipid phase and protein level on meat emulsion stability, texture, and microstructure. *Journal of Food Science* 75: S108–14.

Zeeb, B., Schöck, V., Schmid, N., Majer, L., Herrmann, K., Hinrichs, J., and Weiss, J. (2018). Impact of food structure on the compatibility of heated WPI–pectin-complexes in meat dispersions. *Food & Function* 9: 1647–1656.

Zetzl, A.K., Marangoni, A.G., and Barbut, S. (2012). Mechanical properties of ethylcellulose oleogels and their potential for saturated fat reduction in frankfurters. *Food & Function* 3: 327–337.

Zhang, Z., Xu, G., Ma, M., Yang, J., and Liu, X. (2013). Dietary fiber intake reduces risk for gastric cancer: A meta-analysis. *Gastroenterology* 145: 113–120.e3.

Zhuang, X., Han, M., Kang, Z.-L., Wang, K., Bai, Y., Xu, X-L., and Zhou, G-H. (2016). Effects of the sugarcane dietary fiber and pre-emulsified sesame oil on low-fat meat batter physicochemical property, texture, and microstructure. *Meat Science* 113: 107–115.

4.2

Fat Mimetics in Dairy Products

Ainaz Alizadeh and Mitra Soofi*

Department of Food Science and Technology, Tabriz Branch, Islamic Azad University, Tabriz, Iran
** Corresponding author*

4.2.1 Introduction

Dairy products such as milk, yogurt, cheese, etc. provide crucial nutrients including fat, proteins, and other micro-nutrients and minerals like phosphorous and calcium for our body (Khanal et al., 2017). Although the fat in dairy products is considered the main source of energy, it plays a vital role due to the fat's contribution to the mouthfeel, palatability, lubricity, creaminess as well as other organoleptic and practical features of dairy foods. As well, fat can act as a filler in products like cheese which develops from three-dimensional protein networks (Jooyandeh et al., 2017). Moreover, fat is a major source of fat-soluble vitamins and essential fatty acids in the body. Fat and fat derivatives also lead to the development the flavor along with impacts on the storage stability of dairy foods (Koca and Metin 2004).

Recently, overweight and obesity have been considered one of the high prevalent health concerns in most countries which can lead to the etiology of certain cancers as well as increase the risk factor for coronary heart disease, Type 2 diabetes, and hypertension (Lashkari et al., 2014). In this regard, awareness about the risk of cardiovascular diseases and obesity has led to high consumer demand for the consumption of low fat dairy foods compared to their full-fat counterparts owing to their health benefits. WHO has also provided limiting the consumption of foods containing high amounts of saturated fats, including fatty meat, ice cream, and cheese to maintain a healthy diet (WHO 2015). In this respect, fats and oils, dairy products, meat, snack, and confectionary foods are perceived as the main sources of total fat intake in diets (Omayma and Youssef 2007). Therefore, as the tendency of health-conscious consumers to low- and non-fat dairy products is enhanced, the quality of low fat dairy products should be maintained (Seyed Ahmadi Mamaghani et al. 2020; Alizadeh et al. 2021). However, elimination or reduction of fat from dairy products changes the structure and composition of dairy foods and leads to a poor meltability, rubbery and hard texture in products like cheese and provides poor color and off-flavors as compared to their full-fat counterparts due to perceptible changes in the expected interactions among components (Shafiei et al. 2014). These changes have led to relatively unpalatability of low- or reduced-fat dairy products and decreasing consumer acceptance (Omayma and Youssef 2007).

Hence, the great challenge for the development of low fat dairy products for manufacturers is producing low fat products with comparable characteristics to their full-fat counterparts. One of

Fat Mimetics for Food Applications, First Edition. Edited by Miguel Ângelo Parente Ribeiro Cerqueira and Lorenzo Miguel Pastrana Castro.

the alternative approaches to compensate for fat reduction-related drawbacks is the use of fat replacers which can be incorporated into traditional dairy products as a substitute for all or part of the fat and yield food products with comparable organoleptic properties and consumer acceptability to the full-fat products (Hossain et al. 2021).

4.2.2 The Characteristics of Milk Fat

Milk fat is considered a source of fat-soluble vitamins (E, D, A, and K), essential fatty acids, energy (about 48% of the energy of whole milk), and other health-promoting nutrients (Figure 4.2.1) which shows a remarkable effect on the texture, stability, appearance, flavor and other sensory attributes of dairy products. Milk fat consists of a complex of lipids including esters of fatty acids with glycerol, free sterols, phospholipids, varying amounts of fat-soluble vitamins, and traces of free fatty acids which exist in microscopic globules in an oil-in-water emulsion in milk and a variety of fatty acids (over 400) with different chain lengths such as butyric acid to fatty acids with twenty-six carbon atoms. Therefore, the high proportion of medium-chain (14 carbons in length) and short-chain (4 carbons in length) saturated fatty acids perceive milk fat unique among animal fats (Roufegarinejad et al. 2017). Although fatty acids of milk fat are typically saturated (62%), monounsaturated (30%), and polyunsaturated (4%), with the remaining other minor types of fatty acids (4%), their composition varies according to the season, lactation stage, breed of the cow, feed composition, and geographical region. However, the saturated fatty acids of milk fat, including palmitic, stearic, and myristic acids, can remarkably affect elevating total and low-density lipoprotein cholesterol levels of the body. Too, the effect of individual saturated fatty acids on increasing blood cholesterol levels is different (Alizadeh et al. 2017). The cholesterol content of dairy products is associated with fat content due to its existence in the fat globule membrane of milk fat (Alizadeh and Bahar 2016). So moderation in total fat intake (vegetable and animal origin) is recommended owing to the potential of individual fatty acids in milk fat on increasing the health-related risk factors including coronary heart disease (Alizadeh et al. 2012). Accordingly, numerous studies were investigating the reduction and elimination of milk fat to produce low fat dairy products such as yogurt, cheese, ice cream, etc. with textural and sensorial properties similar to full-fat counterparts (Soofi et al. 2019). Understanding the role of milk fat on different features of dairy foods is important.

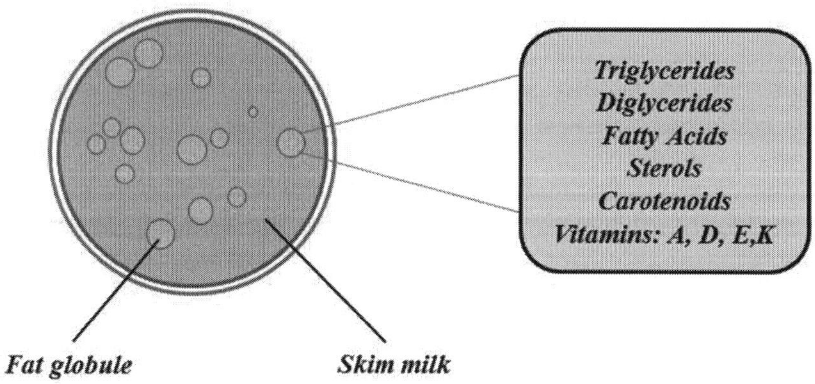

Figure 4.2.1 The composition of milk fat.

4.2.3 The Role of Milk Fat in Dairy Products

The features of dairy products including color, flavor, texture, and functional behavior are affected by the fat content of dairy foods and its significance becomes prominent in any attempt at producing lower-fat counterparts (Badri and Alizadeh 2016). The different effects of milk fat on the features of produced dairy products are described as follows.

4.2.3.1 Milk Fat and Dairy Product's Sensory Characteristics

Enzyme-catalyzed lipolysis of milk fat leads to the release of free fatty acids, which play a main role in dairy products' flavor features. Accordingly, the production of lipases in fermented dairy foods such as cheese as the result of the activity of cheese microbiota during ripening or adding lipases through rennet paste or pre-gastric esterases leads to the decomposition of the milk fat along with the release of free fatty acids (Alizadeh et al. 2009). In this regard, the specificity and activity of the lipase enzymes and milk fatty acid composition play a vital role in the impact of fat on the flavor attributes of products like cheese. Accordingly, fatty acids with shorter chain lengths such as butyric, capric, caproic, and caprylic are the most volatile and impact flavor effectively (Drake et al. 2010). Therefore, reducing the fat content of fermented dairy products has led to changes in the biochemistry and array or proportion of developed volatile compounds during the aging time which subsequently has a significant effect on the flavor profile of the low fat products. For example, low fat cheese contains comparable types of odorants to full-fat cheese; however, the concentrations are different (Alizadeh et al. 2008). Therefore, low fat cheese is characterized by rosy (type of unclean flavor as the result of increases in the phenylethanal content), burnt flavors (attributed to an increase in furanones as the result of deficiency feature of milk fat, sulfur), and brothy flavors as well as lower levels of methanethiol which is an essential precursor to developing the flavor of aged cheese (Johnson et al. 2009). Furthermore, the perceived flavor of dairy foods can be influenced by their polar matrix, which is related to higher levels of fat that impact the detection thresholds of main odorants. Additionally, milk fat imparts a translucent appearance to dairy foods lost by changing the composition of dairy products, including decreasing calcium: protein proportion to soften the dairy foods such as cheese and hold more moisture (Pastorino et al. 2002).

4.2.3.2 Milk Fat and Dairy Products Texture

Texture of the dairy foods is influenced by the fat content of products which show a reinforcing effect on the protein matrix along with elevating their hardness; while letting them easily develop into a smooth mass through exposure to the large-scale deformations such as chewing due to the effect of fat on providing weak points within the protein network. However, the less reinforcing effect of milk fat in dairy products is achieved through raising the temperature as well as its effect in the large-scale deformations such as chewing to develop a smooth mass (Khanal and Bansal 2020). Therefore, as mentioned in different studies, low fat dairy foods are considered as hard and springy, waxier, more fracturable, less sticky, less smooth, less meltable, and chewier compared to their full-fat counterparts which can compensate with the utilization of fat mimetics as a filler particle with comparable characteristics to fat at a phase volume (Johnson et al. 2009).

4.2.3.3 Milk Fat and Dairy Products Melting Properties

Dairy foods are typically classified as foamed emulsions (whipped creams and ice creams), structured emulsions (butter and butter spreads), and coagulated gels (various cheese types). Dairy products vary in microstructure and they all contain a significant proportion of milk fat distributed either in the bulk or dispersed phases which provide the underlying colloidal network of crystalline particles due to the high melting fraction of milk fat. So the macrostructure and sensory properties of dairy foods, including the voluminous body of whipped cream, spreadability of butter, creaminess of ice cream, and spreadability and texture of cheese are affected by milk fat content as well as its high melting fraction. Therefore, the replacement of milk fat is relatively a challenge to produce low fat dairy foods (Patel et al. 2020). The ratio of protein: water and protein: protein interactions and softening effect of fat present in products such as cheese have a remarkable impact on the melting feature of dairy products. Fats' importance has become clearer in cheese products including Mozzarella, especially; when baked and exposed to hot air. In this regard, owing to the contribution of milk fat to the flowability and pliability of Mozzarella, low fat cheese showed a higher scorching when exposed to baking at temperatures between 250°C and 300°C (Wadhwani et al. 2011).

4.2.4 Issues with Low-Fat Dairy Products

Industrial-scale production of low fat dairy products has been started for about two decades. However, inadequate texture and flavor of low fat products compared to full-fat counterparts, in turn, have led to less acceptance of the consumers from the first generation of low fat dairy foods and subsequently slowed down enormous growth in the market of these products (Di Cagno et al. 2014). For example, low fat cheese exhibits bitterness, off-flavor, dry, firm or hard, and grainy texture with higher hardness and springiness, less adhesiveness and cohesiveness, dull color, as well as insufficient melting and mechanical attributes in comparison to full-fat cheese (Shafiei et al. 2014). This phenomenon can be due to the reduced-fat content of low fat dairy foods as well as a higher protein concentration of the milk utilized for the production of low fat products. Therefore, the reduction of fat has led to the larger structural protein matrix per unit cross-sectional area as well as the high amount of protein/casein to the fat which is resulted in the poor stretch and melt properties of low fat cheeses due to the high amount of calcium retention as colloidal calcium. Also, reducing the fat content of dairy products such as cheese has led to the vanishing open-intricate microstructure and a decrease in the production yield (Koca and Metin 2004).

Industrial-scale manufacture of low fat dairy products is still considered a challenge. To overcome texture and flavor defects in light dairy products, processing changesand modern technical approaches, improve their textural, biochemical, and physicochemical features. So several change methodsand various types of fat mimetics, have been suggested which are compensate for milk fat features (Di Cagno et al. 2014).

4.2.5 Fat Mimetics in Dairy Products

Fat replacers are defined as the compounds responsible for providing more physical and sensory features of the fat in food or the substance which imitate the practical features of milk fat completely or partially along with providing lower calories compared to the full-fat products (Soofi et al. 2019). Fat replacers can be classified as fat mimetics and fat substitutes. Fat substitutes include comparable practical and physical features of conventional fat molecules, such as triglycerides. However, fat mimetics can imitate some of the physical and sensorial properties of conventional fat molecules in food which include emulsion-templated networks of proteins and polysaccharides, polymeric networks of ethylcellulose, colloidal and self-assembled fibrillar networks of polar lipid crystalsand solid o/w emulsions of oil trapped within crystallized lamellar mesophases (Badri and Alizadeh 2016; Patel et al. 2020). Fat mimetics cannot replace on a weight-for-weight basis of the fat molecules in foods. Fat mimetics can modify functional, structural, and sensorial attributes of low fat dairy products by increasing moisture content through improving mechanical properties, holding water, and yield. However, they may also produce reduced escalated stickiness and shredability and undesirable flavor in products such as cheese, if used at an enormous level (Johnson 2011). In this regard, the type and features of fat replacers are the two important factors which determines the level of raised functionality in low fat dairy products.

4.2.5.1 Carbohydrate-Based Fat Mimetics

Carbohydrate-based substances can be used as fat mimetics to replace the partially content of fat in dairy products, which include maltodextrins and dextrins, gums (Arabic, guar, κ- and λ-carrageenans, tragacanth), β-glucan, polydextrose, cellulose and starch derivatives, oat flour derivatives, etc. The other carbohydrate-based fat mimetics consist of Novagel™, Litesse®, Slendid™NC-200, N-Lite®, Oatrim, Z-trim, Raftuline®, and Avicel® RCN 30 (Kavas et al. 2004). Carbohydrate-based fat mimetics have health-friendly properties and provide digestible 4 kcal g^{-1} calories for carbohydrates and no calories for complex non-digestible carbohydrates and gums. These compounds can stabilize water existence in the dairy foods in a gel-like matrix and represent a fat replacement function. The utilization of all these aforementioned fat replacers impacts improved viscosity, body, and a creaminess intensity of low fat dairy foods comparable to full-fat products (Omayma and Youssef 2007). On the contrary, these mimetics are not favorable for frying processes. Furthermore, including fat mimetics into the food products has led to increasing water activity of foods which resulted in the elevated potential for microbial growth and shelf-life reduction. too, the kind and quantity of the used fat mimetic can be effectively influenced the flavor profile of the end product (Jooyandeh et al. 2017). Gum-based fat mimetics show no remarkable effects on sensory quality; while starch-based fat mimetics revealed negative effects on flavor intensity of low fat products. In order to eliminate these drawbacks, using the blend of two or more fat replacers' which can balance the performance of each other as well as the other modifications of the procedure is used to prepare low fat dairy products and achieve desirable textural, thickness, bulkiness, and creaminess properties. In this respect, several studies have been conducted about the use of different carbohydrate-based fat mimetics in dairy products to imitate the textural and sensory features of the full-fat counterparts (Khanal et al. 2017).

Functionalities of hydrocolloids to imitate the features of fat in dairy products include the capability to interact with water to develop the texture and improve the stability of the end product during the storage time by keeping thickness and cryoprotection. Gums are high molecular weight carbohydrate hydrocolloids that can enhance the viscosity of dairy food systems at a low concentration and act as gelling agents and stabilizers. Gums including locust bean gum, Arabic gum, xanthan gum, guar gum, carrageenan, and pectin can all be run as fat mimetics individually or combined with other varieties of fat replacers. The strength and rheological attributes of developed gel by including gums as well as its compatibility with other components in the food, and the effects of shearing forces and temperature on the practical features of the gum are considered critical factors in selecting types of gum to use as a fat mimetic (Khanal et al. 2017).

Starch is another carbohydrate-based fat mimetic that consists of glucose molecules and are considered a common polysaccharide carbohydrate. In recent years, many studies have been conducted to change starch and its functionality through using chemicals or enzymes. Owing to the improved function of modified starch such as enhanced thickening and elevated water holding capacity, it shows the capability to use as a fat mimetic in different foods (Shenana 2021). Potato, rice, maize, oat, wheat, and tapioca origin changed starches by physical and/or chemical means are cost-effective, abundant, and can act as fat mimetics by improving the moisture retention and creaminess of dairy foods. However, they show no sufficient stability to acid, freeze/thaw, and heat for some applications and can also mask the flavor of foods in which they are incorporated (Omayma and Youssef 2007). Maltodextrins and Dextrins are non-sweet starch hydrolyzes that usually have dextrose equivalence (DE) of less than 20 and can typically produce by partial hydrolysis of potato, rice, oat, maize, and wheat or tapioca starch through starch fragmentation enzymes and acid. Maltodextrins at a DE range of 1 to 10 show desirable potential as fat replacer which enhances texture, creaminess and gives a fatty mouthfeel to the food products. The combination of maltodextrin gel with solid and liquid fats can form a stable emulsion gel which helps foods to be broken easily into pieces in the mouth and gives a similar taste to fat (Mounsey and O'Riordan 2008). Maltodextrin can improve the low fat dairy product's characteristics by preventing re-crystallization and extending shelf life, enhancing taste, viscosity, consistency, and structure of them and maintaining luster, and original color (Omayma and Youssef 2007). Polydextrose is another carbohydrate-based fat mimetic which is produced by a randomly-bonded polymer of sorbitol, glucose, and phosphoric or citric acid. This compound is commonly used as a bulking agent with a fat-sparing effect that could metabolize partially and provide only 1 kcal g^{-1} (Mounsey and O'Riordan 2008).

Cellulose and cellulose derivatives such as microcrystalline cellulose are also known as fat mimetics which are used individually or combined with other hydrocolloids. A major part of cellulose microcrystals (60–70%) has less than 0.2 mm long which makes them insoluble in water. However, the combination of this compound with other compounds such as gums maintains the insoluble microcrystals in a food matrix as well as mimics fat functions by improving viscosity, body, and mouthfeel of low fat products similar to full-fat counterparts (Lobato-Calleros et al. 2004). As well, Oatrim is another fat mimetic which provides similar sensory features such as mouthfeel to triglycerides. Oatrim is consisting of 5% β-glucan, amylodextrin, and soluble fiber developed by partial enzymatic hydrolysis of oat flour and used to emulate fat features in the food matrix as a texturizing ingredient. Although this compound provides higher stability to high temperature-short time processing and retorting, it shows no acceptable performance for the frying process (Lee et al. 2005).

Fats are normally stabilized as oil-body-like particles or emulsions in dairy products such as milk. So the formation of interactions among proteins, fats, and other colloidal dispersed ingredients in the aqueous phase leads to develop the textural and sensorial features of dairy food. In these products, the elimination of particle-particle interactions as the result of fat reduction is the main reason for the loss of sensory quality of low fat end products. The size and shape of fat particles can be imitated by granular carbohydrate-based fat mimetics which can interact with other colloidal particles or be dispersed individually. Non-granular carbohydrate-based fat mimetics can also imitate the thickening mouthfeel behavior of full-fat products by forming interactions with proteins or other colloidal particulates (Peng and Yao 2017). Moreover, in semi-solid dairy products such as yogurt, heterogeneity of the protein matrix is increased by the presence of fat. So the protein-protein interactions are impeded by the addition of fat in protein networks which resulted in decreasing the elasticity of a system by the formation of a weak structure. In a fat-free system, the formation of a stronger gel network can result from enhancing the density of protein cross-links. So, in products such as yogurt, granular- and non-granular carbohydrate-based fat mimetics can provide thicker and creamier mouthfeel due to disrupting the cross-links of the protein matrix and bringing tenderer and softer textures during chewing and swallowing (Peng and Yao 2017).

4.2.5.2 Protein-Based Fat Mimetics

Protein-based fat mimetics are typically derived from milk, whey, eggs, or vegetables and nutritionally, digested and absorbed as a protein providing 1–4 kcal/g. They are small globular particles that simulate fat globules and provide a sensory quality like fat particles, especially creaminess. Accordingly, one gram of protein-based fat mimetics can be replaced with 3 g of fat in light dairy products and they cannot be used in products subject to frying conditions as the proteins are denatured and lose their functionality. However, they can use as ingredients in foods that undergo less cooking process such as retorting, baking, and ultra-high temperature processing (Omayma and Youssef 2007). Microparticulation is a thermo-mechanical process that involves a high-pressure shearing process of protein with or without heating to form uniform, microscopic spheres analogous to fat particles. As well, heating the proteins has led to the development of a gel structure and enhanced the physicochemical properties of dairy products (Stankey et al. 2017). Microparticulated protein or carbohydrate materials can entrap the water mechanically and provide creaminess and lubricity in the end product. However, they cannot substitute flavor and non-polar carrying properties of the fat. These compounds are perceived as rough in the mouth as the result of the large gel particles perform through heat-coagulation of the proteins that compensate by applying high shearing force during heating of the proteins and the development of small spherical (diameter 0.1–2.0 mm diameter) protein gel particles. These microparticles show no roughness in the mouth and can emulate the taste buds as comparable to fat with a creamy, smooth texture (Dissanayake et al. 2010). Commercial microparticulated proteins used in dairy products are Simplesse® 100 and DairyLo® which have whey particles/aggregates of different sizes (0.5–5 μm). Simplesse® 100 is one of the protein-based fat mimetics which is developed by microparticulated of whey protein concentrate. Whey protein is a good source of bioactive peptides and essential amino acids. Furthermore, protein content in whey affects the emulsion stability of the continuous water phase and the quantity of protein absorbed at the surface of fat globules in dairy products such as ice cream. This compound has been successfully used in frozen dessert products as well as soft cheeses and cheese spreads, yogurt, and sour cream. Simplesse® 100 contributes to the creaminess, viscosity, and opacity of food and provides texture and mouthfeel through inhibiting syneresis and increasing water binding capacity (Romeih et al. 2002). Simplesse® 100 has also been used to prepare low fat cheese for modifying its texture through occlusion of the paracasein as well as binding with water as the result of an enhancement in moisture content in cheese.

4.2.5.3 Lipid-Based Fat Mimetics

One of the most discussed matters among researchers is still classifying the lipid-structuring systems. Based on the literature, Patel et al. (2020) categorized all structured lipids as fat mimetics which are structured using structured biphasic systems, oleogelation using lipidic and nonlipidic gelators, and indirect oil structuring via a solvent exchange and colloid templates (Patel et al. 2020). However, O'Connor and O'Brien (2016) classified fat replacers into two major groups, namely fat mimetics (mimics some of the lipid features without the capability to replace conventional lipids) and fat substitutes (which could fully and directly replace conventional lipids) (O'Connor and O'Brien 2016). Moreover, structured oils were classified as interesterified lipids, organogels, oil bulking systems, and structured emulsions subdivided into hydrogel oil-in-water (O/W) and water-in-oil (W/O) emulsions or organogel O/W and W/O emulsions (Temkov and Mureșan 2021).

4.2.5.3.1 Oleogels as a Possible Fat Mimetics

Dairy products including cheese and butter as well as indulgence products such as ice creams and whipped cream form a part of daily diet and are consumed regularly. These products could be classified as foamed emulsions (ice creams and whipped creams), coagulated gels (various cheese types), and structured emulsions (butter and butter spreads) based on colloid science (Moriano and Alamprese 2017). Although these dairy products vary in microstructures, they have a high proportion of milk fat distributed either in the bulk or dispersed phases responsible for forming the macrostructure and sensorial features of dairy products (including the voluminous body of whipped cream, creaminess of ice cream, the texture of cheese, and spreadability of butter and cheese spreads). Solid or semi-solid vegetable oils application is inevitable for providing the mentioned desirable sensory properties. The plasticity properties of vegetable oils were improved by esterification or the partial hydrogenation process (Adili et al. 2020). However, the content of saturated and trans fatty acids is high in these lipid ingredients and their consumption at high levels has been related to an elevated risk of cardiovascular disease (CVD) and death (Alizadeh et al. 2012; Adili et al. 2020). Due to the functionality of milk fat in dairy products, replacing it with liquid oils without compromising on the product attributes is rather a challenge to replace milk fat. Oleogelation is a novel technic to prepare semi-solid vegetable oils which is the entrapment of the liquid vegetable oil within a three-dimensional oleogel The common oleogels are lipid-based (e.g., plant waxes, fatty alcohols, phospholipids, mono- and diacylglycerols, sorbitan esters, phytosterols) or polymeric oleogelators (Adili et al. 2020). Based on the physicochemical properties of the oleogelator, the physical structure of the vegetable oil exerts a solid-like network without producing trans fatty acid (Aliasl khiabani et al. 2020). Accordingly, wax-based gels had a strong potential for replacing milk fat in different dairy products owing to their abundance, gelling ability, and low cost (Bemer et al. 2016; Aliasl khiabani et al. 2020) The potential of tailoring the physical attributes of gel as well as controlling its behavior, under specific processing conditions, are significant approaches for the food industry (e.g., oil structuring). Gels can be classified as oleogels/organogels, hydrogels (when the gelled liquid phase is water), emulgels (when an emulsion is gelled, a biphasic formulation defined as an emulgel), and bigels (semi-solid formulations, usually prepared by high-speed mixing of oleogel and hydrogel at a certain temperature) based on the polarity of the liquid phase (Martins et al., 2018; Shaikh et al., 2022). Oleogel is known as a thermoreversible viscoelastic semi-solid lipid mixture. Oleogels are usually prepared through direct dispersion of hydrophobic oleogelators such as fatty acids, ethylcellulose, monoglycerides, alcohols, waxes, and sterols (chosen based on their availability, cost-effectiveness, surface activity, thermal stability, self-assembly, as well as the ability to convert liquid oil to solid-like oleogel at low concentration) into lipophilic liquid oils (typically vegetable oil enriched with unsaturated fatty acids) under heating and stirring conditions followed by cooling to generate nucleation and crystal growth, in order to the formation of self-assembled networks (Shakeel et al., 2021). As delivery vehicles, fat mimetics such as oleogels can offer multiple advantages such as i) providing stabilization against recrystallization (due to the high solvency of liquid oil) and preventing oxidation of oily bioactives; ii) controlling and delaying the rate of release of the bioactives; and iii) enhancing bioaccessibility/bioavailability of bioactives because of direct micellization (Cui et al., 2022). However, several factors including oil type or polarity, oleogelator composition or its concentration and type, preparation conditions, and storage conditions can effectively influence on functional features of oleogels (Gutiérrez-Luna et al., 2022).

The possibility of using oleogels as a fat replacer in dairy products such as ice cream and cheese has been studied in earlier literature such as incorporating high-oleic soybean oil/rice bran wax-based oleogel into a formulation of cream cheese products. According to the obtained results during 14 days of storage, no significant difference was shown in the total tocopherol content of oleogel-based cheese compared to that of the control cheese. Furthermore, the content of tocopherol isomers remained unchanged and the content of volatile compounds changed minorly through replacing fat with oleogels in cheese formulation (Park et al., 2018). In another study, incorporating soybean oil with rice bran wax or sunflower wax oleogels as a fat replacer into the formulation of processed cheese samples had led to no significant changes in moisture and fat contents of the samples compared to those of the control counterparts, however, the content of saturated fat was reduced through the inclusion of oleogel (Huang et al., 2018). Sunflower oil oleogel, made with phytosterols and γ-oryzanol, as a milk cream substitute was successfully used to formulate ice cream and resulted in comparable quality attributes to those of the control made with milk cream when the oleogelator concentration of 12 g/100 g oil(Moriano and Alamprese 2017).

Figure 4.2.2 shows an example of an oleogel-based on carnauba wax with acid adipic for use in the food industry. So far, although oleogels show acceptable features as the substitute for saturated fat and trans fatty acids in many foodstuffs including dairy products, the suitability of oleogels and their application as fat mimetics in dairy products need more research.

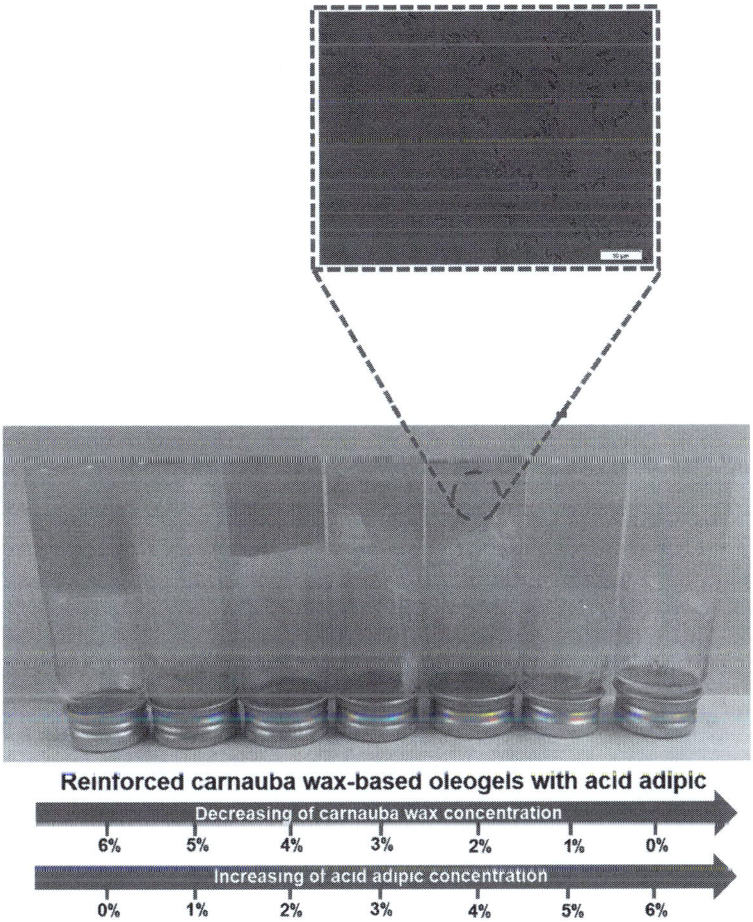

Figure 4.2.2 Prepared carnauba wax/adipic acid oleogel for use in food industry Aliasl khiabani et al. (2020), from Elsevier.

4.2.6 Applications of Fat Mimetics in Different Dairy Products

4.2.6.1 Cheese

Collection of casein micelles and formation of a network structure resulting from hydrophobic interactions take place during the making of rennet-coagulated cheeses (Khanal and Bansal 2020). Fat globules act as inert filler within curd particles due to their hydrophilic membrane which is non-attractive to interactions between the hydrophobic proteins. Too, the effect of fat globules on blocking the protein strand coalescence has led to keeping a small volume of whey in cheese. So the reduction of the fat content of cheese resulted in fewer of these serum reservoirs (Khanal and Bansal 2020).

Fat molecules in cheese provide flow and stretch features as well as a lubricating effect when melted. Low fat (≤10%) cheese is translucent, rubbery, and hard which has a poor flavor, poor meltability, and low performance when cooked (Ipsen 2017). Reducing or lowering the fat content of cheese negatively affects its texture and flavor. Therefore the production of low fat cheese is a major challenge to develop products with comparable characteristics to full-fat counterparts. In order to overcome the defects related to fat reduction and enhance the acceptability of low fat cheese, various suggestions have been proposed (Table 4.2.1). Fat mimetics are polar, water soluble compounds having different chemical structures in contrast to fat which binds moisture, increasing water content, and imparting fat-like features. Utilization of fat mimetics is considered a new approach to producing low fat cheese, regarding their effects on cheese composition, component recoveries, cheese yield, and yield recovery (Alnemr et al., 2016). Fat mimetics provide a lower caloric value and a similar function as the fat in cheese which decreases the consumer's susceptibility to coronary heart diseases. Although fat mimetics inclusion in the low fat cheese improves functional and sensory characteristics of this product by softening the body and improving the texture and yield of the low fat cheese, they cannot fully replace the non-polar functional characteristics of the fat such as its flavor-carrying capacity (Khanal and Bansal 2020). In this regard, Koca and Metin (2004) evaluated the melting, textural and sensorial characteristics of low fat fresh kashar cheeses (70% fat reduction) by utilizing one carbohydrate-based fat replacer (5.0% w/w RaftilinesHP) and two protein-based fat replacers (1.0% w/w Dairy-Lo™ and 1.0% w/w Simplesses® D-100) during the 90 days storage time. Results showed that the cheese samples containing fat mimetics had higher moisture contents compared to the low fat control cheese. Additionally, incorporating fat mimetics had led to increasing cohesiveness and decreasing the gumminess and springiness, hardness, and chewiness of the samples. Moreover, all appearance defects of the low fat cheese improved by using Simplesses® D-100. Although RaftilinesHP and Simplesses® D-100 enhanced the texture, flavor, and overall acceptability of low fat cheese for up to 30 days, the sensory attributes of the low fat samples decreased due to softening, off-flavor and undesirable color of samples containing RaftilinesHP as well as excessive softening and salty flavor of samples containing Simplesses® D-100 up to 60 and 90 days of storage. On the contrary, no significant changes were shown in sensory and textural characteristics of the low fat cheese containing Dairy-Lo™. Among the fat mimetics, RaftilinesHP caused a slight improvement in the meltability of the samples. These results revealed that RaftilinesHP and Simplesses® D-100 can remarkably improve the texture and sensory characteristics of the low fat fresh kashar cheese (Koca and Metin 2004).

B-glucan hydrocolloid composite denoted as "Nutrim", a nutraceutical fat replacer was used to develop 3-month-old low fat cheddar cheeses. In this study, Nutrim I (6.8% fat) and Nutrim II

Table 4.2.1 Summary of studies performed on the application of fat mimetics in cheese.

Food product	Fat mimetic	Major findings	References
Low-fat fresh kashar cheeses	RaftilinesHP, Dairy-Lot and Simplesses® D-100	RaftilinesHP and Simplesses®D-100 can remarkably improve the texture and sensory features of the low-fat fresh kashar cheese.	(Koca and Metin 2004)
Low-fat cheddar cheeses	Nutrim	The melt-flow index, fracturability, and hardness of Nutrim-contained cheese samples were significantly lower than the low-fat control cheeses and the cohesiveness and the elasticity values were comparable in all cheese samples.	(Konuklar et al., 2004)
Low-fat white pickled cheeses	Simplesse® D-100, Dairy-Lo™, Perfectamyl gel MB (changed potato starch), and Satiagel ME4	All the low-fat cheese samples that had fat mimetics had different rheological features with no off-flavor compared to the full-fat cheeses.	(Kavas et al., 2004)
Reduced-fat cheese	Novagel™, Dairy-Lo™, ALACOPALS™	The network structure of low-fat cheese had a carbohydrate-based fat mimetic was more comparable to full-fat cheese than the samples containing protein-based fat mimetics.	(Ma et al., 1997)
Minas fresh cheese	Octenyl succinylated (OS) waxy maize starch	Adding fat mimetics could significantly increase the water holding capacity and the moisture content of the reduced-fat cheese and enhance the overall quality of the reduced-fat Minas fresh cheese.	(Diamantino et al., 2014)
Low-fat white-brined cheeses	Novagelt NC-200 and Simplesses® D-100	The textural properties of the product made with Novagelt resembled comparable to the full-fat cheese. Low-fat cheese products containing Simplesses® showed improving sensory attributes than their full-fat counterpart.	(Romeih et al., 2002).
Low-fat Domiati cheese	Kappa-carrageenan, locust bean, and xanthan gums	Including fat mimetics had led to an increase in the physiochemical features, moisture, and yield of the low-fat cheese samples.	(Alnemr et al., 2016)
Low-fat Mozzarella cheese	Xanthan gum and guar gum	The functionality and yield of low-fat Mozzarella cheese formulated with 0.15% xanthan gum and guar gum were comparable with full-fat counterparts.	(Sattar et al., 2015)
Low-fat Oaxaca cheese	k – and λ -carrageenans	k – carrageenan could improve the melting property of cheese samples. No detrimental changes were seen in sensory attributes of low-fat cheese by incorporating carrageenans in the low cheese.	(Totosaus and Guemes-Vera 2008)
Low-fat brined cheese	Zedu gum	Incorporating fat mimetic caused a softer texture and an increment in moisture content as well as an improvement in low-fat cheese sensory features.	(Baghdadi et al., 2018)

(3.47% fat) cheeses were compared with the low fat control cheese samples. According to the results, the melt-flow index, fracturability, and hardness of Nutrim-contained cheese samples were significantly lower than the low fat control cheeses. As well, the cohesiveness and the elasticity values were comparable in all cheese samples (Konuklar et al., 2004).

Kavas et al. (2004) demonstrated that the total nitrogen and moisture values of the low fat white pickled cheeses elevated significantly compared to the full-fat sample. Although use of fat mimetics (Simplesse® D-100, Dairy-Lo™, Perfectamyl gel MB (modified potato starch), and Satiagel ME 4) showed a negative impact on sensory features of cheese, no remarkable changes were observed in salt, total solids, pH, and acidity values of low fat cheese samples including fat mimetics. According to the results, all the low fat cheese samples that contained fat mimetics had different rheological characteristics with no off-flavor compared to the full-fat cheeses (Kavas et al., 2004).

Ma et al. (1997) produced 60% reduced-fat cheese using three fat mimetics based on carbohydrate (Novagel™ NC200, FMC Corp., Philadelphia, PA), whey-based (Dairy-Lo™, Pfizer, Inc., Groton, CT), and protein-based (ALACOPALS™ 6082, New Zealand Milk Products, Inc., Santa Rosa, CA). According to the results, low fat cheese samples including carbohydrate-based fat mimetics showed higher loss and storage moduli compared to the samples containing two protein-based fat mimetics or low-fat control cheese. However, the storage and loss moduli of low-fat cheese were smaller than their full-fat counterparts. As well, the network structure of low-fat cheese which contained a carbohydrate-based fat mimetic was more comparable to full-fat cheese than the samples that contained protein-based fat mimetics (Ma et al., 1997).

Diamantino et al. (2014) studied the effect of fat reduction and adding octenyl succinylated (OS) waxy maize starch as a fat mimetic on the microstructural, textural, and physicochemical attributes of Minas fresh cheese. According to the results, fat reduction promoted less proteolysis and a denser microstructure and decreased the yield and elevated the hardness of the prepared samples. As well, the addition of fat mimetics could significantly increase the water holding capacity and the moisture content of the reduced-fat cheese and enhance the overall quality of the reduced-fat Minas fresh cheese (Diamantino et al., 2014).

Romeih et al. (2002) investigated the effect of incorporating two commercial hydrocolloid-based fat replacers (0.125% w/w Novagelt NC-200 and1% w/w Simplesses® D-100) on the sensory, proteolysis, composition, and textural features of the low-fat white-brined cheeses (60% fat reduction) throughout aging for 90 days. All the low-fat products showed an increasing trend for the moisture and protein content, but the yield was reduced; however, including the fat replacers improved the moisture and yield values compared to the low-fat product. Also, the extent of lipolysis and proteolysis in the low-fat white-brined cheeses was significantly increased compared to the full-fat cheese samples. The texture and rheological properties of the low-fat samples improved by incorporating fat mimetics, and the textural properties of the product made with Novagelt resembled comparable to the full-fat cheese. Overall, low-fat cheese products containing Simplesses® showed improving sensory attributes than their full-fat counterpart (Romeih et al., 2002).

Alnemr et al. (2016) investigated the effect of hydrocolloids including kappa-carrageenan, locust bean, and xanthan gums in combination as fat mimetics on physicochemical features of low-fat Domiati cheese. Results showed that including fat mimetics had led to an increase in the physiochemical features, moisture, and yield of the low-fat cheese samples. As well low-fat samples had similar sensory attributes to full-fat cheese which are sustained during the 75-day ripening period (Alnemr et al., 2016).

Sattar et al. (2015) reported that fat reduction in cheese negatively affects the functional, sensory, and rheological features of Mozzarella cheese. The potential of different hydrocolloids as fat mimetics was evaluated (xanthan gum and guar gum) by using them at different concentrations in low-fat Mozzarella cheese. According to the results, different concentrations of hydrocolloids impact cheese composition, texture, functionality, sensory features, and yield to varying extents. Additionally, the functionality and yield of low-fat Mozzarella cheese formulated with 0.15% xanthan gum and guar gum were comparable with full-fat counterparts (Sattar et al., 2015).

Totosaus and Guemes-Vera (2008) used k- and λ-carrageenans as the fat mimetic to produce low-fat Oaxaca cheese. According to the results, k–carrageenan could improve the melting property of cheese samples. Generally, no detrimental changes were seen in sensory attributes of low-fat cheese by incorporating carrageenans as the fat mimetics in the low concentrations (Totosaus and Guemes-Vera 2008).

Baghdadi et al. (2018) incorporated Zedu gum (0.1% and 0.25%) as a fat mimetic in low-fat brined cheese. According to the results, incorporating fat mimetic into low-fat cheese caused a softer texture and an increment in moisture content because of the water binding capacity of this hydrocolloid in comparison with the control low-fat cheese. Although the incorporation of Zedu gum into the low-fat cheese matrix caused an improvement in its sensory characteristics, it was still inferior to the sensory properties of full-fat cheese (Baghdadi et al., 2018).

Low-fat cheese without a fat replacer was significantly gummier, chewier, harder, and more elastic; it alsohad lower organoleptic properties and poorer meltability than the full-fat cheese. However, use of fat mimetics in reduced- and low-fat cheeses can affect their characteristics in different ways such as increasing moisture, protein, ash content, and hardness, while their overall quality was acceptable. Fat reduction leads to a denser microstructure with fewer fat globules and less proteolysis during storage. Rheological studies indicated that different types of mimetics provided various effects in improving the protein matrix of low-fat cheeses. For example, a carbohydrate-based mimetic improved the rheological properties but did not simulate the protein matrix of full-fat cheese.

4.2.6.2 Ice Cream

Ice cream is a complex food matrix that has multiple physical phases and eliminating one of its ingredients may affect its physical features as well as multiple sensory attributes that could be important to consumers. According to The Food and Drug Administration, ice cream products with at least 25 and 50% less total fat than regular ice cream (10% fat content) can be labeled as reduced-fat and low-fat (light) ice cream (Liou and Grün 2007). Fat in ice cream contributes to mouthfeel, texture, and flavor, and it serves as a structural element. In ice cream, air bubble surfaces coated with discrete and partially coalesced fat globules present in the matrix as a binding agent and support the cells lined mainly by proteins (Mahdian and Karazhian 2013). The creaminess, texture, flavor, mouthfeel, and sensory features of ice cream are enhanced by the interaction between milk fat and other ingredients. Considering the vital role of fat in the flavor and texture of ice cream, eliminating or reducing fat content in this product led to many defects in the final quality of low-fat ice cream (Shenana 2021). Fat mimetics and flavor enhancers are among the most interesting approaches for researchers to overcome the drawbacks related to fat reduction (Table 4.2.2) and to match the texture and functionality of low-fat ice cream (Shenana 2021). Aime et al., (2001) investigated the effect of incorporating a modified pea starch as the fat mimetic on physicochemical and sensory properties of regular (10%), light (5%), low fat (2.5%), and fat-free vanilla (0.4%) ice creams. Results revealed that low-fat and fat-free ice creams showed lower smoothness, viscosity, and mouthfeel properties compared to their full-fat counterparts. The use of modified starch as a partial fat mimetic in vanilla ice cream was highly acceptable (Aime et al., 2001).

Mahdian and Karazhian (2013) formulated low-fat ice cream using carbohydrate or protein-based fat mimetics to substitute milk fat. Reduced-fat ice cream (5% fat) was developed using inulin

Table 4.2.2 Summary of studies performed on the application of fat mimetics in ice cream.

Food product	Fat mimetic	Major findings	References
Light, low fat, and fat-free vanilla ice creams	Modified pea starch	Using changed starch as a partial fat mimetic in vanilla ice cream was highly acceptable.	(Aime et al., 2001)
Low-fat ice cream	Inulin and milk protein concentrate	No remarkable changes were shown in the hardness of ice cream samples; while overrun values were reduced with increasing milk protein concentrate, and inulin. According to sensory analysis, the highest score for sensory attributes was given to the sample containing 2% inulin.	(Mahdian and Karazhian 2013)
Low-fat ice cream	Simplesse®, and inulin	The products with Simplesse® were firmer compared to other low-fat ice cream samples.	(Hossain et al., 2021)
Low-fat sugar-free ice cream	Maltodextrin and polydextrose	Maltodextrin (3%) and polydextrose (7%) were effective in emulating the mouthfeel of full-fat ice cream and providing bulk to the ice cream without affecting the sensory features and physicochemical features of low-fat ice cream.	(Khan et al., 2018)
Low-fat ice creams	Litesse®, and Simplesse®	Ice creams with Simplesse® and full-fat ice cream were perceived as comparable in sensory attributes, and samples containing the combination of Litesse® and Simplesse® were similar in flavor features to low-fat ice cream and full-fat ice cream in textural features.	(Liou and Grün 2007)
Low-fat ice cream	Inulin, maltodextrin, changed starch, whey protein concentrate powder, and oat powder	The samples containing maltodextrin followed by the samples containing inulin were selected as the best treatment and achieved total panelist scores like the full-fat ice cream.	(Shenana 2021)
Vanilla ice cream	Maltodextrin	Minimal changes were seen in the physical features of low-fat ice creams. Also, no remarkable changes were evaluated in sensory attributes in either fresh or stored low-fat ice cream containing maltodextrin.	(Rolon et al., 2017)

and milk protein concentrate as fat mimetics and the combination of Stab-6924 and Stab-IC80 as commercial stabilizer-emulsifier. Textural, physicochemical, and organoleptic features of low-fat ice cream were examined. According to the results, all the samples were pseudoplastic fluids, and increasing shear rate had led to a decrease in apparent viscosity values. Using milk protein concentrate had also led to elevated viscosity and consistency coefficient of the samples; while the flow behavior index decreased. There were no notable changes associated with the type and amount of stabilizer and fat mimetics in the hardness of ice cream samples; while overrun values were reduced by increasing milk protein concentrate, inulin, and stabilizers' level in the product formulation. Overall, samples containing Stab-IC80 had greater values of melting resistance and overrun than samples with Stab-6924. According to sensory analysis, the highest score for sensory attributes was given to the sample containing 2% inulin (Mahdian and Karazhian 2013).

Hossain et al. (2021) examined the physicochemical and microstructural attributes of low-fat ice cream that had fat mimetics including Simplesse®, and inulin individually or in combination. The first low-fat ice cream samples were treated by pasteurization followed by homogenization (PH), and the second samples were produced by homogenization with subsequent pasteurization (HP). Results showed that the overrun value of PH-treated low-fat ice creams was around 50%, while it increased significantly in the samples formulated with a combination of fat mimetics. However, HP-treated samples showed overrun values of around 40% for all treatments except samples containing inulin. Moreover, both the HP and PH-treated samples showed significant changes in the color intensities as well as the melting behavior. Although both the HP and PH ice cream groups showed a similar firmness, the products with Simplesse® were firmer compared to other low-fat ice cream samples (Hossain et al., 2021).

Khan et al. (2018) formulated desirable low-fat sugar-free ice cream. In this study, RSM (Response Surface Methodology) was used to investigate the effect of predictor variables on textural, sensorial, and melting resistance of low-fat ice cream samples. Maltodextrin (3%) and polydextrose (7%) were effective as fat replacers by providing bulk to the ice cream without affecting the sensory features and physicochemical features of low-fat ice cream. Moreover, produced low-fat ice cream samples were effective in emulating the mouthfeel of full-fat ice cream (Khan et al., 2018).

Liou and Grün (2007) formulated low-fat ice creams containing five strawberry flavor compounds using Litesse® and Simplesse® individually and in combination as fat mimetics. According to the results, ice creams with Simplesse® and full-fat ice cream were perceived as comparable in sensory attributes, and samples containing a combination of Litesse® and Simplesse® were similar in flavor features to low-fat ice cream and full-fat ice cream in textural features. Furthermore, the mouthfeel and flavor profiles of samples that had Simplesse® as a fat mimetic were like full-fat counterparts (Liou and Grün 2007).

Shenana (2021) produced low-fat ice cream by standardized milk fat to 2% to meet consumers' demand for low-energy foods and partial replacement of buffaloes' milk fat with different fat replacers including inulin, maltodextrin, changed starch, whey protein concentrate powder (WPCP), and oat powder. As well, the prepared samples were analyzed for chemical composition, overrun, rheological properties, microbiological quality and sensory features. According to the results, all the treatments were accepted by the panelists except the control of low-fat ice cream (2% fat) which was formulated without using fat replacers. The samples containing maltodextrin followed by the samples containing inulin were selected as the best treatment and achieved total panelist scores like the full-fat ice cream (Shenana 2021).

Rolon et al. (2017) examined the effect of replacing fat with maltodextrin (0, 2, 4, 5, and 8%) in vanilla ice cream formulation (6, 8, 10, 12, and 14% fat) on selected physical properties of ice cream including apparent viscosity, fat destabilization, overrun, melting rate, fat particle size, and hardness as well as consumer acceptability. Tests were also performed after 19 weeks of storage at −18°C to evaluate differences in acceptance because of extended storage time at adverse temperatures. According to the results, decreasing maltodextrin and increasing fat content had led to decreased mix viscosity and increased fat particle size, and fat destabilization significantly. However, no remarkable differences were seen in the acceptability of the fresh samples and ice cream samples stored for 19 weeks. Ice creams formulated with 8% fat and 6% maltodextrin, and 10% fat, and 4% maltodextrin showed a remarkable reduction in sensory attributes after the storage time comparable to changes that occurred in the fresh ice cream with the same fat content. In general, the ice cream samples that had 6 and 12% fat showed no detectable differences among consumers. Although minimal changes were observed in physical characteristics of low-fat ice cream samples that contained maltodextrin as a fat mimetic, no remarkable changes in sensory attributes were evaluated in either fresh or stored low-fat ice cream (Rolon et al., 2017).

The fat content of ice cream has an inverse effect on its melting rate. Fat globule clusters and agglomerates can stabilize air bubbles in the ice cream structure and decrease the melting rate. So the main reason for the higher melting rate of the low-fat ice cream compared to that of its full-fat counterparts can reduce its fat content. The addition of different fat mimetics had various effects on the low-fat ice cream based on their types and features. For example, the hard texture of the ice cream is inversely correlated with the fat content. If an increase in the fat content reduces the ice crystals volume, then the increases in the ice crystals could probably lead to a harder texture. The light ice cream (6% fat) manufacturing process is affected by the melting behavior, air incorporation, and viscosity of the ice cream. However, the microstructural features are also affected by the use of fat mimetics. For example, light ice cream containing fat mimetics had comparable melting behavior and higher overrun, firmness, and viscosity compared to the control full-fat ice cream.

4.2.6.3 Yogurt

Yogurt is a fermented milk product that forms a gel during milk acidification due to the aggregation of destabilized casein micelles along with the formation of a three-dimensional structure. Moreover, whey proteins participate as well as fat globules in the yogurt structure can act as a filler and form the fat globule membrane casein crosslink with the protein matrix (Delavari et al., 2014). The major functional characteristic of yogurt is known as gel formation which is governed by temperature and time of heat pretreatment, milk composition, fermentation temperature, and storage conditions of the final product as well as time, type, and quantity of starter culture used to inoculate the milk (Jalal Aghdasian et al., 2022). However, increasing awareness of the potentially harmful impact of fat on consumers' health has led to a decrease in consumer demand for full-fat yogurt. Although reduced or non-fat yogurt from a health point of view is produced worldwide, the lack of practical features of fat elimination leads to a loss of sensorial quality of low-fat yogurt (Amaya-Llano et al., 2008). The Code of Federal Regulations (FDA) reported that low-fat yogurt has 0.5% to 2% fat; however, non-fat yogurt has less than 0.5% milk fat. In yogurt, milk fat shows the main role in the flavor, texture, and color development of the final product. Therefore, the fat reduction can lead to subsequently reducing the total solids content as well as leading to poor texture, weak body, and increased whey separation (Saleh et al., 2020). Accordingly, many studies (Table 4.2.3) have been conducted about providing the functionality of the missing fat through protein or carbohydrate-based fat mimetics (Nouri et al., 2011). Guven et al. (2005) examined the effect of different concentrations (1, 2, and 3%) of inulin as fat mimetic on the quality of fat-free yogurt products. Results showed that adding over 1% inulin had led to an increasing consistency and whey separation; while pH, titratable acidity, and acetaldehyde levels were not altered by the inclusion of inulin. However, tyrosine and volatile fatty acid levels were negatively affected by inulin addition. Sensory attributes of yogurt samples showed that the yogurt sample that contained 1% inulin was perceived as comparable to control regular fat content yogurt samples (Guven et al., 2005).

The physicochemical, textural, and sensorial features of low-fat strained yogurt (0.5–2% milk fat) that had Dairy-Lo™ and Simplesse® as fat mimetics during the storage time showed that the low-fat sample had higher viscosity, acidity, ash, and color values than the full-fat sample. Additionally, yogurt samples containing Dairy-Lo™ had higher ash, acidity, viscosity, flavor, appearance, and color values compared to the samples that contained Simplesse® (Yazici and Akgun 2004).

Calleros et al. (2004) used microparticulated whey protein, whey protein concentrate (WPC), or the combination of both fat mimetics to determine the effect of fat replacers on flow and creep

Table 4.2.3 Summary of studies performed on the application of fat mimetics in yogurt.

Food product	Fat mimetic	Major findings	References
Fat-free yogurt	Inulin	The yogurt sample that contained 1% inulin was perceived as comparable to control regular fat content yogurt samples.	(Guven et al., 2005)
Low-fat strained yogurt	Dairy-Lo™ and Simplesse®	Yogurt formulated with Dairy-Lo™ had higher physicochemical, textural, and sensorial characteristics compared to the samples containing Simplesse®.	(Yazici and Akgun 2004)
Reduced-fat yogurt	Microparticulated whey protein, whey protein concentrate, or the mixture of them	The yogurt formulated with whey protein concentrate demonstrated comparable flow and viscoelastic characteristics to the full-fat counterpart.	(Lobato-Calleros et al., 2004)
Non-fat yogurt	renneted skim milk	The inclusion of 30% renneted skim milk into the non-fat yogurt samples has provided comparable sensory properties to full-fat yogurt.	(Nouri et al., 2011)
Low-fat stirred yogurt	Hydrolyzed jicama starches	The sensorial and rheological attributes showed the possibility of hydrolyzed jicama starches to produce the low-fat yogurt with desirable functional and sensorial properties.	(Amaya-Llano et al., 2008)
Non-fat yogurt	Protein-based fat mimetic (Dairy-Lo™) or carbohydrate-based fat mimetic (Maltrin)	Both fat mimetics could remarkably improve the rheological properties of yogurts. The samples containing 0.5 and 1% Maltrin were similar in quality attributes to full-fat control yogurt.	(Abd El-Galeel et al., 2017)
Low fat (6%), and reduced-fat (12%) yogurt	Carob bean gum	The sensory and textural characteristics of reduced-fat and low-fat yogurts containing Carob bean gum were significantly higher.	(Aydinol Sonmez and Ozcan 2022)

compliance properties of reduced-fat yogurt samples. The results revealed that yogurt formulated with WPC demonstrated comparable flow and viscoelastic characteristics to the full-fat counterpart (Lobato-Calleros et al., 2004).

Nouri et al. (2011) evaluated the physicochemical, textural, and organoleptic properties of non-fat yogurt samples containing different concentrations of renneted skim milk (RSM). Results revealed that increasing RSM content (50%) had led to reducing total solids in non-fat yogurt samples; while no alterations were seen in the chemical attributes (pH, acidity, total protein, and fat content) of low-fat yogurt samples. Too, the samples including a higher concentration of RSM showed higher a* and b* values; while the highest water holding capacity, viscosity, firmness, L* value, and lowest syneresis were seen in a sample containing 30% RSM. In this study, the authors revealed the greater ability of the yogurts gel network to form the interactions and bind water in the samples containing RSM compared to the control skim yogurt. However, collection of casein in a higher concentration of RSM (higher than 30%) had led to no increase in the L* value, viscosity, and water holding capacity of the samples. According to the sensory attributes, including 30% RSM into the non-fat yogurt samples has provided comparable sensory properties to full-fat yogurt (Nouri et al., 2011).

Amaya-Liano et al. (2008) formulated low-fat stirred yogurt by incorporating hydrolyzed starches. No significant differences were seen in viscosity and pH values of yogurt samples that had hydrolyzed jicama starches as fat mimetic compared to the control sample. Overall, the sensorial attributes showed that producing low-fat yogurt with desirable functional and sensorial properties was possible by using hydrolyzed jicama starches as a fat mimetic (Amaya-Llano et al., 2008).

Abd El-Galeel et al. (2017) investigated quality changes in non-fat yogurt formulated with different levels (0.5 and 1%) of protein-based fat mimetic (Dairy-Lo^{TM}) or carbohydrate-based fat mimetic (Maltrin®). Results suggest incorporating fat mimetics could remarkably improve rheological properties (syneresis and viscosity) of yogurt samples, while no significant changes were shown in the chemical composition of resultant non-fat yogurt. According to the sensorial attributes, the fat replacement had led to an increase in the soluble nitrogenous compounds, formation of acetaldehyde, diacetyl, and total volatile fatty acids (flavor compounds). Although adding both fat mimetics together increased the organoleptic features of yogurt, the effect of Maltrin was more remarkable. Maltrin® could effectively improve the flavor formation and body properties of non-fat yogurt and the samples containing 0.5 and 1% Maltrin were similar in quality attributes to full-fat control yogurt (Abd El-Galeel et al., 2017).

Aydinol Sonmez and Ozcan (2022) formulated the low fat (6%), and reduced-fat (12%) yogurt using Carob bean gum (CBG) as a carbohydrate-based fat mimetic. According to the results, the sensory and textural characteristics (such as firmness and stickiness) of reduced-fat and low-fat yogurts containing CBG were significantly higher. Adding CBG to reduced-fat and low-fat yogurts could effectively improve the sensory attributes and acceptance index of the samples (Aydinol Sonmez and Ozcan 2022).

In conclusion, structure development during yogurt gelation is influenced by both the preheating treatment of the milk and its fat content. However, the reduction of fat in yogurt caused the harder, blander, and more rubbery/chewy texture, color, and melting defects compared to its full-fat counterparts. Accordingly, using different fat mimetics in yogurt formulation could solve the problems associated with enhancing technological properties. Finally, adding fat mimetics to yogurt milk improved the textural and sensorial properties of reduced and low-fat yogurts. Yogurt produced with fat mimetics showed higher apparent viscosity, water holding capacity, firmness, and lower syneresis than control skim yogurt. While this sample had low-fat content, the overall sensory evaluation score was similar to the full-fat yogurt. The final product had low calories, accepted mouthfeel, and was healthier.

4.2.6.4 Dairy Dessert and Beverages

Incorporating fat mimetics could effectively improve the textural properties of reduced-fat dairy desserts and beverages. According to different studies (Table 4.2.4), including fat mimetics in low-fat dairy desserts and beverages decreased instrumental differences in color and rheological traits between low-fat and full-fat samples and made them more akin to full-fat samples in terms of rheological behavior. Nevertheless, to obtain products with different fat content but similar flavor perception, sweetener and aroma concentration should be adjusted. Accordingly, Rodriguez Furlán and Campderrós (2017) investigated the influence of incorporating inulin as a fat mimetic and sucralose and stevia as a sugar substitute on the physicochemical properties of starch-based dairy desserts. The results revealed that fat could be partially replaced by inulin, and a mixture of stevia and sucralose could be used as a sugar substitute to enhance the end product's characteristics. Moreover, including inulin into the fat-reduced dessert sweetened with the 50:50 mixture of sucralose and stevia led to increased stability in the samples; though, syneresis was decreased. Additionally, the inulin-contained low-fat desserts showed an increase in consistency of the samples, which demonstrated the effect of the polysaccharide as a fat mimetic. Overall, the samples prepared by incorporating 50% sucralose, 2.5% inulin, and 50% stevia showed higher sensory characteristics than a commercial reduced-fat dairy dessert (Rodriguez Furlán and Campderrós 2017).

Table 4.2.4 Summary of studies performed on the application of fat mimetics in dairy dessert and beverages.

Food product	Fat mimetic	Major findings	References
Dairy desserts	inulin	The inulin-contained low-fat desserts showed an increase in consistency compared with commercial reduced-fat dairy desserts. As well, the samples prepared by incorporating 50% sucralose, 2.5% inulin, and 50% stevia showed higher sensory characteristics and lower syneresis	(Rodriguez Furlán and Campderrós 2017)
CMC-based dairy beverages	the mixture of 50: 50 long- and short-chain inulin as well as λ-carrageenan	inulin and λ-carrageenan showed remarkable effects in imitation of fat properties in CMC-based dairy beverages	(Arancibia et al., 2011)
Prebiotic dairy dessert	A mixture of long and short-chain inulin	Desserts containing 10% sucrose, 5.5% inulin, and 60 ppm lemon flavor showed similar sensorial attributes (thickness and creaminess) with the sample containing 2.8% fat	(Arcia et al., 2011)
yogurt mousse	long-chain inulin	Formulation containing 2.8% fat, 4.8% inulin, and 1.0% microencapsulated calcium showed rheological and sensory properties comparable to a full-fat, non-fortified yogurt mousse	(Gomez-Betancur et al., 2020)

Arancibia et al. (2011) studied the effects of incorporating the mixture of 50:50 long- and short-chain inulin as well as λ-carrageenan as fat mimetics on textural, physicochemical, and organoleptic attributes of the low-fat dairy beverages formulated with carboxymethyl cellulose. Results revealed the effect of the inulin and λ-carrageenan in imitation of fat properties in CMC-based dairy beverages. However, differences in rheological and color characteristics were distinguishable by consumers (Arancibia et al., 2011).

Arcia et al. (2011) prepared a low-fat prebiotic dairy dessert by incorporating inulin (a mixture of long and short-chain inulin), lemon flavor, and sucrose. The sample developed by incorporating 10% sucrose, 5.5% inulin, and 60 ppm lemon flavor was the optimum sample due to the similar sensorial attributes (thickness and creaminess) with the sample that contained 2.8% fat (Arcia et al., 2011).

Gomez-Betancur et al. (2020) investigated the changes in the overrun, rheological, and sensory attributes of yogurt mousse (fat content of 0.0–8.0%) with the response surface methodology as the result of the fortification with microencapsulated calcium carbonate (0.0–1.0%), and substituting fat with long-chain inulin (0.0–4.8%). The results showed that inulin significantly affected all the rheological responses and creaminess of yogurt mousses. Overall, a formulation containing 2.8% fat, 4.8% inulin, and 1.0% microencapsulated calcium showed rheological and sensory properties comparable to a full-fat, non-fortified yogurt mousse (Gomez-Betancur et al., 2020).

4.2.7 Conclusion

Consumers are attracted to low-fat dairy products because such products help consumers maintain adequate body weight and / or addresshealth-related issues. For this reason, several factors need to be considered during production to make low-fat dairy products with acceptable organoleptic attributes as well as features similar to the low-fat dairy products' full-fat counterparts. Accordingly, various types of fat mimetics are available; however, none of them can emulate all the organoleptic and practical properties of conventional fats. Therefore, using different combinations of fat mimetics may help imitate the functional and sensory attributes of fat as well as decrease differences in the rheological and color properties between full-fat and low-fat dairy foods. This makes it more likely consumers will accept the end product.

References

Abd El-Galeel, A., Atwaa, E., and Abdelwahed, E. (2017). Improving properties of non-fat yoghurt using fat replacers. *Zagazig Journal of Agricultural Research* 44: 583–590.

Adili, L., Roufegarinejad, L., Tabibiazar, M., Hamishehkar, H., and Alizadeh, A. (2020). Development and characterization of reinforced ethyl cellulose based oleogel with adipic acid: Its application in cake and beef burger. *Lwt* 126: 109277.

Aime, D.B., Arntfield, S.D., Malcolmson, L.J., and Ryland, D. (2001). Textural analysis of fat reduced vanilla ice cream products. *Food Research International* 34: 237–246.

Aliasl, A.K., Tabibiazar, M., Roufegarinejad, L., Hamishehkar, H., and Alizadeh, A. (2020). Preparation and characterization of carnauba wax/adipic acid oleogel: a new reinforced oleogel for application in cake and beef burger. *Food Chemistry* 333: 127446.

Alizadeh, A., Aghayi, N., Soofi, M., and Roufegarinejad, L. (2021). Development of synbiotic added sucrose-free mango nectar as a potential substrate for Lactobacillus casei: physicochemical characterisation and consumer acceptability during storage. *Acta Alimentaria* 50: 299–309.

Alizadeh, A. and Bahar, M. (2016). Cholesterol content of Lighvan cheese: A natural probiotic traditional cheese. *Journal of Clinical Gastroenterology* 50: S199–S199.

Alizadeh, A., Ehsani, M.R., Ezzatpanah, H., Mousavi, G., Darabi, M., and Roufegarinejad, L. (2012). Effect of cow and sheep milk diet on plasma lipids in rats. *Milchwissenschaft-Milk Science International* 67: 410.

Alizadeh, A., Ehsani, M.R., and Homayouni, A. (2008). Acetaldehyde production rate in yoghurt made from ultrafiltered skim milk. *Asian Journal of Chemistry* 20: 6529.

Alizadeh, A., Ehsani, M.R., and Rofehgarinejad, L. (2017). Fatty acid profile and cholesterol content of Ghezel sheep milk during lactation period. *Acta Alimentaria* 46: 457–463.

Alizadeh, A., Ehsani, M.R., and Safari, M. (2009). The effects of ultra-filtered milk proteins on chemical and sensory properties of yoghurt. *Iranian Journal of Food Science and Technology* 6: 109–115.

Alnemr, T., Helal, A., Hassan, A., and Elsaadany, K. (2016). Utilizing the functions of hydrocolloids as fat mimetic to enhance the properties of low-fat Domiati cheese. *Journal of Food Processing & Technology* 7.

Amaya-Llano, S.L., Martínez-Alegría, A.L., Zazueta-Morales, J.J., and Martínez-Bustos, F. (2008). Acid thinned jicama and maize starches as fat substitute in stirred yogurt. *LWT - Food Science and Technology* 41: 1274–1281.

Arancibia, C., Costell, E., and Bayarri, S. (2011). Fat replacers in low-fat carboxymethyl cellulose dairy beverages: color, rheology, and consumer perception. *Journal of Dairy Science* 94: 2245–2258.

Arcia, P.L., Costell, E., and Tárrega, A. (2011). Inulin blend as prebiotic and fat replacer in dairy desserts: optimization by response surface methodology. *Journal of Dairy Science* 94: 2192–2200.

Aydinol, S.P. and Ozcan, T. (2022). Assessment of structure and sensory characteristics of reduced fat yoghurt manufactured with carob bean gum polysaccharides. *Food Science and Technology* 42: 1–7.

Badri, M. and Alizadeh, A. (2016). Effect of beta-glucan addition on survival of Lactobacillus acidophilus in low fat stirred probiotic yoghurt. *Journal of Clinical Gastroenterology* 50: S215–S215.

Baghdadi, F., Aminifar, M., Farhoodi, M., and Shojaee, A.A.S. (2018). Study of macromolecular interactions in low-fat brined cheese modified with Zedu gum. *International Journal of Dairy Technology* 71: 382–394.

Bemer, H.L., Limbaugh, M., Cramer, E.D., Harper, W.J., and Maleky, F. (2016). Vegetable organogels incorporation in cream cheese products. *FRIN* 85: 67–75.

Cui, X.T., Saleh, A.S.M., Yang, S., Wang, N., Wang, P., Zhu, M., and Xiao, Z. (2022). Oleogels as animal fat and shortening replacers: Research advances and application challenges. *Food Reviews International* 38: 1–22.

Delavari, M., Pourahmad, R., and Sokutifar, R. (2014). Production of low fat synbiotic yogurt containing lactobacillus plantarum and inulin. *Advances in Environmental Biology* 8: 17–24.

Di Cagno, R., De Pasquale, I., De Angelis, M., Buchin, S., Rizzello, C.G., and Gobbetti, M. (2014). Use of microparticulated whey protein concentrate, exopolysaccharide-producing Streptococcus thermophilus, and adjunct cultures for making low-fat Italian Caciotta-type cheese. *Journal of Dairy Science* 97: 72–84.

Diamantino, V.R., Beraldo, F.A., Sunakozawa, T.N., and Penna, A.L.B. (2014). Effect of octenyl succinylated waxy starch as a fat mimetic on texture, microstructure and physicochemical properties of Minas fresh cheese. *LWT - Food Science and Technology* 56: 356–362.

Dissanayake, M., Kelly, A.L., and Vasiljevic, T. (2010). Gelling properties of microparticulated whey proteins. *Journal of Agricultural and Food Chemistry* 58: 6825–6832.

Drake, M.A., Miracle, R.E., and McMahon, D.J. (2010). Impact of fat reduction on flavor and flavor chemistry of Cheddar cheeses. *Journal of Dairy Science* 93: 5069–5081.

Gomez-Betancur, A.M., Carmona-Tamayo, R., Martínez-Álvarez, O.L., Casanova-Yepes, H., and Torres-Oquendo, J.D. (2020). Effect of fat substitution using long-chain inulin and fortification with microencapsulated calcium in the rheological and sensory properties of yogurt mousse. *Journal of Food Process Engineering* 45: 1–12.

Gutiérrez-Luna, K., Astiasarán, I., and Ansorena, D. (2022). Gels as fat replacers in bakery products: a review. *Critical Reviews in Food Science and Nutrition* 62: 3768–3781.

Guven, M., Yasar, K., Karaca, O.B., and Hayaloglu, A.A. (2005). The effect of inulin as a fat replacer on the quality of set-type low-fat yogurt manufacture. *International Journal of Dairy Technology* 58: 180–184.

Hossain, M.K., Petrov, M., Hensel, O., and Diakité, M. (2021). Microstructure and physicochemical properties of light ice cream: effects of extruded microparticulated whey proteins and process design. *Foods* 10.

Huang, H., Hallinan, R., and Maleky, F. (2018). Comparison of different oleogels in processed cheese products formulation. *International Journal of Food Science & Technology* 53: 2525–2534.

Ipsen, R. (2017). Microparticulated whey proteins for improving dairy product texture. *International Dairy Journal* 67: 73–79.

Jalal Aghdasian, A., Alizadeh, A., and Soofi, M. (2022). Development of iron and vitamin D3 fortified low-fat synbiotic yogurt as a potential substrate for Lactobacillus acidophilus and Bifidobacterium lactis: evaluation of physicochemical and sensory Properties during the storage time. *Journal of Food Measurement and Characterization* 16: 2718–2725.

Johnson, M. (2011). Cheese low-fat and reduced-fat cheese. In: *Encyclopedia of Dairy Sciences* (ed. J.W. Fuquay, P.F. Fox, and P.L.H Mcsweeney), pp. 833–842 United Kingdom: Elsevier Ltd.

Johnson, M.E., Kapoor, R., McMahon, D.J., McCoy, D.R., and Narasimmon, R.G. (2009). Reduction of sodium and fat levels in natural and processed cheeses: scientific and technological aspects. *Comprehensive Reviews in Food Science and Food Safety* 8: 252–268.

Jooyandeh, H., Goudarzi, M., Rostamabadi, H., and Hojjati, M. (2017). Effect of Persian and almond gums as fat replacers on the physicochemical, rheological, and microstructural attributes of low-fat Iranian White cheese. *Food Science and Nutrition* 5: 669–677.

Kavas, G., Oysun, G., Kinik, O., and Uysal, H. (2004). Effect of some fat replacers on chemical, physical and sensory attributes of low-fat white pickled cheese. *Food Chemistry* 88: 381–388.

Khan, S., Rustagi, S., Choudhary, S., Pandey, A., Khan, M.K., and Singh, A. (2018). Sucralose and maltodextrin-An altrernative to low fat sugar free ice-cream. *Bioscience Biotechnology Research Communication* 11: 136–143.

Khanal, B., Bhandari, B., Prakash, S., and Bansal, N. (2017). Effect of sodium alginate addition on physical properties of rennet milk gels. *Food Biophysics* 12.

Khanal, B.K.S. and Bansal, N. (2020). Dairy fat replacement in low-fat cheese (LFC): a review of successful technological interventions. In: *Dairy Fat Products and Functionality* (ed. T. Truong, C. Lopez, B. Bhandari, and S. Prakash), pp. 549–581. Springer: Switzerland.

Koca, N. and Metin, M. (2004). Textural, melting and sensory properties of low-fat fresh kashar cheeses produced by using fat replacers. *International Dairy Journal* 14: 365–373.

Konuklar, G., Inglett, G.E., Warner, K., and Carriere, C.J. (2004). Use of a β-glucan hydrocolloidal suspension in the manufacture of low-fat Cheddar cheeses: textural properties by instrumental methods and sensory panels. *Food Hydrocolloids* 18: 535–545.

Lashkari, H., Madadlou, A., and Alizadeh, M. (2014). Chemical composition and rheology of low-fat Iranian white cheese incorporated with guar gum and gum arabic as fat replacers. *Journal of Food Science and Technology* 51: 2584–2591.

Lee, S., Kim, S., and Inglett, G.E. (2005). Effect of shortening replacement with oatrim on the physical and rheological properties of cakes. *Cereal Chemistry* 82: 120–124.

Liou, B.K. and Grün, I.U. (2007). Effect of fat level on the perception of five flavor chemicals in ice cream with or without fat mimetics by using a descriptive test. *Journal of Food Science* 72.

Lobato-Calleros, C., Martínez-Torrijos, O., Ofelia, S-C., Pérez-Orozco, J.P., and Vernon-Carter, E.J. (2004). Flow and creep compliance properties of reduced-fat yoghurts containing protein-based fat replacers. *International Dairy Journal* 14: 777–782.

Ma, L., Drake, M.A., Barbosa-Cánovas, G.V., and Swanson, B.G. (1997). Rheology of full-fat and low-fat cheddar cheeses as related to type of fat mimetic. *Journal of Food Science* 62: 748–752.

Mahdian, E. and Karazhian, R. (2013). Effects of fat replacers and stabilizers on rheological, physicochemical and sensory properties of reduced-fat ice cream. *Journal of Agricultural Science and Technology* 15: 1163–1174.

Martins, A.J., Vicente, A.A., Cunha, R.L., and Cerqueira, M.A. (2018). Edible oleogels: an opportunity for fat replacement in foods. *Food & Function* 9: 758–773.

Moriano, M.E. and Alamprese, C. (2017). LWT - Food Science and Technology Organogels as novel ingredients for low saturated fat ice creams. *LWT - Food Science and Technology* 86: 371–376.

Mounsey, J.S. and O'Riordan, E.D. (2008). Modification of imitation cheese structure and rheology using pre-gelatinised starches. *European Food Research and Technology* 226: 1039–1046.

Nouri, M., Ezzatpanah, H., and Abbasi, S. (2011). Application of renneted skim milk as a fat mimetics in nonfat yoghurt. *Food and Nutrition Sciences* 2: 541–548.

O'Connor, T. and O'Brien, N. (2016). Fat replacers. In: *Reference Module in Food Science* (ed. G.W. Smithers). Amsterdam: Elsevier.

Omayma, E. and Youssef, M.M. (2007). Fat replacers and their applications in food products: a review. *Journal of Food Science and Technology* 4: 29–44.

Park, C., Bemer, H.L., and Maleky, F. (2018). Oxidative stability of rice bran wax oleogels and an oleogel cream cheese product. *Journal of the American Oil Chemists' Society* 95: 1267–1275.

Pastorino, A.J., Dave, R.I., Oberg, C.J., and McMahon, D.J. (2002). Temperature effect on structure-opacity relationships of nonfat Mozzarella cheese. *Journal of Dairy Science* 85: 2106–2113.

Patel, A.R., Nicholson, R.A., and Marangoni, A.G. (2020). Applications of fat mimetics for the replacement of saturated and hydrogenated fat in food products. *Current Opinion in Food Science* 33: 61–68.

Peng, X. and Yao, Y. (2017). Carbohydrates as fat replacers. *Annual Review of Food Science and Technology* 8: 331–351.

Rodriguez Furlán, L.T. and Campderrós, M.E. (2017). The combined effects of Stevia and sucralose as sugar substitute and inulin as fat mimetic on the physicochemical properties of sugar-free reduced-fat dairy dessert. *International Journal of Gastronomy and Food Science* 10: 16–23.

Rolon, M.L., Bakke, A.J., Coupland, J.N., Hayes, J.E., and Roberts, R.F. (2017). Effect of fat content on the physical properties and consumer acceptability of vanilla ice cream. *Journal of Dairy Science* 100: 5217–5227.

Romeih, E.A., Michaelidou, A., Biliaderis, C.G., and Zerfiridis, G.K. (2002). Low-fat white-brined cheese made from bovine milk and two commercial fat mimetics: chemical, physical and sensory attributes. *International Dairy Journal* 12: 525–540.

Roufegarinejad, L., Ashrafi, R., Alizadeh, A., and Ghaemi, H. (2017). Effect of cream and yogurt butter diets on serum lipid profile in rats. *Journal of Nutritional Sciences and Dietetics* 3: 3–8.

Saleh, A., Mohamed, A.A., Alamri, M.S., Hussain, S., Qasem, A.A., and Ibraheem, M.A. (2020). Effect of different starches on the rheological, sensory and storage attributes of non-fat set yogurt. *Foods* 9.

Sattar, M.U., Sameen, A., Huma, N., and Shahid, M. (2015). Exploit fat mimetic potential of different hydrocolloids in low fat Mozzarella cheese. *Journal of Food and Nutrition Research* 3: 518–525.

Seyed Ahmadi Mamaghani, M.H., Alizadeh, A., and Soofi, M. (2020). Effect of ultrasound treatment on the viability of probiotics and physicochemical properties of synbiotic carrot juice. *Iranian Food Science and Technology Research Journal* 16: 15–25.

Shafiei, Z., Hojjatoleslamy, M., Soha, S., and Shariati, M.A. (2014). The influence of malt extraction adding to uf fresh low fat cheese on its textural properties. *International Journal of Engineering Science* 6: 52–55.

Shaikh, H.M., Anis, A., Poulose, A.M., Madhar, N.A., and Al-Zahrani, S.M. (2022). Development of bigels based on date palm-derived cellulose nanocrystal-reinforced guar gum hydrogel and sesame oil/candelilla wax oleogel as delivery vehicles for moxifloxacin. *Gels* 8: 330.

Shakeel, A., Farooq, U., Gabriele, D., Marangoni, A.G., and Lupi, F.R. (2021). Bigels and multi-component organogels: An overview from rheological perspective. *Food Hydrocolloids* 111: 106190.

Shenana, M.E. (2021). Improving the quality of low-fat ice cream using some fat replacers. *Annals of Agricultural Science, Moshtohor* 59: 463–472.

Soofi, M., Alizadeh, A., and Mousavi Kalajahi, S.E. (2019). Optimization of low fat prebiotic yogurt processing containing inulin at different temperatures and shear stresses Soofi. *Iranian Journal of Food Science and Technology* 16: 109–118.

Stankey, J.A., Lu, Y., Abdalla, A., Govindasamy-Lucey, S., Jaeggi, J.J., Mikkelsen, B.Ø, Pederson, K.T., and Anderson, C.B. (2017). Low-fat Cheddar cheese made using microparticulated whey proteins: effect on yield and cheese quality. *International Journal of Dairy Technology* 70: 481–491.

Temkov, M. and Mureșan, V. (2021). Tailoring the structure of lipids, oleogels and fat replacers by different approaches for solving the trans-fat issue—a review. *Foods* 10.

Totosaus, A. and Guemes-Vera, N. (2008). Effect of κ- and λ-carrageenans as fat-replacers in low-fat Oaxaca cheese. *International Journal of Food Properties* 11: 656–668.

Wadhwani, R., McManus, W.R., and McMahon, D.J. (2011). Improvement in melting and baking properties of low-fat Mozzarella cheese. *Journal of Dairy Science* 94: 1713–1723.

WHO (2015). *Healthy diet. Fact sheet No. 394 [WWW Document].*

Yazici, F. and Akgun, A. (2004). Effect of some protein based fat replacers on physical, chemical, textural, and sensory properties of strained yoghurt. *Journal of Food Engineering* 62: 245–254.

4.3

Margarine and Fat Spreads

Filip Van Bockstaele, Ivana A. Penagos, Kato Rondou, and Koen Dewettinck*

Ghent University, Department of Food Technology, Safety and Health, Food Structure and Function (FS&F) Research Group, Coupure Links 653, Ghent, Belgium
** Corresponding author*

4.3.1 Introduction

Margarine represents a pinnacle of culinary modernity. At the end of the 19th century, driven by population growth and ongoing industrialization, there was a shortage of butter (Blackmore 1960). Undernourishment and high prices led to a quest to produce a cheap, nutritious, and easy-to-produce butter substitute. Thus, in 1869, the French chemist Hippolyte Mège-Mouriès invented margarine in response to a "prize offered by Napoleon III" (Turner 2018). In 1869, Hippolyte Mège-Mouriès completed his experiments and made his new product in a small factory at Poissy outside Paris. He had noticed that the solidified mixture of the ingredients had a pearly appearance, and it is from the Greek words for "pearl" *margaron* and "pearly" *margarites*, that the name "margarine" was derived (Blackmore 1960). In 1869, he patented his process in France and England. In 1871, he sold his knowledge to the Dutch firm of Antoon Jurgens and his British, American, and Prussian patents to various buyers in 1873 and 1874 (Deelstra et al. 2014), leading to an exponential increase in production. The product "margarine" thus celebrated its 150th anniversary in 2019; today margarine contributes to an important share in the global consumption of fats. Despite the scientific innovation behind the production of margarine, and its association with progress and modernity from a food engineering perspective, historically, margarine's status in the public consciousness has been rather less than flattering (Turner 2018). Especially in the early days, margarine was perceived as a low-quality butter imitation and the butter producers did whatever they could to close off their home markets from the imitation product. Today, it is consumed in many countries, especially as spread for bread or as an ingredient in baked products. Furthermore, margarine gained importance as a plant-based food and as a carrier of essential lipid components in the human diet.

4.3.1.1 Definitions and Legislation

Legislation differs from country to country. Here we present information from the Codex Alimentarius, EU and US regulations. The Codex Alimentarius Standard 32–1981 (FAO/WHO 2001) defines margarine (as a food) as a plastic or fluid emulsion, which is mainly of the type water-in-oil (W/O), produced principally from edible fats and oils, which are not mainly derived from milk. The product should include at least 80% fat (w/w) and a maximum of 16% water (w/w). Additives which can be included in margarine, next to water and edible oils and fats, are colorants, flavors, emulsifiers, preservatives, antioxidants, acidity regulators, and anti-foaming agents. Next to margarine, the Codex Alimentarius Standard 135–1981 (FAO/WHO 2001) defines minarine as a

Fat Mimetics for Food Applications, First Edition. Edited by Miguel Ângelo Parente Ribeiro Cerqueira and Lorenzo Miguel Pastrana Castro.

food in the form of a spreadable emulsion, which is mainly of the type W/O, produced principally from water and edible fats and oils not solely derived from milk, and in which the fat content is not less than 39% (w/w) and not more than 41% (w/w).

In the European Union, fat product terminology is regulated by Regulation 1234/2007 (Council of the European Union 2007). Under this regulation, the sales description of a fat product as a solid, malleable emulsion, principally of the W/O type, is determined by the fat origin. More precisely, three subgroups are defined of products containing (1) exclusively milk fats, (2) mainly vegetable and or animal fats (< 3% milk fats), and (3) fat blends (10–80% milk fat). According to this regulation, margarine is defined as the product obtained from vegetable and/or animal fats with a fat content between 80% and 90%. The terminology "fat spreads" is generally used for lower fat contents with three-quarter-fat (60–62% fat) and half-fat margarine (39–41%) as specific sales descriptions.

The US law (Code of Federal Regulations 2021) has regulations for labeling and establishing requirements for margarine. Margarine (or oleomargarine) is the food in plastic form or liquid emulsion, containing no less than 80% fat determined by the method prescribed in the "Official Methods of Analysis of the Association of Official Analytical Chemists," 13th Ed. (1980), section 16.206, "Indirect Method". In short, according to the Code of Federal Regulations (CFR), margarine has four requirements. First, margarines must be produced from edible fats and/or oils, or mixtures of these, whose origin is vegetable, animal, or marine, and that has been affirmed as GRAS (Generally Recognized as Safe) or has been listed as a food additive for this use. They may have small amounts of other lipids naturally present in the fat or oil, such as phosphatides, unsaponifiable constituents, or free fatty acids. Second, margarines may have one or more aqueous phase ingredients present. Water, milk, milk products, and/or suitable edible protein ((modified) whey, albumin, casein, caseinate, vegetable protein) are examples. Third, they must contain vitamin A (not less than 15,000 international units per pound of finished margarine). Fourth, they can contain optional ingredients such as vitamin D, salt, carbohydrate sweeteners, emulsifiers, preservatives, colorants, flavors, acidulants, and alkalizers.

4.3.1.2 Microstructure

Margarine and low-fat spreads are W/O emulsions, comprising a finely dispersed aqueous phase into liquid oil, both of which are embedded and stabilized within a fat crystal network (Arellano et al. 2015). The microstructure of two commercial fat spreads, containing 38 and 25% fat respectively, is illustrated in Figure 4.3.1 by cryo-scanning electron microscopy (cryo-SEM). In cryo-SEM, controlled sublimation of the unbound water allows the underlying microstructure to be exposed. Water droplets appear as craters surrounded by the lipid matrix, containing fragments related to solutes in the water phase (Figure 4.3.1A-B). Water droplets are finely dispersed and a broad distribution in water droplet sizes is present in commercial fat spreads ranging from some μm up to around 30 μm. In case of high water content, such as in low-fat spreads, hydrocolloids may be added to stabilize or even gel the water phase (Figure 4.3.1C-D). Another approach to visualize the microstructure of W/O emulsions is confocal scanning laser microscopy (CSLM). By applying different fluorescent dyes, specific structural components are visible as shown in Figure 4.3.2 in which nile red dyes the fat matrix (usually visualized in red) and fluorescein isothiocyanate (FITC) colors the proteins (usually visualized in green) in the water phase. Fat crystals are primarily responsible for product stabilization by providing a physical barrier at the oil-water interface, preventing coalescence of the water droplets (Frasch-Melnik et al. 2010), and forming a three-dimensional network entrapping liquid oil and water droplets. This is illustrated in Figure 4.3.3 by the cryo-SEM image of

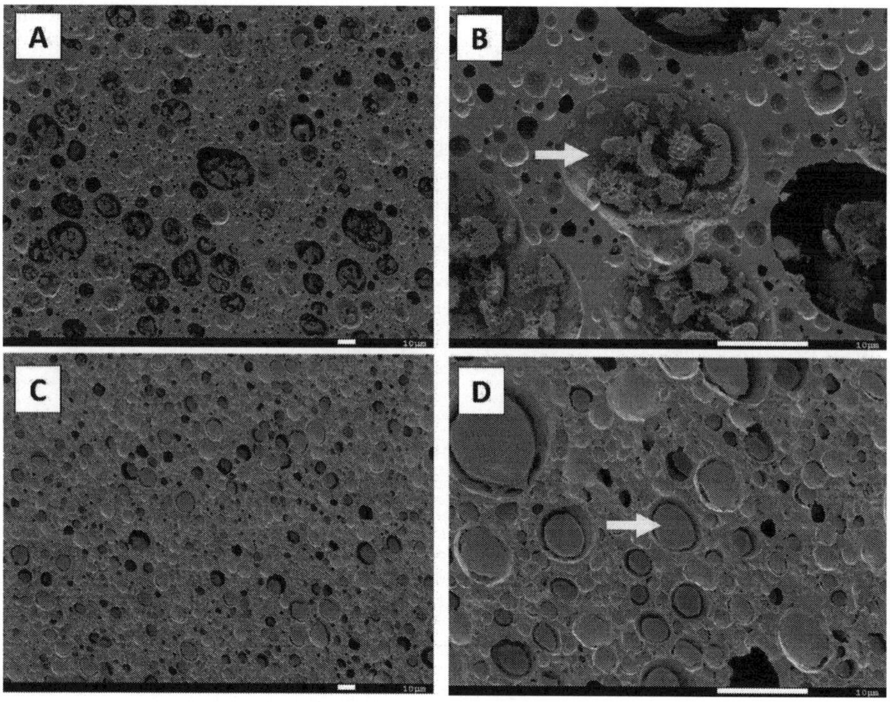

Figure 4.3.1 Cryo-SEM images of commercial fat spreads (A-B: 38% fat, no hydrocolloids in the water phase; C-D: 25% fat, potassium alginate as stabilizer in the water phase) obtained after cryo-planing the frozen sample on a cryo-ultramicrotome with image magnifications at 500x (A-C) and 2500x (B-D). Arrows indicate water droplet content after (partial) sublimation of the water (Courtesy of Vandemoortele Centre, Food Structure and Function Research Group, Ghent University).

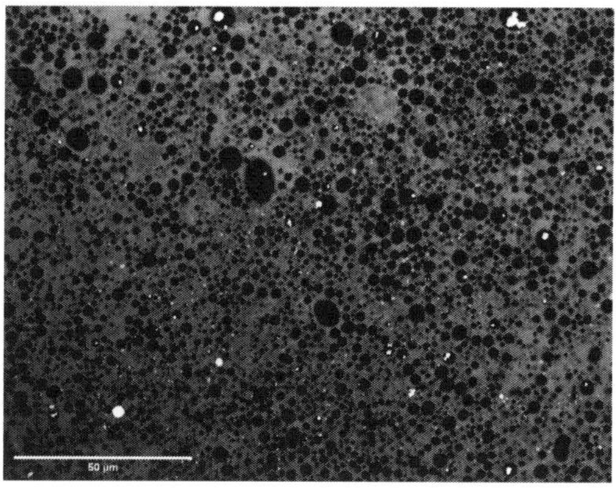

Figure 4.3.2 Microstructure of a commercial fat spread visualized by confocal laser scanning microscopy: gray = fat matrix (nile red dye, conventionally visualized in red), black: water phase (no dye), white: protein aggregates (FITC dye, conventionally visualized in green) (Courtesy of Vandemoortele Centre, Food Structure and Function Research Group, Ghent University).

a de-oiled margarine. A central void formed by the sublimation of a water droplet is visible. The water droplet is surrounded by a crystalline shell, which is further embedded in the aggregated crystal network. The primary crystal aggregates usually have a size in the order of 15–20 μm (Bockisch 1998). This network shows smaller voids from the liquid oil. The formation of the fat crystals at the interface may be mediated with a surfactant (*i.e.*, saturated/unsaturated mono- and diacylglycerols), which may act as a template that promotes the interfacial heterogeneous nucleation of

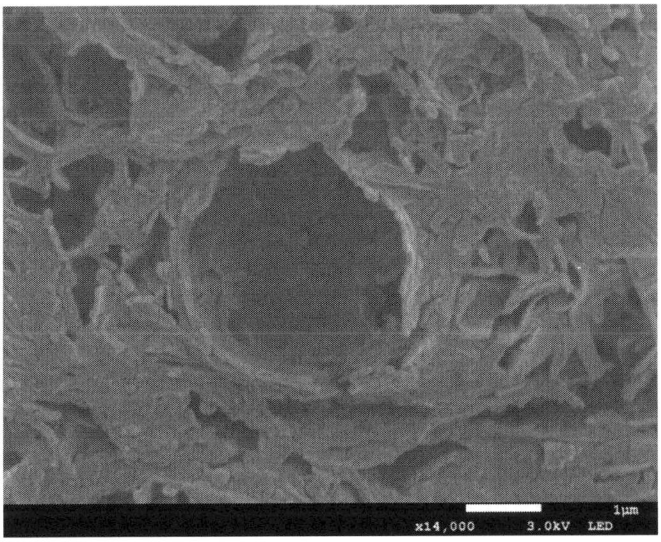

Figure 4.3.3 Microstructure of the fat crystal network in a margarine visualized by cryo-scanning electron microscopy (cryo-SEM). Sample obtained after removal of the liquid oil by solvent extraction, vitrification, fracturing and sublimation (Courtesy of Vandemoortele Centre, Food Structure and Function Research Group, Ghent University).

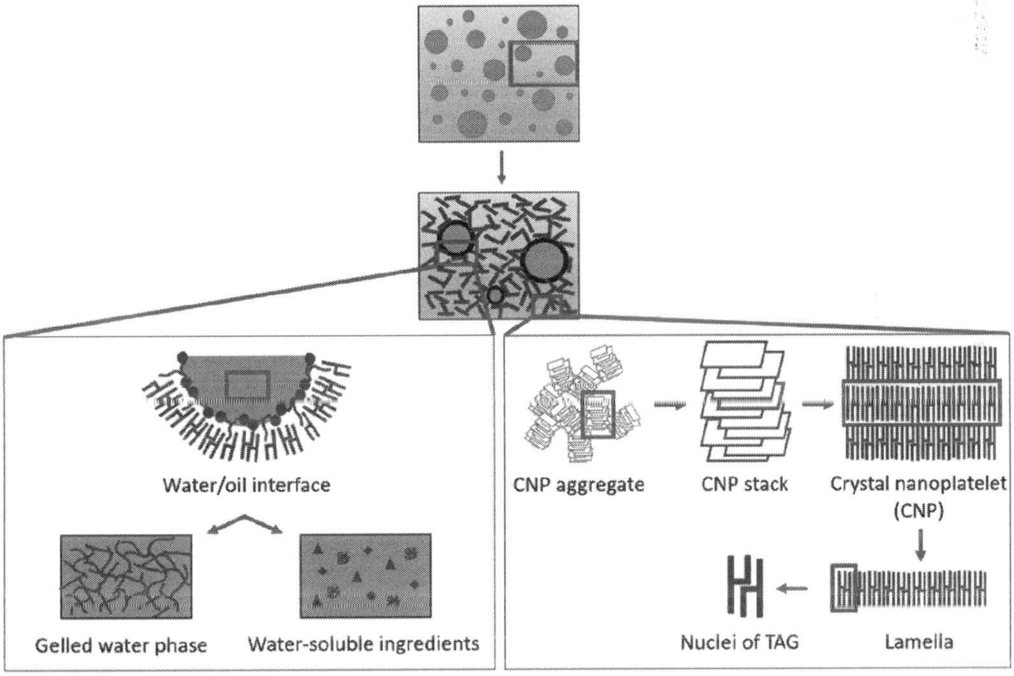

Figure 4.3.4 Schematic drawing of the microstructure of edible semi-solid W/O emulsions (Food Structure and Function Research Group, Ghent University).

the triglycerides (TAGs) present in the lipid phase (Douaire et al. 2014) resulting in the crystalline shell. Eventually, margarine draws its mechanical stability from the crystal network and from the stability of the emulsion. Taking all these components into account, a schematic drawing of the multiscale structure of a margarine/spread is presented in Figure 4.3.4.

4.3.1.3 Formulation

Margarines and fat spreads formulation allows for flexible ingredient and additive implementation, which leads to diverse types (differing in nutritional profile, texture, sensory properties, and techno-functional properties). Margarine has developed into a tailor-made fat product with adjustable properties to fit different industrial and household uses. As discussed, margarines and spreads are W/O emulsions consisting of a continuous lipid phase in which water droplets are dispersed. The lipid phase consists of a fat blend to which the emulsifier and lipophilic additives are added. The water phase has water-soluble ingredients and additives.

The choice of a margarine's fat blend is based on several criteria including the achievement of certain techno-functional properties, nutritional profile, availability, and cost. Taking those criteria into account, the composition of a fat blend can vary broadly and can vary over time within a brand. The blend composition for margarine has changed dramatically over the last century. When initially most margarine fat blends were of animal origin, today, animal fat has disappeared, and edible vegetable oils and fats are dominant in all countries. A fat blend with techno-functional properties is achieved by mixing appropriate (modified) hardstocks and liquid oils. Modification processes used to obtain suitable hardstocks include fractionation, (partial) hydrogenation, and interesterification. However, because of adverse nutritional impacts of trans fatty acids, partial hydrogenation has been largely abandoned as modification technology for margarine hardstock (Korver and Katan 2006). Nowadays, palm oil and its fractions are the most popular hard fats for producing shortenings and margarines. Owning a good balance between saturated and unsaturated fatty acids, this low-cost plant oil forms very stable tiny crystals in long-term storage (Nguyen et al. 2020). Further, some lipophilic additives may be included in the lipid phase such as emulsifiers (e.g., mono- and diglycerides, polyglycerolricinoleate (PGPR)), anti-spattering agents (especially in frying margarines, example lecithin), vitamins (A, D, E), and colorants (e.g., β-carotene).

The water phase, which comprises 10 to 75% of the composition, contains water-soluble ingredients such as milk components, organic acids, salt, flavors, preservatives, and stabilizers. Most margarines and fat spreads have milk solids, which can be understood historically from the attempt to imitate butter. Most milk components carry flavors such as traces of butyric acid, ketones, lactones, and diacetyl. Salt (NaCl, 0.2–3% depending on local habits) decreases microbiological sensitivity and adds flavor. Flavors enhance the taste or mask off-flavors. Organic acids like citric and lactic acid lower the pH value thus improving bacteriological stability and providing a better, fresher taste. In addition, their chelating properties allow for binding metals, such as complexed iron, to reduce the sensitivity to oil oxidation. Preservatives (free benzoic or free sorbic acid (0.05–0.1%)) are especially needed for reduced fat margarines to enhance the microbial shelf life and the safety of the product. Also thickening agents or stabilizers are mainly added to the water phase in low-fat spreads. They bind water or even gel the water phase thus stabilizing the emulsion (cfr. Figure 4.3.1C-D).

4.3.1.4 Processing

The modern production process of margarines/fat spreads is outlined in Figure 4.3.5. The process consists of a series of unit operations which create an emulsion that upon cooling is transformed into a semi-solid, plastic product. First, the lipid phase and the water phase are separately produced. The fat blend containing the hardstock fat and liquid oil is heated above its melting temperature, blended with liquid oil and lipophilic additives (e.g., colorant, antioxidant, and emulsifier) are added. The water phase, including the water-soluble ingredients, is prepared at a similar temperature. Next,

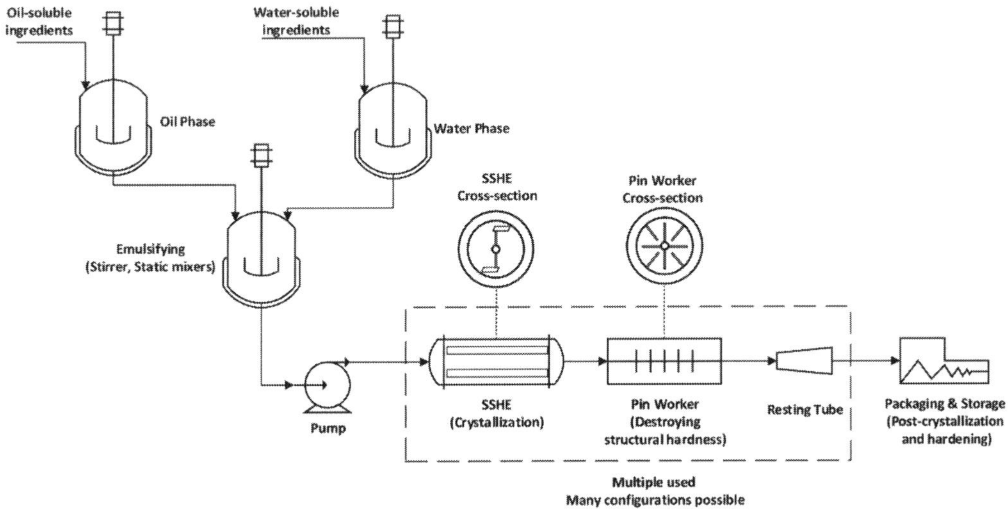

Figure 4.3.5 Schematic overview of the margarine and fat spread production process (Food Structure and Function Research Group, Ghent University).

both phases are blended to create the pre-emulsion. Then, the pre-emulsion is rapidly cooled in a scraped surface heat exchanger (SSHE) creating fat crystals. Processing factors affecting the crystallization process are flow rate, cooling profile and cooling rate, and shear intensity. The sub-cooling degree and high shear rates promote the formation of fat crystals in the α-polymorph, which will grow and transform into the β' polymorph. High shear rates (300–500 s^{-1}) have the effect of breaking the fat crystals into small debris and of mixing them into the bulk lipid phase, where they will act as seed crystals that induce the formation of new fat crystals (*i.e.*, through secondary nucleation) (Arellano et al. 2015). The shear forces also aid in further dispersing and homogenizing the water phase into small droplets. The chilled margarine may move to intermediate crystallizers or waiting tubes for further crystallization or may be mechanically kneaded in a pin-worker. This action will provoke partial fat crystal network disruption to obtain desired texture properties by partially breaking up the fat crystal network. In this step also the polymorphic transition of the fat crystals from α to β' will proceed. Small crystals are needed to obtain a homogeneous and coherent fat crystal network with ideal oil binding and in-mouth melting properties (Arellano et al. 2015). Finally, the margarine or fat spread is portioned or extruded in large blocks and during storage post-crystallization occurs. An optional tempering step may aid post-crystallization.

4.3.1.5 Properties

Semi-solid W/O emulsions are usually formulated and processed to achieve certain textural, sensorial, and shelf-life features. These features largely determine the multiscale properties of the fat crystal network, and are affected by the fat blend composition and its interaction with other parts (emulsifier and dispersed water droplets), processing, and storage conditions. Although many factors impact the fat crystal network properties, solid fat content (SFC) seems correlated with most of the macroscopic features (Arellano et al. 2015). The SFC denotes the quantity of solidified lipid phase at a particular temperature. Figure 4.3.6 illustrates the SFC profiles of three types of commercial

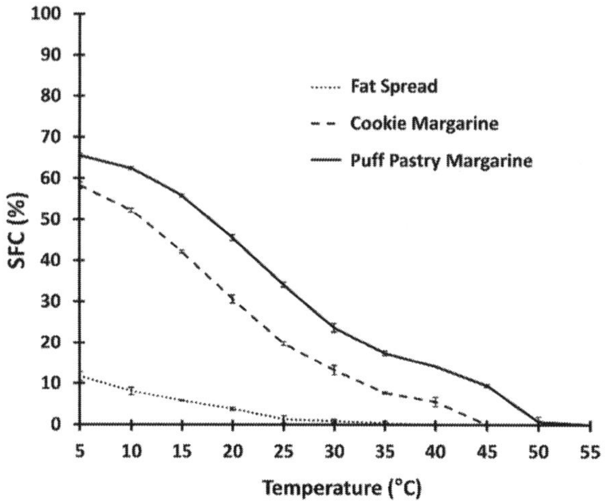

Figure 4.3.6 Solid fat content (SFC) profiles of three types of semi-solid W/O emulsions: low-fat spread, cookie, and puff pastry margarine (Food Structure and Function Research Group, Ghent University).

W/O emulsions. Fat spreads usually have a lower fat content with a limited amount of solid fat, which results in a soft, spreadable, fat product. A minimum amount of solid fat is necessary to prevent oiling off when it is at room temperature. At body temperature, the solid fat should be molten to prevent a waxy mouthfeel. Bakery margarines have much higher levels of solid fat. Cookie margarines should typically have a SFC of 25% at room temperature to aid in air incorporation and to prevent gluten agglomeration (shortening effect). The gradual decrease in SFC, upon being heated, controls the way a cookie spreads during baking (Sciarini et al. 2013). Puff pastry margarines contain the most solid fats at room temperature (SFC around 50%). These margarines should show plasticity when deformed. To allow the homogeneous incorporation in laminated dough systems, a constant consistency is desired with prolonged kneading.

The shelf life of W/O emulsions is limited due to physicochemical or microbial deterioration. For palm oil-based margarines, besides oxidation, the most prevalent physicochemical deterioration in storage is the appearance of granular crystals resulting in a graininess feeling for customers. According to many studies, granular crystals are clusters of fat crystals with a size in the range of 0.1–3 mm, which are visible in products stored at 5–10°C after 3–6 months. Hypotheses on the formation granular crystals have been reviewed by Nguyen et al. (2020). Although the water phase has been finely dispersed, microbial growth can still occur limiting the microbial shelf life. To prevent the growth of microorganisms, different strategies are usually combined such as lowering pH, adding synthetic preservatives, and reducing water droplet size. However, water droplet size also plays an important role in flavor release, so an optimal balance is required.

4.3.2 Alternative Structuring Approaches for Margarines and Fat Spreads

4.3.2.1 Stabilization Mechanisms

W/O emulsions comprise a continuous fat phase and a dispersed water phase, which generally coexist due to the presence of one or more stabilizing components. Since emulsions are not kinetically stable systems and separate over time, stabilization mechanisms slow down emulsion breakdown (caused by coalescence, flocculation, Ostwald ripening, or gravitational separation). Figure 4.3.7 presents an

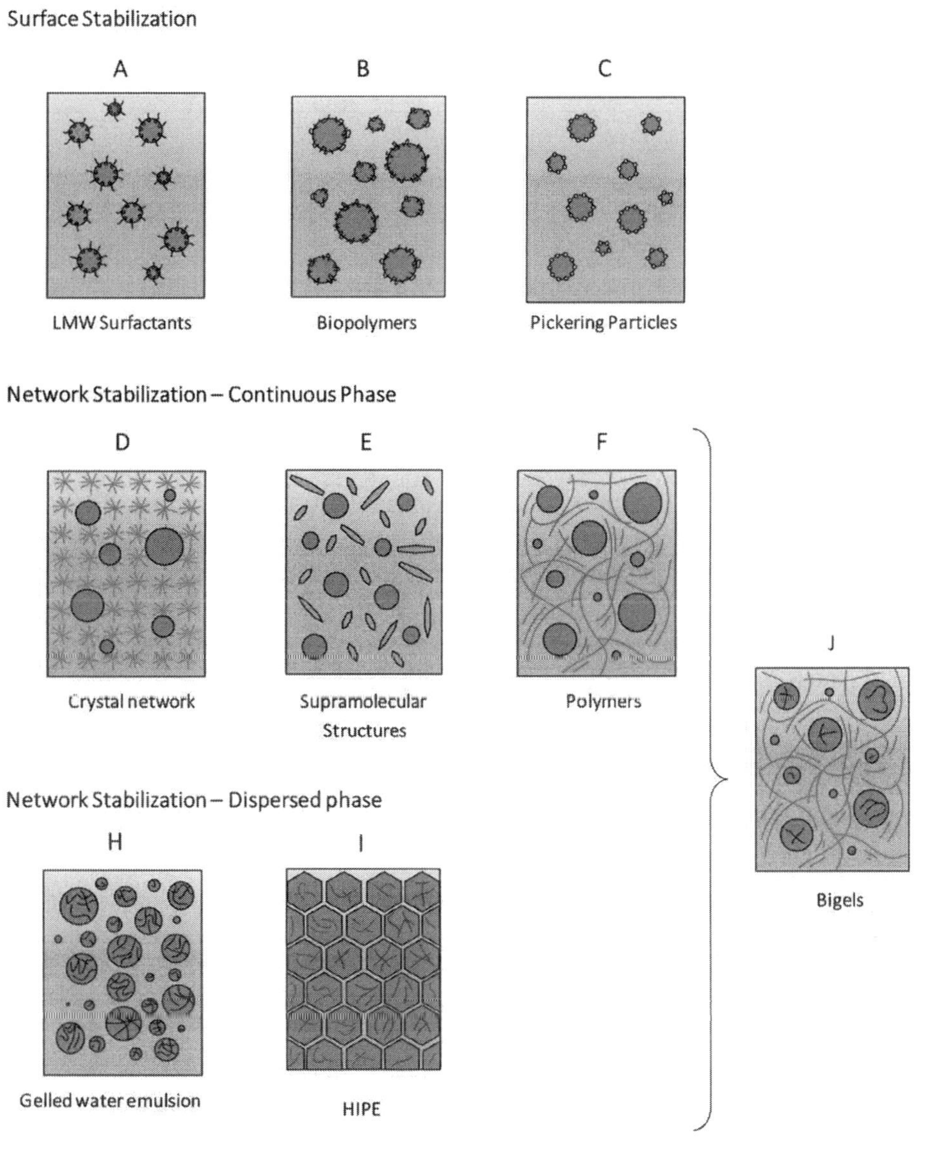

Figure 4.3.7 Schematic overview of possible stabilization mechanisms of W/O Emulsions (LMW = low molecular weight, HIPF = high internal phase emulsion) (Food Structure and Function Research Group, Ghent University).

overview of different stabilization mechanisms, which may be subdivided into two main subgroups: (1) surface stabilization and (2) network stabilization. Network stabilization mechanisms can be further divided according to the phase in which they act (continuous versus dispersed). Realistically, most food products have synergistic stabilization mechanisms, and often combine both surface stabi lization and network stabilization. For example, conventional fat spreads combine surfactants (at the interface) with saturated fats (network formation). Low caloric fat spreads are often gelled emulsions additionally stabilized with surfactants.

By definition, surface stabilization requires the presence of a stabilizing component on the W/O interface. Often, these components will be amphiphilic, so they can interact with both polar and non-polar phases. This is conventionally done with low molecular weight (LMW) emulsifiers (Figure 4.3.7A). Some examples of conventional emulsifiers used in margarine include partial glycerides (e.g., monoacylglycerol and diacylglycerol), lecithin (e.g., soy lecithin), fatty acid polyglycerol esters (e.g., polyglycerol polyricinoleate), sorbitan monostearate, etc.

Biopolymers, such as proteins or polysaccharides, can also stabilize surfaces. Their working mechanism is different. They form an interfacial film, leading to steric or electrostatic repulsion between droplets, thus conferring meta-stability (Figure 4.3.7B) (Berton-Carabin and Schroen 2015). For this, biopolymers can adhere directly to the two-phase boundary or can interact with another surfactant or biopolymer already in the interface. Polysaccharides, due to their non-surface-active properties, often need modifications or are part of a collaborative stabilization process. Proteins are usually amphiphilic and absorb directly into the interface. Although biopolymers are widely studied to stabilize oil-in-water emulsions (O/W), there are little to no examples of proteins or polysaccharides stabilizing W/O emulsions on their own (Zembyla et al. 2020). Combinations of biopolymers and surfactants are more common. A non-extensive list of proteins and polysaccharides used for margarines includes whey proteins, sodium caseinate, lactoferrins, β-globulins, high methoxyl or low methoxyl pectins, xanthan, gelatin, cellulose, κ-carrageenan, etc., which are found in synergy with a surfactant (e.g., PGPR and lecithin) (Zembyla et al. 2020).

Pickering emulsions are another surface stabilization mechanisms that has been of increasing interest over the last few decades (Dickinson 2011; Berton-Carabin and Schroen 2015). Researchers have been so enthusiastic about the topic that some suggest that “we have now entered the neo-Pickering era of colloid science” (Dickinson 2020). In these systems, the molecular emulsifier is replaced by a solid or semi-solid particle, which leads to emulsions with high stability against coalescence (Figure 4.3.7C). In principle, if the particles are subject to change by triggers (e.g., temperature, ionic strength, pH), it is also possible to alter the interface after emulsion formation, creating responsive interfacial properties (Xiao et al. 2016). As most research has been done on O/W emulsions, W/O emulsions have been neglected to a certain extent. Silica and latex are examples of non-bio-derived particles for W/O. Some bio-derived examples include cellulose, lignin, zein, starch, and polyphenols (Zembyla et al. 2020).

Additional surface stabilization systems have been reported, such as bijels (Clegg and Thijssen 2020), nano-emulsions (Akram et al. 2021), and multilayer emulsions (Guzey and McClements 2006). However, their application in the formulation of margarine is rather limited, and, therefore, a more detailed review remains out of the scope of this book chapter.

In contrast to surface stabilized emulsions, network stabilized systems do not rely on surface-active components to achieve meta-stability. Instead, they depend on the formation of sufficiently dense three-dimensional networks that prevent droplet motion and hinder destabilization. Traditionally, in margarine production, saturated fats are used as network stabilizers (Figure 4.3.7D) (Ghosh et al. 2011). Some examples include hydrogenated fats and stearines, which show little to no surface activity. In principle, when present in sufficiently large concentrations, they will provide kinetic stability by entrapping the dispersed phase in a continuous crystalline matrix.

Due to increasing concerns related to saturated fatty acid consumption, alternative network stabilization mechanisms have recently been explored. Studied alternatives may be classified into two main categories: oil structuring techniques and water structuring techniques. Oleogelation is the clearest example of an oil structuring technique (Figure 4.3.7E-F-G). During this process, a liquid oil is transformed into a solid-like material due to the presence of a certain gelling agent. Lipidic gelators are commonly used for this purpose. Examples include waxes, partial glycerides, phytosterols, and fatty alcohols and fatty acids. Although less common, non-lipidic gelators also exist—ethylcellulose is a well-known example (Figure 4.3.7F). These systems will be discussed in further detail in the following sections.

Within oleogelation, other indirect oil structuring techniques have been described. These include emulsion-templated oleogelation, foam-templated oleogelation, and solvent exchange of protein hydrogels (Patel et al. 2020). However, these techniques have rarely been used to produce W/O emulsions and will not be discussed further.

Water structuring techniques stabilize the dispersed water phase. Hydrocolloids are commonly used to gel the water phase, creating the so-called gelled emulsions (Figure 4.3.7G). Starch, high methoxyl or low methoxyl pectins, xanthan, gelatine, cellulose, κ-carrageenan are examples. When the volume fraction of the dispersed phase is above 0.74, a high internal phase emulsion (HIPE) is formed (Figure 4.3.7H). Given the high concentration of water, droplets are closely packed together and deform, separated by a thin film of continuous phase (Patel and Dewettinck 2015). Although gelled emulsions and HIPEs might not strictly be considered fat mimetics, they help reduce the total fat content in food products and thus, help tackle health concerns related to saturated fatty acid consumption. Finally, a synergy between a continuous phase stabilization and dispersed phase stabilization constitutes a bigel system (*i.e.*, two gels).

In the following section, some examples of fat mimetics will be grouped according to their stabilization mechanisms. In each section, the reader will be briefly introduced the fat mimetic and a general overview of the working principle. In every case, examples of W/O emulsions are presented. The reader should be mindful that the classification mechanisms are mere categorizations, which must be interpreted only as guidelines.

4.3.2.2 Network Stabilization – Continuous Phase

4.3.2.2.1 Waxes

Natural waxes have been intensively studied over the last 10 to 15 years as promising food-grade oleogelators. They are of major scientific interest due to their low required concentration, economic feasibility, and food-grade status. When combined with liquid oil, natural waxes behave as efficient gelling agents, as low concentrations are required to form stable oleogels. Although highly dependent on the wax, it has been reported that concentrations as low as 1–1.5% were sufficient to form oleogels of sunflower wax (SFW), candelilla wax (CLW), berry wax (BEW), and beeswax (BZW) (Doan et al. 2015). Since natural waxes are often obtained as by-products of existing agricultural processes, they are considered an economically interesting oil structuring alternative. Since some are already approved as food ingredients for certain applications, they are often considered a promising alternative to conventional saturated fats.

Recent developments in wax oleogelation led to these findings being applied toward formulating water-in-oil emulsions. Several authors reported successful stabilization of water-in-wax oleogel matrixes. As for the stabilization mechanism, depending on the wax, they can provide surface stabilization, continuous phase network stabilization, or a combination of both (Figure 4.3.8). Table 4.3.1 summarizes the most relevant studies on the topic for both model systems and margarine systems.

Although there are differences in the literature, preparing water-in-wax oleogel emulsions can be roughly described as a three-step process. First, the oil phase and the water phase are prepared separately. For this, the waxes are added into the fat phase and heated above the wax's melting

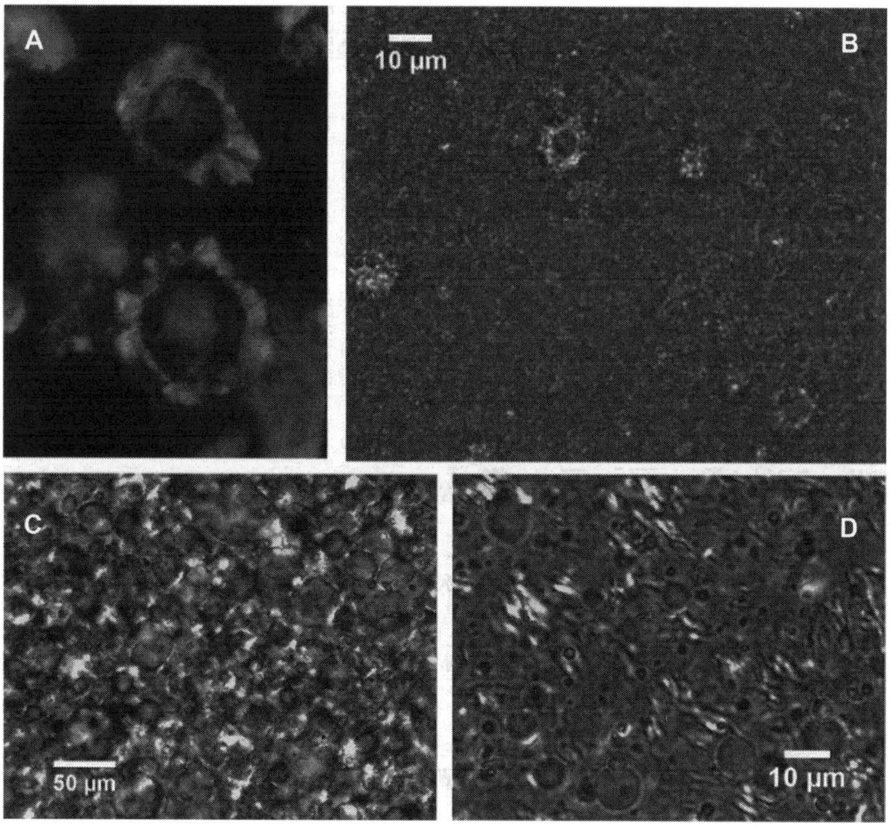

Figure 4.3.8 Polarized light microscopy images of W/O emulsions stabilized with (A) shellac wax, (B) carnauba wax, (C) sunflower wax and (D) beeswax. In shellac wax and carnauba wax, wax crystals were visualized in the W/O interface, displaying some affinity to the water phase. Sunflower wax and beeswax crystals act as network stabilizers by entrapping the water droplets on a three-dimensional network (Food Structure and Function Research Group, Ghent University). The system in A is described in a high level of detail by Patel et al. (2013) from Elsevier; B-D, courtesy of the Vandemoortele Centre.

temperature. If present, surfactants are included in that mixture. In parallel, the water phase, which includes all of the water-soluble ingredients, is heated to a similar temperature. The second step is an emulsification step in which water and oil are combined. In this step, energy is added to the system. This is often done by mixing (e.g., Ultra Turrax, homogenizer, mechanical stirring); however, high-pressure homogenization and high-intensity ultrasound are other alternatives (Toro-Vazquez et al. 2013; da Silva et al. 2019b). Finally, the emulsion may cool down. Fast cooling is most commonly reported as it enhances long-term emulsion stability (Binks and Rocher 2009).

In Table 4.3.1, studies are ordered by complexity. First, so-called "model systems" are presented. These usually include oil, wax, water and possibly, an emulsifier. BZW (Gao et al. 2021), CRW (Binks and Rocher 2009; Szumała and Luty 2016), and shellac wax (SHW) (Patel et al. 2013) are described first. What is interesting about these studies is that the resulting emulsions were stable in the absent surfactants (Figure 4.3.8A-B). Some authors suggest that wax crystals at the interface reduce diffusion or coalescence, granting stability to the system (Ghosh et al. 2011; Patel et al. 2013; Silva et al. 2021b). In contrast, emulsions formulated with CLW and rice bran wax (RBW) are often presented in cooperation

Table 4.3.1 Water-in-oil emulsions stabilized by natural waxes.

Reference	Formulation				Preparation	
	Wax	Fat Phase	Water Phase	Others	Mixing Protocol	Cooling Protocol
Model Systems						
(Gao et al. 2021)	BZW (1–6%)	Soybean oil	Water (10–55%)	–	Dropwise inclusion with magnetic stirring (1,100 rpm, 5 h)	–
(Szumała and Luty 2016)	CRW (5–30%)	Paraffin oil	Water (10–60%)	–	Mechanic Stirring (500 rpm)	Stirring until reached 25°C
(Binks and Rocher 2009)	CRW (0–15%)	Squalene	Water (50%)	–	Hand shaking (2 min) or Ultra Turrax (11,000 rpm, 30 s)	Different protocols tested
(Patel et al. 2013)	SHW (1–6%)	Rapeseed oil	Water (20%)	–	Ultra Turrax (11,000 rpm, 5 min)	Room temperature
(Wijarnprecha et al. 2019)	RBW (3.75–4.6%).	Canola Oil	Water (7.5–25%)	GMS (0.75–2.50%)	Ultra Turrax (9,000 rpm, 3 min)	Magnetic stirring (300 rpm) in ice bath
(Toro-Vazquez et al. 2013)	CLW (0.5–4%)	Safflower oil high in triolein	Water (20%)	MAG (0–0.5%)	High-Pressure Homogenizer (200 bars, 65°C, 3 passes)	Coper coil bypass immersed in water bath (31–33°C)
(da Silva et al. 2019b)	CLW (2%)	HOSO, Hard fat (2%)	Water (5–25%)	MAG (2%)	HIU (diameter tip 3.2 mm, amplitude 216 μm, 3 min) or Ultra Turrax (3 min, 10,000 rpm).	Stored at 5°C
Margarine Systems						
(Öğütcü et al. 2015)	BZW (3.5–4.5%)	Virgin olive oil	Water (8.7–24.2%)	Tween20 Tween80 Salt Citric Acid Xanthan Gum.	Ultra Turrax (40,000 rpm, 5 min)	Room Temperature
(Silva et al. 2021b)	CLW (1.05–1.68%)	Soybean oil, FHPO (hydrogenated palm oil)	Water (39–64%)	MAG (0.99–1.40%) Salt (1.0%) Potassium Sorbate (0.05%)	Ultra Turrax (11,000 rpm, 1 min)	Incubator at 5°C

(Continued)

Table 4.3.1 (Continued)

Reference	Formulation				Preparation	
	Wax	Fat Phase	Water Phase	Others	Mixing Protocol	Cooling Protocol
Model Systems						
(da Silva et al. 2018a)	CLW (2.66%)	HOSO, Interesterified fat (4.91%)	Water (16.65%)	MAG (1.77%) Skim powdered milk (1.8%) Salt (1.5%) Beta-carotene Butter flavor Antioxidant Potassium sorbate	Laboratory scale jacketed reactor (dynamic crystallization)	Cooled jacketed reactor with an anchor blade surface scraper (60 rpm) until 30°C
(Hwang and Winkler-Moser 2020a)	CLW + BZW (3–7%)	Soybean oil	Skimmed milk (19.76%)	MAG/DAG (0.20%) Lecithin (0.10%) Salt (0.10%) Citric acid Calcium disodium EDTA Potassium sorbate	Mechanical Stirring (3,000 rpm, 5 minutes)	Ice bath 30 min
(Hwang et al. 2013)	CLW (1–6%) or RBW (1–6%) or SFW (1–6%)	Soybean oil	Skimmed milk (19.46%)	MAG/DAG (0.20%) Lecithin (0.10%) Salt (0.10%) Citric acid Calcium disodium EDTA Potassium sorbate	Homogenizer (3,000 rpm, 7 min)	Ice bath 30 min
(Merchan Sandoval et al. 2020)	SFW (2–5%)	HOSO	Water	Lecithin (0.12%) Sodium chloride (0.55%) Fat-free milk powder (11%) Calcium disodium EDTA Potassium sorbate	Homogenizer (12,000 rpm, 5 min)	Refrigeration 5°C

BZW: beeswax / CRW: carnauba wax / SHW: shellac wax / RBW: rice bran wax / CLW: candelilla wax / SFW: sunflower wax / HOSO: high oleic sunflower oil / GMS: glycerol monostearate / MAG: monoglycerides / DAG: diglycerides / EDTA: ethylenediaminetetraacetic acid.

with a surface-active compound. For example, GMS was used to prepare RBW W/O emulsions, resulting in the formation of crystalline shells around the water droplets (Wijarnprecha et al. 2019). CLW emulsions were developed using CLW-MAG oleogels. The resulting products had interesting intercrystal interactions (Toro-Vazquez et al. 2013), lower fat contents and greater health appeal (da Silva et al. 2019a).

Findings obtained from model systems are similar irrespective of the wax used. First, long-term stability of emulsions was obtained by several authors. For example, Patel et al. (2013) reported four months of stability for SHW emulsions, while Gao et al. (2021) and Wijarnprecha et al. (2019) reported two month stability for their BZW and RBW emulsions respectively. In terms of rheology, it is well described that increasing amounts of wax leads to an increase in the complex modulus (Patel et al. 2013; Toro-Vazquez et al. 2013; Wijarnprecha et al. 2019; Gao et al. 2021). The influence of the water phase was described as dominant, as viscosity, emulsion elasticity, and firmness were enhanced with increasing amounts of water (Wijarnprecha et al. 2019; Gao et al. 2021). Emulsions that were obtained were described as thermo-sensitive with excellent shear-dependent structure recovery (Patel et al. 2013; Gao et al. 2021).

In contrast to model systems, "margarine-like systems" include more ingredients. Publications in this category try to closely mimic commercial products, replacing the fat phase with a wax oleogel. Therefore, they include water, oil, waxes, and emulsifiers (MAGs, DAGs, lecithin), and some minor components such as food acidifiers (citric acid), preservatives (calcium disodium EDTA, potassium sorbate), and flavor ingredients (salt, milk powder, and butter flavor). Literature suggests that the resulting systems have great potential in terms of shelf life, appearance, and thermal stability. However, taste and firmness need improvement (da Silva et al. 2018a). In addition, since wax-based margarines have melting points higher than conventional margarine, sensory results often show an undesirable residual waxy taste (da Silva et al. 2018b). To overcome these challenges, Hwang and Winkler-Moser (2020b) proposed using eutectic properties. By mixing different waxes at specific concentrations, higher firmness and lower melting points were achieved (Hwang and Winkler-Moser 2020b). However, a deeper understanding on waxes' gelation mechanism is still necessary to elucidate wax interactions with the oil and water phase.

4.3.2.2.2 Mono- and Diglycerides

Mono- and diglycerides (MAGs and DAGs) are naturally present in fats and oils as minor components, or they can be derived from triglycerides (TAGs) (e.g., by hydrolyzation). Similar to the name of triacylglycerols, which specifies the presence of three fatty acid chains, MAGs and DAGs contain respectively one or two fatty acid chains esterified to a glycerol backbone. Some hydroxyl groups of the glycerol molecule remain unesterified, leading to better self-assembling properties compared to TAGs. Therefore, DAGs and MAGs are used as structuring lipids in shortenings, margarines, and emulsions (Pehlivanoğlu et al. 2018).

Partial glycerides show good gelation properties in oil, even at low concentrations. When MAG are dissolved in oil, they can form an inverse lamellar phase when cooled below their gelling temperature. This structure is the same as the α-polymorph (hexagonal packing) and is driven by the arrangement of the glycerol molecules. Further cooling below the crystallization temperature leads to the transformation into orthorhombic packing of the sub-α crystalline phase. The lamellar and the sub-α crystalline phase are only metastable and transform into the more stable β-crystalline phase (Chen and Terentjev 2018; Bin Sintang et al. 2017a; Heymans et al. 2018). Contrarily to MAGs, DAGs can occur in two isomers, namely 1,2-DAG and 1,3-DAG. The former one can crystallize in either the α or β' polymorph while the latter one only crystallizes in the β-polymorph (β1, β2) (Tavernier et al. 2019). Some authors suggest that the use of DAGs as oleogelator may provide more health benefits (Lai 2019; Lee et al. 2020). DAGs might be able to reduce the postprandial fatty acids levels, suppress fat accumulation, and decrease the cholesterol levels (total cholesterol and low density lipoprotein (LDL)-cholesterol). Therefore, DAGs might positively affect human health by acting as an anti-obesity agent (Lee et al. 2020). However, most attention on partial glycerides has been toward MAGs.

The behavior of MAGs changes when water is added to the MAG/oil system. Due to their amphiphilic nature, MAGs migrate to the interface of W/O emulsions. MAGs act as emulsifiers and/or crystallization templates (Tavernier et al. 2019). MAGs are also known to form a liquid crystalline lamellar phase (Lα) when concentrated in water at high temperatures. Upon cooling, the mobility of the surfactant is reduced, resulting in an α-gel with hexagonal packing. The dense packing, although still lamellar, gives short-distance ordering leading to a gel-like structure. This metastable structure transforms into an anhydrous β-crystalline form. This is often called a coagel (Chen and Terentjev 2018).

Altogether, they are versatile ingredients that can result or contribute to different microstructures. In this respect, it is challenging to use MAGs or DAGs oleogel emulsions to obtain a margarine-like product. However, different approaches are used in literature. Contreras-Ramírez et al. (2020) investigated W/O emulsions with MAG (10% w/w) and polyglycerol polyricinoleate (PGPR, 1% w/w) in canola oil and coconut oil. They found that the structure of the emulsions changed as a function of the storage time because of MAG-oil interactions; although, no phase separation occurred. Adding PGPR resulted in a more organized structure. The combination of MAG and PGPR at the interface leads to an increase in electrostatic repulsions so the water droplets are better dispersed. This might prevent the release of water during polymorphic transitions. Contrarily, prolonged storage resulted in the separation of these aggregates into crystalline fibrillar structures (Contreras-Ramírez et al. 2020).

Recently, a new approach to structure oils with partial glycerides has been introduced by Nicholson and Marangoni (2021). This approach is based on glycerolysis of edible liquid oil so partial acylglycerols are produced. Under the conditions used (2% w/w enzyme in different oils), the final mono-, di- and triglyceride concentrations were respectively around 31–35%, 42–50%, and 17–22%, depending on the oil. Hereby, MAGs and DAGs influence the crystallization and melting behavior of the oils. Next to the partial acylglycerols, also small amounts of free fatty acids were present (<2%). This was only measured in peanut oil. Different oils were investigated, and it was found that the onset of crystallization was 20°C higher after the glycerolysis reaction for each oil. The glycerolysis-structured oils could be used to produce margarines. After glycerolysis, tigernut oil was mixed with an aqueous phase (17.1% w/w). Within this aqueous phase, sodium chloride and potassium sorbate were dissolved. This margarine showed minimal phase separation after a storage time of 10 months (5°C), indicating high stability. Spreadability was also analyzed. The glycerolysis-based margarines had a smooth force-deformation curve similar to commercial margarine and butter. Based on these results, using this technique might be useful to produce margarines. However, no sensory evaluation was performed. Contrarily to the margarines produced with hydrogenated oil, this approach can provide structure to edible oils without changing the fatty acid composition. Additionally, DAGs might bring more health benefits (Nicholson and Marangoni 2021).

A third approach is the formulation of a novel O/W emulsions based on partial glycerides. Although microstructurally different from conventional margarines, these systems have the potential to be used as spreadable fats and shortenings (Batte et al. 2007). In these structured O/W emulsions, MAG and co-surfactants form a gel network in which the oil droplets are stabilized by MAG walls. These walls have alternating bilayers of MAG and water (Calligaris et al. 2021). Batte et al. (2007) investigated O/W emulsions with a 4% dimodan/stearic acid mixture (1 g/50 mg; 90% stearic acid, 10% palmitic acid monoglycerides) in canola oil emulsions. The emulsions were produced by heating and mixing the water phase (40%) and the fat phase, followed by a cooling step. They found that the size of the oil droplets was largely influenced by the homogenization rate, the homogenization temperature, and the cooling rate. Adding salt resulted in network breakdown. The added sodium molecules interacted with the co-surfactant (stearic acid) and prevented hydration. This is followed by the collection of MAG, which is macroscopically visible as curdling (Batte et al. 2007). Therefore, it is necessary to have proper hydration.

The effect of the water content on the hydration of structured O/W emulsions was investigated by Calligaris et al. (2021). They found that the water content should at least be 30% (w/w) to ensure proper hydration (Calligaris et al. 2021). Still, the water content of margarines is typically lower, namely 20% (w/w). To mimic the water content of margarines while maintaining stability, a drying step was performed. The emulsions were spread in thin layers and partially dried to achieve the desired 20% water content. The structure of the (partially) dried emulsions behaved more solid-like compared to the initial emulsions. Also, it was impossible to produce emulsions with a lower water content. A certain amount of water was essential for the MAG to create a network that could entrap both the liquid oil and the water (Calligaris et al. 2021).

4.3.2.2.3 Ethylcellulose

Although polysaccharides are mostly used to structure the water phase, some can also gel the lipid phase. Ethylcellulose (EC) is a well-known example. This linear polysaccharide is derived from cellulose. Through chemical modifications, cellulose's hydroxyl-end-groups are replaced with ethyl-end-groups (Zetzl and Marangoni 2012). The resulting polymer is compatible with organic material and is highly flexible, thermoplastic, transparent, non-caloric, tasteless, odorless, inert, and has considerable mechanical strength (Gravelle et al. 2018). Interestingly, it is also approved for use in food (Zetzl and Marangoni 2012). To produce an EC oleogel, oil and EC are first mixed and heated above the glass transition temperature (≈140°C) until the solution clarifies. The mixture is then cooled statically, allowing a physically entangled polymer network to form (Davidovich-Pinhas et al. 2015; Patel and Dewettinck 2016). Because of the high temperatures needed for oleogelation, lipid oxidation is often a concern. In addition, variations in the heating procedure can have a big impact on the mechanical strength of the resulting gel (Gravelle et al. 2012).

Studies of W/O emulsions using EC are rather limited. In 2003, Melzer et al. published their findings on EC W/O emulsions for pharmaceutical applications. Surfactant-free systems made of water, EC and 2-octyldodecanol were investigated. The authors suggest that the EC acts as a polymeric emulsifier and precipitates sediments in the interface of the droplet (building a Pickering emulsion) (Melzer et al. 2003; Silva et al. 2021a). Another study reported the successful stabilization of propylene glycol in medium chain triglycerides and soybean oil with EC (Ceballos et al. 2014). Again, the results suggested that a co-surfactant was not needed to stabilize "cream-like emulsions" for at least a month. However, EC changes orientation with temperature (Silva et al. 2021a). For instance, at higher temperatures, EC is directed toward the oil phase, while at lower temperatures, it forms a film on the inner phase (Silva et al. 2021b). Thus, EC systems could phase inverse if the temperature is not monitored (Melzer et al. 2003). Recently, García-Ortega et al. (2021) published their findings on emulsions composed of water, high oleic safflower oil, EC, and GMS. In this case, GMS acted as a co-surfactant. The authors concluded that the water droplets acted as active fillers on the rheological properties of the system. In addition, the synergy between EC and GMS had a positive stabilizing effect. In summary, although EC W/O emulsions are regarded as novel systems with high potential for SAFA-free formulations, applied research is still lacking for margarine-like systems.

4.3.2.2.4 Phytosterols

Phytosterols occur in the cell membranes of plants and in as minor parts in vegetable oils (typically 0.1–1%) (Bot and Flöter 2018). The structure of phytosterols is similar compared to cholesterol. It consists of a steroid skeleton in which the A-ring is substituted with a hydroxy group and the D-ring with an aliphatic side chain (Bin Sintang et al. 2017a). Phytosterols are added to food products (e.g., vegetable oils and margarine) for their effect in lowering cholesterol levels (Law 2000). An intake of 2–3 g/day might result in a decrease of the LDL-cholesterol levels with 5–15% according to research (Trautwein et al. 2018). Moreover, phytosterols can be used in oleogels to simultaneously reduce the saturated fatty acid content and provide more health benefits (Matheson et al. 2018).

Phytosterols cannot entrap the liquid oil in a gel network, although, it is possible to structure the liquid oil when combined with other molecules (Moschakis et al. 2016). A combination of phytosterols with phytosterol esters (e.g., γ-oryzanol) results in the formation of unique structures able to entrap the liquid oil (Bot and Agterof 2006). Sitosterol and oryzanol will self-assemble in a helical way, so hollow tubules with a diameter around 10 nm are formed. The tubules result from hydrogen bonds between the hydroxyl group and the carbonyl group of respectively sitosterol and oryzanol. These tubules interact with each other to form a fibril network able to entrap the liquid oil. The liquid oil is present in the inside and the outside of the tubule (Matheson et al. 2018).

Several issues occur when water is introduced to these oleogels. Water prevents the fibrillar network from forming by interfering with the formation of hydrogen bonds. Water interacts with the hydroxyl group of sitosterol so the formation of hydrate crystals is promoted, and the gel network is broken down (Matheson et al. 2018). However, some studies report the possibility to form structured emulsions with a mixture of β-sitosterol and γ-oryzanol.

Bot et al. (2011) illustrated the effect of total sterol concentration on the formation of hydrate crystals. Emulsions were prepared with 16% and 32% total sterols (40% β-sitosterol, 60% γ-oryzanol, w/w) in sunflower oil. They found that a low total sterol concentration of 16% resulted in the presence of sitosterol and oryzanol hydrate crystals. On the other hand, emulsions prepared with a high total sterol concentration of 32% were structured with tubules. Their structure was similar compared to the water-free oleogels, however, not identical. Additionally, some sitosterol and oryzanol hydrate crystals were also found (Bot et al. 2011).

To prevent the formation of monohydrate crystals, different approaches can be used. First, the mixtures can be heated above 45°C to induce a transition from the monohydrate to the hemihydrate form (Bot et al. 2011). Second, reducing the water activity by adding salts or sugars can be a useful approach. According to research, the water activity should be below 0.9 to prevent the formation of monohydrates (Sawalha et al. 2012). The tubules in the emulsion prepared with a high salt concentration (>12%) were similar compared to these in the water-free oleogel. Also, the firmness of the emulsion with a salt concentration of 20% was three times higher compared to the emulsion without added salt (Sawalha et al. 2012). A third approach includes using oils with a lower polarity compared to sunflower oil (e.g., decane and limonene). It was found that the low polarity oils showed similar results as the water-free oleogels and the structured emulsion with a salt concentration of 20%. This could be due to the reduced water solubility resulting in a reduced water transport so the formation of monohydrates will be hindered or delayed. Another possibility is the stronger oil-sitosterol interaction, so the formation of monohydrates is energetically less favorable (Sawalha et al. 2012, 2021).

Adding salts or sugar can also have more benefits related to their stability upon storage. Without added salts or sugars, the shape of the water droplets and the structure of the network changed after one week of storage at 5°C. After one week, the tubules were still present, however, also larger structures were detected which might lead to instability (Bot et al. 2011). After a storage period of 29 weeks, the tubular structures were decomposed. Still, the addition of salt in the aqueous phase prolonged the stability of the tubules. A salt concentration of 5% resulted in the presence of tubules until 50 weeks after production. Further decreasing the water content below 0.9 was necessary to maintain the structure up 50 weeks (Sawalha et al. 2012).

4.3.2.2.5 Lecithin

Lecithin is obtained during the refining process of liquid oil, more precisely during the degumming step. Therefore, the origin and the degumming method influences the composition of lecithin. Generally, lecithin is a mixture containing different naturally occurring glycolipids and phospholipids with minor parts, such as triglycerides, free fatty acids, and carbohydrates. The main components are phospholipids which have fatty acids and a (polar) group esterified to a glycerol backbone. Soy phospholipids are associated with the presence of palmitic, stearic, oleic, and linoleic acids (Bodennec et al. 2016; Xie et al. 2021).

Due to their amphiphilic nature, they can be used as surface-active agents in a variety of applications. However, their complex composition might result in a limited functionality (Okuro et al. 2019). Some phospholipids are better in stabilizing W/O emulsions (phosphatidylethanolamine, phosphatidylinositol) while others are better for O/W emulsions (phosphatidylcholine). To improve their functionality, other conditions such as pH, water content, and salt concentration should be chosen carefully. Lecithin can be changed by extracting the desired parts (Joshi et al. 2006).

Lecithin alone has a weak gelling capacity when added to liquid oil. The network comprises reversed hexagonal arrays (H_{II}) that are formed by the lecithin's self-assembling. These arrays further align into cylinders that interact with each other to form a fibrous gel network. Because of its weak gelling capacity, lecithin is mostly used in combined systems, for example with vitamin C (Hashizaki et al. 2012), α-tocopherol (Nikiforidis and Scholten 2014), sorbitan tristearate (Pernetti et al. 2007), and sucrose esters (Bin Sintang et al. 2017b). It was found that adding lecithin resulted in changes in the self-assembly route of other oleogelators so lecithin might act as a crystal morphology modifier (Bin Sintang et al. 2017b; Gaudino et al. 2019).

Adding water resulted in a more structured crystal network, however, above a critical water concentration of 2% w/w, phase separation occurred (Bodennec et al. 2016). Therefore, other structurants should be added to increase the water content. Gaudino et al. (2019) investigated the microstructure and the stability of oleogels and oleogel emulsions produced with lecithin and stearic acid in canola oil. They concluded that the crystal network's formation was different between the oleogel and the oleogel emulsion. The oil entrapment in the oleogel was achieved using reverse micelles-based fibers formed by lecithin in which stearic acid interacts with the non-polar tails and the matrix itself. Contrarily, the oleogel emulsions were primary structured through stearic acid crystals and spherulites. These crystals stabilize the emulsion because of their interaction with water at the interface. Higher stearic acid concentrations led to an increased hardness and a higher oil binding capacity (Gaudino et al. 2019). Another approach is combining lecithin with polyglycerol polyricinoleate (PGPR). This HIPE is characterized by tightly packed droplets which increases the overall viscosity. A ratio of 0.5:1.5 and 1:1 lecithin:PGPR was most stable (Okuro et al. 2019).

4.3.2.3 Network Stabilization – Dispersed Phase

Low-fat spreads are often produced by gelling the dispersed water phase instead of the continuous fat phase. Carbohydrates, polysaccharides, and proteins have been long used as water structuring ingredients in the production of low fat spreads (Figure 4.3.9). As lower amounts of fat are used, the influence

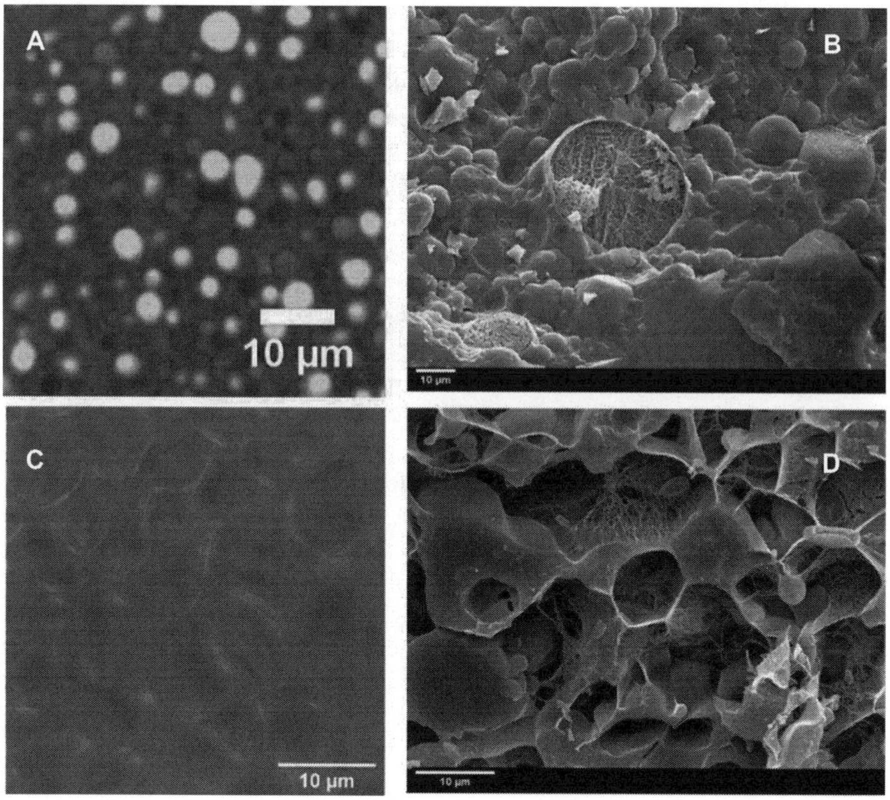

Figure 4.3.9 Examples of dispersed phase networks stabilized emulsions. (A) Confocal laser scanning microscopy, (B) cryo-scanning electron microscopy and (C) optical microscopy of a commercial low-fat spread. (D) Cryo-scanning electron microscopy of an 80% W/O HIPE stabilized by locust bean gum and carrageenan. In (A) and (B), it is possible to see the protein network formed in the dispersed phase. In (C) and (D), the microstructure of HIPE is visible (Courtesy of Vandemoortele Centre, Food Structure and Function Research Group, Ghent University). The system in D is described in a high level of detail by Patel and Dewettinck (2015), from John Wiley & Sons.

of the aqueous phase becomes increasingly important (e.g., HIPEs), leading to poor sensory attributes and shorter shelf-life. That means hydrocolloids are often included to enhance flavor release and stability (inhibit water droplet aggregation) (Clegg et al. 1996). In commercial low-fat spreads, starch and alginates are commonly used, but maltodextrins, caseinates, gums, and carrageenan have also been studied (Clegg et al. 1996; Iqbal et al. 2019; Nasirpour-Tabrizi et al. 2020; Pan et al. 2021).

Proteins are another water phase gelling alternative. Protein functionality depends on their composition and external factors. Generally, these are predominantly hydrophilic, although, some amino acids have hydrophobic side chains (Scharfe and Flöter 2020). Because of this, proteins can self-assemble and form stable colloidal structures in the water phase. Whey protein isolates, soy protein isolates, gelatin, and caseins can gel the water phase (Nascimento et al. 2020; Tomadoni et al. 2020).

Finally, a combination of polysaccharides and proteins can formulate hydrogels. This might lead to better properties due to electrostatic interactions between both parts. However, it is essential that the polysaccharides and the proteins have opposite charges (Tomadoni et al. 2020). Examples are short linear glucan and gelatin (Ge et al. 2018), xanthan gum or carrageenan, and soy protein (Hu et al. 2017).

4.3.2.4 Bigels

Bigels, as the name suggests, are bi-component gel systems. These are often uniform, semi-solid dispersions created by combining two gel phases. Eventually, the resulting system is visually seen as a single gel. The main advantage of bigels over emulsions and emulsion gels is their increased stability, which is attributed to the immobilization of the two mobile phases (Singh et al. 2018). Rheological properties might be optimized in bigels by changing composition and mixing speed (Shakeel et al. 2021). There are different types of bigels depending on the two phases. However, for the scope of this book chapter, only hydrogel-in-oleogel bigels will be reviewed.

Hydrogel-in-oleogel systems are formed by dispersing a structured water phase into an oleogel. Table 4.3.2 summarizes some relevant studies on the topic.

Findings suggest that the synergistic interactions between different gelators help optimizing the rheological properties of the resulting systems. It is commonly reported that moduli values (G′ and G″) are higher multi-component organogels, as compared to mono-component systems (Shakeel et al. 2021). Long-term physical stability and high thermal stability are some of the key benefits of bigel systems (Sagiri et al. 2015; Zheng et al. 2020).

Table 4.3.2 Overview of bigels (hydrogel-in-oleogel) systems.

Reference	Oil Phase	Oil Phase Gelator	Water Phase Gelator	Preparation Method
(Zheng et al. 2020)	Corn Oil	Monoglycerides (20%)	κ-carrageenan (1.5%)	High speed blender (300 rpm, 20 min + 500 rpm, 2 min)
(Martins et al. 2019)	Medium Chain Triglycerides	Beeswax (3–6%)	sodium alginate (2%)	Mechanical Stirring (600 rpm, 45 min)
(Paul et al. 2018)	Rice bran oil	Stearic acid (20%)	Tamarind gum (10%)	Mechanical Stirring (200 rpm)
(Kodela et al. 2017)	Rice bran oil	Stearyl alcohol (12%)	Agar (1%)	Mechanical stirring (200 rpm, 5 min)
(Bollom et al. 2020)	Soy bean oil	Soy lecithin + Stearic acid (20%)	Whey protein (15–25%)	Ultra Turrax (23,000 rpm, 3 min)
(Fasolin et al. 2021)	High oleic sunflower oil	Glycerol monostearate (5–15%)	Gellan gum (1–1.5%)	Mechanical stirring (500–1500 min-1)
(Zhu et al. 2021)	Soybean oil	Glycerol monostearate (1%) + Beeswax (1%)	Gellan gum (0.3%)	Ultra Turrax (10,000 rpm, 1 min)

4.3.3 Conclusion

It can be concluded that different gelling agents can produce margarine-like systems. Model systems for margarine consist of a continuous fat phase and a dispersed water phase (known as a W/O emulsions). These W/O emulsions are formed by solidifying either the fat phase, the water phase, or a combination of both. Most research has been conducted on gelling agents that are able to stabilize the fat continuous network. However, not all of them are fully characterized. The main difficulty is the way their behavior changes in the presence of the water phase. Many gelling agents (e.g., MAG, phytosterols, lecithin) are sensitive to the water content and require proper hydration to form a stable and structured emulsion. Therefore, these systems mostly include co-surfactants and/or other gelators to improve the stability. Comprehensive applied research for margarine-systems is still lacking. Next to the network stabilization, stabilization of the dispersed water phase is often used in low-fat spreads. Within these systems, carbohydrates, polysaccharides, and proteins are successfully used to structure the water phase. Finally, research regarding bi-component gels or bigels showed they have a lot of potential for further application.

Acknowledgments

Vandemoortele Lipids NV is acknowledged for financially supporting the UGent Vandemoortele Centre on “Lipid Science and Technology”. The Hercules foundation is recognized for its financial support in the acquisition of the scanning electron microscope JEOL JSM-7100F equipped with cryo-transfer system Quorum PP3000T (grant number AUGE-09–029). We thank Benny Lewille for the support during the acquisition of cryo-SEM images. The Research Foundation – Flanders (FWO) is acknowledged for providing the SB grant 1SA5321N.

Conflict of Interest Statement

Nothing was declared.

References

Akram, S., Anton, N., Omran, Z., and Vandamme, T. (2021). Water-in-oil nano-emulsions prepared by spontaneous emulsification: New insights on the formulation process. *Pharmaceutics* 13.

Arellano, M., Norton, I.T., and Smith, P. (2015). 10 – specialty oils and fats in margarines and low-fat spreads. In: *Specialty Oils and Fats in Food and Nutrition* (ed. G. Talbot). Woodhead Publishing.

Batte, H.D., Wright, A.J., Rush, J.W., Idziak, S.H.J., and Marangoni, A.G. (2007). Effect of processing conditions on the structure of monostearin–oil–water gels. *Food Research International* 40: 982–988.

Berton-Carabin, C.C. and Schroen, K. (2015). Pickering emulsions for food applications: Background, trends, and challenges. *Annual Review of Food Science and Technology* 6: 263–297.

Binks, B.P. and Rocher, A. (2009). Effects of temperature on water-in-oil emulsions stabilised solely by wax microparticles. *Journal of Colloid and Interface Science* 335: 94–104.

Bin Sintang, M.D., Danthine, S., Brown, A., Van de Walle, D., Patel, A.R., Tavernier, I, Rimaux, T., and Dewettinck, K. (2017a). Phytosterols-induced viscoelasticity of oleogels prepared by using monoglycerides. *Food Research International* 100: 832–840.

Bin Sintang, M.D., Danthine, S., Patel, A.R., Rimaux, T., Van de Walle, D., and Dewettinck, K. (2017b). Mixed surfactant systems of sucrose esters and lecithin as a synergistic approach for oil structuring. *Journal of Colloid and Interface Science* 504: 387–396.

Blackmore, M. (1960). *The Story of Margarine*. London: Unilever Limited.

Bockisch, M. (1998). Fat as or in food. In: *Fats and Oils Handbook* (ed. M. Bockisch). Urbana: AOCS Press.

Bodennec, M., Guo, Q., and Rousseau, D. (2016). Molecular and microstructural characterization of lecithin-based oleogels made with vegetable oil. *RSC Advances* 6: 47373–47381.

Bollom, M.A., Clark, S., and Acevedo, N.C. (2020). Development and characterization of a novel soy lecithin-stearic acid and whey protein concentrate bigel system for potential edible applications. *Food Hydrocolloids* 101.

Bot, A. and Agterof, W.G.M. (2006). Structuring of edible oils by mixtures of γ-oryzanol with β-sitosterol or related phytosterols. *Journal of the American Oil Chemists' Society* 83: 513–521.

Bot, A., Den Adel, R., Regkos, C., Sawalha, H., Venema, P., and Flöter, E. (2011). Structuring in β-sitosterol+γ-oryzanol-based emulsion gels during various stages of a temperature cycle. *Food Hydrocolloids* 25: 639–646.

Bot, A. and Flöter, E. (2018). Chapter 2 – Edible oil oleogels based on self-assembled β-sitosterol+γ-oryzanol tubules. In: *Edible Oleogels*, 2e (ed. A.G. Marangoni and N. Garti). AOCS Press.

Calligaris, S., Plazzotta, S., Barba, L., and Manzocco, L. (2021). Design of roll in margarine analogous by partial drying of monoglyceride-structured emulsions. *European Journal of Lipid Science and Technology* 123: 2000206.

Ceballos, M.R., Brailovsky, V., Bierbrauer, K.L., Cuffini, S.L., Beltramo, D.M., and Bianco, I.D. (2014). Effect of ethylcellulose on the structure and stability of non-aqueous oil based propylene glycol emulsions. *Food Research International* 62: 416–423.

Chen, C.-H. and Terentjev, E.M. (2018). Chapter 5 – monoglycerides in oils. In: *Edible Oleogels*, 2e (ed. A.G. Marangoni and N. Garti). AOCS Press.

Clegg, P.S. and Thijssen, J.H.J. (2020). Introduction to bijels. In: *Bijels: Bicontinuous Particle-stabilized Emulsions* (ed. P.S. Clegg). The Royal Society of Chemistry.

Clegg, S.M., Moore, A.K., and Jones, S.A. (1996). Low-fat margarine spreads as affected by aqueous phase hydrocolloids. *Journal of Food Science* 61: 1073–1079.

Code of Federal Regulations (2021). *Title 21 food and drugs – Chapter 1 – Subchapter B – Part 166: margarine [Online]*. https://www.ecfr.gov/current/title-21/chapter-I/subchapter-B/part-166 (accessed 25 November 2021).

Contreras-Ramírez, J.I., Gallegos-Infante, J.A., Pérez-Martínez, J.D., Dibildox-Alvaredo, E., Rocha-Guzmán, N.E., Moreno-Jiménez, M.R., González-Laredo, R.F., and Rosas-Flores, W. (2020). Influence of vegetable oil, monoglycerides and polyglycerol polyricinoleate into the physical stability of organogel-emulsion (w/o) systems. *SN Applied Sciences* 2: 1343.

Council of the European Union (2007). *Establishing a common organisation of agricultural markets and on specific provisions for certain agricultural products*. [Online]. *Official Journal of the European Union*. https://eur-lex.europa.eu/legal-content/EN/TXT/HTML/?uri=CELEX:32007R1234&from=EN (accessed 25 November 2021).

da Silva, T.L.T., Arellano, D.B., and Martini, S. (2019a). Interactions between candelilla wax and saturated triacylglycerols in oleogels. *Food Research International* 121: 900–909.

da Silva, T.L.T., Arellano, D.B., and Martini, S. (2019b). Use of high-intensity ultrasound to change the physical properties of oleogels and emulsion gels. *Journal of the American Oil Chemists' Society* 96: 681–691.

da Silva, T.L.T., Chaves, K.F., Fernandes, G.D., Rodrigues, J.B., Bolini, H.M.A., and Arellano, D.B. (2018a). Sensory and technological evaluation of margarines with reduced saturated fatty acid contents using oleogel technology. *JAOCS, Journal of the American Oil Chemists' Society* 95: 673–685.

da Silva, T.L.T., Chaves, K.F., Fernandes, G.D., Rodrigues, J.B., Bolini, H.M.A., and Arellano, D.B (2018b). Sensory and technological evaluation of margarines with reduced saturated fatty acid contents using oleogel technology. *Journal of the American Oil Chemists' Society* 95: 673–685.

Davidovich-Pinhas, M., Barbut, S., and Marangoni, A.G. (2015). The gelation of oil using ethyl cellulose. *Carbohydrate Polymers* 117: 869–878.

Deelstra, H.A., Burns, D.T., and Walker, M.J. (2014). The adulteration of food, lessons from the past, with reference to butter, margarine and fraud. *European Food Research and Technology* 239: 725–744.

Dickinson, E. (2011). Mixed biopolymers at interfaces: Competitive adsorption and multilayer structures. *Food Hydrocolloids* 25: 1966–1983.

Dickinson, E. (2020). Advances in food emulsions and foams: Reflections on research in the neo-Pickering era. *Current Opinion in Food Science* 33: 52–60.

Doan, C.D., van de Walle, D., Dewettinck, K., and Patel, A.R. (2015). Evaluating the oil-gelling properties of natural waxes in rice bran oil: Rheological, thermal, and microstructural study. *Journal of the American Oil Chemists' Society* 92: 801–811.

Douaire, M., Di Bari, V., Norton, J.E., Sullo, A., Lillford, P., and Norton, I.T. (2014). Fat crystallisation at oil–water interfaces. *Advances in Colloid and Interface Science* 203: 1–10.

FAO/WHO. 2001. *The codex alimentarius – volume 8: Fats, oils and related products [Online]*. Rome. https://www.fao.org/3/Y2774E/y2774e00.htm#Contents (accessed 25 November 2021).

Fasolin, L.H., Martins, A.J., Cerqueira, M.A., and Vicente, A.A. (2021). Modulating process parameters to change physical properties of bigels for food applications. *Food Structure* 28.

Frasch-Melnik, S., Norton, I.T., and Spyropoulos, F. (2010). Fat-crystal stabilised w/o emulsions for controlled salt release. *Journal of Food Engineering* 98: 437–442.

Gao, Y., Lei, Y., Wu, Y., Liang, H., Li, J., Pei, Y., Li, Y., Li, B., Luo, X., and Liu, S. (2021). Beeswax: a potential self-emulsifying agent for the construction of thermal-sensitive food W/O emulsion. *Food Chemistry* 349: 129203.

García-Ortega, M.L., Toro-Vazquez, J.F., and Ghosh, S. (2021). Development and characterization of structured water-in-oil emulsions with ethyl cellulose oleogels. *Food Research International* 150.

Gaudino, N., Ghazani, S.M., Clark, S., Marangoni, A.G., and Acevedo, N. (2019). Development of lecithin and stearic acid based oleogels and oleogel emulsions for edible semisolid applications. *Food Research International* 116: 79–89.

Ge, S., Li, M., Ji, N., Liu, J., Mil, H., Xiong, L., and Sun, Q. (2018). Preparation of a strong gelatin–short linear glucan nanocomposite hydrogel by an in situ self-assembly process. *Journal of Agricultural and Food Chemistry* 66: 177–186.

Ghosh, S., Tran, T., and Rousseau, D. (2011). Comparison of Pickering and network stabilization in water-in-oil emulsions. *Langmuir* 27: 6589–6597.

Gravelle, A.J., Barbut, S., and Marangoni, A.G. (2012). Ethylcellulose oleogels: manufacturing considerations and effects of oil oxidation. *Food Research International* 48: 578–583.

Gravelle, A.J., Marangoni, A.G., and Davidovich-Pinhas, M. (2018). Ethylcellulose oleogels. *Edible Oleogels* 1.

Guzey, D. and McClements, D.J. (2006). Formation, stability and properties of multilayer emulsions for application in the food industry. *Advances in Colloid and Interface Science* 128–130: 227–248.

Hashizaki, K., Watanabe, N., Imai, M., Taguchi, H., and Salto, Y. (2012). Possibility of vitamin C to induce the formation of lecithin organogel. *Chemistry Letters* 41: 427–429.

Heymans, R., Tavernier, I., Danthine, S., Rimaux, T., Van der Meeren, P., and Dewettinck, K. (2018). Food-grade monoglyceride oil foams: the effect of tempering on foamability, foam stability and rheological properties. *Food & Function* 9: 3143–3154.

Hu, B., Chen, Q., Cai, Q., Fan, Y., Wilde, P.J., Rong, Z., and Zeng, X. (2017). Gelation of soybean protein and polysaccharides delays digestion. *Food Chemistry* 221: 1598–1605.

Hwang, H.-S., Singh, M., Bakota, E.L., Winkler-Moser, J.K., Kim, S., and Liu, S.X. (2013). Margarine from organogels of plant wax and soybean oil. *Journal of the American Oil Chemists' Society* 90: 1705–1712.

Hwang, H.S. and Winkler-Moser, J.K. (2020a). Properties of margarines prepared from soybean oil oleogels with mixtures of candelilla wax and beeswax. *Journal of Food Science* 85: 3293–3302.

Hwang, H.S. and Winkler-Moser, J.K. (2020b). Properties of margarines prepared from soybean oil oleogels with mixtures of candelilla wax and beeswax. *Journal of Food Science* 85: 3293–3302.

Iqbal, S., Xu, Z., Huang, H., and Chen, X.D. (2019). Structuring of water-in-oil emulsions using controlled aggregation of polysaccharide in aqueous phases. *Journal of Food Engineering* 258: 34–44.

Joshi, A., Paratkar, S.G., and Thorat, B.N. (2006). Modification of lecithin by physical, chemical and enzymatic methods. *European Journal of Lipid Science and Technology* 108: 363–373.

Kodela, S.P., Pandey, P.M., Nayak, S.K., Uvanesh, K., Anis, A., and Pal, K. (2017). Novel agar–stearyl alcohol oleogel-based bigels as structured delivery vehicles. *International Journal of Polymeric Materials and Polymeric Biomaterials* 66: 669–678.

Korver, O. and Katan, M.B. (2006). The elimination of trans fats from spreads: How science helped to turn an industry arounda. *Nutrition Reviews* 64: 275–279.

Lai, O.-M. (2019). *Healthful lipids*.

Law, M. (2000). Plant sterol and stanol margarines and health. *Bmj* 320: 861–864.

Lee, W.J., Zhang, Z., Lai, O.M., Tan, C.P., and Wang, Y. (2020). Diacylglycerol in food industry: Synthesis methods, functionalities, health benefits, potential risks and drawbacks. *Trends in Food Science & Technology* 97: 114–125.

Martins, A.J., Silva, P., Maciel, F., Pastrana, L.M., Cunha, R.L., Cerqueira, M.A., and Vicente, A.A. (2019). Hybrid gels: Influence of oleogel/hydrogel ratio on rheological and textural properties. *Food Research International* 116: 1298–1305.

Matheson, A., Dalkas, G., Clegg, P.S., and Euston, S.R. (2018). Phytosterol-based edible oleogels: A novel way of replacing saturated fat in food. *Nutrition Bulletin* 43: 189–194.

Melzer, E., Kreuter, J., and Daniels, R. (2003). Ethylcellulose: A new type of emulsion stabilizer. *European Journal of Pharmaceutics and Biopharmaceutics* 56: 23–27.

Merchan Sandoval, J., Carelli, A., Palla, C., and Baumler, E. (2020). Preparation and characterization of oleogel emulsions: A comparative study between the use of recovered and commercial sunflower waxes as structuring agent. *Journal of Food Science* 85: 2866–2878.

Moschakis, T., Panagiotopoulou, E., and Katsanidis, E. (2016). Sunflower oil organogels and organogel-in-water emulsions (part I): Microstructure and mechanical properties. *LWT - Food Science and Technology* 73: 153–161.

Nascimento, L.G.L., Casanova, F., Silva, N.F.N., de Varvalho Teixeira, A.V.N., and de Carvalho, A.F. (2020). Casein-based hydrogels: A mini-review. *Food Chemistry* 314: 126063.

Nasirpour-Tabrizi, P., Azadmard-Damirchi, S., Hesari, J., Heshmati, M.K., and Savage, G.P. (2020). Production of a spreadable emulsion gel using flaxseed oil in a matrix of hydrocolloids. *Journal of Food Processing and Preservation* 44.

Nguyen, V., Rimaux, T., Truong, V., Dewettinck, K., and Van Bockstaele, F. (2020). Granular crystals in palm oil based shortening/margarine: A review. *Crystal Growth & Design* 20: 1363–1372.

Nicholson, R.A. and Marangoni, A.G. (2021). Lipase-catalyzed glycerolysis extended to the conversion of a variety of edible oils into structural fats. *Current Research in Food Science* 4: 163–174.

Nikiforidis, C.V. and Scholten, E. (2014). Self-assemblies of lecithin and α-tocopherol as gelators of lipid material. *RSC Advances* 4: 2466–2473.

Öğütcü, M., Arifoğlu, N., and Yilmaz, E. (2015). Preparation and characterization of virgin olive oil-beeswax oleogel emulsion products. *Journal of the American Oil Chemists' Society* 92: 459–471.

Okuro, P.K., Gomes, A., Costa, A.L.R., Adame, M.A., and Cunha, R.L. (2019). Formation and stability of W/O-high internal phase emulsions (HIPEs) and derived O/W emulsions stabilized by PGPR and lecithin. *Food Research International* 122: 252–262.

Pan, H., Xu, X., Qian, Z., Cheng, H., Shen, X., Chen, S., and Ye, X. (2021). Xanthan gum-assisted fabrication of stable emulsion-based oleogel structured with gelatin and proanthocyanidins. *Food Hydrocolloids* 115.

Patel, A.R. and Dewettinck, K. (2015). Comparative evaluation of structured oil systems: Shellac oleogel, HPMC oleogel, and HIPE gel. *European Journal of Lipid Science and Technology* 117: 1772–1781.

Patel, A.R. and Dewettinck, K. (2016). Edible oil structuring: an overview and recent updates. *Food and Function* 7: 20–29.

Patel, A.R., Nicholson, R.A., and Marangoni, A.G. (2020). Applications of fat mimetics for the replacement of saturated and hydrogenated fat in food products. *Current Opinion in Food Science* 33: 61–68.

Patel, A.R., Schatteman, D., de Vos, W.H., Lesaffer, A., and Dewettinck, K. (2013). Preparation and rheological characterization of shellac oleogels and oleogel-based emulsions. *Journal of Colloid and Interface Science* 411: 114–121.

Paul, S.R., Qureshi, D., Yogalakshmi, Y., Nayak, S.K., Singh, V.K., Syed, I., Sarkar, P., and Pal, K. (2018). Development of bigels based on stearic acid-rice bran oil oleogels and tamarind gum hydrogels for controlled delivery applications. *Journal of Surfactants and Detergents* 21: 17–29.

Pehlivanoğlu, H., Demirci, M., Toker, O.S., Konar, N., Karasu, S., and Sagdic, O. (2018). Oleogels, a promising structured oil for decreasing saturated fatty acid concentrations: Production and food-based applications. *Critical Reviews in Food Science and Nutrition* 58: 1330–1341.

Pernetti, M., van Malssen, K., Kalnin, D., and Flöter, E. (2007). Structuring edible oil with lecithin and sorbitan tri-stearate. *Food Hydrocolloids* 21: 855–861.

Sagiri, S.S., SinghSingSi, V.K., Kulanthaivel, S., Banerjee, I., Basak, P., Battachrya, M.K., and Pal, K. (2015). Stearate organogel-gelatin hydrogel based bigels: physicochemical, thermal, mechanical characterizations and in vitro drug delivery applications. *Journal of the Mechanical Behavior of Biomedical Materials* 43: 1–17.

Sawalha, H., den Adel, R., Venema, P., Bot, A., Flöter, E., and van der Linden, E. (2012). Organogel-emulsions with mixtures of β-sitosterol and γ-oryzanol: Influence of water activity and type of oil phase on gelling capability. *Journal of Agricultural and Food Chemistry* 60: 3462–3470.

Sawalha, H., Venema, P., Bot, A., Flöter, E., Lan, Y., and van der Linden, E. (2021). Effects of oil type on sterol-based organogels and emulsions. *Food Biophysics* 16: 109–118.

Scharfe, M. and Flöter, E. (2020). Oleogelation: From scientific feasibility to applicability in food products. *European Journal of Lipid Science and Technology* 122: 2000213.

Sciarini, L.S., Van Bockstaele, F., Nusantoro, B., Perez, G.T., and Dewettinck, K. (2013). Properties of sugar-snap cookies as influenced by lauric-based shortenings. *Journal of Cereal Science* 58(2): 234–240. ISSN 0733-5210. doi 10.1016/j.jcs.2013.07.005.

Shakeel, A., Farooq, U., Gabriele, D., Marangoni, A.G., and Lupi, F.R. (2021). Bigels and multi-component organogels: An overview from rheological perspective. *Food Hydrocolloids* 111.

Silva, T.J., Barrera-Arellano, D., and Ribeiro, A.P.B. (2021a). Oleogel-based emulsions: Concepts, structuring agents, and applications in food. *Journal of Food Science* 86: 2785–2801.

Silva, T.J., Fernandes, G.D., Bernardinelli, O.D., da Rosa Silva, E.C., Barrera-Arellano, D., and Ribeiro, A.P.B. (2021b). Organogels in low-fat and high-fat margarine: a study of physical properties and shelf life. *Food Research International* 140: 110036.

Singh, V.K., Qureshi, D., Nayak, S.K., and PAL, K. (2018). Bigels. *Polymeric Gels* 1.

Szumała, P. and Luty, N. (2016). Effect of different crystalline structures on W/O and O/W/O wax emulsion stability. *Colloids and Surfaces A: Physicochemical and Engineering Aspects* 499: 131–140.

Tavernier, I., Moens, K., Heyman, B., Danthine, S., and Dewettinck, K. (2019). Relating crystallization behavior of monoacylglycerols-diacylglycerol mixtures to the strength of their crystalline network in oil. *Food Research International* 120: 504–513.

Tomadoni, B., Capello, C., Valencia, G.A., and Gutiérrez, T.J. (2020). Self-assembled proteins for food applications: A review. *Trends in Food Science & Technology* 101: 1–16.

Toro-Vazquez, J.F., Mauricio-Pérez, R., González-Chávez, M.M., Sánchez-Becerril, M., Ornelas-Paz, J.J., and Pérez-Martínez, J.D. (2013). Physical properties of organogels and water-in-oil emulsions structured by mixtures of candelilla wax and monoglycerides. *Food Research International* 54: 1360–1368.

Trautwein, E.A., Vermeer, M.A., Hiemstra, H., and Ras, R.T. (2018). LDL-cholesterol lowering of plant sterols and stanols-which factors influence their efficacy? *Nutrients* 10: 1262.

Turner, E. (2018). Margarine, mystery and modernity: Margarine and class in literary texts (1880–1945). *Food, Culture & Society* 21: 521–538.

Wijarnprecha, K., de Vries, A., Santiwattana, P., and Sonwai, S. (2019). Microstructure and rheology of oleogel-stabilized water-in-oil emulsions containing crystal-stabilized droplets as active fillers. *LWT - Food Science and Technology* 115.

Xiao, J., Li, Y., and Huang, Q. (2016). Recent advances on food-grade particles stabilized Pickering emulsions: fabrication, characterization and research trends. *Trends in Food Science and Technology* 55: 48–60. doi: 10.1016/j.tifs.2016.05.010.

Xie, M., Yu, Y., and Zhang, L. (2021). Oil-gelling properties of soy lecithin fractions. *Food & Function* 12: 10390–10396.

Zembyla, M., Murray, B.S., and Sarkar, A. (2020). Water-in-oil emulsions stabilized by surfactants, biopolymers and/or particles: A review. *Trends in Food Science & Technology* 104: 49–59.

Zetzl, A.K. and Marangoni, A.G. (2012). Structured oils and fats (organogels) as food ingredient and nutraceutical delivery systems. In: *Encapsulation Technologies and Delivery Systems for Food Ingredients and Nutraceuticals* (ed. N. Garti and D. Julian McClements).

Zheng, H., Mao, L., Cui, M., Liu, J., and Gao, Y. (2020). Development of food-grade bigels based on κ-carrageenan hydrogel and monoglyceride oleogels as carriers for β-carotene: Roles of oleogel fraction. *Food Hydrocolloids* 105.

Zhu, Q., Gao, J., Han, L., Han, K., Wei, W., Wu, T., Li, J., and Zhang, M. (2021). Development and characterization of novel bigels based on monoglyceride-beeswax oleogel and high acyl gellan gum hydrogel for lycopene delivery. *Food Chemistry* 365: 130419.

4.4

Baked Products

Ilkem Demirkesen[1], Suyong Lee[2], and Behic Mert[3]

[1] *Department of Animal Health, Food and Feed Research, General Directorate of Agricultural Research and Policies, Ministry of Agriculture and Forestry, Ankara, Turkey*
[2] *Department of Food Science & Biotechnology and Carbohydrate Bioproduct Research Center, Sejong University, Seoul, Republic of Korea*
[3] *Department of Food Engineering, Middle East Technical University, Ankara, Turkey*

4.4.1 Introduction

A fat replacer is a carbohydrate, protein, or fat-based compound that has the primary function of decreasing calories in food products with no health hazards while offering some or all of the functional properties of fat to food products. Based on their chemical structure and function, fat replacers can be classed as fat substitutes or fat mimetics (Chen et al. 2020). Fat substitutes are ingredients with a chemical structure like fats and comparable physiochemical features, and can replace fat on an equal weight basis. But fat mimetics are compounds with chemical structures distinct from fats but that imitate the organoleptic properties of natural fats and oils in foodstuffs. Thus, fat mimetics are generally used for partial fat replacement. Fat mimetics are like fats in terms of bulk and flavor. They can absorb substantial water content to achieve their functionality. Fat mimetics, also called texturizing agents, can imitate physical and organoleptic properties (Gutiérrez-Luna et al. 2022).

Fat has long been one of the most crucial ingredients in bakery items such as breads, cakes, cookies, and pastries. Fat has several functional roles in foods, one of which is being a solvent for minerals, bioactive chemicals, and flavor carriers. Because it is responsible for the entrapment of air bubbles during the creaming process and the stability of air bubbles during mixing, it also produces a delicate flavor by aerating the dough and creating the sensation of lubrication in the mouth. It is also necessary for bakery items to have a soft and uniform texture. So, it affects the sensory aspects of bakery products including the color, taste, texture, and aroma, which significantly influence overall customer acceptability. Fat also affects the lubrication of gluten particles and the dough's plasticity. This allows air incorporation by covering gluten and starch granules and preventing them from sticking to each other. Another advantage is that its ability to make complexes with amylose delays water transfer into the starch granules, granting bakery products a longer shelf life. Shortening, a blend of solid fat crystals and liquid oil, is a common ingredient in the baking industry. The solid fat components of shortening form a three-dimensional crystal matrix that can provide plasticity by holding the liquid portion of the dough together (Demirkesen and Mert 2020).

The impact of fat on human health has been linked to the quantity of fat consumed, fatty acid composition, and the presence of bioactive micronutrients. In the literature, several works have

Fat Mimetics for Food Applications, First Edition. Edited by Miguel Ângelo Parente Ribeiro Cerqueira and Lorenzo Miguel Pastrana Castro.

emphasized that consuming a diet containing high in saturated and trans fats presents an elevated risk of cardiovascular and coronary heart disease (Demirkesen and Mert 2020). Even though the negative health effects of trans fatty acids (TFAs) include noncommunicable diseases—especially cardiovascular disease, hypertension, obesity, cancer, infertility, and diabetes, there is a lot of uncertainty over the role of SFA (saturated fatty acid) consumption in the development of coronary heart diseases and death (Ari et al. 2021). The reason for contradictory health findings related to saturated fatty acids is because restricting saturated fat could lead to their replacement with carbohydrate, which would result in increased TAGs (triacylglycerols) with decreased HDL (high-density lipoprotein/good cholesterol), both of which are more detrimental health-wise. So, while the debate over the replacement of saturated fatty acids is still ongoing, current systematic studies reflect the impact of replacing saturated fatty acids with polyunsaturated fatty acids (PUFA), with essential fatty acids, and even with monounsaturated fatty acids (MUFA) to decrease the risk of coronary heart disease. Since TFA is an important public health problem, TFA removal from the global food supply is one of WHO's (World Health Organization) primary objectives. The WHO policy brief suggests that total TFA (industrially generated and ruminant) should not exceed 1% of total energy for both adults and children (Ari et al. 2021).

Consumer concern regarding the adverse health consequences overconsuming trans fats as well as consumer demand for fat reduction for a healthy lifestyle and expected lifespan inspired development of a variety of reduced-fat food products. The biggest challenge is to imitate the same product qualities the same qualities that the original solid fats would give. Saturated fatty acids in the form of TAGs are responsible for forming a fat crystal network that structures the fat into a solid-like substance while preserving the fat's liquid oil components and inhibiting exudation (Demirkesen and Mert 2020). Therefore, the aim of this chapter is to provide a comprehensive and concise review of using fat mimetics with emphasis on recent research updates and their potential applications in bakery products.

4.4.2 Fat Mimetics in Bakery Products

Dough or cake batter is an aqueous colloidal dispersion. In bakery products, the "shortening effect" is known as the interaction of lipids with wheat flour macromolecules. Even though shortening effects can be linked both chemically and physically, the physical effects involved in lipid oxidation, such as lubrication, foam formation, hydrogen, and hydrophobic interactions, the delay of carbon dioxide release, and helping heat transfer play a more pronounced role. During bread production, the hydrophilic and hydrophobic components of the dough crosslink and create a three-dimensional structure responsible for the entrapment of air and flour lipids. Entrapment of wheat flour lipids in a three-dimensional gluten matrix lets them increase gas cell capability throughout the bread making process. With the lack of shortening, carbon dioxide is released faster from the dough, so, generally, higher hardness and lower volume are obtained. When shortening is included in dough formulation, the shortening's plasticity allows for the coating of gluten particles, which prevents protein and starch structure adherence. Hence, the lubrication of gluten particles leads to greater expansion during fermentation and bread baking, which provides a softer and aerated texture, higher volume, and longer shelf life for bakery products. Also, the interaction of fat with starch delays retrogradation. The shortening's solid-like elasticity and oil-binding capabilities keep the fat in its liquid TAG phase and prevent it from soaking into the dough. At room temperature, the plastic ranges and a fat fraction of shortenings plays a critical role in bread making. It has been stated that bakery shortening must include at least 20% SFC (solid fat content) at room temperature, while at higher temperatures, such as 40°C, it must have at least 5% SFC (Demirkesen and Mert 2020).

Fat mimetics also have body, lubricity, perceived creaminess or smoothness, absorption, and adsorption effects on the taste buds or tongue; they also have cohesiveness, adhesiveness, and waxiness. Various fat mimetics such as inulin, maltodextrin, polydextrose, and whey protein concentrates have also been used in bakery products. Materials with gel-like structures such as hydrogels, emulgels, and oleogels have also been applied as fat mimetic approaches in the bakery industry. Carbohydrate-based fat mimetics generally consist of dietary fibers from different sources. Since they are often mixed with water, they typically provide only 1 to 2 kcal/g instead of 4 kcal/g, and some (cellulose) have zero calories. They also lend bulk, viscosity, structure, and texture to food products. Commonly, they can resist heat processing, so they can have suitable thermal stability during baking. In bakery goods, the carbohydrate-based fat mimetics commonly used are fibers, gums, inulin, maltodextrins, polydextrose, and starches. Protein-based fat mimetics include lower molecular weight or modified proteins that provide a fat-like structure and mouthfeel. Most protein-based fat mimetics cannot be used at high temperatures due to their coagulation problems, resulting in a decrease in their functionalities. Protein-based fat mimetics provide 1 to 4 kcal/g. The simplest protein-based fat replacement strategy is the addition of milk protein and milk solids to products. Whey protein concentrates and isolates can also be used as protein-based fat mimetics. Microparticulated protein, modified protein concentrates, milk protein hydrolysates, gelatin, soy, and corn proteins can be classified as protein-based fat mimetics (Swanson 2006). Recently, structured solid-like materials called oleogels have provent to have great potential to serve a healthier alternative to bakery shortenings (Gutiérrez-Luna et al. 2022).

4.4.2.1 Fat Mimetics in Bread Formulation

Bread typically has lower fat content (2% fat based on flour weight) compared to other bakery products such as cookies, doughnuts, cakes, etc. However, a high amount of fat is incorporated into formulating certain types of bread, such as sweet pan bread. In bread formulation, shortening is generally the fat used as helps entrap air bubbles into the dough during mixing and helps leaven, tenderize the crumb, add moistness, and improve mouth feel and shelf-life (Balic 2015). In bread formulations, inulin, Simplesse® (microparticulated whey protein), gums such as HPMC (hydroxypropyl methylcellulose) and changed starches have yielded promising results (Balic 2015).

As a protein-based (whey protein) fat replacer, Simplesse® presents both hydrophilic and hydrophobic properties and can act as an emulsifier. Microparticulation is a method that creates a high number of protein microparticles with the size and shape required for a fat-like mouthfeel. Microparticulated protein is usually obtained from whey protein concentrate and from other sources of protein. Simplesse® is obtained from whey protein concentrate by simultaneously heating and shearing, resulting in a particle size of 0.1–10 μm, and these small particles can create a creaminess sensation associated with fat. Simplesse® reportedly provides nutritional advantages and functional features such as the supplementation of limited amino acids like lysine and methionine. The performance of several fat replacers (inulin powder, inulin gel, and Simplesse®) at different quantities (2.5 and 5%) was tested in wheat dough and the final product by O'brien et al. (2003). The addition of inulin increased water absorption. Simplesse® weakened the gluten network by having a more porous structure unable to retain gas while under stress. As a result, lower volume and higher hardness values were obtained from the loaves. Compared to Simplesse® or inulin powder, inulin gel provided better loaf volume, having similar volume yields to the control. Similarly, both inulin powder and Simplesse® increased crumb hardness, but breads with the control fat and inulin gel had similar hardness values. Overall, among different fat replacers, incorporating inulin gel at a concentration of 2.5% was the most effective, presenting similar quality features to the control bread containing fat. In another study, the influence of inulin on the formation and release of white bread volatiles was also studied during baking (Poinot et al. 2010). Inulin enhanced the formation of the bread crust and maillard reaction, and hence, bread baked for a shorter time had similar quality parameters to those of non-enriched bread.

Fibers derived from various sources can replace dietary fat in bakery products. Native or changed starches have been found in preceding studies to be effective as a fat replacer in a variety of food items. Corn, potato, wheat, rice, cassava, sorghum, and barley starches can all be used as fat mimetics. Also, changed starches are often used as emulsifiers to form a stronger dough, which may create liquid films at the gluten-starch interface, resulting in gas retention and increased dough and bread volume. When hydrophobic side chains are added to hydrophilic starch molecules, starch can adsorb at the water-oil interface that stabilizes the emulsion. Esterification of starch with octenyl succinate anhydride (OSA) is an example. In bread formulation, the effectiveness of OSA changed starches (2 and 4%) from two sources (wheat and tapioca) as fat replacers have also been studied. This study revealed that OSA changed tapioca starch had a more significant influence on dough and bread quality compared to OSA changed wheat starch (Balic et al. 2017). The positive influence of enzymatically changed corn starch as a fat replacer on the crumb structure, volume, and sensory properties of bread formulations has also been shown (Scheuer et al. 2014, 2015, 2016).

To meet the necessity of high-quality and highly nutritional gluten free products, using different kinds of fat mimetics from different fibers has also been proposed in healthy gluten-free food products (Mert et al. 2016). These studies reflect the potential of developing fiber-rich gluten-free breads with a suitable amount of fiber (Beran et al. 2018; Rios et al. 2020).

Hydrogels are polymeric substances capable of swelling and keeping a significant fraction of water within their structure. Several hydrogels prepared with different gelling agents have been used as fat replacers in bakery products, including chia, pectin, maltodextrin, etc. Chia seeds (*Salvia hispanica* L.) have been used to substitute eggs and fat in a variety of baking items due to their hydrophilic properties (Borneo et al. 2010; Felisberto et al. 2015; Fernandes and de las Mercedes Salas-Mellado 2017). Mucilage is a polysaccharide that functions as a soluble dietary fiber and that can act as a texturizing and stabilizing agent by forming gels or thickening the continuous phase (Capitani et al. 2013). Chia seeds can absorb water at levels up to 12-fold larger than their own mass. This property gives consistency and makes it a versatile and low-calorie ingredient (Gutiérrez-Luna et al. 2020). Zettel and Hitzmann (2016) conducted a study to define the effects of gel from ground chia as a fat replacer in sweet pan breads. The substitution of chia (*Salvia hispanica* L) gel with up to 50% fat replacement improved the nutritional value of the products, while reducing the total fat in them. As a carbohydrate-based fat replacer, inulin can also form a gel when combined with water, so inulin can also be categorized as a hydrogel. It is also regarded as a prebiotic. Inulin and oligofructose can potentially be used to mimic the functional and sensorial properties of fat and sugar while providing a healthier, lower-calorie alternative. So Rößle et al. (2011) evaluated the effects of both inulin and oligofructose as fat and sugar replacers on the quality parameters of quick breads. The inclusion of inulin and oligofructose increased the crust and crumb color. The presence of inulin and oligofructose in formulations also led to slight increases in crust and crumb hardness. The increase in crumb hardness with the replacement of inulin by fat might be related to the dominant role of fat in the softening of crumb and crust hardness. On the other hand, including inulin and oligofructose significantly increased volume. Results showed that the mixture design successfully decreased both the fat

(10%) and sugar (10%) levels in the formulation. The findings of this study suggest that low-fat and low sugar quick breads can be produced to have potential health benefits of inulin and oligofructose.

Organogelation also called oleogelation is a new, alternative structuring technology that forms structured, solid-like materials by trapping a large volume of liquid oil within a thermo-reversible and three-dimensional gel network. Oil type, oleogelator type, and concentrations are all critical parameters that affect the rheological, thermal, and textural qualities of oleogels and the quality of the finished products. In the literature, sunflower oil, olive oil, hazelnut oil, canola oil, rice bran oil, maize oil, soybean oil, grape seed oil, rapeseed oil, and flaxseed oil have all been used to make oleogels. It has been shown that oil with a higher saturated TGA and a higher melting fatty acid yields a higher viscosity and yields harder oleogels. The gelling ability of gelators also varies depending on fatty acid composition, molecular weight, and acyl chain length, as well as its concentration. The interaction between oleogelators and oils has also favorably impacted the gelling ability of oleogelators. Many types of oleogelators have been used in the literature, including triacylglycerols, diacylglycerols, monoacylglycerols, fatty acids, fatty alcohols, waxes, wax esters, sorbitan monostearate, phospholipids, phytosterols, and mixtures of fatty acids and fatty alcohols, lecithin and sorbitan tristearate, phytosterols, and oryzanol.

Meng et al. (2019) used emulsifier sodium stearoyl lactylate as a gelling agent to form sunflower oil oleogels at different concentrations. Higher oleogelator concentrations created a denser crystalline network, which provided stronger mechanical strength and made it easier to keep the oil phase. Baking test results indicated that the oleogels could represent a possible strategy to design low saturated fat bakery products (Meng et al. 2019).

In the study by Ye et al. (2019), the performance of ethylcellulose oleogels were tested as bakery shortenings. Ethylcellulose oleogels were prepared using the base oil and different types (viscosities) of ethylcellulose (ethylcellulose7 (viscosity 6~9 cP), ethylcellulose20 (viscosity 18~22 cP), ethylcellulose50 (viscosity 45~55 cP), ethylcellulose100 (viscosity 90~110 cP), and ethylcellulose200 (viscosity 180~220 cP)). Various types of emulsifiers (glyceryl monostearate, triglyceryl monostearate, glyceryl distearate, and sorbitan stearate) were also tested to see whether they affected the mechanical properties of the oleogels. Compared to breads produced with conventional shortening, loaves prepared with ethylcellulose100 oleogel and triglyceryl monostearate had a higher specific volume and softer texture. In another study, emulsifier sodium stearoyl lactylate was a gelling agent to form sunflower oil oleogels at different concentrations. Higher concentration of oleogelator gave a denser crystalline network, which provided stronger mechanical strength and enhanced the ability to keep the oil phase. Results of the baking tests indicated that the oleogels could represent a possible strategy to design low saturated fat bakery products (Meng et al. 2019).

Demirkesen and Mert (2020) have postulated that reducing the shortening content of products without changing the quality parameters could be achieved using oleogel/shortening blends. Similarly, Jung et al. (2020) tested butter and oleogel blends in sweet pan bread to reduce saturated fat. The increasing amount of oleogels prepared by rice bran oil and candelilla wax in the blends created a more dominant viscous dough. It has been shown that replacing butter with oleogels at up to 75% by weight produced baked breads with similar specific volumes to the control bread prepared only with butter. Using oleogel for butter provided nutritional superiority by decreasing the saturated fat content from 71.0 to 25.3 g/100 g. This study has shown that using oleogels in sweet pan bread may provide helpful information on the processing performance of oleogels in yeast-leavened baked products, perhaps encouraging the food sector to adopt oleogels in a wider variety of baked goods.

4.4.2.2 Fat Mimetics in Cookie/Biscuit Formulation

Unlike bread, large amounts of sugar and fat are used in biscuit/cookie formulations, but moisture content remains in low proportions. Shortening used in cookie/biscuit formulations should have "plastic" qualities as described by SFC values. In the case of high SFC values, shortening does not have enough oil volume for adequate aeration. Low SFC values will prevent shortening from keeping enough air bubbles until the end of mixing (Demirkesen and Mert 2020). Therefore, both used shortening, and SFC content of the shortening play an important role in the cookie dough and the final product.

Without shortening, the aqueous phase of dough interacts with gluten to generate a cohesive and extensible network; the resulting texture is firmer and/or chewier. In cookies or biscuits, shortening controls gluten development by breaking the continuity of the protein and starch structure, so limited gluten development is obtained. Typically, cookie/biscuit dough has less-elastic properties; it does not shrink after lamination, and hence a crisp or crumbly texture is obtained from the resulting products. Aeration is another important function of shortening in cookies. While the liquid phase of shortening is essential for air incorporation, fat crystals are responsible for air retantion at the end of mixing and during the initial stage of baking (Demirkesen and Mert 2020). Hence, the plasticity of shortening is also critical for air entrapment and for retention during the initial creaming phase, which turns the cookie into a uniform, fine, and tender crumb. Another role of shortening in cookie manufacturing is to create the proper spreading capabilities while also keeping moisture during baking. The lubricating effect of shortening has a noteworthy impact on the final quality of cookies in terms of spreading properties. Generally, better spreading occurs when more shortening is used. The interaction of shortening with flour provides flavor and an overall lubrication sensation to the product (Demirkesen and Mert 2020). Table 4.4.1 shows studies conducted to understand the effect of various fat mimetics on the quality features of cookies and biscuits.

Several studies have shown inulin or a mixture of inulin and oligofructose may be employed as a potential fat mimetic in cookies, biscuits, and their gluten-free counterparts (Table 4.4.1). The effects of inulin (10, 20, 30, and 40%) as a fat replacer on short dough biscuits and their corresponding doughs were studied (Rodríguez-García et al. 2013). Flour was more accessible for hydration at greater fat replacement levels, resulting in substantially harder doughs. Weight loss during baking, water activity, biscuit dimensions, and aeration decreased significantly as fat replacement levels increased (Rodríguez-García et al. 2013). Likewise, the study by Krystyjan et al. (2015) showed that higher inulin replacement levels (30, 40, and 50%) worsen the sensory and textural quality features of biscuits. Devereux et al. (2003) indicated using inulin and oligofructose might be limited by adverse physiological effects when consumed at high levels. Similar to inulin, replacing fat with other fat mimetics such as maltodextrin, polydextrose, etc. provided a harder texture (Sudha et al. 2007; Forker et al. 2012; Chugh et al. 2013). It has been suggested that combining fat mimetics with other ingredients might improve fat-like properties since the additional ingredients are usually intended to reduce the negative effects of the individual fat mimetics. Improvement in texture was noted when maltodextrin was combined with either glycerol monostearate, guar gum, corn fiber, etc. was used along with maltodextrin (Table 4.4.1).

Different types of fat mimetics (Dairytrim (β-glucans), polydextrose (Microparticulated whey proteins), Litesse (improved polydextrose), C*deLight (maltodextrins), Simplesse® (a blend of micropatriculated whey proteins and emulsifiers), Raftiline (inulin), and pectin were compared for cookie quality by Zoulias et al. (2000, 2002a and 2002b). Polydextrose and Dairytrim resulted in harder cookies. C*deLight, Simplesse®, and Raftiline produced softer cookies. All fat mimetics

Table 4.4.1 Overview of research conducted on the application of various fat mimetics in cookies/biscuits.

Fat mimetic type	Product type	Main findings	Reference
Inulin	Biscuit	The effects of inulin (10, 20, 30, and 40%) as a fat replacer on short dough biscuits and their corresponding doughs were tested. At higher fat replacement levels, flour was more available for hydration, leading to significantly harder doughs. Weight loss during baking, water activity, biscuit dimensions, and aeration decreased significantly with the increasing fat replacement levels. The findings of the study suggest that shortening may be partially replaced, up to 20%, with inulin.	Rodríguez-García et al. (2013).
Inulin	Cookie	The partial replacement of fat (0, 25, 50, 75%) by inulin was effective in reducing the percentage of total lipids in the final product without changing moisture, crude fiber, and instrumental color. The 50% fat replacement had similar quality parameters to the control one.	Lourencetti et al. (2013).
Inulin	Gluten-free cookie	The nutritional value of gluten-free cookies was improved with inulin in terms of prebiotics and fiber. Inulin was added at four levels (3.0, 3.5, 4.0, and 4.5%) to the gluten-free cookie formulation. The results indicated that adding 3.0% inulin resulted in the highest protein, fat, and ash content, while 4.0% inulin gave the highest moisture content. The total dietary fiber content increased with the inulin level, but the spread factor decreased with the increasing inulin level.	Maghaydah et al. (2013).
Inulin	Biscuit	Inulin replaced fat in short dough biscuits. Adding inulin at reduced fat content significantly affected the acoustic parameters of short-dough biscuits.	Błońska et al. (2014).
Inulin	Biscuit	The substitution of 20% of fat by inulin did not worsen the sensory evaluation of biscuits and had the least influence on their texture as compared to higher replacement levels (30, 40, and 50%). The decrease in fat and the increase in sugar content with a simultaneous reduction in caloric value were noticed. The soluble dietary fiber increased compared with the control.	Krystyjan et al. (2015).
Soluble fibers (nutriose, inulin, polydextrose), insoluble fibers (bamboo, potato, pea)	Cookie	The influence of different soluble and insoluble fibers on cookie quality were evaluated. Soluble fibers decreased dough consistency, favored cookie spread during baking, and produced wider and thinner cookies, while insoluble fibers increased dough consistency and gave higher moisture, lower spread factor, and higher hardness. The specific length and shape of insoluble fibers added were important, especially for texture.	Mancebo et al. (2018).
Inulin, oligofructose	Cookie, cake and muffin	Inulin and oligofructose were used as fat replacers in cookies, blueberry muffins, carrot cake, chocolate cake, lemon cheesecake, ice cream, and beef sausages. These foods were rated as acceptable by an untrained taste panel but scored consistently lower than their full-fat counterparts. Unlike the controls, texture was more important than flavor in determining the overall acceptability of the low-fat foods. Some have suggested that inulin and oligofructose are readily incorporated into bakery and meat formulations, but their use might be limited by adverse physiological effects when consumed at high levels.	Devereux et al. (2003).

Table 4.4.1 (Continued)

Fat mimetic type	Product type	Main findings	Reference
Maltodextrin, polydextrose, glycerol mono stearate, guar gum	Biscuit	Replacing fat with equal quantities of maltodextrin and polydextrose reduced the dough consistency and dough hardness to some extent. Adding glycerol monostearate and guar gum had a positive effect on dough consistency and hardness. Fat reduction had a negative effect on the texture of the biscuits. However, texture improved significantly when fat was replaced with maltodextrin. Further improvement in texture was noted when either glycerol monostearate or guar gum was used along with maltodextrin.	Sudha et al. (2007).
Maltodextrin, guar gum	Biscuit	Hardness increased with increases in levels of sugar and fat replacers and decreased in fat levels. Increased fat and water levels had an additive effect on the spread ratio, lowering it.	Chugh et al. (2013).
Corn fiber, maltodextrin, lupine extract	Biscuit	The influences of different fat mimetics and their combinations on the quality features and sensory properties of the biscuits were studied. A replacement of 30% baking fat-reduced the total fat content of the biscuits by 27.3–29.8%, depending on the exact composition of the fat replacer. On the other hand, the replacement of 30% fat gave higher firmness and lower volume as compared to control biscuits.	Forker et al. (2012).
Rice starches	Cookie and muffins	Native and chemically changed (acetylated and hydroxypropylated) rice starches were partly substituted for butter in the production of reduced-fat muffins and for shortening in cookies. Both native and changed rice starches at substitution levels of 20% for fat are equally suitable for both reduced-fat muffins and reduced-fat cookie production.	Lee and Puligundla (2016).
Polydextrose Dairytrim, C*deLight, Simplesse®, Raftiline	Cookie	The increase in polydextrose or Dairytrim content resulted in harder cookies, while the increase in C*deLight, Simplesse®, or Raftiline content had the opposite effect. The brittleness of the cookies was increased with the addition of all fat mimetics, but a moderate increase was obtained with C*deLight, Simplesse®, or Raftiline.	Zoulias et al. (2002a).
Litesse, C*deLight, Dairytrim, pectin, Simplesse®	Cookie	Soft-type cookies were prepared with up to 35% fat replacement by Litesse, C*deLight, Dairytrim, pectin, and Simplesse®. All formulations with fat replacement up to 23% showed properties like control cookies except hardness.	Zoulias et al. (2000).
Raftiline, Simplesse®, C*delight	Cookie	Raftiline, Simplesse® or C*delight combined with lactitol or sorbitol in 35% fat-reduced, sugar-free cookies gave comparable hardness and brittleness as compared to control. Polydextrose produced hard and brittle cookie. Except for cookies made with Simplesse® and polydextrose, the diameter increase was minumum. Further fat replacement of 50% was achieved using Raftiline, Simplesse® or C*deLight combined with a blend of lactitol and sorbitol, however the resultant products were hard, brittle and did not expand adequately. Cookies prepared with Simplesse® had the least acceptable flavor, while cookies prepared with C*deLight were rated as the most acceptable by a sensory panel.	Zoulias et al. (2002b).

(Continued)

Table 4.4.1 (Continued)

Fat mimetic type	Product type	Main findings	Reference
Mung bean paste	Cookie	Reducing the fat content of butter by 25% did not affect overall product acceptability, but product acceptability was reduced when higher replacement levels were used.	Adair et al. (2001).
Puree of canned green peas	Cookie	Substitution of up to 75% of egg yolk and margarine with green peas resulted in acceptable cookies with lower fat and calories, as well as higher moisture levels and moistness acceptability compared to full-fat cookies.	Romanchik-Cerpovicz et al. (2019).
Wheat bran	Biscuit	Wheat bran fiber with a bigger particle size was more favorable in terms of the textural properties of the dough and the quality parameters of biscuits, but the fibers with a smaller particle size enhanced the viscoelastic properties of the dough like the control. Although the use of these fibers in the production of low-fat biscuits was suitable in terms of workability of dough, increasing fiber content and reducing fiber size gave harder biscuits with a lower spread ratio.	Erinc et al. (2018).
Microparticulated whey proteins	Cookie	Highly acceptable reduced-calorie biscuits can be produced by using dairy–multigrain composite flour with maltitol and fructooligosaccharides-sucralose (as sweetener) and polydextrose (as fat replacer).	Aggarwal et al. (2016).
Oatrim, avocado puree	Biscuit	The effectiveness of partial replacement (50% level) of butter fat by two fat replacers on the physical, textural, and sensory properties of oatmeal cookies was tested. Compared with the control cookie, both fat-reduced cookies had greater heights, reduced spreads, increased moisture loss, higher water activity, and a tougher and less brittle texture.	Wekwete and Navder (2008).
Trimchoice-5, Kel-Ute 8K, Sledid	Biscuit	Three selected carbohydrate-based fat replacers derived from pectin, gums, and oat bran were tested in biscuit formulation. Fat substitutes provided a higher moisture content and a lighter crust color. Variations in volume and crumb firmness were associated with the fat substitute and level of fat. Crumb size decreased and bitterness was detected in biscuits as fat levels decreased despite the fat substitute.	Conforti et al. (1997).
Inulin, HPMC	Biscuit	Both the level and type of fat replacer influenced the texture features of biscuits and their sensory qualities. Texture became harder with the use of fat replacer. Fat replacement of up to 15% provided acceptable biscuits for consumers. HPMC could be used at higher fat replacement rates while keeping the dough workable, but concentrations higher than 15% cannot be employed if good sensory properties are to be kept.	Laguna et al. (2014).
HOSO, lecithin and β-Glucan	Biscuit	Using lecithin in the products improved biscuit quality and sensory features. β-glucan helped to reduce fat migration, but it had a detrimental impact on the biscuits' sensory quality. Biscuits containing inulin and lecithin, but no fiber additions were similar to the shortening control sample. However, they were different in terms of sensory and mechanical qualities.	Onacik-Gür et al. (2015).
HOSO, inulin	Cookie	Incorporation of inulin as a fat replacer helped to maintain sensory properties. Inulin served as a simple sugar substitute and a source of dietary soluble fiber.	Doménech-Asensi et al. (2016).

Table 4.4.1 (Continued)

Fat mimetic type	Product type	Main findings	Reference
Emulsion filled gel (EFG) based on inulin and extra virgin olive oil (EVOO)	Cookie	Higher volumes, harder textures, higher water availability, and lower fat content were obtained by increasing the EFG in shortbread cookies. Darker color and toasted notes were registered for increasing levels of EFG, in relation to inulin involved in the Maillard reaction.	Paciulli et al. (2020).
EFG based on EVOO and inulin	Biscuit	The replacement of 50% and 100% of butter in shortbread cookies with an EFG based on inulin and EVOO was evaluated. The total substitution of butter with EFG resulted in higher number of small-sized pores and lower level of fracture stress, with poorer sensory properties compared to control cookies. Cookies with 50% butter replacement showed comparable fracture and microstructure properties and were well accepted by consumers.	Giarnetti et al. (2015).
Oleogels	Cookie	Oleogels of candelilla wax with canola oil were evaluated as an alternative to shortening in cookies. In the oleogel cookies, the unsaturated fatty acid levels were increased, up to around 92%, compared to the shortening cookies (47%). The cookies prepared with the oleogels had desirable spreadable properties, and the replacement of shortening with the oleogels produced cookies with soft eating features.	Jang et al. (2015).
Oleogels	Cookie	Higher spreadable values, softer textures and longer biological shelf life were obtained from cookies prepared with hazelnut oil-beeswax and hazelnut oil-sunflower wax oleogels.	Yılmaz and Öğütcü (2015).
Oleogels	Cookie	Two oleogels containing carnauba wax and candelilla wax were evaluated to substitute shortening in cookies. When compared to sunflower oil, oleogels created softer cookies with a superior spread ratio and appearance. However, texture measurements revealed that oleogels could not form short dough and commercial shortening with a greater solid fat content. Cookies made using oleogels had a somewhat tougher texture and a spread ratio larger than those made with commercial shortening. Although using oleogels in short dough cookies improved the quality of the cookies significantly overusing liquid oil, the necessity of further work to match the quality attributes of commercial shortening-containing cookies has been highlighted.	Mert and Demirkesen (2016a).
Oleogels	Cookie	The potential application of candelilla wax containing oleogels for partial replacement of the shortening in cookies was studied. Oleogels were prepared with different amounts of candelilla wax and blended with a commercial bakery shortening. After crystallizing the oleogel/shortening blends by using a pilot-scale crystallization unit, the blends were used in a cookie formulation. When the shortening was substituted with oleogel, the resulting products were softer than liquid oil, but harder than the shortening containing products. Shortening was partially replaced with oleogels, which resulted in considerably better dough and cookie qualities. The findings show that gradually replacing shortening with oleogels might be a good way to reduce saturated fat in short dough products.	Mert and Demirkesen (2016b).

(Continued)

Table 4.4.1 (Continued)

Fat mimetic type	Product type	Main findings	Reference
Oleogels	Cookie	Oleogels were prepared using β-sitosterol and/or monoacylglycerol with crude soybean oil and refined soybean oil. Incorporatingβ-sitosterol and/or monoacylglycerol into crude and refined soybean oil could produce oleogels with solid-like properties. Stronger and firmer oleogels were obtained when refined soybean oil was used. The cookies using oleogels had similar quality features with that using the shortening.	Zhao et al. (2020a).
Oleogels	Cookie	Oleogels prepared from expeller-pressed corn germ oil and rice bran wax were used at different concentrations (3, 5, 7, and 9 wt%). Rice bran wax could form oleogels in both refined and expeller-pressed corn germ oils at a concentration $\geq$ 3 wt%. Although refined corn oil produced a gel stronger than crude corn oil, cookies made with both types of oleogels had similar properties with that of shortenining containing cookies.	Zhao et al. (2020b).
Oleogels	Cookie	Different types of oleogels were prepared by various gelators, including HPMC, monoacylglycerol, SSL, rice bran wax, and beeswax, and their applications in cookies were compared. According to rheological, textural, microstructural, and color measurements, MAG, rice bran wax oleogels had similar values.	Li et al. (2021).

C*delight: Maltodextrins; Dairytrim: β -glucans; HPMC: Hydroxypropyl Methylcellulose; HOSO: High-oleic acid sunflower oil; Kel-Ute 8K: Xanthan gum, guar gum, cellulose gel, sodium stearoyl lactylate, gum Arabic, dextrin, lecithin, and mono- and diglycerides; Litesse: Improved polydextrose; Oatrim: Oat β-glucan amylodextrins; Polydextrose: Microparticulated whey proteins; Raftiline: Inulin; Simplesse®: a blend of micropatriculated whey proteins and emulsifiers; SSL: Sodium stearoyl lactylate; Sledid Fat Replacer: Slendid Pectin Low Esterified; Trimchoice-5: P-glucans and low dextrose equivalent maltodextrins.

enhanced the brittleness of the cookies; however, only a small increase was produced with C*deLight, Simplesse®, or Raftiline. Further fat replacement was achieved using Raftiline, Simplesse®, or C*deLight combined with a blend of lactitol and sorbitol, but the final products were hard and brittle and did not expand properly after baking. Simplesse® cookies were ranked as the least acceptable, while C*deLight cookies were rated as the most acceptable. Thus, among these fat replacers, maltodextrin has the most success in replacing fat in cookie/biscuit products. Inulin has also shown promising results in lowering the fat and energy content of cookie and biscuit products while having the least impact on sensory quality. Other commercial fat replacers such as okra gum, apple sauce, pureed white beans, mung bean paste, pawpaw puree mungo flour, and Lotus root (Saeed et al. 2020a, 2020b) have all been used with some success in baked products.

Additionally, per the literature, gum and gel fat replacers including oatrim, HPMC, guar gum, pectin, and others have also been used extensively. Although they are mostly carbohydrate-based, lipid-based and protein-based gums and gels are also available. Oatrim (oat β-glucan amylodextrins) is as a powder or gel that may also be effectively used as a partial replacement of fat in cookies (Wekwete and Navder 2008). Conforti et al. (1997) compared the influences of three selected carbohydrate-based fat mimetics derived from pectin, gums, and oat bran on the quality parameters of a biscuit formulation. The level and type of fat mimetics affected textural and sensory properties of biscuits (Table 4.4.1). It has been found that the use of fat mimetics provided both higher moisture content and a lighter crust color. Crumb size decreased and bitterness was detected in biscuits as fat levels decreased despite the fat mimetic. The suitability of inulin and HPMC as fat replacers was compared by Laguna et al. (2014). Texture became harder with the increasing use of fat mimetics. HPMC might be used at higher fat replacement levels while keeping the dough workable. However, the authors suggested that concentrations of over 15% fat mimetics cannot be used while maintaining excellent sensory qualities.

The suitability of high oleic sunflower oil (HOSO) with inulin, lecithin, and β-glucan (oil + lecithin, oil + inulin, oil + inulin + lecithin, oil + β-glucan, oil + lecithin + β-glucan, oil) as fat replacers was also tested. Replacement of shortening with oil had detrimental effects on the volume, color, and texture of the biscuits (Onacik-Gür et al. 2015). Using lecithin in the products improved the biscuits' quality by lowering their hardness and increasing their volume. Including lecithin improved sensory features; although β-glucan helped reduce fat migration, it had a detrimental impact on the sensory quality of the biscuits. The findings showed that the biscuits containing inulin and lecithin with no fiber additions (β-glucan) were comparable to the shortening control sample. However, biscuits were different in terms of sensory and textural qualities. Thus, more research is needed to develop oil-based biscuits with the same features as solid fat-based biscuits (Onacik-Gür et al. 2015). In another study, replacing margarine and sunflower oil with HOSO and inulin was studied in four different bakery products. Conversely, in the study by Doménech-Asensi et al. (2016), the incorporation of inulin as a fat replacer helped maintain sensorial properties (Table 4.4.1). The authors have proposed that it is necessary to evaluate inulin levels after baking and control possible losses to understand the role of inulin in these studies.

A combination of emulsion filled gel (EFG) based on inulin and extra virgin olive oil (EVOO) was also tested as a fat mimetic in cookie and biscuit formulations. The increasing amount of EFG gave higher volumes, higher water availability, and lower fat content, but also harder texture and darker color (Paciulli et al. 2020). Replacing 50% of the butter fraction of shortbread with EFG produced a product with potentially healthier properties like the control cookies and were well accepted by consumers. These cookies were characterized by a fat content reduced by 19% and a saturated fat content reduced by 39%. However, the authors stated that further research is needed to overcome sensory problems (Giarnetti et al. 2015) (Table 4.4.1).

Studies conducted on applying oleogels in the bakery industry indicated that they can be good alternatives as fat replacers. The effects of oleogel technology on baked goods were reviewed by Demirkesen and Mert (2020). Previous studies showed the potential of different kind of waxes with various types of oils for the formation of oleogels (Demirkesen and Mert 2020). Jang et al. (2015) obtained desirable spreadable properties and soft eating characteristics from cookies prepared with the oleogels of candelilla wax and canola oil. Similarly, Yılmaz and Öğütcü (2015) obtained higher spreadable values from cookies that were prepared with hazelnut oil-beeswax and hazelnut oil-sunflower wax oleogels than shortening cookies. The result of the higher spreadable characteristics of cookies with the oleogel might be attributed to the lower aeration of cookie dough. On the other hand, shortening-containing dough had higher moisture content that resulted in higher weight and height, but same diameter values, and lower spreadable characteristics in cookies with shortening. The authors also stated that because of fine β′ type crystals in oleogels, cookies prepared with oleogels had significantly lower hardness but higher fracturability values.

Because waxes also include other elements such as extremely long chain hydrocarbons, acylglycerols, sterol esters, and other chemicals in addition to long chain fatty acids, the inclusion of waxes into oil resulted in some increases in the quantity of saturated fatty acids. The percentage of the solid fractions of fats at a certain temperature is known as the solid fat content of edible oils and fats; hence it denotes the differences in consistency and plasticity of products at various temperatures. As mentioned, solid fat is necessary for a well-aerated, soft, lubricated, and viscoelastic dough structure. Fats with relatively high SFC values are utilized during the mixing process in bakery products, and thus, shortenings must have a particular solid fat index when used in cookies, especially at dough mixing temperatures (Demirkesen and Mert 2020). Despite having substantially lower SFC values, oleogels were able to form self-standing structures. This might be related to wax gelation, in which liquid oil is confined to the crystal particle network generated by waxes through a process. Therefore, oleogels, particularly those containing a higher amount of wax, had similar rheological properties to shortening.

Due to their space-filling properties, oleogels have shown to have considerably high elastic properties and viscosities in the linear regime. The rheological properties of oleogels can be significantly altered because of structural breakdown under deformation conditions, and thus, oleogels may quickly lose their solid-like structure as a result of shear forces under dough mixing conditions (Mert and Demirkesen 2016a). As a result, they may become considerably softer than before.

In the literature, there are contradictory findings about the influence of oleogels on hardness values. The potential use for two different oleogels produced from candelilla wax, carnauba wax, and sunflower oil at different levels (2.5 g/100 g and 5 g/100 g wax) in cookie dough as well as the resulting product was also tested (Mert and Demirkesen 2016a). In this study, the extensibility of dough samples was also studied to determine the degree of gluten network formation. The lowest extensibility values were obtained from shortening, indicating a characteristic property of short dough. Nevertheless, small droplets of oil dispersed throughout the dough, which revealed mixing to be ineffective in preventing crosslink formation within the gluten network as well as in aeration activities when liquid oil was used for formulation. Consequently, oleogel had extensibility values between shortening-containing dough and sunflower containing samples. Due to the significant amount of liquid oil in candelilla or carnauba wax oleogels, they were not as effective as shortening in forming short dough. Oleogel-containing dough samples were found to be softer than dough prepared with shortening. This result revealed that oleogels lose their solid-like structure when exposed to shear. As compared to cookies prepared with sunflower oil, oleogel cookies had a softer texture and a higher spread ratio. On the other hand, oleogel cookies have a slightly harder texture but a higher spread ratio than the samples prepared with commercial shortening. Although oleogels could form self-standing structures, rheological measurements revealed that they could quickly lose their solid-like structure when mixed with dough. When the amount of the waxes used in the oleogel was increased, the extensibility values decreased, indicating decreased gluten network formation by crystals, but they were not as effective as solid fat in forming short dough due to their significant amount of liquid oil. This study revealed that using oleogels in cookies was better than using of liquid oil but that further optimization is necessary to match the quality of commercial cookies with shortening.

The same authors also employed oleogel/shortening blends in cookies (Mert and Demirkesen 2016b). Based on their previous findings, they selected the oleogels with canola oil with candelilla wax, and they tested both oleogels at two concentrations (3% and 6% w/w) and the oleogel/shortening blends at different ratios (30:70 and 60:40) as shortening replacements. At an unperturbed state, both oleogels and oleogel/shortening blends formed self-standing structures, and they had comparable rheological properties to shortening.

While oleogel/shortening blends could withstand shear, oleogels could not withstand the application of shear. Rheological measurements showed that a higher amount of shortening in blends improved the shear sensitivity of the samples and gave the dough samples texture properties closer to those of short dough. Using oleogels provided enhancements in the quality features of cookies compared to liquid oil, but further enhancements were obtained when oleogel-shortening blends were used in cookie formulations. So partial replacement of shortening with oleogels offered more acceptable dough and cookie features as compared to total replacement of shortening with oleogels. The authors proposed that gradually replacing shortening with oleogels might be a suitable approach for decreasing saturated fat in short dough products.

The effects of various oleogels made with different types of oils (olive oil, soybean oil, and flaxseed oil) and different waxes (sunflower wax, rice bran wax, beeswax, and candelilla wax) were also tested in cookie formulations (Hwang et al. 2016). It has been discovered that oleoel features, such as firmness and melting behavior, are influenced by both wax and oil types. Among different waxes, sunflower wax and rice bran wax made cookie dough the hardness, but the firmness values of oleogels did not directly reflect the dough hardness. Even though, wax type and oil type affected the properties of oleogels and cookie dough, they did not significantly influence the properties of cookies in terms of spread factor, hardness, and fracturability. The result showed that some cookies prepared with oleogels had similar quality parameters to those prepared with shortening.

In a recent study, Li et al. (2021) evaluated the effects of several types of oleogels prepared by various gelators such as HPMC, MAG (monoacylglycerol), SSL (sodium stearyl lactate), rice bran wax, and beeswax on the quality parameters of cookies. MAG, rice bran wax oloegel, and shortening cookies all displayed comparable rheological, textural, microstructural, and color properties. The cookies prepared with MAG and rice bran wax oleogel had homogeneous porous structure (Figure 4.5.1), an ideal color, and a crispy texture.

As depicted in Figure 4.4.1, SSL and beeswax oleogels had insufficient capacities to hold and stabilize air bubbles. Therefore, the holes in the cross-section of SSL and beeswax oleogel cookies were unevenly distributed as compared to shortening. HPMC oleogels made with both an emulsion and a foam template were not as effective in cookies as the other four oleogels made with low molecular weight gelators; these cookies had almost no pores (Figure 4.5.1). Consequently, in the case of wax-based oleogels, higher content of β' crystal and solid fat content resulted in cookies that were less hard, whereas β' content and solid fat content did not

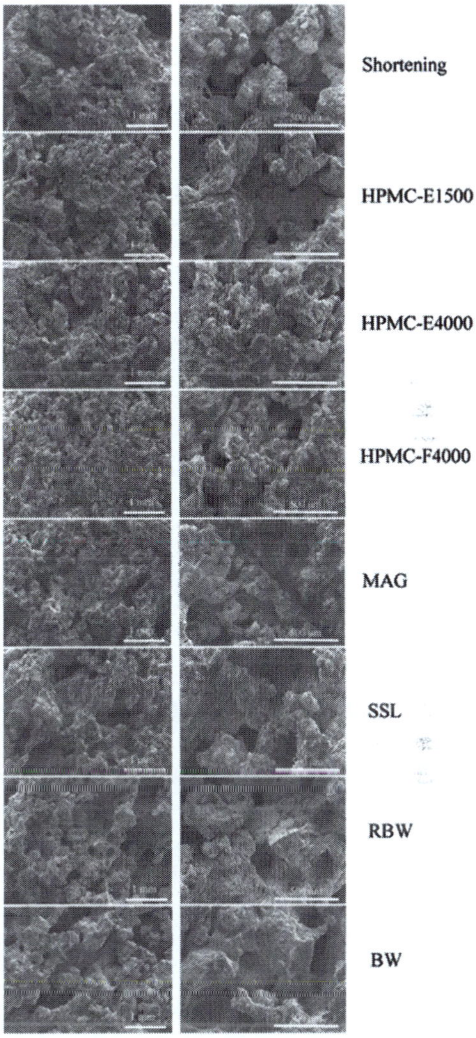

Figure 4.4.1 SEM images of cookies prepared with shortening and oleogels. HPMC-E1500: emulsion templated method, molecular weight 1500; HPMC-E4000: emulsion templated method, molecular weight 4000; HPMC-F4000: foam templated method, molecular weight 4000; MAG: monoglyceride; SSL: sodium stearoyl lactate; RBW: Rice bran wax; BW: bee wax. × 30 (left) and × 100 (right). Reproduced from Li et al. (2021), with permission from Elsevier.

affect the hardness values of cookies prepared with emulsifier-based oleogels. This finding provided a reference for developing oleogels to match shortening's quality in cookies. Waxes seem to be among the most suitable oleogelators to design oleogels that can be used in bakery products.

Generally, cookies with fat reductions of more than 30–40% using different kinds of fat mimetic such as inulin, oatrim, maltodextrin, etc. had higher hardness values and less acceptable sensory properties. Studies conducted on fat mimetics have shown that oleogel technology as a fat replacement strategy allows one to use higher fat replacement levels while maintaining acceptable product quality. The negative effects of removing the fat from a product may be solved with oleogel/shortening blends and partial fat replacement; using oleogel/shortening blends is seemingly the most successful technique for providing higher fat replacement levels while preserving healthy and good product quality.

4.4.2.3 Fat Mimetics in Cake/Muffin Formulation

One of shortening's roles in cake baking is formulating stable emulsions capable of withstanding the heat of baking. Hence, shortening promotes aeration by incorporating small air cells into the plastic phase during mixing. This enables air bubbles to be kept in the continuous phase of the emulsion at an ambient temperature rather than remaining in the aqueous phase. Shortening also helps with aeration throughout the baking process (Demirkesen and Mert 2020).

The fat crystals, which are covered with an interfacial layer of adsorbed protein during mixing, melted during baking, exposing the oil on the inside surface of the cells. This offers an additional interface for expansion. The baking mechanism lets air bubbles move from the fatty phase to the aqueous phase. A fine texture with homogenous pore distribution and a high volume are obtained from cakes (Demirkesen and Mert 2020).

Since the plasticity of shortening incorporates air bubbles during mixing, generally, super glycerinated shortenings containing monoglycerides and diglycerides (having surface activity) are preferred in cake formulations. In addition to plasticity, cake shortening must have a creamier structure. Cake shortening is a w/o type emulsion, and small β' crystals of fat in the continuous phase are essential for trapping air into the batter at a solid fat content. Saturated monoglycerides are chosen for cake manufacture because they tend to form complexes with the amylose part of starch, resulting in a softer texture, higher volume, and prolonged shelf life. Shortening is also responsible for moistness and tenderness. Monoglycerides, lactylated monoglycerides, propylene glycol esters, lecithin, polyglycerol esters, polysorbate 60, and sodium stearoyl lactylate are usually used as emulsifiers for the emulsified shortenings (Demirkesen and Mert 2020).

As presented in Table 4.4.2, the effects of different fat mimetics such as chia seed gel, inulin, oligofructose, maltodextrin, etc. have also been tested in cake/muffin formulations. Like cookie formulations, fat replacements with inulin at high levels gave a firm structure, hence cakes and muffins prepared with inulin instead of shortening were unacceptable by the panelist. But partial replacement levels (50%–70%) resulted in acceptable quality in cakes/muffins (Zahn et al. 2010; Rodríguez-García et al. 2012; Rodríguez-García et al. 2014). Inulin and oligofructose were also tested as fat replacers in cake formulations. Even though they were rated as acceptable by an untrained taste panel, 100% replacement levels resulted in lower scores compared to controls (Devereux et al. 2003).

The overall acceptability, sensory characteristics, functional properties, and nutrient content of cakes made using chia gel as a replacement for oil (25%, 50%, and 75%) or eggs were determined (Borneo et al. 2010). The formulation with 25% chia gel provided acceptable color, taste, and texture, and it also yielded a more nutritious product with acceptable sensory features. Chocolate

Table 4.4.2 An overview of research conducted on the application of various fat mimetics in cakes/muffins.

Fat mimetic type	Product type	Main findings	Reference
Inulin	Cake	The effects of several fat replacement levels (0%, 35%, 50%, 70%, and 100%) by inulin on sponge cake microstructure and physicochemical properties were studied. Oil substitution for inulin significantly reduced batter viscosity, providing heterogeneous bubble size distributions. Cakes with fat replacement of up to 70% had high crumb air cell values; they were softer and rated as acceptable by an untrained sensory panel. In the absence of oil, a less aerated structure after baking was obtained.	Rodríguez-García et al. (2012).
Inulin	Muffin	In a muffin formulation, 50%, 75%, or 100% of the baking fat was replaced with inulin. With increasing amounts of added inulin, product moisture and crumb density increased significantly, but muffin volume was reduced. The replacement of 50% fat in the formulation resulted in muffins that were comparable or slightly higher in crumb firmness.	Zahn et al. (2010).
Inulin and Oligofructose	Cookie, cake, and muffin	Inulin and oligofructose were used as fat replacers in cookies, blueberry muffins, carrot cake, chocolate cake, lemon cheesecake, ice cream, and beef sausages. Even though products were rated as acceptable by an untrained taste panel, 100% replacement levels resulted in lower scores compared to controls.	Devereux et al. (2003).
Chia gel	Cake	Overall acceptability, sensory features, functional properties, and nutrient content of cakes made using chia gel as a replacement for oil (25%, 50%, and 75%) or eggs were determined. 25% chia gel provided acceptable color, taste, and texture. Higher replacements caused a loss of the quality parameters. A 25% replacement of chia gel by oil or eggs in cakes yielded a more nutritious product with acceptable sensory features.	Borneo et al. (2010).
Chia mucilage gel	Cake	No significant improvement in specific volume, symmetry, uniformity, moisture, and water activity of the cakes with the replacement of vegetable fat by chia mucilage gel was seen. The lowest replacement level (25%) was technologically feasible in pound cakes with no significant alterations to their quality parameters.	Felisberto et al. (2015).
Chia seeds mucilage	Cake and bread	Breads and chocolate cakes made with chia mucilage were made with different levels of chia mucilage dried at 50°C or lyophilized as fat. For chocolate cakes, the technological features were improved in the formulation with a fat replacement level of less than 50%, but higher replacement levels deteriorated product quality.	Fernandes and de las Mercedes Salas-Mellado (2017)

(Continued)

Table 4.4.2 (Continued)

Fat mimetic type	Product type	Main findings	Reference
Maltodextrin	Cake	In a fat-reduced muffin, maltodextrin and high-fructose corn sweetener were used in combination to find their synergistic effects. Moisture and water activity values were like those of the full-fat muffin. However, using fat replacer deteriorated the color and texture. Results from both the physical and sensory analyses showed that the combination of maltodextrin gel up to a 75% reduction in shortening and sucrose/HFCS-90 resulted in satisfactory cakes.	Conforti and Archilla (2001)
Maltodextri, inulin, oligofructose, citrus pectin, microparticulated protein	Cake	The effects of different kinds of fat replacers at varying levels on the quality parameters of cake batters and resulting products were also tested. Using fat replacers above 65% increased firmness, elasticity, and decreased crumbliness and volume. The protein-based fat substitute showed the greatest differentiation among the other fat mimetics in terms of its textural properties.	Psimouli and Oreopoulou (2013)
Oleogels	Muffin	Muffin prepared with higher amounts of oleogels exhibited lower viscosity, greater shear-thinning behavior, and a less elastic nature. X-ray tomography showed a lower fragmentation index (*i.e.*, a more connected solid structure) in the oleogel-incorporated muffins, providing more enclosed and isolated air cells. The shortening replacement with oleogels produced muffins with a firmer and springier texture. Based on fatty acid compositions, the ratio of saturated to unsaturated fatty acids was significantly reduced from 2.81 to 0.41.	Lim et al. (2017).
Oleogels	Cake	Cakes prepared using oleogel-based w/o emulsions (20 wt% water) as a shortening alternative showed comparable functionalities (texture and sensory attributes) to the standard cake. But the nonuniform distribution of air in the batter due to the lack of solid fat crystals led to a lower uniformity in cell size or voids for the shellac oleogel emulsion cake. The authors propose investigating the use of emulsifiers to enable enhanced air incorporation and subsequent stabilization during baking as a future study.	Patel et al. (2014a).
Oleogels	Cake	Oleogels (containing >97 wt% of sunflower oil) were prepared using a combination of water-soluble food polymers (methylcellulose and xanthan gum) and evaluated for potential application as a shortening alternative. While the batter properties (air incorporation, rheology, and microstructure) of the oleogel batch were closer to those of the oil batch, the textural properties of the cakes were significantly better than oil and resembled those prepared using shortening and margarine.	Patel et al. (2014b).

Table 4.4.2 (Continued)

Fat mimetic type	Product type	Main findings	Reference
Oleogels	Cake	Natural waxes (rice bran wax, beeswax, and candelilla wax) were used to produce sunflower oil oleogels, and the physicochemical features of oleogels as a shortening replacer in cakes were compared. Cakes prepared with shortening and beeswax oleogels had crumb structure with homogenously distributed fine air cells. The shortening replacement with oleogels reduced the total porosity of the cakes, while there was no significant difference in the specific volume between the control and beeswax cakes. Among all the oleogel cakes, the lowest hardness was obtained from the beeswax samples, and the cakes with candelilla and rice bran wax became harder and chewy in texture. The levels of saturated fatty acids in the cakes containing oleogels were significantly reduced to 14–17%, compared to the control with shortening (58%). Beeswax oleogels were the most effective in producing nutritionally superior cakes with comparable quality attributes to those made with shortening.	Oh et al. (2017).
Oleogels	Cake	Canola oil-carnauba wax oleogels were evaluated as a replacement for shortening in cakes. The shortening replacement with oleogels at up to 50% was effective in maintaining the ability to hold air cells in the cake batters. The increasing levels of shortening replacement with oleogels decreased the volume and increased the firmness of cakes. The tomographic analysis showed the total porosity and fragmentation index were reduced in the oleogel cakes, showing a more connected solid structure. The levels of saturated fatty acids in the cakes containing oleogels were significantly reduced to 13.3%, compared to the control with shortening (74.2%). Using oleogels for shortening by up to 25% produced cakes with lower levels of saturated fatty acids without quality loss.	Kim et al. (2017).
Oleogels	Muffin	HPMC was used to structure sunflower oil into solid-like oleogels and the feasibility of HPMC oleogels as a shortening replacer in muffins was investigated in terms of physical, rheological, and tomographic properties. The specific gravity of batter tended to increase with elevating replacement levels of shortening, but the specific volume of baked muffins was not significantly reduced at shortening replacement levels of up to 50%. According to X-ray micro-computed tomography analysis, muffins prepared with HPMC oleogels had large and homogeneously distributed air cells. The results showed HPMC oleogels could be effective in replacing shortening by up to 50% without significant deterioration in the quality attributes of muffins.	Oh and Lee (2018).

(Continued)

Table 4.4.2 (Continued)

Fat mimetic type	Product type	Main findings	Reference
Oleogels	Cake	According to the sensory features, cakes, including oleogels, had sensory scores close to or greater than the control sample. Results indicated that oleogel formulation significantly influenced the quality features of the end product.	Pehlivanoglu et al. (2018).
Oleogels	Muffin	The effectiveness of using oleogels from high oleic sunflower oil and monoglycerides as solid fat replacers in muffins. The replacement of commercial margarine with the optimized oleogels in muffin formulation resulting in improved spreadability, higher specific volume, identical hardness values, and a more uniform crumb structure. After 10 days of storage, these muffins showed a 50% reduction in oil migration compared to the control muffins.	Giacomozzi et al. (2018).
Oleogels	Cake	Since gluten-free bakery products require a high amount of fat, the fat used in these products is often high in saturated fatty acids. Beeswax containing oleogels for partial replacement of the shortening in gluten-free aerated products was evaluated. Oleogels prepared with beeswax were cocrystallized with a commercial cake shortening in the laboratory scale crystallization unit. Then, the resulting blends were tested in the gluten-free cake formulations. When the beeswax oleogel was used alone, the overrun values of the batter samples decreased, indicating reduced air-holding ability. Product porosity and specific volume of the samples were also diminished with the complete replacement of the shortening with beeswax oleogel. Still, 45%, 30%, and 15% replacement of the shortening with beeswax oleogel gave batter and baked product properties comparable to those of the control products. It has been shown that gradual replacement of shortening with oleogels may be an alternative method for a partial reduction of saturated fat without changing the physical properties of gluten-free aerated products.	Demirkesen and Mert (2019).
Oleogels	Cake	The influence of candelilla wax/canola oil oleogel on the starch digestibility and texture of wheat sponge cake bread was investigated. Candelilla wax/canola oil oleogel/butter blends were made by substituting butter at 0, 25, 50, 75, and 100% of the oleogel. Cake hardness decreased with oleogel substitution but increased cake specific volume. Incorporating oleogel into the *in vitro* starch digestibility, the digestible starch fraction increased from 70% to about 84%.	Alvarez-Ramirez et al. (2020).

Table 4.4.2 (Continued)

Fat mimetic type	Product type	Main findings	Reference
Oleogels	Cake	The potential application of reinforced carnauba wax-based oleogel with adipic acid in the cake and the beef burger was evaluated. Adding adipic acid to carnauba wax-based oleogels formed new intramolecular or intermolecular hydrogen bonds and enhanced the thermal behavior and crystallinity of oleogels. The increase of adipic acid concentration to over 3% of the oleogel formulation significantly improved the strength of oleogels. Cake prepared with the partial substitution of adipic acid in carnauba wax-based oleogels gave an acceptable texture profile, color, and organoleptic features.	Khiabani et al. (2020).
Oleogels	Muffin	The rheological properties and oxidative stability of candelilla wax oleogels fortified with glycerol monostearate and β-carotene, and their applications to muffin as a shortening replacer were evaluated. The interaction between candelilla wax and glycerol monostearate strengthened the oleogel structure. The oleogels with β-carotene showed the lowest peroxide values compared to the other samples. The muffins prepared with oleogels for shortening had greater specific gravity and harder texture, but the specific volume of the shortening and oleogel samples with glycerol monostearate were statistically found to be similar. Muffins with β-carotene oleogels had the highest oxidative stability. Therefore, this study indicated that incorporatingβ-carotene and glycerol monostearate in oleogels positively affected the storage stability of muffins.	Jeong et al. (2021).

cakes were made with different levels of chia mucilage, were dried at 50°C, or were lyophilized as fat (Fernandes and de las Mercedes Salas-Mellado 2017). For chocolate cakes, the technological characteristics were enhanced in the formulation with a fat replacement level of less than 50%, but the higher replacement levels deteriorated product quality (Fernandes and de las Mercedes Salas-Mellado 2017). On the contrary, Felisberto et al. (2015) stated that there was no significant improvement in specific volume, symmetry, uniformity, moisture, and water activity of the cakes with the replacement of vegetable fat by chia mucilage gel. Similar to the findings of Borneo et al. (2010), the lowest replacement level (25%) was found to be technologically feasible in pound cakes with no significant alterations to their quality parameters, and higher replacements caused a loss of the quality parameters (Felisberto et al. 2015).

The influences of different kinds of fat replacers (maltodextrin, inulin (high performance and granulated), oligofructose, citrus pectin, and microparticulated protein) at varying levels (35, 65, and 100%) on quality parameters of cake batters and resulting products were also tested by Psimouli and Oreopoulou (2013). Generally, fat replacement by 35% did not induce significant differences. Above 65% fat replacement the increasing levels of fat replacers resulted in high firmness, elasticity, and lower crumbliness and volume values. At 65% fat replacement level, cakes presented acceptable textural, physical, and sensorial attributes. But cakes prepared with higher fat replacement levels received lower scores on taste and flavor, but at total fat replacement they were evaluated as not acceptable. The protein-based fat substitute showed the greatest differentiation among the other fat mimetics in terms of textural properties (Psimouli and Oreopoulou 2013).

The conducted studies have shown that each fat mimetic functions well when used for low-fat replacement levels. Cakes/muffins prepared with fat mimetics at high levels lead to unacceptable sensory quality. In sensory aspects, consider customer acceptance, which varies based on the kind and percentage of substitution, as well as the reformulated product. More research must explore other plant derived ingredients that can be incorporated either for their structural role or for their bioactive compounds and nutritional value.

But the negative effects of removing all fat from a product may be solved with product development. The feasibility of different types of oleogels as shortening replacers in cakes and muffins was evaluated in several studies (Table 4.4.2). Natural waxes (rice bran wax, beeswax, and candelilla wax) were used to produce sunflower oil oleogels and physicochemical features of oleogels as a shortening replacer in cakes were compared in the study of Oh et al. (2017). Cakes prepared with shortening and beeswax oleogels had crumb structure with homogenously distributed fine air cells. The shortening replacement with oleogels reduced the total porosity of the cakes while there was no significant difference in the specific volume between the control and beeswax cakes. Among all oleogel cakes, the lowest hardness was obtained from cakes prepared with beeswax samples, and the cakes with candelilla and rice bran wax became harder and chewy in texture. The levels of saturated fatty acids in the cakes containing oleogels were significantly reduced to 14–17%, compared to the control with shortening (58%). Beeswax oleogels proved to be the most effective in producing nutritionally superior cakes with comparable quality attributes to the shortening cakes. In another study, the shortening replacement with canola oil-carnauba wax oleogels at up to 50% was effective in holding air cells in the cake batters. However, higher replacement levels decreased the cakes' volume but increased their firmness. Using of oleogels for shortening by 25% provided cakes with lower levels of saturated fatty acids without quality loss (Kim et al. 2017). Oleogel incorporation also positively affected the *in vitro* starch digestibility, as the digestible starch fraction increased from 70% to about 84% (Alvarez-Ramirez et al. 2020).

The potential edible application of shellac oleogels as a shortening alternative for cake preparation was also tested. Cakes prepared using oleogel-based w/o emulsions (20 wt% water) as a shortening alternative indicated comparable functionalities (texture and sensory attributes) to the standard cake (Patel et al. 2014a).

In addition to waxes and shellac gels, hydrocolloids have also been used to form oleogels. In the study of Oh and Lee (2018), the feasibility of HPMC oleogels as a shortening replacer in muffins was investigated in terms of physical, rheological, and tomographic properties. According to X-ray, micro-computed tomographic investigation, air bubbles were large and were heterogeneously dispersed in the muffins prepared with HPMC oleogels (Figure 4.4.2).

The specific gravity of batter tended to increase with increased replacement levels of shortening, but the specific volume of baked muffins was not significantly reduced at shortening replacement levels of up to 50%.

The to apply oleogels from methylcellulose and xanthan gum was evaluated as a shortening alternative in cakes (Patel et al. 2014b). While the batter properties (air incorporation, rheology, and microstructure) of the oleogel batch were closer to the oil batch, the textural properties of cakes were significantly better than oil and resembled more to those prepared using shortening and margarine.

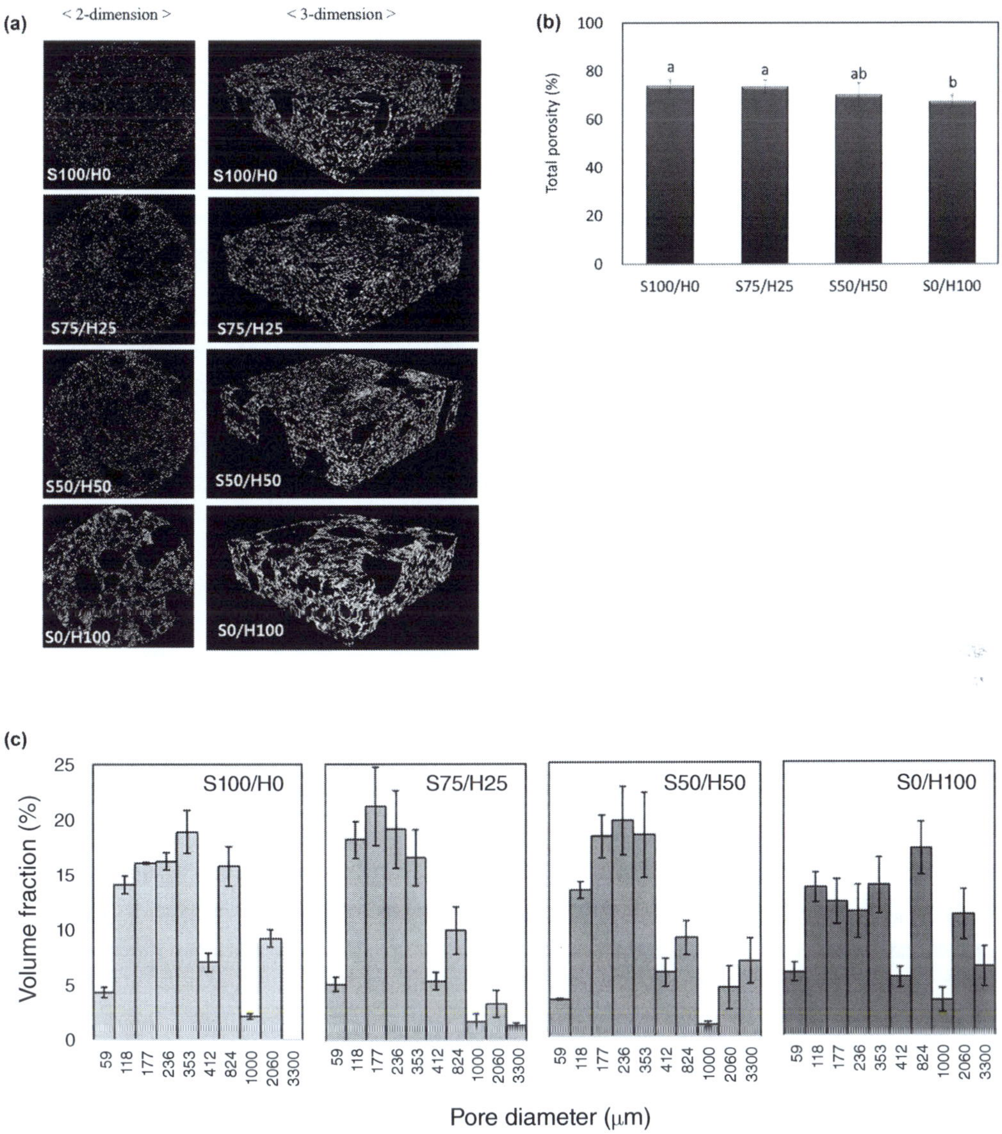

Figure 4.4.2 Micro-CT tomographical analysis of muffins prepared with different levels of shortening (S) replacement with HPMC oleogel (H); 2D and 3D images (a), total porosity (b), and pore size distribution (c). Reproduced from Oh and Lee (2018) with permission from Elsevier.

Gluten-free bakery products often require a high amount of fat to match the quality parameters of their counterparts. In the study by Demirkesen and Mert (2019), beeswax containing oleogels for partial replacement of the shortening in gluten-free aerated products were evaluated. Oleogels made of beeswax were co-crystallized with a commercial cake shortening in the laboratory scale crystallization unit, and the resulting blends were used in the gluten-free cake formulations. The overrun values of the batter samples decreased, indicating a reduced air-holding capacity when only beeswax oleogels were used. Product porosity and specific sample volumes were also reduced with shortening being completely replaced with beeswax oleogels. However, 45%, 30%, and 15% replacement of the shortening with beeswax oleogels provided comparable quality to batter and baked products as the control products. Rheological and textural measurements, microscopy, and bubble size distribution suggest that gradually replacing shortening with oleogels may be an alternative method for a partial reduction of saturated fat without changing the physical properties of gluten-free aerated products.

Among all these fat mimetics, oleogels seem to be the most successful fat mimetics in cakes and muffins, with no alterations in sensory quality even at higher fat replacement levels. But using oleogels at concentrations higher than 50% may cause considerable changes in the physical properties of cakes and muffins, particularly in terms of firmness values, which can make cake and muffin manufacturing challenging. Even while oleogel technology has great potential to be applied to a wide range of bakery products, still some challenges need to be faced. The choice of structuring agents, which can endure mixing and other processing conditions while also presenting the desired functionality, should also be determined. Although the literature shows that oleogels have great potential to be used as shortening replacers in bakery products, further studies on oil structuring are needed to optimize oleogels to match the quality attributes of commercial shortening-containing bakery products.

4.4.3 Conclusion

Fat mimetics also have body, lubricity, perceived creaminess or smoothness, absorption, and adsorption effects on taste buds or tongue, as well as cohesiveness, adhesiveness, and waxiness. Various fat mimetics such as chia seed gel, inulin, maltodextrin, polydextrose, and whey protein concentrates have also been utilized in bakery products. They can mimic the functional and sensorial properties of fat while providing a healthier and lower-calorie alternative. Studies have shown that each fat mimetic (except oleogel) functions well when it is used for low-fat replacement levels. However, higher use levels of these kinds of mimetics worsen the sensory and textural quality characteristics of products. It has been suggested that combining fat mimetics with other ingredients might improve fat-like properties since the additional ingredients are usually intended to reduce the negative effects of the individual fat mimetics. In the case of sensory aspects, it is important to consider customer acceptance, which varies based on the kind and percentage of substitution, as well as the reformulated product. The negative effects of removing all the fat from a product may be solved with product development. From this perspective, among all these fat mimetics, oleogels appear to be the most successful fat mimetics in bakery products, with no alterations in sensory quality even at higher fat replacement levels. On the other hand, it should be noted that utilizing oleogels at concentrations greater than 50% may cause considerable changes in the physical properties of products, particularly in terms of firmnesss. The negative effects of removing the fat from a product may be solved by gradually replacing shortening with oleogels and partial fat replacement by using oleogel/shortening blends seems to be the most suitable technique to provide higher fat replacement levels while preserving healthy and good product quality. Even while oleogel technology has great potential to be applied to a wide range of bakery products, there are still some challenges to be faced, such as the selection of proper structuring agents for each kind of bakery product. Therefore, further oil structuring studies are needed to optimize oleogels to match the quality commercial shortening-containing bakery products' attributes.

References

Adair, M., Knight, S., and Gates, G. (2001). Acceptability of peanut butter cookies prepared using mungbean paste as a fat ingredient substitute. *Journal of the American Dietetic Association* 101 (4): 467–467.

Aggarwal, D., Sabikhi, L., and Kumar, M.S. (2016). Formulation of reduced-calorie biscuits using artificial sweeteners and fat replacer with dairy–multigrain approach. *NFS Journal* 2: 1–7.

Alvarez-Ramirez, J., Vernon-Carter, E.J., Carrera-Tarela, Y. et al. (2020). Effects of candelilla wax/canola oil oleogel on the rheology, texture, thermal properties and in vitro starch digestibility of wheat sponge cake bread. *LWT Food Science and Technology* 130: 109701.

Ari, H.O., Işlek, E., Özatkan, Y., Bilir, K., Demirkesen Mert, I., Özcan Cetin, E.H., Yildrim, H.H., and Erguder, T. (2021). *EVIPNet evidence brief for policy: Reducing the consumption of trans-fats and their negative impacts on health in Turkey*. Copenhagen: WHO Regional Office for Europe. License: CC BY-NC-SA 3.0 IGO.

Balic, R., Miljkovic, T., Ozsisli, B., and Simsek, S. (2017). Utilization of modified wheat and tapioca starches as fat replacements in bread formulation. *Journal of Food Processing and Preservation* 41 (1): e12888.

Balic, R.P. (2015). *Utilization of modified starch as a fat replacement in bread*.

Beran, M., Drahorad, J., Vltavsky, O., Urban, M., Lakenerova, I., Fronek, M., Sova, J., Ondracek, J., Ondrackova, L., Kralova, M., and Fromankova, S. (2018). Pilot-scale production and application of microparticulated plant proteins. *Journal of Nutrition & Food Sciences* 8 (655): 2.

Błońska, A., Marzec, A., and Błaszczyk, A. (2014). Instrumental evaluation of acoustic and mechanical texture properties of short-dough biscuits with different content of fat and inulin. *Journal of Texture Studies* 45 (3): 226–234.

Borneo, R., Aguirre, A., and León, A.E. (2010). Chia (Salvia hispanica L) gel can be used as egg or oil replacer in cake formulations. *Journal of the American Dietetic Association* 110 (6): 946–949.

Capitani, M.I., Ixtaina, V.Y., Nolasco, S.M., and Tomás, M.C. (2013). Microstructure, chemical composition and mucilage exudation of chia (Salvia hispanica L.) nutlets from Argentina. *Journal of the Science of Food and Agriculture* 93 (15): 3856–3862.

Chen, Y., She, Y., Zhang, R., Wang, J., Zhang, Z., and Gou, X. (2020). Use of starch-based fat replacers in foods as a strategy to reduce dietary intake of fat and risk of metabolic diseases. *Food Science and Nutrition* 8 (1): 16–22.

Chugh, B., Singh, G., and Kumbhar, B.K. (2013). Development of low-fat soft dough biscuits using carbohydrate-based fat replacers. *International Journal of Food Science* 2013.

Conforti, F.D. and Archilla, L. (2001). Evaluation of a maltodextrin gel as a partial replacement for fat in a high-ratio white-layer cake. *International Journal of Consumer Studies* 25 (3): 238–245.

Conforti, F.D., Charles, S.A., and Duncan, S.E. (1997). Evaluation of a carbohydrate-based fat replacer in a fat-reduced baking powder biscuit. *Journal of Food Quality* 20 (3): 247–256.

Demirkesen, I. and Mert, B. (2019). Utilization of beeswax oleogel-shortening mixtures in gluten-free bakery products. *Journal of the American Oil Chemists' Society* 96 (5): 545–554.

Demirkesen, I., and Mert, B. (2020). Recent developments of oleogel utilizations in bakery products. *Critical Reviews in Food Science and Nutrition* 60 (14): 2460–2479.

Devereux, H.M., Jones, G.P., McCormack, L., and Hunter, W.C. (2003). Consumer acceptability of low fat foods containing inulin and oligofructose. *Journal of Food Science* 68 (5): 1850–1854.

Doménech-Asensi, G., Merola, N., López-Fernández, A., Ros-Berruezo, G., and Frontela-Saseta, C. (2016). Influence of the reformulation of ingredients in bakery products on healthy characteristics and acceptability of consumers. *International Journal of Food Sciences and Nutrition* 67 (1): 74–82.

Erinc, H., Mert, B., and Tekin, A. (2018). Different sized wheat bran fibers as fat mimetic in biscuits: its effects on dough rheology and biscuit quality. *Journal of Food Science and Technology* 55 (10): 3960–3970.

Felisberto, M.H.F., Wahanik, A.L., Gomes-Ruffi, C.R. et al. (2015). Use of chia (Salvia hispanica L.) mucilage gel to reduce fat in pound cakes. *LWT Food Science and Technology* 63 (2): 1049–1055.

Fernandes, S.S. and de las Mercedes Salas-Mellado, M. (2017). Addition of chia seed mucilage for reduction of fat content in bread and cakes. *Food Chemistry* 227: 237–244.

Forker, A., Zahn, S., and Rohm, H. (2012). A combination of fat replacers enables the production of fat-reduced shortdough biscuits with high-sensory quality. *Food and Bioprocess Technology* 5 (6): 2497–2505.

Giacomozzi, A.S., Carrín, M.E., and Palla, C.A. (2018). Muffins elaborated with optimized monoglycerides oleogels: From solid fat replacer obtention to product quality evaluation. *Journal of Food Science* 83 (6): 1505–1515.

Giarnetti, M., Paradiso, V.M., Caponio, F., Summo, C., and Pasquaione, A. (2015). Fat replacement in shortbread cookies using an emulsion filled gel based on inulin and extra virgin olive oil. *LWT Food Science and Technology* 63 (1): 339–345.

Gutiérrez-Luna, K., Astiasarán, I., and Ansorena, D. (2022). Gels as fat replacers in bakery products: A review. *Critical Reviews in Food Science and Nutrition* 1–14.

Hwang, H.S., Singh, M., and Lee, S. (2016). Properties of cookies made with natural wax–vegetable oil organogels. *Journal of Food Science* 81 (5): 1045–1054.

Jang, A., Bae, W., Hwang, H.S., Lee, H.G., and Lee, S. (2015). Evaluation of canola oil oleogels with candelilla wax as an alternative to shortening in baked goods. *Food Chemistry* 187: 525–529.

Jeong, S., Lee, S., and Oh, I. (2021). Development of antioxidant-fortified oleogel and its application as a solid fat replacer to Muffin. *Foods* 10 (12): 3059.

Jung, D., Oh, I., Lee, J., and Lee, S. (2020). Utilization of butter and oleogel blends in sweet pan bread for saturated fat reduction: Dough rheology and baking performance. *LWT Food Science and Technology* 125: 109194.

Khiabani, A.A., Tabibiazar, M., Roufegarinejad, L., Hamishehkar, H., and Alizadeh, A. (2020). Preparation and characterization of carnauba wax/adipic acid oleogel: A new reinforced oleogel for application in cake and beef burger. *Food Chemistry* 333: 127446.

Kim, J.Y., Lim, J., Lee, J., Hwang, H.S., and Lee, S. (2017). Utilization of oleogels as a replacement for solid fat in aerated baked goods: Physicochemical, rheological, and tomographic characterization. *Journal of Food Science* 82 (2): 445–452.

Krystyjan, M., Gumul, D., Ziobro, R., and Sikora, M. (2015). The effect of inulin as a fat replacement on dough and biscuit properties. *Journal of Food Quality* 38 (5): 305–315.

Laguna, L., Primo-Martín, C., Varela, P., Salvador, A., and Sanz, T. (2014). HPMC and inulin as fat replacers in biscuits: Sensory and instrumental evaluation. *LWT Food Science and Technology* 56 (2): 494–501.

Lee, Y.T. and Puligundla, P. (2016). Characteristics of reduced-fat muffins and cookies with native and modified rice starches. *Emirates Journal of Food and Agriculture* 311–316.

Li, S., Wu, G., Li, X., Jin, Q., Wang, X., and Zhang, H. (2021). Roles of gelator type and gelation technology on texture and sensory properties of cookies prepared with oleogels. *Food Chemistry* 356: 129667.

Lim, J., Jeong, S., Lee, J., Park, S., Lee, J., and Lee, S. (2017). Effect of shortening replacement with oleogels on the rheological and tomographic characteristics of aerated baked goods. *Journal of the Science of Food and Agriculture* 97 (11): 3727–3732.

Lourencetti, R.E., Benossi, L., Marques, D.R., Joia, B.M., and Monteiro, A.R.G. (2013). Development of biscuit type cookie with partial replacement of fat by inulin. *International Journal of Food Sciences and Nutrition* 2 (5): 261–265.

Maghaydah, S., Abdul-Hussain, S., Ajo, R., Obeidat, B., and Tawalbeh, Y. (2013). Enhancing the nutritional value of gluten-free cookies with inulin. *Advance Journal of Food Science and Technology* 5 (7): 866–870.

Mancebo, C.M., Rodríguez, P., Martínez, M.M., and Gómez, M. (2018). Effect of the addition of soluble (nutriose, inulin and polydextrose) and insoluble (bamboo, potato and pea) fibres on the quality of sugar-snap cookies. *International Journal of Food Science and Technology* 53 (1): 129–136.

Meng, Z., Guo, Y., Wang, Y., and Liu, Y. (2019). Oleogels from sodium stearoyl lactylate-based lamellar crystals: Structural characterization and bread application. *Food Chemistry* 292: 134–142.

Mert, B. and Demirkesen, I. (2016a). Evaluation of highly unsaturated oleogels as shortening replacer in a short dough product. *LWT Food Science and Technology* 68: 477–484.

Mert, B. and Demirkesen, I. (2016b). Reducing saturated fat with oleogel/shortening blends in a baked product. *Food Chemistry* 199: 809–816.

Mert, I.D., Sumnu, G., and Sahin, S. (2016). Microstructure of gluten-free baked products. In: *Imaging Technologies and Data Processing for Food Engineers* (ed. N. Sozer), Cham: Springer. 197–242.

O'brien, C.M., Mueller, A., Scannell, A.G.M., and Arendt, E.K. (2003). Evaluation of the effects of fat replacers on the quality of wheat bread. *Journal of Food Engineering* 56 (2–3): 265–267.

Oh, I.K., Amoah, C., Lim, J., Jeong, S., and Lee, S. (2017). Assessing the effectiveness of wax-based sunflower oil oleogels in cakes as a shortening replacer. *LWT Food Science and Technology* 86: 430–437.

Oh, I.K. and Lee, S. (2018). Utilization of foam structured hydroxypropyl methylcellulose for oleogels and their application as a solid fat replacer in muffins. *Food Hydrocolloids* 77: 796–802.

Onacik-Gür, S., Żbikowska, A., and Jaroszewska, A. (2015). Effect of high-oleic sunflower oil and other pro-health ingredients on physical and sensory properties of biscuits. *CYTA Journal of Food* 13 (4): 621–628.

Paciulli, M., Littardi, P., Carini, E., Paradiso, V., Castellino, M., and Chiavaro, E.(2020). Inulin-based emulsion filled gel as fat replacer in shortbread cookies: Effects during storage. *LWT Food Science and Technology* 133: 109888.

Patel, A.R., Cludts, N., Sintang, M.D.B., Lesaffer, A., and Dewetinck, K. (2014b). Edible oleogels based on water soluble food polymers: Preparation, characterization and potential application. *Food and Function* 5 (11): 2833–2841.

Patel, A.R., Rajarethinem, P.S., Grędowska, A., Turhan, O., Lesaffer, A., De Vos, W.H., Van de Walle, D., and Dewettinck, K. (2014a). Edible applications of shellac oleogels: Spreads, chocolate paste and cakes. *Food and Function* 5 (4): 645–652.

Pehlivanoglu, H., Ozulku, G., Yildirim, R.M., Demirci, M., Toker, O.S., and Sagdic, O. (2018). Investigating the usage of unsaturated fatty acid-rich and low-calorie oleogels as a shortening mimetics in cake. *Journal of Food Processing and Preservation* 42 (6): e13621.

Poinot, P., Arvisenet, G., Grua-Priol, J., Fillonneau, C., Le-Bail, A., and Prost, C. (2010). Influence of inulin on bread: Kinetics and physico-chemical indicators of the formation of volatile compounds during baking. *Food Chemistry* 119 (4): 1474–1484.

Psimouli, V. and Oreopoulou, V. (2013). The effect of fat replacers on batter and cake properties. *Journal of Food Science* 78 (10): 1495–1502.

Rios, M.B., Iriondo-DeHond, A., Iriondo-DeHond, M., Herrera, T., Velasco, D., Gómez-Alonso, S., Callejo, M.J., and del Castillo, M.D. (2020). Effect of coffee cascara dietary fiber on the physicochemical, nutritional and sensory properties of a gluten-free bread formulation. *Molecules* 25 (6): 1358.

Rodriguez-Garcia, J., Laguna, L., Puig, A., Salvador, A., Hernando, I (2013). Effect of fat replacement by inulin on textural and structural properties of short dough biscuits. *Food and Bioprocess Technology* 6 (10): 2739–2750.

Rodríguez-García, J., Puig, A., Salvador, A., and Hernando, I. (2012). Optimization of a sponge cake formulation with inulin as fat replacer: structure, physicochemical, and sensory properties. *Journal of Food Science* 77 (2): C189–C197.

Rodríguez-García, J., Salvador, A., and Hernando, I. (2014). Replacing fat and sugar with inulin in cakes: Bubble size distribution, physical and sensory properties. *Food and Bioprocess Technology* 7 (4): 964–974.

Romanchik-Cerpovicz, J.E., Jeffords, M.J., and Onyenwoke, A.C. (2019). College student acceptance of chocolate bar cookies containing puree of canned green peas as a fat-ingredient substitute. *Journal of Culinary Science and Technology* 17 (6): 507–518.

Rößle, C., Ktenioudaki, A., and Gallagher, E. (2011). Inulin and oligofructose as fat and sugar substitutes in quick breads (scones): A mixture design approach. *European Food Research* 233 (1): 167–181.

Saeed, S.M.G., Ali, S.A., Ali, R., Naz, S., Sayeed, S.A., Mobin, L., and Ahmed, R. (2020a). Utilization of Vigna mungo flour as fat mimetic in biscuits: Its impact on antioxidant profile, polyphenolic content, storage stability, and quality attributes. *Legume Science* 2 (4): e58.

Saeed, S.M.G., Tayyaba, S., Ali, S.A., Tayyab, S., Sayeed, S.A. and Ali, R. (2020b). Evaluation of the potential of Lotus root (Nelumbo nucifera) flour as a fat mimetic in biscuits with improved functional and nutritional properties. *CYTA Journal of Food* 18 (1): 624–634.

Scheuer, P.M., Ferreira, J.A.S., Mattioni, B., de Miranda, M.Z., and de Francisco, A. (2015). Optimization of image analysis techniques for quality assessment of whole-wheat breads made with fat replacer. *Food Science and Technology* 35: 133–142.

Scheuer, P.M., Luccio, M.D., Zibetti, A.W., de Miranda, M.Z., and de Francisco, A. (2016). Relationship between instrumental and sensory texture profile of bread loaves made with whole-wheat flour and fat replacer. *Journal of Texture Studies* 47 (1): 14–23.

Scheuer, P.M., Mattioni, B., Barreto, P.L.M., Montenegro, F.M., Gomes-Ruffi, C.R., Biondi, S., Kilpp, M., and de Francisco, A. (2014). Effects of fat replacement on properties of whole wheat bread. *Brazilian Journal of Pharmaceutical Sciences* 50: 703–712.

Sudha, M.L., Srivastava, A.K., Vetrimani, R., and Leelavathi, K. (2007). Fat replacement in soft dough biscuits: Its implications on dough rheology and biscuit quality. *Journal of Food Engineering* 80 (3): 922–930.

Swanson, B.G. (2006). Fat replacers: mimetics and substitutes. In: *Nutraceutical and Specialty Lipids and Their Co-Products* (ed. F. Shahidi), CRC Press. 343–354.

Wekwete, B. and Navder, K.P. (2008). Effects of avocado fruit puree and oatrim as fat replacers on the physical, textural and sensory properties of oatmeal cookies. *Journal of Food Quality* 31 (2): 131–141.

Ye, X., Li, P., Lo, Y.M., Fu, H., and Cao, Y. (2019). Development of novel shortenings structured by ethylcellulose oleogels. *Journal of Food Science* 84 (6): 1456–1464.

Yılmaz, E. and Öğütcü, M. (2015). The texture, sensory properties and stability of cookies prepared with wax oleogels. *Food and Function* 6 (4): 1194–1204.

Zahn, S., Pepke, F., and Rohm, H. (2010). Effect of inulin as a fat replacer on texture and sensory properties of muffins. *International Journal of Food Science and Technology* 45 (12): 2531–2537.

Zettel, V. and Hitzmann, B. (2016). Chia (Salvia hispanica L.) as fat replacer in sweet pan breads. *International Journal of Food Science and Technology* 51 (6): 1425–1432.

Zhao, M., Lan, Y., Cui, L., Monono, E., Rao, J., and Chen, B. (2020a). Physical properties and cookie-making performance of oleogels prepared with crude and refined soybean oil: a comparative study. *Food and Function* 11 (3): 2498–2508.

Zhao, M., Lan, Y., Cui, L., Monano, E., Rao, J., and Chen, B. (2020b). Formation, characterization, and potential food application of rice bran wax oleogels: Expeller-pressed corn germ oil versus refined corn oil. *Food Chemistry* 309: 125704.

Zoulias, E.I., Oreopoulou, V., and Kounalaki, E. (2002b). Effect of fat and sugar replacement on cookie properties. *Journal of the Science of Food and Agriculture* 82 (14): 1637–1644.

Zoulias, E.I., Oreopoulou, V., and Tzia, C. (2000). Effect of fat mimetics on physical, textural and sensory properties of cookies. *International Journal of Food Properties* 3 (3): 385–397.

Zoulias, E.I., Oreopoulou, V., and Tzia, C. (2002a). Textural properties of low-fat cookies containing carbohydrate-or protein-based fat replacers. *Journal of Food Engineering* 55 (4): 337–342.

Section V

Industrial Perspective

5.1

Molecular Gels – Barriers, Advances, and Opportunities

Michael A. Rogers

Associate Professor and Tier II Canada Research Chair in Food Nanotechnology
University of Guelph, Guelph, Ontario, Canada

5.1.1 Introduction

Molecular gels, although first reported in the mid-19th century with the self-assembly of lithium laurate in water (Lipowitz 1841), received little attention until 1991 (Weiss 2014), after which citations per year increased to over 3,000 and have steadily increased since. The key feature differentiating a molecular gel and a polymeric gel is the networking forming compound mass, whereby molecular gels are 0-Dimensional (0D) and must self-assemble via non-covalent interactions before forming 1-Dimensional (1D) fibers that are similar in size and perspective to a polymeric (1D) gelator. More rarely, self-assembly may lead to higher-order aggregates such as 2D platelets (Daniel and Rajasekharan 2003; Dassanayake et al. 2009; Hwang et al. 2013) or 3D spherulites (Rogers et al,. 2009b; Guo et al. 2020b) that result in a gel formation (Weiss 2014); however, the network-liquid interface decrease typically produces less effective gels. Molecular gels are predominantly liquid systems whereby the molecular gelator undergoes microscopic separation, nucleation, and crystal growth resulting in 1D fibers that permeate the entire system via a continuous self-assembled fibrillar network (SAFiN) (3D). Although the resultant networks are dynamic and typically reversible, they must persist for the duration of the analytical experiment to have solid-like rheological properties where the elastic modulus (G′) exceeds the loss or viscous modulus (G″) across frequencies (Flory 1974; Weiss 2014). Molecular self-assembly depends not only on the position, stereochemistry, and type of functional group on the gelator but also on the solvent functional groups (Raynal and Bouteiller 2011; Diehn et al. 2014a; Lan et al. 2014a, 2015).

5.1.2 Reliance Serendipitous Discovery

Presently, researchers rely mainly on serendipitously discovering new molecular gelators, which impedes application across many fields, including food science. The complex interplay between gelator structure and concentration, solvent chemistry, and supersaturation on self-assembly has hindered the rational design of molecular gels. The formation of a sol by heating the gelator solvent mixture above its dissolution temperature precedes self-assembly, after which cooling the sol drives supersaturation starting stochastic nucleation. (Weiss and Terech 2006) Bulk crystallization occurs via macroscopic phase separation, forming visible solids and liquids; however, the supersaturated state causes the gelator molecules to microscopically phase separate and self-assemble via stochastic nucleation events driven by enthalpic forces (Lam et al. 2010). How a sol is cooled (e.g.,

Fat Mimetics for Food Applications, First Edition. Edited by Miguel Ângelo Parente Ribeiro Cerqueira and Lorenzo Miguel Pastrana Castro.

isothermal or non-isothermal cooling rate and crystallization temperature) varies the degree of supersaturation, a time/temperature-dependent parameter that alters crystallization, and, ultimately, the SAFiN morphology (Rogers and Marangoni 2009; Li and Liu 2010). Aggregating molecular gelators into SAFiNs requires a meticulous balance between contrasting solubility parameters and those controlling epitaxial growth. Unlike common nucleation events, molecular gels require highly specific interactions that promote 1D crystal growth; these interactions include hydrogen bonding (Duval et al. 2006; Wang et al. 2013), π-π stacking, electrostatic (Fabri et al. 1998), and van der Waals interactions (Draper and Adams 2017; Chivers and Smith 2019). The consequent fibers formed from 1D growth are described as "crystal-like" and have the same features as polymer chains in polymeric gels.

Following nucleation, for a 3D SAFiN to result in a molecular gel, the 1D crystalline fibers must interact or entangle, forming a network of transient and permanent junctions between fibers that permeate the entire system. Permanent junction zones, resulting from crystallographic mismatches at the solvent-fiber interface of the growing crystals, effectively entrains the apolar liquid in the mesh-like network (Wu et al. 2013; Diehn et al. 2014a). The degree of supercooling or the driving force for crystallization varies crystallographic mismatches, altering fiber length and mesh pore size, which ultimately affects the gelation outcome (Liu and Sawant 2001, 2002; Lam et al. 2009; Wang et al. 2013). Low supersaturation at slow cooling rates or high isothermal crystallization temperatures produces 1D crystals; conversely, high cooling rates and greater supersaturation introduce mismatches with fiber branching and increased crystal dimensionality transitioning from 1D to 3D spherulitic crystal growth patterns. The junction zones and branching of SAFiN fibers are responsible for gel elasticity and rigidity. Crystallographic mismatches form widely branched networks that create more elastic gels that hold more solvent. The junction zones combine 1D fibers into 3D networks that pervade the whole system and entrap the liquid part macroscopically via capillary forces and surface tension.

5.1.3 Solvent Confluence on Gelation Outcome

Any chemical process in a solution, including aggregation or crystal growth, is governed by solution properties (Gu et al. 2004). Solvent affects self-assembly through both solvent – solute interactions and bulk solvent properties such as viscosity and chemical potential, making it near impossible to predict a priori solvent parameters that lead to self-assembly (Lan et al. 2015; Singh et al. 2017; Guo et al. 2020a). This interplay between molecular structure and the thermodynamics of crystalization affecting gelation outcomes is confounded by the diversity of apolar solvents considered, and likely no universal molecular gelator capable of self-assembly in all dilute solvents exits. Presently, the primary barrier is the inability to predict gelation outcomes for solvent gelator combinations that impeded the rational design or identified new self-assembling small molecules. Clearly molecular gelators require a meticulous balance between contrasting parameters of solubility and those controlling epitaxial growth; if the solvent and gelator are too similar, the result is a solution; conversely, if the solvent is too dissimilar, then macroscopic phase separation and precipitation occurs (Raynal and Bouteiller 2011; Gao et al. 2012; Bonnet et al. 2014; Diehn et al. 2014a; Lan et al. 2014b, 2015; De La Peña-Gil et al. 2016; Rogers and Marangoni 2016; Rosa Nunes et al. 2018).

In molecular self-assembly, considering solvent as a macroscopic continuum characterized with a bulk physical property is insufficient as individual solvent molecules interact differently at the surface or with a solute than with another solvent molecule; and potentially, variations exist at different points along a reaction or nucleation coordinate (Kaszynska et al. 2012; De La Peña-Gil et al. 2016). The most informative studies on the effect of solvent on small molecules' ability to self-assemble use Hansen solubility parameters (HSP) derived from Hildebrand solubility parameters (Singh et al. 2017; Rogers 2018; Nasr et al. 2020). HSPs decompose energy of vaporization of a species (*i.e.*, Hildebrand solubility parameter), a measure of its total cohesive energy, into three energetic components: dispersion, dipole-dipole, and hydrogen-bonding interactions plotted on cartesian axis termed Hansen space (Diehn et al. 2014a; Abbott 2020). In theory, the closer a solute and a solvent are in Hansen space, the higher the probability of miscibility. When represented as a point in Hansen space, each solvent clusters based on whether the gelator is soluble or insoluble or forms a SAFiN (Hansen and Smith 2004; Diehn et al. 2014a, 2014b; Lan et al. 2014b; Bonnet et al. 2015; Rosa Nunes et al. 2018; Nasr et al. 2020). For low molecular weight organogelators, the enclosed spheres that cluster gels are compact and globular; for example, the sol sphere (S) encloses most solvents in which the gelator forms a sol while excluding the solvents that give rise to different outcomes.

Food hydrocolloids have been studied and applied for decades while food oleocolloid research remains in its infancy. Further, in the case of edible oils, their diversity of chemical structure gives rise to a range of potential "solvent" HSPs, as does their co-missable molecules (*i.e.*, phospholipids, surfactant, and oxidation products) (De La Peña-Gil et al. 2016). Thus, Hansen space must be applicable to complex solvent mixtures, which is of central importance in finding new food-grade oleogels. The most eloquent effect of solvent HSPs on low molecular weight organogelators was for glucono-pyrenyl molecules in water, tetrahydrofuran and their mixtures (Figure 5.1.1) (Yan et al. 2013). The molecular gelator is insoluble in either THF or water, resulting in a precipitate. Yan et al. applied HSPs to illustrate why glucono-pyrenyl molecules placed in pure water or tetrahydrofuran (THF) are insoluble; yet, when the solvents were mixed at 30:70 or 20:80 ratios of THF to water, they gel. (Yan et al. 2013). Near the center of the Hansen sphere, solvent mixtures solubilized the gelator, and at the far radius, the molecular gel precipitated; however, at intermediate HSPs, the resultant mixture forms a SAFiN and molecular gel upon cooling (Figure 5.1.1) (Yan et al. 2013). This finding directly supports that the aggregation of molecular gelators into SAFiNs requires a meticulous balance between contrasting parameters of solubility and those controlling epitaxial growth.

5.1.3.1 Dissecting Gelator and Solvent Molecular Features that Drive Self-Assembly–A Step Toward Rational Design

To move toward rational design and choice of molecular gelators which enable molecule-by-molecule ordering into nanoscale hierarchical architectures, it must first be understood what features (*i.e.*, molecular synthons) are responsible for self-assembly and ultimately gelation. (Nasr et al. 2020) Although the need to understand the type and position of functional groups that drive assembly is apparent, even the most simple and well-studied gelators (*i.e.*, 1:3, 2:4 dibenzylidene sorbitol (DBS), 12-hydroxystearic acid (12HSA)) are just now being studied (Grahame et al. 2011; Abraham et al. 2012; Rogers et al. 2012a, 2012b; Pal et al. 2013; Chen et al. 2015; Rogers and Weiss 2015; Singh et al. 2017; Nasr et al. 2020), making it a molecular gelator of particular interest (Grassi et al. 2011; Song et al. 2013; Li et al. 2014; Shen et al. 2014; Okesola et al. 2015). The complex interplay between solvent and gelator was recently established for DBS by changing the molecular synthons and examining the Hansen space of gelled solvents (Singh et al. 2017; Nasr et al. 2020). DBS self-assembly is attributed to forming intermolecular hydrogen bonds between free hydroxyl groups and $\pi - \pi$ stacks between aromatic groups and its "butterfly-like" shape

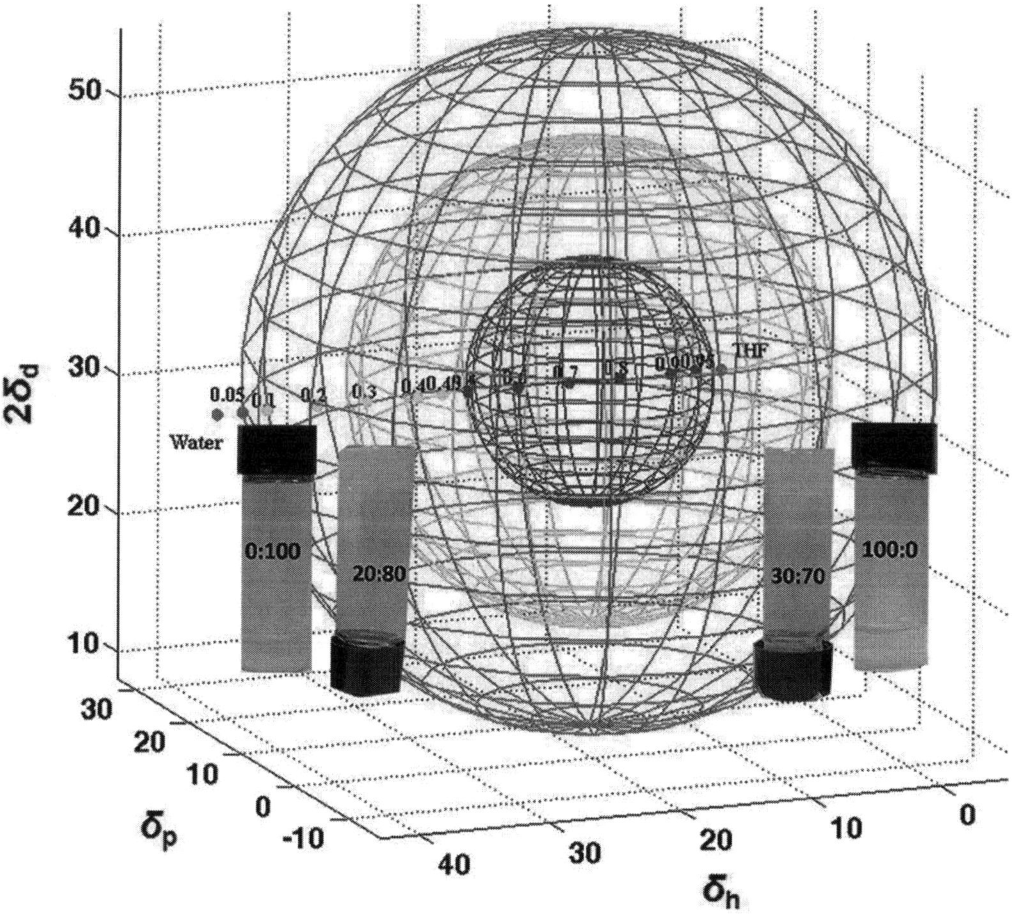

Figure 5.1.1 Hansen Solubility plots for 2.0% Glucono-Pyrenyl Molecules (Pn 7) in liquid mixtures represented in a Hansen space with spheres/shells: Largest (soluble), intermediate-size (gel), and smallest (solution). Yan et al. (2013), with permission from the American Chemical Society.

(Wilder et al. 2003; Basrur et al. 2013). The "butterfly" shape refers to the sorbitol backbone acting as the body while the benzylidene rings act as wings. The hydrophobic benzylidene wings are responsible for DBS being slightly soluble in many organic solvents, and their removal or replacement with cyclohexyl groups reduces the solvents gelled to only carbon tetrachloride (Singh et al. 2017). Clearly, a reduction in gelation capacity is seen when the ability of DBS to π–π stack is eliminated. However, the role of DBSs hydroxyl groups is much more nuanced and illustrates why rational design has eluded scientists.

Five derivatives of DBS, changing the primary and secondary hydroxyl groups, were synthesized and studied in an array of solvents to understand the ability of DBSs to self-assemble (Figure 5.1.2). DBS was the precursor molecule to synthesize derivatives by selectively removing the primary or secondary alcohol or acetylating a hydroxyl group. Irrespective of the modified primary or secondary molecular synthon, it results in fewer solvents gelled (Table 5.1.1) (Nasr et al. 2020). For each derivative, the shift in the gel sphere HSP coordinates did not correspond to the calculated HSP parameters, implying that molecular gelation is not only a function of solubility (Nasr et al. 2020). The importance of combining studies that change molecular synthons and testing the effect

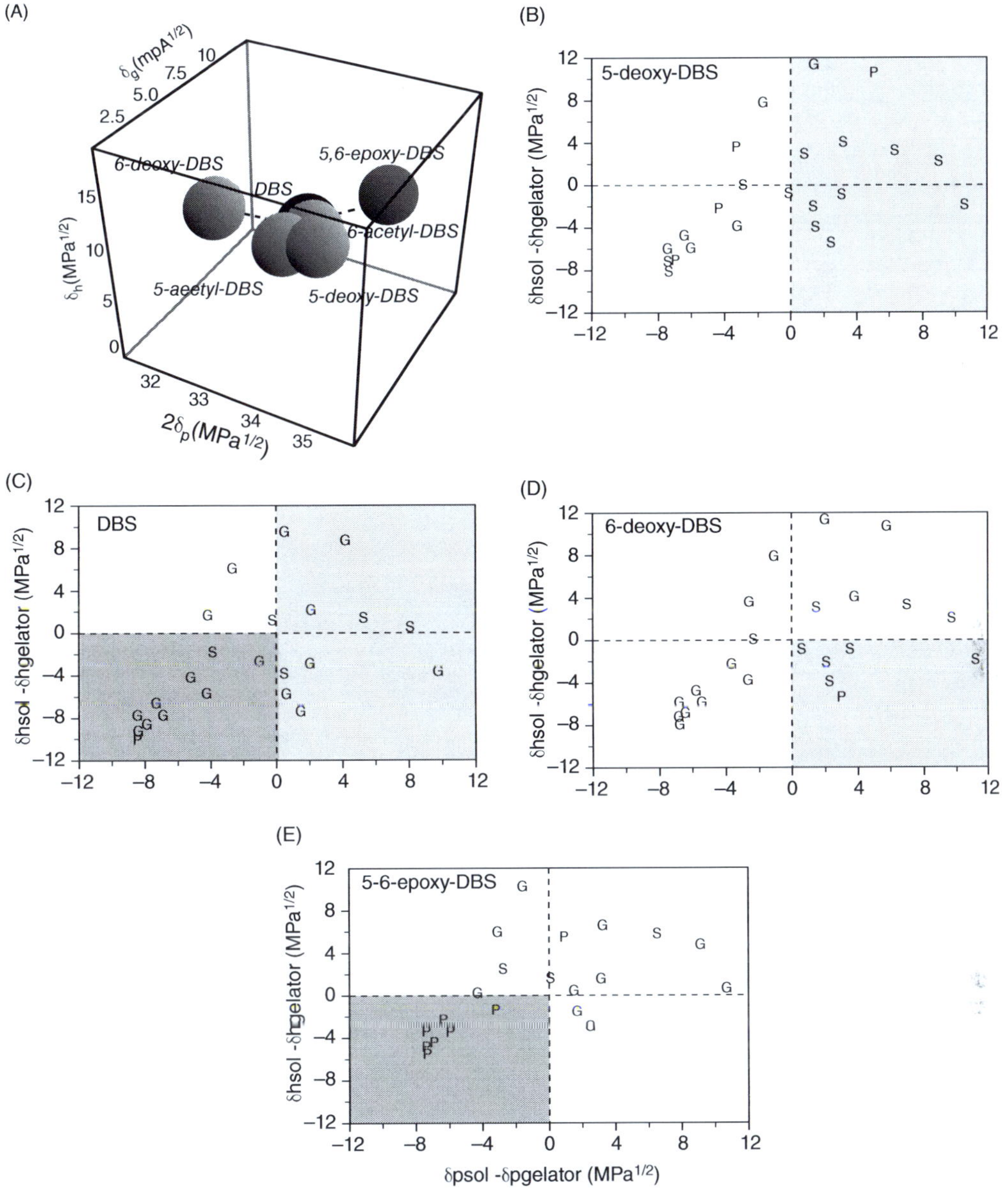

Figure 5.1.2 (A) Gelation spheres for DBS and its derivatives plotted in Hansen space. The size of a sphere is proportional to the calculated radius of the gel sphere. (B–E) δ_p and δ_h distances between gelator and solvent calculated from HSPiP software. Reproduced with permission from Nars et al., 2020 / American Chemical Society, *Journal of Physical Chemistry C*, 2020, 124, 48, 26455–26466.

on gelation must be done in a broad spectrum of solvents; if not, the finding may be misleading. For example, in some solvents, the primary alcohol of DBS was identified as the critical hydrogen bond (Yamasaki et al. 1995). When examined in a breadth of 23 solvents, the number DBS formed molecular gels in was 17; while altering the primary hydroxyl position decreased the solvents gelled to 13 and changes to the secondary hydroxyl group decreased those gelled to 7 (Rogers 2018). Hence it was concluded that changes to the 5-OH had a greater impact on self-assembly.

Table 5.1.1 Outcomes from 5% (w/w) DBS and its derivatives in solvents. Sample vials were inverted for 30 minutes and categorized as 1) (G) gels (defined as no observable flow upon inversion); 2) (S) solutions (remained transparent and flowed); or 3) (P) precipitates (non-gel and non-solution). Reproduced with permission from Nasr et al., 2020 / American Chemical Society, *Journal of Physical Chemistry* C, 2020, 124, 48, 26455–26466.

Solvent	DBS	6-acetyl	5-acetyl	6-deoxy	5-deoxy	5,6-epoxy
Salicylaldehyde	G	G	S	G	S	G
Dimethyl sulfoxide	S	S	S	S	S	G
***o*-Xylene**	G	G	P	G	G	P
Isobutyl alcohol	G	G	G	G	G	G
Triethylene glycol	G	G	S	G	P	P
Butylamine	S	S	S	S	S	S
Hexanoic acid	G	G	G	G	P	G
Dichlorotoluene	G	S	S	P	S	G
Tetramethylurea	S	S	S	S	S	P
Carbon tetrachloride	G	G	G	G	S	P
Benzyl methacrylate	G	P	G	G	G	P
Acetophenone	G	G	S	S	S	G
Chloroform	G	G	S	G	P	G
Triethylamine	G	G	G	G	P	P
Hexane	P	P	P	G	S	P
Benzene	G	G	S	G	G	P
Acetone	G	P	S	S	S	G
Acetonitrile	G	P	S	S	S	G
Pyridine	S	S	S	S	S	G
Toluene	G	G	G	G	G	P
Dichloromethane	G	G	S	S	S	S
Ethanol	G	G	G	G	G	G
Dimethyl formamide	S	S	S	S	S	S

The gel spheres Hansen coordinates for 6-deoxy-DBS and 5-deoxy-DBS shift in the opposite orthogonal orientations relative to DBS. Since the shift in gelled Hansen space did not correspond to the shift in gelator HSP, the solvent must play a more coordinated role in the self-assembly of the molecular gelator. Changes to the primary molecular synthon corresponded to a shift in the position of the gel spheres to lower dispersive HSPs (δ_d) values compared to DBS, but changes to the 5-OH shifted the gelation sphere to higher δ_d values even though the gelators had similar HSPs (Nasr et al. 2020). The reason for the shift is clear when plotting the vector distance between the solvents' and gelators' polar ($\Delta\delta_p$) versus hydrogen bonding ($\Delta\delta_h$) HSPs (Figure 5.1.2). DBS with both primary and secondary alcohols was the only gelator to gel solvents in all four quadrants, meaning it can gel solvents that are more and less polar and with greater or less hydrogen bonding (Figure 5.1.2 C). This finding implies that DBS's unique gelation ability is because of both primary and secondary molecular synthons. When either hydroxyl group is removed, solvents that are more polar and with weaker hydrogen bonding no longer form gels illustrated in the blue quadrants (Figure 5.1.2 B-D). Additionally, removing the secondary hydroxyl impeded self-assembly in the upper right quadrant corresponding to solvents with greater H-bonding and greater polarity. Coupling HSPs and structural changes provide important fundamental insights into the rational design of molecular gelators, removing some reliance on serendipitous discoveries to identify new food-grade gelators.

5.1.3.2 Solvent Complexity of Edible Oils–An Opportunity for Advancement

Commonly used vegetable oils are complex mixtures of triglycerides, monoglycerides, and diglycerides, free fatty acids, and polar lipids, giving rise to a range of potential solvent HSP. Traditionally, HSPs for molecules are calculated using the Hoftyzer and Krevelen methods (Grulke 2005), where each functional group contributes to the dispersive, polar, and hydrogen bonding interactions. For triglycerides, the calculation can be simplified by assuming that fats and oils comprise simple TAG (*i.e.*, trilaurin, tristearin, triolein, etc.) in the same ratios as their fatty acids, it is then possible to estimate δ_d, δ_p, and δ_h simply using the weighted molar average (Equation 5.1.1 and Table 5.1.2) without the need to use the group contributions for each TAG.

Table 5.1.2 HSP (δ_d, δ_p, and δ_h) determined for triglycerides with the HSPiP software using the SMILE system. Reproduced with permission from De La Peña-Gil, A. et al., 2016 / Springer Nature, Food Biophysics, 2016, 11:283–291.

Triacylglycerol (TAG)	δd (Mpa$^{1/2}$)	δp (Mpa$^{1/2}$)	δh (Mpa$^{1/2}$)	δT (Mpa$^{1/2}$)
Tributytin	16.3	3.9	5.7	17.7
Trivalerin	16.6	3.7	5.3	17.8
Tricaproin	16.6	3.4	5	17.7
Trioenanthin	16.3	3.3	4.5	17.2
Tricaprylin	16.2	3.4	4	17
Tripelargonin	16.6	2.6	3.4	17.1
Tricaprin	16.5	2.5	3.3	17
Triundecanoin	16.3	2.5	3.1	16.8
Trilaurin	16.1	2.7	2.8	16.6
Tritridecanoin	16.6	2.1	2.4	16.9
Trimyristin	16.6	2.1	2.5	16.9
Tripentadecanoin	16.4	2.2	2.4	16.7
Tripalmitin	16.2	2.4	2.2	16.5
Tripalmitolein	16.1	2.4	2.8	16.5
Trimargarin	16.9	1.9	2	17.1
Tristearin	16.9	1.9	2.1	17.1
Triolein	16.6	2.1	2.3	16.9

(Continued)

Table 5.1.2 (Continued)

Triacylglycerol (TAG)	δd ($Mpa^{1/2}$)	δp ($Mpa^{1/2}$)	δh ($Mpa^{1/2}$)	δT ($Mpa^{1/2}$)
Trivaccenin	16.7	1.9	2.6	17
Trilinolein	16.5	2	2.9	16.9
Trilinolenin	16.2	2.1	3.2	16.6
Tristearidonin	17	1.5	3.5	17.4
Trinonadecanoin	16.7	2	2.1	16.9
Triarachidin	16.6	2.2	2	16.9
Trieicosenoin	17.3	1.7	2.2	17.5
Triarachidonin	16.6	1.7	3.6	17.1
Trieicosapentaenoin	16.2	1.7	3.8	16.7
Triheneicosanoin	17.5	1.8	1.8	17.7
Tribehenin	17.4	1.8	1.9	17.6
Triclupanodonate	16.7	1.3	3.6	17.1
Tridocosahexaenoin	16.4	1.3	4	16.9
Tritricosanoin	17.3	1.9	2	17.5
Trilignocerin	17.2	2.1	1.9	17.4

$$\delta_d^{mix} = \sum_{i=1}^{i=\infty} \delta_{d_i x_i}, \delta_p^{mix} = \sum_{i=1}^{i=\infty} \delta_{p_i x_i}, \delta_h^{mix} = \sum_{i=1}^{i=\infty} \delta_{h_i x_i} \tag{5.1.1}$$

where δ_d, δ_p, and δ_h are the HSP calculated for the mixed solvent system using Eq. 1 and x_i is the corresponding mass fraction of the individual solvent (De La Peña-Gil et al. 2016; Rogers 2018). Based on these calculations, although there is little variation in the dispersive HSP component (~16 $MPa^{1/2}$), the polar component ranges $1.5 < \delta_p < 2.5$ $MPa^{1/2}$ for polyunsaturated fish oil and medium-chain coconut oil while the hydrogen-bonding component varies between $2.2 < \delta_h < 3.6$ $MPa^{1/2}$ (Table 5.1.3) (De La Peña-Gil et al. 2016). The value for oils can drastically change during storage due to the formation of polar oxidation products and can be tailed by adding co-miscible solvents (*i.e.*, glycerol, DAG oils, FA). Based on these narrow Hansen space ranges for edible oils, a comprehensive report was done on molecular gelators and on where they are by approximating their HSP using the group contributions method (Figure 5.1.3) (Rogers 2018).

Knowing where non-food grade molecular gelators capable of gelling, edible oils living in Hansen space provide a targeted approach to narrow the search for new food-grade gelators (Yan et al. 2013). To observe if a region in Hansen space clusters molecular gelators, a meta-analysis was performed on reported molecular gels that gelled vegetable oil. (Rogers 2018) Most of the reported molecular gelators (85%) found their dispersive HSP ranged between $16.0 < \delta_d < 17.5$ $MPa^{1/2}$; while δ_p was greater than 5.0 $MPa^{1/2}$ for all gelators, and 50% of the gelators were between $7.0 < \delta_p < 11.0$ $MPa^{1/2}$. The gelators had a $\delta_h > 5$ $MPa^{1/2}$, and 45% are between $7.0 < \delta_h < 9.0$ $Mpa^{1/2}$. A narrowed region of Hansen space, derived from non-food grade gelators, provides a framework for rationally designing the next generation of low molecular weight organogelators hoping to find new technologies for the food industry.

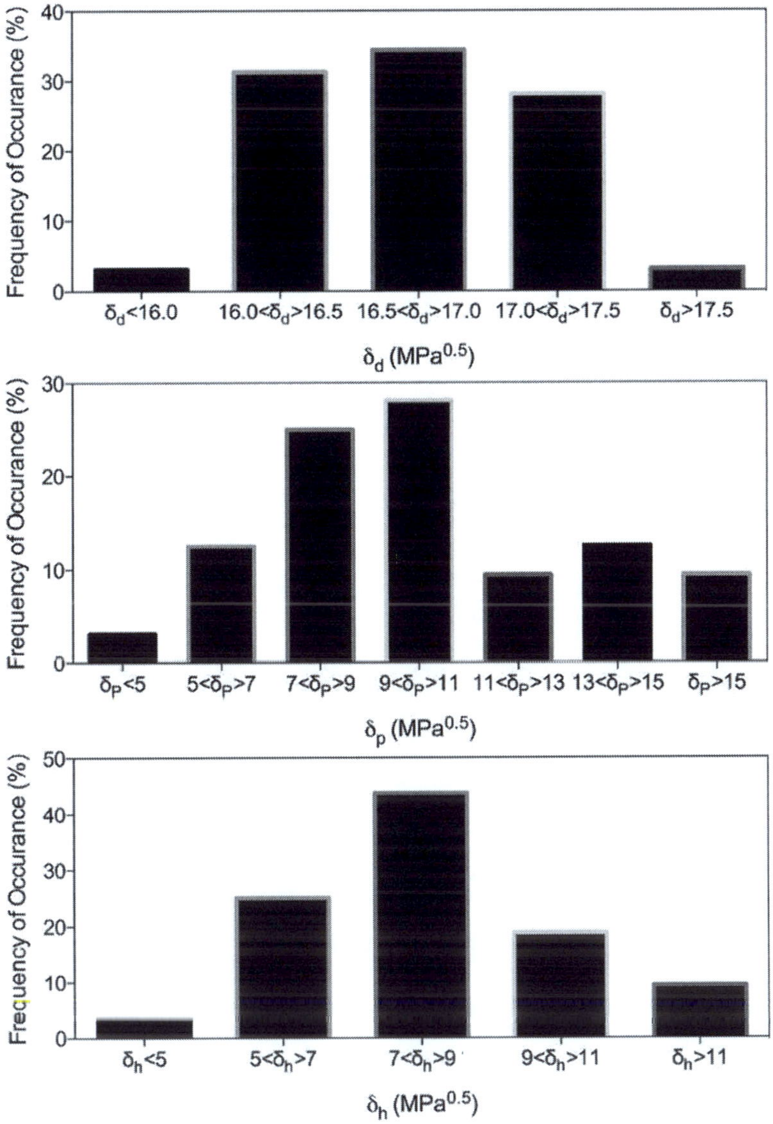

Figure 5.1.3 HSP's frequency of occurrence of reported molecular gelators that gel vegetable oil. Data were obtained for $n = 34$ unique molecular gelators. Reproduced with permission from Rogers, et al. 2018 / John Wiley & Sons, *Journal of the American Oil Chemists Society*, 2018, 95, 393–405.

Table 5.1.3 HSPs (δ_d, δ_p, δ_h) determined for the oils using the HSPiP software and calculated using the Hansen solubility parameters (δ_d, δ_p, δ_h) for symmetrical triglycerides (TAGs) and fatty acid mass fraction. Reproduced with permission from De La Peña-Gil A. et al., 2016 / Springer Nature, Food Biophysics, 2016, 11:283–291.

Sample	TAGs	d_d (MPa$^{1/2}$)	d_p (MPa$^{1/2}$)	d_h (MPa$^{1/2}$)	d_T (MPa$^{1/2}$)
Cocoa Butter	PPP	16.2	2.4	2.2	0.30
	SSS	16.9	1.9	2.1	0.31
	OOO	16.6	2.1	2.3	0.38

(Continued)

Table 5.1.3 (Continued)

Sample	TAGs	d_d (MPa$^{1/2}$)	d_p (MPa$^{1/2}$)	d_h (MPa$^{1/2}$)	d_T (MPa$^{1/2}$)
	LnLnLn	16.5	2.0	2.9	0.02
	HSPs TAGs$_{mixture}$	16.6	2.1	2.2	
	HSPs HSPiP$_{Software}$	16.6	2.1	2.2	
Coconut oil	CaCaCa	16.2	3.4	4.0	0.02
	CCC	16.5	2.5	3.3	0.16
	LaLaLa	16.1	2.7	2.8	0.56
	MMM	16.6	2.1	2.5	0.14
	PPP	16.2	2.4	2.2	0.03
	OOO	16.6	2.1	2.3	0.05
	LLL	16.5	2.0	2.9	0.03
	^{HSPs}TAGsmixture	16.2	2.5	2.8	
	HSPs HSPiP$_{Software}$	16.2	2.5	2.8	
Canola oil	PPP	16.2	2.4	2.2	0.07
	SSS	16.9	1.9	2.1	0.01
	OOO	16.6	2.1	2.3	0.49
	LLL	16.5	2.0	2.9	0.35
	LnLnLn	16.5	2.0	2.9	0.08
	^{HSPs}TAGsmixture	16.5	2.1	2.6	
	HSPs HSPiP$_{Software}$	16.5	2.1	2.6	
Fish oil	PPP	16.2	2.4	2.2	0.03
	SSS	16.9	1.9	2.1	0.03
	PlPlPl	16.1	2.4	2.8	0.04
	OOO	16.6	2.1	2.3	0.04
	EEE	16.2	1.7	3.8	0.13
	DpDpDp	16.7	1.3	3.6	0.33
	DhDhDh	16.4	1.3	4.0	0.40
	^{HSPs}TAGsmixture	16.5	1.5	3.6	
	HSPs HSPiP$_{Software}$	16.5	1.5	3.6	
Soybean oil	PPP	16.2	2.4	2.2	0.05
	SSS	16.9	1.9	2.1	0.03
	OOO	16.6	2.1	2.3	0.27
	LLL	16.5	2.0	2.9	0.64
	LnLnLn	16.2	2.1	3.2	0.02
	^{HSPs}TAGsmixture	16.5	2.0	2.7	
	HSPs HSPiP$_{Software}$	16.5	2.0	2.7	

5.1.4 Emerging Low Molecular Mass Organogelating Technologies

5.1.4.1 Peptide Gelators

Low molecular weight peptide gelators have many structural motifs arising from the complex chemical space associated with amino acid side chain variability (e.g., structural programmability, biocompatibility, availability, and versatile functionality), leading to a gamut of biomaterials (Na et al. 2013; Xue et al. 2013; Parween et al. 2014; Rogers et al. 2016a; Liu et al. 2017; Shaikh et al. 2018). To date, low molecular weight peptide organogelators range from surfactant-like peptides (Vauthey et al. 2002) to aromatic dipeptides (Nguyen et al. 2016) to cyclopeptides (Rogers et al. 2016a; Manchinella and Govindaraju 2017). Peptide molecular gelators originally had the same core recognition motif, diphenylalanine, as Alzheimer's amyloids, which formed discrete and extraordinary stiff nanotubes (Yan et al. 2008).The first reported dipeptide molecular gel was diphenylalanine, which was one of the smallest known molecular gelators. Since then, many other symmetric and asymmetric dipeptides containing just D or L isomers have been reported to gel organic solvents (Manchinella and Govindaraju 2017). Self-assembly of dipeptide gel was initially thought to be driven by the hydrogen bonding of the peptide backbone and π-π interactions between aromatic residues of the peptide (Na et al. 2013); however, π-π interactions are not required as cyclic dileucine forms gels in an array of solvents (Lan et al. 2019).

Cyclic peptides with alternating D and L-amino acids, β residues, or with alternating α and γ residues are planar stacking into β-sheets that result in synthetic nanotube building blocks (Chapman et al. 2012). The amide bond orientation is perpendicular to the ring's axis, which leads to the equatorial protrusion of the amino acid side chains. This orientation allows for the enthalpically driven assemblage of CPs into extended tubular structures (Chapman et al. 2012). By tailoring the chemical structure of the cyclic peptide moieties, the self-assembled supramolecular architectures have various applications, including stimuli-responsive nanomaterials, antibacterial agents, ion channeling and ion sensing (Chapman et al. 2012). The number of amino acids in the cyclic peptide allows sub-nanometer control over the van der Waals internal tubular diameter, which is difficult to achieve in other nanotubes (Chapman et al. 2012). Another major advantage of cyclic peptides that form nanotubes is the ease with which the exterior functionality by simply chaining the amino acid side chain (Chapman et al. 2012).

Up to tripeptides have been tested and effective at self-assembling and gelling certain organic solvents. To illustrate how subtle changes of each molecular synthon that drives SAFiN assembly, both linear dipeptides of diphenylanine and ditryrosine (L-FF and L-LL) gelled similar solvents, yet when the peptide backbone was cyclized between the free amine and carboxylic acid groups cyclized, only C-LL cyclic-dityrosine was a robust gelator, while cyclo-diphenylalanine was did not assemble (Lan et al. 2019). Upon cyclization, the diphenylalanine which drives the self-assembly of linear peptides fails to confer the ability to self-recognize and ultimately form a SAFiN. From this observation, and that cyclic-dileucine is a robust gelator, the core-recognition motif diving self-assembly via non-covalent hydrogen-bonding, is the cyclic peptide backbone (Lan et al. 2019). Several cyclic peptides assemble into nanotubes through β-sheet hydrogen bonding. Cyclic peptide molecular gels require a fine balance between hydrophobicity and hydrophilicity to form longer fibrils that can entangle into a denser network. Although none that are capable of gelling vegetable oil have been found, however, only a small fraction of combinations has been tested

Larger cyclic peptides have also been reported to form SAFiNs, including naturally-occurring LOB3 (a.k.a. cyclolinopeptide A), extracted from Linum usitatissimum L. (flaxseed) (Rogers et al.

2016b). LOB3 is a nine amino acid peptide (NαC-(Ile-Leu-Val-Pro-Pro-Phe- Phe-Leu-Ile) that when it is covalently linked between the free amine and carboxylic acid group forms crystalline fibrillar aggregates; conversely, the linear peptide derivates could not assemble, which points to the peptide backbone as the molecular synthon driving assembly (Shaikh et al. 2018). Cyclic peptides designed to have d-leucine have been shown to help promote self-assembly via increased hydrophobic interactions (Chapman et al. 2012). The rational design of functional nanotubes based on cyclic peptides requires an understanding of both the ring size and chemical functionality, and the chemical diversity of amino acid building blocks introduces a large variation into the molecular synthons driving self-recognition and assembly into SAFiNs (Chapman et al. 2012). This likely has the largest potential to be a disruptive technology pushing these biomaterials again to the forefront of edible gel technologies.

5.1.4.2 Sugar Gelators

One of the first reported food-grade gelators was D-sorbitol; unfortunately, its ability to form SAFiNS includes only ethanol and its mixtures with water at higher concentrations (Grassi et al. 2011). At low concentration, D-sorbitol gel formation requires sonication during cooling, likely to promote nucleation over crystal growth as its absence results in macroscopically phase-separated liquid and gel' beads' (Grassi et al. 2011). What is remarkable is that a simple structural addition of one benzylidene ring to monosaccharides turns a poor gelator into a "supergelator", which forms extraordinarily high aspect ratio fibers in a breadth of solvents at concentrations as low as 0.03–0.05 wt.% (Luboradzki et al. 2000; Gronwald and Shinkai 2001a, 2001b). The addition of two benzylidene rings forms 1:3,2:4-dibenzylidene sorbitol (DBS), the most widely studied gelators due to their ability to self-assemble and gel an array of solvents (Okesola et al. 2015). It would imply that the aromatic groups determine gelation ability (Singh et al. 2017). As mentioned in section 5.3.1, the primary and secondary hydroxyl groups play a more nuanced yet often essential role in determining gel outcome (Nasr et al. 2020).

Although many reports on sugar-appended gelators exist, few can gel edible oil; medium-chain dialkanoates of low-calorie sugars (sugar alcohol dioctanoates) form multilayered nanostructures. (Jadhav et al. 2010) The small lamellar-crystals aggregate and form crystallographic mismatches branching the network, forming a coherent network in a wide range of canola, soybean, and grapeseed oils and gelled at low concentrations (1 − 3 wt %/v) (Jadhav et al. 2010). Even minor changes to the sugar group drastically alter gelation; for example, mannitol gels near 1 wt% while sorbitol derivatives require 3 wt% structurant (Jadhav et al. 2010). Sugar-based (e.g., sucrose, maltose, lactose, and trehalose) diesters synthesized using the lipase B from Candida antarctica (CALB) were transesterified in acetone with either vinyl stearate or vinyl butyrate as the hydrophobic ester donors (John et al. 2006). Only trehalose resulted in gel formation during transesterification (John et al. 2006). The edible oils (e.g., olive, canola, soybean, and grapeseed) were selected because they vary in amounts of unsaturation fatty acids; olive and canola are high in monounsaturated; soybean oil is high in mono and di-unsaturated FA while grapeseed oils are predominantly di- and tri unsaturated fatty (Jadhav et al. 2010). Irrespective of the oil selected, it has little effect on the gelation outcome or rheological properties.

5.1.4.3 Lipids Gelators

Hydrogenated castor oil derivates (Wright and Marangoni 2006; Rogers et al. 2008, 2009a; Co and Marangoni 2013) and waxes (Toro-Vazquez et al. 2010; Jang et al. 2015; Lim et al. 2016) were two of the first food-grade gelators; however, their application has not progressed into edible applications. The challenge of hydroxylated fatty acids, such as 12-hydroxystearic acid (12HSA), is the high purity requirement of a single geometric and positional isomer. For example, both enantiopure isomers (D or L) of 12HSA gel at 1 wt% form high aspect ratio fibers effectively entraining the solvent phase, while a 1:1 D- to L-HSA requires 2 wt% and no longer form fibers (Grahame et al. 2011). For fiber formation, enantio-enriched mixtures of 12HSA must exceed 90% of one isomer to self-assemble effectively (Grahame et al. 2011). Racemic mixtures of 12HSA form platelet-like crystals and the molecular arrangement are single, in-plane, hydrogen-bonded acyclic dimers, preventing longitudinal growth and limiting the polar groups' ability to phase separate during nucleation (Grahame et al. 2011). Enantiopure 12HSA SAFiNs have high aspect ratio fibers because the molecules are organized as head-to-head dimers, and the 12-hydroxyl groups helped with the H-bonding network along the axis transverse to the longitudinal growth. The enantio-enriched D:L ratios lower than 80:20 present equal cyclic and acyclic dimers; conversely D-12HSA, and 95:05 and 90:10 DL-12HSA have significantly more cyclic than acyclic dimers.(Grahame et al. 2011)

Regarding enantiopurity, the position of the molecular synthon, the hydroxyl group of 12HSA alters the molecular amphiphilicity and the ability to self-assemble. If the two polar groups of 12HSA (e.g., the carboxylic acid and the hydroxyl group) are near each other (2HSA and 3HSA), it prevents gelation. (Abraham et al. 2012) In the case of 2HSA, it does not form a molecular dimer, as indicated by FT-IR, and growth along the secondary axis is inhibited as the secondary hydroxyl group does not form intermolecular H-bonds; 3-hydroxystearic acid (3HSA) forms an acyclic dimer, but the secondary hydroxyl groups cannot hydrogen bond, preventing the crystal structure from growing along the secondary axis. (Abraham et al. 2012) Carboxylic cyclic dimers form for 6, 8, 10, 12, and 14HSA isomers and form hydrogen bonding along the secondary axis resulting in SAFiN formation.

Soft gels form with various waxes in vegetable oil that self-assemble into small microcrystalline platelets of colloidal dimensions aggregating into a 3D network entrapping the liquid oil. Plant waxes shown to form vegetable oil gels have been extracted from candelilla (*Euphorbia cerifera*), carnauba (*Copernicia cerifera*), sugarcane (*Saccharum officinarum*), sunflower (*Helianthus annuus*), and rice (*Oryza sativa*) bran (Buitimea-Cantúa et al. 2021). Waxes differ in their proportions of n-alkanes, fatty acids, fatty alcohols, wax esters, ketones and sterols, and their gelling behavior is governed by the ester chain length (typically C24 and C28 carboxylic acids and C32 and C34 straight-chained primary alcohols). The differences in chemical composition cause variations in their melting point, crystallization behaviors, and, ultimately, oil structuring ability (Mandu et al. 2020). Candelilla wax has > 50% n-alkanes between 29–33 carbons and structures safflower oil at 1% wax (Toro-Vazquez et al. 2010). The wax gel elasticity increases with gelator concentration displaying similar textural properties to commercially available spreads; tailoring the physicochemical properties of organogels toward specific applications is possible (Silva et al. 2021). Recent advances in understanding how refining and extracting waxes alters gelation have provided a new avenue of biorefinery potential to derive more functional gelators. Three commercially available food-grade carnauba waxes (e.g., type I (Refined), type III (Filtered), and type IV (Crude)) showed the waxes refined and filtered waxes structured edible oil at lower concentrations compared to the crude wax (Buitimea-Cantúa et al. 2021). These differences induced by the refining and filtration processes favor the structuring properties of carnauba wax

Due to the amphiphilicity of monoglycerides (MAGs), they self-assemble into various structures (*i.e.*, lamellar, micellar, cubic, and hexagonal mesophases) in both edible oils and water (Ojijo et al. 2004). Below 1.3 wt.% MAGs form an entangled network resulting in a soft gel with its elasticity being concentration and frequency-dependent; however, monopalmitate and monostearate form reverse bilayers exhibiting gel-like behavior > 1.3% (Ojijo et al. 2004). Water and oil emulsified with monoglycerides result in the L_α liquid-crystalline lamellar) surrounding individual oils droplets forming an effective fat replacer (Marangoni et al. 2007). Cooling oil-in-water emulsions with MAGs induces the L_α liquid-crystalline lamellar) to crystallize efficiently at the droplet interface, entraining oil. Aging the gel causes a transition induced by the lamella expelling water to a β-gel which imparts functionality and solid-like properties emulating a fat crystal. From a nutritional standpoint, this fat-like material has no trans fats and may be designed with as little as 4% saturated fat. The primary benefit of the β-gel is the physiological lower blood TAGs, free fatty acids, and insulin after an acute intake of the oil-water–MAG gel versus an oil-in-water emulsion (Marangoni et al. 2007).

In addition to MAGs, many other polar lipids structure edible oils. The hydrophobic nonionic surfactant sorbitan monostearate (Span 60) forms opaque, thermoreversible gels in canola, sesame, and olive oils at 10 wt% (Murdan et al. 1999). Span 60 has a hydrophilic-lipophilic balance (HLB) of ~ 4.7 that favors water-in-oil emulsions; however, cooling decreases sorbitan monostearate solubility, and the molecules self-assemble into toroidal inverse rod-shaped tubules (Murdan et al. 1999). Polysorbates, also known as Tweens, change the Span 60 rod-like tubules, improving organogel stability and lowering critical gelator concentration (Murdan et al. 1999). Lecithin, another polar lipid, also forms a jelly-like organogel with little water consisting of a coherent network of entangled reverse micelles between 2.0–2.5 nm in the radius and hundreds to thousands of microns in length only form in highly pure soy or egg lecithin (e.g., > 95% phosphatidylcholine) (Schurtenberger et al. 1990; Capitani et al. 1996). Lecithins' network of entangled reverse micelles is similar to the sorbitan monostearates' network; however, lecithin, a small amount of non-polar liquid must be added to form the tubules. This narrow concentration range of polar solvent is the primary disadvantage to lecithin organogels and is what makes it difficult to start in complex food systems.

Recently two antioxidants were found to gel in sunflower oil, (2 E)-3-Phenylprop-2-enoic (cinnamic acid) and 4′-Hydroxy-3′,5′-dimethoxyacetophenone (acetosyringone) and are approved as additives (Pommella et al. 2020). A requirement for their gelation is the controlled crystallization rate; at faster cooling rates, the cinnamic acid-based oleogels do not show the same rigid platelets network produced under low supersaturation conditions (Pommella et al. 2020). At low cinnamic acid concentrations, the organogels have long ribbons, and increasing concentration reduces the crystallite size. Acetosyringone gels have fibers transitioning from jammed spherulites to entangled rods as concentration increases and is attributed to the disappearance of branching points or secondary nucleation defects. Acetosyringone and cinnamic acid are plant-derived, and the latter is known for its anti-oxidant, -diabetic, -inflammatory, and -cancer activities (Pommella et al. 2020). The plethora of phenolic anti-oxidants has been hardly investigated when this chapter was prepared. Numerous others will probably form SAFiN as they no longer rely only on van der Waals interactions, and these molecular synthons coordinate through hydrogen bonds or π-stacking interactions (Pommella et al. 2020). The variety in phenolic structure will alter solubility at high temperature and crystal habit (Pommella et al. 2020), giving rise to potential unique applications. Furthermore, adding phenolic compounds improves the nutritional value of formulated foods by reducing saturated fat and the conferred health benefits and oxidative stability of the anti-oxidant (Pommella et al. 2020).

5.1.4.4 Mixed System Gelators

Synergistic interactions between gelators in multi-component organogels have higher elasticity and other beneficial properties than their corresponding mono-component systems (Shakeel et al.

2021). Thus, novel systems can be formulated, varying the ratio of the structuring components to tune the rheological properties based on their end-use application. Also, the combination of a network forming molecule and inert filler or bulking agent improved stability and increased storage modulus compared to the alone self-assembling structurant. Using combined structurants such as long-chain fatty acids and fatty alcohols (Gandolfo et al. 2004; Callau et al. 2020), β-sitosterol and γ-oryzanol (Bot et al. 2008, 2012; den Adel et al. 2010; AlHassawi and Rogers 2013; Sawalha et al. 2013), and ceramides and lecithin (Guo et al. 2020a, 2020b) all effectively form SAFiNs in vegetable oil. The latter two systems are of particular interest due to the bioactive properties of β-sitosterol, γ-oryzanol and ceramides. For example, phytosterols are shown to lower blood cholesterol levels and contribute to satiety, making these attractive alternatives to colloidal fat crystal networks (Hughes et al. 2009). Ceramides have several beneficial physiological effects, including regulating cells' growth, senescence, and apoptosis and changing cholesterol profiles (Zheng et al. 2006). Additionally, ceramides are the primary sphingolipid in the skin's stratum corneum and are responsible for moisture retention; they are a common water-retaining agent in skincare products (Engberg et al. 2020).

In the case of β-sitosterol and γ-oryzanol, neither self-assemble sufficiently to entrain oil; however, at a 60:40 wt% γ-oryzanol/β-sitosterol ratio (or a 1:1 M ratios), SAFiNs and strong oleogels are formed from hollow tubules at concentrations as low as 2% (Bot et al. 2008, 2012). The ternary phase diagram (TPD) for γ-oryzanol, β-sitosterol, and canola oil (Figure 5.1.4)

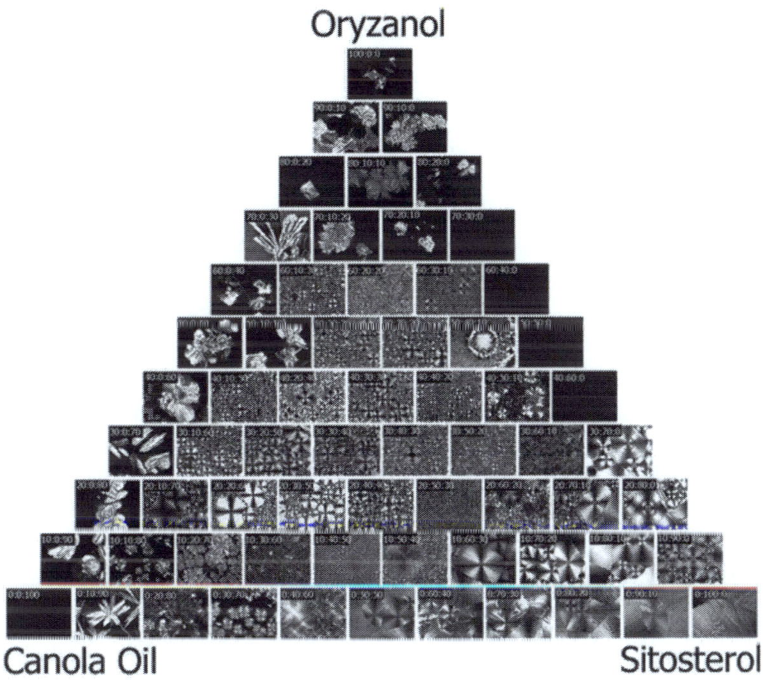

Figure 5.1.4 Ternary phase diagram (TPD) consisting of microscopy images for each tested sample. Gray scale groupings are based on similarities in polymorphic d-spacing patterns. The composition of each point, in the upper right corner of each micrograph, is expressed as the γ-oryzanol:β-sitosterol:canola oil ratios. Inset outlines the borders of the different regions of the ternary phase diagram and the corresponding figure which describes the physical characterization. Rogers et al., 2013, Reproduced with permission from John Wiley & Sons, *Journal of the American Oil Chemists Society*, 2013, 90, 1533–1540.

exhibit many polymorphic forms, melting points, and morphological features (AlHassawi & Rogers, 2013). γ-oryzanol ratios between 30% and 100%, β-sitosterol levels between 0% and 20%, and canola oil between 0% and 70% formed mixed crystals that required large supercooling to start nucleation resulting in rapid crystallization and many small spherulitic crystals (AlHassawi and Rogers 2013). A large central region (e.g., 20%–70% γ-oryzanol, 60% – 10% β-sitosterol, and 20% – 60% canola oil) form a single crystalline phase. The TPD shows various polymorphic phases and microstructural elements formed simply by altering the ratio of γ-oryzanol, β-sitosterol and canola oil. Systematic studies show the molecular synthon requirements for β-sitosterol and γ-oryzanol to form a SAFiN are listed in decreasing order of importance: synergistic hydrogen bonding, a ring system, and alkyl residue (Bot and Agterof 2006). Other plant sterols such as cholesterol, cholestenol, ergosterol, and stigmasterol form co-crystals with γ-oryzanol resulting in molecular gelation (Bot et al. 2012). Variations in the chemical structures are reflected in the SAFiN tubule diameter and wall thickness (Bot et al. 2012). In sunflower oil, β-sitosterol and γ-oryzanol oleogels are transparent and varying the sterol type, and concentration affects the mechanical and rheological properties of the oleogel (Panagiotopoulou et al. 2016).

Different γ-oryzanol and β-sitosterol ratios form a variety of crystal morphologies, and the resultant gels vary in lipid emulsification efficiency and free fatty acid release rate during lipolysis (Dong et al. 2020). During simulated digestion, the oil emulsions increase particle size more extensively than oleogel emulsions, and the free fatty acid release rate correlates with the emulsification efficiency (Dong et al. 2020). The self-assembled structures within these oleogels physically obstruct interactions between lipase and lipid, lowering the digestion rate. The gel strength, crystal morphology, and polymorphism have important roles in lipid digestion, including emulsification efficiency and lipolysis rate. These findings imply that oleogels can control lipid digestibility and bioavailability of lipid-soluble bioactives.

Ceramides have sphingosine bound to a fatty acid via an amide bond and differ based on the fatty acid chain length. Short-chain ceramides such as C2 ceramide are effective individual gelators of canola oil, while long-chain ceramides (C18) and mixed ceramides (from egg yolk) are less effective gelators (Rogers et al. 2009b). The shorter chain ceramides form large aspect ratio fibers, while hybrid system form a small maltese cross, spherulitic crystals (Rogers et al. 2009b). Recently, adding lecithin to these systems has been shown to change gelation outcomes (Dong et al. 2020; Guo et al. 2020a, 2020b). Lecithin is not an effective gelator; however, it has been reported to change the crystal morphology of gelators and other non-gelators, improving gelation capacity (Shchipunov and Shumilina 1995; Tamura and Ichikawa 1997; Murdan et al. 1999; Guo et al. 2020a). Mixing lecithin (L) and ceramides (C) at different molar ratios in sunflower oil led to varied gelation behaviors (Guo et al. 2020b). The crystal morphology of gels changed with gelator ratio; needle-like crystals and small rosette crystals were seen in 3:2 and 1:1, while the other ratios formed spherulites, which differed from the monocomponent gel structures (10L0C and 0L10C) (Figure 5.1.5) (Guo et al. 2020b). Stronger gels form when the ratio of lecithin exceeds ceramide, which is not expected as lecithin is a less effective gelator compared to ceramide III (Guo et al. 2020b). At 3:2 and 1:1 lecithin: ceramide III ratios, stronger gels formed with greater oil binding capacity (up to 99.84%) while ratios 4:1, 7:3, 2:3 and 3:7 formed soft gels (Guo et al. 2020b). Lecithin and ceramides co-assemble near an equal molar ratio, and excess lecithin or ceramide self-sort to combine with the co-assembled fibers by lateral association, forming different underlying microstructures.

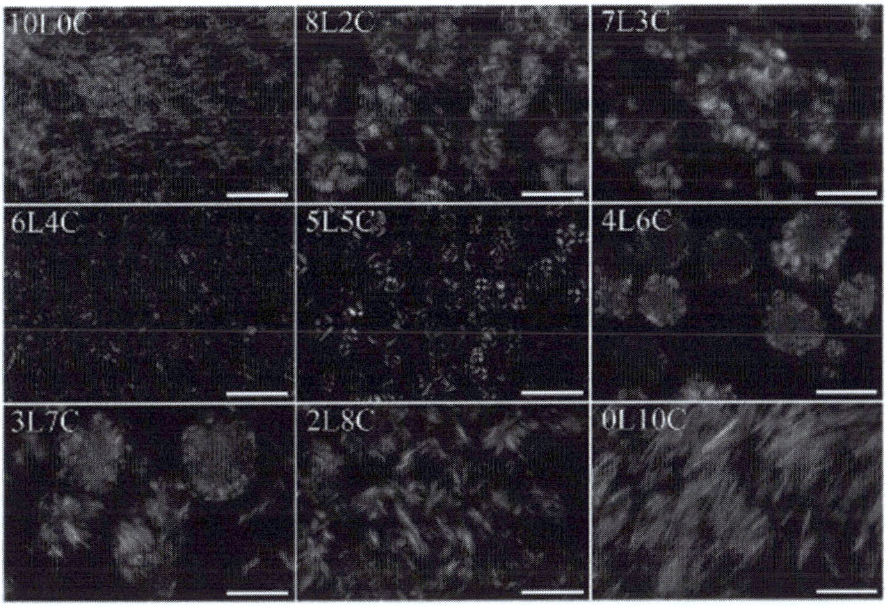

Figure 5.1.5 Polarized light microscopy images of ratios of lecithin and ceramide in sunflower oil. Scale bar is 20 μm. Guo et al. (2020), Reproduced with permission from Royal Society of Chemistry.

5.1.5 Polymeric Gelation

Even though polymers do not require monomers to self-assemble before forming junction zones responsible for the coherent network, there are far fewer known polymeric organogelators than molecular gelators (Suzuki and Hanabusa 2010). Non-polar liquids transformed into soft solids, using the 3D network of polymers, limits oil via surface and capillary forces without affecting the chemical properties of the oil. To form a supramolecular 3D network of polymers, they must form interactions with adjacent polymers. They may form crosslinks via 1) conformational changes, 2) adding a crosslinking agent, or 3) physical non-covalent interactions between segments (Suzuki and Hanabusa 2010). The primary polymers reported to date capable of forming organogels are poly(ethylene glycol)s, polycarbonates, polyesters, polycaprolactones, polyolefins (Suzuki and Hanabusa 2010). In the case of stereoregular polymers (e.g., poly(methyl methacrylate) and poly(ester)s), they rely on conformation changes exposing crosslinking sites between helices to induce organogelation. Polymer organogelators composed of L-amino acids have a good organogelation ability for many solvents due to the hydrogen bonding formation between peptide backbones inducing the formation of α-helices and β-sheets, which are further stabilized with π–π stacking interactions when aromatic amino acids are present (Suzuki and Hanabusa 2010).

Due to the hydrophilic nature of most polymers, they are less suitable to structure apolar liquids, which is why most oleogeling polymers require a surfactant to form a coherent network (Scholten 2019). Added surfactants increase the dispersibility of the polymer and are shown to improve both ethylcellulose and chitin oleogelation of vegetable oil (Scholten 2019). To overcome the poor solubility of the polymer, another approach is to use structured oil-in-water emulsions or foam templates, after which water can be evaporated or undergo solvent exchange (Scholten 2019).

5.1.5.1 Emerging Polymeric Organogelating Technologies

The first edible polymeric oleogel was made using ethylcellulose (EC), which is a linear polymer of 1,4-β-D-glucose units with ethoxy substitutions at carbons 2, 3, or 6 (Laredo et al. 2011). EC is commercially available across a molecular weight range characterized by their viscosity in a 5% solution of EC in 80% toluene and 20% ethanol; whereby ECs with viscosities from 4–100 cP have substitution of ~ 2.5, which gives them the solubility in the vegetable oils to form a gel (Laredo et al. 2011). EC becomes soluble in organic oils above its glass transition temperature (~140 °C) when the degree of ethoxy substitutions reaches 2.3–2.6. Oleogels form after EC is heated in vegetable oil partly solubilized above its glass transition temperature, and upon cooling, microscopic phase separation, nucleation and self-assembly occur (Dey et al. 2011). The EC in various vegetable oils is plasticized by edible surfactants (*i.e.*, glycerol mono-oleate), modulating EC solubility. Increasing levels of unsaturation in the oil increase the mechanical strength of the gels when all other parameters are controlled; with canola oil, rich in oleic acid (18:1), the gel was much softer than in flaxseed oil rich in linolenic acid (18:3) (Zetzl et al. 2012). Differences in hardness are attributed to the fatty acids packing within the EC gel network; more dense oil, such as flaxseed, pack more tightly in the EC network, creating a stronger gel (Zetzl et al. 2012). Gel hardness increases with larger EC molecular weight because of more polymer-polymer junction zones. Gel hardness is not linearly correlated to the EC molecular weight; instead, it follows an exponential or power-law function and differs for each vegetable oil (Rogers et al. 2014).

Another interesting system worth mentioning, although slightly outside the norm of what is thought of for "molecular gels" and "organogels," is chitin and chitin nanocrystals packed systems (Nikiforidis and Scholten 2015). The particle size of chitin in sunflower oil was ~ 70 μm, and these large aggregates reduce the contact points between polymers, decreasing gel strength. The degree of aggregation plays an important role in the network formation and subsequent behavior of chitin organogels; since the number of contact points (e.g., junction zones) between the structural elements determines the coherence of the gel network and the final gel features, using pure chitin should be avoided. Crude chitin in triglyceride oil does not form an organogel as substantial aggregation occurs; to reduce the macroscopic aggregation, either a co-surfactant (Span 60 or phosphatidylcholine) is added, or an acid hydrolyzed chitin is used (Nikiforidis and Scholten 2015). Acid hydrolysis converted the long and flexible chitin strands into rigid and short chitin nanoparticles. The acid hydrolysis of purified chitin had nanoparticles from 200 to 500 nm in length and 10 to 15 nm in width (Goodrich and Winter 2007). The gel strength is highly influenced by the surfactant, as is the temperature sensitivity of the gels. Surfactants added to chitin organogels plasticize the polymer, reducing interpolymer friction and molecular rigidity. The most elastic gels were produced using acid-hydrolyzed chitin nanocrystals, co-surfactants, in vegetable oil at concentrations between 1–10 wt%, which suggests efficient packing of the particles (Nikiforidis and Scholten 2015). The most interesting result is these gels are compatible with considerable water, up to 25 wt%; and adding water increases gel strength (e.g., storage modulus). The increase in elasticity with added water is attributed to water absorption by chitin, and during hydration, swelling of the chitin increases the effective filler volume fraction (Nikiforidis and Scholten 2015).

Also, within polymeric organogels are foam templated gels. First, a porous foam containing a water-soluble cellulose derivative, hydroxypropyl methylcellulose (HPMC), approved for pharmaceutical and food applications, is freeze-dried into a cryogen (Patel et al. 2013). The porous cryogel exhibits excellent oil sorption, absorbing close to 100 times its weight in oil. The elasticity of the gel and the rheological characterization of HPMC foam templated organogels had comparable or, sometimes, superior to organogels made from ethylcellulose or waxes (Patel et al. 2013). Remarkably, these organogels had no observable oiling-out or polymer sedimentation at 5°C for over 3 months

(Patel et al. 2013). Adding a copolymer (xanthan gum) to HPMC improves the organogelation and entrainment of the continuous oil phase (Jiang et al. 2021). The cryogels had 97% porosity and absorbed 72 times their weight of low saturated oils at 0.8 wt% HPMC. Adding xanthan to HPMC formed a hydrogen bonding bi-continuous network, which improved the structural recovery of the organogel, an important part of spreads (Jiang et al. 2021). More copolymers (e.g., κ-carrageenan and flaxseed gum) have shown their ability to improve HPMC cyrogels with higher hardness and oil holding capacity (>98%) (Jiang et al. 2022). Oil absorption curves indicated that flaxseed gum and guar gum-enhanced cryogels provided the fastest oil absorption rate due to bigger pores. The absorption model was fitted well with all experimental data. Physical and mechanical properties of cryogels were positively related to the rheological property and oil bonding capacity of oleogels. κ-carrageenan and flaxseed gum were more suitable as structural enhancers to improve hydroxypropyl-methylcellulose-based cryogels for preparing oleogels to replace plastic fats in foods.

A similar approach has been shown using the cryogel rice bran protein (Wei et al. 2022). The foam first prepared at pH 3.0 showed the highest foam stability; however, the corresponding cryogel was less efficient at absorbing oil due to the smaller pores and higher surface tension to overcome during wetting. Cryogels prepared from foams at pH 7, 9, and 11 resulted in oleogels with G′ reaching 1.5×10^5 Pa and yield stress anear 1000 Pa (Wei et al. 2022). Interestingly, the final oleogel strength inversely correlated to oil absorption rate and the gels from RBP foam templates showed mechanical properties like shortening, providing a potential alternative to hard-stock fat. The essential part that all these studies show is that the properties of the organogel are highly inherited from the cryogel, and the properties of the cryogel are highly inherited from the foam properties (Jiang et al. 2021). Thus, advances made over the last 100 years for hydrocolloids may have applicability to foam templated organogels which could accelerate the rational design of the precursor formulation biopolymer blend, pH other physiochemical properties identified.

5.1.6 Conclusion

The emergence of biorefinery converts biomass of renewable feedstock using the associated infrastructure employed to produce chemicals and materials analogous to petroleum refining and is of particular interest in the hope of new food-grade gelators emerging (John et al. 2010). The creation of purified, naturally-occurring small molecules from food waste may serve as substrates to rationally design new molecular gelators. Examples already include new amphiphiles with amygdalin, vitamin C, cardanol, arjunolic acid, and trehalose that are molecular gelators. (John et al. 2010) Compared to the field of hydrocolloids and "hydro" gelation, organogel research is still very much in its infancy; many platforms, have been established with divergent physical properties. As such, the application of oil gels extends beyond simple trans and saturated fat replacement and is a new platform for culinary design with application in molecular gastronomy, military rations, and targeted bioactive delivery.

References

Abbott, S. (2020). Solubility, similarity, and compatibility: A general-purpose theory for the formulator. *Current Opinion in Colloid & Interface Science* 48: 65–76.

Abraham, S., Lan, Y., Lam, R.S.H., Grahame, D.A.S., Kim, J.J.H., Weiss, R.G., and Rogers, M.A. (2012). Influence of positional isomers on the macroscale and nanoscale architectures of aggregates of racemic hydroxyoctadecanoic acids in their molecular gel, dispersion, and solid states. *Langmuir* 28: 4955–4964.

AlHassawi, F. and Rogers, M.A. (2013). Ternary phase diagram of B-sitosterol, G-oryzanol and canola oil. *Journal of the American Oil Chemists Society* 90: 1533–1540.

Basrur, V.R., Guo, J., Wang, C., and Raghavan, S.R. (2013). Synergistic gelation of silica nanoparticles and a sorbitol-based molecular gelator to yield highly-conductive free-standing gel electrolytes veidhes. *ACS Applied Materials & Interfaces* 5: 262–267.

Bonnet, J., Suissa, G., Raynal, M., and Bouteiller, L. (2014). Organogel formation rationalized by Hansen solubility parameters: Dos and don'ts. *Soft Matter* 10: 3154–3160.

Bonnet, J., Suissa, G., Raynal, M., and Bouteiller, L. (2015). Organogel formation rationalized by Hansen solubility parameters: influence of gelator structure. *Soft Matter* 11 (11): 2308–2312.

Bot, A. and Agterof, W.G.M. (2006). Structuring of edible oils by mixtures of γ-Oryzanol with β-sitosterol or related phytosterols. *Journal of the American Oil Chemical Society* 83: 513–521.

Bot, A., den Adel, R., and Roijers, E.C. (2008). Fibrils of gamma-Oryzanol plus beta-sitosterol in edible oil organogels. *Journal of the American Oil Chemists Society* 85 (12): 1127–1134.

Bot, A., Gilbert, E.P., Bouwman, W.G., Sawalha, H., den Adel, R., Garamus, V.M., Venema, P., van der Linden,E., and Flöter, E. (2012). Elucidation of density profile of self-assembled sitosterol + oryzanol tubules with small-angle neutron scattering. *Faraday Discussion* 158: 223–238.

Buitimea-Cantúa, G.V., Serna-Saldívar, S.O., Pérez-Carrillo, E., Silva, T.J., Barrera-Arellano, D., and Buitimea-Cantúa, N.E. (2021). Effect of quality of carnauba wax (Copernica cerífera) on microstructure, textural, and rheological properties of soybean oil-based organogels. *LWT* 136: 110267.

Callau, M., Sow-Kébé, K., Nicolas-Morgantini, L., and Fameau, A.L. (2020). Effect of the ratio between behenyl alcohol and behenic acid on the oleogel properties. *Journal of Colloid and Interface Science* 560: 874–884.

Capitani, D., Segre, A.L., Dreher, F., and Anzeigen, A. (1996). Multinuclear NMR investigation of phosphatidylcholine organogels. *The Journal of Physical Chemistry* 100 (37): 15211–15217.

Chapman, R., Danial, M., Koh, M.L., Jolliffe, K.A., and Perrier, S. (2012). Design and properties of functional nanotubes from the self-assembly of cyclic peptide templates. *Chemical Society Reviews* 41 (18): 6023–6041.

Chen, L., Corradini, M.G., and Rogers, M.A. (2015). Self-assembly of 12-hydroxystearic acid molecular gels in mixed solvent systems rationalized using hansen solubility parameters. *Colloid & Polymer Science* 293: 975–983.

Chivers, P.R.A. and Smith, D.K. (2019). Shaping and structuring supramolecular gels. *Nature Reviews Materials* 4 (7): 463–478.

Co, E. and Marangoni, A.G. (2013). The formation of a 12Hydroxystearic acid/vegetable oil organogel under shear and thermal fields. *Journal of the American Oil Chemical Society* 90: 529–544.

Daniel, J. and Rajasekharan, R. (2003). Organogelation of plant oils and hydrocarbons by long-chain saturated FA, fatty alcohols, wax esters, and dicarboxylic acids. *Journal of the American Oil Chemists' Society* 80 (5): 417–421.

Dassanayake, L.S.K., Kodali, D.R., Ueno, S., and Sato, K. (2009). Physical properties of rice bran wax in bulk and organogels. *Journal of the American Oil Chemical Society* 86: 1163–1173.

De La Peña-Gil, A., Toro-Vazques, J.F., and Rogers, M.A. (2016). Simplifying hansen solubility parameters for complex edible fats and oils. *Food Biophysics* 11: 283–291.

den Adel, R., Heussen, P.C.M., and Bot, A. (2010) Effect of water on self-assembled tubules in β-sitosterol + γ-oryzanol-based organogels. *Journal of Physics: Conference Series*, 247, 012025.

Dey, T., Kim, D.A., and Marangoni, A.G. (2011). Edible oleogels: Structure and health implications. In: *Ethylcellulose oleogels* (ed. A.G. Marangoni and N. Garti), 295–311. Urbana, IL: AOCS Press.

Diehn, K.K., Oh, H., Hashemipour, R., Weiss, R.G., and Raghavan, S.R. (2014a). Insights into organogelation and its kinetics from hansen solubility parameters. Toward a priori predictions of molecular gelation. *Soft Matter* 10 (15): 2632–2640.

Diehn, K.K., Oh, H., Hashemipour, R., Weiss, R.G., and Raghavan, S.R (2014b). The physical chemistry of molecular gels: Insights into self-assembly-based organogelation and its kinetics via hansen solubility parameters. *Soft Matter* 10: 2632–2640.

Dong, L., Lv, M., Gao, X., Zhang, L., Rogers, M., Cao, Y., and Lan, Y. (2020). In vitro gastrointestinal digestibility of phytosterol oleogels: Influence of self-assembled microstructures on emulsification efficiency and lipase activity. *Food & Function* 11 (11): 9503–9513.

Draper, E.R. and Adams, D.J. (2017). Low-molecular-weight gels: the state of the art. *Chem* 3 (3): 390–410.

Duval, F.P., van Duynhoven, J.P.M., and Bot, A. (2006). Practical implications of the phasecompositional assessment of lipid-based food products by time-domain NMR. *Journal of the American Oil Chemists Society* 83: 905–912.

Engberg, O., Kováčik, A., Pullmannová, P., Juhasčik, M., Opálka, L., Huster, D., and Vávrová, K. (2020). The sphingosine and acyl chains of ceramide [NS] show very different structure and dynamics that challenge our understanding of the skin barrier. *Angewandte Chemie International Edition* 59 (40): 17383–17387.

Fabri, D., Guan, J., and Cesàro, A. (1998). Crystallisation and melting behaviour of poly (3-hydroxybutyrate) in dilute solution: Towards an understanding of physical gels. *Thermochimica Acta* 321 (1): 3–16.

Flory, P.J. (1974). Introductory lecture. *Faraday Discussions* 57: 7–18.

Gandolfo, F., Bot, A., and Floter, E. (2004). Structuring of edible oils by long-chain FA, fatty alcohols, and their mixtures. *Journal of the American Oil Chemists' Society* 81: 1–6.

Gao, J., Wu, S., and Rogers, M.A. (2012). Harnessing Hansen solubility parameters to predict organogel formation. *Journal of Materials Chemistry* 22: 12651–12658.

Goodrich, J.D. and Winter, W.T. (2007). Alpha-chitin nanocrystals prepared from shrimp shells and their specific surface area measurement. *Biomacromolecules* 8 (1): 252–257.

Grahame, D.A.S., Olauson, C., Lam, R.S.H., Pedersen, T., Borondics, F., Abraham, S., Weiss, R.G., and Rogers, M.A. (2011). Influence of chirality on the modes of self-assembly of 12-hydroxystearic acid in molecular gels of mineral oil. *Soft Matter* 7: 7359–7365.

Grassi, S., Carretti, E., Dei, L., Branham, C.W., Kahr, B., and Weiss, R.G. (2011). D-Sorbitol, a structurally simple, low molecular-mass gelator. *New Journal of Chemistry* 35: 445–452.

Gronwald, O. and Shinkai, S. (2001a). 'Bifunctional' sugar-integrated gelators for organic solvents and water—on the role of nitro-substituents in 1-O-methyl-4,6-O-(nitrobenzylidene)-monosaccharides for the improvement of gelation ability. *Journal Chemical Society Perkin Trans 2* 2001, 1933–1937.

Gronwald, O. and Shinkai, S. (2001b). Sugar-integrated gelators of organic solvents. *Chemistry – A European Journal* 7 (20): 4328–4334.

Grulke, E.A. (ed) (2005). *Solubility Parameter Values*, 4e. New York: John Wiley & Sonds.

Gu, C.-H., Li, H., Gandhi, R.B., and Raghavan, K. (2004). Grouping solvents by statistical analysis of solvent property parameters: implication to polymorph screening. *International Journal of Pharmaceutics* 283: 117–125.

Guo, S., Lv, M., Chen, Y., Hou, T., Zhang, Y., Huang, Z., Cao, Y., Rogers, M., and Lan, Y. (2020a). Engineering water-induced ceramide/lecithin oleogels: Understanding the influence of water added upon pre-and post-nucleation. *Food & Function* 11: 2048–2057.

Guo, S., Song, M., Gao, X., Dong, L., Hou, T., Lin, X., Tan, W., Cao, Y., Rogers, M., and Lan, Y. (2020b). Assembly pattern of multicomponent supramolecular oleogel composed of ceramide and lecithin in sunflower oil: Self-assembly or self-sorting? *Food & Function* 11 (9): 7651–7660.

Hansen, C. and Smith, A.L. (2004). Using Hansen solubility parameters to correlate solubility of C60 fullerene in organic solvents and in polymers. *Carbon* 42: 1591–1597.

Hughes, N.E., Marangoni, A.G., Wright, A.J., Rogers, M., and Rush, J. (2009). Potential food applications of edible oil organogels. *Trends in Food Science & Technology* 20 (10): 470–480.

Hwang, H.-S., Singh, M., Bakota, E.L., Winkler-Moser, J.K., Kim, S., and Liu, S.X (2013). Margarine from organogels of plant wax and soybean oil. *Journal of the American Oil Chemists Society* 90: 1705–1712.

Jadhav, R.S., Vemula, P.K., Kumar, R., Raghavan, S.R., and John, G. (2010). Sugar-derived phase-selective molecular gelators as model solidifiers for oil spills. *Angewandte Chemie* 49: 7695–7698.

Jang, A., Bae, W., Hwang, H.S., Lee, H.G., and Lee, S. (2015). Evaluation of canola oil oleogels with candelilla wax as an alternative to shortening in baked goods. *Food Chemistry* 187: 525–529.

Jiang, Q., Du, L., Li, S., Liu, Y., and Meng, Z. (2021). Polysaccharide-stabilized aqueous foams to fabricate highly oil-absorbing cryogels: Application and formation process for preparation of edible oleogels. *Food Hydrocolloids* 120: 106901.

Jiang, Q., Li, P., Ji, M., Du, L., Li, S., Liu, Y., and Meng, Z. (2022). Synergetic effects of water-soluble polysaccharides for intensifying performances of oleogels fabricated by oil-absorbing cryogels. *Food Chemistry* 372: 131357.

John, G., Vijai Shankar, B., Jadhav, S.R., and Vemula, P.K. (2010). Biorefinery: a design tool for molecular gelators. *Langmuir* 26 (23): 17843–17851.

John, G., Zhu, G., Li, J., and Dordick, J.S. (2006). Enzymatically derived sugar-containing self-assembled organogels with nanostructured morphologies. *Angewandte Chemie International Edition* 45 (29): 4772–4775.

Kaszynska, J., Lapinski, A., Bielejewski, M., Luboradzki, R., and Tritt-Goc, J. (2012). On the relation between the solvent parameters and the physical properties of methyl-4,6-O-benzylidene-α-d-glucopyranoside organogels. *Tetrahedron* 68: 3803–3810.

Lam, R., Quaroni, L., Pederson, T., and Rogers, M.A. (2010). A molecular insight into the nature of crystallographic mismatches in self-assembled fibrillar networks under non-isothermal crystallization conditions. *Soft Matter* 6: 404–408.

Lam, R., Rogers, M.A., and Marangoni, A.G. (2009). Thermo-mechanical method for the determination of the fractal dimension of fat crystal networks. *Journal of Thermal Analysis and Calorimetry* 98 (1): 7–12.

Lan, Y., Corradini, M.G., Liu, X., May, T.E., Borondics, F., Weiss, R.G., and Rogers, M.A. (2014a). Comparing and correlating solubility parameters governing the self-assembly of molecular gels using 1, 3:2,4-dibenzylidenesorbitol as the gelator. *Langmuir* 30 (47): 14128–14142.

Lan, Y., Corradini, M.G., and Rogers, M.A. (2014b). Do molecular gelators cluster in hansen space? *Crystal Growth & Design* 14: 4811–4818.

Lan, Y., Corradini, M.G., Weiss, R.G., Raghavan, S.R., and Rogers, M.A. (2015). To gel or not to gel: Correlating molecular gelation with solvent parameters. *Chemical Society Reviews* 44: 6035–6058.

Lan, Y., Lv, M., Guo, S., Nasr, P., Ladizhansky, V., Vaz, R., Corradini, M.G., Hou, T., Ghazani, S.M., Marnangoni, A., and Rogers, M.A. (2019). Molecular motifs encoding self-assembly of peptide fibers into molecular gels. *Soft Matter* 15 (45): 9205–9214.

Laredo, T., Barbut, S., and Marangoni, A.G. (2011). Molecular interactions of polymer oleogelation. *Soft Matter* 7 (6): 2734–2743.

Li, J., Fan, K., Guan, X., Yu, Y., and Song, J. (2014). Self-assembly mechanism of 1, 3:2,4-di(3,4-dichlorobenzylidene)-d-sorbitoland control of the supramolecular chirality. *Langmuir* 30 (44): 13422–13429.

Li, J.L. and Liu, X.Y. (2010). Architecture of supramolecular soft functional materials: From understanding to micro-/nanoscale engineering. *Advanced Functional Materials* 20: 3196–3216.

Lim, J., Hwang, H.S., and Lee, S. (2016). Oil-structuring characterization of natural waxes in canola oil oleogels: Rheological, thermal, and oxidative properties. *Applied Biological Chemistry* 60: 17–22.

Lipowitz, A. (1841). Versuche und Resultate über die Löslichkeit der Harnsäure. *Justus Liebigs Annalen der Chemie* 38 (3): 348–355.

Liu, X., Fei, J., Wang, A., Cui, W., Zhu, P., and Li, J. (2017). Transformation of dipeptide-based organogels into chiral crystals by cryogenic treatment. *Angewandte Chemie International Edition* 56 (10): 2660–2663.

Liu, X.Y. and Sawant, P.D. (2001). Formation kinetics of fractal nanofiber networks in organogels. *Applied Physics Letters* 19: 3518–3520.

Liu, X.Y. and Sawant, P.D. (2002). Determination of the fractal characteristic of nanofiber-network formation in supramolecular materials. *CHEMPHYSCHEM* 4: 374–377.

Luboradzki, R., Gronwald, O., Ikeda, A., and Shinkai, S. (2000). Sugar-integrated "supergelators" which can form organogels with 0.03–0.05% [g mL−1]. *Chemistry Letters* 29 (10): 1148–1149.

Manchinella, S. and Govindaraju, T. (2017). Molecular self-assembly of cyclic dipeptide derivatives and their applications. *ChemPlusChem* 82: 88–106.

Mandu, C.C., Barrera-Arellano, D., Santana, M.H.A., and Fernandes, G.D. (2020). Waxes used as structuring agents for food organogels: A review. *Grasas y Aceites* 71 (1): e344.

Marangoni, A.G., Idziak, S.H.J., Vega, C., Batte, Heidi, Ollivon, M., Jantzi, P.S., and Rush, J.W.E. (2007) Encapsulation-stucturing of edible oil attenuates acute elevation of blood lipids and insulin in humans. *Soft Matter* 3 (2): 183–187.

Murdan, S., Gregoriadis, G., and Florence, A.T. (1999). Novel sorbitan monostearate organogels. *Journal of Pharmaceutical Sciences* 88 (6): 608–614.

Na, N., Mu, X., Liu, Q., Wen, J., Wang, F., and Ouyang, J. (2013). Self-assembly of diphenylalanine peptides into microtubes with "turn on" fluorescence using an aggregation-induced emission molecule. *Chemical Communications* 49 (86): 10076.

Nasr, P., Corradini, M.G., Hill, J., Read, S.T., Rosendahl, S.M., Weiss, R.G., Auzanneau, F.I., and Rogers, M.A. (2020). Hansen solubility parameters clarify the role of the primary and secondary hydroxyl groups on the remarkable self-assembly of 1:3,2:4-dibenzylidene sorbitol. *The Journal of Physical Chemistry C* 124 (48): 26455–26466.

Nguyen, V., Zhu, R., Jenkins, K., and Yang, R. (2016). Self assembly of diphenylalanine peptide with controlled polarization for power generation. *Nature Communiations* 7: 13566.

Nikiforidis, C.V. and Scholten, E. (2015). Polymer organogelation with chitin and chitin nanocrystals. *RSC Advances* 5 (47): 37789–37799.

Ojijo, N.K.O., Neeman, I., Eger, S., and Shimoni, E. (2004). Effects of monoglyceride content, cooling rate and shear on the rheological properties of olive oil/monoglyceride gel networks. *Journal of the Science of Food and Agriculture* 84 (12): 1585–1593.

Okesola, B.O., Vieira, V.M.P., Cornwell, D.J., Whitelaw, N.K., and Smith, D.K. (2015). 1,3:2,4-Dibenzylidene-D-sorbitol (DBS) and its derivatives – efficient, versatile and industrially relevant low-molecular-weight gelators with over 100 years of history and a bright future. *Soft Matter* 11: 4768–4787.

Pal, A., Abraham, S., Rogers, M.A., Dey, J., and Weiss, R.G. (2013). Comparison of dipolar, H-bonding, and dispersive interactions on gelation efficiency of positional isomers of keto and hydroxy substituted octadecanoic acids. *Langmuir* 29: 6467–6475.

Panagiotopoulou, E., Moschakis, T., and Katsanidis, E. (2016). Sunflower oil organogels and organogel-in-water emulsions (part II): Implementation in frankfurter sausages. *Lwt – Food Science and Technology* 73: 351–356.

Parween, S., Misra, A., Ramakumar, S., and Chauhan, V.S. (2014). Self-assembled dipeptide nanotubes constituted by flexible β-phenylalanine and conformationally constrained α,β-dehydrophenylalanine residues as drug delivery system. *Journal of Materials Chemistry B* 2 (20): 3096.

Patel, A.R., Schatteman, D., Lesaffer, A., and Dewettinck, K. (2013). A foam-templated approach for fabricating organogels using a water-soluble polymer. *RSC Advances* 3 (45): 22900–22903.

Pommella, A., Mathonnat, M., and In, M. (2020). Texturing edible oil with crystals of phenolic compounds: Platelets versus rods. *Journal of Food Engineering* 283: 110039.

Raynal, M. and Bouteiller, L. (2011). Organogel formation rationalized by hansen solubility parameters. *Chemical Communication* 47: 8271–8273.

Rogers, M.A. (2018). Hansen solubility parameters as a tool in the quest for new edible oleogels. *Journal of the American Oil Chemists' Society* 95 (4): 393–405.

Rogers, M.A., Abraham, S., Bodondics, F., and Weiss, R.G. (2012a). Influence of the hydroxyl position in racemic hydroxyoctadecanoic acids on the crystallization kinetics and activation energies of gels and dispersions in mineral oil. *Crystal Growth & Design* 12: 5497–5504.

Rogers, M.A., Abraham, S., Bodondics, F., and Weiss, R.G. (2012b). Positional isomers of hydroxyoctadecanoic acid molecular gels and dispersions influence crystallization kinetics and activation energies. *Crystal Growth & Design* 12: 5497–5504.

Rogers, M.A., Feng, Q., Ladizhansky, V., Good, D.B., Smith, A.K., Corridini, M., Grahame, D.A.S., Bryska, B.C., Jadhav, P.D., Sammynaiken, S., Lim, L.T., Guild, B., Shim, Y.Y., Burnett, P.G., and Reaney, M.J.T. (2016a). Self-assembled supramolecular fibrillar networks comprised of a cyclic peptide–LOB3. *RCS Advances* 6: 40765–40776.

Rogers, M.A., Feng, Q., Ladizhansky, V., Good, D.B., Smith, A.K., Corridini, M., Grahame, D.A.S., Bryska, B.C., Jadhav, P.D., Sammynaiken, S., Lim, L.T., Guild, B., Shim, Y.Y., Burnett, P.G., and Reaney, M.J.T. (2016b). Self-assembled fibrillar networks comprised of a naturally-occurring cyclic peptide—LOB3. *RSC Advances* 6 (47): 40765–40776.

Rogers, M.A. and Marangoni, A.G. (2009). Solvent-modulated nucleation and crystallization kinetics of 12-hydroxystearic acid: A nonisothermal approach. *Langmuir* 25: 8556–8566.

Rogers, M.A. and Marangoni, A.G. (2016). Kinetics of 12-hydroxyoctadecanoic acid SAFiN crystallization rationalized using hansen solubility parameters. *Langmuir* 32 (48): 12833–12841.

Rogers, M.A., Strober, T., Bot, A., Toro-Vazquez, J.F., Stortz, T., and Marangoni, A.G. (2014). Edible oleogels in molecular gastronomy. *International Journal of Gastronomy and Food Science* 2 (1): 22–31.

Rogers, M.A. and Weiss, R.G. (2015). Alkane-based molecular gelators and the structures and properties of their gels. *New Journal of Chemistry* 39: 785–799.

Rogers, M.A., Wright, A.J., and Marangoni, A.G. (2008). Crystalline stability of self-assembled fibrillar networks of 12-hydroxystearic acid in edible oils. *Food Research International* 41 (10): 1026–1034.

Rogers, M.A., Wright, A.J., and Marangoni, A.G. (2009a). Nanostructuring fiber morphology and solvent inclusions in 12-hydroxystearic acid/canola oil organogels (vol 14, pg 33, 2009). *Current Opinion in Colloid & Interface Science* 14 (3): 223-223.

Rogers, M.A., Wright, A.J., and Marangoni, A.G. (2009b). Oil organogels: the fat of the future? *Soft Matter* 5 (8): 1594–1596.

Rosa Nunes, D., Raynal, M., Isare, B., Albouy, P.A., and Bouteiller, L. (2018). Organogel formation rationalized by Hansen solubility parameters: improved methodology. *Soft Matter* 14 (23): 4805–4809.

Sawalha, H., Margry, G., den Adel, R., Venema, P., Bot, A., Flöter, E., and van der Linden, E. (2013). The influence of the type of oil phase on the self-assembly process of γ-oryzanol + β-sitosterol tubules in organogel systems.*European Journal of Lipid Science and Technology* 115: 295–300.

Scholten, E. (2019). Edible oleogels: How suitable are proteins as a structurant? *Current Opinion in Food Science* 27: 36–42.

Schurtenberger, P., Scartazzini, R., Magid, L.J., Leser, M.E., and Luisi, P.L. (1990). Structural and dynamic properties of polymer-like reverse micelles. *The Journal of Physical Chemistry* 94 (9): 3695–3701.

Shaikh, H., Rho, J.Y., Macdougall, L.J., Gurnani, P., Lunn, A.M., Yang, J., Husband, S., Mansfield, E.D.H., Peltier, R., and Perrier, S. (2018). Hydrogel and organogel formation by hierarchical self-assembly of cyclic peptides nanotubes. *Chemistry – A European Journal* 24 (71): 19066–19074.

Shakeel, A., Farooq, U., Gabriele, D., Marangoni, A.G., and Lupi, F.R. (2021). Bigels and multi-component organogels: An overview from rheological perspective. *Food Hydrocolloids* 111: 106190.

Shchipunov, Y.A. and Shumilina, E.V. (1995). Lecithin bridging by hydrogen bonds in the organogel. *Materials Science and Engineering: C* 3 (1): 43–50.

Shen, H., Niu, L., Fan, K., Li, J., Guan, X., Song, J. (2014). Application of solubility parameters in 1, 3:2,4-bis(3,4-dimethylbenzylidene) sorbitolorganogel in binary organic mixtures. *Langmuir* 30: 9176–9182.

Silva, P.M., Martins, A.J., Fasolin, L.H., and Vicente, A.A. (2021). Modulation and characterization of wax-based olive oil organogels in view of their application in the food industry. *Gels* 7 (1): 12.

Singh, A., Auzanneau, F. I., Corradini, M.G., Grover, G., Weiss, R.G., and Rogers, M.A. (2017). Molecular nuances governing the self-assembly of 1,3:2,4-dibenzylidene-d-sorbitol. *Langmuir* 33 (41): 10907–10916.

Song, J., Sun, H., Sun, S., and Feng, R. (2013). Synthesis and gel properties of sorbitol derivative gelators. *Transactions of Tianjin University* 19 (5): 319–325.

Suzuki, M. and Hanabusa, K. (2010). Polymer organogelators that make supramolecular organogels through physical cross-linking and self-assembly. *Chemical Society Reviews* 39 (2): 455–463.

Tamura, T. and Ichikawa, M. (1997). Effect of lecithin on organogel formation of 12-hydroxystearic acid. *Journal of the American Oil Chemists Society* 74 (5): 491–495.

Toro-Vazquez, J.F., Morales-Rueda, J., Mallia, V.A., and Weiss, R.G. (2010). Relationship between molecular structure and thermo-mechanical properties of candelilla wax and amides derived from (R)-12-hydroxystearic acid as gelators of safflower oil. *Food Biophysics* 5 (3): 193–202.

Vauthey, S., Santoso, S., Gong, H., Watson, N., and Zhang, S. (2002). Molecular self-assembly of surfactant-like peptides to form nanotubes and nanovesicles. *Proceedings of the National Academy of Sciences* 99 (8): 5355–5360.

Wang, R.Y., Wang, P., Li, J.L., Yuan, B., Liu, Y, Li, L., and Liu, X.Y. (2013). From kinetic-structure analysis to engineering crystalline fiber networks in soft materials. *Physical Chemistry Chemical Physics* 15 (9): 3313–3319.

Wei, F., Lu, M., Li, J., Xiao, J., Rogers, M.A., Cao, Y., and Lan, Y. (2022). Construction of foam-templated oleogels based on rice bran protein. *Food Hydrocolloids* 124: 107245.

Weiss, R.G. (2014). The past, present, and future of molecular gels. What is the status of the field, and where is it going? *Journal of the American Chemcial Society* 136: 7519–7530.

Weiss, R.G. and Terech, P. (2006). Introduction. In: *Molecular Gels: Materials with Self-Assembled Fibrillar Networks* (ed. R.G. Weiss and P. Terech), 1–13. Dordrecht The Netherlands: Springer.

Wilder, E.A., Spontak, R.J., and Hall, C.K. (2003). The molecular structure and intermolecular interactions of 1, 3:2,4-dibenzylidene-D-sorbitol. *Molecular Physics* 101: 3017–3027.

Wright, A.J. and Marangoni, A.G. (2006). Formation, structure, and rheological properties of ricinelaidic acid-vegetable oil organogels. *Journal of the American Oil Chemists' Society* 83 (6): 497–503.

Wu, S., Gao, J., Emge, T., and Rogers, M.A. (2013). Solvent induced polymorphic nanoscale transitions for 12-hydroxyoctadecanoic acid molecular gels. *Crystal Growth & Design* 13: 1360–1366.

Xue, P., Xu, Q., Gong, P., Quan, C., Ren, A., Zhang, Y., and Lu, R. (2013). Fibrous film of a two-component organogel as a sensor to detect and discriminate organic amines. *Chemical Communications* 49 (52): 5838–5840.

Yamasaki, S., Ohashi, Y., Tsutsumi, H., and Tsujii, K. (1995). The aggregated higher-structure of 1, 3:2,4-di-O-benzylidene-D-sorbitolin organic gels. *Bulletin of the Chemical Society Japan* 68: 146–151.

Yan, N., Xu, Z., Diehn, K.K., Raghavan, S.R., Fang, Y., and Weiss, R.G. (2013). How do liquid mixtures solubilize insoluble gelators? Self-assembly properties of pyrenyl-linker-glucono gelators in tetrahydrofuran-water mixtures. *Journal of the American Chemical Society* 135: 8989–8999.

Yan, X., Cui, Y., He, Q., Wang, K., and Li, J. (2008). Organogels based on self-assembly of diphenylalanine peptide and their application to immobilize quantum dots. *Chemistry of Materials* 20 (4): 1522–1526.

Zetzl, A.K., Marangoni, A.G., and Barbut, S. (2012). Mechanical properties of ethylcellulose oleogels and their potential for saturated fat reduction in frankfurters. *Food & Function* 3: 327–337.

Zheng, W., Kollmeyer, J., Symolon, H., Momin, A., Munter, E., Wang, E., Kelly, S., Allegood, J.C., Liu, Y., Peng, Q., Ramaraju, H., Sullards, M.C., Cabot, M., and Merrill Jr., A.H. (2006). Ceramides and other bioactive sphingolipid backbones in health and disease: Lipidomic analysis, metabolism and roles in membrane structure, dynamics, signaling and autophagy. *Biochimica et Biophysica Acta (BBA) – Biomembranes* 1758 (12): 1864–1884.

5.2

Research and Development Toward the Commercialization of Fat Mimetics

Miguel Ângelo Parente Ribeiro Cerqueira, Buse N. Gürbüz, and Lorenzo Miguel Pastrana Castro*

INL–International Iberian Nanotechnology Laboratory, Braga, Portugal
Corresponding author

5.2.1 Introduction

In the last five years, food industry research and development have been boosted by different stakeholders' understanding of the agri-food system's relevance to human well-being and world sustainability. For example, overall food production must be increased by more than 50% to feed 10 billion people by 2050, but the agri-food system contributes to 30% of greenhouse gas emissions and 70% of freshwater withdrawals. In addition, it is reported that gobally, 1/4 of all food is lost or wasted, which is a contradiction if we consider that malnutrition affects around 3.5 billion people. Simultaneously, unhealthy lifestyles increase the worldwide numbers of overweight (~2 billion) and obese (>500 million) (WEF 2016; WRI 2018). Another challenge of the agri-food sector is digitilizing the food value chain to increase consumer engagement. For these reasons, there is an urgent need for increased private and public investments in research and innovation to make the food system more sustainable, resilient, healthier (also improving food safety), trustworthy (securing food authenticity and traceability), and fair.

In recent years, some reports and publications tried to track the R&D investment in the agri-food industry and how it has evolved in different countries, from the public to the private sector. The food system is one of the largest global value chains connecting producers to consumers worldwide. The food industry is the European Union's biggest manufacturing sector in terms of jobs (24 million people) and gross value added (6% of Europe's GVA). But, a recent report from JRC mentioned that in the EU, the agro-food industry is usually considered an industry with relatively low-intensity R&D and that the share of output attributable to R&D is around 0.27%, which is lower than R&D in the automobile (5.5%), software (10.6%), and pharmaceutical (13.1%) industries (Commission et al., 2018). In 2016, Pardey et al. showed that in 2011, 5% of worldwide R&D investment (public and private) was in food and agriculture. These values were supported by OECD reports that showed that the public investment in agricultural R&D value-added is less than 3% when measured by the government budget allocation for R&D (OECD 2019a). Private funding is difficult to track in many countries, however, some have suggested that private spending on food and agricultural R&D increased worldwide. A recent report mentioned that in 2019, the private R&D investment of food and drink companies was higher in the US, EU, Japan, and Switzerland, which is justifiable given these countries are the headquarters of big companies (FoodDrinkEurope 2021).

One strategy to overcome low R&D investment is to change the innovation models and the strategies for the R&D. Those strategies need to consider the agri-food sector complexity, which is one

Fat Mimetics for Food Applications, First Edition. Edited by Miguel Ângelo Parente Ribeiro Cerqueira and Lorenzo Miguel Pastrana Castro.

of the challenges for starting R&D projects in the sector. The agri-food sector has substantial interconnectedness and interdependency of various elements ranging from primary production, processing, packaging, and distribution to consumption and waste management, which in most cases is impossible to consider in a single project. Countries have been applying strategic plans where different stakeholders participate, defining long term goals. Some countries, such as Canada and the Netherlands, facilitate discussion and cooperation throughout the innovation process by developing formal institutions to manage those processes, which encourages dialogue and cooperation.

5.2.2 Research & Development in Fat Mimetics

As already presented in earlier chapters, fat mimetics can replace partially hydrogenated oils and saturated fats with other healthy ingredients in foods while avoiding using unsustainable oils and promoting a more sustainable food system. This has driven several stakeholders (companies and research organizations) to develop R&D on fat mimetics for food applications. Table 5.2.1 presents R&D projects from the last 20 years focused on fat mimetics, in which information is public.

Based on the project's summaries, different strategies, mainly the ones using carbohydrates, proteins, and lipids to develop fat mimetics, have been explored. Some projects were focused on fat mimetics based on carbohydrates and proteins (e.g., *Alternative sweeteners, fat mimetics, anti-staling agents and novel food gums* and *Kinetic trapping: a novel, energy-efficient approach to designing protein-based fat replacers* and *Fat replacement system from natural plant materials for improving food nutrition, quality, and reducing costs, PALM-ALT; MIPRO*), and others in emulsions and oleogels (e.g., *SAT-FAT-FREE, Novel Strategies For The Functional Replacement Of Trans And Saturated Fats In Finished Food Products, Plant Particle-based Hybrid Bicontinuous Oleogels, BetterFat4Meat, hOLIVEcream*, and *Architecture of colloidal delivery systems: what is the role of structure on the digestibility?*).

The projects mostly focus on replacing and reducing fats in foods where trans fatty acids and saturated fats are the problems being addressed. However, two projects discuss the possibility of enhanced health benefits by adding olive oil (*Preparation of nutritionally healthy margarines using olive pomace oil as a substitute for saturated fats* and *hOLIVEcream*), which can be one of the unique advantages of using fat mimetics (*i.e.*, tailoring the fatty acids profile of food products). Another point explored in some projects is the control of lipids digestion. Those strategies can control the absorption of lipids while maintaining foods' organoleptic properties (POLIMEATSI, *Architecture of colloidal delivery systems: what is the role of structure on the digestibility*).

5.2.3 Patents and Commercial Products

Another important asset among companies is intellectual property rights (IPRs). IPRs are an important incentive to invest in innovation since it lets companies protect their investment. By using IP, the companies that are rights-holders can exclude competitors during a limited period or, in the case of open innovation approaches, they can promote access and sharing. One challenge for IPR regulations is managing it when the company wants to incentivize public-private partnerships, with public investment, without compromising knowledge sharing and further innovation (OECD 2019).

Table 5.2.1 Funded projects related to fat mimetics.

Country or Region / Year	Name of the project	Budget	Website (accessed on 12th April 2022)
USA (2004–2009)	Alternative sweeteners, fat mimetics, anti-staling agents and novel food gums	–	https://reeis.usda.gov/web/crisprojectpages/0088724-alternative-sweeteners-fat-mimetics-anti-staling-agents-and-novel-food-gums.html
Ireland (2009–2010)	Kinetic trapping: a novel, energy-efficient approach to designing protein-based fat replacers		https://www.teagasc.ie/media/website/publications/2010/Kinetic-trapping-a-novel-energy-efficient-approach-to-designing-protein-based-fat-replacers_6041.pdf
Scotland (2012–2015)	Foaming and fat replacer ingredients	42 363 £	https://gow.epsrc.ukri.org/NGBOViewGrant.aspx?GrantRef=EP/J501712/1
European Commission (2012–2016)	TeRiFiQ: Combining technologies to achieve significant binary reductions in sodium, fat and sugar content in everyday foods while optimizing their nutritional quality	2 994 363 €	https://cordis.europa.eu/project/id/289397
USA (2013–2015)	Fat replacement system from natural plant materials for improving food nutrition quality, and reducing costs	–	https://reeis.usda.gov/web/crisprojectpages/1000375-fat-replacement-system-from-natural-plant-materials-for-improving-food-nutrition-quality-and-reducing-costs.html
Belgium (2013–2017)	SAT-FAT-FREE – Emulsion-templated structuring of liquid oil for future development of saturated fat-free food products	100 000 €	https://cordis.europa.eu/project/id/618157
Canada (2014–2017)	Novel Strategies For The Functional Replacement Of Trans And Saturated Fats In Finished Food Products	–	https://foodmap.lib.uoguelph.ca/project/2255
Spain (2014–2021)	Study of strategies for stabilizing oils in matrices for the reformulation of healthier meat products. Changes in technological properties and in vitro digestion processes	25 940 €	https://www.csic.es/en/node/415770
Netherlands (2018–2020)	Plant Particle-based Hybrid Bicontinuous Oleogels	165 5989 €	https://cordis.europa.eu/project/id/798917
European Commission (2019-)	OLEOGELS – healthy and sustainable replacement for saturated fat and palm oil in spreads	–	https://www.eitfood.eu/projects/oleogels-healthy-and-sustainable-replacement-for-saturated-fat-and-palm-oil-in-spreads
Portugal (2019–2022)	BetterFat4Meat: Development of Structured fat to use in meat products in substitution of animal fat	1 246 691 €	https://inl.int/funded-project

(Continued)

Table 5.2.1 (Continued)

Country or Region / Year	Name of the project	Budget	Website (accessed on 12th April 2022)
Spain (2020–2023)	POLIMEATSI: Biopolymer-based emulsions with controlled lipid digestibility for developing functional meat products as a strategy to manage Type 3 diabetes mellitus	–	https://digital.csic.es/dc/investigacionImagenes/infoTotal.jsp?id=375
Spain (2020–2023)	Preparation of nutritionally healthy margarines using olive pomace oil as a substitute for saturated fats	44 000 €	https://www.csic.es/es/investigacion/proyectos-de-investigacion/elaboracion-de-margarinas-nutricionalmente-saludables
Scotland (2020–2022)	PALM-ALT: Novel palm fat replacer for the food industry (phase 1 and phase 2)	39 384 £ and 643 855 £	https://gtr.ukri.org/projects?ref=81448 https://gtr.ukri.org/projects?ref=10004518
Germany (2020–2023)	MIPRO	–	https://www.ivv.fraunhofer.de/en/food/plant-proteins/mipro.html
Portugal (2021–2023)	hOLIVEcream – Healthy olive oil-based creams enriched with berries for application in bakery and pastry	500 077 €	https://inl.int/funded-project
Brazil (2021–2026)	Architecture of colloidal delivery systems: what is the role of structure on the digestibility?	–	https://bv.fapesp.br/en/auxilios/108164/architecture-of-colloidal-delivery-systems-what-is-the-role-of-structure-on-the-digestibility

Several companies and research institutes have been protecting their findings in the fat mimetics area. A search for patents on fat mimetics showed over 90 patents related to fat mimetics since the beginning of the 21st century. The number of patents has increased over the last five years. Table 5.2.2 presents some patents on the topic of fat mimetics. A detailed analysis of the patents shows a clear trend in using oleogels and lipid-based systems for replacing, where the fat mimetics that use no oil as an ingredient represent around 15% of all analyzed patents.

Among the commercial products, different types of fat mimetics can be found. For example, there are available on the market carbohydrate-based, such as JELUCEL® (https://www.jelu-werk.com) and Novagel® (https://www.dupontnutritionandbiosciences.com), protein-based such as Dairy Lo™ (https://www.lactalisingredients.ca) and SIMPLESSE® (https://www.cpkelco.com), and the lipid-based, in which monoacylglycerols are used, such as Coasun™ (https://www.coasun.com and Trancendim® (https://www.corbion.com).

Table 5.2.2 Patents on fat mimetics (search in the United States Patent and Trademark Office, in World Intellectual Property Organization and Google Patents).

			Legal Status		
Objective	**Materials**	**Reference**	**Active**	**Pending**	**Abandoned**
Margarine/ Fat substitute in bakery	Vegetable oil, protein, and water	One step procedure for producing a protein oleogel		AU2020286152A1 (Scholten 2022b) EP3975752A1 (Scholten 2022a) CA3141790A1 (Scholten 2020b) WO2020242313A1 (Scholten 2020a) SG11202113239VA (Scholten 2021) BR112021023956A2 (Scholten 2022c) KR1020220018505A (Scholten 2022d) IL288422D0 (Elke 2022)	
Mimic animal fat	Fat, vegetable oil, protein, water, hydrocolloid	Oleogel	NL2026242B1 (Oscar Jan et al., 2022)	WO2022031172A1 (Camilleri et al., 2022)	
Replacement of solid and semi-solid fat	Soluble and insoluble fiber, vegetable oil, water	Fiber-based oleogel		WO2022060289A1 (Phoon Pui 2022)	
Margarine / Replacement of fat	Vegetable oil, water, compound of Formula (I) "oxalamide"	Composition comprising oxalamide gelators and vegetable oil	EP3897560A1 (Sijakovic Vujicic et al., 2021)	CA3120396A1 (Sijakovic Vujicic et al., 2020b) US20220023179A1 (Sijakovic Vujicic et al., 2022) WO2020125926A1 (Sijakovic Vujicic et al., 2020a)	

(Continued)

Table 5.2.2 (Continued)

Objective	Materials	Reference	Legal Status: Active	Legal Status: Pending	Legal Status: Abandoned
Replacement of fat in chocolate (Mars)	Edible oil, ethyl cellulose, surfactant (e.g., sorbitan monostearate (SMS))	Chocolate compositions containing ethylcellulose oleogel and preparation methods	US11241021B2 (Marangoni Alejandro 2022) US20190261644A1 (Marangoni Alejandro 2019) RU2524097C2 (Marangoni Alekhandro 2014a) MX2011013353A (Marangoni Alejandro 2012b) ES2464368T3 (Marangoni 2014) EP2440067B1 (Gregorio Marangoni Alejandro 2014a) EP2440067A1 (Marangoni Alejandro 2012d) CN107594047A (Marangoni Alejandro 2018) CN102548425B (A·G·马兰戈尼 2017) CN102548425A (Gregorio Marangoni 2012a) CA2763645C (Marangoni Alejandro 2017a) CA2763645A1 (Marangoni Alejandro 2010a) BRPI1009679B1 (Alejandro 2017a) BRPI1009679A2 (Gregorio 2015a) AU2010258360B2 (Marangoni Alejandro 2015a) AU2010258360A1 (Marangoni Alejandro 2012c)	EG26866A (Gregorio Marangoni Alejandro 2014b) WO2010143067A1 (Marangoni Alejandro 2010b) HK1169923A1 (Marangoni Alejandro 2013a)	US20120183651A1 (Marangoni Alejandro 2012a)

Margarine / Replacement of fat	Vegetable oil, natural waxes (e.g., waxes from sunflower, candelilla, etc.)	Wax oleogels as fat substitutes	CA3106795C (Marangoni et al., 2021c) CA3106795A1 (Marangoni et al., 2021c)	WO2021/046642a1 (Marangoni et al., 2021b)	US20210161166A1 (Marangoni et al., 2021a)
Margarine / Replacement of fat (Unilever)	Fat powder, lecithin, water, edible oil, protein, thickener (e g., starch or gums like alginate)	Process for preparing emulsifier-free edible fat-continuous emulsions	AU2014334012B2 (Adel Rudi et al., 2016) AU2014334012A1 (Adel Rudi et al., 2016) BR112016007493B1 (Abraham et al., 2021) BR112016007493A2 (Abraham et al., 2017) CA2925300C (Adel Rudi et al., 2021) CA2925300A1 (Adel Rudi et al., 2015) EP3054776B1 (Adel Rudi et al., 2018) EP3054776A1 (Adel Rudi et al., 2016) MX2016004474A (Georg Christian et al., 2016)	PL3054776T3 (Adel Rudi et al., 2019a) HUE041672T2 (Adel Rudi et al. 2019b) WO2015052037A1 (Adel Rudi et al., 2015)	US20160213018A1 (Den Adel et al., 2016)
Margarine / Replacement of fat	Edible oil, ethyl cellulose	Method of preparing an edible oleogel	BR112014027719B1 (Bradley et al., 2020) BR112014027719A2 (Bradley et al., 2017) EP2866577B1 (Ergun et al., 2019) EP2866577A1 (Ergun et al., 2015a) KR102308971B1 (에르건 et al., 2021) KR20150034208A (에르건 et al., 2015)	CN107616241A (Ergun et al., 2018) KR1020200052996A (Ergun et al,. 2020) JP2015521846A（ロジャ・アーガン et al. 2015) WO2014004018A1 (Ergun et al., 2014)	CN104582496A (Ergun et al., 2015b) US20150157038A1 (Ergun et al., 2015c)

(Continued)

Table 5.2.2 (Continued)

			Legal Status		
Objective	Materials	Reference	Active	Pending	Abandoned
Margarine	Vegetable oil, walnut protein polysaccharide composite, dandelion root polysaccharide	Walnut protein polysaccharide composite oleogel		CN112293746A (Yan et al., 2021)	
Replacement of fat	Sodium alginate, water, vegetable oil, beeswax	Semi-finished confectionery product based on hybrid gel	RU2753975C1 (Nepovinnykh Nataliya et al., 2021)		
Replacement of fat	Rice bran protein, saline, vegetable oil	Rice bran protein oleogel as well as preparation method and application thereof	CN113261594B (兰雅淇 et al., 2022) CN113261594A (Lan et al., 2021)		
Replacement of fat	Wax (eg: beeswax, rice bran wax), emulsifier (monoglycerides, sucrose, etc.), vegetable oil, solution of vitamin C	Preparation method of vitamin C loaded oleogel and shortening blend as well as product and application of vitamin C loaded oleogel and shortening blend		CN112931633A (Li et al., 2021)	
Replacement of fat	Ethyl cellulose, vegetable oil, dispersant agent (eg: fatty acids such as oleic or stearic acid), water	Ethyl cellulose oleogel dispersion	JP6883588B2 (デイヴイツト・エル・マロトキー et al., 2021) JP2019513169A (デイヴイツト・エル・マロトキー et al., 2019)	EP3435777A1 (Malotky David et al., 2019a) US20190075810A1 (Malotky David et al., 2019b) WO2017172594A1 (Malotky David et al., 2017)	CN108712863A (Malotky David et al., 2018a) KR1020180131565A (Malotky David et al., 2018b) BR112018069122A2 (David et al., 2019)
Oleogel	Soybean oil, rice bran wax, water	Preparation method of soybean oil-based oleogel		CN112889935A (He et al., 2021)	

Replacement of fat	Soy protein isolated, vegetable oil, thickener (eg: xanthan gum, carrageenan, etc.), gelling agent (magnesium sulfate, etc.), cross-linking agent (e.g.: glutamine transaminase)	Preparation method of oleogel for replacing hydrogenated vegetable oil		CN113100298A (Lin et al., 2021)
Partial replacement of coconut fat in chocolate	Canola oil, chitosan, carrageenan, water, soy lecithin	Polymer-based oleogel and related use		BR102020002731A2 (AndrÉS Silvestre Gallegos et al., 2021) WO2021155455A1 (Dupas Hubinger et al., 2021)
Replacement of fat	Gelling agents (eg: beeswax, ethyl cellulose, etc.), vegetable oil, monoglycerides, water	Oleogel-based emulsion with stable interface crystallization as well as preparation method and application of oleogel-based emulsion		CN113678899A (Zhang et al., 2021)
Replacement of fat	Vegetable oil, gelatin, polysaccharides (eg: xanthan gum, etc.), polyphenol (eg: tannin)	Oleogel rich in unsaturated fatty acid and preparation method and application thereof	CN107950684B (仇超颖 et al., 2021) CN107950684A (Qiu et al., 2018)	
Replacement of fat for frying	Vegetable oil, waxes (eg: sunflower wax)	Oil-wax compositions and uses of oil-wax compositions for cooking food articles		AR120631A1 (Di Bari et al., 2022) WO2021064453A1 (Sarney Andrew et al., 2021)
Replacement of fat for chocolate	Gelling agents (eg: monoglycerides, glycerol monoestreate, waxes, etc.), edible oil	Method to produce oleogels		US20210127720A1 (Valoppi et al., 2021)

(Continued)

Table 5.2.2 (Continued)

			Legal Status		
Objective	**Materials**	**Reference**	**Active**	**Pending**	**Abandoned**
Replacement of fat	Gelling agent (eg: gelatin, hydroxypropylmethylcellulose, etc.), water, vegetable oil, oil-soluble polysaccharide (ethyl cellulose and chitin), oil-soluble micromolecule gel (eg: monoglycerides, etc.)	Method for preparing phase-change adjustable emulsion gel-type fat substitute and application		CN113383947A (Meng et al., 2021)	
Margarine / Replacement of fat	Vegetable oil, emulsifier (e.g.: tween 20, lecithin, etc.), water, diglyceride solid lipid nanoparticles	Preparation method of water-in-oil emulsion gel based on diglyceride solid lipid nanoparticles		CN112868816A (Qiu et al., 2021)	
Replacement of fat	Agar-agar, water, vegetable oil, protein or surfactant (e.g.: carrageenan, etc.)	Method for obtaining oleogels	ES2738923B2 (LÓPez Rubio et al. 2020) ES2738923A8 (LÓPez Rubio et al., 2020) ES2738923A1 (LÓPez Rubio et al., 2020b)	WO2020021152A1 (LÓPez Rubio et al., 2020a) WO2021019117A1 (LÓPez Rubio et al., 2021)	
Replacement of fat	Wax (e.g.: beeswax, sunflower wax, etc.), structuring agent (e.g.: fatty acids, etc.), vegetable oil	Composition of organogels/oils based on combinations of waxes and vegetable oils		BR102019027238A2 (Noadia Genuario et al., 2021)	

Replacement of fat	Water, edible oil, emulsifier (e.g., monoglycerides, etc.), sugar and/or sugar alcohols (e.g., sorbitol, xylitol, etc.)	Structured oil-in-water emulsion and food product comprising the same		WO2020218965A1 (Demeurisse and Marangoni 2020) EP3958684A1 (Demeurisse and Marangoni 2022b) CN114144068A (J 德米尤瑞斯 and A 马兰戈尼 2022) US20220167638A1 (Demeurisse and Marangoni 2022a) AR118761A1 (Marangoni and Demeurisse 2021) UY38675A (Demeurisse and Marangoni 2021)	
Replacement of fat	Vegetable oil, gelling agent (e.g., monoglycerides, waxes, etc.), water, polysaccharide (e.g., xanthan gum, agar, alginate, etc.)	Preparation method and application of bicontinuous emulsion gel		CN111700264A (Liu et al., 2020)	
Margarine / Replacement of fat	Vegetable oil, water, solid fat, phytosterol	Phytosterol composition in food		CN112004415A (Daels and Foubert 2020c) AU2018387328A1 (Daels and Foubert 2020b) CA3086654A1 (Daels and Foubert 2019a) EP3727006A1 (Daels and Foubert 2020a) RU2020123991A3 (ДАЕЛС and ФАУБЕРТ 2022) WO2019122227A1 (Daels and Foubert 2019b)	US20200315200A1 (Daels and Foubert 2020d)
Replacement of fat	Composite gel (monostearyl glyceryl oleate and beta-sitosterol/ gamma-oryzanol), palm oil	Oryzanol/sitosterol changed palm oil oleogel and preparation method thereof		CN111743012A (Mou et al., 2020)	

(Continued)

Table 5.2.2 (Continued)

			Legal Status		
Objective	**Materials**	**Reference**	**Active**	**Pending**	**Abandoned**
Margarine / Replacement of fat	Olive oil, coconut butter, lecithin, microcrystalline cellulose, stabilizing agent (eg: mono e diglycerides)	Olive oil-based, solid vegetable fatty product substitute for butter or margarine and use thereof		CA3130858A1 (Ortiz Garcia Juan et al., 2020) ES2787974A1 (Ortiz Garcia 2020) EP3957181A1 (Ortiz Garcia Juan et al., 2022a) WO2020212633A1 (Ortiz Garcia et al., 2020) CL2021002596A1 (Ortiz Garcia Juan et al., 2022c) PE20220382A1 (Ortiz Garcia Juan et al., 2022b)	
Replacement of fat	Ethyl cellulose, edible oil, stearic acid	Oleogel with stearic acid	EP3515200B1 (Ergun and Meunier David 2020a) EP3515200A1 (Ergun and Meunier David 2019a)	CN109788774A (Ergun and Meunier David 2019b) US20200015493A1 (Ergun and Meunier David 2020b) WO2018057384A1 (Ergun and Meunier David 2018)	
Replacement of fat in margarine	Vegetable oil, oryzanol, sterol, cinnamic acid, monoglyceride	A novel artificial cream and its preparation method and application	CN106720458B (肖志刚 et al., 2020) CN106720458A (Xiao et al., 2017)		
Replacement of fat	Vegetable oil, wax (e.g., polycosanol), protein, water	A antioxidant and edible oleogel and preparation method thereof	CN109588497B (李松林 and 林静 2020) CN109588497A (Li and Lin 2019)	WO2020133652A1 (Li et al., 2020)	
Replacement of fat	Ethyl cellulose, vegetable oil, stearic acid, saturated fat	Solid composition containing oleogel		CN109788773A (Ergun et al., 2019a) EP3515196A1 (Ergun et al., 2019b) WO2018057383A1 (Ergun et al., 2018)	US20190307145A1 (Ergun et al., 2019c)

Replacement of fat	Edible oil, ethyl cellulose	Method of making oleogel	EP3435776B1 (Hild et al., 2020) EP3435776A1 (Hild et al., 2019a) JP6876063B2 (アレクサンドラ・ヒルド et al., 2021) JP2019513011A (アレクサンドラ・ヒルド et al., 2019)	WO2017172590A1 (Hild et al., 2017)	CN108777972A (Hild et al., 2018) KR1020180128002A (Hild et al., 2018) US20190075811A1 (Hild et al., 2019b) BR112018069041A2 (Alexandra et al., 2019)
Replacement of fat	Rice bran protein, water, phenolic compounds (catechin), corn oil, pectin	Method for preparing antioxidant stable oleogel through electrochemical treatment and polyphenol and polysaccharide changed rice bran protein		CN110623138A (Yu et al., 2019)	
Margarine/ Replacement of fat	Surfactant (e.g.: monoglyceride, etc.), vegetable oil	A food-grade novel antibacterial oleogel rich in polyunsaturated fatty acid and preparation method of it based on vegetable oil	CN109527120B (赵海珍 et al., 2022) CN109527120A (Zhao et al., 2019)		
Replacement of fat	Vegetable oil, gelling agent (e.g., lecithin, phospholipids, etc.), fat	Thermoreversible edible oleogel and method for producing same	EP3387909B1 (Pérez Gallardo et al., 2021) US10874115B2 (Perez Gallardo et al., 2020) ES2909381T3 (Perez Gallardo et al., 2022) EP3387909A4 (Pérez Gallardo et al., 2019) US2018/0352826A1 (Perez Gallardo et al., 2018) EP3387909A1 (Pérez Gallardo et al., 2018)	MX2015016900A (Alfonso Pérez et al., 2017) CA3007962A1 (Perez Gallardo et al., 2017) WO2017099574A1 (Pérez Gallardo et al., 2017)	

(Continued)

Table 5.2.2 (Continued)

			Legal Status		
Objective	**Materials**	**Reference**	**Active**	**Pending**	**Abandoned**
Replacement of fat (Mars)	Ethyl cellulose, edible oil, surfactant (e.g.: tween 80, tween 60, glycerol monostearate, etc.)	Polymer gelation of oils	HK1169922A1 (Marangoni Alejandro 2013b) US8940354B2 (Marangoni Alejandro 2015b) AU2010258359B2 (Marangoni Alejandro 2015c) AU2010258359A1 (Marangoni Alejandro 2012h) BRPI1009677B8 (Alejandro Gregorio 2018) BRPI1009677B1 (Alejandro 2017b) BRPI1009677A2 (Gregorio 2015b) CA2763644C (Marangoni Alejandro 2017b) CN102510725B (Gregorio Marangoni Alejandro 2014d) CN102510725A (Gregorio Marangoni 2012b) EP2440066B1 (Marangoni Alejandro 2013c) EP2440066A1 (Marangoni Alejandro 2012g) ES2437394T3 (Gregorio Marangoni Alejandro 2014c) RU2527027C2 (Marangoni Alekhandro 2014b) US8940354B2 (Marangoni Alejandro 2015b) US20120183663A1 (Marangoni Alejandro 2012f) MX2011013355A (Marangoni Alejandro 2012e)	EG26592A (Grogorio Marangoni Alejandro 2014) WO2010143066A1 (Marangoni Alejandro 2010c)	

Creme/ Replacement of fat	Water, vegetable oil, surfactant (e.g.: monoglycerides, etc.)	Method to make fat substitute and/or fat-imitator compounds		IT201600069429A1 (Calligaris et al., 2018b) WO2018007399A1 (Calligaris et al., 2018a)	
Replacement of fat	Vegetable oil, ethyl cellulose	Low fat laminated dough and pastry	ES2669614T3 (Ergun et al., 2018) EP2956007B1 (Ergun et al., 2018) JP6126301B2 (ロジャ・エルグン et al., 2017) JP2016514486A (ロジャ・エルグン, ブラッドリー・エス・トムソン et al., 2016) EP 2956007 A1 (Ergun et al., 2015b)	WO2014172135A1 (Ergun et al., 2014)	BR112015025714A2 (Bradley et al., 2017) CN105120670A (Ergun et al., 2015a) US20160021898A1 (Ergun et al., 2016) KR20160002929A (Ergun et al., 2016)
Creme/ Replacement of fat	Olive oil, vegetable fat, water, emulsifier (e.g.: lecithin, etc.), vegetable fiber, organic acid (e.g.: citric acid, etc.)	Solid creamy food formulation based on olive oil, process for its preparation and uses thereof		ITUB20154616A1 (Pompili Ferrari et al., 2017a) EP3361876A1 (Pompili Ferrari et al., 2018a) WO2017/064737A1 (Pompili Ferrari et al., 2017b)	AU2016339162A1 (Pompili Ferrari et al., 2018b) US20180295851A1 (Pompili Ferrari et al., 2018c)
Replacement of fat (Nestec)	Gelling agent (e.g.: diacylglycerol, etc.) edible oil and/or saturated fat	Structured oil compositions	AU2014267411B2 (Kornel Nagy 2017) AU2014/267411A1 (Nagy 2015a) ES2623293T3 (Kornél Nagy 2017a) EP3001789A1 (Kornél Nagy 2016) EP3001789B1 (Kornél Nagy 2017b)	CL2015003366A1 (KornÉI Nagy 2016)	BR112015028696A2 (Kornél 2017) CA2912425A1 (Kornel Nagy 2014) CN105208872A (Nagy 2015b) EP3001789B1 (Kornél Nagy 2017b) JP2016523523A (ナジー 2016) RU2015153876A (НАГИ 2017) US20160073654A1 (Kornel Nagy 2016) WO2014184118A1 (Kornél Nagy 2014)

(Continued)

Table 5.2.2 (Continued)

			Legal Status		
Objective	**Materials**	**Reference**	**Active**	**Pending**	**Abandoned**
Creme (Unilever)	Fat powder, olive oil, soybean oil, water	Process for preparing a fat slurry comprising olive oil and for preparing a spread with said slurry	EP3393265B1 (äGrebenkämper et al., 2019a) EP3393265A1 (Grebenkämper et al., 2018) ZA201802881B (Kai et al., 2019a) TR201904940T4 (Kai et al., 2019b)	CA3004505A1 (Grebenkamper et al., 2017) MX2018007635A (Rogier Antoine Floris et al., 2018) US20180368434A1 (Grebenkämper et al., 2018) PL3393265T3 (Grebenkämper et al., 2019b) HUE043408T2 (Grebenkämper et al., 2019) WO2017108275A1 (Grebenkämper et al., 2017)	
Margarine / Replacement of fat	Ethyl cellulose, edible oil	Process for preparing an oleogel	BR112015028312B1 (David et al., 2020) BR112015028312A2 (David et al., 2017) CN105228461B (R·埃尔贡 et al., 2019) CN105228461A (Ergun et al., 2016c) EP3007565B1 (Ergun et al., 2018) EP3007565A1 (Ergun et al., 2016b) JP6412115B2 (ロジャ・エルグン et al., 2018) JP2016519956A (ロジャ・エルグン, ロバート・ビー・アッペル, et al., 2016) KR102254288B1 (에르군 et al., 2021) KR20160030892A (Ergun et al., 2016d) US9655376B2 (Ergun et al., 2017) US20160081374A1 (Ergun et al., 2016a)	WO2014193667A1 (Ergun et al., 2014)	

Replacement of fat	Edible oil, nano-crystalline cellulose compound (Pickering emulsion)	Edible oil gel and preparation method thereof			CN105994697A (Sui et al., 2016)
Replacement of fat (Archer Daniels Midland Company)	Vegetable oil, gelling agent (e.g., lecithin, etc.), water	Food compositions comprising organogels	AU2009313928B2 (Baseeth Shireen and Sebree Bruce 2015b) AU2009/313928A1 (Baseeth Shireen and Sebree Bruce 2010a) JP6073552B2 (バシース,シリーン,エス. & セブリー,ブルース,アール 2017) KR101685218B1 (바시쓰 and 세브리 2016) JP5827709B2 (バシース,シリーン,エス. And セブリー,ブルース,アール 2015) ZA201104382B (Baseeth Shireen and Sebree Bruce 2012a) MX2011005076A (Baseeth Shireen and Sebree Bruce 2011a) KR20110083738A (Baseeth Shireen and Sebree Bruce 2011c)	WO2010057007A1 (Baseeth Shireen and Sebree Bruce 2010b)	BRPI0915271A2 (Sebree Bruce and Baseeth Shireen 2016) CA2743675A1 (Baseeth Shireen and Sebree Bruce 2010c) CN103349285A (Baseeth Shireen and Sebree Bruce 2013) CN105532922A (Baseeth Shireen and Sebree Bruce 2016) EP2355803A1 (Baseeth Shireen and Sebree Bruce 2011b) EP2711001A3 (Baseeth Shireen and Sebree Bruce 2014) EP2711001A2 (Baseeth Shireen and Sebree Bruce 2014) EP2355803A4 (Baseeth Shireen and Sebree Bruce 2012b) US20110250299A1 (Baseeth Shireen and Sebree Bruce 2011d) WO2010/057007A1 (Baseeth Shireen and Sebree Bruce 2010b) AU2015227472A1 (Baseeth Shireen and Sebree Bruce 2015a) CO6361902A2 (Shireen Steve and Bruce 2012)
Replacement of fat	Edible oil, gelling agent (anhydrohexitol)	Structuring and gelling agents		WO2015095472A1 (Stensrud et al., 2015)	US20160324178A1 (Stensrud et al., 2016)

(Continued)

Table 5.2.2 (Continued)

			Legal Status		
Objective	**Materials**	**Reference**	**Active**	**Pending**	**Abandoned**
Margarine (Unilever)	Fat powder, sunflower oil, lecithin, emulsifier (e.g., triglycerides, etc.), β-Carotene, water, potassium sorbate, sodium chloride	Process for preparing a spreadable dispersion	CN101137293B (Kempen et al., 2011) CN101137293A (Eckhard and Hindrik 2008) AT535152T (Barendse Sandra et al. 2011b) AU2006215829B2 (Barendse Sandra et al., 2009) AU2006215830B2 (Potman Ronald et al., 2010) AU2006215829A1 (Barendse Sandra et al., 2006) CA2598401C (Barendse Sandra et al., 2013) CA2598401A1 (Kempen Gijsbert Michiel et al., 2006) EP1865786B2 (P. Barendse Sandra et al., 2015a) EP1865786B1 (Barendse Sandra et al., 2011a) EP1865786A2 (Barendse Sandra et al., 2007) US8147895B2 (Petronella Barendse Sandra et al., 2012a) US8927045B2 (Barendse Sandra et al., 2015) US9661864B2 (Barendse Sandra et al., 2017)	PL1865786T5 (P. Barendse Sandra et al., 2015b) PL1865786T3 (P. Barendse Sandra et al., 2012) WO2006087091A3 (Barendse Sandra et al., 2006) WO2006087093A1 (Garbolino and Huizinga 2006e) WO2006087090A1 (Garbolino and Huizinga 2006b) WO2006087091A2 (Barendse Sandra et al., 2006)	CN101119640A (Chiara and Hindrik 2008b) CN101119638A (Chiara and Hindrik 2008c) AU2006215828B2 (Garbolino and Huizinga 2010) AU2006215831B2 (Huizinga and Garbolino 2009) AU2006215828A1 (Garbolino and Huizinga 2006a) AU2006215831A1 (Huizinga and Garbolino 2006) BRPI0606341A2 (Garbolino and Huizinga 2009b) BRPI0606354A2 (Janssen Johannes Jozef et al., 2009) BRPI0606356A2 (Garbolino and Huizinga 2009c) CA2595281A1 (Garbolino and Huizinga 2006d) CA2597478A1 (Garbolino and Huizinga 2006c)

			RU2377782C2 (Barendse Sandra et al., 2010) RU2379906C2 (Janssen Jokhannes Jozef et al., 2010) US20150118384A1 (Petronella Barendse Sandra et al., 2015) US20120263860A1 (Petronella Barendse Sandra et al., 2012b) US20090110801A1 (Barendse Sandra et al., 2009) ZA200706749B (Petronella Barendse et al., 2008) ZA200706244B (Chiara and Hindrik 2008d) ZA200706742B (Chiara and Hindrik 2008a)		EP1850676A1 (Garbolino and Huizinga 2007a) EP1858341A1 (Garbolino and Huizinga 2007b) US20080193628A1 (Garbolino and Huizinga 2008) US20090136645A1 (Garbolino and Huizinga 2009a) RU2384071C2 (Garbolino and Khejzinga 2010b) RU2379905C2 (Garbolino and Khejzinga 2010a) BRPI0606545A2 (Barendse Sandra et al. 2009)
Margarine (Unilever)	Fat powder, sunflower oil, lecithin, emulsifier (e.g., triglycerides, etc.), β-Carotene, water, potassium sorbate, sodium chloride	Fat granules	EP1885192B8 (Janssen Johannes et al., 2015) EP1885192B1 (Janssen Johannes et al., 2015) EP1885192A2 (Janssen Johannes et al., 2008) AU2006215830A1 (Potman Ronald et al., 2006) CA2595381C (Janssen Johannes et al., 2013) CA2595381A1 (Potman Ronald et al., 2006) US8124152B2 (Janssen Johannes Jozef et al., 2012) CN101119639B (Hogervorst Wim et al., 2011) US20080317917A1 (Janssen Johannes Jozef et al., 2008) ZA200706748B (Maria Janssen Johannes et al., 2008)	PL1885192T3 (Janssen et al., 2015) WO2006087092A3 (Janssen Johannes et al., 2006) WO2006087092A2 (Janssen Johannes et al., 2006)	

(Continued)

Table 5.2.2 (Continued)

			Legal Status		
Objective	**Materials**	**Reference**	**Active**	**Pending**	**Abandoned**
Replacement of fat	Edible oil, protein (e.g., β – lactoglobulin, etc.), emulsifier (e.g.: sorbitol esters, etc.), water	Oleogel	MX2012008439A (Clement et al., 2012)	WO2011089171A1 (Savin et al., 2011b)	EP2347658A1 (Savin et al., 2011) EP2525665A1 (Savin et al., 2012c) AU2011208707B2 (Savin et al., 2014a) AU2011208707A1 (Savin et al., 2012a) BR112012018153A2 (Savin et al., 2015) CA2787545A1 (Savin et al., 2011a) CN102834018A (Savin et al., 2012b) RU2012135483A (CABEH et al., 2014) US8796342B2 (Savin et al., 2014b) US20120289610A1 (Savin et al., 2012d)
Margarine / Replacement of fat (Unilever)	Vegetable oil, structuring agent (e.g., triglycerides, etc.), water	Edible dispersions comprising oil and structuring agent	DE602006018858D1 (Boer et al., 2011c) DK1841327T3 (Boer et al., 2011d) DK2266413T3 (Boer et al., 2015d) EP1841327B1 (Boer et al., 2010) EP2266413B1 (Boer et al., 2015c) EP2266413A1 (Boer et al., 2010) EP1841327A1 (Boer et al., 2007) ES2355741T3 (Boer et al., 2011a) ES2553627T3 (Boer et al., 2015a) ZA200705709B (Brigitta et al., 2009)	PL2266413T3 (Boer et al., 2015b) PL1841327T3 (Boer et al., 2011) BRPI0606151A2 (Boer et al., 2009) WO2006079445A1 (Boer et al., 2006)	AT491339T (Boer et al., 2011b) US20080118624A1 (Boer et al., 2008)

Replacement of fat	Microcrystalline cellulose, distilled water, hydrophilic colloids (e.g., xanthan, pectin, chitosan, etc.), edible oil (e.g.: sunflower oil, olive oil, etc.)	A preparation method of high dietary-fiber oleogel			CN107296260A (Liu et al., 2017)
Replacement of fat	Vegetable oil (e.g., sunflower, olive, palm oil, etc.), protein (e.g.: globular proteins such as hemoglobin, etc.), surfactant (e.g.: non-ionic, cationic, etc.), gelling agent (e.g.: natural gums, pectin, gelatin, etc.), water	Protein-stabilized oleogels		WO2016062685A1 (Scholten and De Vries 2016a)	EP3011836A1 (Scholten and De Vries 2016b)
Replacement of fat	Canola oil, candelilla wax	Bakery product with canola gel	KR101759502B1 (이수용 et al., 2017) KR20160115195A (Lee Su et al., 2016)		
Replacement of fat	Ethyl cellulose, peanut oil	Preparation method of vegetable oil gel fat substitute and application of it in sausage		CN113367200A (Liu et al., 2021)	
Oleogel	Protein (e.g : soybean, corn, collagen, whey protein, etc.), fish oil,	Preparation method of fish oil gel		CN112617200A (Ren et al., 2021)	

(Continued)

Table 5.2.2 (Continued)

			Legal Status		
Objective	**Materials**	**Reference**	**Active**	**Pending**	**Abandoned**
Replacement of fat	Vegetable oil (e.g.: high oleic canola, or sunflower oil), emulsifying agent (e.g.: monoglyceride, diglyceride, etc.)	Trans-fatty acid free shortening		WO2006053097A1 (Miller 2006d)	US20060105092A1 (Miller 2006c) AU2005304449A1 (Miller 2006b) BRPI0517483A (Miller 2008) CA2587403A1 (Miller 2006a) CN101098630A (Miranda 2008) EP1819234A1 (Miller 2007a) MX2007005684A (Miller 2007b) RU2007121722A (МИЛЛЕР 2008) NO20072482L (Miller 2007d) KR1020070085885A (Miller 2007c)
Replacement of fat	Yolk, phytosterol (e.g.: cupreol, stigmasterol, etc.), polysaccharide (e.g.: gellan, xanthan gum, etc.), low saturated liquid oil (e.g.: walnut oil, beta carotene oil, etc.), phosphate (e.g.: sodium phosphate, sodium tripolyphosphate, etc.), carbonate (e.g.: sodium bicarbonate and/or sodium carbonate), edible salt, white sugar, water	Egg yolk-phytosterol-polysaccharide composite emulsion gel and preparation method thereof		CN110089694A (Yang et al., 2019)	

Replacement of fat	Vegetable oil (e.g.: sunflower, canola oil, etc.), natural wax (e.g.: beeswax, candelilla wax, etc.), hydrocolloid (e.g.: guar gum. Carrageenan, etc.), water, tallow	Vegetable fat composition with oleogel and hydrocolloid	KR102138833B1 (이수용 et al., 2020) KR20200050165A (Lee Su et al., 2020)	
Replacement of fat	Water, sodium citrate, millet protein, polyvinylpyrrolidone, phosphate buffer solution, tremella ergosterol	Preparation method of fat substitute and edible rhizomorph food conditioning food	CN112369596B (史德芳 et al., 2022) CN112369596A (Shi et al., 2021)	
Replacement of fat	Edible oil (e.g.: soybean, sunflower, olive oil, etc.), phosphate buffer and/or water, pea/mung bean protein, edible gum (guar, gum carrageenan, etc.)	Method for preparing healthy fat substitute based on pea/mung bean protein		CN113812609A (Meng et al., 2021)
Replacement of fat	Whey protein isolate powder, deionized water, rapeseed oil, flaxseed gum	Method for preparing fat substitute by regulating and controlling oil drop aggregation in emulsion through flaxseed gum	CN109938328B (刘骞 et al., 2022) CN109938328A (Q.Liu et al., 2019)	
Replacement of fat	Natural wax (e.g., beeswax, chinese insect wax, etc.), vegetable oil (soybean, sunflower oil, etc.), nanocellulose suspension (nanocellulose fibers, crystals, etc.), water	Oil-in-water gel Pickering emulsion as well as preparation method and application thereof		CN113647608A (Qi et al., 2021)

(Continued)

Table 5.2.2 (Continued)

			Legal Status		
Objective	Materials	Reference	Active	Pending	Abandoned
Replacement of fat	Medium-long chain diglyceride/sitosterol (e.g.: caprylic, capric acid, etc.), vegetable oil (e.g.: soybean, rapeseed oil, etc.)	Method for preparing ultra-stable temperature-sensitive nutritional oil foam by molecular assembly and application		CN110934196A (Qiu et al., 2020)	
Fat analog/ Replacement of fat	Curdlan, carrageenan, vegetable oil, water	Special fat mimetic for meat product and preparation method of fat mimetic			CN106615220A (Zhang et al., 2017)
Fat analog/ Replacement of fat	Konjac glucomannan, carrageenan, oat bran extract, calcium hydroxide, olive oil, water	Fat mimic and application of it in Sichuan style sausages	CN105901744A (Yang et al., 2016) CN105901744B (杨勇 et al., 2020)		
Fat analog/ Replacement of fat	Konjaku flour, curdlan, water, corn oil, β-glucan	Method for preparing blocky fat stimulant with konjaku flour and curdlan as base materials		CN109730273A (Q Liu et al., 2019)	
Replacement of fat	Curdlan, water, vegetable oil (e.g., soybean oil, etc.)	Preparation method for animal fat substitute adopting curdlan as matrix	CN103125615B (Zhang et al., 2014) CN103125615A (Zhang et al., 2013)		
Fat substitute / Replacement of fat	Vegetable oil (e.g., corn, palm oil, etc.), konjuka flour, carrageenan, ovalbumin, phosphate, salt, meat flavor, water	Simulated fat and method for preparing same	CN101558890B (Aiju et al., 2011) CN101558890A (Aiju et al., 2009)		

Fat substitute / Replacement of fat	Water, curdlan, thickener (e.g.: starch, potato starch, etc.), albumen powder class nutrients (e.g.: soybean protein isolate, gluten, etc.), vegetable oil (e.g.: soybean, rapeseed oil, etc.), plant fiber (e.g.: wheat, soybean fiber, etc.), yam flour, wolfberry fruit powder, water retention agent (e.g.: sodium phosphate, etc.), antioxidant (e.g.: bamboo leaf antioxidant, etc.)	Novel health fat substitute and preparation method and application thereof		CN109123629A (J. Liu et al., 2019)	
Fat analog / Replacement of fat	Water caltrop starch, phosphate buffer, distilled water	Preparation method of fat mimics using water chestnut starch and konjak refined powder as matrix			CN104770640B (丁保淼 2017) CN104770640A (Ding 2015)
Replacement of fat	Water, konjac powder, carrageenan, sodium carbonate	Compounded gel used as fat substitute, and sausage made of compounded gel			CN107259442A (An 2017)
Fat mimic/ Replacement of fat	Konjac glucomannan, gellan gum, maltodextrin, soybean protein isolate, cassava changed starch, pig bone essence, water	Fat mimic with Harbin red sausage flavor and preparation method thereof		CN111084332A (Xu et al., 2020)	
Fat mimetic/ replacement of fat	Protein powder (soybean protein isolate), vegetable oil (palm oil), konjac glucomannan, carrageenan, sodium carbonate, water	Pig back fat simulant and preparation method thereof		CN113080433A (Wang et al., 2021)	
Replacement of fat	Curdlan, gelatin, solid and liquid oil (e.g.: rapeseed, rice salad oil, etc.)	Solid oil and fat-like composition		HK1208781A1 (Tetsuya 2016) WO2014020734A1 (Yoshimoto 2014)	CN104507323A (Yoshimoto 2015)

(Continued)

Table 5.2.2 (Continued)

			Legal Status		
Objective	**Materials**	**Reference**	**Active**	**Pending**	**Abandoned**
Replacement of fat	Soluble dietary fiber (e.g.: soybean, etc.) edible gelatin, glutamine transaminase, anthocyanidin.	Low-fat meatball with novel complex gum as fat substitute and preparing method		CN110063460A (Niu et al., 2019)	
Fat substitute/ Replacement of fat	Konjac glucomannan, sodium alginate, calcium hydroxide, purified water, pork	Pig fat substitute and production method thereof	CN104939100B (李正纪 et al., 2018) CN104939100A (Li et al., 2015)		
Fat analog/ Replacement of fat	Konjac glucomannan, carrageenan, oat bran extract, calcium hydroxide, olive oil, water	Fat substitute and application of it in low-fat fermented sausages			CN105851901A (Yang et al., 2016)
Fat substitute/ Replacement of fat	Sodium alginate, calcium salts (e.g.: calcium carbonate, calcium lactate). chelating agent (e.g.: calgon, sodium phosphate), emulsifying agent (e.g.: casein sodium, xanthans, konjac glucomannan), lipid matter (e.g.: pig back fat, vegetable oil)	Preparation method of fat substitute for fermented sausage	CN102871141B (Xu et al., 2014) CN102871141A (Xu et al., 2013)		
Fat substitute/ Replacement of fat	Oligosaccharide (e.g.: fructose, glucose, etc.), oligodextran, water, emulsifier (e.g.: lecithin, etc.), hydrocolloids (e.g.: carrageenans, pectins, etc.)	Fat replacers and filling materials		EP2575498A1 (Dendooven Els Ginette et al., 2013a) WO2011140212A1 (Dendooven Els Ginette et al., 2011)	BR112012029519A2 (Dendooven Els Ginette et al., 2015) US20130052300A1 (Dendooven Els Ginette et al., 2013b) US20170027209A1 (Dendooven Els Ginette et al., 2017)

Fat substitute/ Replacement of fat	Inulin, cassia gum, water	Use of fat substitute compositions comprising inulin and cassia gum	EP2755492B1 (Veen and Budemann 2015) EP2755492A1 (Veen and Budemann 2014)		WO2013037739A1 (Veen and Budemann 2013)
Fat substitute/ Replacement of fat	Xanthan gum, arabic gum, water, citrus pectin	Composite gel fat substitute and low-fat egg tart filling prepared from same	CN109288030B (王兆梅 et al., 2022) CN109288030A (Wang et al., 2019)		
Fat substitute/ Replacement of fat	Sodium alginate, calcium dehydrate, glucono-δ-lactone, alginic acid, diol ester, maltodextrin, fucoidan, fucoxanthin, vegetable oil, water	Preparation of fat substitution and its application in meatballs		CN109619537A (Lin et al., 2019)	

5.2.4 Conclusion

Productivity is one of the main goals of the agri-food innovation system. Still, climate changes, human health, and food security have demanded new strategies from different stakeholders to overcome the recent challenges of the agri-food system. One way to face them is through the joint collaboration of multidisciplinary teams from different sectors. This investment in agri-food R&D yielded several new developments in recent years. While public funding has not increased significantly in the last several years, private funding has been increasing. This increase is mainly related to big companies' investments; however, there still exists the need for R&D investment from small and medium companies. One strategy is the support of public funding; for example, in some EU countries, R&D investment is focused on in SMEs over big corporations. Besides the investment, it is also needed help in regulatory issues and IP protection, where the complexity and lack of knowledge in those SMEs' companies lead to a low success of the new R&D findings. In the fat mimetics area, it was possible to get some public investment through funded projects, but one of the exciting conclusions was the high number of patents and the products already in the marketplace, which undoubtedly reflect R&D investment.

Acknowledgements

The authors acknowledge the financial support of the projects: hOLIVEcream – Healthy olive oil-based creams enriched with berries for application in bakery and pastry, with the reference n.º NORTE-01–0247-FEDER-046947, co-funded by Norte 2020 – North-Regional Operational Programme under the PORTUGAL 2020 Partnership Agreement, through the European Regional Development Fund (ERDF); NUTRI4ICECARE: Ice cream with incorporation of compounds of nutritional interest (NORTE-01-0247-FEDER-045380), cofinanced by Norte 2020 and the European Union, through the European Regional Development Fund (ERDF). and BetterFat4Meat: Development of structured fat for use in meat products by replacement of animal fat (POCI-01–0247- FEDER-039718). cofinanced by Compete 2020, Portugal 2020 and the European Union, through the European Regional Development Fund (ERDF).

References

马兰戈尼, A·G. (2017). *含有乙基纤维素的巧克力组合物*(CN Patent No. CN 102548425 B). https://lens.org/000-370-546-452-550.

Adel, R.D., Dol, G.C., Grebenkämper, K. Leenhouts, A., Potman, R.P., Smit-Kingma, E., and Tio, F.F. (2015). *Process for preparing emulsifier-free edible fat-continuous emulsions* (AU Patent No. CA AU 2925300 A1). https://lens.org/068-241-921-254-051.

Adel, R.D., Dol G.C., Grebenkämper, K., Leenhouts, A., Potman, R.P., Smit-Kingma, I.E., and Tio, F.F. et al. (2016). *Process for preparing emulsifier-free edible fat-continuous emulsions* (AU Patent No. AU 2014/334012 B2). https://lens.org/121-863-955-805-233.

Adel, R.D., Dol, G.C., Grebenkämper, K., Leenhouts, A., Potman, R.P., Smit-Kingma, I.E., and Tio, F.F.. (2021). *Process for preparing non-lecithin emulsifier-free edible fat-continuous emulsions* (CA Patent No. CA 2925300 C). https://lens.org/069-745-093-063-262.

Adel, R.D., Dol G.C., Grebenkämper, K., Leenhouts, A., Potman, R.P., Smit-Kingma, I.E., and Tio, F.F. (2015). *Process for preparing emulsifier-free edible fat-continuous emulsions* (WO Patent No. WO 2015/052037 A1). https://lens.org/100-345-417-045-618.

Adel, R.D., Dol G.C., Grebenkämper, K., Leenhouts, A., Potman, R.P., Smit-Kingma, I.E., and Tio, F.F. (2016). *Process for preparing emulsifier-free edible fat-continuous emulsions* (EP Patent No. EP 3054776 A1). https://lens.org/027-702-145-258-296.

Adel, R.D., Dol G.C., Grebenkämper, K., Leenhouts, A., Potman, R.P., Smit-Kingma, I.E., and Tio, F.F. (2018). *Process for preparing emulsifier-free edible fat-continuous emulsions* (EP Patent No. EP 3054776 B1). https://lens.org/042-277-685-303-069.

Adel, R.D., Dol G.C., Grebenkämper, K., Leenhouts, A., Potman, R.P., Smit-Kingma, I.E., and Tio, F.F (2019a). *process for preparing emulsifier-free edible fat-continuous emulsions* (PL Patent No. PL 3054776 T3). https://lens.org/189-074-976-574-310.

Adel, R.D., Dol, G.C., Grebenkämper, K., Leenhouts, A., Potman, R.P., Smit-Kingma, I.E., and Tio, F.F. (2019b). *Process for preparing emulsifier-free edible fat-continuous emulsions* (HU Patent No. HU E041672 T2). https://lens.org/136-312-755-985-720.

Adel Rudi, D. E. N., Dol Georg, C., Grebenkamper, K. A. I., Leenhouts, A., Potman Ronald, P., Smit-Kingma Irene, E., & Tio Farley, F. (2016). *Process for preparing emulsifier-free edible fat-continuous emulsions* (AU Patent No. AU 2014/334012 A1). https://lens.org/070-871-451-675-22X.

Aiju, G.U.O., Wenhua, C., Zhiqiang, N.I.E., and Chao, Q.U. (2009). *Simulated fat and method for preparing same* (CN Patent No. CN 101558890 A). https://lens.org/198-222-316-605-087.

Aiju, G.U.O., Wenhua, C., Zhiqiang, N.I.E., and Chao, Q.U. (2011). *Simulated fat and method for preparing same* (CN Patent No. CN 101558890 B). https://lens.org/146-915-070-999-535.

Alejandro, G. (2017a). *composição de chocolate contendo etilcelulose* (BR Patent No. BR PI1009679 B1). https://lens.org/078-679-414-876-63X.

Alejandro, G. (2017b). *gelificação de polímero de óleos* (BR Patent No. BR PI1009677 B1). https://lens.org/160-025-334-860-178.

Alejandro Gregorio, M. (2018). *gelificação de polímero de óleos* (BR Patent No. BR PI1009677 B8). https://lens.org/006-042-592-538-959.

Alexandra, H., Britta, H.-K., Matthias, K., and Roja, E. (2019). *método para fazer um oleogel de etilcelulose, e, óleogel de etilcelulose* (BR Patent No. BR 112018069041 A2). https://lens.org/191-405-024-548-356.

Alfonso, PérezP.G., Blanca, A.G.C., and Gregorio, J.D.H.R. (2017). *Thermoreversible edible oleogel and method for producing the same* (MX Patent No. MX 2015016900 A). https://lens.org/001-406-818-839-029.

An, P. (2017). *Compounded gel used as fat substitute, and sausage made of compounded gel* (CN Patent No. CN 107259442 A). https://lens.org/064-967-188-133-253.

Andrés, S.G.S., Miriam, D.H., Priscilla, E., and Renata, S.R. (2021). *oleogéis de base polimérica e uso* (BR Patent No. BR 102020002731 A2). https://lens.org/148-919-957-238-324.

Barendse S.P., Flöter, E., Hogervorst W.T., Van Horsen, D.S.H., Huizinga, H., and Van Kempen Gijsbert Michiel, P. (2010). *Food dispersion of easily spread consistence and method of its production* (RU Patent No. RU 2377782 C2). https://lens.org/002-733-168-819-425.

Barendse S.P., Flöter, E., Hogervorst W.T., Van Horsen, D.S.H., Huizinga, H., and Van Kempen Gijsbert Michiel, P. (2006). *Process for the preparation of a spreadable dispersion* (WO Patent No. WO 2006/087091 A2). https://lens.org/087-879-478-024-180.

Barendse S.P., Flöter, E., Hogervorst W.T., Van Horsen, D.S.H., Huizinga, H., and Van Kempen Gijsbert Michiel, P. (2013). *Process for the preparation of a spreadable dispersion* (CA Patent No. CA 2598401 C). https://lens.org/078-933-225-917-704.

Barendse S.P., Flöter, E., Hogervorst W.T., Van Horsen, D.S.H., Huizinga, H., and Van Kempen Gijsbert Michiel, P.. (2007). *Process for the preparation of a spreadable dispersion* (EP Patent No. EP 1865786 A2). https://lens.org/048-601-279-914-075.

Barendse S.P., Flöter, E., Hogervorst W.T., Van Horsen, D.S.H., Huizinga, H., and Van Kempen Gijsbert Michiel, P. (2011a). *Process for the preparation of a spreadable dispersion* (EP Patent No. EP 1865786 B1). https://lens.org/045-495-594-268-442.

Barendse S.P., Flöter, E., Hogervorst W.T., Van Horsen, D.S.H., Huizinga, H., and Van Kempen Gijsbert Michiel, P. (2011b). *Verfahren zur herstellung einer streichbaren dispersion* (AT Patent No. AT 535152 T). https://lens.org/069-612-894-520-022.

Barendse S.P., Flöter, E., Hogervorst W.T., Van Horsen, D.S.H., Huizinga, H., and Van Kempen Gijsbert Michiel, P. (2009). *dispersão comestivel espalhável e seu processo de preparação* (BR Patent No. BR PI0606545 A2). https://lens.org/100-706-798-258-158.

Barendse S.P., Flöter, E., Hogervorst W.T., Van Horsen, D.S.H., Huizinga, H., and Van Kempen Gijsbert Michiel, P. (2012a). *Process for the preparation of a spreadable dispersion* (US Patent No. US 8147895 B2). https://lens.org/035-593-746-677-592.

Barendse S.P., Flöter, E., Hogervorst W.T., Van Horsen, D.S.H., Huizinga, H., and Van Kempen Gijsbert Michiel, P. (2012b). *Process for the preparation of a spreadable dispersion* (US Patent No. US 2012/0263860 A1). https://lens.org/005-972-855-251-795.

Barendse S.P., Flöter, E., Hogervorst W.T., Van Horsen, D.S.H., Huzinga, H., Van Kempen Gijsbert Michiel, P.Van Horsen, D.S.H., (2012). *Process for the preparation of a spreadable dispersion* (PL Patent No. PL 1865786 T3). https://lens.org/137-685-960-577-441.

Barendse S.P., Flöter, E., Hogervorst W.T., Van Horsen, D.S.H., Huzinga, H., Van Kempen Gijsbert Michiel, P. (2015a). *Process for the preparation of a spreadable dispersion* (EP Patent No. EP 1865786 B2). https://lens.org/183-956-009-567-525.

Barendse S.P., Flöter, E., Hogervorst W.T., Van Horsen, D.S.H., Huzinga, H., Van Kempen Gijsbert Michiel, P. (2015b). *Process for the preparation of a spreadable dispersion* (PL Patent No. PL 1865786 T5). https://lens.org/154-496-918-360-613.

Barendse S.P., Flöter, E., Hogervorst W.T., Van Horsen, D.S.H., Huzinga, H., Van Kempen Gijsbert Michiel, P. (2009). *Process for the preparation of a spreadable dispersion* (US Patent No. US 2009/0110801 A1). https://lens.org/185-973-237-529-050.

Barendse S.P., Flöter, E., Hogervorst W.T., Van Horsen, D.S.H., Huzinga, H., Van Kempen Gijsbert Michiel, P. (2015). *Process for the Preparation of a Spreadable Dispersion* (US Patent No. US 2015/0118384 A1). https://lens.org/189-831-210-960-23X.

Barendse S.P., Flöter, E., Hogervorst W.T., Van Horsen, D.S.H., Huzinga, H., Van Kempen Gijsbert Michiel, P. (2017). *Process for the preparation of a spreadable dispersion* (US Patent No. US 9661864 B2). https://lens.org/104-428-697-194-042.

Barendse S.P., Van Horsen, D.S.H.,. Hogervorst, W.T., Huzinga, H., Flöter, E., and Van Kempen Gijsbert Michiel, P. (2006). *Process for the preparation of a spreadable dispersion* (AU Patent No. AU 2006/215829 A1). https://lens.org/089-086-951-922-879.

Barendse Sandra, P., Floeter, E., Hogervorst Wim, T., Horsen Dirk Simon Hendriks, V. A. N., Huizinga, H., & Kempen Gijsbert Michiel, P. (2006). *Process for the preparation of a spreadable dispersion* (WO Patent No. WO 2006/087091 A3). https://lens.org/147-759-728-898-445.

Barendse S.P., Van Horsen, D.S.H.,. Hogervorst, W.T., Huzinga, H., Flöter, E., and Van Kempen Gijsbert Michiel, P. (2009). *Process for the preparation of a spreadable dispersion* (AU Patent No. AU 2006/215829 B2). https://lens.org/146-528-310-832-09X.

Barendse Sandra, P., Floter, E., Hogervorst Wim, T., Van Horsen Dirk Simon, H., Huizinga, H., & Van Kempen Gijsbert Michiel, P. (2015). *Process for the preparation of a spreadable dispersion* (US Patent No. US 8927045 B2). https://lens.org/092-248-852-331-202.

Baseeth, S.S. and Sebree, B.R. (2010a). *Organogel compositions and processes for producing* (AU Patent No. AU 2009/313928 A1). https://lens.org/068-809-113-314-158.

Baseeth, S.S. and Sebree, B.R. (2010b). *Organogel compositions and processes for producing* (WO Patent No. WO 2010/057007 A1). https://lens.org/108-086-779-982-905.

Baseeth, S.S. and Sebree, B.R. (2010c). *Organogel compositions and processes for producing* (CA Patent No. CA 2743675 A1). https://lens.org/017-017-421-159-828.

Baseeth, S.S. and Sebree, B.R. (2011a). *Organogel compositions and processes for producing* (MX Patent No. MX 2011005076 A). https://lens.org/194-917-093-837-256.

Baseeth, S.S. and Sebree, B.R. (2011b). *Organogel compositions and processes for producing* (EP Patent No. EP 2355803 A1). https://lens.org/156-247-231-421-649.

Baseeth, S.S. and Sebree, B.R. (2011c). *Organogel compositions and processes for producing* (KR Patent No. KR 20110083738 A). https://lens.org/133-127-342-068-197.

Baseeth, S.S. and Sebree, B.R. (2011d). *Organogel compositions and processes for producing* (US Patent No. US 2011/0250299 A1). https://lens.org/182-545-844-225-868.

Baseeth, S.S. and Sebree, B.R. (2012a). *Organogel compositions and processes for producing* (ZA Patent No. ZA 201104382 B). https://lens.org/168-050-864-072-510.

Baseeth, S.S. and Sebree, B.R. (2012b). *Organogel compositions and processes for producing* (EP Patent No. EP 2355803 A4). https://lens.org/116-717-183-955-621.

Baseeth, S.S. and Sebree, B.R. (2013). *Rganogel compositions and processes for producing* (CN Patent No. CN 103349285 A). https://lens.org/107-788-949-989-465.

Baseeth, S.S. and Sebree, B.R. (2014). *Organogel compositions and processes for producing* (EP Patent No. EP 2711001 A2). https://lens.org/058-675-845-181-561.

Baseeth, S.S. and Sebree, B.R. (2015a). *Organogel compositions and processes for producing* (AU Patent No. AU 2015/227472 A1). https://lens.org/109-610-700-406-484.

Baseeth, S.S. and Sebree, B.R. (2015b). *Organogel compositions and processes for producing* (AU Patent No. AU 2009/313928 B2). https://lens.org/104-176-249-975-218.

Baseeth, S.S. and Sebree, B.R. (2016). *Organogel compositions and processes for producing* (CN Patent No. CN 105532922 A). https://lens.org/143-694-643-222-972.

Boer, B., Flöter, E., and Van Kempen Gijsbert Michiel, P. (2009). *dispersão entornável comestìvel e processo para a preparação de uma dispersão contendo gordura comestìvel* (BR Patent No. BR PI0606151 A2). https://lens.org/167-056-289-734-362.

Boer, B., Flöter, E., and Van Kempen Gijsbert Michiel, P. (2010). *Edible dispersions comprising oil and structuring agent* (EP Patent No. EP 2266413 A1). https://lens.org/175-834-793-686-572.

Boer, B., Flöter, E., and Van Kempen Gijsbert Michiel, P. (2015a). *Dispersiones comestibles que comprenden aceite y un agente estructurante* (ES Patent No. ES 2553627 T3). https://lens.org/077-274-677-586-851.

Boer, B., Flöter, E., and Van Kempen Gijsbert Michiel, P. (2015b). *Edible dispersions comprising oil and structuring agent* (PL Patent No. PL 2266413 T3). https://lens.org/148-670-360-066-397.

Boer, B., Flöter, E., and Van Kempen Gijsbert Michiel, P. (2015c). *Edible dispersions comprising oil and structuring agent* (EP Patent No. EP 2266413 B1). https://lens.org/031-111-612-928-149.

Boer, B., Flöter, E., and Van Kempen Gijsbert Michiel, P. (2015d). *Spiselige dispersioner omfattende olie og strukturmiddel* (DK Patent No. DK 2266413 T3). https://lens.org/018-334-804-385-131.

Boer, B., Flöter, E., and Van Kempen Gijsbert Michiel, P. (2006). *Edible dispersions comprising oil and structuring agent* (WO Patent No. WO 2006/079445 A1). https://lens.org/126-784-753-980-566.

Boer, B., Flöter, E., and Van Kempen Gijsbert Michiel, P. (2007). *Edible dispersions comprising oil and structuring agent* (EP Patent No. EP 1841327 A1). https://lens.org/033-720-821-000-36X.

Boer, B., Flöter, E., and Van Kempen Gijsbert Michiel, P. (2010). *Edible dispersions comprising oil and structuring agent* (EP Patent No. EP 1841327 B1). https://lens.org/110-249-485-211-254.

Boer, B., Flöter, E., and Van Kempen Gijsbert Michiel, P. (2011a). *Dispersiones comestibles que comprenden aceite y un agente estructurante* (ES Patent No. ES 2355741 T3). https://lens.org/162-158-201-522-468.

Boer, B., Flöter, E., and Van Kempen Gijsbert Michiel, P. (2011b). *Essbare dispersionen mit öl und strukturmittel* (AT Patent No. AT 491339 T). https://lens.org/041-209-885-809-952.

Boer, B., Flöter, E., and Van Kempen Gijsbert Michiel, P. (2011c). *Essbare dispersionen mit öl und strukturmittel* (DE Patent No. DE 602006018858 D1). https://lens.org/091-198-855-562-672.

Boer, B., Flöter, E., and Van Kempen Gijsbert Mi Michiel, P. (2011d). *Spiselige dispersioner indeholdende olie og strukturmiddel* (DK Patent No. DK 1841327 T3). https://lens.org/050-546-648-464-579.

Boer, B., Flöter, E., and Van Kempen Gijsbert Michiel, P. (2008). *Edible dispersions comprising oil and structuring agent* (US Patent No. US 2008/0118624 A1). https://lens.org/059-723-886-167-686.

Boer, B., Flöter, E., and Van Kempen Gijsbert, M.P. (2011). *Edible dispersions comprising oil and structuring agent* (PL Patent No. PL 1841327 T3). https://lens.org/070-875-256-645-517.

Boer, B., Flöter, E., and Van Kempen Gijsbert Michiel, P. (2009). *Edible dispersions comprising oil and structuring agent* (ZA Patent No. ZA 200705709 B). https://lens.org/028-460-987-585-178.

Calligaris, S., Manzocco, L., and Plazzotta, S. (2018a). *Method to make fat-substitute and/or fat-imitator compounds* (WO Patent No. WO 2018/007399 A1). https://lens.org/131-484-840-790-425.

Calligaris, S., Manzocco, L., and Plazzotta, S. (2018b). *Metodo per la realizzazione di composti sostitutivi e/o imitativi dei grassi* (IT Patent No. IT 201600069429 A1). https://lens.org/154-685-930-941-624.

Camilleri, M., Hoogland Oscar, J., and Franken Wouter, M. (2022). *Oleogel* (WO Patent No. WO 2022/031172 A1). https://lens.org/054-035-211-075-192.

Chiara, G. and Hindrik, H. (2008a). *Granules comprising sterol* (ZA Patent No. ZA 200706742 B). https://lens.org/162-569-809-249-742.

Chiara, G. and Hindrik, H. (2008b). *Granules comprising sterol* (CN Patent No. CN 101119640 A). https://lens.org/095-317-013-262-360.

Chiara, G. and Hindrik, H. (2008c). *Process for the preparation of a spreadable dispersion comprising sterol* (CN Patent No. CN 101119638 A). https://lens.org/195-464-057-664-121.

Chiara, G. and Hindrik, H. (2008d). *Process for the preparations of a spreadable dispersion comprising sterol* (ZA Patent No. ZA 200706244 B). https://lens.org/180-277-965-430-832.

Clement, V., Savin, G., Jung, J.-M., Mezzenga, R., Erwin, L.M. (2012). *Oil gel* (MX Patent No. MX 2012008439 A). https://lens.org/069-025-264-548-075.

Daels, E.V.A. and Foubert, I. (2019a). *Phytosterol compositions in food products* (CA Patent No. CA 3086654 A1). https://lens.org/120-128-918-823-988.

Daels, E.V.A. and Foubert, I. (2019b). *Phytosterol compositions in food products* (WO Patent No. WO 2019/122227 A1). https://lens.org/074-732-304-061-637.

Daels, E.V.A. and Foubert, I. (2020a). Phytosterol compositions in food products (EP Patent No. EP 3727006 A1). https://lens.org/126-205-683-560-574.

Daels, E.V.A. and Foubert, I. (2020b). *Phytosterol compositions in food products* (AU Patent No. AU 2018/387328 A1). https://lens.org/044-117-021-701-267.

Daels, E.V.A. and Foubert, I. (2020c). *Phytosterol compositions in food products* (CN Patent No. CN 112004415 A). https://lens.org/108-117-380-216-853.

Daels, E.V.A. and Foubert, I. (2020d). *Phytosterol compositions in food products* (US Patent No. US 2020/0315200 A1). https://lens.org/068-611-171-085-193.

David, L.M., Robert, B.A., and Roja, E. (2017). *processo para preparar um oleogel* (BR Patent No. BR 112015028312 A2). https://lens.org/123-351-048-776-663.

David, L.M., Robert, B.A., and Roja, E. (2019). *dispersão aquosa* (BR Patent No. BR 112018069122 A2). https://lens.org/153-347-335-728-038.

David, L.M., Robert, B.A., and Roja, E. (2020). *processo para preparar um oleogel* (BR Patent No. BR 112015028312 B1). https://lens.org/127-262-619-312-362.

Demeurisse, J. and Marangoni, A. (2020). *Structured oil-in-water emulsion and food product comprising the same* (WO Patent No. WO 2020/218965 A1). https://lens.org/066-334-029-764-915.

Demeurisse, J. and Marangoni, A. (2021). *Emulsión estructurada de aceite en agua y producto alimenticio que la comprende* (UY Patent No. UY 38675 A). https://lens.org/046-800-745-379-495.

Demeurisse, J. and Marangoni, A. (2022a). *Structured oil-in-water emulsion and food product comprising the same* (US Patent No. US 2022/0167638 A1). https://lens.org/165-212-422-789-709.

Demeurisse, J. and Marangoni, A. (2022b). *Structured oil-in-water emulsion and food product comprising the same* (EP Patent No. EP 3958684 A1). https://lens.org/107-722-039-824-395.

Den Adel, R., Dol, G.C., Grebenkämper, K.., Leenhouts, A., Potman, R.P., Smit-Kingma, I.E., and Tio, F.F. (2016). *Process for preparing emulsifier-free edible fat-continuous emulsions* (US Patent No. US 2016/0213018 A1). https://lens.org/092-809-319-915-557.

Dendooven Els Ginette, A., Nguyen Van, S.A.U., Van-Nieuwenhuyze, L., and Vercauteren Ronny Leontina, M. (2011). *Fat replacers and filling materials* (WO Patent No. WO 2011/140212 A1). https://lens.org/024-471-590-695-772.

Dendooven, E.G.A., Nguyen V.S., Van-Nieuwenhuyze, L., and Vercauteren, R.L.M. (2013a). *Fat replacers and filling materials* (EP Patent No. EP 2575498 A1). https://lens.org/079-693-864-892-60X.

Dendooven, E.G.A., Nguyen V.S., Van-Nieuwenhuyze, L., and Vercauteren, R.L.M. (2013b). *Fat replacers and filling materials* (US Patent No. US 2013/0052300 A1). https://lens.org/031-435-597-210-913.

Dendooven, E.G.A., Nguyen V.S., Van-Nieuwenhuyze, L., and Vercauteren, R.L.M. (2017). *Fat replacers and filling materials* (US Patent No. US 2017/0027209 A1). https://lens.org/139-096-045-941-841.

Dendooven, E.G.A., Nguyen V.S., Van-Nieuwenhuyze, L., and Vercauteren, R.L.M. (2015). *substituidores de gordura e materiais de enchimento* (BR Patent No. BR 112012029519 A2). https://lens.org/064-007-817-699-033.

Di Bari, V., Foster, T., Toleman, R., and Sarney, A.M. (2022). *Composiciones de cera oleosa, procesos para cocinar artículos alimenticios y artículos alimenticios cocinados* (AR Patent No. AR 120631 A1). https://lens.org/002-429-904-235-780.

Ding, B. (2015). *Preparation method of fat mimics using water chestnut starch and konjak refined powder as matrix* (CN Patent No. CN 104770640 A). https://lens.org/009-321-635-353-075.

Dol, G.C., Leenhouts, A., Potman, R.Po., Tio, F.F., Adel, R.D., Smit Kingma, I.E., and Grebenkämper, K. (2016). Process for preparing emulsifier-free edible fat-continuous emulsions (MX Patent No. MX 2016004474 A). https://lens.org/156-299-020-869-405.

Dupas, H.M., Silvestre, G.S.A., Santos, R.R., and Efraim, P. (2021). *Polymer-based oleogel and related use* (WO Patent No. WO 2021/155455 A1). https://lens.org/167-782-710-101-064.

Elke, S. (2022). *One step procedure for producing a protein oleogel* (IL Patent No. IL 288422 D0). https://lens.org/057-499-012-642-945.

Ergun, R., Appell, R.B., and Malotky, D.L. (2014). *Process for preparing an oleogel* (WO Patent No. WO 2014/193667 A1). https://lens.org/076-864-360-471-54X.

Ergun, R., Appell, R.B., and Malotky, D.L. (2016a). *Process for preparing an oleogel* (US Patent No. US 2016/0081374 A1). https://lens.org/142-871-308-402-00X.

Ergun, R., Appell, R.B., and Malotky, D.L. (2016b). *Process for preparing an oleogel* (EP Patent No. EP 3007565 A1). https://lens.org/112-990-132-914-389.

Ergun, R., Appell, R.B., and Malotky, D.L. (2016c). *Process for preparing an oleogel* (CN Patent No. CN 105228461 A). https://lens.org/049-827-310-273-420.

Ergun, R., Appell, R.B., and Malotky, D.L. (2016d). *Process for preparing an oleogel* (KR Patent No. KR 20160030892 A). https://lens.org/140-779-872-730-78X.

Ergun, R., Appell, R.B., and Malotky, D.L. (2017). *Process for preparing an oleogel* (US Patent No. US 9655376 B2). https://lens.org/088-221-825-782-425.

Ergun, R., Appell, R.B., and Malotky, D.L. (2018). *Process for preparing an oleogel* (EP Patent No. EP 3007565 B1). https://lens.org/144-900-914-548-247.

Ergun, R. and Meunier, DD.M. (2018). *Oleogel with stearic acid* (WO Patent No. WO 2018/057384 A1). https://lens.org/114-681-735-153-535.

Ergun, R. and Meunier, D.M. (2019a). *Oleogel with stearic acid* (EP Patent No. EP 3515200 A1). https://lens.org/047-884-186-615-283.

Ergun, R. and Meunier, D.M. (2019b). *Oleogel with stearic acid* (CN Patent No. CN 109788774 A). https://lens.org/118-476-896-173-097.

Ergun, R. and Meunier, D.M. (2020a). *Oleogel with stearic acid* (EP Patent No. EP 3515200 B1). https://lens.org/013-836-967-912-985.

Ergun, R. and Meunier, D.M. (2020b). *Oleogel with stearic acid* (US Patent No. US 2020/0015493 A1). https://lens.org/134-865-865-939-52X.

Ergun, R., Syed, A., and Adden, A. (2018). *Solid composition containing oleogel* (WO Patent No. WO 2018/057383 A1). https://lens.org/111-591-648-527-799.

Ergun, R., Syed, A., and Adden, A. (2019a). *Solid composition containing oleogel* (CN Patent No. CN 109788773 A). https://lens.org/011-879-816-946-376.

Ergun, R., Syed, A., and Adden, A. (2019b). *Solid composition containing oleogel* (EP Patent No. EP 3515196 A1). https://lens.org/002-421-998-925-929.

Ergun, R., Syed, A., and Adden, A. (2019c). *Solid composition containing oleogel* (US Patent No. US 2019/0307145 A1). https://lens.org/088-672-293-520-768.

Ergun, R., Thomson, B.S., and Huebner-Keese, K.B. (2016). *Low fat laminated dough and pastry* (KR Patent No. KR 20160002929 A). https://lens.org/162-990-335-598-13X.

Ergun, R., Thomson, B.S., and Huebner-Keese, B. (2018). *Low fat laminated dough and pastry* (EP Patent No. EP 2956007 B1). https://lens.org/113-415-685-194-583.

Ergun, R., Thomson, B. S., and Huebner-Keese, B. (2014). *Low fat laminated dough and pastry* (WO Patent No. WO 2014/172135 A1). https://lens.org/147-768-642-053-69X.

Ergun, R., Thomson, B. S., and Huebner-Keese, B. (2015a). *Low fat laminated dough and pastry* (CN Patent No. CN 105120670 A). https://lens.org/079-082-983-928-019.

Ergun, R., Thomson, B.S., and Huebner-Keese, B. (2015b). *Low fat laminated dough and pastry* (EP Patent No. EP 2956007 A1). https://lens.org/011-755-356-044-536.

Ergun, R., Thomson, B. S., and Huebner-Keese, B. (2016). *Low fat laminated dough and pastry* (US Patent No. US 2016/0021898 A1). https://lens.org/092-345-102-233-886.

Ergun, R., Thomson, B.S., and Huebner-Keese, B. (2018). *Masa y pastel estratificado de bajo contenido en grasa* (ES Patent No. ES 2669614 T3). https://lens.org/137-342-762-874-977.

Ergun, R., Thomson, B. S., Wallick, D.E., and Meunier, D.M. (2014). *Method of preparing an edible oleogel* (WO Patent No. WO 2014/004018 A1). https://lens.org/023-087-155-060-876.

Ergun, R., Thomson, B. S., Wallick, D.E., and Meunier, D.M. (2015a). *Method of preparing an edible oleogel* (EP Patent No. EP 2866577 A1). https://lens.org/155-145-018-583-573.

Ergun, R., Thomson, B. S., Wallick, D.E., and Meunier, D.M. (2015b). *Method of preparing an edible oleogel* (CN Patent No. CN 104582496 A). https://lens.org/057-270-057-717-116.

Ergun, R., Thomson, B. S., Wallick, D.E., and Meunier, D.M. (2015c). *Method of preparing an edible oleogel* (US Patent No. US 2015/0157038 A1). https://lens.org/186-965-693-272-937.

Ergun, R., Thomson, B. S., Wallick, D.E., and Meunier, D.M. (2018). *Method of preparing edible oleogel* (CN Patent No. CN 107616241 A). https://lens.org/015-528-592-222-619.

Ergun, R., Thomson, B. S., Wallick, D.E., and Meunier, D.M. (2019). *Method of preparing an edible oleogel* (EP Patent No. EP 2866577 B1). https://lens.org/023-481-561-442-605.

Ergun, R., Thomson, B. S., Wallick, D.E., and Meunier, D.M. (2020). *Method of preparing an edible oleogel* (KR Patent No. KR 20200052996 A). https://lens.org/027-869-866-585-697.

European Commission, Joint Research Centre, Gomez y Paloma, S., Garzón, D.V., Ciaian, P. (2018). *The impact of private R&D on the performance of food-processing firms: evidence from Europe, Japan and North America: task 4.4 report*. Publications Office. https://doi.org/doi/10.2788/85468.

Flöter, E., and Huzinga, H. (2008). *Process for the preparation of a spreadable dispersion* (CN Patent No. CN 101137293 A). https://lens.org/197-177-152-730-95X.

FoodDrinkEurope (2021). *Data & trends of the European food and drink industry* https://www.fooddrinkeurope.eu/wp-content/uploads/2021/11/FoodDrinkEurope-Data-Trends-2021-digital.pdf (accessed on 20th April 2022).

Garbolino, C. and Huizinga, H. (2006a). *Granules comprising sterol* (AU Patent No. AU 2006/215828 A1). https://lens.org/176-653-117-429-179.

Garbolino, C. and Huizinga, H. (2006b). *Granules comprising sterol* (WO Patent No. WO 2006/087090 A1). https://lens.org/131-285-643-759-769.

Garbolino, C. and Huizinga, H. (2006c). *Granules comprising sterol* (CA Patent No. CA 2597478 A1). https://lens.org/175-443-718-842-055.

Garbolino, C. and Huizinga, H. (2006d). *Process for the preparation of a spreadable dispersion comprising sterol* (CA Patent No. CA 2595281 A1). https://lens.org/094-224-913-510-375.

Garbolino, C. and Huizinga, H. (2006e). *Process for the preparation of a spreadable dispersion comprising sterol* (WO Patent No. WO 2006/087093 A1). https://lens.org/132-469-819-432-851.

Garbolino, C. and Huizinga, H. (2007a). *Granules comprising sterol* (EP Patent No. EP 1850676 A1). https://lens.org/112-504-288-376-533.

Garbolino, C. and Huizinga, H. (2007b). *Process for the preparation of a spreadable dispersion comprising sterol* (EP Patent No. EP 1858341 A1). https://lens.org/000-275-511-297-815.

Garbolino, C. and Huizinga, H. (2008). *Process for the preparation of a spreadable dispersion comprising sterol* (US Patent No. US 2008/0193628 A1). https://lens.org/163-592-375-273-287.

Garbolino, C. and Huizinga, H. (2009a). *Granules Comprising Sterol* (US Patent No. US 2009/0136645 A1). https://lens.org/009-846-192-434-862.

Garbolino, C. and Huizinga, H. (2009b). *grânulo, processo para a preparação de uma dispersão comestivél, cubo ou tablete, processo para a preparação de cubos ou tabletes e tempero particulado ou concentrado* (BR Patent No. BR PI0606341 A2). https://lens.org/039-375-805-418-559.

Garbolino, C. and Huizinga, H. (2009c). *processo para a preparação de uma dispersão comestivel espalhável, dispersão comestivel espalhável e emulsão espalhável de baixo teor de gordura* (BR Patent No. BR PI0606356 A2). https://lens.org/137-732-841-505-719.

Garbolino, C. and Huizinga, H. (2010). *Granules comprising sterol* (AU Patent No. AU 2006/215828 B2). https://lens.org/188-526-357-561-439.

Garbolino, C. and Khejzinga, K. (2010a). *Fat granules containing sterol and method of production of food dispersion with them* (RU Patent No. RU 2379905 C2). https://lens.org/082-926-233-509-46X.

Garbolino, C. and Khejzinga, K. (2010b). *Food dispersion of spreadable consistency (versions), method of its preparation and low-fat emulsion spread* (RU Patent No. RU 2384071 C2). https://lens.org/131-424-230-932-305.

Grebenkämper, K., Leenhouts, A., and The Rogier Antoine Floris (2019). *Process for preparing a fat slurry comprising olive oil and for preparing a spread with said slurry* (HU Patent No. HU E043408 T2). https://lens.org/096-306-716-514-918.

Grebenkämper, K., Leenhouts, A., and The Rogier Antoine Floris. (2017). *Process for preparing a fat slurry comprising olive oil and for preparing a spread with said slurry* (CA Patent No. CA 3004505 A1). https://lens.org/184-876-333-021-686.

Grebenkämper, K., Leenhouts, A., and The Rogier Antoine Floris. (2017). *Process for preparing a fat slurry comprising olive oil and for preparing a spread with said slurry* (WO Patent No. WO 2017/108275 A1). https://lens.org/132-849-401-272-728.

Grebenkämper, K., Leenhouts, A., and The Rogier Antoine Floris. (2018). *Process for preparing a fat slurry comprising olive oil and for preparing a spread with said slurry* (US Patent No. US 2018/0368434 A1). https://lens.org/080-361-308-943-255.

Grebenkämper, K., Leenhouts, A., and The Rogier Antoine Floris. (2018). *Process for preparing a fat slurry comprising olive oil and for preparing a spread with said slurry* (EP Patent No. EP 3393265 A1). https://lens.org/027-867-910-637-820.

Grebenkämper, K., Leenhouts, A., and The Rogier Antoine Floris. (2019a). *Process for preparing a fat slurry comprising olive oil and for preparing a spread with said slurry* (EP Patent No. EP 3393265 B1). https://lens.org/148-846-397-610-532.

Grebenkämper, K., Leenhouts, AL., and The Rogier Antoine Floris. (2019a). *Process for preparing a fat slurry comprising olive oil and for preparing a spread with said slurry* (ZA Patent No. ZA 201802881 B). https://lens.org/175-803-397-676-354.

Grebenkämper, K., Leenhouts, AL., and The Rogier Antoine Floris. (2019b). *Zeytinyağı içeren bir katı yağ bulamacı hazırlanmasına ve söz konusu bulamacı içeren bir sürülebilir ürün hazırlanmasına yönelik işlem* (TR Patent No. TR 201904940 T4). https://lens.org/153-865-620-795-491.

Grebenkämper, K., Leenhouts, A., and The Rogier Antoine Floris. (2019b). *Process for preparing a fat slurry comprising olive oil and for preparing a spread with said slurry* (PL Patent No. PL 3393265 T3). https://lens.org/199-094-455-294-445.

Gregorio, A. (2015a). *Composição de chocolate contendo etilcelulose* (BR Patent No. BR PI1009679 A2). https://lens.org/037-882-055-885-273.

Gregorio, A. (2015b). *Gelificação de polímero de óleos* (BR Patent No. BR PI1009677 A2). https://lens.org/115-176-680-401-069.

Gregorio, M.A. (2012a). *Chocolate compositions containing ethylcellulose* (CN Patent No. CN 102548425 A). https://lens.org/150-298-725-937-241.

Gregorio, M.A. (2012b). *Polymer gelation of oils* (CN Patent No. CN 102510725 A). https://lens.org/039-663-273-350-43X.

He, R., Wang, Z., Ju, X., Chen, C., Yuan, J., Wang, L. (2021). *Preparation method of soybean oil-based oleogel* (CN Patent No. CN 112889935 A). https://lens.org/120-963-923-329-510.

Hild, A., Knarr, M., Huebner-Keese, B., and Ergun, R. (2018). *올레오겔의 제조방법* (KR Patent No. KR 20180128002 A). https://lens.org/145-978-057-546-779.

Hild, A., Knarr, M., Huebner-Keese, B., and Ergun, R. (2017). *Method of making oleogel* (WO Patent No. WO 2017/172590 A1). https://lens.org/026-192-419-047-207.

Hild, A., Knarr, M., Huebner-Keese, B., and Ergun, R. (2018). *Method of making oleogel* (CN Patent No. CN 108777972 A). https://lens.org/081-626-720-554-021.

Hild, A., Knarr, M., Huebner-Keese, B., and Ergun, R. (2019a). *Method of making oleogel* (EP Patent No. EP 3435776 A1). https://lens.org/156-305-703-942-42X.

Hild, A., Knarr, M., Huebner-Keese, B., and Ergun, R. (2019b). *Method of making oleogel* (US Patent No. US 2019/0075811 A1). https://lens.org/033-154-222-134-493.

Hild, A., Knarr, M., Huebner-Keese, B., and Ergun, R. (2020). *Method of making oleogel* (EP Patent No. EP 3435776 B1). https://lens.org/097-925-292-629-04X.

Hogervorst, W.T., Verbeek, J.H.T., Janssen, J.M., and Potman, R.P. (2011). *Fat granules* (CN Patent No. CN 101119639 B). https://lens.org/074-222-224-546-124.

Huizinga, H. and Garbolino, C. (2006). *Process for the preparation of a spreadable dispersion comprising sterol* (AU Patent No. AU 2006/215831 A1). https://lens.org/195-479-867-102-990.

Huizinga, H. and Garbolino, C. (2009). *Process for the preparation of a spreadable dispersion comprising sterol* (AU Patent No. AU 2006/215831 B2). https://lens.org/189-402-123-404-651.

Janssen, J., Hogervorst, W.I.M., Potman, R., and Verbeek, J.A.N. (2015). *Fat granules* (PL Patent No. PL 1885192 T3). https://lens.org/132-156-258-270-129.

Janssen, J.J.M., Hogervorst, W.T., Potman, R.P., and Verbeek, J.H.T. (2006). *Fat granules* (WO Patent No. WO 2006/087092 A2). https://lens.org/037-539-150-060-271.

Janssen, J.J.M., Hogervorst, W.T., Potman, R.P., and Verbeek, J.H.T. (2008). *Fat granules* (EP Patent No. EP 1885192 A2). https://lens.org/104-286-641-286-932.

Janssen, J.J.M., Hogervorst, W.T., Potman, R.P., and Verbeek, J.H.T. (2013). *Fat granules* (CA Patent No. CA 2595381 C). https://lens.org/180-233-176-650-459.

Janssen, J.J.M., Hogervorst, W.T., Potman, R.P., and Verbeek, J.H.T. (2015). *Fat granules* (EP Patent No. EP 1885192 B1). https://lens.org/043-964-057-209-953.

Janssen, J.J.M., Hogervorst, W.T., Potman, R.P., and Verbeek, J.H.T. (2008). *Fat granules* (US Patent No. US 2008/0317917 A1). https://lens.org/155-531-231-335-240.

Janssen, J.J.M., Hogervorst, W.T., Potman, R.P., and Verbeek, J.H.T. (2009). *grânulo, processo para a preparação de uma dispersão comestivel, cubo ou tablete, processo para a preparação de cubos ou tabletes de caldo de sopa ou de temperos e tempero particulado ou concentrado* (BR Patent No. BR PI0606354 A2). https://lens.org/167-535-484-629-363.

Janssen, J.J.M., Hogervorst, W.T., Potman, R.P., and Verbeek, J.H.T. (2012). *Fat granules* (US Patent No. US 8124152 B2). https://lens.org/196-790-706-043-158.

Janssen, J.J.M., Khogervorst, V.T., Potman, R.P., and Verbik, J.K. T. (2010). *Fat granules and method of production of food dispersion with them* (RU Patent No. RU 2379906 C2). https://lens.org/118-231-730-314-51X.

Kempen, G.M.P., Hindrik, H., Barendse, S.P., Horsen, D.S.H.V, Hogervorst, W.T., and Flöter, E. (2006). *Process for the preparation of a spreadable dispersion* (CA Patent No. CA 2598401 A1). https://lens.org/174-535-179-978-94X.

Kempen, G.M.P.V., Barendse, S.P., Flöter, E., Horsen, D.S.H.V., Hogervorst, W.T., and Hindrik, H.et al. (2011). *Process for the preparation of a spreadable dispersion* (CN Patent No. CN 101137293 B). https://lens.org/047-786-322-030-660.

Kornél, N. (2017). *composições de óleo estruturadas* (BR Patent No. BR 112015028696 A2). https://lens.org/101-766-711-768-173.

Lan, Y., Zheng, S., Wei, F., Dong, L., Zheng, R., and Hu, B. (2021). *Rice bran protein oleogel as well as preparation method and application thereof* (CN Patent No. CN 113261594 A). https://lens.org/097-893-725-221-282.

Lee, S.Y., Bae, W.S., Kim, H.J.I., a Reum, J., and Hoon, K.S. (2016). *bakery product with canola gel* (KR Patent No. KR 20160115195 A). https://lens.org/064-938-238-805-351.

Lee S.Y., Park, S.K., Oh, I.K., and Jung, D.E. (2020). *Vegetable fat composition with oleogel and hydrocolloid* (KR Patent No. KR 20200050165 A). https://lens.org/164-226-811-712-239.

Leenhouts, A., Tio, F.F., Dol, G. C., Smit-Kingma, I.E., Grebenkämper, K., Potman, RP., and Adel, R.D. (2017). *processo para preparação de emulsões lipídicas contínuas comestíveis sem emulsificante e emulsao continua lipidica comestivel* (BR Patent No. BR 112016007493 A2). https://lens.org/116-954-853-368-303.

Leenhouts, A., Tio, F.F., Dol, G.C., Smit-Kingma, E.S., Grebenkämper, K., Potman, RP., and Adel, R.D. (2021). *processo para preparação de emulsoes lipidicas contínuas comestíveis sem emulsificante e emulsão continua lipídica comestível* (BR Patent No. BR 112016007493 B1). https://lens.org/186-625-913-198-193.

Li, S., Li, M., Yang, Y., Chen, X, Ye, H., and Bai, Q. (2020). *Edible antioxidant oil gel and preparation method therefor* (WO Patent No. WO 2020/133652 A1). https://lens.org/159-817-297-574-091.

Li, S. and Lin, J. (2019). *Antioxidative edible oil gel and preparation method thereof* (CN Patent No. CN 109588497 A). https://lens.org/005-392-769-725-682.

Li, Y., Zhu, S., and Shang, J. (2021). *Preparation method of vitamin C loaded oleogel and shortening blend as well as product and application of vitamin C loaded oleogel and shortening blend* (CN Patent No. CN 112931633 A). https://lens.org/013-881-185-992-792.

Li, Z., He, X., and Gong, Q. (2015). *Pig fat substitute and production method thereof* (CN Patent No. CN 104939100 A). https://lens.org/066-847-323-612-713.

Lin, C., Dong, X., Wang, B.. Xin, Z., Zhao, L. (2019). *Preparation of fat substitution and its application in meatballs* (CN Patent No. CN 109619537 A). https://lens.org/084-759-542-726-73X.

Lin, W., Wang, C., Wang, C., and Wang, S. (2021). *Preparation method of oil gel for replacing hydrogenated vegetable oil* (CN Patent No. CN 113100298 A). https://lens.org/097-684-577-887-654.

Liu, J., Chang, Z., Bu, G., and Xie, X. (2019). *Novel health fat substitute and preparation method and application thereof* (CN Patent No. CN 109123629 A). https://lens.org/041-514-773-293-710.

Liu, Q., Jia, X., Kong, B., Cao, C., and Zhao, S.. (2019). *Method for preparing fat substitutes by regulating and controlling oil droplet gathering in emulsion through flaxseed gum* (CN Patent No. CN 109938328 A). https://lens.org/152-020-266-124-052.

Liu, Q., Kong, B., Jia, X., Cao, C., and Zhao, S. (2019). *Method for preparing blocky fat stimulant with konjaku flour and curdlan as base materials* (CN Patent No. CN 109730273 A). https://lens.org/057-499-034-199-171.

Liu, S., Zhang, X., Wu, X.I.A., and He, K. (2017). *Preparation method of high-dietary-fiber oleogel* (CN Patent No. CN 107296260 A). https://lens.org/078-887-376-279-565.

Liu, Y., Liu, C., Zheng, Z., Xo, C., Shi, Y.. (2020). *Preparation method and application of bicontinuous emulsion gel* (CN Patent No. CN 111700264 A). https://lens.org/134-169-022-173-870.

Liu, Y.I., Chen, J.I.E., Bi, J., and Li, X. (2021). *Preparation of vegetable oil gel fat substitute and application thereof in sausage* (CN Patent No. CN 113367200 A). https://lens.org/040-645-059-938-28X.

López, R.A., Martínez, S.M., Fontes, C., and Fabra Rovira, M.J. (2020a). *Method for obtaining oleogels* (WO Patent No. WO 2020/021152 A1). https://lens.org/036-314-156-477-12X.

López, R.A., Martínez, S.M., Fontes, C., and Fabra Rovira, M.J. (2020b). *Oleogel obtaining procedure (Machine-translation by Google Translate, not legally binding)* (ES Patent No. ES 2738923 A1). https://lens.org/188-325-738-944-862.

López, R.A., Martínez, S.M., Fontes, C., and Fabra Rovira, M.J. (2020). *Procedimiento de obtencion de oleogeles* (ES Patent No. ES 2738923 A8). https://lens.org/181-176-623-313-912.

López, R.A., Martínez, S.M., Fontes, C., and Fabra Rovira, M.J. (2021). *Method for obtaining oleogels* (WO Patent No. WO 2021/019117 A1). https://lens.org/051-006-481-117-995.

López Rubio, A., Martínez Sanz, M., Fontes Candia, C., & Fabra Rovira María, J. (2020). *Method for obtaining oleogels* (ES Patent No. ES 2738923 B2). https://lens.org/176-692-700-893-034

Malotky, D.L., Appell R.B., and Ergun, R. (2017). *Ethyl cellulose oleogel dispersion* (WO Patent No. WO 2017/172594 A1). https://lens.org/020-239-528-492-69X.

Malotky, D.L., Appell R.B., and Ergun, R. (2018a). *Ethyl cellulose oleogel dispersion* (CN Patent No. CN 108712863 A). https://lens.org/181-707-716-632-998.

Malotky, D.L., Appell R.B., and Ergun, R. (2018b). *에틸 셀룰로오스 올레오겔 분산액*(KR Patent No. KR 20180131565 A). https://lens.org/165-713-007-431-182.

Malotky, D.L., Appell R.B., and Ergun, R. (2019a). *Ethyl cellulose oleogel dispersion* (EP Patent No. EP 3435777 A1). https://lens.org/099-542-300-254-212.

Malotky, D.L., Appell R.B., and Ergun, R. (2019b). *Ethyl cellulose oleogel dispersion* (US Patent No. US 2019/0075810 A1). https://lens.org/054-756-423-054-908.

Marangoni, A. (2014). *Composiciones de chocolate que contienen oleogel de etilcelulosa y métodos de preparación* (ES Patent No. ES 2464368 T3). https://lens.org/131-951-770-214-835.

Marangoni, A. and Demeurisse, J. (2021). *Emulsión estructurada de aceite en agua y producto alimenticio que la comprende* (AR Patent No. AR 118761 A1). https://lens.org/072-504-873-067-628.

Marangoni, A., Kranis, N., and Ghazani, S. (2021a). *Wax oleogels as fat substitutes* (US Patent No. US 2021/0161166 A1). https://lens.org/111-636-239-865-822.

Marangoni, A., Kranis, N., and Ghazani, S. (2021b). *Wax oleogels as fat substitutes* (WO Patent No. WO 2021/046642 A1). https://lens.org/063-863-861-993-112.

Marangoni, A., Kranis, N., and Ghazani, S. (2021c). *Wax oleogels as fat substitutes* (CA Patent No. CA 3106795 A1). https://lens.org/151-150-962-798-834.

Marangoni A.G. (2010a). *Chocolate compositions containing ethylcellulose* (CA Patent No. CA 2763645 A1). https://lens.org/147-338-252-794-45X.

Marangoni A.G. (2010b). *Chocolate compositions containing ethylcellulose* (WO Patent No. WO 2010/143067 A1).https://lens.org/062-555-318-691-568.

Marangoni A.G. (2010c). *Polymer gelation of oils* (WO Patent No. WO 2010/143066 A1). https://lens.org/066-053-691-956-435.

Marangoni A.G. (2012a). *Chocolate compositions containing ethylcellulose* (US Patent No. US 2012/0183651 A1). https://lens.org/178-452-720-139-91X.

Marangoni A.G. (2012b). *Chocolate compositions containing ethylcellulose* (MX Patent No. MX 2011013353 A). https://lens.org/090-710-441-668-593.

Marangoni A.G. (2012c). *Chocolate compositions containing ethylcellulose* (AU Patent No. AU 2010/258360 A1). https://lens.org/019-610-474-044-629.

Marangoni A.G. (2012d). *Chocolate compositions containing ethylcellulose oleogel and methods for preparing* (EP Patent No. EP 2440067 A1). https://lens.org/061-624-906-916-276.

Marangoni A.G. (2012e). *Polymer gelation of oils* (MX Patent No. MX 2011013355 A). https://lens.org/051-513-753-531-791.

Marangoni A.G. (2012f). *Polymer gelation of oils* (US Patent No. US 2012/0183663 A1). https://lens.org/156-374-465-834-640.

Marangoni A.G. (2012g). *Polymer gelation of oils* (EP Patent No. EP 2440066 A1). https://lens.org/081-061-270-225-38X.

Marangoni A.G. (2012h). *Polymer gelation of oils* (AU Patent No. AU 2010/258359 A1). https://lens.org/029-341-039-059-391.

Marangoni A.G. (2013a). *Chocolate compositions containing ethylcellulose oleogel and methods for preparing* (HK Patent No. HK 1169923 A1). https://lens.org/082-057-713-888-970.

Marangoni A.G. (2013b). *Polymer gelation of oils* (HK Patent No. HK 1169922 A1). https://lens.org/143-081-793-496-475.

Marangoni A.G. (2013c). *Polymer gelation of oils* (EP Patent No. EP 2440066 B1). https://lens.org/040-853-872-036-920.

Marangoni A.G. (2014). *Polymer gelation of oils* (EG Patent No. EG 26592 A). https://lens.org/018-059-146-485-785.

Marangoni A.G. (2014a). *Chocolate compositions containing ethylcellulose oleogel and methods for preparing* (EP Patent No. EP 2440067 B1). https://lens.org/101-947-018-615-286.

Marangoni A.G. (2014b). *Chocolate compositions containing ethylcellulose oleogel and methods for preparing* (EG Patent No. EG 26866 A). https://lens.org/072-487-484-860-903.

Marangoni A.G. (2014c). *Gelificación de polímeros de aceites* (ES Patent No. ES 2437394 T3). https://lens.org/046-703-054-804-881.

Marangoni, A.G. (2014d). *Polymer gelation of oils* (CN Patent No. CN 102510725 B). https://lens.org/179-709-289-431-688.

Marangoni A.G. (2015a). *Chocolate compositions containing ethylcellulose* (AU Patent No. AU 2010/258360 B2). https://lens.org/102-073-600-590-346.

Marangoni A.G. (2015b). *Polymer gelation of oils* (US Patent No. US 8940354 B2). https://lens.org/100-432-644-214-296.

Marangoni A.G. (2015c). *Polymer gelation of oils* (AU Patent No. AU 2010/258359 B2). https://lens.org/103-481-547-769-695.

Marangoni A.G. (2017a). *Chocolate compositions containing ethylcellulose* (CA Patent No. CA 2763645 C). https://lens.org/119-024-675-560-703.

Marangoni A.G. (2017b). *Polymer gelation of oils* (CA Patent No. CA 2763644 C). https://lens.org/158-915-046-152-217.

Marangoni A.G. (2018). *Heat-resisting chocolate composition* (CN Patent No. CN 107594047 A). https://lens.org/077-685-001-318-546.

Marangoni A.G. (2019). *Chocolate compositions containing ethylcellulose* (US Patent No. US 2019/0261644 A1). https://lens.org/189-360-735-388-418.

Marangoni A.G. (2022). *Chocolate compositions containing ethylcellulose* (US Patent No. US 11241021 B2). https://lens.org/063-067-495-523-69X.

Marangoni A.G. (2014a). *Chocolate compositions containing ethyl cellulose* (RU Patent No. RU 2524097 C2). https://lens.org/074-982-282-664-921.

Marangoni A.G. (2014b). *Polymer jellying of oils* (RU Patent No. RU 2527027 C2). https://lens.org/015-342-353-397-889.

Maria, J.J.J., Theodorus, H.W., Potman, R.P., and Theodoor, V.Jan, H. (2008). *Fat Granules* (ZA Patent No. ZA 200706748 B). https://lens.org/166-116-587-477-144.

Meng, Z., Liu, Y., and Jiang, Q. (2021). *Method for preparing phase-change-adjustable emulsion gel type fat substitute and application* (CN Patent No. CN 113383947 A). https://lens.org/162-815-466-799-005.

Meng, Z., Liu, Y., and Li, S. (2021). *Method for preparing healthy fat substitute based on pea/mung bean protein* (CN Patent No. CN 113812609 A). https://lens.org/196-085-421-554-161.

Miller, M. (2006a). *Trans-fatty acid free shortening* (CA Patent No. CA 2587403 A1). https://lens.org/187-306-545-186-882.

Miller, M. (2006b). *Trans-fatty acid free shortening* (AU Patent No. AU 2005/304449 A1). https://lens.org/115-486-965-019-094.

Miller, M. (2006c). *Trans-fatty acid free shortening* (US Patent No. US 2006/0105092 A1). https://lens.org/128-333-556-170-337.

Miller, M. (2006d). *Trans-fatty acid free shortening* (WO Patent No. WO 2006/053097 A1). https://lens.org/093-477-489-908-286.

Miller, M. (2007a). *Trans-fatty acid free shortening* (EP Patent No. EP 1819234 A1). https://lens.org/197-229-354-242-374.

Miller, M. (2007b). *Trans-fatty acid free shortening* (MX Patent No. MX 2007005684 A). https://lens.org/164-709-436-171-165.

Miller, M. (2007c). *Trans-fatty acid free shortening* (KR Patent No. KR 20070085885 A). https://lens.org/173-949-607-053-622.

Miller, M. (2007d). *Trans-fettsyrefritt fett* (NO Patent No. NO 20072482 L). https://lens.org/026-637-777-278-493.

Miller, M. (2008). *composição de gordura, produto alimentìcio, e, método para formar uma composição de gordura* (BR Patent No. BR PI0517483 A). https://lens.org/168-021-289-676-701.

Miranda, M. (2008). *Trans-fatty acid free shortening* (CN Patent No. CN 101098630 A). https://lens.org/178-330-808-295-321.

Mou, Y., Liu, Z., Zhang, M., Zhang, Q., Liu, W., Tian, H., Yu, X., Chen, Q., Zhang, L., Zhou, B., Li, R., Sun Z., Zhu, L., Gao, F., Xu, Y., Lu, B., Zhang, Y. (2020). *Oryzanol/sitosterol modified palm oil oleogel and preparation method thereof* (CN Patent No. CN 111743012 A). https://lens.org/074-542-912-646-201.

Nagy, K. (2014). *Structured oil compositions* (WO Patent No. WO 2014/184118 A1). https://lens.org/078-572-166-706-251.

Nagy, K. (2014). *Structured oil compositions* (CA Patent No. CA 2912425 A1). https://lens.org/198-913-887-280-469.

Nagy, K. (2015a). *Structured oil compositions* (AU Patent No. AU 2014/267411 A1). https://lens.org/157-488-717-414-418.

Nagy, K. (2015b). *Structured oil compositions* (CN Patent No. CN 105208872 A). https://lens.org/032-397-342-269-755.

Nagy, K. (2016). *Composiciones de aceites estructuradas* (CL Patent No. CL 2015003366 A1). https://lens.org/183-883-577-729-233.

Nagy, K. (2016). *Structured oil compositions* (US Patent No. US 2016/0073654 A1). https://lens.org/006-306-875-969-990.

Nagy, K. (2016). *Structured oil compositions* (EP Patent No. EP 3001789 A1). https://lens.org/128-351-205-717-512.

Nagy, K. (2017). *Structured oil compositions* (AU Patent No. AU 2014/267411 B2). https://lens.org/050-869-655-223-585.

Nagy, K. (2017a). *Composiciones de aceite estructurado* (ES Patent No. ES 2623293 T3). https://lens.org/161-420-874-133-226.

Nagy, K. (2017b). *Structured oil compositions* (EP Patent No. EP 3001789 B1). https://lens.org/167-365-669-221-939.

Nepovinnykh, N.V., Kutsenkova, V.S., Utesheva, M.S., Eganekhzad, S.A.S.A., Faezyan, A.K., Khesarinedzhad, M.M., Baratiangkhorgi, Z.E. (2021). *Semi-finished confectionery product based on hybrid gel* (RU Patent No. RU 2753975 C1). https://lens.org/125-935-020-954-28X.

Niu, Y., Fang, H., Xia, Q., Gu, M., Lu, W. (2019). *Low-fat meatball with novel complex gum as fat substitute and preparing method* (CN Patent No. CN 110063460 A). https://lens.org/044-908-924-302-858.

Noadia Genuario, B., Paula Kiyomi, O., and Rosiane Lopes Da, C. (2021). *composição de organogéis/oleogéis baseados em combinações de ceras e óleos vegetais* (BR Patent No. BR 102019027238 A2). https://lens.org/112-471-318-608-879.

OECD. (2019). *Innovation, Productivity and Sustainability in Food and Agriculture: Main Findings from Country Reviews and Policy Lessons*, https://doi.org/10.1787/c9c4ec1d-en.

OECD. (2019a). *Food and Agricultural Reviews*, OECD Publishing, Paris, https://doi.org/10.1787/c9c4ec1d-en.

Ortiz, G.J.M., Ortiz, G.S., and Blanco, D.V. (2020). *Olive oil-based, solid vegetable fatty product substitute for butter or margarine and use thereof* (CA Patent No. CA 3130858 A1). https://lens.org/004-596-551-089-802.

Ortiz, G.J.M., Ortiz, G.S., and Blanco, D.V. (2022a). *Olive oil-based solid vegetable fat product* (EP Patent No. EP 3957181 A1). https://lens.org/097-143-757-791-914.

Ortiz, G.J.M., Ortiz, G.S., and Blanco, D.V. (2022b). *Producto graso vegetal solido basado en aceite de oliva sustituto de la mantequilla o la margarina y utilizacion del mismo* (PE Patent No. PE 20220382 A1). https://lens.org/102-501-036-403-228.

Ortiz, G.J.M., Ortiz, G.S., and Blanco, D.V. (2022c). *Producto graso vegetal sólido basado en aceite de oliva sustituto de la mantequilla o la margarina y utilización del mismo* (CL Patent No. CL 2021002596 A1). https://lens.org/069-441-412-223-110.

Ortiz, G.S. (2020). *Solid vegetable fat product based on olive oil substitute for butter or margarine and the use of the same (Machine-translation by Google Translate, not legally binding)* (ES Patent No. ES 2787974 A1). https://lens.org/058-839-207-984-27X.

Ortiz, G.J.M., Ortiz, G.S., and Blanco, D.V. (2020). *Olive oil-based solid vegetable fat product* (WO Patent No. WO 2020/212633 A1). https://lens.org/069-507-056-777-342.

Oscar, J.H., Wouter, M.F., and Michel, C. (2022). *Oleogel* (NL Patent No. NL 2026242 B1). https://lens.org/089-172-391-885-571.

Pardey, P.G., Chan-Kang, C., Dehmer, S.P., and Beddow, J.M. (2016). Agricultural R&D is on the move. *Nature* 537 (7620): 301–303. doi: 10.1038/537301a.

Pérez, G.A., Granados, C.B.A., and De Haene, R.G.J. (2017). *Edible and thermoreversible oleogel and method for preparation thereof* (CA Patent No. CA 3007962 A1). https://lens.org/179-047-272-709-970.

Pérez, G.A., Granados, C.B.A., and De Haene, R.G.J. (2018). *Edible and thermoreversible oleogel and method for preparation thereof* (US Patent No. US 2018/0352826 A1). https://lens.org/092-295-363-000-359.

Pérez, G.A., Granados, C.B.A., and De Haene, R.G.J.. (2020). *Edible and thermoreversible oleogel and method for preparation thereof* (US Patent No. US 10874115 B2). https://lens.org/079-858-492-360-524.

Pérez, G.A., Granados, C.B.A., and De Haene, R.G.J.Pérez (2022). *Oleogel comestible y termorreversible y procedimiento para elaborarlo* (ES Patent No. ES 2909381 T3). https://lens.org/107-224-251-478-627.

Pérez, G.A., Granados, C.B.A., and De Haene, R.G.J.Pérez (2017). *Thermoreversible edible oleogel and method for producing same* (WO Patent No. WO 2017/099574 A1). https://lens.org/168-738-560-645-303.

Pérez, G.A., Granados, C.B.A., and De Haene, R.G.J.Pérez (2018). *Thermoreversible edible oleogel and method for producing same* (EP Patent No. EP 3387909 A1). https://lens.org/008-020-421-135-990.

Pérez, G.A., Granados, C.B.A., and De Haene, R.G.J.Pérez (2019). *Thermoreversible edible oleogel and method for producing same* (EP Patent No. EP 3387909 A4). https://lens.org/150-569-669-883-294.

Pérez, G.A., Granados, C.B.A., and De Haene, R.G.J.Pérez (2021). *Thermoreversible edible oleogel and method for producing same* (EP Patent No. EP 3387909 B1). https://lens.org/000-864-121-229-249.

Barendse, S.P., Flöter, E., Hogervorst, W.T., Van Horsen, D.S.H., Huizinga, H. Van Kempen, G.M.P. (2008). *Process for the preparation of a spreadable dispersion* (ZA Patent No. ZA 200706749 B). https://lens.org/195-829-750-338-489.

Phoon, P., Y.E.U. (2022). *Fibre-based oleogel* (WO Patent No. WO 2022/060289 A1). https://lens.org/149-304-117-379-568.

Pompili, F.F., Rossi, E., Luneia, R., Rossi, M., Ascenzi, F. (2017a). *Formulazione alimentare solida cremosa a base di olio di oliva, procedimento per la sua preparazione e usi* (IT Patent No. IT UB20154616 A1). https://lens.org/008-963-957-382-547.

Pompili, F.F., Rossi, E., Luneia, R., Rossi, M., Ascenzi, F. (2017b). *Solid creamy food formulation based on olive oil, process for its preparation and uses thereof* (WO Patent No. WO 2017/064737 A1). https://lens.org/052-941-507-325-789.

Pompili, F.F., Rossi, E., Luneia, R., Rossi, M., Ascenzi, F. (2018a). *Solid creamy food formulation based on olive oil, process for its preparation and uses thereof* (EP Patent No. EP 3361876 A1). https://lens.org/001-552-133-015-540.

Pompili, F.F., Rossi, E., Luneia, R., Rossi, M., Ascenzi, F. (2018b). *Solid creamy food formulation based on olive oil, process for its preparation and uses thereof* (AU Patent No. AU 2016/339162 A1). https://lens.org/181-954-563-159-279.

Pompili, F.F.,Rossi, E., Luneia, R., Rossi, M., Ascenzi, F. (2018c). *Solid creamy food formulation based on olive oil, process for its preparation and uses thereof* (US Patent No. US 2018/0295851 A1). https://lens.org/196-262-790-042-449.

Potman, R.P., Verbeek, J.H.T., Hogervorst, W.T., and Janssen J.J.M. (2006). *Fat granules* (CA Patent No. CA 2595381 A1). https://lens.org/108-365-910-027-252.

Potman, R.P., Verbeek, J.H.T., Janssen, J.J.M., and Hogervorst, W.T. (2006). *Fat granules* (AU Patent No. AU 2006/215830 A1). https://lens.org/073-503-082-180-685.

Potman, R.P., Verbeek, J.H.T., Janssen, J.J.M., and Hogervorst, W.T. (2010). *Fat granules* (AU Patent No. AU 2006/215830 B2). https://lens.org/120-826-450-846-348.

Qi, W., Zhang, Z., Shu, Y., Rao, W., Liang, T., Wang, H., and Yang, Q. (2021). *Oil-in-water gel Pickering emulsion as well as preparation method and application thereof* (CN Patent No. CN 113647608 A). https://lens.org/070-208-552-095-514.

Qiu, C., Lei, M., and Wang, Y. (2020). *Method for preparing ultra-stable temperature-sensitive nutrient oil foams through molecular assembly and application* (CN Patent No. CN 110934196 A). https://lens.org/106-294-159-049-747.

Qiu, C., Li, G., and Wang, Y. (2021). *Preparation method of water in oil emulsion gel based on diglyceride solid lipid nanoparticles* (CN Patent No. CN 112868816 A). https://lens.org/031-863-298-313-972.

Qiu, C., Wang, Y., and Huang, Y.U. (2018). *Oil gel rich in unsaturated fatty acid and preparation method and application thereof* (CN Patent No. CN 107950684 A). https://lens.org/023-988-400-214-348.

Ren, Z., Shi, L., and Weng, W. (2021). *Preparation method of fish oil gel* (CN Patent No. CN 112617200 A). https://lens.org/187-349-824-692-77X.

Rogier, Antoine Floris, The, Kai, G., and Leenhouts, A. (2018). *Process for preparing a fat slurry comprising olive oil and for preparing a spread with said slurry* (MX Patent No. MX 2018007635 A). https://lens.org/166-641-268-091-733.

Sarney Andrew, M., Tolcman, R., Foster, T., and Di Bari, V. (2021). *Oil-wax compositions and uses of oil-wax compositions for cooking food articles* (WO Patent No. WO 2021/064453 A1). https://lens.org/111-165-347-724-164.

Savin, G., Clement, V., Jing, J.-M., Mezzenga, R., Leser, M.E.. (2015). *Gel de óleo* (BR Patent No. BR 112012018153 A2). https://lens.org/139-494-788-403-554.

Savin, G., Clement, V., Jing, J.-M., Mezzenga, R., Leser, M.E. (2011a). *Oil gel* (CA Patent No. CA 2787545 A1). https://lens.org/051-560-332-621-07X.

Savin, G., Clement, V., Jing, J.-M., Mezzenga, R., Leser, M.E.. (2011b). *Oil gel* (WO Patent No. WO 2011/089171 A1). https://lens.org/028-172-830-524-653.

Savin, G., Clement, V., Jing, J.-M., Mezzenga, R., Leser, M.E.. (2012a). *Oil gel* (AU Patent No. AU 2011/208707 A1). https://lens.org/161-941-450-246-079.

Savin, G., Clement, V., Jing, J.-M., Mezzenga, R., Leser, M.E. (2012b). *Oil gel* (CN Patent No. CN 102834018 A). https://lens.org/053-464-645-080-799.

Savin, G., Clement, V., Jing, J.-M., Mezzenga, R., Leser, M.E. (2012c). *Oil gel* (EP Patent No. EP 2525665 A1). https://lens.org/099-191-752-908-115.

Savin, G., Clement, V., Jing, J.-M., Mezzenga, R., Leser, M.E. (2012d). *Oil gel* (US Patent No. US 2012/0289610 A1). https://lens.org/142-186-559-391-282.

Savin, G., Clement, V., Jing, J.-M., Mezzenga, R., Leser, M.E. (2014a). *Oil gel* (AU Patent No. AU 2011/208707 B2). https://lens.org/177-724-459-998-517.

Savin, G., Clement, V., Jing, J.-M., Mezzenga, R., Leser, M.E. (2014b). *Oil gel* (US Patent No. US 8796342 B2). https://lens.org/190-732-601-849-450.

Savin, G., Clement, V., Jing, J.-M., Mezzenga, R., Leser, M.E. (2011). *Oil gel* (EP Patent No. EP 2347658 A1). https://lens.org/010-160-155-550-601.

Scholten, E. (2020a). *One step procedure for producing a protein oleogel* (WO Patent No. WO 2020/242313 A1). https://lens.org/138-380-342-131-070.

Scholten, E. (2020b). *One step procedure for producing a protein oleogel* (CA Patent No. CA 3141790 A1). https://lens.org/108-721-467-911-059.

Scholten, E. (2021). *One step procedure for producing a protein oleogel* (SG Patent No. SG 11202113239V A). https://lens.org/098-559-586-486-75X.

Scholten, E. (2022a). *One step procedure for producing a protein oleogel* (EP Patent No. EP 3975752 A1). https://lens.org/020-016-050-953-099.

Scholten, E. (2022b). *One step procedure for producing a protein oleogel* (AU Patent No. AU 2020/286152 A1). https://lens.org/080-488-016-862-03X.

Scholten, E. (2022c). *Processo de etapa única para produzir um oleogel de proteína* (BR Patent No. BR 112021023956 A2). https://lens.org/071-098-973-692-204.

Scholten, E. (2022d). *단백질 올레오겔을 제조하는 단일 단계 방법*(KR Patent No. KR 20220018505 A). https://lens.org/069-163-968-802-372.

Scholten, E. and De Vries, A. (2016a). *Protein-stabilised oleogels* (WO Patent No. WO 2016/062685 A1). https://lens.org/161-642-994-821-81X.

Scholten, E. and De Vries, A. (2016b). *Protein-stabilised oleogels* (EP Patent No. EP 3011836 A1). https://lens.org/082-248-191-401-599.

Sebree, B.R. and Baseeth, S.S. (2016). *composição de organo-gel fosfolipídico estruturado termorreversível, uso da composição de organo-gel fosfolipídico estruturado termorreversível, processo para a obtenção de um produto cristalino líquido e método de carregar organo-gel fosfolipídico estruturado termorreversível* (BR Patent No. BR PI0915271 A2). https://lens.org/131-224-876-796-81X.

Shi, D., Gao, H., Fan, X., Yin, C., Yao, F., Shen, W., Feng, C., Meng, J., Chen, C., and Ye, S. (2021). *Fat substitute and preparation method of edible mushroom vegetarian diet prepared food* (CN Patent No. CN 112369596 A). https://lens.org/033-135-616-692-486.

Shireen, S.B. and Bruce, R.S. (2012). *Composiciones de organogel y procesos para su producción* (CO Patent No. CO 6361902 A2). https://lens.org/068-174-385-898-592.

Sijakovic, V.N., Jeric, I., Suc, S.J., and Radosevic, P. (2020a). *Composition comprising oxalamide gelators and vegetable oil* (WO Patent No. WO 2020/125926 A1). https://lens.org/104-713-352-177-055.

Sijakovic, V.N., Jeric, I., Suc, S.J., and Radosevic, P. (2020b). *Composition comprising oxalamide gelators and vegetable oil* (CA Patent No. CA 3120396 A1). https://lens.org/116-866-136-526-671.

Sijakovic, V.N., Jeric, I., Suc, S.J., and Radosevic, P. (2021). *Composition comprising oxalamide gelators and vegetable oil* (EP Patent No. EP 3897560 A1). https://lens.org/075-535-306-028-308.

Sijakovic, V.N., Jeric, I., Suc, S.J., and Radosevic, P. (2022). *Composition comprising oxalamide gelators and vegetable oil* (US Patent No. US 2022/0023179 A1). https://lens.org/139-487-633-993-806.

Stensrud, K., Baseeth, S., Jadhav, S., Lee, J.I., and Wicklund, L. (2016). *Structuring and Gelling Agents* (US Patent No. US 2016/0324178 A1). https://lens.org/056-414-842-554-213.

Stensrud, K., Baseeth, S., Jadhav, S.R., Lee, J.I., and Wicklund, L.Lee, J (2015). *Structuring and gelling agents* (WO Patent No. WO 2015/095472 A1). https://lens.org/172-245-466-169-776.

Sui, X., Jiang, Y., Mao, Z., Xu, H., Zhong, Y., and Zhang, L. (2016). *Edible oil gel and preparation method thereof* (CN Patent No. CN 105994697 A). https://lens.org/040-710-210-972-594.

Tetsuya, Y. (2016). *Solid oil and fat-like composition* (HK Patent No. HK 1208781 A1). https://lens.org/105-424-517-992-919.

Thomson, B.S., Huebner-Keese, B., and Ergun, R. (2017). *artigo e processo* (BR Patent No. BR 112015025714 A2). https://lens.org/041-273-171-780-832.

Thomson, B.S., Wallick, D.E., Meunier, D.M., and Ergun, R. (2017). *método para preparar um gel óleo comestível, gel óleo comestível, uso do gel óleo comestível, produto alimentício e uso de produto alimentício* (BR Patent No. BR 112014027719 A2). https://lens.org/060-413-190-666-542.

Thomson, B.S., Wallick, D.E., Meunier, D.M., and Ergun, R. (2020). *método para preparar um gel de óleo comestível, gel de óleo comestível e uso do gel de óleo comestível* (BR Patent No. BR 112014027719 B1). https://lens.org/067-541-575-539-828.

Valoppi, F., Salmi, A.R.I., and Haeggström, E. (2021). *Method to produce oleogels* (US Patent No. US 2021/0127720 A1). https://lens.org/150-918-878-952-64X.

Veen, M. and Budemann, A. (2013). *Fat substitute compositions comprising inulin and cassia gum* (WO Patent No. WO 2013/037739 A1). https://lens.org/154-241-929-497-733.

Veen, M. and Budemann, A. (2014). *Use of fat substitute compositions comprising inulin and cassia gum* (EP Patent No. EP 2755492 A1). https://lens.org/147-340-824-533-97X.

Veen, M. and Budemann, A. (2015). *Use of fat substitute compositions comprising inulin and cassia gum* (EP Patent No. EP 2755492 B1). https://lens.org/030-428-399-614-077.

Wang, L., Gao, Y., Li, B.., Li, J., Jiang, S., and Hao, L. (2021). *Pig back fat simulant and preparation method thereof* (CN Patent No. CN 113080433 A). https://lens.org/188-838-632-937-553.

Wang, Z., Luo, S., Lin, Z., Zheng, M., Pan, Z., and Fang, J. (2019). *Composite gel fat substitute and low fat egg tart filling prepared therefrom* (CN Patent No. CN 109288030 A). https://lens.org/068-428-081-722-499.

WEF (World Economic Forum) (2016). Food security and why it matters. https://www.weforum.org/agenda/2016/01/food-security-and-why-it-matters (accessed on 3rd April 2022).

WRI (Word Resources Institute) (2018). How to sustainably feed 10 billion people by 2050, in 21 charts. https://www.wri.org/insights/how-sustainably-feed-10-billion-people-2050-21-charts (accessed on 3rd April 2022).

Xiao, Z., Li, X., Wang, P., Wang, Q., Wu, Y., Li, F., Yang, S., Li, Z., Yang, Q., and Shi, C. (2017). *Novel margarine and preparation method and applications thereof* (CN Patent No. CN 106720458 A). https://lens.org/100-084-118-902-920.

Xu, B., Cai, K., Wu, X., Li, X., and Chen, BL. (2020). *Fat mimic with Harbin red sausage flavor and preparation method thereof* (CN Patent No. CN 111084332 A). https://lens.org/022-242-485-559-523.

Xu, X., Zou, B., Yang, N., Jin, Z., Zhou, X., Tian, Y., Wang, J., Jiao, A., Zhao, J., and Xie, Z. (2013). *Preparation method of fat substitute for fermented sausage* (CN Patent No. CN 102871141 A). https://lens.org/082-353-142-840-792.

Xu, X., Zou, B., Yang, N., Jin, Z., Zhou, X., Tian, Y., Wang, J., Jiao, A., Zhao, J., and Xie, Z. (2014). *Preparation method of fat substitute for fermented sausage* (CN Patent No. CN 102871141 B). https://lens.org/166-064-249-786-364.

Yan, S., Wang, Q., Yin, L., Yang, L., and Yang, J. (2021). *Walnut protein polysaccharide composite oleogel* (CN Patent No. CN 112293746 A). https://lens.org/107 362 563 820 693.

Yang, Y., Li, J., Chang, C., Su, Y., and Gu, L.Su, Y. (2019). *Egg yolk-phytosterol-polysaccharide composite emulsion gel and preparation method thereof* (CN Patent No. CN 110089694 A). https://lens.org/176-061-880-483-866.

Yang, Y., Yang, M., Zhang, N., Qing, D., Sun, X., Li, J., Li, Y., Li, C., Zhang, T., Cai, J., Li, M., Feng, C., Li, S., and Li, C. (2016). *Fat mimic and application thereof in Sichuan style sausages* (CN Patent No. CN 105901744 A). https://lens.org/114-447-333-570-234.

Yang, Y., Yang, M., Zhang, N., Qing, D., Sun, X., Li, J., Li, Y., Li, C., Zhang, T., Cai, J., Li, M., Feng, C., Li, S., and Li, C. (2016). *Fat substitute and application thereof in low-fat fermented sausages* (CN Patent No. CN 105851901 A). https://lens.org/014-340-639-842-436.

Yoshimoto, T. (2014). *Solid oil and fat-like composition* (WO Patent No. WO 2014/020734 A1). https://lens.org/010-681-855-166-962.

Yoshimoto, T. (2015). *Solid oil and fat-like composition* (CN Patent No. CN 104507323 A). https://lens.org/087-332-287-921-381.

Yu, D., Li, D., Wang, T., Wang, X., Li, T., Zhang, X., Zhang, H., Yang, F., Sun, L., Zhang, D., Jiang, L., and Wang, L. (2019). *Method for preparing antioxidant stable oleogel through electrochemical treatment and polyphenol and polysaccharide modified rice bran protein* (CN Patent No. CN 110623138 A). https://lens.org/022-525-764-257-333.

Zhang, J., Lan, Y., Liao, Z., and Hu, B. (2021). *Oleogel-based emulsion with stable interface crystallization as well as preparation method and application of oleogel-based emulsion* (CN Patent No. CN 113678899 A). https://lens.org/058-701-437-830-434.

Zhang, M., Li, Q., Liu, R., and Wu, T. (2017). *Special fat mimetic for meat product and preparation method of fat mimetic* (CN Patent No. CN 106615220 A). https://lens.org/104-654-705-926-721.

Zhang, Q., Huang, J., Zhou, W., Fan, L., Wu, C., Qiao, J., and Chen, Y. (2013). *Preparation method for animal fat substitute adopting curdlan as matrix* (CN Patent No. CN 103125615 A). https://lens.org/171-375-451-556-801.

Zhang, Q., Huang, J., Zhou, W., Fan, L., Wu, C., Qiao, J., and Chen, Y. (2014). *Preparation method for animal fat substitute adopting curdlan as matrix* (CN Patent No. CN 103125615 B). https://lens.org/032-230-504-617-19X.

Zhao, H., Liu, P., Fan, H., Lu, Z., Lyu, F., Bie, X., and Zhang, C.. (2019). *Vegetable oil-based food grade novel antibacterial oil gel rich in polyunsaturated fatty acids, and preparation method thereof* (CN Patent No. CN 109527120 A). https://lens.org/008-944-902-467-195.

ДАЕЛС, Э. and ФАУБЕРТ, И. (2022). *КОМПОЗИЦИИ НА ОСНОВЕ ФИТОСТЕРИНОВ В ПИЩЕВЫХ ПРОДУКТАХ* (RU Patent No. RU 2020123991 A). https://lens.org/031-838-489-753-765.

МИЛЛЕР, М. (2008). *ШОРТЕНИНГ, НЕ СОДЕРЖАЩИЙ ТРАНСИЗОМЕРОВ ЖИРНЫХ КИСЛОТ* (RU Patent No. RU 2007121722 A). https://lens.org/172-593-334-616-889.

НАГИ, К. (2017). *СТРУКТУРИРОВАННЫЕ КОМПОЗИЦИИ МАСЕЛ* (RU Patent No. RU 2015153876 A). https://lens.org/028-484-174-420-33X.

САВЕН, Г., КЛЕМАН, В., ЦЗЮН, Ц.-М. et al. (2014). *МАСЛЯНЫЙ ГЕЛЬ* (RU Patent No. RU 2012135483 A). https://lens.org/123-187-510-278-831.

アレクサンドラ・ヒルド, マティアス・クナール, ブリッタ・ヒユーブナー–キーゼ, & ロジャ・アーガン. (2019). *オレオゲルの製造方法* (JP Patent No. JP 2019513011 A). https://lens.org/163-229-610-871-118.

アレクサンドラ・ヒルド, マティアス・クナール, ブリッタ・ヒユーブナー–キーゼ, & ロジャ・アーガン. (2021). *オレオゲルの製造方法* (JP Patent No. JP 6876063 B2). https://lens.org/050-315-936-403-078.

デイヴィット・エル・マロトキー, ロバート・ビー・アツベル, & ロジャ・アルグン. (2019). *エチルセルロースオレオゲル分散体* (JP Patent No. JP 2019513169 A). https://lens.org/082-224-950-189-319.

デイヴィット・エル・マロトキー, ロバート・ビー・アツベル, & ロジャ・アルグン. (2021). *エチルセルロースオレオゲル分散体* (JP Patent No. JP 6883588 B2). https://lens.org/147-481-414-407-398.

ナジー, コ. (2016). *構造化油組成物* (JP Patent No. JP 2016523523 A). https://lens.org/172-153-474-184-513.

バシース, シリーン, エス., & セブリー, ブルース, アール. (2015). *オルガノゲル組成物及び製造方法* (JP Patent No. JP 5827709 B2). https://lens.org/016-152-871-371-522.

バシース, シリーン, エス., & セブリー, ブルース, アール. (2017). *オルガノゲル組成物及び製造方法* (JP Patent No. JP 6073552 B2). https://lens.org/040-268-771-259-490.

ロジャ・アーガン, ブラッドリー・エス・トムソン, デイヴィット・イー・ウオリック, & デイヴィット・エム・ミユニエル. (2015). *食用オレオゲルの製造方法* (JP Patent No. JP 2015521846 A). https://lens.org/026-653-292-193-643.

ロジャ・エルグン, ブラッドリー・エス・トムソン, & ブリッタ・ヒユーブナー–キーゼ. (2016). *低脂肪積層生地及び低脂肪積層菓子* (JP Patent No. JP 2016514486 A). https://lens.org/078-883-520-651-360.

ロジャ・エルグン, ブラッドリー・エス・トムソン, & ブリッタ・ヒユーブナー–キーゼ. (2017). *低脂肪積層生地及び低脂肪積層菓子* (JP Patent No. JP 6126301 B2). https://lens.org/195-142-360-577-491.

ロジャ・エルグン, ロバート・ビー・アツペル, & デイヴィット・エル・マロトキー. (2016). *オレオゲルを調製するための方法* (JP Patent No. JP 2016519956 A). https://lens.org/017-143-339-531-982.

ロジャ・エルグン, ロバート・ビー・アツペル, & デイヴィット・エル・マロトキー. (2018). *オレオゲルを調製するための方法* (JP Patent No. JP 6412115 B2). https://lens.org/050-962-155-122-987.

바시쓰, 쉬.에. and 세브리, 브.알. (2016). *Organogel compositions and processes for producing* (KR Patent No. KR 101685218 B1). https://lens.org/076-102-174-134-717.

에르건, 로., 톰슨 브래들리, 에., 월릭 데이비드, 이., and 메니에 데이비드, 엠. (2015). *Method of preparing an edible oleogel* (KR Patent No. KR 20150034208 A). https://lens.org/172-835-484-367-549.

에르건, 로., 톰슨 브래들리, 에., 월릭 데이비드, 이., and 메니에 데이비드, 엠. (2021). *Method of preparing an edible oleogel* (KR Patent No. KR 102308971 B1). https://lens.org/127-063-767-321-98X.

에르군, 로., 에펠, 로. 비., & 마로키, 데. 엘. (2021). *Process for preparing an oleogel* (KR Patent No. KR 102254288 B1). https://lens.org/176-642-684-262-498.

이수용, 박성권, 오임경, & 정다은. (2020). *Vegetable fat composition with oleogel and hydrocolloid* (KR Patent No. KR 102138833 B1). https://lens.org/018-382-609-907-519.

이수용, 배우성, 김현지, 장아름, & 고상훈. (2017). *bakery product with canola gel* (KR Patent No. KR 101759502 B1). https://lens.org/146-893-422-918-723.

丁保淼. (2017). *种菱角淀粉-魔芋精粉为基质的脂肪模拟物的制备方法* (CN Patent No. CN 104770640 B). https://lens.org/162-197-083-063-855.

仇超颖, 汪勇, & 黄玉. (2021). *一种富含不饱和脂肪酸的油凝胶及其制备方法与应用* (CN Patent No. CN 107950684 B). https://lens.org/163-319-733-961-067.

兰雅淇, 郑松柏, 魏飞龙, 董路路, 郑汝婷, & 胡冰洁. (2022). *一种米糠蛋白油凝胶及其制备方法和应用* (CN Patent No. CN 113261594 B). https://lens.org/190-485-912-262-117.

刘骞, 曹传爱, 孔保华, 贾惜文, & 赵钟イ. (2022). *一种亚麻籽胶调控乳状液中油滴聚集制备脂肪替代物的方法* (CN Patent No. CN 109938328 B). https://lens.org/057-408-146-131-992.

史德芳, 高虹, 范秀芝, 殷朝敏, 姚芬, 沈汪洋, ... 叶爽. (2022). *一种脂肪替代物及食用菌素食调理食品的制备方法* (CN Patent No. CN 112369596 B). https://lens.org/094 889 301 858 940.

埃尔贡, R., 阿佩尔, R.·.B., and 马洛特基, D.·.L. (2019). *制备油凝胶的方法* (CN Patent No. CN 105228461 B). https://lens.org/007-526-626-838-51X.

德米尤瑞斯, J. and 马兰戈尼, A. (2022). *水包油结构乳液和包含该乳液的食品* (CN Patent No. CN 114144068 A). https://lens.org/109-501-132-611-638.

李松林, & 林静. (2020). *一种抗氧化食用油凝胶及其制备方法* (CN Patent No. CN 109588497 B). https://lens.org/197-387-142-878-585.

李正纪, 何兴春, & 巩桥. (2018). *种猪肥膘 替代物及其制作方法*(CN Patent No. CN 104939100 B). https://lens.org/011-819-722-956-181.

杨勇, 杨敏, 张楠, 卿丹丹, 孙霞, 李静, ... 李诚. (2020). *一种脂肪模拟 物及其在川式香肠 中的应 用*(CN Patent No. CN 105901744 B). https://lens.org/117-729-178-053-599.

王兆梅, 罗司丹, 林梓茵, 郑明坤, 潘振辉, & 方嘉沁. (2022). *一种复 合凝胶 脂肪替代物及由其制备的低脂蛋挞馅料*(CN Patent No. CN 109288030 B). https://lens.org/145-094-060-254-060.

肖志刚, 李雪, 王鹏, 王庆峰, 武英华, 李凡姝, ... 时超. (2020). *一种 新型人造奶 油及其制备 方法和应 用*(CN Patent No. CN 106720458 B). https://lens.org/157-406-310-399-031.

赵海珍, 刘盼盼, 范海潇, 陆兆新, 吕凤霞, 别小妹, & 张充. (2022). *一种基于植物油的富含多不饱 和脂肪酸的食品级 抑菌油凝胶 及其制备 方法*(CN Patent No. CN 109527120 B). https://lens.org/174-674-698-823-309.

Index

Note: Page number followed by "*f*" denotes figures and "*t*" denotes tables, respectively.

Fat Mimetics for Food Applications, First Edition. Edited by Miguel Ângelo Parente Ribeiro Cerqueira and Lorenzo Miguel Pastrana Castro.

d

e

i

k

l

p

t

x

y

z

Plates

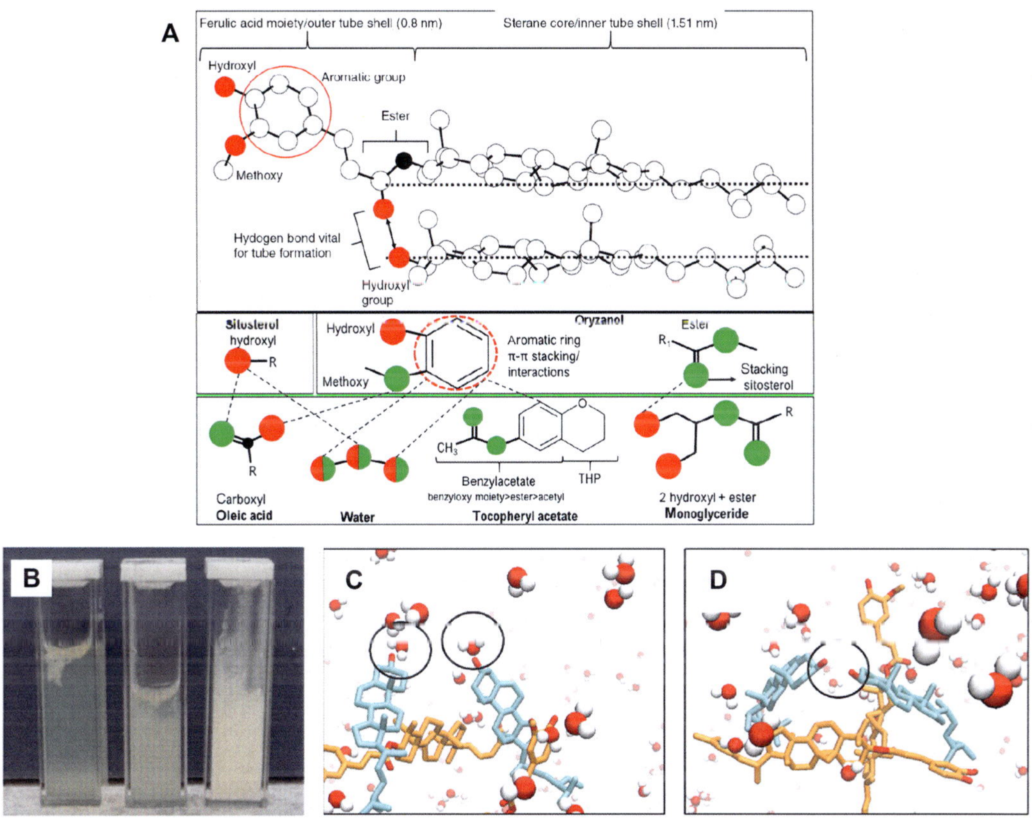

Plate 1 A) Top: Stacking of γ-oryzanol and β-sitosterol, red indicates main interactions points that contribute to the tube's stability or to form inter-tubule interactions. Bottom: Potential interactions of additives' functional groups with functional groups of sitosterol and oryzanol. Red circles: hydrogen bond donor. Green circles: hydrogen bond acceptor. Red dotted line: $\pi-\pi$ interactions and aromatic stacking, graphic does not aim to provide complete overview of all possible interactions, B) sunflower oil sterol-based oleogel organogel (left), a gel from the same batch with added glycerol droplets (center), and an emulsion-based oleogel from the initial batch with added (right); C) and D) show the hydrogen bond network between the hydroxyl group of sitosterols and water molecules and the hydrogen bond between sitosterols. Sitosterol is shown with cyan carbons and the hydrogen bonds are in circles. A) Reproduced from Scharfe, M. et al. 2021 / with permission of John Wiley & Sons; B, C, and D) Reproduced from Matheson et al. (2018), with permission from Royal Society of Chemistry.

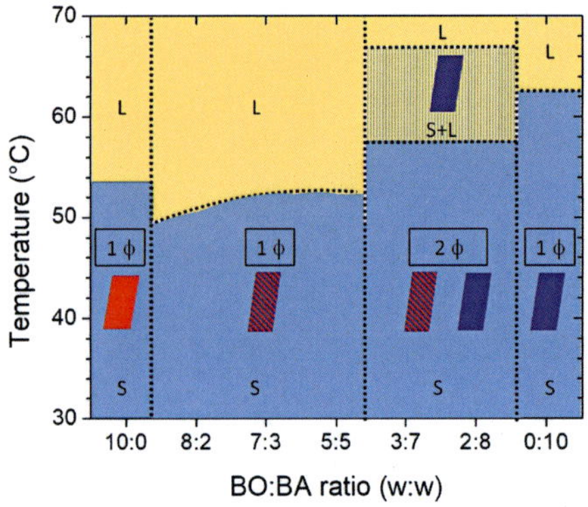

Plate 2 Schematic phase diagram for BO/BA oleogelator system in oil. One phase region is indicated by 1ϕ, and the two phase region is indicated by 2ϕ. S corresponds to solid and L to liquid. Platelets crystals are drawn: pure BO in red, mixed BO/BA crystals in red and blue dashed, and pure BA in blue. Reproduced from (Callau et al. 2021) with Permission from Society of Cosmetic Chemists.

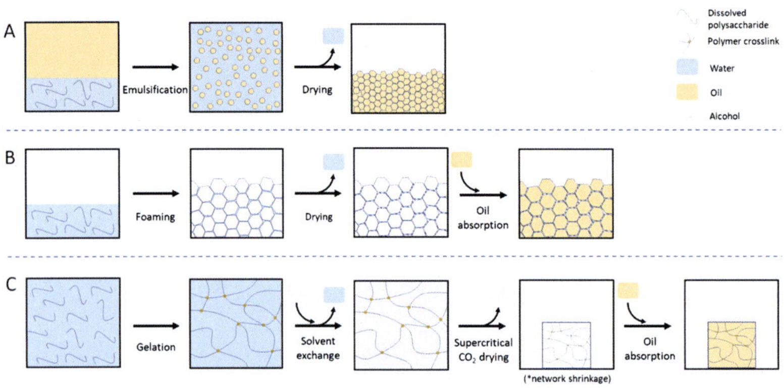

Plate 3 Schematic representation outlining the standard workflows for preparing polysaccharide-based oleogels using indirect methods; emulsion-templating (A), foam-templating (B), and hydrogel-templating via supercritical CO_2 drying after substitution with an intermediate solvent (C).

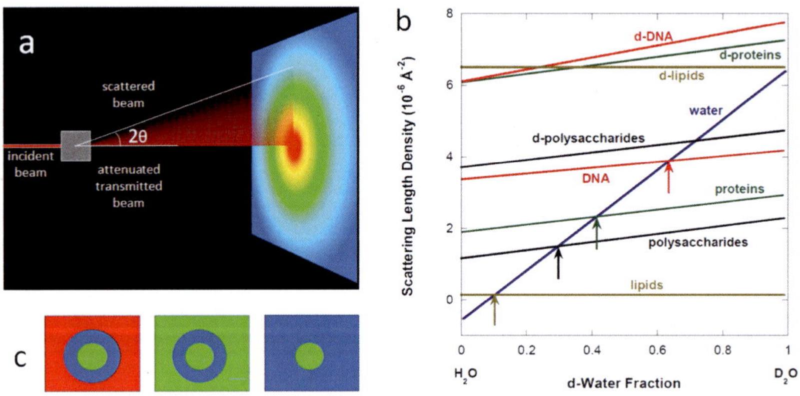

Plate 4 (a) Scattering geometry for a SAXS or SANS measurement; (b) Neutron scattering length densities for common food-based materials (Hammouda 2008); (c) Core-shell nanostructure in solution showing the possibility of selectively contrast matching either phase through changing solvent $C_{16}H_{34}/C_{16}D_{34}$ composition: by matching the SLD of the solvent to core, only the shell is visible, while matching the solvent to the shell yields information only on the core. Reproduced from Gilbert 2019 / with permission of Elsevier.

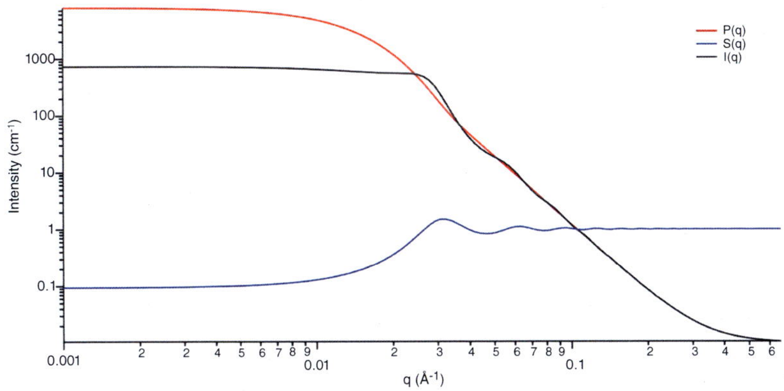

Plate 5 SAS from a 30% volume fraction, 100 Å particles with 30% size polydispersity with a contrast of 5.0×10^{-6} Å^{-2} and background of 0.01 cm^{-1}. The resultant intensity is the convolution of P(q) and S(q); note the effect of the structure factor, S(q) on reducing the intensity at low q due to the repulsive hard sphere interaction and generates a broad peak-like feature between q = 0.02 and 0.03 Å^{-1}.

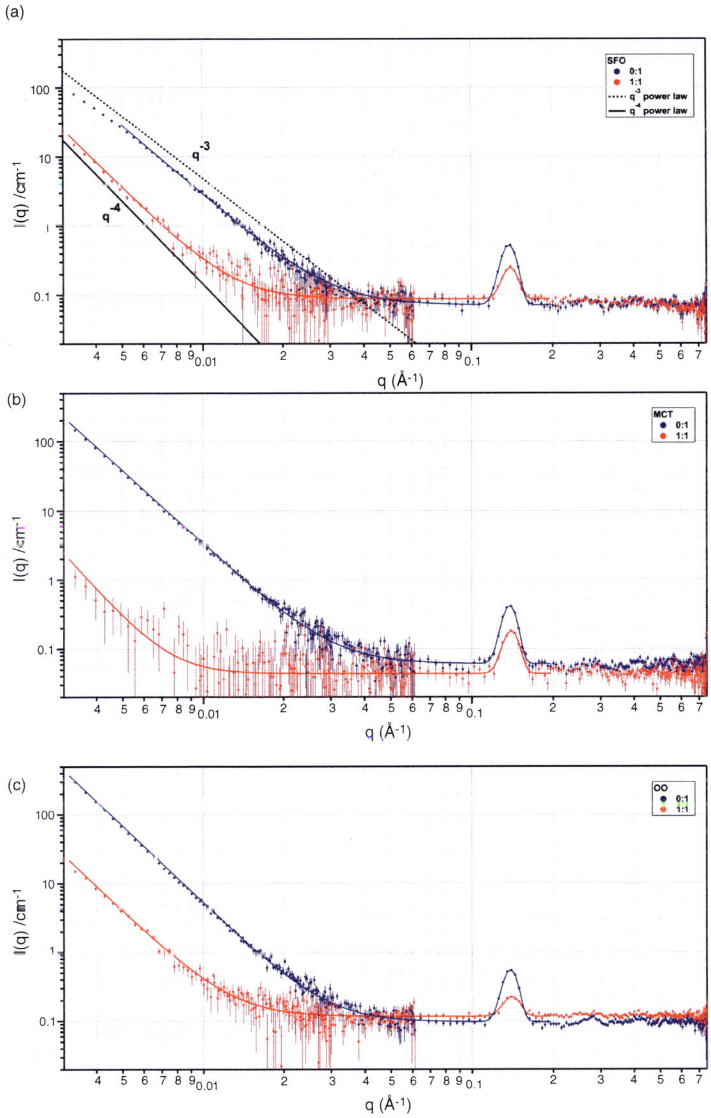

Plate 6 SANS from 1:1 and 0:1 oleogels: (a) SFO (with q^{-4} and q^{-3} power-law scattering superimposed); (b) MCT and (c) OO. Gels have been prepared with TAG oils (hydrogenated) and D2O (deuterated). Fits using power law plus Gaussian peak model are shown as solid lines. Reproduced from Cornet et al. 2021 / with permission of Elsevier.

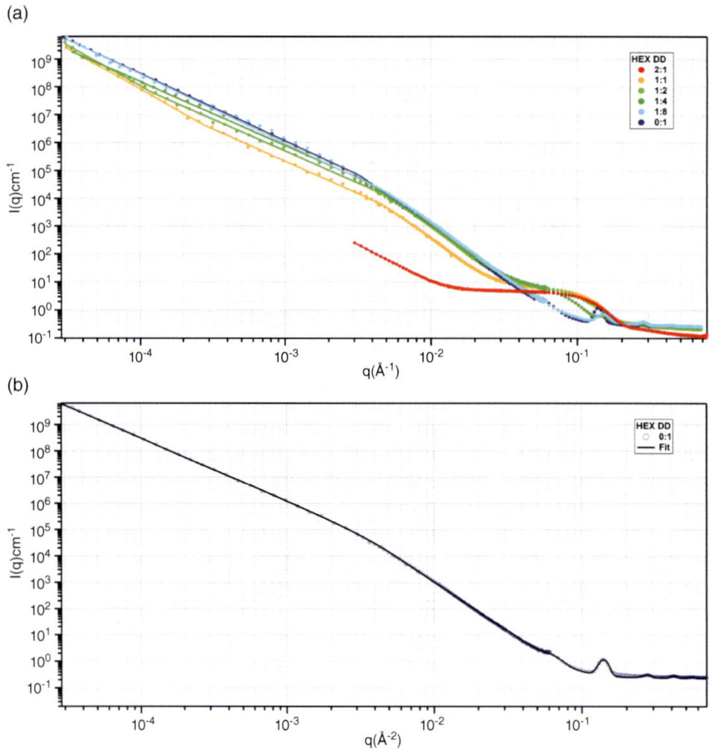

Plate 7 (a) Combined USANS and SANS from 16 wt% samples prepared with deuterated hexadecane (*i.e.*, DD contrast) including fits to single, double or triple power law; (b) 0:1 data with fit to Guinier-Porod model plus five Gaussian peak function. Reproduced from Cornet et al. 2021 / with permission of Elsevier.

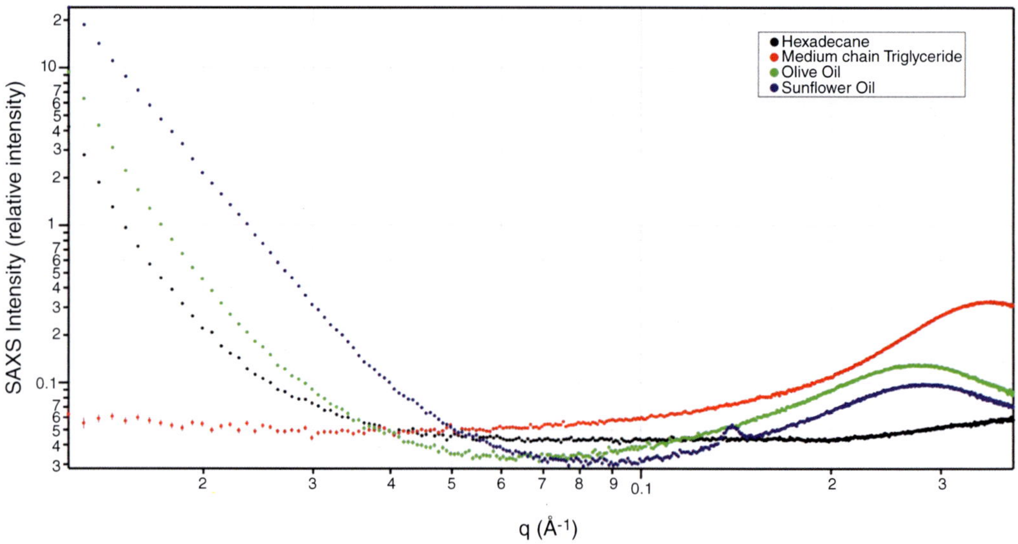

Plate 8 SAXS from typical edible oils along with apolar solvent, hexadecane. The weak peak in sunflower oil at $q \approx 0.14$ Å^{-1} arises from a crystalline impurity.

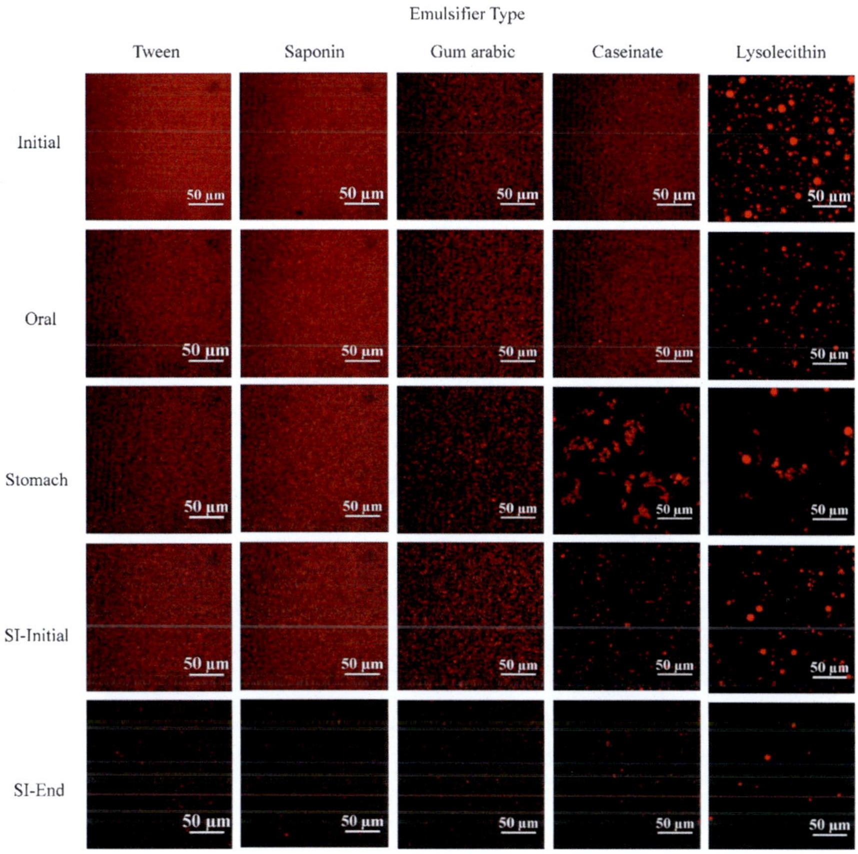

Plate 9 The effect of different emulsifier type on the confocal photos of corn oil-in-water emulsions during *in vitro* gastrointestinal digestion Tan et al. (2020), from Elsevier.

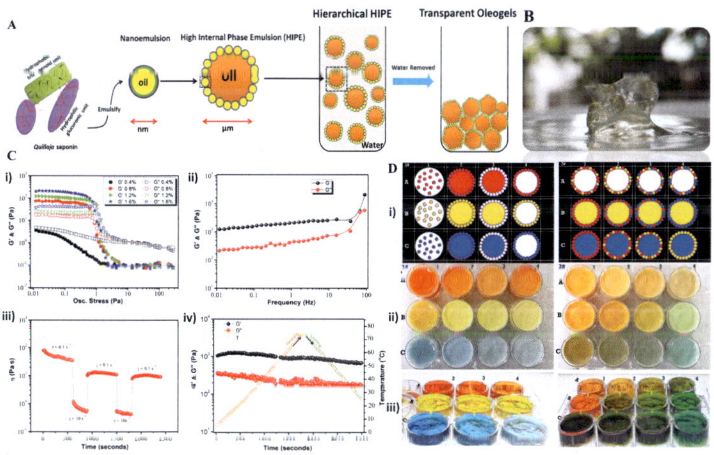

Plate 10 (A) Schematic for the formation of hierarchical high internal phase emulsions and oleogels from *quillaja* saponin; (B) Appearance of the oleogels obtained by drying the high internal phase emulsions (ϕ = 0.75) formed at 1.2% nanodroplets concentration; (C) Amplitude sweeps, frequency sweeps, flow measurements followed in time against alternating low (0.1 s^{-1}) and high (10 s^{-1}) shear rates and G' and G" versus time at varying temperatures; (D) Schematic of the position of the colorants in the hierarchical architecture (i) and representative image of the brightly colored emulsions (ii) and oleogels (iii) (Chen et al., 2017b, from the Royal Society of Chemistry).

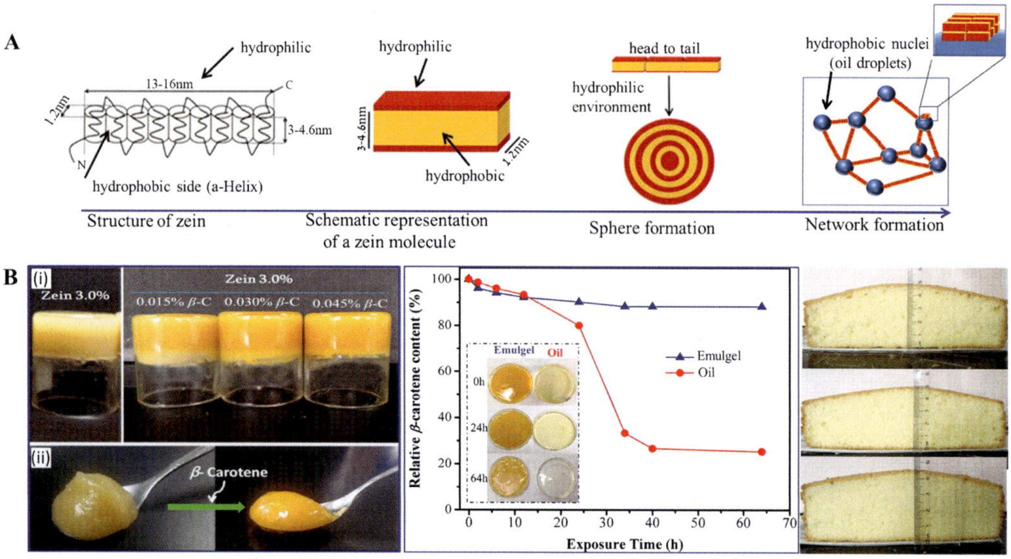

Plate 11 (A) Structure and schematic representation of a zein molecule, spherical particles, and emulsion gels formation by zein self-assembly in hydrophilic and hydrophobic environments (de Vries et al. (2014) / EPL Association). (B) Physical appearance (i) of zein-based emulsion gels incorporated varying concentrations of β-carotene; Panel (ii) presents the photograph of zein-based emulsion gels enriched with β-carotene (3.0% zein, 0.030% β-carotene) (Chen et al., (2016c), from Elsevier).

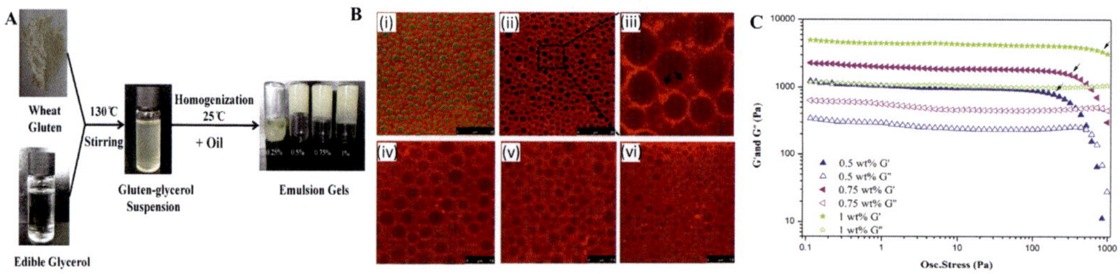

Plate 12 Schematic representation (A) of the fabrication process of the wheat gluten-based emulsion gels and their microstructure characterized by confocal laser scanning microscopy (B) and rheology and stress rheology (C) of emulsion gels Liu et al. (2016) / with permission from Elsevier).

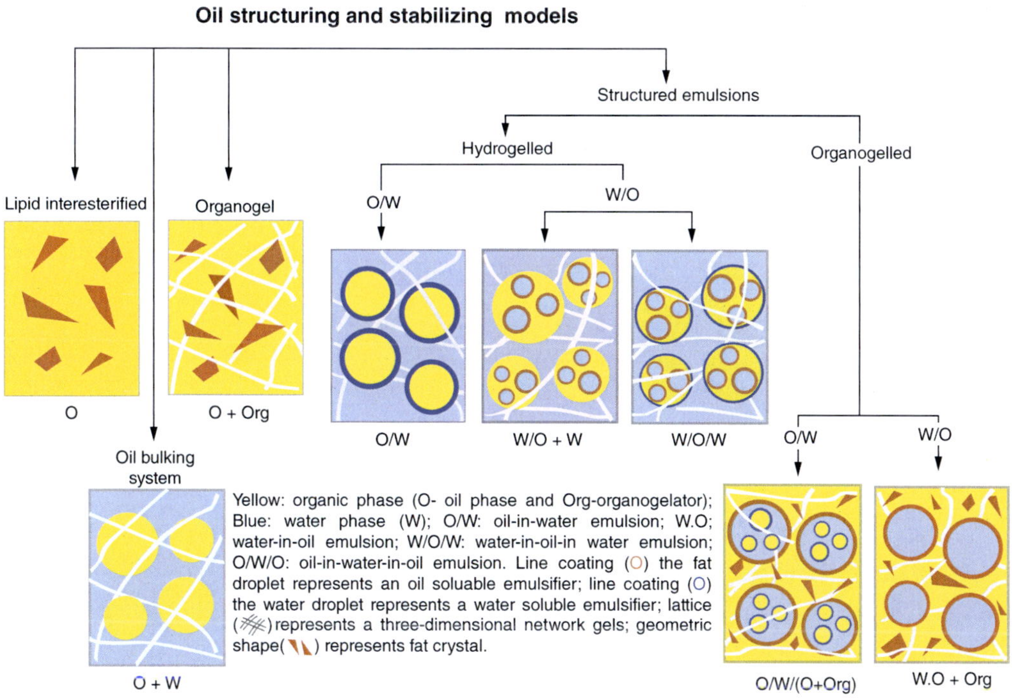

Plate 13 Schematic presentation of structural arrangement of the structuring/stabilizing systems of edible oils in solid-like materials suitable as meat fat alternatives. Reprinted with permission from Jimenez-Colmenero et al., 2015.

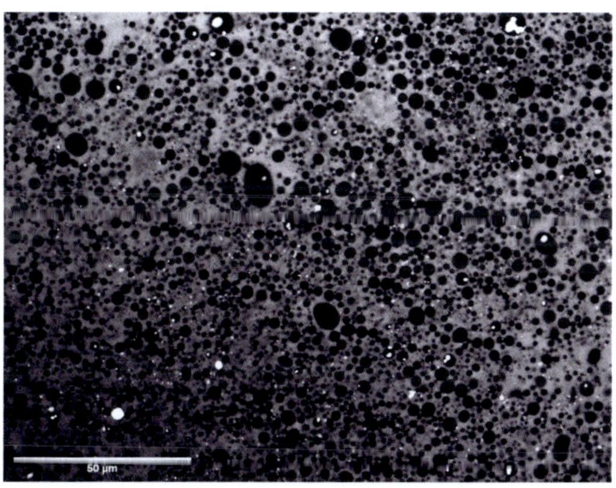

Plate 14 Microstructure of a commercial fat spread visualized by confocal laser scanning microscopy: gray = fat matrix (nile red dye, conventionally visualized in red), black: water phase (no dye), white: protein aggregates (FITC dye, conventionally visualized in green) (Courtesy of Vandemoortele Centre, Food Structure and Function Research Group, Ghent University).

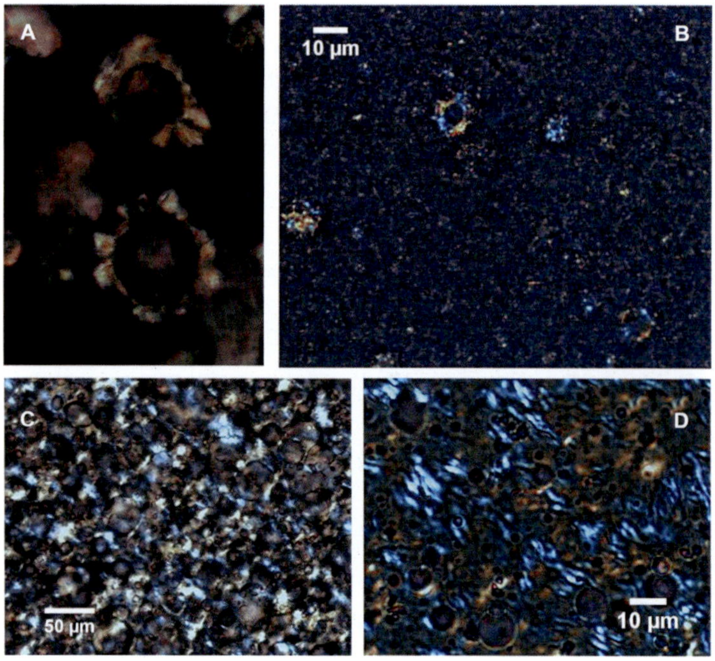

Plate 15 Polarized light microscopy images of W/O emulsions stabilized with (A) shellac wax, (B) carnauba wax, (C) sunflower wax and (D) beeswax. In shellac wax and carnauba wax, wax crystals were visualized in the W/O interface, displaying some affinity to the water phase. Sunflower wax and beeswax crystals act as network stabilizers by entrapping the water droplets on a three-dimensional network (Food Structure and Function Research Group, Ghent University). The system in A is described in a high level of detail by Patel et al. (2013) from Elsevier; B-D, courtesy of the Vandemoortele Centre.

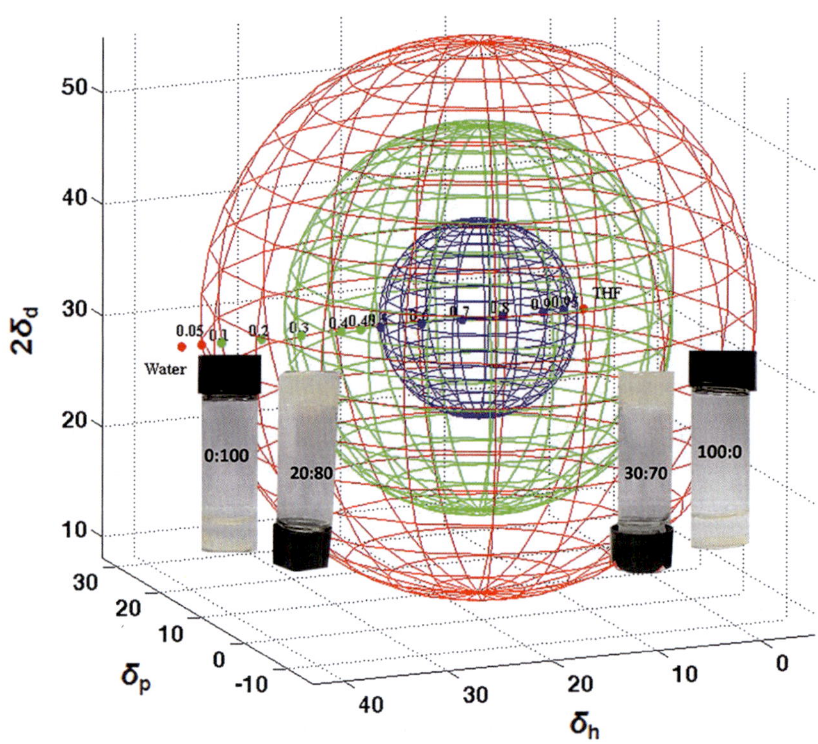

Plate 16 Hansen Solubility plots for 2.0% Glucono-Pyrenyl Molecules (Pn 7) in liquid mixtures represented in a Hansen space with spheres/shells: Largest (soluble), intermediate-size (gel), and smallest (solution). Yan et al. (2013), with permission from the American Chemical Society.